Progress in Mathematical Physics
Volume 34

Clifford Algebras

Applications to Mathematics, Physics, and Engineering

Rafał Abłamowicz

Editor

Birkhäuser

Boston • Basel • Berlin

Rafał Abłamowicz
Department of Mathematics, Box 5054
Tennessee Technological University
Cookeville, TN 38505
U.S.A.
rablamowicz@tntech.edu
http://www.math.tntech.edu/rafal/

Library of Congress Cataloging-in-Publication Data

A CIP catalogue record for this book is available from the Library of Congress,
Washington D.C., USA.

AMS Subject Classifications: Primary: 30G35, 15A66, 15A69, 15A75, 53C27, 58B32, 58B34, 68U10, 81R25,
81R50; Secondary: 05E10, 05E15, 11E81, 11E88, 11F03, 11F41, 13E10, 14D21, 15-04, 15A33, 16W30,
16W50, 17A35, 17B, 17B10, 17B35, 17B37, 17B45, 19D55, 20C05, 20C30, 20G15, 20G42, 22E25, 30G25,
31C12, 31C20, 32A36, 32A40, 32A50, 32W05, 35F15, 35H10, 35K15, 35Q40, 35S05, 42B30, 42B35,
43A80, 46H30, 46L87, 47A13, 47A60, 51N25, 53A30, 53B20, 53C07, 53C15, 53C25, 53C30, 53C50,
58C35, 58J40, 58J60, 65D17, 65K05, 65M60, 65R10, 78A45, 78A46, 81P15, 81Q05, 81T10, 81T13, 81T75,
81V22, 83A05, 83C05, 83C40, 83C57, 83C60

ISBN-13: 978-1-4612-7393-6   e-ISBN-13: 978-1-4612-2044-2
DOI: 10.1007/978-1-4612-2044-2

©2004 Birkhäuser Boston   *Birkhäuser*
Softcover reprint of the hardcover in 1st edition 2004

Typeset by the editor in $\mathcal{A}_{\mathcal{M}}\mathcal{S}$-LATEX.

9 8 7 6 5 4 3 2 1   SPIN 10955864

Birkhäuser is part of *Springer Science+Business Media*

www.birkhauser.com

Contents

Preface

This volume contains a selection of invited papers, often in expanded form, that are based on presentations made at the 6th Conference on Clifford Algebras and their Applications in Mathematical Physics, May 20–25, 2002, in Cookeville, Tennessee [1]. The organizers of the conference, Rafał Abłamowicz (Tennessee Technological University), and John Ryan (University of Arkansas) grouped all conference presentations into five sessions in Clifford analysis, geometry, mathematical structures, physics, and applications in engineering. Thus it seemed natural to organize the book into five parts under the same titles.

The 6th Conference on Clifford Algebras continued a 16-year sequence of international conferences devoted to the mathematical aspects of Clifford algebras and their varied applications in mathematical physics, and, more recently, in cybernetics, robotics, image processing and engineering. Previous meetings took place at the University of Kent, Canterbury, U.K., 1985; University of Montpellier, Montpellier, France, 1989; University of Gent, Gent, Belgium, 1993; University of Aachen, Germany, 1996; and Ixtapa, Mexico, 1999. Three edited volumes appeared after the Ixtapa conference, published by Birkhäuser, Boston [2–4].

Chapter 1, in this volume, is devoted exclusively to topics from Clifford analysis and range from the Morera problem, inverse scattering associated with the Schrödinger equation, through discrete Stokes equations in the plane, a symmetric functional calculus, Poincaré series, to differential operators in Lipschitz domains, Paley–Wiener theorems and Shannon sampling, Bergman projections, and quaternionic calculus for a class of boundary value problems. Among geometry topics, not so visibly present at previous conferences that gave rise to Chapter 2, are spin structures and Clifford bundles, eigenvalue problems for Dirac and Rarita–Schwinger operators, differential forms on conformal manifolds, connection and torsion, Bochner identities on Riemannian manifolds, and noncommutative geometry in physics. Chapter 3 is devoted to mathematical structures such as Grassmann algebras, Lie superalgebras, Grassmann–Hopf algebras, symplectic Clifford algebras and graded central algebras. Applications in physics, collected in Chapter 4, cover a wide range of topics from classical mechanics to general relativity, twistor and octonionic methods, electromagnetism and gravity, elementary particle physics, noncommutative physics, Dirac's equation, quantum spheres, and the Standard Model. Chapter 5 includes papers on Clifford geometric algebras and designing co-processors in computer engineering, applications in the description of an image space using Cayley–Klein geometry, and pose estimation. Below the reader will find brief introductions to all chapters in this volume.

While the papers collected in this volume require that the reader possess a solid knowledge of appropriate background material, as they lead to the most current research topics, the fundamentals of this background were presented prior to the 6th Conference in the form of six lectures. The lectures were delivered by prominent specialists in the field (including Pertti Lounesto to whom this volume is dedicated). They were aimed at graduate students and newcomers to the field. These lectures will appear in a separate volume from Birkhäuser in 2003 [6].

Papers from regular sessions at the 6th Conference included here were selected by the conference session organizers who oversaw the refereeing process:

o Clifford Analysis: Marius Mitrea (University of Missouri-Columbia, Columbia, Missouri), and Mircea Martin (Baker University, Baldwin City, Kansas);

o Geometry: Tom Branson (University of Iowa, Iowa City, Iowa), and Ugo Bruzzo (International School for Advanced Studies, Trieste, Italy);

o Mathematical Structures: Ludwik Dabrowski (SISSA, Trieste, Italy), and Bertfried Fauser (Universität Konstanz, Konstanz, Germany);

o Physics: William Baylis (University of Windsor, Windsor, Canada), and Giovanni Landi (Università di Trieste, Trieste, Italy);

o Applications: Jon Selig (South Bank University, London, England), and Gerald Sommer (Christian-Albrechts-Universität Kiel, Kiel, Germany).

Papers presented as plenary talks were selected by the Editor. The main aim of this volume, which guided the selection process, was to collect the best papers in each of the five representative areas ranging from pure mathematical theory to a wide array of applications in physics and computer engineering. Following the spirit of the conference, this volume unlike previously published volumes, includes a separate chapter on geometry.

Dedication
I am dedicating this book to the memory of Pertti Lounesto, friend and colleague, whose plenary lecture at the 6th Conference proved to be his last major contribution to the field of Clifford algebras – the field he loved so much.

PART I. CLIFFORD ANALYSIS

Berenstein, Chang and Eby: In this paper the authors discuss the Morera problem for functions f taking values in a Clifford algebra $C\ell_n$. As a replacement for the Morera theorem from the complex plane $\mathbb{C}$, which gives necessary and sufficient conditions for a continuous function f in an open subset Ω of $\mathbb{C}$ to be holomorphic in Ω, they formulate integral conditions whose vanishing on a Lipschitz boundary of every bounded domain secures that f be left regular. They establish a connection, similar to the one from the Euclidean space, between the

Pompeiu and Morera problems in Clifford analysis. Finally, the Morera theorem under Möbius transformations of a disk and the moment problem in the context of Clifford analysis is considered. Then, the authors discuss functions defined on the Heisenberg group $\mathbf{H}^n$ and investigate conditions under which these functions are CR-functions. In particular, they study moment and conjugate moment conditions that, if satisfied by functions in $L^2(\mathbf{H}^n)$, $L^p(\mathbf{H}^n)$, and eventually $L^\infty(\mathbf{H}^n)$, guarantee that these functions are CR-functions or conjugate CR-functions, respectively.

Bernstein: The author applies Clifford analysis methods to a study of the inverse scattering problem for an n-dimensional Schrödinger-type equation. In particular, she relies on a Borel–Pompeiu formula in $\mathbb{C}^n$ and a representation of holomorphic functions from complex Clifford analysis. This Clifford algebra setting is particularly useful in explaining the so-called "compatibility conditions for inverse scattering". A method of reconstructing the potential from a scattering data by quadratures with the help of the Borel–Pompeiu formula is provided. It is applied to a regularized Schrödinger equation in $\mathbb{C}^n$ and to time-dependent and the time-independent Schrödinger equation in $\mathbb{R}^n$. In fact, for a data T fulfilling a certain operator equation determined by an operator $\overline{\mathcal{D}}$ defined by the author in analogy to the $\bar{\partial}$ operator, the compatibility conditions, and decaying properly as $|k| \to \infty$, the scattering potential is computed by taking the inverse Fourier transform of the scalar part of the Borel–Pompeiu formula that involves T.

Gürlebeck and Hommel: The authors apply the theory of discrete analytic functions to the solution of the Dirichlet problem for the Stokes and Navier–Stokes equations. Finite difference operators are used to approximate the Cauchy–Riemann operator. A discrete T-operator and a discrete Cauchy integral are studied. In particular, authors discuss that the T-operator is a right-inverse of the discrete Cauchy–Riemann operator. A discrete Borel–Pompeiu formula is established and discrete Bergman projections are defined via a suitable orthogonal decomposition of the underlying l_2-space. Adapted finite difference schemes are used to approximate solutions to the boundary value problems.

Jefferies: The author discusses a symmetric functional calculus $f \mapsto f(A)$ for bounded monogenic functions f defined in sectors in $\mathbb{R}^{n+1}$ that take values in the Clifford algebra $C\ell(\mathbb{C}^n)$. Here, $A = (A_1, \ldots, A_n)$ is a finite sequence of closed linear operators, whose real combinations have spectra contained in a fixed sector of $\mathbb{C}$. If f has decay at zero and infinity, $f(A)$ is defined even if $A_1, \ldots, A_n$ do not commute. In the case when $n = 1$, the author obtains the usual Riesz–Dunford functional calculus for a single operator. In particular, if $A = (A_1, \ldots, A_n)$ commute and satisfy certain square functions estimates, the author uses a recently proven bijection between bounded monogenic functions defined on a sector in $\mathbb{R}^{n+1}$ and bounded holomorphic functions defined on a corresponding sector in $\mathbb{C}^n$ to obtain a functional calculus $f \mapsto f(A)$ for bounded holomorphic functions f defined in a sector in $\mathbb{C}^n$. The method is applicable to the Dirac operator on a Lipschitz surface Σ in $\mathbb{R}^{n+1}$ and leads to another proof of the boundedness of the Cauchy integral operator on Σ.

Krausshar: This paper is a continuation of a series of earlier papers by the same author in which began developing a theory of monogenic and polymonogenic hypercomplex-valued modular forms using methods of Clifford analysis. Here, the author provides nontrivial examples of such forms related to discrete subgroups of the Vahlen group. This is accomplished by constructing nontrivial monogenic and polymonogenic Eisenstein and Poincaré series.

Marmolejo-Olea and Mitrea: The authors study a large class of Dirac-like operators $D : C^\infty(M, \mathcal{E}) \to C^\infty(M, \mathcal{F})$ where $\mathcal{E}$, $\mathcal{F} \to M$ are two smooth, Hermitian vector bundles on a smooth, compact, boundaryless, orientable, Riemannian manifold M. They provide a variety of examples of such operators, both for real and complex manifolds M. The main goal is to show that there is a rather rich function theory associated with these operators, derivable by means of real-variable techniques. The starting point is a generalization of the classical Pompeiu formula for some suitably defined Cauchy type integral operators. Later, they study the interior regularity and boundary behavior of these Cauchy type operators, and prove results related to Hardy spaces associated with such operators in Lipschitz domains. Then they specialize their results by discussing separately Clifford monogenic functions and octonionic analysis as manifestations of a general, unifying function theory. Finally, they observe that many of the versions of the classical Bochner–Martinelli–Koppelman formula currently scattered in the literature are, in fact, particular cases of their generalized Pompeiu formula.

Qian: The author discusses Paley–Wiener theorems for entire functions in one and several complex variables, and their generalization to left-monogenic functions on $\mathbb{R}_1^n = \{x = x_0 + \underline{x} \mid x_0 \in \mathbb{R}, \underline{x} \in \mathbb{R}^n\}$. The $\mathbb{R}^n$ version uses Alan McIntosh's generalized exponential function $e(x, \underline{\xi})$. The author remarks that a proof of the classical theorem in $\mathbb{C}$ can be adapted to a proof of the generalized version due to the availability of Laurent series, Fourier multipliers and convolution integrals of Clifford monogenic functions. An analysis of both proofs follows. As an application, the Clifford monogenic sinc function is studied. A theorem is stated giving an extension of Shannon sampling to functions that are left-Clifford entire with exponential bounds of the Paley–Wiener type, and whose restriction to $\mathbb{R}^n$ belongs to $L^2(\mathbb{R}^n)$. A complete proof of the theorem is given.

Ren and Malonek: The authors prove a theorem that gives conditions when the Bergman projection operator P_α from the Banach space $L^p(\mathbb{B}, C\ell_{0,n}, dV_\alpha)$ onto a subspace of monogenic functions $A^p(\mathbb{B}, C\ell_{0,n}, dV_\alpha)$ is continuous. The proof follows a study of the Clifford algebra valued Bergman kernel.

Spr össig: Methods of quaternionic operator calculus developed earlier by the author in cooperation with K. Gürlebeck for the solution of time-independent boundary value problems are extended here to a study of Galpern–Sobolev equations with a variable dispersive term. The time derivative is approximated by forward finite differences. The author discusses stability and convergence of the time-discretization method.

PART II. GEOMETRY

Bartocci and Jardim: The authors study a generalization of the Nahm transform for instantons on minimal resolutions of $\mathbb{R}^4/G$ where G is a finite subgroup of $SU(2)$. Their approach allows them to define the Nahm transform on hyper-Kähler ALE 4-manifolds. Unlike the usual Nahm transform that associates instantons on $T^n \times \mathbb{R}^{4-n}$, $n = 1, 2, 3, 4$ to (possibly singular) solutions of the dimensionally reduced anti-self-duality equations on $\hat{T}^n$, the torus dual to T^n, the Nahm transformed instanton connection has no singularities. Under certain conditions described by the authors this new Nahm transform is invertible. Differential geometric properties of this transform are discussed.

Grantcharov: The author studies the hyper-Kähler geometry with torsion. The geometry of a connection with totally skew-symmetric torsion and holonomy in $Sp(n)$ is referred to as an HKT-geometry. He finds that a hyper-Hermitian manifold admits an HKT-connection if and only if for each complex structure there is a holomorphic $(0, 2)$-form. This allows him to study the potential theory for the HKT-geometry. In particular, he constructs a family of inhomogeneous HKT-structures on compact manifolds including $S^1 \times S^{4n+3}$. Then he discusses the reduction theory for HKT-structures. He eventually shows local existence of HKT metrics on any hypercomplex manifold and points out that global existence in general does not hold.

Homma: The author studies the enveloping algebra and Casimir elements of $\mathfrak{so}(n)$. He gives relations among them corresponding to Bochner identities. He then defines the principal symbols of gradients called *the Clifford homomorphisms* and relates them to the enveloping algebra through conformal weights. He calculates the eigenvalues of Casimir elements. In the last section, Homma gives all the Bochner identities for the gradients explicitly. Examples of geometric first-order differential operators which can be realized as gradients on Riemannian manifolds include the exterior derivative, its adjoint, and the conformal Killing operator and on spin manifolds, the Dirac operator, the twistor operator and the Rarita–Schwinger operator.

Hong: The author gives explicit formulas for the eigenvalues of Dirac and Rarita–Schwinger operators on $M = S^1 \times S^{n-1}$ where S^1 is endowed with a Lorentzian or Riemannian metric and a nontrivial spin structure, while S^{n-1} (n even) carries Riemannian metric and a standard spin structure.

Ugalde: Author presents partial results on differential forms associated to even-dimensional compact conformal manifolds. He associates to any pseudodifferential operator S of order 0 and acting on sections of a bundle B on a compact manifold without a boundary M, a differential form $\Omega_{n,S}(f, h)$ of order n acting on $C^\infty(M) \times C^\infty(M)$. The form is uniquely given as the Wodzicki 1-density of the product of commutators and can be expressed in terms of the total symbol of S. The author discusses a family of differential forms $\Omega_6(f, g)$ of order 6 with f, $h \in C^\infty(M)$ on a 6-dimensional, conformal, oriented, and compact manifold M without boundary. He shows that the integral $\int_M f_0 \Omega_6(f_1, f_2)$ defines a Hochschild 2-cocycle over the algebra $C^\infty(M)$. Conformally flat and non-flat cases are considered.

Várilly: The author discusses connections between noncommutative geometry and quantum physics. As examples that have influenced development of mathematical theory, he brings the Standard Model (SM) of fundamental interactions, noncommutative quantum field theory and Feynman graph combinatorics. The paper begins with a review of reconstruction of the SM due to Connes and Lott, and mentions the Mainz–Marseille approach to the SM. Then the author discusses spectral triples and their role in constructing spin manifolds. He reviews the spectral action principle introduced in order to include gravity in the early SM model and then moves on to discuss noncommutative field theory. Mathematics and physics are closely linked: "Seiberg–Witten map" and gauge theory, Moyal product and NC theory, phase space approach to quantum mechanics, noncommutative spaces and quantum gravity, Hopf algebras and QFT calculations. Extensive references are included.

PART III. MATHEMATICAL STRUCTURES

Brini, Regonati and Teolis: The authors consider supersymmetric letterplace algebras $Super[\mathcal{L}|\mathcal{P}]$ over a pair of $\mathbb{Z}_2$-graded alphabets $\mathcal{L}$ and $\mathcal{P}$ of letters and places. These algebras can be viewed as bimodules under the action of general linear Lie superalgebras $pl(\mathcal{L})$ and $pl(\mathcal{P})$ (and their enveloping algebras) associated to the alphabets $\mathcal{L}$ and $\mathcal{P}$. Following A. Capelli's idea, they do so by embedding $Super[\mathcal{L}|\mathcal{P}]$ into a larger "virtual letterplace algebra", obtained by adjoining new "virtual symbols" to the former. They describe how "virtual operators" may be used in place of operators from the universal enveloping algebras $U(pl(\mathcal{L}))$ and $U(pl(\mathcal{P}))$. As an example of this process they treat generalized Capelli operators as products of virtual polarizations. They extend these ideas to the so-called "symmetrized bitableaux" that give a basis for $Super[\mathcal{L}|\mathcal{P}]$, and "Young–Capelli symmetrizers" that act on these bitableaux. They give several complete decomposition theorems for these modules and algebras. As an application, they consider Berele–Regev representation theory of the symmetric group S_n over $\mathbb{Z}_2$-graded tensor spaces $T^n[V_0 \oplus V_1]$ built from $\mathbb{Z}_2$-graded finite dimensional vector spaces.

Eastwood: Various filtered and graded algebras arise, like the Clifford algebra, as quotients of the tensor algebra over a finite-dimensional real or complex vector space V, the latter possibly equipped with a nondegenerate symmetric or skew bilinear form. Examples presented include the exterior (Grassmann) algebra, the symmetric algebra, the symplectic Clifford algebra (also known as Weyl algebra), the enveloping algebra of a (real or complex) Lie algebra, and Laplace algebra that arises from symmetries of the Laplacian. These algebras can be recovered via a general construction that involves the tensor algebra over an irreducible representation W of a semisimple Lie algebra $\mathfrak{g}$ and suitably chosen homomorphisms of $\mathfrak{g}$-modules: $\Phi : W \otimes W \to \mathbb{R}$ or $\mathbb{C}$ and $\Psi : W \otimes W \to W$. In general, three classes of algebras can be obtained that are determined by the symmetric product, skew product or Cartan product. The author discusses three conjectures about the Cartan and tensor products of irreducible representations of $\mathfrak{g}$ and shows how they apply to the general construction and the specific cases. In an appendix he gives an outline of his talk "Symmetry and Differential Invariants" that was presented at the 6th Conference and that gave a motivation for the presented study.

Fauser: A modern categorical approach is necessary to properly formulate grade-free product formulas for almost all algebra and coalgebra products in Grassmann, Grassmann–Cayley, and Clifford algebras. That is, formulas that apply to general multivector polynomials instead of only generators or homogeneous elements. General Rota–Stein's cliffordization process—a categorically formulated generalization of Chevalley deformation—yields a combinatorial, instead of Chevalley's recursive, computationally efficient formula for the most important of these products—the Clifford product. Starting from a Grassmann–Hopf algebra, the author derives by categorial duality a self-dual Grassmann–Cayley double algebra with formulas for meet and join, comeet and cojoin, left/right contraction and co-contraction, Clifford and co-Clifford products. Chevalley formulas are then shown to be a specialization of this general approach while the co-Chevalley formulas are new. The method is based on algebraic invariant theory. Implementations of these products by mostly combinatorial algorithms prove to be computationally effective.

Hahn: This is a survey of the theory of quadratic forms q over a commutative and associative ring R with unity (often a field of characteristics not 2 (because there are integral results as well)), the theory of quadratic modules (M, q) (eventually free modules with finite bases), and the theory of Clifford algebras $C(M)$ of quadratic modules. A classification theorem of Clifford algebras as central simple algebras is given. The author shows that any finite-dimensional central simple algebra A with involution over a field R is "Brauer equivalent" to a Clifford algebra $C(M)$ for some quadratic space M of even rank and discriminant 1. He then moves on to discuss analogous results when R is a global field, or an arithmetic Dedekind domain. For quadratic forms over a field F of characteristics not 2, the author reviews the role of the Witt ring $W(F)$ of F. He shows that three maps dim, disc, and Cliff are sufficient to classify quadratic spaces when F is a local field, while a total signature map sig is needed when F is an algebraic number field. Then, he moves on to discuss the spin groups $\mathbf{Spin}(M)$ as the central extensions of the commutator subgroup of the orthogonal group $O(M)$. Generalized even Clifford algebras are Clifford algebras $C(A, \sigma, f)$, where A is a finite-dimensional central simple algebra over a field F, char $F \neq 2$, and (σ, f) is a quadratic pair on A. These algebras are defined as factor algebras of the tensor algebra $T(A)$. The associated spin groups are central in the classification theory of algebraic groups. Finally, in historical retrospective, the author discusses the role that Clifford algebras play in Adams' solution to the existence problem of the largest number $\rho(N + 1) - 1$ of linearly independent vector fields on the sphere S^N where $\rho(N + 1)$ is the Radon–Hurwitz number of $N + 1$.

Helmstetter: In this first paper the author compares (orthogonal) Clifford algebras and symplectic Clifford algebras, also called "Weyl algebras". Following Chevalley, one can introduce the Clifford product on the exterior algebra $\bigwedge(M)$ of a vector space M equipped with a bilinear form $\beta : M \times M \to K$, (here $K = \mathbb{R}$ or $\mathbb{C}$, and the form β is assumed to have a nondegenerate symmetric part ϕ). Analogously, when the dimension dim $M = m$ is even, M can be equipped with a bilinear form γ that has a nondegenerate skewsymmetric part ψ, and the finite-dimensional exterior algebra may be replaced with the infinite-dimensional symmetric algebra $S(M)$. Indeed it is possible to introduce a γ-dependent Weyl product on the vector space $S(M)$ so that $S(M)$ equipped with this new product becomes isomorphic to the Weyl algebra of (M, ψ), also referred to as symplectic Clifford algebra. The author then moves on to state the Lipschitz Theorem for the orthogonal Clifford algebras. The theorem relates an inner automorphism of $\bigwedge(M, \beta)$, which is generated by an exterior exponential of an element of $\bigwedge^2(M)$ and that leaves M invariant, with an "induced-by-it" orthogonal transformation of M without an eigenvalue of 1. In the main part of this paper the author explains difficulties and misconceptions related to extending the Lipschitz theorem to symplectic Clifford algebras: the need for a proper enlargement $\bar{S}(M)$ of $S(M)$ and the need to realize that Weyl multiplication of elements of $\bar{S}(M)$ often fails due to divergent infinite sums. In order to construct a "local" symplectic Lipschitz group, the author proves a symplectic counterpart of the Lipschitz theorem. Then, he presents ideas as to how to introduce a "global" symplectic Lipschitz group.

Helmstetter: It is well known that Clifford algebras $C\ell(M, Q)$ over a vector space M with a nondegenerate quadratic form Q are examples of graded central simple (G.C.S.) algebras provided with an involution called the "reversion". In general, finite-dimensional G.C.S. algebras A over a field K are isomorphic to matrix algebras (with matrices of suitable sizes) over some graded central division algebra D. Thus, these algebras can be classified by the isomorphy classes of D which form the Brauer–Wall group $BW(K)$ under a twisted tensor product $\hat{\otimes}$ as the group multiplication. The author studies G.C.S. algebras $A = A_0 \oplus A_1$

over a field K of characteristic $\neq 2$, provided with an involution ρ such that $\rho(A_j) = A_j$ for $j = 0, 1$. Using a "complex divided trace" of ρ, which is an element of the group R_8 of the eighth roots of 1 in $\mathbb{C}$, he obtains a classification of these involutions. Then, the pairs (A, ρ) can be classified by suitable pairs $(b, r) \in BW(K) \times R_8$. As an example, he gives a classification of such pairs over $\mathbb{R}$ and, as a particular case, he recovers the classification of the Clifford algebras $C\ell(p, q)$. The last section of this paper contains two specializations of a bijective mapping $A \otimes Z^g(A, M) \to M$ derived from a graded bimodule M over A and the graded centralizer $Z^g(A, M)$ of A in M.

Marks: The author describes a binary labeling scheme for powers of generators γ_μ $(\mu = 0, 1, \ldots, n - 1)$ in the Clifford algebra $C\ell(p, q)$ $(n = p + q)$ in terms of prescripts, postscripts, subscripts, and superscripts. While the subscripted (superscripted) indices denote covariant (contravariant) generators, prescripted (postscripted) indices denote monotonically decreasing (increasing) order of the generators. Thus, a typical (Clifford) basis monomial in a preferred labeling will look like $_m e = {}_{m_{n-1} \cdots m_\mu \cdots m_0} e = (\gamma_{n-1})^{m_{n-1}} \ldots (\gamma_\mu)^{m_\mu} \ldots (\gamma_0)^{m_0}$. One advantage of this approach is that the powers m_μ can be treated as bits of a binary number $m = \sum_{\mu=0}^{n-1} m_\mu 2^\mu$. Then, the grade of the monomial is just the sum of the bits, while the actions of reversion, grade involution, conjugation, and duality transformations on $_m e$ can be expressed as simple binary operations on the labels. An important formula is, of course, Clifford multiplication of two basis monomials $_l e$ and $_m e$, which is similar to a formula due to P.-E. Hagmark and P. Lounesto that uses Walsh functions and is implemented in CLICAL.

Schmeikal: Traditionally, in Euclidean spaces $\mathbb{R}^n$ with a positive definite metric, transpositions of base units are carried out by Weyl reflections. Those reflections represent the lattice of root spaces A_n connected with the classical Lie algebras $su(n+1, \mathbb{C})$ and the other real forms of $sl(n, \mathbb{C})$. Schmeikal develops that concept further by defining and investigating transposition involutions of Clifford algebras generated by spaces with indefinite metric. He applies that tool to the real Clifford algebra $C\ell_{3,1}$ of the Minkowski spacetime to derive the normal real forms of the algebra $su(3, \mathbb{C})$ and its Lie group. It becomes evident that a definite algebra representation can cover different forms of a group. The original quark multiplet found by Gell-Mann and Zweig is reconstructed. The generators of the inner symmetries of strong interaction turn out as Lorentz transformations of inhomogeneous differential forms. The six copies of the normal real $SU(3)$-form act on six color spaces which are isomorphic with the fourfold ring of real numbers $^4\mathbb{R}$. The flavor $SU(3)$ is an isotropy (sub)group at the point of a fixed lepton, the most prominent element of a color space. Those spaces are the appropriate spinor spaces for fermions with baryon number $\frac{1}{3}$. Possibly this concept opens new mathematical schemes to treat the internal symmetries of particle physics. This can also lead to a new perspective on the physical meaning of these symmetries.

PART IV. PHYSICS

Baylis: The aim of this paper is to use the spinorial formalism of the Clifford algebra of "classical physics" to explore connections between the classical and quantum theories. In particular, the author uses his paravector representation of spacetime in $C\ell_3$ to discuss spin-$\frac{1}{2}$ systems. One advantage of selecting $C\ell_3$ as the main algebra is that reversion (called "reversal" by the author) in $C\ell_3$ coincides with hermitian conjugation since the orthonormal basis vectors (generators of $C\ell_3$) are represented by hermitian matrices (normally there is no such connection). This gives many classical expressions their corresponding quantum form. The author discusses spin transformations of paravectors, eigenspinors and their time evolution in accelerating and rotating systemS, classical de Broglie waves, electromagnetic gauge potential, spin and g-factor, magnetic moments, Dirac's equation, and the Stern–Gerlach experiment.

Bonechi, Ciccoli and Tarlini: This paper discusses noncommutative geometry of Podleś 2-sphere and its generalization, the 4-sphere, proposed earlier by the authors. In particular, they analyze local "quantum charts" that can be used to define local trivialization for the monopole and instanton vector bundles.

Budinich: The author discusses "simple" spinors of É. Cartan (renamed as "pure" by C. Chevalley) and shows how the Cartan's equations defining them may be identified as the spinor-field equations in ten-dimensional spacetime $\mathbb{R}^{1,9}$ after restriction to $\mathbb{R}^{1,3}$. In this way the author obtains equations of motion for fermion multiplets where charges and internal symmetry groups originate from the known correlation of Clifford algebras with the three complex division algebras: complex numbers ($U(1)$ of charges), quaternions (isospin $SU(2)$, and electroweak $SU(2)_L$) and octonions ($SU(3)$ flavor and color). Furthermore in this framework, momentum spaces, bilinear in simple or pure spinorsn, are compact and also several problematic aspects of recent elementary particle theoretical physics might obtain simple geometrical explanations.

Chen, Nester and Tung: This paper addresses problems associated with a proper mathematical formulation of gravitational energy-momentum using spinors. An important achievement in this field (described in this paper) is Witten's spinorial positive energy proof, which can best be understood in terms of the associated Hamiltonian density. The latter is a candidate for a "truly physical" energy-momentum density for a gravitational field. The authors discuss the role of the Hamiltonian boundary terms, the issue of positivity, (quasi)localization, and the various roles the spinor field can play. Using a certain spinor-curvature identity they recover Witten's proof and the Dougan–Mason quasilocalization. A similar identity involving $SU(2)$ spinors gives an alternate reparametrization of the Hamiltonian, allowing the authors to provide an alternate positive energy proof and quasilocalization. Yet another identity allows for the inclusion of the spinor field into a "quadratic spinor Lagrangian" as a dynamic physical field. Although they find that spinors give nice results for energy-momentum, they note a serious limitation of this approach to gravitational energy-momentum based on spinor Hamiltonians: it seems that it cannot give all of the other physically conserved quantities of asymptotically flat spacetimes.

Daviau: The author discusses a chiral relativistic wave equation for neutrinos with a nonzero mass term. Unlike in the classical complex formalism, where the Dirac equation is formulated in the Clifford algebra $C\ell_3$ of the space, two new vector $K_{(1)} = \phi\sigma_1\phi^\dagger$, $K_{(2)} = \phi\sigma_2\phi^\dagger$ and two new bivector $S_{(1)} = \phi\sigma_1\overline{\phi}$, $S_{(2)} = \phi\sigma_2\overline{\phi}$ Lorentz-covariant tensorial densities appear that are related to Pauli's σ_1 and σ_2 directions. They are not electric gauge invariant. Here $\phi \in C\ell_3$ is a general multi-vector that the author eventually expresses in terms of two Weyl spinors ξ, η. With the $K_{(j)}$ and $S_{(j)}$ tensors, the author has 36 densities, and not only 16 as in the classical approach. Then, the author undertakes a detailed study of a chiral wave equation with mass and charge terms in the spacetime algebra setting. He solves this equation for the chiral wave for the hydrogen atom using H. Krüger's method.

Dray and Manogue: This paper is a continuation of earlier work by these authors on providing an octonionic description of the massless Dirac equation in $(9 + 1)$ Minkowski space. Using 2-component octonionic spinors they previously showed that three generations of leptons combine into such a description. The three generations live in three distinct quaternionic subalgebras of the octonions whose intersection is a complex subalgebra generated by a preferred octonionic unit. This paper begins with a review of the earlier work. The authors then extend their method to a 3-component formalism used by J. Schray for the superparticle. They hope to formulate a description for particles in the language of a special exceptional Jordan algebra of 3×3 octonionic Hermitian matrices (here called "Jordan matrices"). These matrices satisfy the usual characteristic equation of order 3. Of interest are those Jordan matrices $\mathcal{V}$ that under the Jordan product "$\circ$" (symmetric commutator) are primitive idempotents of trace 1 (Moufang plane). The authors exploit the fact that each such matrix can be dyadically factored as $\mathcal{V} = vv^\dagger$ where v is a 3-component octonionic column vector with quaternionic components. They then consider the Jordan eigenvalue problem $\mathcal{A} \circ \mathcal{V} = \lambda\mathcal{V}$ for Jordan matrices $\mathcal{A}$ and $\mathcal{V}$.

Lasenby, Doran and Arcaute: The authors apply spacetime algebra (STA) formalism to problems in electromagnetism, gravitation and multiparticle quantum systems. As examples of applications they discuss solutions to full Maxwell equations in the presence of a conductor, a spectrum of bound states created by a black hole, conformal models of space and spacetime, and multiparticle systems. They give a detailed study of a minimally-coupled relativistic wave equation for fermions in a spherically-symmetric black hole. Since this equation reduces to a free-field equation with a single interaction term H_I that lacks Hermiticity, energies of the gravitational bound states contain an imaginary component that causes states to decay. They display sample energy spectra of $S_{\frac{1}{2}}$ and $2P_{\frac{3}{2}}$ states. Their goal is to include multiparticle effects in the quantum physics of the black hole. Then, they move on to conformal models of Euclidean geometry and their applications in computer graphics. In particular, they discuss a stereographic representation of $\mathbb{R}^n$ versus a projective one. The latter approach is especially conveniently implemented via blades in a geometric algebra. Conformal geometry enters with rotors that, initially, give rotations and translations (Euclidean transformations) about the origin in the Euclidean spaces, and then can be used for rotations about any point. The last part of the paper is devoted to applications of a geometric algebra of the $4n$-dimensional relativistic configuration space (called the "multiparticle spacetime algebra" or MTSA for short) in entanglement and multiparticle wave equations. At the end, the authors return to the conformal geometry in the context of R. Penrose's twistors treated as elements of STA and classically expressed in terms of one dotted and one undotted Pauli spinors. Authors recall that a physical point in the spacetime can be (classically) constructed in terms of antisymmetric bilinear forms in twistorial components[1] and show how this can be implemented in MSTA.

Majid: The author explores the idea that spacetime itself may have noncommutative coordinates and gives three such models based on Lie algebras viewed as "noncommutative spacetime". Other examples of noncommutative spaces discussed are the famous "noncommutative torus", θ-spaces, and, unusually, Clifford algebras themselves with the familiar γ-matrices viewed as noncommutative coordinates. In all these cases "quantization" may be achieved by means of a Moyal-product type cocycle deformation from a commutative space, in the Clifford algebra case as quantization of the finite lattice $(\mathbb{Z}_2)^n$. As well as several mathematical results, of which the main one is a full account of how differential forms on such noncommutative spaces can be obtained by a cocycle twist of the classical differential forms, the article also contains a fully worked example in which noncommutative geometry is used to quantize $U(1)$ Yang–Mills theory on a finite lattice of 4 points in a path integral approach.

Owczarek: This paper is a survey of difficulties that need to be overcome on the way to unify S. L. Woronowicz's approach to the noncommutative geometry on quantum groups with A. Connes' noncommutative geometry. One problem is to provide a proper description of spectral triples on quantum groups. Another is to find a way to define Dirac operators on quantum homogeneous spaces. The author outlines difficulties of defining the Dirac operator with Clifford algebras. He then moves on to compare several (similar) definitions of quantum homogeneous spaces and related problems, e.g., problems with differential calculi on such

---

[1]*Editor's comment:* There exist two additional projection formulas that give a spacetime point in terms of symmetric (bilinear) and hermitian forms, respectively, in two twistors. The original idea goes back to J. Rzewuski (Wrocław, Poland) while the actual formulas, derived entirely using coordinate-free geometric algebra formalism, appeared in [10]. For another view on the Penrose's projection, see J. Rzewuski [11].

spaces. He reviews results by P. Podleś on this issue. As an example, he discusses the quantum a 2-sphere of Podleś as a homogeneous space of the quantum $SU(2)$ of Woronowicz and shows that a natural Dirac operator on such 2-sphere does not satisfy one of Connes' axioms. At the end he gives an outlook on possible future developments and concludes that axioms of Connes should be modified.

Pozo and Parra: An "r-fold Clifford algebra" $^{r}C\ell_{p,q}$ of r-fold multivectors is defined by the authors by tensoring r Clifford algebras $C\ell_{p,q}$. Grassmann and Clifford products, grade involution, reversion, conjugation, and the Hodge duality can be extended to $^{r}C\ell_{p,q}$ by applying them to each tensor slot. They contrast $^{r}C\ell_{p,q}$ with a multiparticle geometric algebra (MPGA). Then, they discuss the ∇ differential operator on r-fold multivectors and apply their formalism to superenergy tensors, which are important objects in General Relativity.

Trayling and Baylis: The authors review a recent approach to the standard model in terms of the geometric algebra $C\ell_7$. The Dirac algebra is then embedded in $C\ell_7$. In particular, they demonstrate how the exact gauge symmetries of the minimal standard model are related to a Poincaré group in a linear space with four extra spacelike dimensions. In this model, the gauge symmetries arise from the new algebra as unique symmetry groups of the current. The authors show how spinors used to represent all fermions of a single generation can be combined into one algebraic spinor of $C\ell_7$; how the currents can be computed from such spinors, and how contributions from individual spinors can be identified. They further discuss the rotational symmetries of these spinors and derive from them the gauge symmetries of the standard model. Finally, they show how a minimal Higgs field with the correct weak hypercharge assignment emerges in this approach.

PART V. APPLICATIONS IN ENGINEERING

Gebken, Perwass and Sommer: Due to expanding uses of (orthogonal) Clifford algebras $C\ell_{p,q}$ in applied fields, i.e., robotics, computer vision, image processing, in which coordinate-free computations with multivectors (blades) replace traditional matrix computations, increasing computational efficiency with these algebras is of paramount importance. The authors present a first-ever design of a Clifford (geometric) algebra co-processor and discuss its implementation on a Field Programmable Gate Array (FPGA). The design allows for soft changing the signature (p, q), the dimension $n = p + q$, and the numerical accuracy. High calculation speeds are claimed via a pipeline architecture. They also review briefly existing software packages, both numeric and symbolic, for computations with Clifford algebras.[2] Internally, the co-processor represents multivectors as

---

[2]*Editor's comment:* See also [4, 5] and, more recently [6, Appendix] (and references therein) for an extensive review of various software packages for these algebras.

lists of basis Grassmann monomials with associated scalar factors. The geometric product of two basis monomials is computed directly rather than retrieved from pre-computed multiplication tables, as this is computationally superior and more efficient. The authors then present their hardware design of the co-processor. The main algebraic operation implemented in the co-processor is, of course, the geometric product in $C\ell_{p,q}$. Others are: the inner and the outer products, addition, subtraction, and operations between two multivectors. This presents demands on the arithmetic logic unit (ALU) that the authors discuss. Then, they present technical aspects of FPGA, how the direct computations are performed on a chip, and then go on to the basis blade pipeline architecture.[3]

Koenderink: The author begins with a premise that image processing techniques, for example, in medical imaging alone, are "inconsistent" with basic principles of physics because images are surfaces in $\mathbb{R}^3$ (called "image space") where two dimensions give a localization of a pixel on a surface, while the third dimension serves as the "intensity dimension". As such, the latter is not interchangeable (commensurable) with the first two. The author proposes to look at the intensity dimension z as a (full) affine line (fiber) attached to the picture (Euclidean) plane (x, y). This construction yields a trivial fiber bundle in which sections are images, or, more precisely, Monge parameterizations of the "intensity surface". Since the author considers images to be invariant under action of a group of movements and similarities, it is natural now that he postulates the image space to be a homogeneous space. This rules out the Euclidean group as the group of proper motions. A pixel then is the base point of a fiber while the latter is just a line parallel to the intensity dimension. The result is that a bundle of parallel lines is preserved by a unique three-dimensional homogeneous space, one of the twenty seven Cayley–Klein geometries that has a single isotropic dimension. An eight-parameter group G_8 of similarities acts on the image space preserving the fibers. Two of the eight parameters are moduli of similarities of the first and the second kind, with the remaining six parameters needed for a group of proper motions G_6. Similarities of the image space, upon projection onto the image plane, give the Euclidean similarities in that plane. Then, in order to understand the G_8 similarities, the author gives a detailed description of how one-parameter subgroups of G_8 act in the image space: some appear like rotations, and some like translations in the picture plane. Others act trivially in that plane while acting as, for example, "gamma transformations" in the fibers. Since the Cayley–Klein geometry is dependent on one isotropic vector and the image plane is the Euclidean plane, one obtains, in

---

[3]*Editor's comment:* In Clifford algebras $C\ell(B)$ of an arbitrary bilinear form B (not covered by this hardware design), computational complexity is of many orders higher than in $C\ell_{p,q}$. A substantial improvement in speed of symbolic computations in $C\ell(B)$ has been achieved by using non-iterative Rota–Stein cliffordization based on Hopf algebra theory (see [8, 9] and the help pages of the "Bigebra" package [7] for further literature). It would be of great benefit to quantum physics to implement the Clifford product from $C\ell(B)$ via a hardware design similar to the one presented in this chapter for $C\ell_{p,q}$.

fact, a Clifford algebra degenerate in one dimension $C\ell_{2,0,1}$. The author describes consequences of this approach to the image space and pixels, explains why the similarities of the image space have two moduli, and how dual numbers (a suitable subalgebra of the Clifford algebra) are related to the so-called "normal planes". He discusses the geometry of the latter and then moves on to discuss the geometry of the entire image space. Finally, he describes the differential geometry aspects of images and points out the relevance of his approach in the image presentation business.

Rosenhahn and Sommer: Pose estimation is a task of determining a position and an orientation of a 3D object in a reference camera coordinate system from its observed 2D image. Since mathematical structures involved are the projective plane, projective space, kinematic space and Euclidean space, the authors use a conformal geometric algebra (CGA), as it contains all geometric algebras modeling these spaces as subalgebras. In this chapter the modeled objects are cycloidal curves (circles rolling on circles or lines). As a (CGA) they use a Clifford algebra $C\ell_{4,1}$ (denoted here as $\mathcal{G}_{4,1}$) over the vector space $\mathbb{R}^{4,1} = \mathbb{R}^3 \oplus \mathbb{R}^{1,1}$ that contains an identified hyperbolic plane $\mathbb{R}^{1,1}$ with a null basis. The latter is needed to identify a point at infinity and the origin. Products of Clifford exponentials of bivectors in $C\ell_{4,1}$ (rotors and translators) called "motors" give rigid transformations (elements of $SE(3)$) in $\mathbb{R}^3$. The authors show how cycloidal curves can be generated in CGA by means of twists and discuss how pose estimation of these curves can be accomplished in that algebra. In particular, they investigate 2-twist and 3-twist curves. The chapter ends with a display of experimental results on pose estimations of convex and non-convex objects.

Acknowledgments: The Editor would like to acknowledge and express appreciation for a generous support provided by the National Science Foundation, Award 0201303. The Award made it possible to bring an unprecedented number of researchers, postdocs, graduate and undergraduate students to the 6th Conference. Additional funding, for which he is grateful, came from the College of Arts and Sciences, the Center for Manufacturing Research, and the Provost Office at Tennessee Technological University; the Graduate School at the University of Arkansas in Fayetteville, and the College of Arts and Sciences at George Mason University.

Thanks are due to all session organizers for selecting papers from their sessions for inclusion in this volume and for overseeing their refereeing process.

Furthermore, the Editor is grateful to all invited speakers and contributors for submitting their works, and to all referees who have contributed to maintaining quality of this volume through their constructive criticism.

Since editing such a book could not be done without valuable help from our publisher, I would like to express my gratitude to Ann Kostant, Executive Editor/Mathematics and Physics, Birkhäuser, for her unremitting support to the Clifford community through encouragement and enthusiasm in bringing collected works in the area of Clifford algebras and their applications to the scientific community at large and to Elizabeth Loew for help with TEX-ing. Big thanks are due to Charmaine McMurry, TTU Mathematics Department, for her help with proofreading the entire book, and to Alexander Shibakov for his expert handling of eps files.

REFERENCES

[1] The 6th Conference on Clifford Algebras, May 20–25, 2002, Home page URL: http://math.tntech.edu/rafal/cookeville/cookeville.html.

[2] R. Abłamowicz and B. Fauser, Eds., *Clifford Algebras and their Applications in Mathematical Physics*, Vol. 1: Algebra and Physics, Birkhäuser, Boston, 2000.

[3] J. Ryan and W. Sprössig, Eds., *Clifford Algebras and their Applications in Mathematical Physics*, Vol. 2: Clifford Analysis, Birkhäuser, Boston, 2000.

[4] E. Bayro-Corrochano and G. Sobczyk, Eds., *Geometric Algebra: A Geometric Approach to Computer Vision, Neural and Quantum Computing, Robotics and engineering*, Birkhäuser, Boston, 2001.

[5] R. Abłamowicz, P. Lounesto, and J. Parra, Eds., *Clifford Algebras with Numeric and Symbolic Computations*, Birkhäuser, Boston, 1996.

[6] R. Abłamowicz and G. Sobczyk, Eds., *Lectures on Clifford Geometric Algebras and Applications*, Birkhäuser, Boston, 2003.

[7] R. Abłamowicz and B. Fauser, Clifford and Grassmann Hopf algebras via the BIGEBRA package for Maple, paper presented at ACA'2002: 8th International Conference on Applications of Computer Algebra, June 25–28, 2002, Volos, Greece (Submitted to *Journal of Symbolic Computation*, Special Issue on Applications of Computer Algebra).

[8] G.-C. Rota and J. A. Stein, Plethystic Hopf algebras, *Proc. Natl. Acad. Sci. USA* **91**, 1994:13057–13061.

[9] B. Fauser, *A Treatise on Quantum Clifford Algebras*, Konstanz, 2002, Habilitationsschrift, arXiv:math.QA/0202059.

[10] J. Rzewuski, Z. Oziewicz, R. Abłamowicz, Clifford algebra approach to twistors, *J. Math. Phys.* **23**, 1982, 231–242.

[11] J. Rzewuski, On the projections of the representation spaces of the symmetry group on the Minkowski space, *J. Math. Phys.* **23** (9), 1982, 1573–1576.

Rafał Abłamowicz
Cookeville, Tennessee, U.S.A.
October 9, 2003

In Memory of Professor Pertti Lounesto

PART I. CLIFFORD ANALYSIS

1

The Morera Problem in Clifford Algebras and the Heisenberg Group

Carlos A. Berenstein, Der-Chen Chang, and Wayne M. Eby

ABSTRACT In an open subset Ω of the complex plane $\mathbb{C}$ the Morera theorem gives a simple looking necessary and sufficient condition for a continuous function f to be holomorphic in Ω. Namely the vanishing of all the integrals $\int_\gamma f(z)\, dz$, where γ is an arbitrary Jordan curve in Ω whose interior also lies in Ω. The Morera problem consists in finding relatively small families Γ of Jordan curves such that the vanishing of the corresponding integrals still ensures that the conclusion of Morera's theorem still holds. In this lecture we discuss a number of recent results obtained in the case one considers functions defined in two fairly different settings, the Clifford algebras and the Heisenberg group.

Keywords: Morera theorem, Pompeiu problem, Clifford algebras, Heisenberg group

1.1   Introduction

Let us summarize briefly what is known about the Morera problem in the case of the complex plane $\mathbb{C}$. We refer to the excellent survey article by L. Zalcman [26]. (See also [27].) The theorem of Morera asserts that if f is a continuous function in $\mathbb{C}$, then f is an entire function, i.e., $f \in H(\mathbb{C})$, if and only if for every Jordan curve γ,

$$\int_\gamma f(z)\, dz = 0. \qquad (1.1)$$

The Morera problem consists in finding "small" families Γ of Jordan curves such that if (1.1) holds only for $\gamma \in \Gamma$, one can still conclude that $f \in H(\mathbb{C})$.

Let $\Gamma = \{\gamma\}$ be a family of Jordan curves in the plane $\mathbb{C}$ and assume that Γ is invariant under the action of the group $\mathcal{A}$ of all affine, orientation, and area

AMS Subject Classification: 22E25, 32A40, 32A50, 35H10, 42B30, 43A80.

This manuscript is an expanded version of the lecture given by the first author at the 6th conference on Clifford Algebras, Tenn. Tech. Univ., May 20–25, 2002. The research was partially supported by NSF Grant DMS 0070044.

preserving transformations of $\mathbb{C}$, and let f be a C^1 function that we are testing for holomorphicity. Then, by Stokes' formula

$$\int_\gamma f(z)\, dz = \int_{\text{int}\gamma} \frac{\partial f}{\partial \bar{z}}(z)\, d\bar{z} \wedge dz. \tag{1.2}$$

So that if f passes the test, i.e., the left-hand side of (1.2) vanishes for all $\gamma \in \Gamma$, then the corresponding area integrals of the continuous function $g = \frac{\partial f}{\partial \bar{z}}$ vanishes also.

Assume that Γ is generated over the group $\mathcal{A}$ by a single curve γ_0, and let $\Omega_0 = \text{int}(\gamma_0)$ be the corresponding Jordan domain. Then it is elementary that what one needs to verify is the Pompeiu property. That is, if $g \in C(\mathbb{R}^2)$ and satisfies

$$\int_{\sigma(\Omega_0)} g\, dx\, dy = 0 \quad \text{for all} \quad \sigma \in \mathcal{A}, \tag{1.3}$$

then $g \equiv 0$.

In fact the Pompeiu property for Ω_0 and the Morera property for $\partial\Omega_0$ are really equivalent. What is really useful to decide whether $\partial\Omega_0$ has the Morera property or not is the following equivalence with Schiffer's overdetermined boundary value problem, as defined in the following proposition.

Proposition 1. *Let $C = \overline{\Omega}$, where Ω is a bounded open set, C^c is connected and ∂C is (at least) Lipschitz. Then Ω fails to have the Pompeiu property if and only if there is an eigenvalue $\alpha > 0$ and a function u on Ω satisfying the overdetermined Neumann problem*

$$\begin{cases} \Delta u + \alpha u = 0, & \text{in} \quad \Omega \\ u = 1, \quad \frac{\partial u}{\partial n} = 0, & \text{on} \quad \partial\Omega. \end{cases} \tag{1.4}$$

See for instance [4, 5]. The following corollary allows us to find a plethora of regions satisfying the Pompeiu property.

Corollary 1. *Let Ω be as above. Then if Ω fails to have the Pompeiu property, it must be that $\partial\Omega$ is real analytic.*

For instance, any triangle, rectangle, or polygonal region has the Pompeiu property. Similarly, a simply connected domain with an infinitely flat point in the boundary will also have the Pompeiu property. It is easy to see that in the case where Ω_0 is a disk, then there is an infinite number of eigenvalues for the overdetermined Neumann problem; simply look for those that correspond to radial eigenfunctions. Therefore a disk does not have the Pompeiu property. Let us also remark that one can prove that if there are infinitely many overdetermined eigenvalues, then Ω_0 is necessarily a disk [4]. The question of whether disks are the only domains for which there is at least one such eigenvalue, sometimes called Schiffer's conjecture, remains open.

Although a single disk, with its translations, cannot help detect whether a continuous function in $\mathbb{C}$ is identically zero if all its averages over the disks of that radius vanish, one can decide this question using two different disks as long as the quotient of the radii is not in a certain countably infinite set.

Theorem 1 (Two Radii Theorem [24]). *Let $f \in C(\mathbb{R}^2)$ and let $r_1, r_2 > 0$. Suppose these radii satisfy the condition*

$$\frac{r_1}{r_2} \notin \mathcal{Q}(J_n(x)) = \{\frac{r}{s} : J_1(r\lambda) = J_1(s\lambda) = 0 \quad \text{for some} \quad \lambda \in \mathbb{C}\}.$$

Then $f = 0$ if and only if, for $r = r_1$ and r_2, one has

$$\int_{D_r(\mathbf{y})} f(\mathbf{x}) \, dm(\mathbf{x}) = 0 \quad \text{for all} \quad \mathbf{y} \in \mathbb{R}^2.$$

The condition on the quotients of the two radii is equivalent to the condition that the Fourier transforms of the measures associated to the two disks have no common zeros. The non-existence of common zeros of the Fourier transforms of the associated measures is a necessary condition for the Pompeiu property, though not in general sufficient. Note that the disk is the only shape for which integrals over the rotations of that shape do not provide any extra information. When S belongs to a large class of domains, including those with a corner, it has been demonstrated that the collection $\{\widehat{\mu_{\sigma(S)}}\}_{\sigma \in \text{rotations}}$ does not possess any common zero [16]. However it is intriguing that disks are a special case, for the following reasons. This collection then degenerates to one element, $\widehat{\mu_S}$, and it is known that for any domain S, $\widehat{\mu_S}$ will have zeros [16]. Thus we see why two disks are required for the Pompeiu property and have explained the source of the conditions for the radii in terms of zeros of the associated Fourier transforms. Note that circles, taken with Lebesgue measure, present a situation comparable to disks. Later a different method of avoiding the common zeros of the Fourier transforms of the associated measures will appear: the use of moments.

It is natural to ask whether one can study the Pompeiu and the Morera problems on a local region rather than requiring information on the whole plane. That is, suppose that f is a continuous function in the open unit disk $\mathbf{D}$, Ω is a Jordan domain in $\mathbf{D}$. Then we would like to test whether f is zero (respectively holomorphic) based on the vanishing of integrals of f over all copies of Ω that remain within the disk $\mathbf{D}$. Clearly, one cannot consider translations only because one will then really only test points near the center of $\mathbf{D}$. The natural thing is to consider all Möbius transformations of $\mathbf{D}$, that is the group $\mathcal{M} \simeq SU(1,1)$ instead of the group $\mathcal{A}$, as done above. For instance one obtains a result similar to the above for the Morera problem.

Theorem 2. *Let $\Omega \subset \mathbf{D}$ be a Jordan domain of class $C^{2,\epsilon}$ for some $\epsilon > 0$ and suppose that the Jordan curve $\Gamma = \partial\Omega$ is not real analytic. Assume $f \in C(\mathbf{D})$ satisfies*

$$\int_\Gamma f(\sigma(z)) \, dz = 0 \quad \text{for every} \quad \sigma \in \mathcal{M}.$$

Then f is holomorphic on $\mathbf{D}$.

There is also a Pompeiu type theorem in this setting; one only needs to replace the Lebesgue measure by the $\mathcal{M}$-invariant measure in $\mathbf{D}$,

$$d\mu = \frac{d\bar{z} \wedge dz}{1 - |z|^2}.$$

Investigating the Pompeiu problem on bounded open subsets of the ambient space, there is a related property, called the local Pompeiu problem, in which only a subset of the group (Euclidean or Möbius) is used. This question is discussed, for instance, in [11, 12]. In fact, similar ideas allow one to prove two-radii Pompeiu theorems for all non-compact irreducible symmetric spaces of rank 1 [14]. In the same spirit, it is natural to consider these questions for variants of the Cauchy–Riemann equations. One such variant appears in Clifford analysis, related to the study of the Dirac operator; another appears in the context of the boundary Cauchy–Riemann operator, that is, in the Heisenberg groups. It turns out that when one studies the "natural" Morera problem in Clifford analysis, it is easily reduced to the Euclidean case, as noted in [22].

On the other hand, for a holomorphic function f in the complex plane, we clearly have that for any $r > 0$ and all $z_0 \in \mathbb{C}$,

$$\frac{1}{2\pi} \int_{-\pi}^{\pi} f(z_0 + re^{i\theta})e^{in\theta}\, d\theta = 0$$

for any $n \geq 0$, not just $n = 0$. In other words, all the non-negative moments of f vanish on the circle. The converse is also correct. In [25], Zalcman investigated whether the converse of this result holds when only a few of the moments are considered and obtained the following theorem.

Theorem 3 (Two Moments Theorem [25]). *Let $f \in L^1_{\mathrm{loc}}(\mathbb{R}^2)$ and let $r > 0$ be fixed. Suppose there exist integers m, n such that for almost all $w \in \mathbb{C}$,*

$$\int_0^{2\pi} f(w + re^{i\theta})e^{in\theta}\, d\theta = 0, \qquad \int_0^{2\pi} f(w + re^{i\theta})e^{im\theta}\, d\theta = 0.$$

Then

(a) *if $0 \leq n < m$, f agrees almost everywhere with a solution of*

$$\left(\frac{\partial}{\partial \bar{z}}\right)^n f = 0;$$

(b) *if $0 \geq n > m$, f agrees almost everywhere with a solution of*

$$\left(\frac{\partial}{\partial z}\right)^{|n|} f = 0;$$

(c) *if $n > 0 > m$, $m \neq -n$, f agrees almost everywhere with a solution of the pair of equations $\left(\frac{\partial}{\partial \bar{z}}\right)^n f = 0$, $\left(\frac{\partial}{\partial z}\right)^{|m|} f = 0$. Thus in this case f is (essentially) a polynomial.*

For instance, the case $1 = n < m$ characterizes when f is a holomorphic function. Similarly, $0 = n < m$ characterizes when $f \equiv 0$.

It is important to discuss the concept of homogeneity in relation to these moments. This concept is directly related to the moments themselves, as well as the conclusions which may be drawn from them. For example, the two moments condition in Theorem 3 is equivalent to the condition that, for all $w \in \mathbb{C}$,

$$\int_{|z|=r} z^m f(w+z)\, d\sigma(z) = 0 \quad \text{and} \quad \int_{|z|=r} z^n f(w+z)\, d\sigma(z) = 0$$

when $n, m > 0$, letting σ be area measure on the circle in $\mathbb{C}$. Or, if both $n, m < 0$, the conjugate moments turn up in the statement:

$$\int_{|z|=r} \bar{z}^m f(w+z)\, d\sigma(z) = 0 \quad \text{and} \quad \int_{|z|=r} \bar{z}^n f(w+z)\, d\sigma(z) = 0.$$

Note the distribution given by

$$< f, T_m > = \int_{|z|=r} z^m f(z)\, d\sigma(z)$$

is of homogeneity m, while $\bar{T}_m$, given by

$$< f, \bar{T}_m > = \int_{|z|=r} \bar{z}^m f(z)\, d\sigma(z),$$

is of homogeneity $-m$. Note the conclusion of the theorem matches the homogeneity of the smaller moment. When considering the moments on circles in $\mathbb{C}$ represented by the distributions T_m and T_n, the Fourier transforms are

$$\hat{\mu}_1(\xi) = \xi^m J_m(r|\xi|) \quad \text{and} \quad \hat{\mu}_2(\xi) = \xi^n J_n(r|\xi|).$$

When $|n| \neq |m|$, say $|n| > |m|$, the Bessel functions $J_m(r|\xi|)$ and $J_n(r|\xi|)$ have no common zeros. Therefore the only common zero between $\hat{\mu}_1$ and $\hat{\mu}_2$ is $\xi = 0$, with a multiplicity of m. These facts lead to the conclusion $(\frac{\partial}{\partial \bar{z}})^m f = 0$. The connection between common zeros, moments, homogeneity, and the Pompeiu problem on spheres in $\mathbb{C}$ has now been described.

Notice that when the lower moment is $m = 0$, then the two moments theorem gives the conclusion $f = 0$, similarly to the two radii theorem. However the two moments theorem has one advantage, namely this theorem is valid for any sphere of one fixed radius. In contrast, the two radius theorem does not work for many pairs of radii. In fact, the set of bad radii is dense. Further, by adjusting the lower moment, the two moments theorem can be utilized to detect more subtle properties of f : namely,

$$\left(\frac{\partial}{\partial \bar{z}}\right)^m f = 0 \quad \text{or} \quad \left(\frac{\partial}{\partial z}\right)^{|m|} f = 0.$$

It is of considerable interest that differential properties, such as these just mentioned, may be tested by use of integral conditions. In closing let us remind the reader that there are other examples of theorems using moments and yielding differential relations for a function. Zalcman has an earlier version of a theorem involving moments [24], and instead of getting the results by use of two moments, it uses one moment on two spheres of distinct radii satisfying certain conditions comparable to those of the two radii theorem. This theme of moments and differential conditions will be explored further in the coming sections.

1.2 The Morera problem and moment problem in Clifford analysis

In this section we present an overview of results on the moment problem in Clifford analysis first presented in [21]. Before discussing these problems in the Clifford analysis setting, it is necessary to introduce the Clifford algebra $C\ell_n$ and present results relevant to the Morera problem, such as the version of the Morera theorem which appears in Clifford analysis. This material is adapted from [22], and further information may be found in the references [15, 20]. Let us first recall that the Clifford algebra $C\ell_n$ is generated from $\mathbb{R}^n$ by the basis elements $\mathbf{e}_1, \ldots, \mathbf{e}_n$ of $\mathbb{R}^n$ and the anti-commutation relations

$$\mathbf{e}_i\mathbf{e}_j + \mathbf{e}_j\mathbf{e}_i = -2\delta_{ij}.$$

The algebra $C\ell_n$ will have basis elements

$$1, \mathbf{e}_1, \ldots, \mathbf{e}_n, \ldots, \mathbf{e}_{j_1} \cdots \mathbf{e}_{j_r}, \ldots, \mathbf{e}_1 \cdots \mathbf{e}_n$$

for all possible collections of indices $1 \leq j_1 < \cdots < j_r \leq n$ and $1 \leq r \leq n$. An element of $C\ell_n$ may then be written in terms of the basis elements as

$$a = \sum_\alpha a_\alpha \mathbf{e}_\alpha$$

where $a_\alpha \in \mathbb{R}$ and $\mathbf{e}_\alpha = \mathbf{e}_{j_1} \cdots \mathbf{e}_{j_r}$ corresponding to $\alpha = (\mathbf{e}_{j_1}, \ldots, \mathbf{e}_{j_r})$.

There are two involutions which are defined by their actions on the basis elements. These are given by

$$\sim : C\ell_n \to C\ell_n : \mathbf{e}_{j_1} \cdots \mathbf{e}_{j_r} \to \mathbf{e}_{j_r} \cdots \mathbf{e}_{j_1},$$

where $\tilde{a}$ is written for $\sim(a)$, and

$$- : C\ell_n \to C\ell_n : \mathbf{e}_{j_1} \cdots \mathbf{e}_{j_r} \to (-1)^r \mathbf{e}_{j_r} \cdots \mathbf{e}_{j_1},$$

where $\bar{a}$ will be written for $-(a)$.

The Clifford algebra $C\ell_n$ becomes a Hilbert space and a Banach algebra with the inner product given by

$$< a, b > = \sum_\alpha a_\alpha b_\alpha$$

for any $a, b \in C\ell_n$. The space of functions to be considered will be the space of smooth $C\ell_n$-valued functions $\mathcal{E}(\mathbb{R}^n, C\ell_n)$. This space is also a $C\ell_n$ module under pointwise multiplication. More generally there is the corresponding space of continuous functions $C(\mathbb{R}^n, C\ell_n)$.

The notion of left and right regular functions replaces the concept of analytic functions as used for complex space. To define these we need to define the following differential operator known as the Dirac operator.

$$D = \sum_{i=1}^{n} \mathbf{e}_i \frac{\partial}{\partial x_i}.$$

A function $f \in C^1(\mathbb{R}^n, C\ell_n)$ is called *left regular* if

$$Df = \sum_{i=1}^{n} \mathbf{e}_i \frac{\partial f}{\partial x_i} = \sum_{i=1}^{n} \sum_{\alpha} \mathbf{e}_i \mathbf{e}_\alpha \frac{\partial f_\alpha}{\partial x_i} = 0,$$

and similarly $g \in C^1(\mathbb{R}^n, C\ell_n)$ is called *right regular* if

$$gD = \sum_{i=1}^{n} \frac{\partial g}{\partial x_i} \mathbf{e}_i = \sum_{i=1}^{n} \sum_{\alpha} \mathbf{e}_\alpha \mathbf{e}_i \frac{\partial g_\alpha}{\partial x_i} = 0.$$

Note that the involution $\sim$ yields $\widetilde{Df} = -\tilde{f}D$ so that f being left regular is equivalent to $\tilde{f}$ being right regular.

Note that $D^2 = -\triangle$, so that regular functions are harmonic. Also consider the function

$$G(x) = \frac{1}{\omega_n} \frac{-x}{\|x\|^n} = \frac{1}{\omega_n} \frac{x^{-1}}{\|x\|^{n+2}},$$

where ω_n is the surface area of the unit sphere in $\mathbb{R}^n$. This function is both left and right regular, and furthermore it takes the place of the Cauchy kernel. This observation leads to the following theorems [22].

Theorem 4. *Let M be a bounded domain with Lipschitz boundary. Then for $f \in C^1(U, C\ell_n)$ and $x \in M$,*

$$f(\mathbf{x}) = \int_{\partial M} G(\mathbf{y} - \mathbf{x}) n(\mathbf{y}) f(\mathbf{y}) \, dS(\mathbf{y}) - \int_M G(\mathbf{y} - \mathbf{x}) Df(\mathbf{y}) \, dv(\mathbf{y}).$$

The Cauchy integral formula for the Clifford algebras is a consequence of this identity, as the second term on the right-hand side vanishes when f is left regular. The following takes the place of the Morera theorem.

Theorem 5. *If f is a Clifford algebra-valued continuous function on the domain U such that*

$$\int_{\partial M} n(\mathbf{y}) f(\mathbf{y}) \, dS(\mathbf{y}) = 0$$

for every bounded domain M in U with Lipschitz boundary, then f is left regular.

With the Morera theorem in place, it is desirable to present what form the Morera problem takes in Clifford analysis. The following result [22] establishes that the relationship between the Pompeiu and Morera problems in Clifford analysis is the same as it is in Euclidean space. Let $\{S_j\}$ be a collection of Jordan surfaces and let $M_j = \overline{\mathrm{int}\, S_j}$.

Theorem 6. $\{S_j\}$ *has the Morera property in* $C\ell_n$ *if and only if* $\{M_j\}$ *has the Pompeiu property in* $\mathbb{R}^n$.

Therefore the surfaces which possess the Morera property are the same whether in the underlying Euclidean space or in the setting of Clifford analysis.

It is possible also to consider the Morera problem with respect to Möbius transformations for Clifford analysis. They are the functions ϕ acting on $x \in C\ell_n$ by

$$\phi(x) = (ax + b)(cx + d)^{-1},$$

where $a, b, c, d \in C\ell_n$ and are assumed to satisfy the two conditions:

(i) $a\tilde{c}, c\tilde{d}, d\tilde{b}$, and $d\tilde{a} \in \mathbb{R}^n$,

(ii) $a\tilde{d} - b\tilde{c} = \pm 1$,

and the transformation ϕ defines then a Möbius transformation formula over $\mathbb{R}^n \cup \{\infty\}$. The corresponding 2×2 matrices, $\left(\begin{smallmatrix} a & b \\ c & d \end{smallmatrix}\right)$, are called the Vahlen matrices.

The following theorem [23] then describes the change of variable for these Möbius transformations in Clifford analysis.

Theorem 7. *Suppose that* $\mathbf{y} = \phi(\mathbf{x}) = (a\mathbf{x} + b)(c\mathbf{x} + d)^{-1}$ *is a Möbius transformation and that* f *and* g *are Clifford algebra-valued functions. If* S *is a closed, bounded and oriented surface, then*

$$\int_S g(\mathbf{y})n(\mathbf{y})f(\mathbf{y})\, dD(\mathbf{y})$$

$$= \int_{\phi^{-1}(S)} g(\phi(\mathbf{x}))\widetilde{J(\phi,\mathbf{x})}n(\mathbf{x})J(\phi,\mathbf{x})f(\phi(\mathbf{x}))\, dS(\mathbf{x}),$$

where

$$J(\phi,\mathbf{x}) = \frac{\widetilde{c\mathbf{x}+d}}{\|c\mathbf{x}+d\|^n}.$$

Finally it is possible to consider the Morera theorem under Möbius transformations of the disk in the setting of Clifford analysis.

Theorem 8 ([22]). *Let* f *be a continuous Clifford algebra-valued function defined in the unit ball* $\mathbf{B}$ *in* $\mathbb{R}^n$ *and let* S *be a Jordan surface in* $\mathbf{B}$. *If*

$$\int_S \widetilde{J(\phi,\mathbf{x})}n(\mathbf{x})J(\phi,\mathbf{x})f((\phi(\mathbf{x}))\, dS(\mathbf{x}) = 0,$$

for every $\phi \in \mathcal{M}$, where $\mathcal{M}$ is the group of Möbius transformations of the ball, then f is left regular if and only if $\mathcal{M} = \overline{\text{int } S}$ has the Pompeiu property with respect to $\mathcal{M}$.

Finally the moment problem will be considered in the context of Clifford analysis. A set of moment conditions is presented which characterize those continuous functions which are left regular.

Theorem 9 ([22]). *Let $f : \mathbb{R}^n \to C\ell_n$ be a continuous Clifford algebra-valued function. Let $r > 0$ be fixed. If for each $\mathbf{x} \in \mathbb{R}^n$,*

$$\int_{\partial B(\mathbf{x},r)} n(\mathbf{y})f(\mathbf{y})\, dS(\mathbf{y}) = 0,$$

and

$$\int_{\partial B(\mathbf{x},r)} P_i(\mathbf{y} - \mathbf{x})n(\mathbf{y})f(\mathbf{y})\, dS(\mathbf{y}) = 0,$$

for

$$P_i(\mathbf{x}) = \mathbf{e}_1 \mathbf{x}_i + \mathbf{e}_i \mathbf{x}_1, \quad i = 2, \ldots, n,$$

then f is a left regular function.

1.3 The moment problem in the Heisenberg group

Let us look at some of the recently established results for the moment problem in the setting of the Heisenberg group. But first, as a reference point, it would be useful to describe the known results about the Pompeiu and Morera problems on the Heisenberg group which do not involve moment conditions. A good reference both for these results and for the methods used on the Heisenberg group is the book [10].

Formally the Heisenberg group $\mathbf{H}^n$ may be defined by the collection of points $\{[\mathbf{z}, t] \in \mathbb{C}^n \times \mathbb{R}\}$ under the group law

$$[\mathbf{z}, t] \cdot [\mathbf{w}, s] = [\mathbf{z} + \mathbf{w}, t + s + 2 \operatorname{Im}(\mathbf{z} \cdot \bar{\mathbf{w}})].$$

However, it is easier to visualize by identifying it with a manifold in $\mathbb{C}^{n-1}$ (of real codimension 1). This manifold $\{(\mathbf{z}, z_{n+1}) \in \mathbb{C}^{n+1} : \operatorname{Im} z_{n+1} = |\mathbf{z}|^2\}$, which is the boundary of the Siegel generalized upper half space

$$\Omega_n = \{(\mathbf{z}, z_{n+1}) \in \mathbb{C}^{n+1} : \operatorname{Im} z_{n+1} > |\mathbf{z}|^2\}.$$

There is a natural group action of $\mathbf{H}^n$ on $\partial\Omega_n$ giving an identification between these two objects, both as spaces of points and as groups. The map that identifies one to the other is given by

$$\Phi : \mathbf{H}^n \to \partial\Omega_n, \qquad [\mathbf{z}, t] \mapsto (\mathbf{z}, t + i|\mathbf{z}|^2).$$

Thus, there is an effective method to study $\mathbf{H}^n$ as the boundary of the Siegel generalized upper half space $\partial\Omega_n$. The set of left-invariant complex vector fields $\bar{Z}_j, Z_j$, and T, defined by

$$\bar{Z}_j = \frac{\partial}{\partial\bar{z}_j} - iz_j\frac{\partial}{\partial t}, \quad Z_j = \frac{\partial}{\partial z_j} + i\bar{z}_j\frac{\partial}{\partial t}, \quad T = \frac{\partial}{\partial t}$$

forms a complex basis for the Lie algebra $\mathfrak{h}_n$ of left-invariant vector fields on $\mathbf{H}^n$. These vector fields have the following commutation relations:

$$[\bar{Z}_j, Z_j] = 2i\delta_{jk}T,$$

and all the other commutators among $\bar{Z}_j, Z_j$ and T vanish. These vector fields endow the space $\mathbf{H}^n$ with a CR structure. At some places in the paper we will use the notation $\bar{\mathbb{Z}}$ to represent $\bar{Z}_1 \cdots \bar{Z}_n$, so that $\bar{\mathbb{Z}}f$ represents successive application to the function f of the vector fields $\bar{Z}_1$ out to $\bar{Z}_n$. Similarly, $\bar{\mathbb{Z}}^m$ represents $\bar{Z}_1^{m_1} \cdots \bar{Z}_n^{m_n}$.

A class of functions on $\mathbf{H}^n$ known as CR functions play a role comparable to the holomorphic functions on the Euclidean space $\mathbb{C}^n$. Similar to the Cauchy–Riemann equations $\frac{\partial}{\partial\bar{z}_j}f = 0$ for $j = 1, \ldots, n$ satisfied by holomorphic functions on $\mathbb{C}^n$, a function f on $\mathbf{H}^n$ is said to be a CR function when it satisfies the equations $\bar{Z}_jf = 0$ for $j = 1, \ldots, n$. CR functions are closely related to holomorphic functions as can be seen in the following result.

Proposition 2. *A function f on $\mathbf{H}^n$ is a CR function if and only if there exists a function F holomorphic on the upper half space and continuous to the boundary, $F \in H(\Omega_n) \cap C(\overline{\Omega_n})$ such that $f \circ \Phi = F|_{\partial\Omega_n}$.*

This relation could be stated as: CR functions are the boundary values of functions holomorphic on the upper half space $\Omega_n \subset \mathbb{C}^{n+1}$. The Morera and moment problems on $\mathbf{H}^n$ will deal primarily with CR functions and means of characterizing them by the vanishing of certain integrals. Similarly, the moment problem can be used to characterize functions satisfying certain differential relations, which are now expressed in terms of the complex vector fields $\bar{Z}_j$ and Z_j.

On the Heisenberg group the function spaces being considered become more of a matter of importance. The first result proven in this direction was a Morera type result and was proved for the function space $L^2(\mathbf{H}^n)$.

Theorem 10 ([2]). *Let $f \in C^1(\mathbf{H}^n) \cap L^2(\mathbf{H}^n)$. Then f is a CR function if and only if f satisfies the following integral conditions: For all $\mathbf{g} \in \mathbf{H}^n$ and $k = 1, \ldots, n$,*

$$\int_{|\mathbf{z}|=\rho} L_{\mathbf{g}}f(\mathbf{z}, 0)\omega_k(\mathbf{z}) = 0$$

with $\omega_k(\mathbf{z}) = dz_1 \wedge \cdots \wedge dz_n \wedge d\bar{z}_1 \wedge \cdots \wedge \widehat{d\bar{z}_k} \wedge \cdots \wedge d\bar{z}_n$.

The methods here are based on decomposition into a Laguerre series. The methods are much more complicated than the corresponding results for L^2 functions in Euclidean space, which require only the Paley–Wiener theorem. This result requires only one sphere of a single radius, but this is no different from what happens in Euclidean space when the function space is also restricted to L^2. Spheres of suitably chosen radii are again required when considering the function space $L^\infty(\mathbf{H}^n)$. Note that the conditions now require avoidance of zeros of Laguerre functions, as well as zeros of Bessel functions.

Theorem 11 ([2]). *Let f be a bounded continuous function on $\mathbf{H}^n$ which satisfies for every $\mathbf{g} \in \mathbf{H}^n$,*

$$\int_{S(r)} (L_{\mathbf{g}} f)(\mathbf{z}, 0)\, d\sigma_r(\mathbf{z}) = 0$$

for two radii r_1, r_2. Here σ_r is the area measure of $S_n(r)$. Assume that the above radii r_1 and r_2 satisfy the following two conditions:

1. *$\left(\frac{r_1}{r_2}\right)^2 \notin \cup_{\nu \in \mathbf{Z}_+} Q(L_\nu^{(n-1)})$,*

2. *$\frac{r_1}{r_2} \notin Q\left(\frac{J_{n-1}(t)}{t^{n-1}}\right)$.*

Then $f = 0$. Conversely, if one of the conditions (1) or (2) fails to hold, then there is a bounded continuous function $f \neq 0$ on $\mathbf{H}^n$ satisfying the integral conditions.

Remark 1. *A Morera version of this same theorem was proved in [1].*

Let us now consider the more recent results about moment conditions. In moving to a multi-dimensional setting, it is important to establish what kinds of moments should be used to generalize those from $\mathbb{C}$ where the moments $e^{im\theta} d\theta$ and $e^{-im\theta} d\theta$ correspond to terms $z^{m-1} dz$ and $\bar{z}^{m-1} dz$. In $\mathbb{C}^n$ the moments will be monomials $\mathbf{z}^{\mathbf{m}} = z_1^{m_1} \cdots z_n^{m_n}$ of homogeneity m and $\bar{\mathbf{z}}^{\mathbf{m}} = \bar{z}_1^{m_1} \cdots \bar{z}_n^{m_n}$ of homogeneity $-$m, corresponding to the negative moments. These lead to conclusions comparable to the one-dimensional setting.

Theorem 12 ([8]). *Given a function $f \in L^1_{\mathrm{loc}}(\mathbb{C}^n)$ satisfying*

$$\int_{|\mathbf{z}|=r} f(\mathbf{w} + \mathbf{z}) \mathbf{z}^{\mathbf{m}}\, d\sigma(\mathbf{z}) = 0$$

for all $\mathbf{w} \in \mathbb{C}^n$, and for $j = 1, \ldots, n$ and $1 \leq a_j \in \mathbb{Z}_+$,

$$\int_{|\mathbf{z}|=r} f(\mathbf{w} + \mathbf{z}) \mathbf{z}^{\mathbf{m} + a_j \mathbf{e}_j}\, d\sigma(\mathbf{z}) = 0.$$

We may then conclude

$$\left(\frac{\partial^{|\mathbf{m}|}}{\partial \bar{z}_1^{m_1} \cdots \partial \bar{z}_n^{m_n}}\right) f = 0.$$

In particular, for the integral conditions

$$\int_{|\mathbf{z}|=r} f(\mathbf{w} + \mathbf{z}) z_j\, d\sigma(\mathbf{z}) = 0, \qquad \int_{|\mathbf{z}|=r} f(\mathbf{w} + \mathbf{z}) z_j^{a_j}\, d\sigma(\mathbf{z}) = 0$$

for $j = 1, \ldots, n$ and a_j as before, we conclude that f is a holomorphic function, i.e.,

$$\frac{\partial}{\partial \bar{z}_1} f = 0, \ldots, \frac{\partial}{\partial \bar{z}_n} f = 0.$$

It is also interesting to consider this kind of moment defined for a more general radial measure than the area measure on the sphere. The following result is in that direction. First, let us recall the Euclidean Laplacian $\triangle$ can be written

$$\triangle = 4 \sum_{j=1}^{n} \frac{\partial^2}{\partial z_j \partial \bar{z}_j}.$$

Denote $S(n, |\mathbf{m}|, \mu)$ as the set of quotients of (nonzero) zeros of

$$F_{\mu,n,|\mathbf{m}|}(z) = \sum_{k=0}^{\infty} \frac{(-1)^k}{k!(k+\ell)!} A_k (-\tfrac{1}{2}z)^{2k+\ell}$$

where $\ell = n + |\mathbf{m}| - 1$ and

$$A_k = \int t^{2k+2\ell+1} \, d\mu(t).$$

Now we may state the following result.

Theorem 13 ([17]). *Let $f \in L^1_{\mathrm{loc}}(\mathbb{C}^n)$ and suppose there exist distinct positive real numbers r_1, r_2 such that*

$$\int_0^\infty \int_{|\mathbf{z}|=\rho} \mathbf{z}^{\mathbf{m}} f(\mathbf{w} + \mathbf{z}) \, d\sigma(\mathbf{z}) \, d\mu_r(\rho) = 0 \qquad (3.1)$$

for almost every $\mathbf{w} \in \mathbb{C}^n$ and $r = r_1, r_2$. Here $d\sigma$ represents surface measure on the sphere. If $\frac{r_1}{r_2} \notin S(n, |\mathbf{m}|, \mu)$, then f agrees almost everywhere with a function g satisfying

$$\left(\frac{\partial^{|\mathbf{m}|}}{\partial(\bar{\mathbf{z}}, \mathbf{z})^{\mathbf{m}}} \triangle^s \right) g = 0.$$

Remark 2. *In special cases, such as the area measure on the sphere or disk, the functions $F_{\mu,n,|\mathbf{m}|}$ reduce to Bessel functions, whose zero sets are thoroughly understood. In these cases it is possible to study criteria which involve only one radius and multiple moments, such as the two preceeding theorems do.*

Returning to the Heisenberg group and the study of moment conditions, let us consider the setting of $L^2(\mathbf{H}^n)$. There is a modification of the integral conditions considered above which allow us to conclude that a function is a CR function. In fact, moment conditions turn out to hold also. Further, with the introduction of conjugate moments (and corresponding differential forms) arises the possibility to conclude that a function is a conjugate CR function, that is, it satisfies the equations $Z_j f = 0$ for $j = 1, \ldots, n$.

Theorem 14 ([8]). *Let $f \in C^1(\mathbf{H}^n) \cap L^2(\mathbf{H}^n)$, and let $\mathbf{m} \in (\mathbb{Z}_+)^n$. Then f is a CR function if and only if f satisfies the following integral conditions: For all $\mathbf{g} \in \mathbf{H}^n$ and $k = 1, \ldots, n$*

$$\int_{|\mathbf{z}|=\rho} \mathbf{z}^\mathbf{m} L_\mathbf{g} f(\mathbf{z}, 0) \omega_k(\mathbf{z}) = 0$$

with $\omega_k(\mathbf{z}) = dz_1 \wedge \cdots \wedge dz_n \wedge d\bar{z}_1 \wedge \cdots \wedge \widehat{d\bar{z}_k} \wedge \cdots \wedge d\bar{z}_n$.

In the case where $-\mathbf{m} \in (\mathbb{Z}_+)^n$, then f is a conjugate-CR function (it satisfies the equations $Z_j f = 0$ for $j = 1, \ldots, n$) if and only if f satisfies the following integral conditions: For all $\mathbf{g} \in \mathbf{H}^n$ and $k = 1, \ldots, n$:

$$\int_{|\mathbf{z}|=\rho} \bar{\mathbf{z}}^\mathbf{m} L_\mathbf{g} f(\mathbf{z}, 0) \omega_k(\bar{\mathbf{z}}) = 0$$

with $\omega_k(\bar{\mathbf{z}}) = dz_1 \wedge \cdots \wedge \widehat{dz_k} \wedge \cdots \wedge dz_n \wedge d\bar{z}_1 \wedge \cdots \wedge d\bar{z}_n$.

Beyond expanding the integral conditions to include moments, it is also possible to consider more general radial measures, corresponding to the exploration of general radial measures in Theorem 13 for $\mathbb{C}^n$. In shifting from the differential forms of the above theorem to certain radial measures, we shift from the realm of Morera to the realm of Pompeiu. Other Heisenberg integral conditions involving radial measures include

(1) $\int_{|\mathbf{z}|=r} \mathbf{z}^\mathbf{m} L_\mathbf{g} f(\mathbf{z}, 0) \, d\sigma(\mathbf{z}) = 0$,

(2) $\int_{|\mathbf{z}|<r} \mathbf{z}^\mathbf{m} L_\mathbf{g} f(\mathbf{z}, 0) \, d\mu(\mathbf{z}) = 0$, also written as

$$\int_0^r \left(\int_{|\mathbf{z}|=s} \mathbf{z}^\mathbf{m} L_\mathbf{g} f(\mathbf{z}, 0) \, d\sigma(\mathbf{z}) \right) s^{2n-1} \, ds = 0,$$

(3) $\int_{r_1 < |\mathbf{z}| < r_2} \mathbf{z}^\mathbf{m} L_\mathbf{g} f(\mathbf{z}, 0) \, d\mu(\mathbf{z}) = 0$, also written as

$$\int_{r_1}^{r_2} \left(\int_{|\mathbf{z}|=s} \mathbf{z}^\mathbf{m} L_\mathbf{g} f(\mathbf{z}, 0) \, d\sigma(\mathbf{z}) \right) s^{2n-1} \, ds = 0,$$

(4) $\int_{|\mathbf{z}|=r} \mathbf{z}^\mathbf{m} L_\mathbf{g} f(\mathbf{z}, 0) \, d\sigma(\mathbf{z}) = |\Sigma_r^{2n-1}| f(\mathbf{g})$.

The above integral conditions all consider area measures on a given radial surface, with the last one describing a mean value. The following two conditions introduce radial polynomial weights onto the measures.

(5) $\int_{|\mathbf{z}|<r} \mathbf{z}^\mathbf{m} L_\mathbf{g} f(\mathbf{z}, 0) p(|\mathbf{z}|^2) \, d\mu(\mathbf{z}) = 0$, also written as

$$\int_0^r \left(\int_{|\mathbf{z}|=s} \mathbf{z}^\mathbf{m} L_\mathbf{g} f(\mathbf{z}, 0) \, d\sigma(\mathbf{z}) \right) p(s^2) s^{2n-1} \, ds = 0,$$

(6) $\int_{r_1<|\mathbf{z}|<r_2} \mathbf{z}^{\mathbf{m}} L_{\mathbf{g}} f(\mathbf{z},0) p(|\mathbf{z}|^2)\, d\mu(\mathbf{z}) = 0$, also written as

$$\int_{r_1}^{r_2} \left(\int_{|\mathbf{z}|=s} \mathbf{z}^{\mathbf{m}} L_{\mathbf{g}} f(\mathbf{z},0)\, d\sigma(\mathbf{z}) \right) p(s^2) s^{2n-1}\, ds = 0.$$

Theorem 15 ([17]). *Choose any one of the sets of integral conditions (1)–(6) and denote these integral conditions as (*). Let $f \in C^1(\mathbf{H}^n) \cap L^2(\mathbf{H}^n)$, and let $\mathbf{m} \in (\mathbb{Z}_+)^n$. Then $f \equiv 0$ if and only if f satisfies the integral conditions (*).*

Before considering the similar problem in the space $L^\infty(\mathbf{H}^n)$, we pause to consider the spaces $L^p(\mathbf{H}^n)$ for $2 < p < \infty$. We already know from Theorem 11 that two radii are required for the space L^∞. Recall that associated to the sphere $\{(\mathbf{z},0) \in \mathbf{H}^n : |\mathbf{z}| = r\}$ is an area measure σ_r. Observe the following convolution relationship in $\mathbf{H}^n$ between the function $\psi_{\nu,\nu}^\lambda$ and the measure σ_r. The T^n-spherical Laguerre function $\psi_{\nu,\nu}^\lambda$ is defined as

$$\psi_{\nu,\nu}^\lambda(\mathbf{z},t) = e^{2\pi i \lambda t} e^{-2\pi |\lambda| |\mathbf{z}|^2} \prod_{j=1}^n L_{\nu_j}^{(0)}(4\pi |\lambda| |z_j|^2) \quad (\lambda,\nu) \in \mathbb{R}^* \times (\mathbb{Z}_+)^n.$$

The convolution relation is given by

$$\left(\psi_{\nu,\nu}^\lambda(\mathbf{z},t) * \sigma_r(\mathbf{z}) \right)(\mathbf{w}) = c \psi_{\nu,\nu}^\lambda(r,0) \cdot \psi_{\nu,\nu}^\lambda(\mathbf{w},t).$$

By choosing λ_0 so that $\psi_{\nu,\nu}^{\lambda_0}(r,0) = 0$, the convolution becomes

$$\psi_{\nu,\nu}^{\lambda_0}(\mathbf{z},t) * \sigma_r(\mathbf{z}) = 0.$$

In this situation we may say that $\psi_{\nu,\nu}^{\lambda_0}$ is mean-periodic with respect to the measure σ_r. As a consequence of this relation, and noting that $\psi_{\nu,\nu}^{\lambda_0}(\mathbf{z},t) \in L^\infty(\mathbf{H}^n)$, we may conclude a single sphere fails the Pompeiu property in $L^\infty(\mathbf{H}^n)$. Similar results are valid for a sphere with a single moment factor. The measures associated with these moments on the sphere are written as

$$< \phi, \sigma_r^{\mathbf{m}} > = \int_{|\mathbf{z}|=r} \mathbf{z}^{\mathbf{m}} \phi(\mathbf{z})\, d\sigma(\mathbf{z}).$$

Then the convolution becomes

$$\left(\psi_{\nu,\nu}^\lambda(\mathbf{z},t) * \sigma_r^{\mathbf{m}}(\mathbf{z}) \right)(\mathbf{w}) = f(\lambda,r) \cdot \psi_{\nu+\mathbf{m},\nu}^\lambda(\mathbf{w},t).$$

Thus, we can choose λ_0 so that $f(\lambda_0, r) = 0$. In effect this yields a mean-periodicity relation, as before

$$\psi_{\nu,\nu}^{\lambda_0}(\mathbf{z},t) * \sigma_r^{\mathbf{m}}(\mathbf{z}) = 0.$$

So when moving to consider moments in $L^\infty(\mathbf{H}^n)$, it will not be enough to consider only one sphere with a moment.

In both cases, whether with or without a moment factor, it is important to ask at what point it is necessary to move from one sphere to two. The following theorems answer that question, and interestingly, a main point in the proofs consists of the above convolution relations.

Theorem 16 ([6]). *Let $r > 0$ and let $f \in L^p(\mathbf{H}^n)$ where $1 \le p < \infty$. Then $f = 0$ on $\mathbf{H}^n$ if and only if*

$$\int_{|\mathbf{z}|=r} L_{\mathbf{g}} f(\mathbf{z}, 0) \, d\sigma(\mathbf{z}) = 0.$$

The proof here is based on the Abel summability of the decomposition [18]:

$$f = \lim_{r \to 1^-} \sum_{|\mathbf{k}|=0}^{\infty} r^{|\mathbf{k}|} f * \psi_{\mathbf{k}, \mathbf{k}}^{\lambda}$$

in L^p- norm, for all $f \in L^p(\mathbf{H}^n)$. This summation formula is used in conjunction with the relation

$$\left(\psi_{\nu, \nu}^{\lambda}(\mathbf{z}, t) * \sigma_r(\mathbf{z}) \right)(\mathbf{w}) = \psi_{\nu, \nu}^{\lambda}(r, 0) \cdot \psi_{\nu, \nu}^{\lambda}(\mathbf{w}, t).$$

Since the $\psi_{\nu, \nu}^{\lambda}$ reappears after convolution with σ_r, the same form of the summation formula appears in the convolution $f * \sigma_r$, which is known to be zero. The summation formula is built from spectral projections, and it is then possible to show that each spectral projection is zero so that the function itself is zero.

A newer result including moment factors is the following

Theorem 17 ([19]). *Let $f \in L^p(\mathbf{H}^n)$ for $1 \le p < \infty$, and suppose $r > 0$. Then $f = 0$ on $\mathbf{H}^n$ if and only if*

$$\int_{|\mathbf{z}|=r} \mathbf{z}^{\mathbf{m}} L_{\mathbf{g}} f(\mathbf{z}, 0) \, d\sigma(\mathbf{z}) = 0 \quad \text{for all} \quad \mathbf{g} \in \mathbf{H}^n.$$

Similarly for the conjugate moment we may conclude $f = 0$ on $\mathbf{H}^n$ if and only if

$$\int_{|\mathbf{z}|=r} \bar{\mathbf{z}}^{\mathbf{m}} L_{\mathbf{g}} f(\mathbf{z}, 0) \, d\sigma(\mathbf{z}) = 0 \quad \text{for all} \quad \mathbf{g} \in \mathbf{H}^n.$$

The proof follows the same outline as that of the previous theorem, but there are some new difficulties in that the relation

$$\left(\psi_{\nu, \nu}^{\lambda}(\mathbf{z}, t) * \sigma_r^{\mathbf{m}}(\mathbf{z}) \right)(\mathbf{w}) = f(\lambda, r) \cdot \psi_{\nu+\mathbf{m}, \nu}^{\lambda}(\mathbf{w}, t)$$

introduces a shift in the indices. There is an additional problem with the Abel summation formula which is no longer valid for these functions $\psi_{\nu+\mathbf{m}, \nu}^{\lambda}$ in place of the $\psi_{\nu, \nu}^{\lambda}$. However this is remedied by considering convolution with the area measure on the sphere weighted by the conjugate moment. This additional convolution shifts the index back to ground level and allows the proof to proceed.

Finally, when we consider bounded functions in $\mathbf{H}^n$, the need for conditions on two radii returns. Furthermore, at this level the moments introduce differential conditions into the conclusions, as they had in the Euclidean case. All of the remaining results are from [19] and are based on methods displayed in [1] and further described in [10]. First consider a moment problem with spheres of two distinct radii.

Theorem 18. *Let* $f \in C(\mathbf{H}^n) \cap L^\infty(\mathbf{H}^n)$ *and suppose that for all* $\mathbf{g} \in \mathbf{H}^n$,

$$\int_{S(r_i)} \mathbf{z}^m L_\mathbf{g} f(\mathbf{z}, 0) \, d\sigma(\mathbf{z}) = 0 \quad for \quad i = 1, 2.$$

Assume that r_1 *and* r_2 *satisfy the following properties:*

(1) $(\frac{r_1}{r_2})^2 \notin \mathcal{Q}(L_{|\nu|}^{(|m|+n-1)}(x))$ *for all* $\nu \in (\mathbb{Z}_+)^n$,

(2) $\frac{r_1}{r_2} \notin \mathcal{Q}(J_{n+|m|-1}) = \{\frac{s_1}{s_2} : J_{n+|m|-1}(s_1) = J_{n+|m|-1}(s_2) = 0\}$.

Then we may conclude $(\bar{\mathbf{Z}}^{2m} f)(\mathbf{z}, t) = 0$.

Note that while the differential conditions are comparable to those for $\mathbb{C}^n$, there is a change from m to 2m. It is believed that this change is due to a limitation in the method, and further work is currently underway to establish the same results for m. The preceeding result can be specialized to a specific set of moments corresponding to CR functions, although a few additional steps are required in the proof.

Theorem 19. *Let* $f \in C(\mathbf{H}^n) \cap L^\infty(\mathbf{H}^n)$ *and suppose that for all* $\mathbf{g} \in \mathbf{H}^n$,

$$\int_{S(r_i)} z_1 L_\mathbf{g} f(\mathbf{z}, 0) \, d\sigma(\mathbf{z}) = \cdots = \int_{S(r_i)} z_n L_\mathbf{g} f(\mathbf{z}, 0) \, d\sigma(\mathbf{z}) = 0 \quad for \quad i = 1, 2.$$

Assume that r_1 *and* r_2 *satisfy the following properties:*

(1) $(\frac{r_1}{r_2})^2 \notin \mathcal{Q}(L_{|\nu|}^{(n)}(x))$ *for all* $\nu \in (\mathbb{Z}_+)^n, \lambda \in \mathbb{R}^*$,

(2) $\frac{r_1}{r_2} \notin \mathcal{Q}(J_n) = \{\frac{s_1}{s_2} : J_n(s_1) = J_n(s_2) = 0\}$.

Then we may conclude f *is a CR function.*

One of the advantages of the two moments theorem was the elimination of the requirement for two radii satisfying special conditions. The following results use $n + 1$ moments, as was the case for the moment theorem for $\mathbb{C}^n$. However the conclusions here are weakened to $\bar{\mathbf{Z}}^{2m} f$, as it happened in the previous theorems for $\mathbf{H}^n$.

Theorem 20. *Let* $f \in C(\mathbf{H}^n) \cap L^\infty(\mathbf{H}^n)$ *and suppose that for all* $\mathbf{g} \in \mathbf{H}^n$,

$$\int_{S(r)} \mathbf{z}^m L_\mathbf{g} f(\mathbf{z}, 0) \, d\sigma(\mathbf{z}) = 0$$

and

$$\int_{S(r)} \mathbf{z}^{\mathbf{m}+\mathbf{e}_j} L_{\mathbf{g}} f(\mathbf{z}, 0) \, d\sigma(\mathbf{z}) = 0 \quad for \quad j = 1, \ldots, n.$$

Then we may conclude $(\bar{\mathbf{Z}}^{2\mathbf{m}} f)(\mathbf{z}, t)$ *is a CR function.*

Once again there is a specific test for CR functions, requiring $n + 1$ specific moments.

Corollary 2. *Let* $f \in C(\mathbf{H}^n) \cap L^\infty(\mathbf{H}^n)$ *and suppose that for all* $\mathbf{g} \in \mathbf{H}^n$,

$$\int_{S(r)} L_{\mathbf{g}} f(\mathbf{z}, 0) \, d\sigma(\mathbf{z}) = 0$$

and

$$\int_{S(r)} z_j L_{\mathbf{g}} f(\mathbf{z}, 0) \, d\sigma(\mathbf{z}) = 0 \quad for \quad j = 1, \ldots, n.$$

Then we may conclude f *is a CR function.*

1.4   Additional questions

Many possibilities arise when considering what may still be done in any of these directions. For the moment problem in Clifford analysis, note that so far only one set of moments has been developed. It is interesting to consider whether certain other sets of moments could be used to obtain conclusions regarding other differential equations involving the Dirac operator. Likewise, it should be possible to consider the moments on more general radial surfaces.

In the Clifford algebra setting, work has been done on the Morera problem when the surface of integration is a general Jordan surface γ rather than only a disk or sphere [22]. It would be interesting to determine whether the same result holds for the Heisenberg group, though at present, the methods used are not well-adapted to non-radial surfaces. In the Clifford algebras it was also possible to consider Möbius transforms and the Morera theorem which appears in this setting. Of course, it would be interesting to investigate the same kind of issues in the Heisenberg setting.

Finally, there are many other results known in the Euclidean case, and it is of considerable interest to find out whether an extension of them to the Heisenberg group or to Clifford algebras is possible. For instance, could the kind of local theorems described in the paper [11] be valid if investigated in these settings? More particularly, for the Heisenberg group it would be nice to be able to investigate whether any result comparable to Proposition 1 could be established. This kind of result may help to determine whether the Pompeiu or Morera properties hold in the Heisenberg group for a larger variety of submanifolds.

REFERENCES

[1] M. Agranovsky, C. Berenstein, and D.C. Chang, Morera theorem for holomorphic H^p spaces in the Heisenberg group, *J. Reine Agnew. Math.* **443** (1993), 49–89.

[2] M. Agranovsky, C. Berenstein, D.C. Chang, and D. Pascuas, A Morera type theorem for L^2 functions in the Heisenberg group, *J. Analyse Math.* **57** (1992), 281–296.

[3] M. Agranovsky, C. Berenstein, D.C. Chang, and D. Pascuas, Injectivity of the Pompeiu transform in the Heisenberg group, *J. Analyse Math.* **63** (1994), 131–173.

[4] C. Berenstein, An inverse spectral theorem and its applications to the Pompeiu problem, *J. Analyse Math.* **37** (1980), 128–144.

[5] C. Berenstein, On the converse to Pompeiu's problem, *Notas e communicaçoes de Matematica* **73** (1976), Universidad Federal de Pernambuco, Recife, Brazil. (Reprinted as the technical report ISR-TR97-22 from ISR, University of Maryland.)

[6] C. Berenstein and D.C. Chang, Variations on the Pompeiu problem, *Proceedings of Hayama Symposium on Several Complex Variables*, eds. T. Ohsawa and J. Noguchi, (2000), pp. 9–26.

[7] C. Berenstein, D.C. Chang, and W. Eby, L^p results for the Pompeiu problem with moments on the Heisenberg group, (preprint).

[8] C. Berenstein, D.C. Chang, W. Eby, L. Zalcman, Moment versions of the Morera problem in $\mathbb{C}^n$ and $\mathbf{H}^n$, *Adv. in Appl. Math.*, to appear.

[9] C. Berenstein, D.C. Chang, D. Pascuas, L. Zalcman, Variations on the theorem of Morera, *Contemporary Mathematics*, Volume **137**, 1992, 63–78.

[10] C. Berenstein, D.C. Chang, and J. Tie, *Laguerre Calculus and its Applications in the Heisenberg Group*, AMS/IP Series in Advanced Mathematics #**22**, International Press, Cambridge, Massachusetts, 2001.

[11] C. Berenstein and R. Gay, A local version of the two circles theorem, *Israel J. Math.* **55** (1986), 267–288.

[12] C. Berenstein and R. Gay, Le probléme de Pompeiu local, *J. Analyse Math.* **52** (1989), 133–166.

[13] C. Berenstein and D. Pascuas, Morera and mean value type theorems, *Israel J. Math.* **86** (1994), 61–106.

[14] C. Berenstein and L. Zalcman, Pompeiu's problem on symmetric spaces, *Comment. Math. Helvetici* **55** (1980), 593–621.

[15] F. Brackx, R. Delanghe, and F. Sommen, *Clifford Analysis*, Pitman Research Notes in Mathematics **76**, 1982.

[16] L. Brown, B. Schreiber, and B.A. Taylor, Spectral synthesis and the Pompeiu problem, *Ann. Inst. Fourier*, Grenoble, **23**, 3 (1973), 125–154.

[17] D.C. Chang and W. Eby, Moment conditions for Pompeiu problem extended to general radial surfaces, to appear in *Microlocal Analysis and Complex Fourier Analysis*, eds. K. Fujita and T. Kawai, pp. 44–66.

[18] D.C. Chang and J. Tie, Estimates for spectral projection operators of the sub-Laplacian on the Heisenberg group, *J. Analyse Math.* **71** (1997), 315–347.

[19] W. Eby, *Moment Version of the Pompeiu Problem on the Heisenberg Group*, Ph.D. Dissertation, Univ. of MD, 2002.

[20] J. Gilbert and M. Murray *Clifford Algebras and Dirac Operators in Harmonic Analysis*, Cambridge University Press, 1991.

[21] E. Marmolejo Olea, *Morera Type Problems in Clifford Analysis*, Ph.D. Dissertation, Univ. of MD, 1999.

[22] E. Marmolejo Olea, Morera type problems in Clifford analysis, *Rev. Mat. Iberoam.* **17** (2001), 559–585.

[23] J. Ryan, Conformally covariant operators in Clifford analysis, *Proc. London Math. Soc.* **64** (1992), 70–94.

[24] L. Zalcman, Analyticity and the Pompeiu problem, *Arch. Rat. Mech. Anal.* **47** (1972), 237–254.

[25] L. Zalcman, Mean values and differential equations, *Israel J. Math.* **14** (1973), 339–352.

[26] L. Zalcman, Offbeat integral geometry, *Amer. Math. Monthly* **87** (1980), 161–175.

[27] L. Zalcman, an updated bibliography of the Pompeiu and Morera problems is available from him by direct request at zalcman@macs.biu.ac.il.

Carlos A. Berenstein
Department of Mathematics
University of Maryland
College Park, MD 20742
E-mail: Carlos@math.umd.edu

Der-Chen Chang
Department of Mathematics
Georgetown University
Washington, D.C. 20057
E-mail: chang@math.georgetown.edu

Wayne M. Eby[1]
Department of Mathematics
255 Hurley Building
Notre Dame, IN 46556
E-mail: weby@nd.edu

Submitted: March 4, 2003.

[1]New address from September 2003: Department of Mathematics, Temple University, Philadelphia, PA 19122, E-mail: eby@math.temple.edu.

2

Multidimensional Inverse Scattering Associated with the Schrödinger Equation

Swanhild Bernstein

ABSTRACT There is a deep connection between the theory of several complex variables and complex Clifford analysis. We will use a Borel–Pompeiu formula in $\mathbb{C}^n$ and the representation of holomorphic functions obtained in the context of Clifford analysis to study the inverse scattering problem for an n-dimensional Schrödinger-type equation. Equations are found for reconstructing the potential from scattering data purely by quadratures. The solution also helps elucidate the problem of characterizing admissible scattering data. Especially we do not need a "miraculous condition".

Keywords: Clifford analysis, multidimensional inverse scattering transform

2.1   Introduction

2.1.1   Opening remarks

The connection between nonlinear differential equations and certain linear spectral problems has motivated the study of various inverse problems. As it is shown in [2] and [3] by recasting the notion of scattering data as $\bar{\partial}$-data, it is possible to generalize the theory to higher dimensions. The use of the one-dimensional generalized Cauchy-formula in [1] which leads to n different reconstruction formulas causes the so-called "miraculous condition", i.e., the inverse data must be of that kind that all reconstruction formulas give one and the same potential. The $\bar{\partial}$ approach, which goes back to Beals and Coifman [2, 3] as well as Nachman and Ablowitz [8], does not need such a condition; neither does our Clifford analysis based approach. A first attempt to use Clifford analysis was made by Dix [6]. His approach replaces "holomorphic function" by "monogenic function" which is equivalent to the replacement of the $\bar{\partial}$-operator by the Dirac-operator D. But the inverse scattering problem is deeply connected with complex analysis which cannot be replaced. Therefore complex Clifford analysis, which is connected with

AMS Subject Classification: 30G35, 78A46.

(real) Clifford analysis as well as complex analysis in $\mathbb{C}$ and $\mathbb{C}^n$, is a good basis for studying inverse scattering problems. The product in Clifford analysis combines the scalar (or inner) product and the exterior (or wedge) product for vectors. For vectors $\mathbf{a}, \mathbf{b}$ the Clifford product $\mathbf{a}\,\mathbf{b} = -\mathbf{a} \cdot \mathbf{b} + \mathbf{a} \wedge \mathbf{b}$. This combination explains in an easy way the crucial role of compatibility conditions for the inverse scattering problem. The heart of the Clifford analysis approach is a generalized Cauchy–Green formula also called the Borel–Pompeiu formula which in fact contains the Martinelli–Bochner formula as well as the compatibility conditions. Essentially, this formula will be used for reconstruction purposes and the conditions under which the reconstruction is correct. We give a rigorous treatment of a regularized Schrödinger equation using a Borel–Pompeiu formula in $\mathbb{C}^n$ as well as the time-dependent and the time-independent Schrödinger equation where we use the Borel–Pompeiu formula in $\mathbb{R}^n$.

It should be emphasized that we do purely formal manipulations to demonstrate the way our method can be used. What we must have in mind is that the operator identities are valid only on certain domains of functions. This problem should be discussed elsewhere.

2.1.2   Real and complex Clifford analysis

We want to start with the real Clifford algebra $C\ell_{0,n}$ which is generated by the elements $\mathbf{e}_1, \mathbf{e}_2, \ldots, \mathbf{e}_n$. These generating elements fulfill the *non*-commutative multiplication rule

$$\mathbf{e}_i\mathbf{e}_j + \mathbf{e}_j\mathbf{e}_i = -2\delta_{ij}e_0, \quad i, j = 1, 2, \ldots, n,$$

the unit element of this algebra is denoted by e_0 and commutes with all generating elements, i.e., $e_0\mathbf{e}_i = \mathbf{e}_i e_0$, obviously $e_0^2 = 1$. Thus we can identify $e_0 \equiv 1$. An arbitrary element of the algebra is now a linear combination of all possible products of generating elements, more exactly, $a = \sum_\alpha a_\alpha \mathbf{e}_\alpha$, where $\mathbf{e}_\alpha = \mathbf{e}_{\alpha_1\alpha_2\ldots\alpha_k} = \mathbf{e}_{\alpha_1}\mathbf{e}_{\alpha_2}\ldots\mathbf{e}_{\alpha_k}$ and $1 \leq \alpha_1 < \alpha_2 < \ldots < \alpha_k \leq n$, $\alpha_j \in \{1, 2, \ldots, n\}$. We identify $\mathbf{e}_\emptyset = e_0$. The part $Sc(a) = a_0 e_0 = a_0$ is called the *scalar-part* and $Vec(a) = a - Sc(a)$ is called the *vector-part* of a.

We will identify elements of the Euclidean space with elements of the Clifford algebra in the following way:

$$\mathbb{R}^n \ni (x_1, x_2, \ldots, x_n) = \mathbf{x} = \sum_{j=1}^n x_j\mathbf{e}_j \in C\ell_{0,n}.$$

Especially for $\mathbf{a}, \mathbf{b} \in \mathbb{R}^n$ we have

$$\mathbf{a}\,\mathbf{b} = -\mathbf{a} \cdot \mathbf{b} + \mathbf{a} \wedge \mathbf{b},$$

where $\mathbf{a} \cdot \mathbf{b} = \mathbf{b} \cdot \mathbf{a} = \sum_{j=1}^n a_j b_j$ denotes the usual scalar product of two vectors and $\mathbf{a} \wedge \mathbf{b} = -\mathbf{b} \wedge \mathbf{a} = \frac{1}{2}(\mathbf{a}\mathbf{b} - \mathbf{b}\mathbf{a})$ denotes the wedge or exterior product. We introduce a first–order differential operator, the Dirac operator, which plays the role of the $\bar{\partial}$ operator in complex analysis:

$$D = \sum_{j=1}^{n} \mathbf{e}_j \frac{\partial}{\partial x_j}.$$

Functions f with $Df = 0$ will be called *(left)-monogenic*. Due to

$$DD = \left(\sum_{j=1}^{n} \mathbf{e}_j \frac{\partial}{\partial x_j} \right) \left(\sum_{j=1}^{n} \mathbf{e}_j \frac{\partial}{\partial x_j} \right) = -\sum_{j=1}^{n} \frac{\partial^2}{\partial x_j^2} = -\Delta$$

monogenic functions are harmonic. A fundamental solution of D is

$$\frac{-1}{\omega_{n-1}} \sum_{j=1}^{n} \frac{x_j \mathbf{e}_j}{|\mathbf{x}|^n} = \frac{-1}{\omega_{n-1}} \frac{\mathbf{x}}{|\mathbf{x}|^n},$$

where ω_{n-1} denotes the surface area of the unit sphere in $\mathbb{R}^n$. Using the fundamental solution we are able to state a (general) Cauchy formula.

Theorem 1. *Let G be a bounded domain in $\mathbb{R}^n$ with boundary Γ. Then*

$$\frac{-1}{\omega_{n-1}} \int_\Gamma \frac{(\mathbf{x} - \mathbf{y})}{|\mathbf{x} - \mathbf{y}|^n} \mathbf{n}(\mathbf{y}) u(\mathbf{y}) \, d\Gamma_\mathbf{y}$$

$$+ \frac{1}{\omega_{n-1}} \int_G \frac{(\mathbf{x} - \mathbf{y})}{|\mathbf{x} - \mathbf{y}|^n} (Du)(\mathbf{y}) \, d\mathbf{y} = \left\{ \begin{array}{ll} u(\mathbf{x}), & \mathbf{x} \in G \\ 0, & \mathbf{x} \in \mathbb{R}^n \backslash \{0\} \end{array} \right. .$$

In case of $G \equiv \mathbb{R}^n$ we have

$$\frac{1}{\omega_{n-1}} \int_{\mathbb{R}^n} \frac{(\mathbf{x} - \mathbf{y})}{|\mathbf{x} - \mathbf{y}|^n} (Du)(\mathbf{y}) \, d\mathbf{y} = u(\mathbf{x}) - u(\infty), \tag{1.1}$$

where $u(\infty) = \lim\limits_{|\mathbf{x}| \to \infty} u(\mathbf{x})$.

Suppose $u(\mathbf{x})$ is a **scalar-valued function** and equation (1.1) is equivalent to the system:

$$u(\infty) - \frac{1}{\omega_{n-1}} \int_{\mathbb{R}^n} \frac{(\mathbf{x} - \mathbf{y}) \cdot (Du)(\mathbf{y})}{|\mathbf{x} - \mathbf{y}|^n} \, d\mathbf{y} = u(\mathbf{x}), \tag{1.2}$$

$$\text{and} \quad \frac{1}{\omega_{n-1}} \int_{\mathbb{R}^n} \frac{(\mathbf{x} - \mathbf{y}) \wedge (Du)(\mathbf{y})}{|\mathbf{x} - \mathbf{y}|^n} \, d\mathbf{y} = 0, \tag{1.3}$$

where (1.2) gives the reconstruction formula and (1.3) describes conditions which have to be fulfilled for the reconstruction. Especially, (1.3) is equivalent to the compatibility conditions. We have

$$
0 = \frac{1}{\omega_{n-1}} \int_{\mathbb{R}^n} \frac{(\mathbf{x} - \mathbf{y}) \wedge (Du)(\mathbf{y})}{|\mathbf{x} - \mathbf{y}|^n} \, dy
$$

$$
= \frac{-1}{(n-2)\omega_{n-1}} \int_{\mathbb{R}^n} D\left(\frac{1}{|\mathbf{x} - \mathbf{y}|^{n-2}}\right) \wedge (Du)(\mathbf{y}) \, dy
$$

$$
= \frac{-1}{(n-2)\omega_{n-1}} \int_{\mathbb{R}^n} \frac{1}{|\mathbf{x} - \mathbf{y}|^{n-2}} \sum_{j=1, j<k}^n \left(\frac{\partial^2 u}{\partial x_j \partial x_k} - \frac{\partial^2 u}{\partial x_k \partial x_j}\right) \mathbf{e}_j \mathbf{e}_k \, dy
$$

and thus

$$
\frac{\partial^2 u}{\partial x_j \partial x_k} = \frac{\partial^2 u}{\partial x_k \partial x_j}, \quad \forall j, k, \ j \neq k.
$$

For our considerations we will also need complex Clifford analysis. We introduce an analogue to the $\bar{\partial}$ operator

$$
\overline{\mathcal{D}} = D_{\mathbf{x}} + i D_{\mathbf{y}},
$$

where $D_{\mathbf{x}} = \sum_{j=1}^n \mathbf{e}_j \frac{\partial}{\partial x_j}$ and $D_{\mathbf{y}} = \sum_{j=1}^n \mathbf{e}_j \frac{\partial}{\partial y_j}$. Then a holomorphic function $f(\mathbf{x}, \mathbf{y})$ in $\mathbb{C}^n$ fulfills the relation $\overline{\mathcal{D}} f = 0$ because

$$
\left(\frac{\partial}{\partial x_j} + i \frac{\partial}{\partial y_j}\right) f(\mathbf{x}, \mathbf{y}) = 0, \quad j = 1, 2, \ldots, n.
$$

We will identify

$$
(\mathbf{x}, \mathbf{y}) \in \mathbb{R}^{2n} \sim \mathbf{x} + i\mathbf{y} \in \mathbb{C}^n
$$
$$
\sim (x, y) \in Cl_{0,n}(\mathbb{R}) \times Cl_{0,n}(\mathbb{R})
$$
$$
\sim x + iy \in Cl_{0,n}(\mathbb{C}).
$$

This allows us to combine methods from real, complex and Clifford analysis. More information about Clifford analysis can be found in [5] and [7]. The connection between (complex) Clifford analysis and the theory of several complex variables is investigated in [9].

2.2   The $\overline{\mathcal{D}}$-operator

Even though the operator $\overline{\mathcal{D}}$ seems only to be a new "structure", we will see that the view of holomorphic functions in several complex variables analysis will be considerably improved. The operator $\overline{\mathcal{D}}$ is not an elliptic operator but if we restrict our consideration to scalar, i.e., complex-valued, functions, we are able to obtain a generalized Cauchy formula in $\mathbb{C}^n$ (also called Borel–Pompeiu formula). The proof is given in [4]. We have

Theorem 2. *Let $G \in \mathbb{R}^{2n}$ be a bounded strongly Lipschitz domain with boundary Γ and exterior unit normal $\mathbf{n} = \mathbf{n_x} + i\mathbf{n_y}$ (we identify $\mathbb{R}^{2n}$ with $\mathbb{C}^n$, thus our normal has a real part in the $\mathbf{x}$-direction and an imaginary part in the $\mathbf{y}$-direction) defined almost everywhere on Γ and $m \in C^1(\overline{G})$. Then we have for $(\mathbf{x}, \mathbf{y}) \in G$,*

$$m(\mathbf{x}, \mathbf{y}) - \frac{1}{\omega_{2n-1}} \int_G \frac{(\overline{\mathcal{D}}m)(\xi, \eta)(\xi - \mathbf{x}) - i(\eta - \mathbf{y})(\overline{\mathcal{D}}m)(\xi, \eta)}{(|\xi - \mathbf{x}|^2 + |\eta - \mathbf{y}|^2)^n} \, d\xi \, d\eta$$

$$= -\frac{1}{\omega_{2n-1}} \int_\Gamma m(\xi, \eta) \frac{[\mathbf{n}(\xi, \eta)(\xi - \mathbf{x}) - i(\eta - \mathbf{y})\mathbf{n}(\xi, \eta)]}{(|\xi - \mathbf{x}|^2 + |\eta - \mathbf{y}|^2)^n} \, d\Gamma,$$

where $(\mathbf{n_\xi} + i\mathbf{n_\eta}) \, d\Gamma = \mathbf{n} \, d\Gamma$. And in case of $G \equiv \mathbb{R}^n$ we have with

$$m(\infty) = \lim_{|\mathbf{x}|^2 + |\mathbf{y}|^2 \to \infty} m(\mathbf{x}, \mathbf{y}),$$

$$m(\infty) + \frac{1}{\omega_{2n-1}} \int_{\mathbb{R}^{2n}} \frac{(\overline{\mathcal{D}}m)(\xi, \eta)(\xi - \mathbf{x}) - i(\eta - \mathbf{y})(\overline{\mathcal{D}}m)(\xi, \eta)}{(|\xi - \mathbf{x}|^2 + |\eta - \mathbf{y}|^2)^n} \, d\xi \, d\eta$$

$$= m(\mathbf{x}, \mathbf{y}). \quad (2.1)$$

In fact (2.1) is a system of equations where the scalar part (2.2) gives the reconstruction formula itself and the vector part (2.3) gives the conditions under which the reconstruction can be done, i.e., (2.1) is equivalent to

$$m(\infty) +$$

$$\frac{1}{\omega_{2n-1}} \int_{\mathbb{R}^{2n}} \frac{(\overline{\mathcal{D}}m)(\xi, \eta) \cdot [(\xi - \mathbf{x}) - i(\eta - \mathbf{y})]}{(|\xi - \mathbf{x}|^2 + |\eta - \mathbf{y}|^2)^n} \, d\xi \, d\eta = m(\mathbf{x}, \mathbf{y}) \quad (2.2)$$

and

$$\frac{1}{\omega_{2n-1}} \int_{\mathbb{R}^{2n}} \frac{(\overline{\mathcal{D}}m)(\xi, \eta) \wedge [(\xi - \mathbf{x}) + i(\eta - \mathbf{y})]}{(|\xi - \mathbf{x}|^2 + |\eta - \mathbf{y}|^2)^n} \, d\xi \, d\eta = 0. \quad (2.3)$$

Moreover the conditions (2.3) can be rewritten as the compatibility conditions

$$\frac{\partial}{\partial \overline{k_l}} \frac{\partial}{\partial \overline{k_j}} m = \frac{\partial}{\partial \overline{k_j}} \frac{\partial}{\partial \overline{k_l}} m, \quad l \neq j.$$

2.3   History and principle of the Inverse Scattering Transform

The historical background of the problem is deeply connected with the connection of overdetermined linear systems and non-linear equations as it will be explained in the following example.

$$u_{xx}(x, t, z) + [q(x, t) + z^2] u(x, t, z) = 0,$$

$$u_t(x, t, z) = [z^2 + s(x, t)] u_x(x, t, z) - [iz^3 + r(x, t)] u(x, t, z).$$

Here z is a real or complex parameter; the coefficients q, r, s vanish rapidly as $|x| \rightarrow \infty$. Cross-differentiating, equating coefficients of powers of z, leads to the *compatibility* or *integrability conditions* linking the coefficients, which can be resolved for r and s in terms of q leaving the *Korteweg–de Vries equation*

$$q_t + \tfrac{3}{2}q_{xxx} + \tfrac{1}{4}qq_x = 0.$$

With the asymptotic condition $u(x, t, z) \sim e^{ixz}$ as $x \rightarrow -\infty$, u has the form

$$u(x, t, z) \sim a(z, t)e^{ixz} + b(z, t)e^{-ixz} \text{ as } x \rightarrow +\infty$$

and the time-evolution of the coefficients a and b is given by

$$a_t(z, t) = 0, \qquad b_t(z, t) = -2iz^3 b(z, t).$$

Thus: The scattering transform $q \mapsto (a, b)$ from the Schrödinger potential q to the asymptotic data (a, b) linearizes the KdV evolution.

This behavior is not exclusive for the KdV equation. Similar results are known for the following equations:

(i) Modified Korteweg–de Vries equation

$$u_t - 6u^2 u_x + u_{xxx} = 0.$$

(ii) Cylindrical Korteweg–de Vries equation

$$u_t + 6uu_x + u_{xxx} + \tfrac{1}{2t}u = 0.$$

(iii) Klein–Gordon equation

$$u_{xt} = e^{2u} - e^{-u}.$$

(iv) Sine–Gordon equation
$$u_{xt} = \sin u.$$

(v) Boussinesq equation

$$u_{tt} - u_{xx} + (u^2)_{xx} \pm u_{xxxx} = 0.$$

(vi) Non-linear Schrödinger equation

$$iu_t + u_{xx} \pm 2|u|^2 u = 0.$$

All these equations have the common properties that the non-linear equation is an integrability condition for the coefficients of an overdetermined linear system. If solutions of that linear system are normalized by their asymptotic behavior, then one can expect the scattering map—the relations satisfied by the asymptotics—to linearize the associated evolution. The analytic problem is to find a suitable normalization for solutions and to determine the scattering map and its inverse. To treat the analytic problem we consider the *direct problem* for a linear equation depending holomorphically on a complex parameter z :

$$P_z u = q(x)u,$$

where P_z is a differential operator in x. We look for Jost-type solutions with some kind of boundary condition:

$$u(x, z) = m(x, z)e_z(x), \qquad P_z e_x = 0, \qquad \lim_x m(x, z) = 1.$$

Then the corresponding equation for m is:

$$P_z m = q(x)m(x, z).$$

Let G_z be a fundamental solution to P_z and rewrite our problem as an integral equation of Lippmann–Schwinger type:

$$m(\,\cdot\,, z) = 1 + G_z[q\, m(\,\cdot\,, z)]$$

which has the formal solution

$$m(\,\cdot\,, z) = (I - G_z q)^{-1} 1.$$

Now the *inverse problem* consists in determining the $\bar{\partial}$-data

$$\bar{\partial}m = [\bar{\partial}, (I - G_z q)^{-1}]1 = (I - G_z q)^{-1}[\bar{\partial}, G_z]q(I - G_z q)^{-1}1$$

$$= (I - G_z q)^{-1}[\bar{\partial}, G_z]q\, m. \qquad (3.1)$$

Because P_z depends holomorphically on z, $\bar{\partial}m$ satisfies the same differential equation. If the $m(\,\cdot\,, z)$ are a complete set of solutions, then $\bar{\partial}m$ should be expressible as a combination of the m, i.e.,

$$\bar{\partial}m = Tm, \qquad (3.2)$$

where T is a multiplication or integral operator which depends (non-linearly) on q. In this case T should be computable from (3.1). With the asymptotic condition $m(x, z) \to 1$ as $z \to \infty$ and the *Cauchy transform* C we get

$$m = 1 + CTm \qquad \text{or, respectively,} \qquad m = (I - CT)^{-1}1.$$

T commutes with P_z and with left multiplication by functions of x. This ensures in practice, that T is determined by a functions of z alone.

Now, *define* functions m by $m = (I - CT)^{-1}1$; we get

$$P_z m = [P_z, (I - CT)^{-1}]1 = (I - CT)^{-1}[P_z, CT](I - CT)^{-1}1$$

$$= (I - CT)^{-1}[P_z, C]Tm.$$

Because P_z depends in an affine way on z, from the form of C we see that the commutator $[P_z, C]$ maps to functions which are independent of z. Thus if we set

$$q(x) = [P_z, C]Tm$$

we get

$$P_z m = (I - CT)^{-1}q = (I - CT)^{-1}[q1] = q(I - CT)^{-1}1 = qm.$$

Thus on a formal level the $\bar{\partial}$-approach provides an invertible correspondence between the operators $P_z - q$ and T.

How can this procedure be generalized to higher dimensions? Again we start with $P_{\bar{z}}u = q(\vec{x})u$ where $P_{\bar{z}}$ is a partial differential operator in $\vec{x}$ and $z \in \mathbb{C}^n$ is again equivalent to $P_{\bar{z}}m = q(\vec{x})m$ for Jost-type solutions

$$u(\vec{x}, \vec{z}) = m(\vec{x}, \vec{z})e_{\bar{z}}(\vec{x})$$

for $\vec{x} = (x_1, x_2, \ldots, x_n)$. Then with m, also $\bar{\partial}m$ is a solution. Again we can conclude if the solutions m form a complete set, then the solution $\bar{\partial}m$ should be expressible as

$$\bar{\partial}m = Tm \Leftrightarrow m - 1 = \bar{\partial}m = CTm \Leftrightarrow m = (I - CT)^{-1}1.$$

If T commutes with $P_{\bar{z}}$ and with left multiplication by functions of $\vec{x}$, we are able to define $m = (I - CT)^{-1}1$ and $q(\vec{x}) = [P_{\bar{z}}, C]Tm$ in the same way as before and we obtain $P_{\bar{z}}m = qm$.

What we need is an appropriate analogue to the Cauchy transform in complex analysis which is given by the Cauchy transform in complex Clifford analysis. In the next section we will demonstrate how the procedure works in higher dimensions.

2.4   Inverse scattering associated with a multidimensional generalized Schrödinger equation

In our considerations we follow [1]. We look for Jost-type eigenfunctions m of the form $ve^{i\mathbf{k}\cdot\mathbf{x}+|\mathbf{k}|^2t/\sigma}$ where $\mathbf{k} \in \mathbb{C}^n$, $\mathbf{k} \cdot \mathbf{x} = \sum_{j=1}^{n}(k_{R_j} + ik_{I_j})x_j$; then m fulfills

$$\sigma m_t + \Delta m + 2i\mathbf{k} \cdot \nabla m - um = 0. \tag{4.1}$$

According to [8] we use the Green's function

$$G(t, \mathbf{x}, \mathbf{k}) = \frac{1}{(2\pi)^{n+1}} \int_{\mathbb{R}^{n+1}} \frac{e^{i(t\tau + \mathbf{x} \cdot \xi)}}{i\sigma\tau - |\xi|^2 - 2\mathbf{k} \cdot \xi} \, d\xi \, d\tau. \tag{4.2}$$

A solution m of (4.1) can be obtained by solving the integral equation

$$m = 1 + \tilde{G}(um), \tag{4.3}$$

where $\tilde{G}$ denotes the convolution operator with kernel $G(t - t', x - x', k)$. We assume that (4.3) has no homogeneous solutions. Differentiating the integral equation for m with respect to $\bar{k}_j$ gives another solution of (4.1), which can be expressed nonlocally in terms of m using the symmetry property:

$$e^{-i\beta(t, \mathbf{x}, \mathbf{k}_R, \xi)} G(t, \mathbf{x}, \mathbf{k}_R, \mathbf{k}_I) = G(t, \mathbf{x}, \xi, \mathbf{k}_I) \tag{4.4}$$

$$\text{on} \quad s(\xi) = \left| \xi + \frac{\sigma_I}{\sigma_R} \mathbf{k}_I \right|^2 - \left| \mathbf{k}_R + \frac{\sigma_I}{\sigma_R} \mathbf{k}_I \right|^2 = 0$$

with $\beta(t, \mathbf{x}, \mathbf{k}_R, \mathbf{k}_I, \xi) = \left(\mathbf{x} + 2\frac{t}{\sigma_R} \mathbf{k}_I \right) \cdot (\xi - \mathbf{k}_R)$. We obtain the $\bar{\partial}_j$ derivative

$$\frac{\partial m}{\partial \bar{k}_j}(t, \mathbf{x}, \mathbf{k}) =$$

$$\frac{-(2\pi)^{-n}}{|\sigma_R|} \int_{\mathbb{R}^n} e^{i\beta(t, \mathbf{x}, \mathbf{k}, \xi)} (\xi_j - k_{R_j}) \delta(s(\xi)) T(\mathbf{k}_R, \mathbf{k}_I, \xi) m(t, \mathbf{x}, \xi, \mathbf{k}_I) \, d\xi \tag{4.5}$$

where the scattering data are found to be

$$T(\mathbf{k}_R, \mathbf{k}_I, \xi) = \int_{\mathbb{R}^{n+1}} e^{-i\beta(t, \mathbf{x}, \mathbf{k}, \xi)} u(t, \mathbf{x}) m(t, \mathbf{x}, \mathbf{k}) \, dt \, d\mathbf{x} \tag{4.6}$$

which defines the scattering data T. Here, following [1] we have used the $\bar{\partial}$ method introduced in [2]. Now, our improvement comes by using Clifford analysis. Applying the general Cauchy formula of Theorem 2 together with $m \to 1$ as $|k| \to \infty$ we get

$$m(t, \mathbf{x}, \mathbf{k}_R, \mathbf{k}_I) = 1 +$$

$$\frac{1}{\omega_{2n-1}} \int_{\mathbb{R}^{2n}} \frac{(\overline{\mathcal{D}}m)(\mathbf{k}_R', \mathbf{k}_I')(\mathbf{k}_R' - \mathbf{k}_R) - i(\mathbf{k}_I' - \mathbf{k}_I)(\overline{\mathcal{D}}m)(\mathbf{k}_R, \mathbf{k}_I')}{(|\mathbf{k}_R' - \mathbf{k}_R|^2 + |\mathbf{k}_I' - \mathbf{k}_I|^2)^n} \, d\mathbf{k}_R' \, d\mathbf{k}_I'.$$

Using Theorem 2 we get a new motivation to follow the idea of M.J. Ablowitz and A.I. Nachman to derive the "characterization equations" from the compatibility conditions

$$\frac{\partial}{\partial \bar{k}_l} \frac{\partial}{\partial \bar{k}_j} m = \frac{\partial}{\partial \bar{k}_j} \frac{\partial}{\partial \bar{k}_l} m, \quad l \neq j.$$

Following [1, 8] we get a non-trivial non-linear equation for T :

$$\mathcal{L}_{ij}[T] = (\xi_j - k_{R_j})\left(\frac{\partial T}{\partial \overline{k}_i} + \frac{1}{2}\frac{\partial T}{\partial \xi_i}\right) - (\xi_i - k_{R_i})\left(\frac{\partial T}{\partial \overline{k}_j} + \frac{1}{2}\frac{\partial T}{\partial \xi_j}\right)$$

$$= \int_{\mathbb{R}^n} [(\eta_j - k_{R_j})(\xi_i - \eta_i) - (\eta_i - k_{R_i})(\xi_j - \eta_j)]\delta(s(\eta))$$

$$\times T(\mathbf{k}_R, \mathbf{k}_I, \eta)T(\mathbf{k}_R, \mathbf{k}_I, \xi)\,d\xi = \mathcal{N}_{ij}[T]\,.$$

To avoid redundancies we consider only $i = 1$ and introduce new variables $(z, w, w_0) \in \mathbb{C}^{n-1} \times \mathbb{R}^n \times \mathbb{R}$, where $z = (z_2, \dots, z_n)$:

$$k_{R_1} = \sum_{j=2}^{n} w_j z_{R_j} - \frac{1}{2}w_1 - \frac{1}{2}\frac{\sigma_I w_0 w_1}{w^2}, \qquad k_{R_j} = -w_1 z_{R_j} - \frac{1}{2}w_j - \frac{1}{2}\frac{\sigma_I w_0 w_j}{w^2},$$

$$k_{I_1} = \sum_{j=2}^{n} w_j z_{I_j} + \frac{1}{2}\frac{\sigma_R w_0 w_1}{w^2}, \qquad k_{I_j} = -w_1 z_{I_j} + \frac{1}{2}\frac{\sigma_R w_0 w_j}{w^2},$$

$$\xi_1 = \sum_{j=2}^{n} w_j z_{R_j} + \frac{1}{2}w_1 - \frac{1}{2}\frac{\sigma_I w_0 w_1}{w^2}, \qquad \xi_j = -w_1 z_{R_j} + \frac{1}{2}w_j - \frac{1}{2}\frac{\sigma_I w_0 w_j}{w^2}.$$

If $w_1 \neq 0$, there exists a one-to-one mapping $(\mathbf{k}_R, \mathbf{k}_I, \xi) \rightarrow (\mathbf{z}, \mathbf{w}, w_0)$ such that

$$\mathbf{w} = \xi - \mathbf{k}_R, \quad w_0 = \frac{2}{\sigma_R}\mathbf{k}_I \cdot (\xi - \mathbf{k}_R), \quad \frac{\partial}{\partial \overline{z}_j} = \mathcal{L}_{1j},$$

and, for $i = 1, j = 2, \dots, n$, we obtain

$$\frac{\partial T}{\partial \overline{z}_j} = \mathcal{N}_{1j}[T](\mathbf{z}, \mathbf{w}, w_0).$$

Applying the $\overline{\mathcal{D}}$ operator of Clifford analysis we are able to write

$$\overline{\mathcal{D}}T = \mathcal{N}_1[T](\mathbf{z}, \mathbf{w}, w_0), \tag{4.7}$$

where

$$\mathcal{N}_1[T](\mathbf{z}, \mathbf{w}, w_0) = \sum_{j=2}^{n} \mathcal{N}_{1j}[T](\mathbf{z}, \mathbf{w}, w_0)\mathbf{e}_j$$

$$\text{and} \quad \overline{\mathcal{D}}T = \sum_{j=2}^{n} \mathbf{e}_j\left(\frac{\partial}{\partial x_j} + i\frac{\partial}{\partial y_j}\right)T.$$

Using Theorem 2 we have that if $T(\mathbf{z}, \mathbf{w}, w_0)$ fulfill the compatibility conditions, which is equivalent to a vanishing vector-part in the Borel–Pompeiu formula, we can reconstruct the potential by using the scalar-part of the Borel–Pompeiu formula:

$$I[T](\mathbf{z}, \mathbf{w}, w_0) := T(\mathbf{z}, \mathbf{w}, w_0)$$

$$-\frac{1}{\omega_{2n-3}} \int_{\mathbb{R}^{2n-2}} \frac{\mathcal{N}_1[T](\zeta, \mathbf{w}, w_0) \cdot [(\zeta_R - \mathbf{z}_R) - i(\zeta_I - \mathbf{z}_I)]}{(|\zeta_R - \mathbf{z}_R|^2 + |\zeta_I - \mathbf{z}_I|^2)^{n-1}} \, d\zeta_R \, d\zeta_I$$

$$= T(\infty, \mathbf{w}, w_0) = \hat{u}(w_0, \mathbf{w}). \quad (4.8)$$

Due to the fact that $|\mathbf{z}| \to \infty$, also $|\mathbf{k}| \to \infty$ and

$$m(t, \mathbf{x}, \mathbf{k}_R, \mathbf{k}_I) = 1 + \mathcal{O}(|\mathbf{k}|^{-1}).$$

Thus we have $m \to 1$ as $|\mathbf{k}| \to \infty$,

$$\lim_{|\mathbf{k}| \to \infty} T(\mathbf{k}_R, \mathbf{k}_I, \xi)$$

$$= \lim_{|\mathbf{k}| \to \infty} \int_{\mathbb{R}^{n+1}} e^{-i\beta(t, \mathbf{x}, \mathbf{k}_R, \mathbf{k}_I, \xi)} u(t, \mathbf{x}) m(t, \mathbf{x}, \mathbf{k}_R, \mathbf{k}_I) \, d\mathbf{x} \, dt$$

$$= \lim_{|\mathbf{k}| \to \infty} \int_{\mathbb{R}^{n+1}} e^{-i(\mathbf{x} \cdot \mathbf{w} + t w_0)} u(t, \mathbf{x}) m(t, \mathbf{x}, \mathbf{k}_R, \mathbf{k}_I) \, d\mathbf{x} \, dt$$

$$= T(\infty, \mathbf{w}, w_0) = \hat{u}(w_0, \mathbf{w}).$$

Here $\hat{u}$ denotes the Fourier transform of u. Thus we have

Theorem 3. *If the homogenous equation of (4.3) has no solutions and the data T fulfill (4.7) as well as the compatibility conditions and decay properly if $|k| \to \infty$, then the potential u can be computed by taking the inverse Fourier transform of $I[T]$ (see (4.8)).*

REFERENCES

[1] M.J. Ablowitz and P.A. Clarkson, *Solitons, Nonlinear Evolution Equations and Inverse Scattering*, London Mathematical Society Lecture Note Series **149** (Cambridge: Cambridge University Press), 1991.

[2] R. Beals and R.R. Coifman, Multidimensional inverse scattering and nonlinear partial differential equations, *Microlocal Analysis* (Proc. Symp. Pure Math. **43**) (Providence, RI: AMS) (1985), 45–70.

[3] R. Beals and R.R. Coifman, Linear spectral problems, non-linear equations and the $\bar{\partial}$-method, *Inverse Problems* **5** (1989), pp. 87–130.

[4] S. Bernstein, A Borel–Pompeiu formula in $\mathbb{C}^n$ and its application to inverse scattering theory, in: *Clifford Algebras and their Applications in Mathematical Physics*, Vol. 2 (Progress in Physics **19**) Eds. W. Sprössig, J. Ryan, Birkhäuser, Boston, 2000, pp. 117–134.

[5] F. Brackx, R. Delanghe and F. Sommen, *Clifford Analysis*, Pitman Research Notes in Mathematics **76** (Boston: Pitman), 1982.

[6] D.B. Dix, Applications of Clifford analysis to inverse scattering for the linear hierarchy in several space dimensions, in: *Clifford Algebras in Analysis and Related Topics*, Studies in Advanced Mathematics, Ed. J. Ryan, CRC Press, Boca Raton, 1996, 261–284.

[7] K. Gürlebeck and W. Sprössig, *Quaternionic Analysis and Elliptic Boundary Value Problems*, International Series of Numerical Mathematics (ISNM) Vol. 89, Birkhäuser, Basel, 1990.

[8] A.I. Nachman and M.J. Ablowitz, A multidimensional inverse-scattering method, *Stud. Appl. Math.* **71** (1984), 243–250.

[9] R. Rocha-Chavez, M.V. Shapiro and F. Sommen, *Integral Theorems for Functions and Differential Forms in* $\mathbb{C}^m$, CRC Research Notes in Mathematics **428**, Chapman & Hall/CRC, Boston, 2001.

Swanhild Bernstein
Mathematische Optimierung
Bauhaus-Universität Weimar
Coudraystr. 13B
D-99421 Weimar
Weimar
E-mail: swanhild.bernstein@bauing.uni-weimar.de

Submitted: August 31, 2002.

3

On Discrete Stokes and Navier–Stokes Equations in the Plane

Klaus Gürlebeck and Angela Hommel

ABSTRACT The main goal of the paper is to apply the theory of discrete analytic functions to the solution of Dirichlet problems for the Stokes and Navier–Stokes equations, respectively. The Cauchy–Riemann operator will be approximated by certain finite difference operators. Approximations of the classical T-operator as well as for the Bergman projections are constructed in such a way that the algebraic properties of the operators from complex function theory remain valid. This is used to approximate the solutions to the boundary value problems by adapted finite difference schemes.

Keywords: finite difference Cauchy–Riemann operators, discrete function theory, boundary value problems

3.1   Introduction

The idea to calculate solutions to boundary value problems by finite difference approximations is one of the major approaches to numerical methods. There are different lines to construct adapted methods. Conservation laws have to be preserved in the discrete model, variational principles are reflected by some finite difference schemes, and last but not least approximation (truncation error) and numerical stability play important roles. Beginning in the 1940s first attempts were made to consider solutions of finite difference approximations to the Cauchy–Riemann equations as a class of discrete holomorphic (discrete analytic) functions and to establish a discrete function theory ([2, 4, 5, 11]). The main problems were the definition of a discrete analogue to the Cauchy integral (and the equivalent definition of the discrete analytic functions by this discrete Cauchy integral) as well as the property that discrete analytic functions do not form an algebra with the usual complex multiplication. Another problem was to find a factorization of the 2-dimensional real Laplacian by two (adjoint) discrete Cauchy–Riemann operators. So, for a long time there was no essential progress in discrete function theories in

AMS Subject Classification: 30G25, 31C20.

the described sense.

A series of works in Clifford analysis has shown that a commutative algebra is not necessary to adapt a lot of function theoretic methods to the solution of boundary value problems and to find representation formulas of the solution. A survey and a collection of examples is for instance contained in [10] and [14]. Also in [10], and later in [9] first ideas for the development and the application of discrete function theories are presented. This work was inspired by analogous ideas to develop a discrete potential theory ([3, 15, 17]). A first serious problem in the three dimensional case is to find discrete Cauchy–Riemann operators that factorize exactly the standard seven-point Laplacian. In [9] this problem was solved and the approximate factorization defined in [10] was improved. For the definition of the discrete T-operator and a discrete Cauchy integral we need a fundamental solution of the discrete Cauchy–Riemann operator. Using the factorization of the Laplacian one can work based on the fundamental solution of the Laplacian. This fundamental solution is well studied (see e.g., [1, 6–8, 13, 16, 18–20]). An additional problem in the two-dimensional case is related to the logarithmic behaviour of the fundamental solution at infinity. This asymptotic behaviour complicates the error and norm estimates in exterior domains and requires a careful definition of Cauchy integral and T-operator. We have to prove analogous algebraic properties of the operators. In detail we discuss in Section 2 that the discrete T-operator is a right-inverse of the introduced finite difference Cauchy–Riemann operator. A discrete Borel–Pompeiu formula is discussed and a suitable orthogonal decomposition of the l_2-space is the basis to define discrete Bergman projections. Applying these tools with analogous properties as in the continuous case, we discuss finite difference approximations of the steady state Stokes and Navier–Stokes equations in Sections 3.3 and 3.4.

3.2   Elements of discrete function theory in the plane

Let $\mathbb{R}^2$ be the Euclidean space with the unit vectors e_i, $i = 1, 2$ and G be a bounded domain. We consider functions $f(mh) = (f_0(mh), f_1(mh))$ for fixed mesh width $h > 0$ on the lattice

$$\mathbb{R}_h^2 = \{mh = (m_1h, m_2h) : m_i \in \mathbb{Z}\}$$

and denote the discrete domain by $G_h = (G \cap \mathbb{R}_h^2)$. Using the notation

$$K = \{k_1(1,0), \ k_2 = (0,1), \ k_3 = (-1,0), \ k_4 = (0,-1)\}$$

we define the discrete boundary

$$\gamma_h^- = \{rh \in \mathbb{R}_h^2 \setminus G_h : \ \exists k_i \text{ with } (r + k_i)h \in G_h, \ i = 1, \ldots, 4\}.$$

Often the boundary γ_h^- is split into the parts

$$\gamma_{hi}^- = \{rh \in \gamma_h^- : (r + k_i)h \in G_h\}, \ i = 1, \ldots, 4.$$

At first we present an orthogonal decomposition of the space $l_2(G_h)$ of functions $w(mh) = (w_0(mh), w_1(mh))$ and $v(mh) = (v_0(mh), v_1(mh))$ with the scalar product

$$< w, v > = \sum_{mh \in G_h} h^2 \begin{pmatrix} w_0(mh) \\ w_1(mh) \end{pmatrix}^T \begin{pmatrix} v_0(mh) \\ v_1(mh) \end{pmatrix}.$$

Outside of G_h we extend our functions by zero. Then the finite difference operators $D_h^{\pm j}$ are defined as

$$D_h^j = h^{-1}(U_j - I), \quad U_j w_i(mh) = w_i(mh + he_j),$$
$$D_h^{-j} = h^{-1}(I - U_{-j}), \quad U_{-j} w_i(mh) = w_i(mh - he_j),$$

for $j \in \{1, 2\}$ and $i \in \{0, 1\}$. If the boundary conditions

$$w(rh) = v(rh) = (0, 0), \quad \forall rh \in \gamma_h^-,$$

are satisfied, then we introduce the space $\overset{\circ}{w}_2^1(G_h)$ with the norm

$$\|f\|_{\overset{\circ}{w}_2^1(G_h)} = \left(\|f\|_{l_2(G_h)}^2 + \sum_{j=1}^2 \|D_h^{\pm j} f(mh)\|_{l_2(G_h)}^2 \right)^{\frac{1}{2}}.$$

Theorem 1. *The space $l_2(G_h)$ permits the orthogonal decomposition*

$$l_2(G_h) = \ker D_{h,M}^1(G_h) \oplus D_{h,M}^2(\overset{\circ}{w}_2^1(G_h)),$$

where

$$D_{h,M}^1 = \begin{pmatrix} D_h^{-2} & D_h^1 \\ -D_h^{-1} & D_h^2 \end{pmatrix} \quad and \quad D_{h,M}^2 = \begin{pmatrix} D_h^2 & -D_h^1 \\ D_h^{-1} & D_h^{-2} \end{pmatrix}.$$

Proof. Let $w(mh) \in \overset{\circ}{w}_2^1(G_h)$ and $v(mh) \in D_{h,M}^2(\overset{\circ}{w}_2^1(G_h))$. At first we prove

$$D_{h,M}^2(\overset{\circ}{w}_2^1(G_h))^\perp \subset \ker D_{h,M}^1(G_h).$$

Suppose that

$$< w, v >= 0 \quad \text{for all} \quad v(mh) \in D_{h,M}^2(\overset{\circ}{w}_2^1(G_h)).$$

Then, for

$$s(mh) = (s_0(mh), s_1(mh)) \in \overset{\circ}{w}_2^1(G_h)$$

we get that $< w, I_N D_{h,M}^2 s >= 0$, where N denotes the number of mesh points $mh \in G_h$ and I_N denotes the $N \times N$ identity matrix. If we write the scalar product in the form

$$0 = \sum_{mh \in G_h} h^2 [w_0(mh) D_h^2 s_0(mh) + w_1(mh) D_h^{-1} s_0(mh)$$

$$+ (-w_0(mh) D_h^1 s_1(mh) + w_1(mh) D_h^{-2} s_1(mh))]$$

and change the index of summation, then we obtain for functions $w, s \in \overset{\circ}{w}{}^1_2(G_h)$:

$$\sum_{mh \in G_h} h^2 w_i(mh) \left[D_h^{\mp j} s_k(mh) \right] = - \sum_{mh \in G_h} h^2 \left[D_h^{\pm j} w_i(mh) \right] s_k(mh) \quad (2.1)$$

with $j \in \{1, 2\}$ and $i, k \in \{0, 1\}$. We conclude

$$
\begin{aligned}
0 = &< w, I_N D^2_{h,M} s > \\
= &- \sum_{mh \in G_h} h^2 [D_h^{-2} w_0(mh) + D_h^1 w_1(mh)) s_0(mh) \\
& + (-D_h^{-1} w_0(mh) + D_h^2 w_1(mh)) s_1(mh)] = - < I_N D^1_{h,M} w, s > .
\end{aligned}
$$

Because the equality should be fulfilled for all $s(mh) \in \overset{\circ}{w}{}^1_2(G_h)$, it follows that $w(mh) \in \ker D^1_{h,M}(G_h)$. We prove now that also

$$\ker D^1_{h,M}(G_h) \subset D^2_{h,M}(\overset{\circ}{w}{}^1_2(G_h))^\perp.$$

For $w(mh) \in \ker D^1_{h,M}(G_h)$ we obtain analogously

$$< w, v > = < w, I_N D^2_{h,M} s > = - < I_N D^1_{h,M} w, s > = 0.$$

From both parts of the proof we get that

$$\ker D^1_{h,M}(G_h) = D^2_{h,M}(\overset{\circ}{w}{}^1_2(G_h))^\perp.$$

We remark that the operators $D^1_{h,M}$ and $D^2_{h,M}$ approximate the Cauchy–Riemann operators

$$D^1 = (-i)\left(\frac{\partial}{\partial x_1} + i \frac{\partial}{\partial x_2} \right) \quad \text{and} \quad D^2 = i \left(\frac{\partial}{\partial x_1} - i \frac{\partial}{\partial x_2} \right)$$

with $x = (x_1, x_2) \in \mathbb{R}^2$ and $i^2 = -1$. Consequently, functions belonging to the kernel of $D^1_{h,M}$ will be called *discrete analytic functions*. In analogy to the continuous case these operators have the property

$$D^1_{h,M} D^2_{h,M} = D^2_{h,M} D^1_{h,M} = I_2 \Delta_h,$$

where I_2 denotes the 2×2 identity matrix and $\Delta_h = D_h^1 D_h^{-1} + D_h^2 D_h^{-2}$ is a discrete Laplacian.

Lemma 1. *Let* $w, s \in \overset{\circ}{w}{}^1_2(G_h)$. *Then we have*

$$< w, I_N D^2_{h,M} s > = - < I_N D^1_{h,M} w, s > \qquad \text{and}$$
$$< w, I_N D^1_{h,M} s > = - < I_N D^2_{h,M} w, s > .$$

The first property is already shown in the proof of Theorem 1. Using formula (2.1) the second property can be proved analogously.

Let

$$\delta_h(mh) = \begin{cases} h^{-2}, & \text{for } mh = (0,0); \\ 0, & \text{for } mh \neq (0,0); \end{cases}$$

and $E_h^1(mh)$ with

$$\begin{pmatrix} D_h^{-2} & D_h^1 \\ -D_h^{-1} & D_h^2 \end{pmatrix} \begin{pmatrix} E_{h11}^1(mh) & E_{h12}^1(mh) \\ E_{h21}^1(mh) & E_{h22}^1(mh) \end{pmatrix} = \begin{pmatrix} \delta_h(mh) & 0 \\ 0 & \delta_h(mh) \end{pmatrix}$$

be the discrete fundamental solution of the operator $D_{h,M}^1$. We define a discrete analogue to the complex T-operator

$$(T_h^1[f_0, f_1])(mh) = (\,(T_{h1}^1[f_0, f_1])(mh), (T_{h2}^1[f_0, f_1])(mh)\,).$$

The components of T_h^1 are represented by

$$(T_{hk}^1[f_0, f_1])(mh) = (T_{hk}^{1,G}[f_0, f_1])(mh) + (T_{hk}^{1,\gamma_h^-}[f_0, f_1])(mh)$$

with

$$(T_{hk}^{1,G}[f_0, f_1])(mh) = \sum_{lh \in G_h} h^2 \begin{pmatrix} E_{hk1}^1(mh - lh) \\ E_{hk2}^1(mh - lh) \end{pmatrix}^T \begin{pmatrix} f_0(lh) \\ f_1(lh) \end{pmatrix}$$

and

$$(T_{hk}^{1,\gamma_h^-}[f_0, f_1])(mh) = \sum_{lh \in \gamma_{h1}^- \cup \gamma_{h4}^- \cup \Gamma_{14}} h^2 \begin{pmatrix} E_{hk1}^1(mh - lh) \\ E_{hk2}^1(mh - lh) \end{pmatrix}^T \begin{pmatrix} f_0(lh) \\ 0 \end{pmatrix}$$

$$+ \sum_{lh \in \gamma_{h2}^- \cup \gamma_{h3}^- \cup \Gamma_{23}} h^2 \begin{pmatrix} E_{hk1}^1(mh - lh) \\ E_{hk2}^1(mh - lh) \end{pmatrix}^T \begin{pmatrix} 0 \\ f_1(lh) \end{pmatrix}$$

for $k = 1, 2$. In the last formula inner corners are counted in the union of special parts of γ_h^- only once. In addition the special outer corners

$$\Gamma_{14} = \{lh \in \mathbb{R}_h^2 \setminus (G_h \cup \gamma_h^-) : (l + k_4)h \in \gamma_{h1}^- \quad \text{and} \quad (l + k_1)h \in \gamma_{h4}^-\}$$

and

$$\Gamma_{23} = \{lh \in \mathbb{R}_h^2 \setminus (G_h \cup \gamma_h^-) : (l + k_3)h \in \gamma_{h2}^- \quad \text{and} \quad (l + k_2)h \in \gamma_{h3}^-\}$$

are considered where $f(mh)$ is defined by zero. Therefore, these points are only important if discrete tangential derivatives are used.

Lemma 2. *For an arbitrary function f it holds that*

$$D_{h,M}^1 (T_h^1[f_0, f_1])(mh) = f(mh) \text{ for all } mh \in G_h\,.$$

The proof of this lemma is published in [12]. We remark that we can find a similar structure for $T_h^2[f_0, f_1](mh)$ such that

$$D_{h,M}^2 \left(T_h^2[f_0, f_1]\right)(mh) = f(mh).$$

Based on Lemma 2 a discrete Borel–Pompeiu formula can be established. We describe normal unit vectors by using the homeomorphism between complex numbers $a + ib$ and the matrices $\begin{pmatrix} a & -b \\ b & a \end{pmatrix}$ and write $\begin{pmatrix} n_1^i & n_2^i \\ n_3^i & n_4^i \end{pmatrix}$ on γ_{hi}^-, $i = 1, \ldots, 4$ with

$$\begin{pmatrix} n_1^1 & n_2^1 \\ n_3^1 & n_4^1 \end{pmatrix} = -\begin{pmatrix} n_1^3 & n_2^3 \\ n_3^3 & n_4^3 \end{pmatrix} = \begin{pmatrix} -1 & 0 \\ 0 & -1 \end{pmatrix} \quad \text{and}$$

$$\begin{pmatrix} n_1^2 & n_2^2 \\ n_3^2 & n_4^2 \end{pmatrix} = -\begin{pmatrix} n_1^4 & n_2^4 \\ n_3^4 & n_4^4 \end{pmatrix} = \begin{pmatrix} 0 & 1 \\ -1 & 0 \end{pmatrix}.$$

Lemma 3. *Let $k = 1, 2$. Then we have a discrete Borel–Pompeiu formula*

$$(T_{hk}^1[D_h^{-2}f_0 + D_h^1 f_1, -D_h^{-1}f_0 + D_h^2 f_1])(mh)$$
$$+ (F_{hk}^1[f_0, f_1])(mh) = f_k(mh)$$

with

$$f_1(mh) = \begin{cases} f_0(mh), & mh \in (G_h \cup \gamma_{h1}^- \cup \gamma_{h2}^-); \\ 0, & mh \notin (G_h \cup \gamma_{h1}^- \cup \gamma_{h2}^-); \end{cases}$$

$$f_2(mh) = \begin{cases} f_1(mh), & mh \in (G_h \cup \gamma_{h3}^- \cup \gamma_{h4}^-); \\ 0, & mh \notin (G_h \cup \gamma_{h3}^- \cup \gamma_{h4}^-); \end{cases}$$

and

$$(F_{hk}^1[f_0, f_1])(mh) = (F_{hk}^{1,\gamma_h^-}[f_0, f_1])(mh) + (F_{hk}^{1,\gamma_I^-}[f_0, f_1])(mh),$$

where

$$F_{hk}^{1,\gamma_h^-}[f_0, f_1])(mh)$$
$$= i \sum_{j=1}^4 \sum_{lh \in \gamma_{hj}^-} h \begin{pmatrix} E_{hk1}^1(mh - lh) \\ E_{hk2}^1(mh - lh)^T \end{pmatrix} \begin{pmatrix} n_1^j & n_2^j \\ n_3^j & n_4^j \end{pmatrix} \begin{pmatrix} f_0(lh) \\ f_-(lh) \end{pmatrix}$$

and

$$(F_{hk}^{1,\gamma_I^-}[f_0, f_1])(mh) =$$

$$- i \sum_{lh \in \gamma_{h1}^- \cap \gamma_{h4}^-} h \begin{pmatrix} E_{hk1}^1(mh - lh) \\ 0 \end{pmatrix}^T \begin{pmatrix} n_1^1 + n_1^4 & n_2^1 + n_2^4 \\ n_3^1 + n_3^4 & n_4^1 + n_4^4 \end{pmatrix} \begin{pmatrix} f_0(lh) \\ f_1(lh) \end{pmatrix}$$

$$- i \sum_{lh \in \gamma_{h2}^- \cap \gamma_{h3}^-} h \begin{pmatrix} 0 \\ E_{hk2}^1(mh - lh) \end{pmatrix}^T \begin{pmatrix} n_1^2 + n_1^3 & n_2^2 + n_2^3 \\ n_3^2 + n_3^3 & n_4^2 + n_4^3 \end{pmatrix} \begin{pmatrix} f_0(lh) \\ f_1(lh) \end{pmatrix}$$

$$- i \sum_{lh \in \gamma_{h1}^- \cap \gamma_{h2}^-} h \begin{pmatrix} E_{hk1}^1(mh - lh) \\ E_{hk2}^1(mh - lh) \end{pmatrix}^T \begin{pmatrix} n_1^1 + n_1^2 & n_2^1 + n_2^2 \\ n_3^1 + n_3^2 & n_4^1 + n_4^2 \end{pmatrix} \begin{pmatrix} f_0(lh) \\ 0 \end{pmatrix}$$

$$- i \sum_{lh \in \gamma_{h3}^- \cap \gamma_{h4}^-} h \begin{pmatrix} E_{hk1}^1(mh - lh) \\ E_{hk2}^1(mh - lh) \end{pmatrix}^T \begin{pmatrix} n_1^3 + n_1^4 & n_2^3 + n_2^4 \\ n_3^3 + n_3^4 & n_4^3 + n_4^4 \end{pmatrix} \begin{pmatrix} 0 \\ f_1(lh) \end{pmatrix}.$$

This discrete Borel–Pompeiu formula was already published in [9]. Obviously, we get a discrete analogue of the Cauchy integral formula if

$$D_{h,M}^1 f(mh) = (0, 0)$$

for all $mh \in G_h$. In a similar way a discrete Borel–Pompeiu formula with respect to $D_{h,M}^2$ can be obtained.

We define finite difference approximations to the gradient and divergence as follows:

$$[\text{grad}_h^+ w_0](mh) = \begin{pmatrix} D_h^1 w_0(mh) \\ D_h^2 w_0(mh) \end{pmatrix},$$

$$\text{div}_h^- s(mh) = D_h^{-1} s_0(mh) + D_h^{-2} s_1(mh).$$

Lemma 4. *For $w_0, s \in w_2^1(G_h)$ with $w_0(rh) = 0$ on $\gamma_{h3}^- \cup \gamma_{h4}^-$, $s_0(rh) = 0$ on γ_{h1}^-, and $s_1(rh) = 0$ on γ_{h2}^- we have*

$$< I_N \text{grad}_h^+ w_0, s > = - \sum_{mh \in G_h} h^2 w_0(mh) \, \text{div}_h^- s(mh).$$

Proof. Using the boundary conditions for w_0 and s we obtain in analogy to formula (2.1):

$$< I_N \text{grad}_h^+ w_0, s > = \sum_{mh \in G_h} h^2 [(D_h^1 w_0(mh)) s_0(mh)$$

$$+ (D_h^2 w_0(mh)) s_1(mh)]$$

$$= - \sum_{mh \in G_h} h^2 w_0(mh) \, \text{div}_h^- s(mh).$$

In the following we denote by P_h^+ and Q_h^+ the orthoprojectors onto

$$\ker D_{h,M}^1(G_h) \quad \text{and} \quad D_{h,M}^2(\overset{\circ}{w}_2^1(G_h)),$$

respectively, that are defined by the orthogonal decomposition of the space $l_2(G_h)$ from Theorem 1. These projectors can be represented by $N \times N$ matrices.

Lemma 5. *The orthoprojectors P_h^+ and Q_h^+ have the properties*

$$D_{h,M}^1(P_h^+[w_0, w_1])(mh) = (0,0), \quad \forall\, mh \in G_h,$$

and

$$\begin{aligned} D_{h,M}^1(Q_h^+[w_0, w_1])(mh) &= D_{h,M}^1((I_N - P_h^+)[w_0, w_1])(mh) \\ &= D_{h,M}^1 w(mh). \end{aligned}$$

The proof immediately follows from the definition of the orthoprojectors.

3.3   Discrete Stokes problem

As a discrete version of the Stokes problem we consider the boundary value problem

$$-\Delta_h u_0(mh) + \frac{1}{\mu} D_h^1 p(mh) = \frac{\varrho}{\mu} f_0(mh), \quad \forall\, mh \in G_h,$$

$$-\Delta_h u_1(mh) + \frac{1}{\mu} D_h^2 p(mh) = \frac{\varrho}{\mu} f_1(mh), \quad \forall\, mh \in G_h,$$

$$D_h^{-1} u_0(mh) + D_h^{-2} u_1(mh) = 0, \qquad \forall\, mh \in G_h,$$

$$u(rh) = (u_0(rh), u_1(rh)) = (0,0), \qquad \forall\, rh \in \gamma_h^-,$$

where ϱ denotes the density, μ the viscosity, p the pressure, f_0 and f_1 the vector components of the external forces and u_0 and u_1 the components of the velocity of the fluid. To prepare the proof of an existence result for the solutions of this discrete boundary value problem we prove at first some decompositions.

Lemma 6. *For functions*

$$u(mh) = (u_0(mh), u_1(mh)) \in \overset{\circ}{w}_2^1(G_h) \cap \ker \operatorname{div}_h^-$$

and $p(mh) \in l_2(G_h)$ we have $< I_N D_{h,M}^2 u, Q_h^+[0,p] > = 0$.

Proof. From Lemma 1, Lemma 5 and the property $\operatorname{div}_h^- u(mh) = 0$ we conclude

$$< I_N D_{h,M}^2 u, Q_h^+[0,p] >$$

$$= < I_N D_{h,M}^2 u, (I_N - P_h^+)[0,p] >$$

$$= \sum_{mh \in G_h} h^2 (\operatorname{div}_h^- u(mh))\, p(mh) + < u, I_N D_{h,M}^1 P_h^+[0,p] > = 0.$$

The proof of the next theorem is related to Lemma 6.

Theorem 2. *For each* $f(mh) = (f_0(mh), f_1(mh)) \in l_2(G_h)$ *functions*

$$u(mh) = (u_0(mh), u_1(mh)) \in \overset{\circ}{w}{}^1_2(G_h) \cap \ker \mathrm{div}^-_h \quad and \quad p(mh) \in l_2(G_h)$$

exist such that

$$\frac{\varrho}{\mu}(Q^+_h T^1_h[f_0, f_1])(mh) = -D^2_{h,M}u(mh) + \frac{1}{\mu}(Q^+_h[0, p])(mh), \qquad (3.1)$$

where T^1_h *is the right inverse operator to the operator* $I_N D^1_{h,M}$.

Proof. Obviously $D^2_{h,M}u(mh) \in \mathrm{im}\, Q^+_h$ and $(Q^+_h[0, p])(mh) \in \mathrm{im}\, Q^+_h$. In Lemma 6 the orthogonality of both expressions with respect to the scalar product is proved. A further orthogonal subspace can not exist, if from

$$< -I_N D^2_{h,M}u, Q^+_h T^1_h[f_0, f_1] > = 0 \qquad (3.2)$$

for all $u(mh) \in \overset{\circ}{w}{}^1_2(G_h) \cap \ker \mathrm{div}^-_h$ and

$$< Q^+_h[0, p], Q^+_h T^1_h[f_0, f_1] > = 0 \qquad (3.3)$$

for all $p(mh) \in l_2(G_h)$ it follows that

$$f(mh) = (f_0(mh), f_1(mh)) = (0, 0).$$

At first we obtain from Lemma 1 and Lemma 5

$$< -I_N D^2_{h,M}u, Q^+_h T^1_h[f_0, f_1] > \, = \, < u, I_N D^1_{h,M} Q^+_h T^1_h[f_0, f_1] >$$

$$= < u, I_N D^1_{h,M} T^1_h[f_0, f_1] >$$

$$= < u, f > .$$

Based on the property $\ker \mathrm{div}^-_h = (\mathrm{im\, grad}^+_h)^\perp$, from Lemma 4, we can find a function $g_0(mh)$ with $g_0(rh) = 0$ on $\gamma^-_{h3} \cup \gamma^-_{h4}$, such that from $f_0(mh) = D^1_h g_0(mh)$ and $f_1(mh) = D^2_h g_0(mh)$ for all $mh \in G_h$ and $u(mh) \in \overset{\circ}{w}{}^1_2(G_h) \cap \ker \mathrm{div}^-_h$ it follows that

$$< u, f > \, = \, < u, I_N \mathrm{grad}^+_h g_0 > = - \sum_{mh \in G_h} h^2 \, (\mathrm{div}^-_h u(mh)) \, g_0(mh) = 0.$$

We assumed that equation (3.3) is fulfilled for any function $p(mh) \in l_2(G_h)$ and consider now the special case $p(mh) = g_0(mh)$. Based on the Borel–Pompeiu formula from Lemma 3 we obtain

$$0 = < Q^+_h[0, g_0], Q^+_h T^1_h[f_0, f_1] >$$

$$= < Q^+_h[0, g_0], Q^+_h T^1_h I_N D^1_{h,M}[0, g_0] >$$

$$= < Q^+_h[0, g_0], Q^+_h[0, g_0] - Q^+_h F^1_h[0, g_0] > \qquad (3.4)$$

with

$$(F_h^1[0, g_0])(mh) = ((F_{h1}^1[0, g_0])(mh), (F_{h2}^1[0, g_0])(mh)).$$

The operator F_h^1 maps the boundary values to ker $D_{h,M}^1$ and due to the orthogonal decomposition of the space $l_2(G_h)$ in Theorem 1 we obtain $Q_h^+ F_h^1[0, g_0] = I_N(0, 0)$. Finally, from equation (3.4) it follows that

$$\|Q_h^+[0, g_0]\|_{l_2(G_h)} = 0.$$

Based on this result and the properties of the orthoprojectors from Lemma 5 we conclude

$$f(mh) = [\text{grad}_h^+ g_0](mh) = (D_{h,M}^1[0, g_0])(mh)$$
$$= D_{h,M}^1(Q_h^+[0, g_0])(mh) = (0, 0).$$

Applying the operator T_h^2 in equation (3.1) we get, with the help of the Borel–Pompeiu formula, the representation

$$\frac{\varrho}{\mu}(T_h^2 Q_h^+ T_h^1[f_0, f_1])(mh) = -T_h^2 I_N D_{h,M}^2 u(mh) + \frac{1}{\mu}(T_h^2 Q_h^+[0, p])(mh)$$
$$= -u(mh) + \frac{1}{\mu}(T_h^2 Q_h^+[0, p])(mh).$$

Using the operators $D_{h,M}^1$ and $D_{h,M}^2$ and their property

$$D_{h,M}^1 D_{h,M}^2 = I_2 \Delta_h$$

we can show now that (u, p) is a solution of the discrete Stokes problem. In the next theorem we prove the uniqueness of the solution.

Theorem 3. *The boundary value problem*

$$-I_2 \Delta_h u(mh) + \frac{1}{\mu}[\text{grad}_h^+ p](mh) = \frac{\varrho}{\mu} f(mh), \quad \forall mh \in G_h,$$

$$D_h^{-1} u_0(mh) + D_h^{-2} u_1(mh) = 0, \quad \forall mh \in G_h,$$

$$u(rh) = (u_0(rh), u_1(rh)) = (0, 0), \quad \forall rh \in \gamma_h^-$$

has for each right-hand side $f(mh) = (f_0(mh), f_1(mh)) \in l_2(G_h)$ *a unique solution* $u(mh) = (u_0(mh), u_1(mh))$. *The pressure*

$$p(mh) \in l_2(G_h \cup \gamma_{h3}^- \cup \gamma_{h4}^-)$$

is unique up to a constant.

Proof. From Lemma 6 and Theorem 2 it follows that

$$\frac{\varrho^2}{\mu^2}\|Q_h^+ T_h^1[f_0, f_1]\|_{l_2(G_h)}^2 = \|I_N D_{h,M}^2 u\|_{l_2(G_h)}^2 + \frac{1}{\mu^2}\|Q_h^+[0, p]\|_{l_2(G_h)}^2. \quad (3.5)$$

We assume that two solutions $u^1(mh) = (u_0^1(mh), u_1^1(mh))$ and $u^2(mh) = (u_0^2(mh), u_1^2(mh))$ exist. The difference

$$u^*(mh) = u^1(mh) - u^2(mh)$$

is a solution of the above boundary value problem with $f(mh) = (0,0)$ in the right-hand side. Without any restriction we can define $f(rh) = (0,0)$ for all $rh \in \gamma_h^-$. Then the left-hand side of equation (3.5) is zero and we can follow $I_N D_{h,M}^2 u^*(mh) = (0,0)$. Applying the right inverse operator T_h^2 and the discrete Borel–Pompeiu formula we get

$$u^*(mh) = (0,0)$$

because $F_h^2[u_0^*, u_1^*] = I_N(0,0)$. We assume now that two solutions

$$\{u(mh); p_1(mh)\} \quad \text{and} \quad \{u(mh); p_2(mh)\}$$

of the boundary value problem exist. Then for

$$p^*(mh) = p_1(mh) - p_2(mh)$$

it follows from equation (3.5) that $Q_h^+[0, p^*] = I_N(0,0)$.

That means $(0, p^*(mh))$ is in the kernel of the operator Q_h^+, and therefore in the kernel of the operator $I_N D_{h,M}^1$. Consequently we have

$$D_h^1 p^*(mh) = 0 \quad \text{and} \quad D_h^2 p^*(mh) = 0$$

for all $mh \in G_h$ and it then follows that $p^*(mh) = \text{const.}$ ☐

We consider now the discrete boundary value problem

$$-I_2 \Delta_h u(mh) + \frac{1}{\mu}[\text{grad}_h^+ p](mh) = \frac{\varrho}{\mu} f(mh), \quad \forall mh \in G_h,$$

$$D_h^{-1} u_0(mh) + D_h^{-2} u_1(mh) = \varphi(mh), \quad \forall mh \in G_h, \quad (3.6)$$

$$u(rh) = \psi(rh) = (\psi_0(rh), \psi_1(rh)), \quad \forall rh \in \gamma_h^-,$$

where $\varphi(mh)$ is a measure for the compressibility of the fluid and $\psi(rh)$ describes the velocity on the boundary. In the case $\psi(rh) = (0,0)$ we have adhesion. This system is only solvable if the necessary condition

$$\sum_{mh \in G_h} \varphi(mh)h^2 + \sum_{rh \in \gamma_{h3}^-} D_h^{-1} u_0(rh)h^2 + \sum_{rh \in \gamma_{h4}^-} D_h^{-2} u_1(rh)h^2$$

$$= \sum_{i=1}^4 \sum_{rh \in \gamma_{hi}^-} \begin{pmatrix} \alpha_0^i \\ \alpha_1^i \end{pmatrix} \begin{pmatrix} \psi_0(rh) \\ \psi_1(rh) \end{pmatrix} h$$

is fulfilled. In this formula $\begin{pmatrix} \alpha_0^i \\ \alpha_1^i \end{pmatrix}$ are the normal unit vectors on γ_{hi}^-, $i = 1, \ldots, 4$.
In detail we have

$$\begin{pmatrix} \alpha_0^1 \\ \alpha_1^1 \end{pmatrix} = \begin{pmatrix} \alpha_0^3 \\ \alpha_1^3 \end{pmatrix} = \begin{pmatrix} 1 \\ 0 \end{pmatrix} \quad \text{and} \quad -\begin{pmatrix} \alpha_0^2 \\ \alpha_1^2 \end{pmatrix} = \begin{pmatrix} \alpha_0^4 \\ \alpha_1^4 \end{pmatrix} = \begin{pmatrix} 0 \\ 1 \end{pmatrix}.$$

This necessary condition results from

$$\sum_{mh \in G_h} \varphi(mh)h^2 + \sum_{rh \in \gamma_{h3}^-} D_h^{-1} u_0(rh)h^2 + \sum_{rh \in \gamma_{h4}^-} D_h^{-2} u_1(rh)h^2$$

$$= \sum_{mh \in G_h \cup \gamma_{h3}^-} D_h^{-1} u_0(mh)h^2 + \sum_{mh \in G_h \cup \gamma_{h4}^-} D_h^{-2} u_1(mh)h^2$$

$$= \sum_{mh \in G_h \cup \gamma_{h3}^-} u_0(mh)h - \sum_{mh \in G_h \cup \gamma_{h1}^-} u_0(mh)h$$

$$\qquad + \sum_{mh \in G_h \cup \gamma_{h4}^-} u_1(mh)h - \sum_{mh \in G_h \cup \gamma_{h2}^-} u_1(mh)h$$

$$= \sum_{mh \in \gamma_{h3}^-} \begin{pmatrix} 1 \\ 0 \end{pmatrix} \begin{pmatrix} u_0(mh) \\ u_1(mh) \end{pmatrix} h + \sum_{mh \in \gamma_{h1}^-} \begin{pmatrix} -1 \\ 0 \end{pmatrix} \begin{pmatrix} u_0(mh) \\ u_1(mh) \end{pmatrix} h$$

$$\qquad + \sum_{mh \in \gamma_{h4}^-} \begin{pmatrix} 0 \\ 1 \end{pmatrix} \begin{pmatrix} u_0(mh) \\ u_1(mh) \end{pmatrix} h + \sum_{mh \in \gamma_{h2}^-} \begin{pmatrix} 0 \\ -1 \end{pmatrix} \begin{pmatrix} u_0(mh) \\ u_1(mh) \end{pmatrix} h.$$

Theorem 4. *The boundary value problem*

$$-I_2 \Delta_h u(mh) + \frac{1}{\mu} [\text{grad}_h^+ p](mh) = \frac{\varrho}{\mu} f(mh), \quad \forall mh \in G_h,$$

$$D_h^{-1} u_0(mh) + D_h^{-2} u_1(mh) = \varphi(mh), \quad \forall mh \in G_h, \qquad (3.7)$$

$$(u_0(rh), u_1(rh)) = (T_h^2[0, \varphi])(rh), \quad \forall rh \in \gamma_a^-,$$

has for $f(mh) = (f_0(mh), f_1(mh)) \in l_2(G_h)$ *and* $\varphi(mh) \in l_2(G_h)$ *with* $\varphi(rh) = 0$ *for all* $rh \in \gamma_h^-$ *a unique solution* $\{u(mh); p(mh)\}$ *in the sense of Theorem 3*.

Proof. The uniqueness of the solution in the case $\varphi(mh) = 0$ was already proved in Theorem 3. We denote this solution with $\{u_h(mh); p_h(mh)\}$ and describe the solution of the problem (3.7) by the addition of a special solution. Using the ansatz

$$u(mh) = u_h(mh) + (T_h^2[0, \varphi])(mh), \quad \forall mh \in G_h \cup \gamma_h^-,$$

$$p(mh) = p_h(mh) + \mu \varphi(mh), \quad \forall mh \in G_h$$

it follows that

$$- I_2 \Delta_h u(mh)$$

$$+ \frac{1}{\mu}[\mathrm{grad}_h^+ p](mh) - I_2 \Delta_h u_h(mh) - D_{h,M}^1 D_{h,M}^2 (T_h^2[0,\varphi])(mh)$$

$$+ \frac{1}{\mu}[\mathrm{grad}_h^+ p_h](mh) + [\mathrm{grad}_h^+ \varphi](mh)$$

$$= \frac{\varrho}{\mu} f(mh) - (D_{h,M}^1[0,\varphi])(mh) + [\mathrm{grad}_h^+ \varphi](mh) = \frac{\varrho}{\mu} f(mh).$$

Because T_h^2 is right inverse to $I_N D_{h,M}^2$ the equalities

$$\mathrm{div}_h^- u(mh) = \mathrm{div}_h^- (T_h^2[0,\varphi])(mh) = \varphi(mh)$$

and $u(rh) = (T_h^2[0,\varphi])(rh)$ are satisfied.

In the following the right-hand side $f(mh)$ and the boundary values $\psi(rh)$ of the system (3.6) are given. We construct a solution of the problem using the ansatz

$$u(mh) = u_h(mh) + u_s(mh), \quad p(mh) = p_h(mh);$$

here $\{u_h(mh); p_h(mh)\}$ denotes the unique solution of the discrete boundary value problem considered in Theorem 3. With this ansatz it follows from the first and last equation of the problem (3.6) that the components $u_{s0}(mh)$ and $u_{s1}(mh)$ of $u_s(mh)$ are solutions of the Dirichlet problems

$$-\Delta_h u_{si}(mh) = 0, \qquad \forall\, mh \in G_h,$$
$$u_{si}(rh) = \psi_i(rh), \quad \forall\, rh \in \gamma_h^-,\ i = 0, 1.$$

A uniqueness theorem for the solution of Dirichlet problems is proved in [13]. We describe the solution of $u_{s0}(mh)$ and $u_{s1}(mh)$ using our operator calculus. From Lemma 5 we obtain

$$D_{h,M}^2 u_s(mh) = (P_h^+ I_N D_{h,M}^2 u_s)(mh)$$

because u_s is discrete harmonic. Applying the right inverse operator T_h^2 we get by the help of the Borel–Pompeiu formula the representation

$$u_s(mh) = (F_h^2[\psi_0, \psi_1])(mh) + (T_h^2 P_h^+ I_N D_{h,M}^2 u_s)(mh).$$

Based on

$$D_{h,M}^2 (F_h^2[\psi_0, \psi_1])(mh) = (0,0)$$

and

$$D_{h,M}^2 (T_h^2 P_h^+ I_N D_{h,M}^2 u_s)(mh) = (P_h^+ I_N D_{h,M}^2 u_s)(mh),$$

the right-hand side of the second equation in (3.6) is determined by

$$
\begin{aligned}
\varphi(mh) &= D_h^{-1} u_{s0}(mh) + D_h^{-2} u_{s1}(mh) \\
&= \operatorname{div}_h^-(F_h^2[\psi_0, \psi_1])(mh) + \operatorname{div}_h^-(T_h^2 P_h^+ I_N D_{h,M}^2 u_s)(mh) \\
&= (P_{h2}^+ I_N D_{h,M}^2 u_s)(mh),
\end{aligned}
$$

where P_{h2}^+ denotes the second component of the orthoprojector P_h^+ at each mesh point mh. As a summary of these results we state the following theorem:

Theorem 5. *Let* $u_{s0}(mh)$ *and* $u_{s1}(mh)$ *be the solutions of the Dirichlet problems*

$$
\begin{aligned}
-\Delta_h u_{si}(mh) &= 0, & \forall\, mh \in G_h, \\
u_{si}(rh) &= \psi_i(rh), & \forall\, rh \in \gamma_h^-, \quad i = 0, 1.
\end{aligned}
$$

The boundary value problem

$$
\begin{aligned}
-I_2 \Delta_h u(mh) + \frac{1}{\mu}[\operatorname{grad}_h^+ p](mh) &= \frac{\varrho}{\mu} f(mh), & \forall mh \in G_h, \\
D_h^{-1} u_0(mh) + D_h^{-2} u_1(mh) &= (P_{h2}^+ I_N D_{h,M}^2 u_s)(mh), & \forall mh \in G_h, \\
u(rh) &= \psi(rh), & \forall\, rh \in \gamma_h^-
\end{aligned}
$$

has a unique solution in the sense of Theorem 3 and can be described by

$$
\begin{aligned}
u(mh) &= u_h(mh) + (F_h^2[\psi_0, \psi_1])(mh) + (T_h^2 P_h^+ I_N D_{h,M}^2 u_s)(mh), \\
p(mh) &= p_h(mh).
\end{aligned}
$$

3.4   Discrete Navier–Stokes equations

Based on the Stokes problem we will present a possibility to solve the discrete Navier–Stokes equations

$$
-\Delta_h u_0(mh) + \frac{1}{\mu} D_h^1 p(mh)
$$

$$
+ \frac{\varrho}{\mu}\left(u_0(mh) D_h^{-1} u_0(mh) + u_1(mh) D_h^{-2} u_0(mh) \right) = \frac{\varrho}{\mu} f_0(mh),
$$

$$
-\Delta_h u_1(mh) + \frac{1}{\mu} D_h^2 p(mh)
$$

$$
+ \frac{\varrho}{\mu}\left(u_0(mh) D_h^{-1} u_1(mh) + u_1(mh) D_h^{-2} u_1(mh) \right) = \frac{\varrho}{\mu} f_1(mh),
$$

$$D_h^{-1} u_0(mh) + D_h^{-2} u_1(mh) = 0,$$

$$u(rh) = (0,0) \tag{4.1}$$

for all $mh \in G_h$ and $rh \in \gamma_h^-$. Therefore the problem is rewritten. To simplify the expressions we use the notation

$$M_{h,1}(mh) = \frac{\varrho}{\mu}\left(u_0(mh) D_h^{-1} u_0(mh) + u_1(mh) D_h^{-2} u_0(mh) \right) - \frac{\varrho}{\mu} f_0(mh),$$

$$M_{h,2}(mh) = \frac{\varrho}{\mu}\left(u_0(mh) D_h^{-1} u_1(mh) + u_1(mh) D_h^{-2} u_1(mh) \right) - \frac{\varrho}{\mu} f_1(mh).$$

Theorem 6. *The boundary value problem (4.1) is equivalent to the problem*

$$u(mh) = (T_h^2 Q_h^+ T_h^1 [M_{h,1}, M_{h,2}])(mh) + \frac{1}{\mu}(T_h^2 Q_h^+ [0, p])(mh)$$

$$- (Q_{h2}^+ T_h^1 [M_{h,1}, M_{h,2}])(mh) = \frac{1}{\mu}(Q_{h2}^+ [0, p])(mh) \tag{4.2}$$

where Q_{h2}^+ denotes the second component of Q_h^+ at each mesh point $mh \in G_h$.

Proof. Let $\{u, p\}$ be a solution of the problem (4.2). It follows that

$$- I_2 \Delta_h u(mh)$$

$$= -D_{h,M}^1 D_{h,M}^2 \left[(T_h^2 Q_h^+ T_h^1 [M_{h,1}, M_{h,2}])(mh) + \frac{1}{\mu}(T_h^2 Q_h^+ [0, p])(mh) \right]$$

$$= -D_{h,M}^1 \left[(Q_h^+ T_h^1 [M_{h,1}, M_{h,2}])(mh) + \frac{1}{\mu}(Q_h^+ [0, p])(mh) \right]$$

$$= -M_h(mh) - \frac{1}{\mu}[\mathrm{grad}_h^+ p](mh)$$

with $M_h(mh) = (M_{h,1}(mh), M_{h,2}(mh))$. In addition we have

$$\mathrm{div}_h^- u(mh) = \mathrm{div}_h^- [(T_h^2 Q_h^+ T_h^1 [M_{h,1}, M_{h,2}])(mh) + \frac{1}{\mu}(T_h^2 Q_h^+ [0, p])(mh)]$$

$$= (Q_{h2}^+ T_h^1 [M_{h,1}, M_{h,2}])(mh) + \frac{1}{\mu}(Q_{h2}^+ [0, p])(mh) = 0.$$

The boundary condition of problem (4.1) is satisfied because, based on the property of the orthoprojector Q_h^+, the solution of the problem (4.2) has the representation

$$(T_h^2 I_N D_{h,M}^2 s)(mh) = s(mh) - (F_h^2 [s_0, s_1])(mh)$$

with $s(mh) \in \overset{\circ}{w}{}_2^1(G_h)$. Let now $\{u, p\}$ be a solution of the problem (4.1). With the help of the Borel–Pompeiu formula we obtain

$$(T_h^1 I_N \Delta_h u)(mh) = (T_h^1 I_N D_{h,M}^1 D_{h,M}^2 u)(mh)$$
$$= D_{h,M}^2 u(mh) - (F_h^1 [D_{h,M}^2 u])(mh).$$

Using the orthoprojector Q_h^+ we get

$$(Q_h^+ T_h^1 I_N \Delta_h u)(mh) = (Q_h^+ I_N D_{h,M}^2 u)(mh) - (Q_h^+ F_h^1 [D_{h,M}^2 u])(mh)$$
$$= D_{h,M}^2 u(mh).$$

Otherwise it follows from (4.1) that

$$(Q_h^+ T_h^1 I_N \Delta_h u)(mh) = (Q_h^+ T_h^1 [M_{h,1}, M_{h,2}])(mh)$$
$$+ \frac{1}{\mu}(Q_h^+ T_h^1 I_N D_{h,M}^1 [0, p])(mh)$$
$$= (Q_h^+ T_h^1 [M_{h,1}, M_{h,2}])(mh) + \frac{1}{\mu}(Q_h^+ [0, p])(mh).$$

From both equations we obtain

$$D_{h,M}^2 u(mh) = (Q_h^+ T_h^1 [M_{h,1}, M_{h,2}])(mh) + \frac{1}{\mu}(Q_h^+ [0, p])(mh).$$

Applying the operator T_h^2 and again the discrete Borel–Pompeiu formula, the first equation of (4.2) follows. The second equation results from

$$0 = \text{div}_h^- u(mh)$$
$$= \text{div}_h^- [(T_h^2 Q_h^+ T_h^1 [M_{h,1}, M_{h,2}])(mh) + \frac{1}{\mu}(T_h^2 Q_h^+ [0, p])(mh)]$$
$$= (Q_{h2}^+ T_h^1 [M_{h,1}, M_{h,2}])(mh) + \frac{1}{\mu}(Q_{h2}^+ [0, p])(mh).$$

Based on the problem (4.2) the following iteration procedure can be used to calculate the solution of the Navier–Stokes equations (4.1):

$$u^n(mh) = (T_h^2 Q_h^+ T_h^1 [M_{h,1}^{n-1}, M_{h,2}^{n-1}])(mh) + \frac{1}{\mu}(T_h^2 Q_h^+ [0, p^n])(mh),$$

$$- (Q_{h2}^+ T_h^1 [M_{h,1}^{n-1}, M_{h,2}^{n-1}])(mh) = \frac{1}{\mu}(Q_{h2}^+ [0, p^n])(mh)$$

with

$$M_{h,j}^{n-1}(mh) = \frac{\varrho}{\mu}\left(u_0^{n-1}(mh)D_h^{-1}u_{j-1}^{n-1}(mh) + u_1^{n-1}(mh)D_h^{-2}u_{j-1}^{n-1}(mh)\right)$$
$$- \frac{\varrho}{\mu} f_{j-1}(mh)$$

for $j = 1, 2$,

$$(u_0^0(mh), u_1^0(mh)) \in \overset{\circ}{w}_2^1(G_h) \cap \ker \mathrm{div}_h^-$$

and $n = 1, 2, 3, \ldots$.

In this way the solution of the boundary value problem (4.1) is reduced to the solution (u^n, p^n) of a Stokes problem for $n = 1, 2, 3, \ldots$ and the uniqueness of this Stokes problem is proved in Theorem 3. We have to prove that the iteration procedure converges to the solution of the problem (4.1). The following two lemmas prepare this proof.

Lemma 7. *For* $1 < p < 2$ *and* $q < \frac{2p}{2-p}$ *or,* $p = 2$ *and* $q < \infty$, *the operator* $T_h^2 : l_p(G_h) \rightarrow l_q(G_h)$ *is bounded.*

Proof. Let $p < q$. At first we estimate the expression $|(T_{h,k}^2[f_0, f_1])(mh)|$ for $k = 1, 2$. Because T_h^2 acts only on functions that are defined in G_h we set $f(rh) = (0, 0)$ for all $rh \in \gamma_h^-$. Based on the property

$$|E_{hkj}^2(mh - lh)| \le C_1(|mh - lh| + h)^{-1} \quad \text{with} \quad k, j \in \{1, 2\},$$

(see [7, 21]) we get

$$|(T_{h,k}^2[f_0, f_1])(mh)| = \left| \sum_{lh \in G_h} h^2 \begin{pmatrix} E_{hk1}^2(mh - lh) \\ E_{hk2}^2(mh - lh) \end{pmatrix}^T \begin{pmatrix} f_0(lh) \\ f_1(lh) \end{pmatrix} \right|$$

$$\le C_1 \sum_{lh \in G_h} h^2(|mh - lh| + h)^{-1}(|f_0(lh)| + |f_1(lh)|)$$

$$\le 2 C_1 \sum_{lh \in G_h} h^2(|mh - lh| + h)^{-1}|f(lh)|.$$

For $1 < p \le 2$ it follows $\frac{2}{q} > 1 - \frac{2}{p^*}$ with $\frac{1}{p^*} + \frac{1}{p} = 1$ and therefore, $1 = \frac{2}{q} + \frac{2}{p^*} - 2\beta$ with $\beta > 0$. We get the equation

$$\sum_{lh \in G_h} h^2(|mh - lh| + h)^{-1}|f(lh)|$$

$$= \sum_{lh \in G_h} h^2 \left(|f(lh)|^{\frac{p}{q}}(|mh - lh| + h)^{-\frac{2}{q}+\beta} \right) \left(|f(lh)|^{p\left(\frac{1}{p} - \frac{1}{q}\right)} \right)$$

$$\times \left((|mh - lh| + h)^{-\frac{2}{p^*}+\beta} \right).$$

We set

$$\alpha_1 = \frac{1}{q}, \quad \alpha_2 = \frac{1}{p} - \frac{1}{q}, \quad \text{and} \quad \alpha_3 = \frac{1}{p^*},$$

such that $\alpha_1 + \alpha_2 + \alpha_3 = 1$ and apply the Hölder inequality to the product of the three functions. This leads to the estimate

$$\sum_{lh \in G_h} h^2(|mh - lh| + h)^{-1}|f(lh)| \leq \left(\sum_{lh \in G_h} h^2 |f(lh)|^p(|mh-lh|+h)^{-2+\beta q} \right)^{\frac{1}{q}}$$

$$\times \left(\sum_{lh \in G_h} h^2 |f(lh)|^p \right)^{\frac{1}{p}-\frac{1}{q}} \left(\sum_{lh \in G_h} h^2(|mh-lh|+h)^{-2+\beta p^*} \right)^{\frac{1}{p^*}}.$$

While we have for the second factor $\left(\sum_{lh \in G_h} h^2 |f(lh)|^p \right)^{\frac{1}{p}-\frac{1}{q}} = \|f\|_{l_p(G_h)}^{1-\frac{p}{q}}$, we can estimate the third factor in the form

$$\left(\sum_{lh \in G_h} h^2(|mh-lh|+h)^{-2+\beta p^*} \right)^{\frac{1}{p^*}} \leq \left(\frac{C_2 \, (\text{diam } G_h)^{\beta p^*}}{\beta p^*} \right)^{\frac{1}{p^*}}.$$

In order to simplify the notation we define

$$K(p^*, q, G_h) = C_2 \frac{(\text{diam } G_h)^{\beta p^*}}{\beta p^*} = C_2 \frac{(\text{diam } G_h)^{\frac{p^*}{q}+1-\frac{p^*}{2}}}{\frac{p^*}{q} + 1 - \frac{p^*}{2}}$$

and obtain

$$|(T_{h,k}^2[f_0, f_1])(mh)|$$

$$\leq 2C_1(K(p^*, q, G_h))^{\frac{1}{p^*}} \|f\|_{l_p(G_h)}^{1-\frac{p}{q}} \left(\sum_{lh \in G_h} h^2 |f(lh)|^p(|mh-lh|+h)^{-2+\beta q} \right)^{\frac{1}{q}}.$$

Using the Hölder inequality again, we conclude

$$\|T_h^2\|_{l_q(G_h)}^q = \sum_{mh \in G_h} h^2(|T_{h,1}^2|^2 + |T_{h,2}^2|^2)^{\frac{q}{2}}$$

$$\leq (2\sqrt{2}\,C_1)^q (K(p^*, q, G_h))^{\frac{q}{p^*}} \|f\|_{l_p(G_h)}^{q-p} \sum_{lh \in G_h} h^2 |f(lh)|^p \, K(q, p^*, G_h).$$

Finally we obtain

$$\|T_h^2\|_{l_q(G_h)} \leq 2\sqrt{2}\,C_1 \, (K(p^*, q, G_h))^{\frac{1}{p^*}} (K(q, p^*, G_h))^{\frac{1}{q}} \|f\|_{l_p(G_h)}.$$

We remark that the operator T_h^1 has the same property. Important are two special cases, which immediately follow from Lemma 7.

Corollary 1. *The operator* $T_h^2 : l_p(G_h) \to l_2(G_h)$ *is bounded for* $1 < p \leq 2$, *while the operator* $T_h^2 : l_2(G_h) \to l_q(G_h)$ *is bounded for all* $q < \infty$.

In the next lemma we prove the boundedness of T_h^2 in the Sobolev norm

$$\|v\|_{w_2^1(G_h)} = \|v\|_{l_2(G_h)}$$
$$+ \left(\sum_{i=1}^{2} [\|D_h^{-i} v_0\|_{l_2(G_h \cup \gamma_{h3}^- \cup \gamma_{h4}^-)}^2 + \|D_h^{-i} v_1\|_{l_2(G_h \cup \gamma_{h3}^- \cup \gamma_{h4}^-)}^2] \right)^{\frac{1}{2}}.$$

Lemma 8. *The operator* $T_h^2 : l_2(G_h) \cap im \, Q_h^+ \to w_2^1(G_h)$ *is bounded.*

Proof. For each

$$u(mh) \in l_2(G_h) \cap im \, Q_h^+$$

there exists a function $v(mh) \in \overset{\circ}{w}_2^1(G_h)$ with $u(mh) = D_{h,M}^2 \, v(mh)$. From the Borel–Pompeiu formula we get $v(mh) = T_h^2 \, u(mh)$. Based on equation (2.1), the first Green's formula, which is proved in [13], and Poincaré's inequality we obtain

$$\|u\|_{l_2(G_h)}^2 = \|D_{h,M}^2 \, v\|_{l_2(G_h)}^2 = \, < I_N \, D_{h,M}^2 v, I_N \, D_{h,M}^2 v >$$

$$= \, < -I_N \, D_{h,M}^1 \, D_{h,M}^2 v, v >$$

$$= \sum_{mh \in G_h} h^2 [(-\Delta_h \, v_0(mh)) \, v_0(mh) + (-\Delta_h \, v_1(mh)) \, v_1(mh)]$$

$$= \sum_{mh \in G_h \cup \gamma_{h3}^- \cup \gamma_{h4}^-} \sum_{i=1}^{2} h^2 \, [(D_h^{-i} v_0(mh))^2 + (D_h^{-i} v_1(mh))^2]$$

$$\geq C_3^{-1} \, (\|v_0\|_{l_2(G_h)}^2 + \|v_1\|_{l_2(G_h)}^2) = C_3^{-1} \, \|v\|_{l_2(G_h)}^2,$$

where the constant C_3 is related to the smallest eigenvalue of the discrete Laplace operator. From this estimate we conclude

$$\|T_h^2 u\|_{w_2^1(G_h)} = \|v\|_{w_2^1(G_h)}$$

$$= \|v\|_{l_2(G_h)}$$

$$+ \left(\sum_{i=1}^{2} [\|D_h^{-i} v_0\|_{l_2(G_h \cup \gamma_{h3}^- \cup \gamma_{h4}^-)}^2 + \|D_h^{-i} v_1\|_{l_2(G_h \cup \gamma_{h3}^- \cup \gamma_{h4}^-)}^2] \right)^{\frac{1}{2}}$$

$$\leq (C_3^{\frac{1}{2}} + 1) \, \|u\|_{l_2(G_h)}.$$

Theorem 7. *Let* $\|f\|_{l_p(G_h)} \leq C_4 \left(\frac{\mu}{16KC_4 \varrho} \right)^2, 1 < p \leq 2$. *For any function*

$$u^0(mh) \in \overset{\circ}{w}_2^1(G_h) \cap ker \, div_h^-$$

with

$$\|u^0\|_{w_2^1(G_h)} \le \frac{\mu}{16KC_4\varrho} + \Omega \quad and \quad \Omega = \sqrt{\left(\frac{\mu}{16KC_4\varrho}\right)^2 - \frac{\|f\|_{l_p(G_h)}}{C_4}},$$

the above iteration procedure converges to the solution of the Navier–Stokes equations (4.1).

Proof. Let $[\mathrm{grad}_h^- u_i^n](mh) = \begin{pmatrix} D_h^{-1}u_i^n(mh) \\ D_h^{-2}u_i^n(mh) \end{pmatrix}$, for $i = 0, 1$ and

$$K = \|T_h^2\|_{[l_2(G_h) \cap im Q_h^+, w_2^1(G_h)]} \|T_h^1\|_{[l_p(G_h), l_2(G_h)]}$$

based on Lemma 7 and Lemma 8. In the following we consider the difference $u^n(mh) - u^{n-1}(mh)$.

Replacing $(f_0(mh), f_1(mh))$ by $(M_{h,1}(mh), M_{h,2}(mh))$ we obtain in analogy to (3.5)

$$\frac{1}{\mu}\|Q_h^+[0, p^n - p^{n-1}]\|_{l_2(G_h) \cap im Q_h^+}$$
$$\le \|Q_h^+ T_h^1[M_{h,1}^{n-1} - M_{h,1}^{n-2}, M_{h,2}^{n-1} - M_{h,2}^{n-2}]\|_{l_2(G_h) \cap im Q_h^+}.$$

Using the property $\|Q_h^+\|_{[l_2(G_h), l_2(G_h) \cap im Q_h^+]} = 1$ of the orthoprojector Q_h^+, it follows from the iteration procedure that

$$\|u^n - u^{n-1}\|_{w_2^1(G_h)}$$
$$\le \|T_h^2 Q_h^+ T_h^1[M_{h,1}^{n-1} - M_{h,1}^{n-2}, M_{h,2}^{n-1} - M_{h,2}^{n-2}]\|_{w_2^1(G_h)}$$
$$\qquad + \frac{1}{\mu}\|T_h^2 Q_h^+[0, p^n - p^{n-1}]\|_{w_2^1(G_h)}$$
$$\le 2\|T_h^2\|_{[l_2(G_h) \cap im Q_h^+, w_2^1(G_h)]} \|Q_h^+\|_{[l_2(G_h), l_2(G_h) \cap im Q_h^+]} \|T_h^1\|_{[l_p(G_h), l_2(G_h)]}$$
$$\qquad \times \|M_{h,1}^{n-1} - M_{h,1}^{n-2}, M_{h,2}^{n-1} - M_{h,2}^{n-2}\|_{l_p(G_h)}$$
$$\le 2K\|M_{h,1}^{n-1} - M_{h,1}^{n-2}, M_{h,2}^{n-1} - M_{h,2}^{n-2}\|_{l_p(G_h)}.$$

Using the inequality $a^2 + b^2 \le (|a| + |b|)^2$ and the Minkowski inequality we obtain

$$\|M_{h,1}^{n-1} - M_{h,1}^{n-2}, M_{h,2}^{n-1} - M_{h,2}^{n-2}\|_{l_p(G_h)}$$

$$= \frac{\varrho}{\mu} \left(\sum_{mh \in G_h} h^2 \left[(u^{n-1} \operatorname{grad}_h^- u_0^{n-1} - u^{n-2} \operatorname{grad}_h^- u_0^{n-2})^2 \right. \right.$$

$$\left. \left. + (u^{n-1} \operatorname{grad}_h^- u_1^{n-1} - u^{n-2} \operatorname{grad}_h^- u_1^{n-2})^2 \right]^{\frac{p}{2}} \right)^{\frac{1}{p}}$$

$$\leq \frac{\varrho}{\mu} \left(\sum_{mh \in G_h} h^2 \left[|u^{n-1} \operatorname{grad}_h^- u_0^{n-1} - u^{n-2} \operatorname{grad}_h^- u_0^{n-2}| \right. \right.$$

$$\left. \left. + |u^{n-1} \operatorname{grad}_h^- u_1^{n-1} - u^{n-2} \operatorname{grad}_h^- u_1^{n-2}| \right]^p \right)^{\frac{1}{p}}$$

$$\leq \frac{\varrho}{\mu} \sum_{j=0}^{1} \|u^{n-1} \operatorname{grad}_h^- u_j^{n-1} - u^{n-2} \operatorname{grad}_h^- u_j^{n-2}\|_{l_p(G_h)}$$

$$\leq \frac{\varrho}{\mu} \sum_{j=0}^{1} \left[\|(u^{n-1} - u^{n-2}) \operatorname{grad}_h^- u_j^{n-1}\|_{l_p(G_h)} \right.$$

$$\left. + \|u^{n-2} \operatorname{grad}_h^- (u_j^{n-1} - u_j^{n-2})\|_{l_p(G_h)} \right].$$

Both norms on the right-hand side have the structure $\|w \operatorname{grad}_h^- v_j\|_{l_p(G_h)}$. From Hölder's inequality it follows that

$$\|w \operatorname{grad}_h^- v_j\|_{l_p(G_h)} \leq \|w\|_{l_{pq_1}(G_h)} \|\operatorname{grad}_h^- v_j\|_{l_{pp_1}(G_h)}$$

with $\frac{1}{p_1} + \frac{1}{q_1} = 1$. If we choose $pp_1 = 2$, then we obtain $q := pq_1 = \frac{2p}{2-p}$ and $p = \frac{2q}{2+q}$. Based on the definition of the discrete Sobolev norm we get for $j = 0, 1$ the estimate

$$\|w \operatorname{grad}_h^- v_j\|_{l_p(G_h)} \leq \|w\|_{l_q(G_h)} \left(\sum_{mh \in G_h} h^2 [(D_h^{-1} v_j)^2 + (D_h^{-2} v_j)^2] \right)^{\frac{1}{2}}$$

$$= \|w\|_{l_q(G_h)} \left(\sum_{i=1}^{2} \|D_h^{-i} v_j\|_{l_2(G_h)}^2 \right)^{\frac{1}{2}}$$

$$\leq \|w\|_{l_q(G_h)} \|v\|_{w_2^1(G_h)}.$$

For functions $w \in \overset{\circ}{w}_2^1(G_h)$ we show now $\|w\|_{l_q(G_h)} \leq C_4 \|w\|_{w_2^1(G_h)}$. Because of the special choice of the boundary values, we obtain from the Borel–Pompeiu formula $\|w\|_{l_q(G_h)} = \|T_h^2 D_{h,M}^2 w\|_{l_q(G_h)}$ with

$$D_{h,M}^2 w \in l_2(G_h) \cap \operatorname{im} Q_h^+.$$

From Corollary 1 and the equivalent formulations in the proof of Lemma 8 we can conclude

$$\|T_h^2 D_{h,M}^2 w\|_{l_q(G_h)} \le C_4 \|D_{h,M}^2 w\|_{l_2(G_h)} \le C_4 \|w\|_{w_2^1(G_h)}.$$

On these estimates the result

$$\|u^n - u^{n-1}\|_{w_2^1(G_h)}$$

$$\le 4\,K\,C_4 \frac{\varrho}{\mu} \|u^{n-1} - u^{n-2}\|_{w_2^1(G_h)} \left(\|u^{n-1}\|_{w_2^1(G_h)} + \|u^{n-2}\|_{w_2^1(G_h)} \right)$$

is based and with the notation

$$L_n = 4\,K\,C_4 \frac{\varrho}{\mu} (\|u^{n-1}\|_{w_2^1(G_h)} + \|u^{n-2}\|_{w_2^1(G_h)})$$

we obtain

$$\|u^n - u^{n-1}\|_{w_2^1(G_h)} \le L_n \|u^{n-1} - u^{n-2}\|_{w_2^1(G_h)}.$$

We assume now that the inequality

$$0 \le \|u^{n-1}\|_{w_2^1(G_h)} \le \frac{\mu}{16KC_4\varrho} + \Omega$$

holds with $\Omega = \sqrt{\left(\frac{\mu}{16KC_4\varrho} \right)^2 - \frac{\|f\|_{l_p(G_h)}}{C_4}}$. Under the assumption

$$\|f\|_{l_p(G_h)} \le C_4 \left(\frac{\mu}{16KC_4\varrho} \right)^2,$$

it follows from the iteration procedure, in analogy to the above steps, that

$$\|u^n\|_{w_2^1(G_h)} \le 2K \frac{\varrho}{\mu} \left(\sum_{mh \in G_h} h^2 \left[(u^{n-1} \operatorname{grad}_h^- u_0^{n-1} - f_0)^2 \right. \right.$$

$$\left. \left. + (u^{n-1} \operatorname{grad}_h^- u_1^{n-1} - f_1)^2 \right]^{\frac{p}{2}} \right)^{\frac{1}{p}}$$

$$\le 2K \frac{\varrho}{\mu} \sum_{j=0}^{1} \left[\|u^{n-1} \operatorname{grad}_h^- u_j^{n-1}\|_{l_p(G_h)} + \|f\|_{l_p(G_h)} \right]$$

$$\le 4KC_4 \frac{\varrho}{\mu} \|u^{n-1}\|_{w_2^1(G_h)}^2 + 4K \frac{\varrho}{\mu} \|f\|_{l_p(G_h)}$$

$$\le 4KC_4 \frac{\varrho}{\mu} \left[\left(\frac{\mu}{16KC_4\varrho} + \Omega \right)^2 + \frac{\|f\|_{l_p(G_h)}}{C_4} \right]$$

$$= 4KC_4 \frac{\varrho}{\mu} \left[\left(\frac{\mu}{16KC_4\varrho} \right)^2 + \frac{\mu\Omega}{8KC_4\varrho} + (\Omega)^2 + \frac{\|f\|_{l_p(G_h)}}{C_4} \right]$$

$$\le \frac{\mu}{64KC_4\varrho} + \frac{\Omega}{2} + 4KC_4 \frac{\varrho}{\mu} \left(\frac{\mu}{16KC_4\varrho} \right)^2 \le \frac{1}{2} \left(\frac{\mu}{16KC_4\varrho} + \Omega \right).$$

Because of the inequality

$$L_{n+1} = 4\,K\,C_4 \frac{\varrho}{\mu} \left(\|u^n\|_{w_2^1(G_h)} + \|u^{n-1}\|_{w_2^1(G_h)} \right)$$

$$\leq 4\,K\,C_4 \frac{\varrho}{\mu} \frac{3}{2} \left(\frac{\mu}{16KC_4\varrho} + \Omega \right)$$

$$\leq 6\,K\,C_4 \frac{\varrho}{\mu} \left(2\,\frac{\mu}{16KC_4\varrho} \right) = \frac{3}{4} < 1$$

for all $n \geq 1$, the assertion of the theorem follows from Banach's fixed-point theorem.

REFERENCES

[1] Boor, C., and Höllig, K. ; Riemenschneider, S.: Fundamental solutions for multivariate difference equations, *Amer. J. Math.* **111** (1989), 403–415.

[2] Deeter, C. R., and Lord, M. E.: Further theory of operational calculus on discrete analytic functions, *J. Math. Anal. Appl.* **26** (1969), 92–113.

[3] Duffin, R.J.: Discrete potential theory, *Duke Math. J.* **20** (1953), 233–251.

[4] Duffin, R. J.: Basic properties of discrete analytic functions, *Duke Math. J.* **23** (1956), 335–363.

[5] Ferrand, J.: Fonctions préharmonique et fonctions préholomorphes, *Bulletin des Sciences Mathématiques*, sec. series, vol. **68** (1944), 152–180.

[6] Gürlebeck, K.: Zur Berechnung von Fundamentallösungen diskreter Cauchy–Riemann–Operatoren. *ZAMM* **74** (1994) 6, T625–T627.

[7] Gürlebeck, K., and Hommel, A.: Finite difference Cauchy–Riemann operators and their fundamental solutions in the complex case. *Operator Theory*, vol. **142**, 101–115.

[8] Gürlebeck, K., and Hommel, A.: On fundamental solutions of the heat conduction difference operator, *Z. Anal. Anw.* **13** (1994), No. 3, 425–441.

[9] Gürlebeck, K., and Hommel, A.: On finite difference Dirac operators and their fundamental solutions, *Advances in Applied Clifford Algebras* **11** (S2) (2001), 89–106.

[10] Gürlebeck, K., and Sprößig, W.: *Quaternionic and Clifford Calculus for Engineers and Physicists*, John Wiley &. Sons, Chichester, 1997.

[11] Hayabara, S.: Operational calculus on the discrete analytic functions, *Math. Japon.* **11** (1966), 35–65.

[12] Hommel, A.: Construction of a right inverse operator to the discrete Cauchy–Riemann operator, *Progress in Analysis*, Proceedings of the 3rd International ISAAC Congress, Vol. I, H. Begehr, R. Gilbert, M.W. Wong, Eds., World Scientific, New Jersey, London, Singapore, Hong Hong, 2003.

[13] Hommel, A.: Fundamentallösungen partieller Differenzenoperatoren und die Lösung diskreter Randwertprobleme mit Hilfe von Differenzenpotentialen. Dissertation, Bauhaus–Universität Weimar, 1998.

[14] Kravchenko, V.V., and Shapiro, M.V., *Integral Representations for Spatial Models of Mathematical Physics*, Research Notes in Mathematics **351**, Pitman Advanced Publishing Program, London, 1996.

[15] Maeda, F.-Y., Murakami, A., and Yamasaki, M.: Discrete initial value problems and discrete parabolic potential theory, *Hiroshima Math. J.* **21** (1991), 285–299.

[16] Pfeifer, E., and Rauhöft, A.: Über Grundlösungen von Differenzenoperatoren. *Z. Anal. Anw.* **3** (1984), 227–236.

[17] Ryabenkij, V. S.: *The Method of Difference Potentials for Some Problems of Continuum Mechanics* (Russian). Moscow, Nauka, 1987.

[18] Sobolev, S. L.: Über die Eindeutigkeit der Lösung von Differenzengleichungen des elliptischen Typs. *Doklady Akad. Nauk SSSR* **87** (1952), No. 2, 179–182 (Russian)

[19] Sobolev, S. L.: Über eine Differenzengleichung. *Doklady Akad. Nauk SSSR* **87** (1952), No. 3, 341–343 (Russian)

[20] Stummel, F.: Elliptische Differenzenoperatoren unter Dirichletrandbedingungen. *Math. Z.* **97** (1967), 169–211.

[21] Thomée, V.: Discrete interior Schauder estimates for elliptic difference operators. *SIAM J. Numer. Anal.* **5** (1968), 626–645.

Angela Hommel
Institut für Mathematik/Physik
Bauhaus-Universität Weimar
Coudraystr. 13 B
99423 Weimar, Germany
E-mail: angela.hommel@bauing.uni-weimar.de

Klaus Gürlebeck
Institut für Mathematik/Physik
Bauhaus-Universität Weimar
Coudraystr. 13 B
99423 Weimar, Germany
E-mail: guerlebe@fossi.uni-weimar.de

Submitted: October 7, 2002; Revised: February 28, 2003.

4

A Symmetric Functional Calculus for Systems of Operators of Type ω

Brian Jefferies

ABSTRACT For a system $A = (A_1, \ldots, A_n)$ of linear operators whose real linear combinations have spectra contained in a fixed sector in $\mathbb{C}$ and satisfy resolvent bounds there, functions $f(A)$ of the system A of operators can be formed for monogenic functions f having decay at zero and infinity in a corresponding sector in $\mathbb{R}^{n+1}$. In the case that the operators $A_1, \ldots, A_n$ commute with each other and satisfy square function estimates in Hilbert space, the correspondence between bounded monogenic functions defined in a sector in $\mathbb{R}^{n+1}$ and bounded holomorphic functions defined in a sector in $\mathbb{C}^n$ is used to define the functional calculus $f \mapsto f(A)$ for bounded holomorphic functions f in a sector of $\mathbb{C}^n$. The treatment includes the Dirac operator on a Lipschitz surface in $\mathbb{R}^{n+1}$.

Keywords: functional calculus, plane wave formula, Dirac operator

4.1   Introduction

For a finite system $A = (A_1, \ldots, A_n)$ of bounded linear operators acting on a Banach space X, functions $f(A)$ of the n-tuple A can be formed via the Cauchy integral formula

$$f(A) = \int_{\partial\Omega} G_x(A)\mathbf{n}(x)f(x)\,d\mu(x), \qquad (1.1)$$

just under the assumption that the spectrum $\sigma(\langle A, \xi \rangle)$ of the operator

$$\langle A, \xi \rangle := \sum_{j=1}^{n} A_j \xi_j$$

is a subset of $\mathbb{R}$ for every $\xi \in \mathbb{R}^n$ [6]. The Cauchy kernel $x \mapsto G_x(A)$ is defined outside a distinguished subset $\gamma(A)$ of $\mathbb{R}^n$, the open set $\Omega \subset \mathbb{R}^{n+1}$ contains $\gamma(A)$ and has a smooth oriented boundary $\partial\Omega$ with volume measure μ and outward unit normal $\mathbf{n}(x)$ at each point x of the boundary. The function f is defined in a neighbourhood of the closure $\bar{\Omega}$ in $\mathbb{R}^{n+1}$, takes values in the Clifford algebra $C\ell(\mathbb{C}^n)$

AMS Subject Classification: 47A60, 46H30, 30G35, 47A13.

over $\mathbb{C}^n$ and is left monogenic in the sense of Clifford analysis. In particular, the bounded linear operator $f(A) : X \to X$ is defined for any real analytic function $f : U \to \mathbb{C}$ defined in a neighbourhood U of $\gamma(A)$ in $\mathbb{R}^n$, simply by replacing f in formula (1.1) by its left monogenic extension to an open set in $\mathbb{R}^{n+1}$ [6]. For a polynomial p in n real variables, $p(A)$ is the operator formed by substituting symmetric products in the bounded linear operators $A_1, \ldots, A_n$ for the monomial expressions in p that is, we have a *symmetric* functional calculus in the n operators $A_1, \ldots, A_n$. In the case $n = 1$, we obtain the usual Riesz–Dunford functional calculus for a single operator. By analogy with spectral theory in one operator, the set $\gamma(A)$ is called the *monogenic spectrum* of the n-tuple $A = (A_1, \ldots, A_n)$.

The key ingredient of this approach is the Cauchy kernel $G_x(A)$. In [6], it is defined by the plane wave formula

$$G_x(A) = \frac{(n-1)!}{2} \left(\frac{i}{2\pi} \right)^n \operatorname{sgn}(x_0)^{n-1}$$

$$\times \int_{S^{n-1}} (\mathbf{e}_0 + is) \left(\langle \vec{x}, s \rangle I - \langle A, s \rangle - x_0 s I \right)^{-n} ds \quad (1.2)$$

for all $x = x_0 \mathbf{e}_0 + \vec{x}$ with x_0 a nonzero real number and $\vec{x} \in \mathbb{R}^n$. Here S^{n-1} is the unit $(n-1)$-sphere in $\mathbb{R}^n$, ds is surface measure and the inverse power $\left(\langle \vec{x}I - A, s \rangle - x_0 s \right)^{-n}$ is taken in the Clifford module $\mathcal{L}(X) \otimes C_{(n)}$. The spectral reality condition

$$\sigma(\langle A, \xi \rangle) \subset \mathbb{R}, \quad \text{for all} \quad \xi \in \mathbb{R}^n, \quad (1.3)$$

ensures the invertibility of $(\langle \vec{x}I - A, s \rangle - x_0 s)$ for all $x_0 \neq 0$ and $s \in S^{n-1}$ by the spectral mapping theorem. The Cauchy kernel $G_x(A)$ given by formula (1.2) coincides with the convergent series expansion of $G_x(A)$ for large $|x|$.

If we now pass to unbounded operators, then a similar analysis holds if we retain the spectral reality condition (1.3), provided that we suitably account for operator domains.

The question of forming functions of noncommuting systems of operators arises in quantum physics [14] and the connection with Clifford analysis is already apparent in [9]. Explicit calculations can be made with two hermitian matrices [7]. The application of ideas from Clifford analysis to Feynman's operational calculus is pursued in joint work of the author with G.W. Johnson [5].

The examination of functional calculi for n-tuples of operators, and formula (1.2) in particular, was also motivated by the study of the commuting n-tuple $D_\Sigma = (D_1, \ldots, D_n)$ of differentiation operators on a Lipschitz surface Σ in $\mathbb{R}^{n+1}$, see [13]. In the case that Σ is just the flat surface $\mathbb{R}^n$, the operators

$$D_j = \frac{1}{i} \frac{\partial}{\partial x_j}, \quad j = 1, \ldots, n,$$

commute with each other and are selfadjoint, otherwise, the unbounded operators $D_j, j = 1, \ldots, n$, have spectra $\sigma(D_j)$ contained in a complex sector

$$S_\omega(\mathbb{C}) = \{ z \in \mathbb{C} : z \neq 0, \ |\arg(z)| \leq \omega \}$$

with an angle ω depending on the variation of the directions normal to the surface Σ.

The existence and properties of the H^∞-functional calculus for the commuting n-tuple D_Σ are now well–understood, see for example [13]. Here we show how a functional calculus for an n-tuple A of unbounded operators of type ω can be constructed directly and to apply this to the problem of forming bounded linear operators $f(D_\Sigma)$ acting on $L^p(\Sigma)$ when f is a bounded holomorphic function defined on a suitable sector in $\mathbb{C}^n$. The construction of an H^∞-functional calculus for D_Σ has applications to the solution of irregular boundary value problems. The spectral reality condition (1.3) needs to be replaced by a condition where the spectra are contained in a fixed sector of angle ω in the complex plane in order to treat systems of operators such as D_Σ. The convergence of the integral on the right–hand side of the plane wave formula (1.2) for the Cauchy kernel is then at issue.

In Section 4.2, it is shown how formula (1.2) for the Cauchy kernel associated with the system A of sectorial operators still works if the spectral reality condition (1.3) is replaced by a sectoriality condition with the appropriate resolvent bounds. The system D_Σ of commuting sectorial operators described above is of this type. By this means functions $f(A)$ of the operators A can be formed, provided that f is left monogenic in a sector in $\mathbb{R}^{n+1}$ and satisfies suitable decay estimates at 0 and ∞, in a fashion similar to the case of a single operator of type ω [12]. Because $G_x(A)$ is only defined for x outside a sector in $\mathbb{R}^{n+1}$, the monogenic spectrum $\gamma(A)$ is now contained in that sector in $\mathbb{R}^{n+1}$. Recall that under condition (1.3), $\gamma(A)$ is a subset of $\mathbb{R}^n \equiv \mathbb{R}^n \times \{0\}$.

A function $f(D_\Sigma)$ of the system D_Σ has a natural interpretation as a multiplier operator acting on L^p-spaces of functions defined on the Lipschitz surface Σ as well as a singular convolution operator, so the multiplier f should be a bounded holomorphic function defined on a sector in $\mathbb{C}^n$ [13], rather than a bounded monogenic function defined in a sector in $\mathbb{R}^{n+1}$.

In this work, we use the recently proven bijection [3] between bounded monogenic functions defined on a sector in $\mathbb{R}^{n+1}$ and bounded holomorphic functions defined on a sector in $\mathbb{C}^n$ to resolve this apparent dilemma about what is the appropriate function space for the symmetric functional calculus. The association between the two function spaces is via the Cauchy–Kowaleski extension to a sector in $\mathbb{R}^{n+1}$ of the restriction of the holomorphic function to $\mathbb{R}^n \setminus \{0\}$.

The bijection between bounded monogenic functions and bounded holomorphic functions in sectors is described in Section 4.3. In Section 4.4, this enables us to form functions $f(A)$ of a commuting system A of operators acting in a Hilbert space when the single operator $\sum_{j=1}^n A_j \mathbf{e}_j$ satisfies square function estimates and f is bounded and holomorphic in a sector.

Some notation and facts from Clifford analysis [1, 2] follow.

Let $C\ell(\mathbb{C}^n)$ be the *Clifford algebra* generated over the field $\mathbb{C}$ by the standard basis vectors $\mathbf{e}_0, \mathbf{e}_1, \ldots, \mathbf{e}_n$ of $\mathbb{R}^{n+1}$ with conjugation $u \mapsto \bar{u}$. The *generalized*

Cauchy–Riemann operator is given by

$$D = \sum_{j=0}^{n} \mathbf{e}_j \frac{\partial}{\partial x_j}.$$

Let $U \subset \mathbb{R}^{n+1}$ be an open set. A function $f : U \to C\ell(\mathbb{C}^n)$ is called *left monogenic* if $Df = 0$ in U and *right monogenic* if $fD = 0$ in U. The *Cauchy kernel* is given by

$$G_x(y) = \frac{1}{\sigma_n} \frac{x-y}{|x-y|^{n+1}}, \quad x, y \in \mathbb{R}^{n+1}, x \neq y, \tag{1.4}$$

with $\sigma_n = 2\pi^{\frac{n+1}{2}}/\Gamma\left(\frac{n+1}{2}\right)$ the volume of the unit n-sphere in $\mathbb{R}^{n+1}$. So, given a left monogenic function $f : U \to C\ell(\mathbb{C}^n)$ defined in an open subset U of $\mathbb{R}^{n+1}$ and an open subset Ω of U such that the closure $\overline{\Omega}$ of Ω is contained in U, and the boundary $\partial\Omega$ of Ω is a smooth oriented n-manifold, then the Cauchy integral formula

$$f(y) = \int_{\partial\Omega} G_x(y)\mathbf{n}(x)f(x)\,d\mu(x), \quad y \in \Omega$$

is valid. Here $\mathbf{n}(x)$ is the outward unit normal at $x \in \partial\Omega$ and μ is the volume measure of the oriented manifold $\partial\Omega$. An element $x = (x_0, x_1, \ldots, x_n)$ of $\mathbb{R}^{n+1}$ will often be written as $x = x_0\mathbf{e}_0 + \vec{x}$ with $\vec{x} = \sum_{j=1}^{n} x_j\mathbf{e}_j$.

4.2   The plane wave decomposition

Let $A = (A_1, \ldots, A_n)$ be an n-tuple of densely defined linear operators

$$A_j : \mathcal{D}(A_j) \to X$$

acting in X such that $\cap_{j=1}^{n} \mathcal{D}(A_j)$ is dense in X. The space $\mathcal{L}_{(n)}(X_{(n)})$ of left module homomorphisms of $X_{(n)} = X \otimes C\ell(\mathbb{C}^n)$ is identified with $\mathcal{L}(X) \otimes C\ell(\mathbb{C}^n)$ in the natural way and becomes a right Banach module under the uniform operator topology.

If we take formula (1.2) as the definition of the Cauchy kernel $G_x(A)$, then we need to establish the convergence of the integral

$$\int_{S^{n-1}} (\mathbf{e}_0 + is)\left(\langle \vec{x}I - A, s\rangle - x_0 sI\right)^{-n} ds$$

for particular values of $x = x_0\mathbf{e}_0 + \vec{x} \in \mathbb{R}^{n+1}$. Now

$$\left(\langle \vec{x}I - A, s\rangle - x_0 sI\right)^{-1} = \left(\langle \vec{x}I - A, s\rangle + x_0 sI\right)\left(\langle \vec{x}I - A, s\rangle^2 + x_0^2 I\right)^{-1}$$

if $0 \notin \sigma\left(\langle \vec{x}I - A, s\rangle^2 + x_0^2\right)$. Thus, we need to ensure the appropriate uniform operator bounds for

$$\left(\langle \vec{x}I - A, s\rangle^2 + x_0^2 I\right)^{-1}, \quad s \in S^{n-1}$$

as $x = x_0 \mathbf{e}_0 + \vec{x}$ ranges over a subset of $\mathbb{R}^{n+1}$. In the case that $\sigma(\langle A, s \rangle) \subset \mathbb{R}$ and $(\lambda I - \langle A, s \rangle)^{-1}$ is suitably bounded for all $s \in S^{n-1}$ and $\lambda \in \mathbb{C} \setminus \mathbb{R}$, then $G_{x_0 \mathbf{e}_0 + \vec{x}}(A)$ is defined for all $x_0 \neq 0$. First, for $0 < \nu < \frac{1}{2}\pi$, set

$$S_\nu(\mathbb{C}) = \{z \in \mathbb{C} : |\arg z| \leq \nu\} \cup \{0\}, \quad S_\nu^\circ(\mathbb{C}) = \{z \in \mathbb{C} : |\arg z| < \nu\}.$$

The set of $s \in S^{n-1}$ with nonzero coordinates s_j for every $j = 1, \ldots, n$ is denoted by S_0^{n-1}. Then S_0^{n-1} is a dense open subset of S^{n-1} with full surface measure.

Definition 1. *Let $A = (A_1, \ldots, A_n)$ be an n-tuple of closed, densely defined linear operators $A_j : \mathcal{D}(A_j) \to X$ acting in X such that $\cap_{j=1}^n \mathcal{D}(A_j)$ is dense in X and let $0 \leq \omega < \frac{\pi}{2}$. The operator*

$$\langle A, s \rangle : \cap_{j=1}^n \mathcal{D}(A_j) \to X$$

is defined by

$$\langle A, s \rangle u = \sum_{j=1}^n s_j (A_j u), \quad \forall s \in S^{n-1}, \, u \in \cap_{j=1}^n \mathcal{D}(A_j).$$

Then A is said to be uniformly of type ω if $\sigma(\langle A, s \rangle) \subset S_\omega(\mathbb{C})$ for all $s \in S_0^{n-1}$ and for each $\nu > \omega$, there exists $C_\nu > 0$ such that

$$\|(zI - \langle A, s \rangle)^{-1}\| \leq C_\nu |z|^{-1}, \quad z \notin S_\nu(\mathbb{C}), \, s \in S_0^{n-1}. \tag{2.1}$$

If $n = 1$, the A is just said to be of type ω.

It follows that $s \mapsto \langle A, s \rangle$ is continuous on S_0^{n-1} in the sense of strong resolvent convergence [8, Theorem VIII.1.5]. The subset S_0^{n-1} of S^{n-1} is used here simply because $\cap_{j=1}^n \mathcal{D}(A_j)$ may be strictly contained in $\mathcal{D}(A_k)$ for $k = 1, \ldots, n$.

Now suppose that equation (2.1) is satisfied and let $z = \langle \vec{x}, s \rangle + ix_0$. Then $z \notin S_\nu^\circ(\mathbb{C})$ means that $|x_0| \geq \tan \nu |\langle \vec{x}, s \rangle|$.

First, let

$$N_\nu = \{x \in \mathbb{R}^{n+1} : x = x_0 \mathbf{e}_0 + \vec{x}, \, |x_0| \geq \tan \nu |\vec{x}| \},$$

$$S_\nu(\mathbb{R}^{n+1}) = \{x \in \mathbb{R}^{n+1} : x = x_0 \mathbf{e}_0 + \vec{x}, \, |x_0| \leq \tan \nu |\vec{x}| \},$$

$$S_\nu^\circ(\mathbb{R}^{n+1}) = \{x \in \mathbb{R}^{n+1} : x = x_0 \mathbf{e}_0 + \vec{x}, \, |x_0| < \tan \nu |\vec{x}| \}.$$

Note that if $x_0 \mathbf{e}_0 + \vec{x} \in N_\nu$, then $z = \langle \vec{x}, s \rangle + ix_0 \notin S_\nu^\circ(\mathbb{C})$ for every $s \in S^{n-1}$, because $|x_0| \geq \tan \nu |\vec{x}| \geq \tan \nu |\langle \vec{x}, s \rangle|$.

The proof of the following result is straightforward and appears in [4].

Lemma 1. *Let $\omega < \nu < \frac{1}{2}\pi$. Suppose that the n-tuple A of linear operators is uniformly of type ω. Then for all $x_0 \mathbf{e}_0 + \vec{x} \in N_\nu$, the integral*

$$\int_{S^{n-1}} \left\| ((\vec{x} I - A, s) - x_0 s I)^{-n} \right\|_{\mathcal{L}_{(n)}(X_{(n)})} \, ds$$

converges and satisfies the bound

$$\int_{S^{n-1}} \left\| (\langle \vec{x}I - A, s \rangle - x_0 sI)^{-n} \right\|_{\mathcal{L}_{(n)}(X_{(n)})} ds \le \frac{C'_\nu}{|x_0|^n}.$$

Thus, if A is uniformly of type ω, then

$$x_0 \mathbf{e}_0 + \vec{x} \mapsto G_{x_0 \mathbf{e}_0 + \vec{x}}(A) \tag{2.2}$$

is defined by the plane wave formula (1.2) for all $x_0 \mathbf{e}_0 + \vec{x} \in N_\nu$ with $\omega < \nu < \frac{1}{2}\pi$. Standard arguments ensure that (2.2) is both left and right monogenic as an element of $\mathcal{L}_{(n)}(X_{(n)})$. If we denote by $\gamma(A) \subset \mathbb{R}^{n+1}$ the set of all singularities of function (2.2), then $\gamma(A) \subseteq S_\omega(\mathbb{R}^{n+1})$.

Suppose that $\omega < \nu < \frac{1}{2}\pi$, $0 < s < n$ and f is a left monogenic function defined on $S_\nu^\circ(\mathbb{R}^{n+1})$ such that for every $0 < \nu' < \theta < \nu$ there exists $C_{\theta,\nu'} > 0$ such that

$$|f(x)| \le C_{\theta,\nu'} \frac{|x|^s}{1 + |x|^{2s}}, \quad x \in S_\theta^\circ(\mathbb{R}^{n+1}) \cap N_{\nu'}. \tag{2.3}$$

According to Lemma 1, for every $\omega < \nu' < \theta < \nu$, we have

$$\|G_x(A)\|.|f(x)| \le C'_{\theta,\nu'} \frac{|x|^s}{|x_0|^n (1 + |x|^{2s})}, \quad x = x_0 \mathbf{e}_0 + \vec{x}$$

for all $x \in S_\theta^\circ(\mathbb{R}^{n+1}) \cap N_{\nu'}$. Now if $\omega < \theta < \nu$ and

$$H_\theta = \{x \in \mathbb{R}^{n+1} : x = x_0 \mathbf{e}_0 + \vec{x}, \; |x_0|/|x| = \tan\theta\} \subset S_\nu^\circ(\mathbb{R}^{n+1}) \tag{2.4}$$

it follows that

$$\|G_x(A)\|.|f(x)| = O(1/|x|^{n-s}) \quad \text{as} \quad x \to 0 \quad \text{in} \quad H_\theta.$$

Hence, $x \mapsto G_x(A)\mathbf{n}(x)f(x)$ is locally integrable at zero with respect to n-dimensional surface measure μ on H_θ. Similarly,

$$\|G_x(A)\|.|f(x)| = O(1/|x|^{n+s}) \quad \text{as} \quad |x| \to \infty \quad \text{in} \quad H_\theta,$$

so $x \mapsto G_x(A)\mathbf{n}(x)f(x)$ is integrable with respect to n-dimensional surface measure on H_θ. Therefore, we define the element $f(A)$ of the module $\mathcal{L}_{(n)}(X_{(n)})$ by the formula

$$f(A) = \int_{H_\theta} G_x(A)\mathbf{n}(x)f(x) \, d\mu(x). \tag{2.5}$$

If $\psi : \mathbb{R}^n \setminus \{0\} \to \mathbb{C}$ has a two-sided monogenic extension $\tilde{\psi}$ to $S_\nu^\circ(\mathbb{R}^{n+1})$ that satisfies the bound (2.3) for all $0 < \theta < \nu$, then $\tilde{\psi}(A)$ is written just as $\psi(A)$. Formula (2.5) does just what we would expect in the noncommuting situation. For example, let p be a polynomial of degree n with

$$p(0) = 0 \quad \text{and} \quad b_\lambda(z) = p(z)(\lambda - z)^{-n-1}$$

for some $\lambda \notin S_\nu^\circ(\mathbb{C})$. Let $\xi \in \mathbb{R}^n$ and set

$$\phi_{\lambda,\xi}(\vec{x}) = b_\lambda(\langle \vec{x}, \xi \rangle)$$

for all $\vec{x} \in \mathbb{R}^n$. Denote the two-sided monogenic extension of $\phi_{\lambda,\xi}$ to $S_\nu^\circ(\mathbb{R}^{n+1})$ by $\tilde{\phi}_{\lambda,\xi}$. Then $\tilde{\phi}_{\lambda,\xi}$ has decay at zero and infinity and we have

$$\phi_{\lambda,\xi}(A) = \tilde{\phi}_{\lambda,\xi}(A) = p(\langle A, \xi \rangle)(\lambda I - \langle A, \xi \rangle)^{-n-1}$$

is a bounded linear operator. More generally, the next result shows that formula (2.5) gives the right result when for a special class of functions f, there is a representation of the bounded linear operator $f(A)$ by an alternative formula, namely, the usual Riesz–Dunford functional calculus. Set $D_r = \{z \in \mathbb{C} : |z| < r\}$ for each $r > 0$.

Theorem 1. *Let $0 < \omega < \nu < \frac{1}{2}\pi$ and suppose that $A = (A_1, \ldots, A_n)$ is an n-tuple of operators uniformly of type ω with the property that for each $\omega < \omega' < \nu$, there exist n-tuples $A^{(k)} = \left(A_1^{(k)}, \ldots, A_n^{(k)}\right)$, $k = 1, 2, \ldots$, of bounded linear operators such that, with ω' replacing ω, the bound (2.1) is satisfied uniformly for $k = 1, 2, \ldots$ and*

$$(zI - \langle A^{(k)}, s \rangle)^{-1} \longrightarrow (zI - \langle A, s \rangle)^{-1}, \quad z \notin S_\nu(\mathbb{C}), \; s \in S_0^{n-1}, \quad (2.6)$$

as $k \to \infty$.

Suppose that $0 < s < 1$, $0 < \omega < \nu$, $r > 0$ and $\phi : S_\nu^\circ(\mathbb{C}) \cup D_r \to \mathbb{C}$ is a holomorphic function satisfying $\phi(0) = 0$, such that for each $0 < \nu' < \nu$, there exists $C_{\nu'} > 0$ with

$$|\phi(z)| \le \frac{C_{\nu'}}{1 + |z|^s}, \quad z \in S_{\nu'}^\circ(\mathbb{C}).$$

Let $\xi \in \mathbb{R}^n$ be a vector with nonzero components. Then the function

$$\vec{x} \longmapsto \phi(\langle \vec{x}, \xi \rangle), \quad \vec{x} \in \mathbb{R}^n \setminus \{0\},$$

has a two-sided monogenic extension f to $S_\nu^\circ(\mathbb{R}^{n+1})$ satisfying the bounds (2.3) and the operator $f(A) \in \mathcal{L}(X)$ given by formula (2.5) has the representation

$$f(A) = \phi(\langle A, \xi \rangle) = \frac{1}{2\pi i} \int_C (\lambda I - \langle A, \xi \rangle)^{-1} \phi(\lambda) \, d\lambda. \quad (2.7)$$

The contour C can be taken to be $\{z \in \mathbb{C} : |\operatorname{Im}(z)| = \tan\theta |\operatorname{Re}(z)| \}$, with $0 < \omega < \theta < \nu$.

Proof. We can suppose that $\xi \in S_0^{n-1}$ by scaling. It is routine to check that the two-sided monogenic function f is given by the Cauchy integral formula

$$f(x) = \frac{1}{2\pi i} \int_C (z - \langle \vec{x}, \xi \rangle + x_0\xi)^{-1} \phi(z) \, dz, \quad x \in S_{\nu'}^\circ(\mathbb{R}^{n+1}),$$

for all $x = x_0 e_0 + \vec{x} \in S^\circ_{\nu'}(\mathbb{R}^{n+1})$ and all $0 < \nu' < \theta$. Furthermore, as noted in [13, Section 5.2], the element $i\xi$ of the Clifford algebra $C\ell(\mathbb{C}^n)$ has the spectral decomposition

$$i\xi = \chi_+(\xi)|\xi| - \chi_-(\xi)|\xi|$$

with respect to the projections

$$\chi_\pm = \frac{1}{2}\left(1 \pm \frac{i\xi}{|\xi|}\right),$$

so that f also has the representation

$$f(x) = \phi(\langle\vec{x},\xi\rangle + ix_0|\xi|)\chi_+(\xi) + \phi(\langle\vec{x},\xi\rangle - ix_0|\xi|)\chi_-(\xi),$$

for all $x = x_0 e_0 + \vec{x} \in S^\circ_\nu(\mathbb{R}^{n+1})$. Hence, f satisfies the bounds (2.3). Note that $\langle\vec{x},\xi\rangle = 0$ if $\vec{x}$ is orthogonal to ξ, but by assumption in (2.3), $|x_0| \geq \tan\nu'|\vec{x}|$ for $\nu' > 0$.

Now suppose that $\omega' < \theta < \nu$. We first observe that

$$\int_{H_\theta} G_x(\tau A^{(k)})\mathbf{n}(x)(\zeta - \langle\vec{x},\xi\rangle + x_0\xi)^{-1}\,d\mu(x) = (\zeta I - \tau\langle A^{(k)},\xi\rangle)^{-1} \quad (2.8)$$

for every $\zeta \in \mathbb{C} \setminus S_\nu(\mathbb{C})$ and $\tau \in \mathbb{R}$. The (improper) integral is over the cone

$$H_\theta = \{x_0 e_0 + \vec{x} \in \mathbb{R}^{n+1} : |x_0| = \tan\theta|\vec{x}|\},$$

which can be deformed by Cauchy's theorem in Clifford analysis to the integral over a ball B_r of radius $r > \|A^{(k)}\|$ centred at zero in $\mathbb{R}^{n+1}$ when $|\zeta|$ is large enough. Both sides of equation (2.8) are holomorphic for τ in a neighbourhood of zero in $\mathbb{C}$ and have the same Taylor series about zero, because

$$\frac{d^j}{d\tau^j}\int_{H_\theta} G_x(\tau A^{(k)})\mathbf{n}(x)(\zeta - \langle\vec{x},\xi\rangle + x_0\xi)^{-1}\,d\mu(x)$$

$$= \int_{H_\theta} [\langle -A^{(k)},\nabla_{\vec{x}}\rangle^j G_x(\tau A^{(k)})]\mathbf{n}(x)(\zeta - \langle\vec{x},\xi\rangle + x_0\xi)^{-1}\,d\mu(x)$$

$$= \int_{\mathbb{R}^n \pm \epsilon e_0} \langle -A^{(k)},\nabla_{\vec{x}}\rangle^j G_x(\tau A^{(k)})\mathbf{n}(x)(\zeta - \langle\vec{x},\xi\rangle + x_0\xi)^{-1}\,d\mu(x)$$

$$= \int_{\mathbb{R}^n \pm \epsilon e_0} G_x(\tau A^{(k)})\mathbf{n}(x)\langle A^{(k)},\nabla_{\vec{x}}\rangle^j(\zeta - \langle\vec{x},\xi\rangle + x_0\xi)^{-1}\,d\mu(x)$$

$$\longrightarrow j!\zeta^{-j-1}\langle A^{(k)},\xi\rangle^j \quad \text{as } \tau \to 0.$$

Here we have used the observation, apparent from the plane wave formula (1.2), that for each $k = 1, 2, \ldots$, the Cauchy kernel $(x,\tau) \mapsto G_x(\tau A^{(k)})$ satisfies the operator equation

$$\frac{\partial}{\partial\tau}G_x(\tau A^{(k)}) + \langle A^{(k)},\nabla_{\vec{x}}\rangle G_x(\tau A^{(k)}) = 0$$

for all $x \in N_\nu$. Then the equality (2.8) is obtained for all $\zeta \in \mathbb{C} \setminus S_\nu(\mathbb{C})$ by analytic continuation of both sides. Next, the cone H_θ can be deformed to an n-dimensional surface consisting of part of the sphere of radius $0 < r' < r$ together with $H_\theta \setminus B_{r'}$ and the contour C can be deformed so as to avoid the disk $D_{r'}$. Then it follows that the equalities

$$f(A^{(k)}) = \int_{H_\theta} G_x(A^{(k)}) \mathbf{n}(x) f(x) \, d\mu(x)$$

$$= \frac{1}{2\pi i} \int_{H_\theta} G_x(A^{(k)}) \mathbf{n}(x) \left[\int_C (\zeta - \langle \vec{x}, \xi \rangle + x_0 \xi)^{-1} \phi(\zeta) \, d\zeta \right] d\mu(x)$$

$$= \frac{1}{2\pi i} \int_C \left[\int_{H_\theta} G_x(A^{(k)}) \mathbf{n}(x) (\zeta - \langle \vec{x}, \xi \rangle + x_0 \xi)^{-1} \, d\mu(x) \right] \phi(\zeta) \, d\zeta$$

$$= \frac{1}{2\pi i} \int_C (\zeta I - \langle A^{(k)}, \xi \rangle)^{-1} \phi(\zeta) \, d\zeta$$

$$= \phi(\langle A^{(k)}, \xi \rangle)$$

hold. Now the assumption (2.6) of strong resolvent convergence guarantees that

$$\phi(\langle A^{(k)}, \xi \rangle) \longrightarrow \phi(\langle A, \xi \rangle)$$

in the strong operator topology as $k \to \infty$, so the convergence of the integrals

$$f(A^{(k)}) = \int_{H_\theta} G_x(A^{(k)}) \mathbf{n}(x) f(x) \, d\mu(x)$$

to $f(A)$ as $k \to \infty$ remains to be established. According to Lemma 1, we can find $C > 0$ such that

$$\|G_x(A^{(k)})\|_{\mathcal{L}_{(n)}(X_{(n)})} \le \frac{C}{|x_0|^n}$$

for all $x \in H_\theta$ and $k = 1, 2, \ldots$. Moreover, the plane wave formula (1.2) and strong resolvent convergence (2.6) ensures that $G_x(A^{(k)}) \to G_x(A)$ in the strong operator topology for each nonzero $x \in H_\theta$. The monogenic function f has decay at zero and infinity, so dominated convergence ensures that $f(A^{(k)}) \to f(A)$ as $k \to \infty$. $\qquad \square$

The approximation (2.6) by bounded operators is somewhat simple-minded. A more sophisticated approximation should yield the result for general systems of operators uniformly of type ω. In order to form functions $f(A)$ of the system A of operators for a class of monogenic functions f larger than those which satisfy a bound like (2.3), a greater understanding of function theory in the sector $S_\omega(\mathbb{R}^{n+1})$ is needed. To this end, the simple system $A = \zeta = (\zeta_1, \ldots, \zeta_n) \in \mathbb{C}^n$ of multiplication operators in the algebra $C\ell(\mathbb{C}^n)$ is considered in the next section.

4.3   Function theory in sectors

For the particular example of the n-tuple D_Σ of differentiation operators on a Lipschitz surface Σ mentioned in the Introduction, the theory described in [10] associates a bounded linear operator $f(D_\Sigma)$ with a Fourier multiplier f, this being a suitable holomorphic function of n complex variables defined in a sector in $\mathbb{C}^n$ associated with D_Σ. By means of formula (2.5), we have seen how to associate a bounded linear operator $f(D_\Sigma)$ with a left monogenic function f with suitable decay in a sector, once we verify that D_Σ satisfies the assumptions of Definition 1.

This suggests that we can associate a bounded holomorphic function of n complex variables in a sector in $\mathbb{C}^n$ with a left monogenic function f with suitable decay at zero and infinity in a sector in $\mathbb{R}^{n+1}$. The association is by analytic continuation from $\mathbb{R}^n \setminus \{0\}$ to a sector in $\mathbb{C}^n$. The precise formulation of this correspondence and its consequences is formulated in this section in Theorem 2. The proof is somewhat technical and will appear elsewhere [3]. In the next section, Theorem 2 and some additional estimates are used to extend the functional calculus defined by means of formula (2.5) to all bounded holomorphic functions defined in a sector $\mathbb{C}^n$, at least in the case when A is an n-tuple of commuting operators.

Let $0 < \nu < \frac{1}{2}\pi$ and let $H_\ell^\infty(S_\nu^\circ(\mathbb{R}^{n+1}))$ denote the set of all left monogenic functions

$$f : S_\nu^\circ(\mathbb{R}^{n+1}) \to C\ell(\mathbb{C}^n)$$

that are uniformly bounded on every subsector $S_{\nu'}^\circ(\mathbb{R}^{n+1})$, $0 < \nu' < \nu$. Endowed with the topology of uniform convergence on subsectors $S_{\nu'}^\circ(\mathbb{R}^{n+1})$, $0 < \nu' < \nu$, the topological vector space $H_\ell^\infty(S_\nu^\circ(\mathbb{R}^{n+1}))$ is a Fréchet space. The analogous space for right monogenic functions is written as $H_r^\infty(S_\nu^\circ(\mathbb{R}^{n+1}))$.

For any $f \in H_\ell^\infty(S_\nu^\circ(\mathbb{R}^{n+1}))$, the restriction

$$f_0 : \mathbb{R}^n \setminus \{0\} \to C\ell(\mathbb{C}^n)$$

of f to $\mathbb{R}^n \setminus \{0\}$ is real analytic and so it has a complex analytic extension $\tilde{f}$ to some open neighbourhood U_ν of $\mathbb{R}^n \setminus \{0\}$ in $\mathbb{C}^n \setminus \{0\}$. The argument above concerning Fourier multiplier operators on Lipschitz surfaces suggests that we can take U_ν to be some open sector in $\mathbb{C}^n$ on which $\tilde{f}$ is uniformly bounded on subsectors. The sectors in $\mathbb{C}^n$ are described as follows.

For $x \in \mathbb{R}^{n+1}$ and n complex variables $\zeta = (\zeta_1, \ldots, \zeta_n)$, the Cauchy kernel $G_x(\zeta)$ is understood to be the maximal analytic continuation of the Cauchy kernel $\vec{y} \mapsto G_x(\vec{y})$, $\vec{y} \in \mathbb{R}^n$, $\vec{y} \neq x$, given by formula (1.4). More precisely, let

$$|x - \zeta|_\mathbb{C}^2 = x_0^2 + \sum_{j=1}^n (x_j - \zeta_j)^2$$

and denote the positive square root of $|x - \zeta|_\mathbb{C}^2$ by $|x - \zeta|_\mathbb{C}$. In the case that x actually lies in $\{0\} \times \mathbb{R}^n \equiv \mathbb{R}^n$, the function $\zeta \mapsto |x - \zeta|_\mathbb{C}$ coincides with the analytic extension of the modulus function $\xi \mapsto |x - \xi|$, $\xi \in \mathbb{R}^n \setminus \{x\}$. Then

$$G_x(\zeta) = \frac{1}{\sigma_n} \frac{\overline{x} + \zeta}{|x - \zeta|_{\mathbb{C}}^{n+1}}, \quad x \in \mathbb{R}^{n+1}. \tag{3.1}$$

Here we take $|x - \zeta|_{\mathbb{C}}^2 \notin (-\infty, 0]$ if n is an even integer and $|x - \zeta|_{\mathbb{C}}^2 \neq 0$ if n is an odd integer. If the dimension n is even, then $|x - \zeta|_{\mathbb{C}}^{n+1}$ has a discontinuity as the complex number $|x - \zeta|_{\mathbb{C}}^2$ moves across $(-\infty, 0]$ in $\mathbb{C}$. If n is odd, we only require the denominator $|x - \zeta|_{\mathbb{C}}^{n+1}$ to be nonzero. Then the function $(x, \zeta) \mapsto G_x(\zeta)$ is two-sided monogenic in x and holomorphic in ζ on its given domain.

According to the point of view mentioned in the Introduction, for fixed $\zeta \in \mathbb{C}^n$, the set of singularities of the Cauchy kernel $G_x(\zeta)$ as a function of $x \in \mathbb{R}^{n+1}$ is called the *monogenic spectrum* of $\zeta \in \mathbb{C}^n$ and denoted by $\gamma(\zeta)$. In fact, we are just considering $\zeta \in \mathbb{C}^n$ as an n-tuple of multiplication operators in the Clifford algebra $C\ell(\mathbb{C}^n)$ and taking the natural definition of the Cauchy kernel $x \mapsto G_x(\zeta)$. We do not need the plane wave decomposition here—the Cauchy kernel $G_x(A)$ is defined by the L^∞-functional calculus for any commuting family $A = (A_1, \ldots, A_n)$ of normal operators in a Hilbert space.

To examine the subset $\gamma(\zeta)$ of $\mathbb{R}^n$ more closely, write $\zeta = \xi + i\eta$ for $\xi, \eta \in \mathbb{R}^n$ and $x = x_0 e_0 + \vec{x}$ for $x_0 \in \mathbb{R}$ and $\vec{x} \in \mathbb{R}^n$. Then

$$|x - \zeta|_{\mathbb{C}}^2 = x_0^2 + \sum_{j=1}^{n}(x_j - \zeta_j)^2 = x_0^2 + |\vec{x} - \xi|^2 - |\eta|^2 - 2i\langle \vec{x} - \xi, \eta \rangle. \tag{3.2}$$

Thus, $|x - \zeta|_{\mathbb{C}}^2$ belongs to $(-\infty, 0]$ if and only if x lies in the intersection hyperplane $\langle \vec{x} - \xi, \eta \rangle = 0$ passing through ξ and with normal η, and the ball $x_0^2 + |\vec{x} - \xi|^2 \leq |\eta|^2$ centred at ξ with radius $|\eta|$. If n is even, then

$$\gamma(\zeta) =$$
$$\{x = x_0 e_0 + \vec{x} \in \mathbb{R}^{n+1} : \langle \vec{x} - \xi, \eta \rangle = 0, \; x_0^2 + |\vec{x} - \xi|^2 \leq |\eta|^2 \}. \tag{3.3}$$

and if n is odd, then

$$\gamma(\zeta) =$$
$$\{x = x_0 e_0 + \vec{x} \in \mathbb{R}^{n+1} : \langle \vec{x} - \xi, \eta \rangle = 0, \; x_0^2 + |\vec{x} - \xi|^2 = |\eta|^2 \}. \tag{3.4}$$

In particular, if $\text{Im}(\zeta) = 0$, then $\gamma(\zeta) = \{\zeta\} \subset \mathbb{R}^n$.

Now suppose that $f \in H_\ell^\infty(S_\nu^o(\mathbb{R}^{n+1}))$. The Cauchy integral formula gives

$$\tilde{f}(\zeta) = \int_{\partial\Omega} G_x(\zeta) \mathbf{n}(x) f(x) \, d\mu(x) \tag{3.5}$$

for a bounded open neighbourhood Ω of $\gamma(\zeta)$ with smooth oriented boundary $\partial\Omega$, outward unit normal $\mathbf{n}(x)$ at $x \in \partial\Omega$ and surface measure μ. The integral does not depend on Ω because $G_x(\zeta)$ is right monogenic in x and f is left monogenic [1,

Corollary 9.3]. Differentiating under the integral sign shows that $\tilde{f}$ is holomorphic on the open subset

$$\{\gamma \in \mathbb{C}^n : \gamma(\zeta) \subset S_\nu^\circ(\mathbb{R}^{n+1})\} \quad \text{of} \quad \mathbb{C}^n,$$

which we now describe [4, Proposition 2.1].

Proposition 1. *Let* $\zeta \in \mathbb{C}^n \setminus \{0\}$ *and* $0 < \omega < \frac{1}{2}\pi$. *Then* $\gamma(\zeta) \subset S_\omega(\mathbb{R}^{n+1})$ *if and only if*

$$|\zeta|_{\mathbb{C}}^2 \neq (-\infty, 0] \quad \text{and} \quad |\mathrm{Im}(\zeta)| \leq \mathrm{Re}(|\zeta|_{\mathbb{C}}) \tan \omega. \tag{3.6}$$

For each $0 < \omega < \frac{1}{2}\pi$, let $S_\omega(\mathbb{C}^n)$ denote the set of all $\zeta \in \mathbb{C}^n$ satisfying condition (3.6) and let $S_\omega^\circ(\mathbb{C}^n)$ be its interior. The sector $S_\omega(\mathbb{C}^n)$ arose in [10] as the set of $\zeta \in \mathbb{C}^n$ for which the exponential functions

$$e_+(x, \zeta) = e^{i\langle \vec{x}, \zeta \rangle} e^{-x_0|\zeta|_{\mathbb{C}}} \chi_+(\zeta), \quad x = x_0 e_0 + \vec{x},$$

have a decay at infinity for all $x \in \mathbb{R}^{n+1}$ with $\langle x, m \rangle > 0$ and all unit vectors $m = m_0 e_0 + \vec{m} \in \mathbb{R}^{n+1}$ satisfying $m_0 \geq \cot \omega |\vec{m}|$.

For each $0 < \nu < \frac{1}{2}\pi$, let $H^\infty(S_\nu^\circ(\mathbb{C}^n))$ denote the set of all holomorphic functions

$$f : S_\nu^\circ(\mathbb{C}^n) \to C\ell(\mathbb{R}^n)$$

which are uniformly bounded on every subsector

$$S_{\nu'}^\circ(\mathbb{C}^n), \quad 0 < \nu' < \nu. \tag{3.7}$$

Endowed with the topology of uniform convergence on subsectors (3.7) the topological vector space $H^\infty(S_\nu^\circ(\mathbb{C}^n))$ is a Fréchet space and a (nonabelian) Fréchet algebra under pointwise multiplication. The subalgebra of $\mathbb{C}$-valued functions is, of course, an abelian Fréchet algebra. The proof of the following result is given in [3].

Theorem 2. *The mapping* $f \mapsto \tilde{f}$ *given by the Cauchy integral formula (3.5) is an isomorphism between the Fréchet spaces* $H_\ell^\infty(S_\nu^\circ(\mathbb{R}^{n+1}))$ *and* $H^\infty(S_\nu^\circ(\mathbb{C}^n))$.

Given two functions $f, g \in H^\infty(S_\nu^\circ(\mathbb{R}^{n+1}))$, the restrictions

$$f_0 : \mathbb{R}^n \setminus \{0\} \to C\ell(\mathbb{C}^n) \quad \text{and} \quad g_0 : \mathbb{R}^n \setminus \{0\} \to C\ell(\mathbb{C}^n)$$

of f and g to $\mathbb{R}^n \setminus \{0\}$ are real analytic, so their product $f_0 g_0$ has a left monogenic extension $f \cdot_\ell g$ to an open neighbourhood of $\mathbb{R}^n \setminus \{0\}$ in $\mathbb{R}^{n+1}$. The analogous product for right monogenic functions is written as $f \cdot_r g$.

Corollary 1. *For each* $f, g \in H_\ell^\infty(S_\nu^\circ(\mathbb{R}^{n+1}))$, *the function* $f \cdot_\ell g$ *has a left monogenic extension to* $S_\nu^\circ(\mathbb{R}^{n+1})$, *denoted by the same symbol, and* $\tilde{f} \cdot_\ell g \in H_\ell^\infty(S_\nu^\circ(\mathbb{R}^{n+1}))$. *Moreover,* $H_\ell^\infty(S_\nu^\circ(\mathbb{R}^{n+1}))$ *is a Fréchet algebra with the product*

$$(f, g) \mapsto f \cdot_\ell g.$$

The mapping $f \mapsto \tilde{f}$ *given by the Cauchy integral formula (3.5) is an isomorphism of the Fréchet algebras* $H_\ell^\infty(S_\nu^\circ(\mathbb{R}^{n+1}))$ *and* $H^\infty(S_\nu^\circ(\mathbb{C}^n))$.

If the restrictions f_0, g_0 of f and g take values in $\mathbb{C}$ rather than $C\ell(\mathbb{R}^n)$, then both f and g are two-sided monogenic and

$$\widetilde{f \cdot_r g} = \widetilde{f \cdot_\ell g} = \tilde{f}.\tilde{g} = \tilde{g}.\tilde{f}.$$

Hence, $f \cdot_\ell g = g \cdot_\ell f$ and the subalgebra of $H_\ell^\infty(S_\nu^\circ(\mathbb{R}^{n+1}))$ consisting of such functions is abelian.

4.4   Joint spectral theory of operators of type ω

Suppose that $T : \mathcal{D}(T) \to \mathcal{H}$ is a single densely defined linear operator acting in the Hilbert space $\mathcal{H}$. If $0 \le \omega < \frac{1}{2}\pi$ is a number for which T is of type ω (see Definition 1), then the one–dimensional version of formula (2.5) gives a bounded linear operator $f(A)$ defined by

$$f(T) = \frac{1}{2\pi i} \int_C (\lambda I - T)^{-1} f(\lambda)\, d\lambda \tag{4.1}$$

for any function f satisfying the bounds (2.3) in $S_\nu(\mathbb{C})$ in the case $n = 1$. The contour C can be taken to be

$$\{ z \in \mathbb{C} : |\operatorname{Im}(z)| = \tan\theta|\operatorname{Re}(z)| \},$$

with $\omega < \theta < \nu$. The operator T of type ω is said to have a *bounded H^∞-functional calculus* if for each $\omega < \nu < \frac{1}{2}\pi$, there exists an algebra homomorphism $f \longmapsto f(T)$ from $H^\infty(S_\nu^\circ(\mathbb{C}))$ to $\mathcal{L}(\mathcal{H})$ agreeing with (4.1) and a positive number C_ν such that

$$\|f(T)\| \le C_\nu \|f\|_\infty \quad \text{for all} \quad f \in H^\infty(S_\nu^\circ(\mathbb{C})).$$

The following result is from [11, Theorem 6.2.2]

Theorem 3. *Suppose that T is a one-to-one operator of type ω in $\mathcal{H}$. Then T has a bounded H^∞-functional calculus if and only if for every $\omega < \nu < \frac{1}{2}\pi$, there exists $c_\nu > 0$ such that T and its adjoint T^* satisfy the square function estimates*

$$\int_0^\infty \|\psi_t(T)u\|^2 \frac{dt}{t} \le c_\nu \|u\|^2, \quad u \in \mathcal{H}, \tag{4.2}$$

$$\int_0^\infty \|\psi_t(T^*)u\|^2 \frac{dt}{t} \le c_\nu \|u\|^2, \quad u \in \mathcal{H}, \tag{4.3}$$

for some function $\psi \in H^\infty(S_\nu^\circ(\mathbb{C}))$, which satisfies

$$\int_0^\infty \psi^3(t) \frac{dt}{t} = \int_0^\infty \psi^3(-t) \frac{dt}{t} = 1, \text{ and} \tag{4.4}$$

$$|\psi(z)| \le C_\nu \frac{|z|^s}{1 + |z|^{2s}}, \quad z \in S_\nu^\circ(\mathbb{C}), \tag{4.5}$$

for some $s > 0$. Here $\psi_t(z) = \psi(tz)$ for $z \in S_\nu^\circ(\mathbb{C})$.

We now use formula (2.5) to generalise this result to n-tuples of commuting operators acting in a Hilbert space $\mathcal{H}$.

Theorem 4. *Let $A = (A_1, \ldots, A_n)$ be an n-tuple of densely defined commuting linear operators*

$$A_j : \mathcal{D}(A_j) \to \mathcal{H}$$

acting in a Hilbert space $\mathcal{H}$ such that $\cap_{j=1}^n \mathcal{D}(A_j)$ is dense in $\mathcal{H}$. Suppose that $0 \leq \omega < \frac{1}{2}\pi$ and A is uniformly of type ω.

If $T = i(A_1\mathbf{e}_1 + \cdots + A_n\mathbf{e}_n)$ is a one-to-one operator of type ω acting in $\mathcal{H}_{(n)}$ and T has an H^∞-functional calculus, then the n-tuple A has a bounded H^∞-functional calculus on $S_\nu^\circ(\mathbb{C}^n)$ for any $\omega < \nu < \frac{1}{2}\pi$, that is, there exists a homomorphism $b \longmapsto b(A)$, $b \in H^\infty(S_\nu^\circ(\mathbb{C}^n))$, from $H^\infty(S_\nu^\circ(\mathbb{C}^n))$ into $\mathcal{L}_{(n)}(\mathcal{H}_{(n)})$ and there exists $C_\nu > 0$ such that

$$\|b(A)\| \leq C_\nu \|b\|_\infty \quad \text{for all} \quad b \in H^\infty(S_\nu^\circ(\mathbb{C}^n)).$$

Moreover, if f is the unique two-sided monogenic function defined on $S_\nu^\circ(\mathbb{R}^{n+1})$ such that $b = \tilde{f}$, as in Theorem 2, and f satisfies the bound (2.3), then $b(A) = f(A)$ is given by formula (2.5).

Proof. By assumption, the operator T has an H^∞-functional calculus, so there exists a function $\psi \in H^\infty(S_\nu^\circ(\mathbb{C}))$ satisfying conditions (4.2) – (4.5). Our aim is to define $b(A)$ for $b \in H^\infty(S_\nu^\circ(\mathbb{C}^n))$, by the formula

$$(b(A)u, v) = \int_0^\infty \left((b\phi_t)(A)\psi_t(T)u, \psi_t(T)^*v \right) \frac{dt}{t} \tag{4.6}$$

for all $u, v \in \mathcal{H}_{(n)}$. The function $\phi : S_\nu^\circ(\mathbb{C}^n) \to \mathbb{C}$ is constructed from ψ by setting

$$\phi(\zeta) = \psi^2\{i\zeta\} = \psi^2(|\zeta|_c)\chi_+(\zeta) + \psi^2(-|\zeta|_c)\chi_-(\zeta),$$

for all $\zeta \in S_\nu^\circ(\mathbb{C}^n)$. Then for a particular choice of the function ψ [3], the holomorphic function ϕ_t defined for $t > 0$ by $\phi_t(\zeta) = \phi(t\zeta)$ has the property that

$$\vec{x} \longmapsto b(\vec{x})\phi_t(\vec{x}), \quad \vec{x} \in \mathbb{R}^n \setminus \{0\}, \tag{4.7}$$

has a left (and right) monogenic extension $b \cdot_\ell \phi_t$ to $S_\nu^\circ(\mathbb{R}^{n+1})$ with decay at zero and infinity [3]. Hence $(b\phi_t)(A) := (b \cdot_\ell \phi_t)(A)$ is defined by formula (2.5) and satisfies

$$\sup_{t>0} \|(b \cdot_\ell \phi_t)(A)\| \leq C\|b\|_\infty .$$

By normalising ψ so that

$$\int_0^\infty \psi^4(t) \frac{dt}{t} = 1,$$

the desired functional calculus $b \longmapsto b(A)$, $b \in H^\infty(S_\nu^\circ(\mathbb{C}^n))$, is obtained. $\square$

Remark. (i) In [13], it is shown that the n-tuple D_Σ of differentiation operators on a Lipschitz surface $\Sigma \subset \mathbb{R}^{n+1}$ mentioned in the Introduction satisfies the conditions of Theorem 4. In particular $\langle D_\Sigma, s \rangle$ is the differentiation operator on the Lipschitz graph determined by the slice of Σ in the direction $s \in S^{n-1}$, so for some $0 \leq \omega < \frac{1}{2}\pi$, the n-tuple D_Σ is uniformly of type ω.

(ii) A result similar to Theorem 4 is obtained in [11, Theorem 6.4.3] using the idea that A generates a bounded monogenic semigroup rather than the assumption that A is uniformly of type ω. I do not know the connection between the two concepts.

REFERENCES

[1] F. Brackx, R. Delanghe and F. Sommen, *Clifford Analysis*, Research Notes in Mathematics 76, Pitman, Boston/London/Melbourne, 1982.

[2] R. Delanghe, F. Sommen and V. Soucek, *Clifford Algebras and Spinor Valued Functions: A Function Theory for the Dirac Operator*, Kluwer, Dordrecht, 1992.

[3] B. Jefferies, Function theory in sectors, To appear in *Studia Math.*.

[4] B. Jefferies, A symmetric functional calculus for noncommuting systems of sectorial operators, Proc. International Conf. on Harmonic Analysis and Related Topics, Macquarie Univ., Jan. 14–18, 2002, *Proceedings of the Centre for Mathematics and Its Applications* **41** (2003), pp. 68–83.

[5] B. Jefferies and G.W. Johnson, Feynman's operational calculi for noncommuting operators: the monogenic calculus, *Advances in Applied Clifford Algebras* **11** (1) (2001), 239–264.

[6] B. Jefferies, A. McIntosh and J. Picton-Warlow, The monogenic functional calculus, *Studia Math.* **136** (1999), 99–119.

[7] B. Jefferies and B. Straub, Lacunas in the support of the Weyl calculus for two hermitian matrices, *J. Austral. Math. Soc.* **75** (2003), 85–124.

[8] T. Kato, *Perturbation Theory for Linear Operators*, 2nd Ed., Springer-Verlag, Berlin/Heidelberg/New York, 1980.

[9] V. V. Kisil and E. Ramírez de Arellano, The Riesz-Clifford functional calculus for non-commuting operators and quantum field theory, *Math. Methods Appl. Sci.* **19** (1996), 593–605.

[10] C. Li, A. McIntosh, T. Qian, Clifford algebras, Fourier transforms and singular convolution operators on Lipschitz surfaces, *Rev. Mat. Iberoamericana* **10** (1994), 665–721.

[11] C. Li, A. McIntosh, Clifford algebras and H_∞ functional calculi of commuting operators, *Clifford algebras in analysis and related topics* (Fayetteville, AR, 1993), pp. 89–101, Stud. Adv. Math., CRC, Boca Raton, FL, 1996.

[12] A. McIntosh, Operators which have an H_∞-functional calculus, in: Miniconference on Operator Theory and Partial Differential Equations 1986, pp. 212–222, *Proc. Centre for Mathematical Analysis* **14**, ANU, Canberra, 1986.

[13] A. McIntosh, Clifford algebras, Fourier theory, singular integrals, and harmonic functions on Lipschitz domains, *Clifford algebras in analysis and related topics* (Fayetteville, AR, 1993), pp. 33–87, Stud. Adv. Math., CRC, Boca Raton, FL, 1996.

[14] H. Weyl, *The Theory of Groups and Quantum Mechanics*, Methuen, London, 1931; reprinted by Dover Publications, New York, 1950.

Brian Jefferies
School of Mathematics
UNSW
Kensington
NSW 2052, Australia
E-mail: b.jefferies@unsw.edu.au

Submitted: September 13, 2002.

5

Poincaré Series in Clifford Analysis

Rolf Sören Krausshar

ABSTRACT In this paper we deal with Clifford-valued generalizations of several families of classical complex-analytic Eisenstein series and Poincaré series for discrete subgroups of Vahlen's group in the framework of Clifford analysis.

Keywords: Eisenstein series, Poincaré series, Clifford analysis, hypercomplex modular groups, automorphic forms

5.1   Introduction

One very central and important part of complex analysis is the classical theory of elliptic modular forms which deals with holomorphic functions on the upper half-plane $H^+(\mathbb{C}) := \{z \in \mathbb{C};\ \mathrm{Im}(z) > 0\}$ that satisfy for all $z \in H^+(\mathbb{C})$ the transformation law

$$f(z) = (f|_k M)(z), \quad \text{for all} \quad M = \begin{pmatrix} a & b \\ c & d \end{pmatrix} \in SL(2, \mathbb{Z}), \qquad (1.1)$$

where $(f|_k M)(z) := (cz + d)^{-k} f\left(\frac{az+b}{cz+d}\right)$. One way to construct examples of holomorphic functions satisfying (1.1) is to start with a holomorphic generating function $\tilde{f} : H^+(\mathbb{C}) \to \mathbb{C}$ that is bounded on $H^+(\mathbb{C})$ and that is furthermore already totally invariant under the translation group $\mathcal{T}$, the group generated by the matrix $\begin{pmatrix} 1 & 1 \\ 0 & 1 \end{pmatrix}$ inducing the translation $z \mapsto z + 1$. If we sum the expressions $(\tilde{f}|_k M)(z)$ over a complete system of representatives of right cosets of $SL(2, \mathbb{Z})$ modulo the translation invariance group $\mathcal{T}$, that is,

$$f(z) := \sum_{M:\mathcal{T}\backslash SL(2,\mathbb{Z})} (\tilde{f}|_k M)(z), \qquad (1.2)$$

then (provided k is sufficiently large, so that this series converges) the limit function f has the requested properties. These function series are often called Poincaré series in a wider sense. The simplest non-trivial examples can be obtained by

---

AMS Subject Classification: 30G35, 11F03, 11F41.

putting $\tilde{f} \equiv 1$, thus yielding—up to a normalization factor—the classical Eisenstein series

$$G_m(z) = \sum_{(c,d)\in\mathbb{Z}^2\setminus\{(0,0)\}} \frac{1}{(cz+d)^m}, \quad m \geq 4 \qquad (1.3)$$

which is one of the *prototypes* for elliptic modular forms. It should be mentioned that the Eisenstein series (1.3) were generated originally from the Laurent coefficients of Weierstrass' $\wp$-function.

Already in the first half of the twentieth century an extensive theory of higher–dimensional (but complex-valued) extensions of the theory of classical elliptic modular forms was developed in the framework of several complex variables theory, starting with Blumenthal [1] in 1904, continuing with Siegel and Maass in the 1930s and early 1940s [24, 31].

One of the first authors who considered automorphic forms in a hypercomplex variable was Fueter. In 1927 Fueter constructed in his paper [8] automorphic forms and functions related to Picard's group in a three-dimensional hypercomplex variable. However, the functions treated in his paper are not in kernels of Dirac type operators.

In 1949 Maass proceeded to introduce in [25] a further different kind of automorphic forms in a hypercomplex variable. The class of automorphic forms introduced by him consists of complex-valued non-analytic eigenfunctions of the higher–dimensional hyperbolic Laplace–Beltrami operator.

Many authors extended these fundamental higher–dimensional theories from a large variety of viewpoints. Just to give a few examples, see works of Elstrodt, Grunewald and Mennicke, [5, 6], Freitag and Hermann [7], Krieg [19–21], O. Richter and H. Skogman [26, 27] which are some of many authors that provided important contributions in this direction.

If one looks for analogous extensions within the framework of Clifford analysis and its regularity concepts, then, as far as we know, one primarily finds only very few contributions to higher–dimensional versions of the doubly periodic holomorphic Weierstrass functions as e.g., [4, 9–11, 28] before we started in the period of 1998 – 2002 to fill this gap step by step, and to develop systematically a theory of monogenic and polymonogenic hypercomplex-valued modular forms in the setting of real and complexified Clifford analysis (see e.g., [13–18]).

In this paper we explain how a Poincaré summation process can be applied to some special generating functions from Clifford analysis in order to construct non-trivial monogenic and polymonogenic Eisenstein and Poincaré series, thus providing many non-trivial examples of hypercomplex-valued modular forms related to discrete subgroups of Vahlen's group for the half-space model as well as for the unit ball model within the Clifford analysis setting. We further discuss the problem of the construction of cusp forms in the setting of two monogenic vector variables in which a certain biregular variant of the exponential function plays a crucial role.

5.2   Notation and preliminaries

For details about Clifford algebras and basic Clifford analysis see e.g., [3]. Let $e_1, e_2, \ldots, e_n$ be the standard basis of the Euclidean space $\mathbb{R}^{0,n}$ and $C\ell_{0,n}$ be the associated real Clifford algebra in which $e_i e_j + e_j e_i = -2\delta_{ij}$ holds, δ_{ij} standing for the Kronecker function. We write $e_A := e_{i_1} \cdots e_{i_k}$ where $A = \{i_1, \ldots, i_k\} \subseteq \{1, \ldots, n\}$, and in particular $e_\emptyset = e_0 = 1$.

By $< a, b > := Sc(a\bar{b})$ we denote the usual scalar product between two Clifford numbers $a, b \in C\ell_{0,n}$. The Clifford norm of an $a = \sum_A a_A e_A \in C\ell_{0,n}$ is then $\|a\| = (\sum_A |a_A|^2)^{\frac{1}{2}}$. The reversion is defined by $(ab)^* = b^* a^*$ and $e_i^* = e_i$ for all $i = 1, \ldots, n$ and the conjugation by $\overline{ab} = \bar{b}\,\bar{a}$ and $\overline{e_i} = -e_i$ for all $i = 1, \ldots, n$. For a natural $p \in \{1, \ldots, n-1\}$,

$$\Gamma_p := \left\langle \underbrace{\begin{pmatrix} 1 & e_1 \\ 0 & 1 \end{pmatrix}}_{=:T_1}, \ldots, \underbrace{\begin{pmatrix} 1 & e_p \\ 0 & 1 \end{pmatrix}}_{=:T_p}, \underbrace{\begin{pmatrix} 0 & -1 \\ 1 & 0 \end{pmatrix}}_{=:Q} \right\rangle \tag{2.1}$$

stands for the hypercomplex modular group (compare with [6]) which acts discontinuously on the upper half-space

$$H^+(\mathbb{R}^n) := \{z = \mathbf{x} + x_n e_n : \mathbf{x} \in \mathbb{R}^{n-1}, x_n > 0\} \tag{2.2}$$

via its associated Möbius transformation

$$M < z >:= (az + b)(cz + d)^{-1} = (zc^* + d^*)^{-1}(za^* + b^*)$$

where $M = \begin{pmatrix} a & b \\ c & d \end{pmatrix} \in \Gamma_p$. The subgroup generated by the first p translation matrices $T_1, \ldots, T_p$ will be denoted by $\mathcal{T}_p$ throughout this paper. The Dirac operator in $\mathbb{R}^{0,n}$ will be denoted by $D_z := \sum_{i=1}^{n} e_i \frac{\partial}{\partial x_i}$. Clifford-valued functions, defined in open subsets of $\mathbb{R}^n$, that are annihilated from the left by the Dirac operator are called left monogenic and functions annihilated from the right are consequently called right monogenic. More generally, one speaks of left (resp. right) k-monogenic functions, if $D^k f = 0$ or $f D^k = 0$, where D^k means the k-th iterate of the Dirac operator. Note that the Dirac operator factorizes the Laplacian viz: $D^2 = -\Delta_{\mathbb{R}^n}$. If n is even, then $\operatorname{Ker} D\Delta^{n/2-1}$ is the class of the Fueter–Sce solutions (see e.g., [22, 30]). Clifford-valued functions of two vector variables that are left monogenic in the first one and right monogenic in the other one are called biregular. For the fundamental solution of D and its partial derivatives (generalized negative powers) we write

$$q_0(z) = \frac{\bar{z}}{\|z\|^n}, \quad q_{\mathbf{m}}(z) := \frac{\partial^{|m|}}{\partial z^{\mathbf{m}}} q_0(z), \tag{2.3}$$

where $\mathbf{m} = (m_1, \ldots, m_n) \in \mathbb{N}_0^n$ using the usual multi-index notation as in [16].

5.3   The monogenic Weierstrass $\wp$-function and associated Eisenstein series

The Eisenstein series (1.3) were originally introduced by means of the Laurent coefficients of the classical Weierstrass $\wp$-function associated to a lattice of the form $\Omega = \mathbb{Z}\tau + \mathbb{Z}$, $\tau \in H^+(\mathbb{C})$ (method of generating functions). Without loss of generality one can restrict to such types of lattices, since every general lattice can be transformed into such a one by a simple rotation.

In this chapter we discuss a generalization of (1.3) that is obtained when we apply the same method on the higher–dimensional monogenic variants of the Weierstrass $\wp$-function. Contributions to higher–dimensional generalizations of the Weierstrass $\wp$-function in the Clifford analysis setting can be found e.g., in [4, 9–11, 28] and more recently in [12–14, 23]. We recall:

Definition 1. *Suppose that $\Omega_n := \mathbb{Z}\omega_1 + \cdots + \mathbb{Z}\omega_n$ is a non-degenerate n-dimensional lattice in $\mathbb{R}^n$. Let $\tau(i)$ be the multi-index $\mathbf{m} := (m_1, \ldots, m_n)$ with $m_j = \delta_{ij}$, for $j = 1, \ldots, n$. The associated generalized monogenic $\wp$-function is then defined as*

$$\wp(z) := q_{\tau(i)}(z) + \sum_{\omega \in \Omega_n \setminus \{0\}} \left[q_{\tau(i)}(z + \omega) - q_{\tau(i)}(\omega) \right]. \tag{3.1}$$

As in the complex plane, let us restrict to lattices of the form $\Omega_{n-1} + \mathbb{Z}\tau$ where $\tau \in H^+(\mathbb{R}^n)$ and where Ω_{n-1} is an $(n-1)$-dimensional lattice from $\mathrm{span}_\mathbb{R}\{e_1, \ldots, e_{n-1}\}$. Notice that every general lattice can be transformed into such a one by a simple rotation. Generalizing the classical idea, the Laurent coefficients generate thus the following function series (when regarding τ as a hypercomplex variable of $H^+(\mathbb{R}^n)$) as a generalization of (1.3):

Definition 2. *(cf. [13, 15])*
For a multi-index $\mathbf{m} \in \mathbb{N}_0^n$ of the form $(m_1, \ldots, m_{n-1}, 0)$ of odd length, i.e., with $|\mathbf{m}| = \sum_{i=1}^{n-1} m_i \equiv 1 \pmod{2}$, the monogenic Eisenstein series associated to the generalized monogenic Weierstrass $\wp$-function is defined by

$$G_{\mathbf{m}}(z) := \sum_{(\alpha, \omega) \in \mathbb{Z} \times \Omega_{n-1} \setminus \{(0,0)\}} q_{\mathbf{m}}(\alpha z + \omega), \quad z \in H^+(\mathbb{R}^n), \; |\mathbf{m}| \geq 3, \tag{3.2}$$

where Ω_{n-1} denotes a non-degenerate lattice in $\mathrm{span}_\mathbb{R}\{e_1, \ldots, e_{n-1}\}$.

One can restrict to indices with $m_n = 0$ as a consequence of the Cauchy–Riemann system. The series $G_{\mathbf{m}}$ is absolutely convergent and thus monogenic on $H^+(\mathbb{R}^n)$ whenever $|\mathbf{m}| \geq 3$ (cf. [15]). Additionally we have

$$G_{\mathbf{m}}(T < z >) = G_{\mathbf{m}}(z)$$

for all $T \in \mathcal{T}_{n-1}$ so that it can be represented by a normally convergent Fourier series on $H^+(\mathbb{R}^n)$. Writing $z \in H^+(\mathbb{R}^n)$ in the form $z = \mathbf{x} + x_n e_n$ where

$\mathbf{x} \in \text{span}_{\mathbb{R}}\{\mathbf{e}_1, \ldots, \mathbf{e}_{n-1}\}$ and $x_n > 0$ then we get, restricting without loss of generality to an orthonormal lattice, that its Fourier expansion on $H^+(\mathbb{R}^n)$ reads:

$$G_{\mathbf{m}}(z) = 2\zeta_M^{\Omega_{n-1}}(\mathbf{m})$$

$$+ A_n(2\pi i)^{|\mathbf{m}|} \sum_{\mathbf{s} \in \mathbb{Z}^{n-1}\backslash\{0\}} \sigma_{\mathbf{m}}(\mathbf{s})(i\mathbf{e}_n + \frac{\mathbf{s}}{\|\mathbf{s}\|})e^{2\pi i <\mathbf{s},\mathbf{x}>}e^{-2\pi\|\mathbf{s}\|x_n}.$$

Here, A_n is the surface of the unit ball in $\mathbb{R}^n$, $\zeta_M^{\Omega_{n-1}}$ the generalized Riemann zeta function from [13] (see [2, 15] for more details) and

$$\sigma_{\mathbf{m}}(\mathbf{s}) = \sum_{\mathbf{r}|\mathbf{s}} \mathbf{r}^{\mathbf{m}}$$

where $\mathbf{r}|\mathbf{s}$ means that there is an $\alpha \in \mathbb{N}$ such that $\alpha\mathbf{r} = \mathbf{s}$.

The proof can be done in a similar way as we presented in [13, 15] in the paravector formalism. The main difference is that the Fourier transform of q_0 in the vector formalism has the different form

$$(i\mathbf{e}_n + \frac{\mathbf{s}}{\|\mathbf{s}\|})e^{2\pi i <\mathbf{s},\mathbf{x}>}e^{-2\pi\|\mathbf{s}\|x_n}.$$

All the other technical steps can be adapted easily.

One observes a very close similarity between the form of the Fourier expansion of the classical Eisenstein series (1.3) and that of the higher–dimensional variant defined in (3.2). As a consequence of the monogenicity, the monogenic plane wave function appears as a generalization of the classical exponential function. Infinitely many Fourier coefficients do not vanish, so that the series $G_{\mathbf{m}}$ are actually non-trivial functions.

Summarizing: from the structure of the Fourier expansion of the series $G_{\mathbf{m}}$ and from their connection to the Laurent coefficients of the generalized $\wp$-function, the series (3.2) provide canonical generalizations of (1.3) in the Clifford analysis setting. However, one important aspect is not yet included in this generalization. Although the set of singularities of (3.2), i.e., $Q\mathbf{e}_1 + \cdots + Q\mathbf{e}_{n-1}$ is totally invariant under the full hypercomplex modular group Γ_{n-1} (like their complex counterparts), on the half-space we do only observe the invariance under the subgroup $\mathcal{T}_p$ and not a quasi-invariance under the whole group Γ_{n-1}.

Nevertheless, as we shall see in the next section, the functions (3.2) serve as useful building blocks for generating non-trivial families of monogenic and polymonogenic vector-valued or Clifford-valued modular forms with respect to larger hypercomplex modular groups. This provides in a certain sense still an analogy to the complex case.

5.4   Poincaré series in Clifford analysis

From the conformal invariance formula for the iterated Dirac operators D^h (see for example [29]) it follows that if $\tilde{f} : H^+(\mathbb{R}^n) \to Cl_{0n}$ is a left monogenic function, then

$$D_z^l[J_h(cz+d)^* \tilde{f}(M < z >)] = 0$$

for every positive integer l with $l \geq h$. Here

$$J_h(cz+d) := (cz+d)/\|cz+d\|^{n-h+1}$$

whenever h is odd, and

$$J_h(cz+d) := \|cz+d\|^{h-n}$$

whenever h is an even positive integer. Starting now with an appropriate monogenic generating function $\tilde{f}$ from Clifford analysis that is totally invariant under the action of the translation group T_p, and summing then expressions of the form

$$(\tilde{f}|_h M)(z) := J_h(cz+d)^* \tilde{f}(M < z >)$$

over a complete set of representatives of right cosets in Γ_p modulo T_p, will lead to several classes of non-trivial Clifford-valued modular forms with respect to the whole group Γ_p within the function classes Ker D^l for all positive integers l satisfying $l \geq h$ whenever h is an even positive integer. We are going to explain this now in detail. In the sequel the notation $M : T_p\backslash\Gamma_p$ means that M runs through a system of representatives $\mathcal{R}_p$ of the right cosets of Γ_p with respect to T_p, i.e., $\cup_{M \in \mathcal{R}_p} T_p M = \Gamma_p$ and $T_p M \neq T_p N$ for $M, N \in \mathcal{R}_p$ with $M \neq N$. We showed in [16]:

Construction Theorem 1. *Let $n \in \mathbb{N}, h \in 2\,\mathbb{N}, h < n, p \in \{1,\ldots,n-1\}$ and $p < n - h - 1$. Let $\tilde{f} : H^+(\mathbb{R}^n) \to Cl_{0n}$ be a bounded and left monogenic function on $H^+(\mathbb{R}^n)$ that is totally invariant under the translation group T_p. Then*

$$f(z) = \sum_{M:T_p\backslash\Gamma_p} (\tilde{f}|_h M)(z), \quad z \in H^+(\mathbb{R}^n) \tag{4.1}$$

is a Clifford-valued function which is bounded in any compact subset of $H^+(\mathbb{R}^n)$. Moreover, it satisfies on the whole upper half-space $D^l f(z) = 0$ for all $l \geq h$ and $f(z) = (f|_h M)(z)$ for all $M \in \Gamma_p$.

Short sketch of the proof. As $\tilde{f}$ is bounded on $H^+(\mathbb{R}^n)$ it suffices to make clear that $\sum_{M:T_p\backslash\Gamma_p} \|cz+d\|^{h-n}$ is normally convergent on $H^+(\mathbb{R}^n)$ for $p < n-h-1$. To this end, let $\varepsilon > 0$ and consider the vertical strip

$$\mathcal{V}_\epsilon(H^+(\mathbb{R}^n)) := \left\{ z = \mathbf{x} + x_n \mathbf{e}_n \in H^+(\mathbb{R}^n), \|\mathbf{x}\| \leq \frac{1}{\varepsilon}, x_n \geq \varepsilon \right\}. \tag{4.2}$$

With a standard compactification argument one concludes that one can find for every $\varepsilon > 0$ a real $\rho > 0$ such that

$$\|cz+d\| \geq \rho\|c\mathbf{e}_n + d\|$$

for all $z \in \mathcal{V}_\epsilon(H^+(\mathbb{R}^n))$ and for all $M \in \Gamma_p$. The series

$$\sum_{M:\mathcal{T}_p\backslash\Gamma_p} \|c\mathbf{e}_n + d\|^{-\alpha}$$

has the same convergence abscissa as the series

$$\sum_{M:\mathcal{T}_p\backslash\Gamma_p} \|c\mathbf{e}_{p+1} + d\|^{-\alpha}$$

which is – according to [6] – precisely $p + 1$. Thus the series (4.1) converges normally whenever $p < n - h - 1$ so that f is in Ker D^l whenever $l \geq h$ as a consequence of Weierstrass convergence theorem. To prove the automorphic behavior it suffices to verify it for the generators of the group. This can be done by applying direct rearrangement arguments in the series and using furthermore the property $J_h(ab) = J_h(b)J_h(a)$ for elements a, b from the Clifford group. For the detailed proof, see [16].

By means of the Eisenstein series introduced in the previous section we can generate a couple of non-trivial vector-valued examples:

Theorem. *Let* $n \in \mathbb{N}, h \in 2\,\mathbb{N}, h < n,\ p \in \{1, \ldots, n-1\}, p < n - h - 1$ *and suppose that* $\mathbf{m} \in \mathbb{N}_0^{n-1}$ *is a multi-index where* $|\mathbf{m}| \geq 3$ *is an odd number such that* $\zeta_M^{\Omega_{n-1}}(\mathbf{m}) \neq 0$. *Then, putting* $\tilde{G}_\mathbf{m}(z) := G_\mathbf{m}(z + \mathbf{e}_n)$, *the series*

$$E_{h,\mathbf{m}}(z) = \sum_{M:\mathcal{T}_p\backslash\Gamma_p} (\tilde{G}_\mathbf{m}|_h M)(z) \tag{4.3}$$

does not vanish and represents a non-trivial transcendental vector-valued automorphic form with respect to Γ_p *of weight* $(n - h)$ *on the upper half-space in the classes* Ker D^l *for every positive integer* l *with* $l \geq h$.

Notice, $(\tilde{G}_\mathbf{m}|_h M)(z)$ stands for $J_h(cz+d)G_\mathbf{m}(M < z > +\mathbf{e}_n)$. To verify the non-triviality, consider $\lim_{x_n \to \infty} E_{h,\mathbf{m}}(z)$ which equals

$$\sum_{M:\mathcal{T}_p\backslash\Gamma_p, c_M=0} \|d\|^{h-n} \lim_{x_n \to \infty} G_\mathbf{m}(a[\mathbf{x} + x_n\mathbf{e}_n]d^{-1} + bd^{-1} + \mathbf{e}_n)$$

$$= 2^{p+1} \lim_{x_n \to \infty} G_\mathbf{m}(z) = 2^{p+1}\zeta_M^{\Omega_{n-1}}(\mathbf{m}).$$

It suffices to show that there are indices $\mathbf{m}$ for which $\zeta_M^{\Omega_{n-1}}(\mathbf{m}) \neq 0$. To this end notice that $\zeta_M^{\Omega_{n-1}}(\mathbf{m})$ ($|\mathbf{m}|$ odd) are the Laurent coefficients of the series

$$\epsilon_\mathbf{r}(z) = \sum_{\omega \in \Omega_{n-1}} q_\mathbf{r}(z + \omega) \quad |\mathbf{r}| \geq 1.$$

If $\zeta_M^{\Omega_{n-1}}(\mathbf{m})$ vanished for all $\mathbf{m}$, then one would obtain that $\epsilon_\mathbf{r}(z) \not\equiv q_\mathbf{r}(z)$ within a whole ball which is a contradiction.

As a concrete example consider the class of Clifford-valued functions in $\text{Ker } D\Delta^4$ defined in $\mathbb{R}^{10}$ (Fueter–Sce equation). Let us consider the group Γ_2 which acts discontinuously on $H^+(\mathbb{R}^{10})$. If we take a multi-index $\mathbf{m} \in \mathbb{N}_0^9$ of odd length $|\mathbf{m}| \geq 3$ for which $\zeta_M^{\Omega_{n-1}}(\mathbf{m}) \neq 0$, then the families of function series $E_{2,\mathbf{m}}$, $E_{4,\mathbf{m}}$ and $E_{6,\mathbf{m}}$ provide non-trivial examples of vector-valued hypercomplex modular forms with respect to Γ_2 in the function class of holomorphic Cliffordian functions when expressing them in the paravector formalism.

Notice that if we substituted $G_\mathbf{m}$ by a general $(n-1)$-fold periodic function $\tilde{f} \in \text{Ker } D\Delta^4$, then in general $\tilde{f}|_{2\alpha}M < z >$ would not be anymore an element of $\text{Ker } D\Delta^4$, since α has to be chosen out of the set $\{1, 2, 3\}$ in order to preserve the convergence of the associated function series.

Notice that the construction (4.1) generates only the zero function whenever h is odd. See [16] for details. More generally, whenever we have a group that includes the negative identity matrix, then all functions satisfying

$$f(z) = (f|_h M)(z)$$

for all M from that group vanish identically, whenever h is odd.

If we want to construct non-trivial h-monogenic functions that satisfy the transformation law $f(z) = (f|_h M)(z)$ for odd h, then we need to restrict on smaller groups. Just to give one very simple example of h-monogenic functions with such a transformation behavior involving odd h, let us take for G the transversion group

$$\mathcal{W}_p := \left\langle V_1 := \begin{pmatrix} 1 & 0 \\ e_1 & 1 \end{pmatrix}, \ldots, V_p := \begin{pmatrix} 1 & 0 \\ e_p & 1 \end{pmatrix} \right\rangle \tag{4.4}$$

where $p \leq n - 1$. This group does not contain the negative identity matrix. All matrices from this group induce transformations of the form

$$V(z) = z(vz + 1)^{-1}$$

with a $v \in \mathbb{Z}^p$. The expressions $V(z)$ do not have singularities in $H^+(\mathbb{R}^n)$ and it is not difficult to show (compare with [17]) that the series

$$v_h(z) := \sum_{M \in \mathcal{W}_p} J_h(cz + d) \tag{4.5}$$

converges normally on $H^+(\mathbb{R}^n)$ whenever $p < n - h$. Furthermore, a computation leads to $\lim\limits_{x_n \to \infty} v_h(e_n x_n) = 1$, providing us with the non-triviality argument of this series. This example is indeed one of the simplest ones, since the transversion group is conjugated to the translation group $\mathcal{T}_p$.

One possibility to construct also non-trivial modular forms in $\text{Ker } D^l$ $(l \geq h)$ where h is odd (in particular for $h = 1$), with respect to the whole modular group Γ_p and even with respect to the full modular group Γ_{n-1}, is to consider *two* weight factors (one from the left, another one from the right) and to introduce a second auxiliary variable. We exploited this idea recently in [18] in order to construct monogenic and polymonogenic Hilbert modular forms on Cartesian products of the upper half-space.

In what follows we focus on the construction of non-trivial monogenic and polymonogenic modular forms on

$$H_2^+(\mathbb{R}^n) = H^+(\mathbb{R}^n) \oplus H^+(\mathbb{R}^n)$$

that transform

$$f(z, w) = (f\|_{h,l}M)(z, w) \quad \text{for all } M \in \Gamma_p \tag{4.6}$$

and whose restriction to the diagonal $(z = w)$ is still a non-constant C^∞-automorphic form satisfying

$$f(z, z) = (f\|_{h,l}M)(z, z)$$

for all $M \in \Gamma_p$. Here

$$(f\|_{h,l}M)(z, w) = J_h(cz + d)^* f(M < z >, M < w >)\overline{J_l(wc^* + d^*)}.$$

The following construction theorem provides many non-trivial examples:

Construction Theorem 2. *Let $p < 2n - (h + l) - 1$ and let $h + l \equiv 0 \pmod{2}$. Suppose $\tilde{f} : H_2^+(\mathbb{R}^n) \to Cl_{0n}$ is a bounded function that satisfies for all $(z, w) \in H_2^+(\mathbb{R}^n)$ the relation*

$$D_z^h[\tilde{f}(z, w)] = [\tilde{f}(z, w)]D_w^l = 0$$

and furthermore

$$\tilde{f}(T < z >, T < w >) = \tilde{f}(z, w)$$

for all $T \in \mathcal{T}_p$. Then

$$f(z, w) := \sum_{M:\mathcal{T}_p\backslash\Gamma_p} (\tilde{f}\|_{h,l}M)(z, w) \tag{4.7}$$

is left h-monogenic in the variable z and right l-monogenic in w and satisfies for all $(z, w) \in H_2^+(\mathbb{R}^n)$,

$$f(z, w) = (f\|_{h,l}M)(z, w), \quad \text{for all } M \in \Gamma_p. \tag{4.8}$$

Sketch of the proof (for details see [18]). Let $\mathcal{K}$ be a compact subset of $H_2^+(\mathbb{R}^n)$. Choose an $\varepsilon > 0$ such that

$$\mathcal{V}_\varepsilon(H^+(\mathbb{R}^n)) \times \mathcal{V}_\varepsilon(H^+(\mathbb{R}^n))$$

covers $\mathcal{K}$ where $\mathcal{V}_\varepsilon(H^+(\mathbb{R}^n))$ is the vertical strip domain defined in (4.2). With the argumentation of the convergence proof of the first construction theorem, we know that there is a real $\rho > 0$ such that for all $M \in \Gamma_p$:

$$\|J_h(cz + d)\| \le \rho^{h-n}\|ce_n + d\|^{h-n} \tag{4.9}$$

for all z that belong to $\mathcal{V}_\varepsilon(H^+(\mathbb{R}^n))$. Since $\tilde{f}$ is bounded, we obtain that for every $(z,w) \in \mathcal{K}$:

$$\left\| \sum_{M:\mathcal{T}_p\backslash\Gamma_p} \|J_h(cz+d)^* \tilde{f}(M<z>, M<w>)\overline{J_l(wc^*+d^*)}\right\|$$

$$\leq \tilde{L}\rho^{h+l-2n} \sum_{M:\mathcal{T}_p\backslash\Gamma_p} \frac{1}{\|ce_n+d\|^{n-h}} \frac{1}{\|e_nc^*+d^*\|^{n-l}}$$

$$\leq L \sum_{M:\mathcal{T}_p\backslash\Gamma_p} \frac{1}{\|ce_n+d\|^{2n-h-l}},$$

where $\tilde{L}$ and L are non-negative real constants. The series in the previous line is absolutely convergent if and only if $p < 2n - h - l - 1$ (see [6, 16]). From Weierstrass' convergence theorem and the quasi-conformal invariance formulae for the iterated Dirac operators from [29], we infer that the limit function f is left h-monogenic in z and right l-monogenic in w. To show the automorphy relation it is again sufficient to verify the relation for the generators of the group. With respect to the inversion we observe that

$$f(-z^{-1}, -w^{-1}) = \sum_{M:\mathcal{T}_p\backslash\Gamma_p} \Big[J_h(-cz^{-1}+d)^*$$

$$\times \tilde{f}\Big((-az^{-1}+b)(-cz^{-1}+d)^{-1}, (-w^{-1}c^*+d^*)^{-1}(-w^{-1}a^*+b^*)\Big)$$

$$\times \overline{J_l(-w^{-1}c^*+d^*)}\Big]$$

$$= \sum_{M:\mathcal{T}_p\backslash\Gamma_p} \Big[J_h(-z^{-1})^{-1*} J_h(c-dz)^* \times$$

$$\tilde{f}\Big((a-bu)(c-du)^{-1}, (c^*-wd^*)^{-1}(a^*-wb^*)\Big) \overline{J_l(c^*-wd^*)}\, \overline{J_l(-w^{-1})^{-1}}\Big]$$

$$= J_h(-z^{-1})^{-1*} f(z,w)\overline{J_l(-w^{-1})^{-1}}.$$

By a similar rearrangement procedure we can show that the automorphy relation holds for simultaneous translations $(z,w) \mapsto (z+m, w+m)$ where $m = \mathbf{e}_i$ for $i = 1, \ldots, p$.

We get even convergence for $p = n - 1$ whenever $h + l \leq n - 1$. Even for the monogenic case, i.e., $h = l = 1$, we get non-trivial Clifford-valued modular forms for the full modular group provided we are in a space of dimension $n \geq 3$. The simplest examples are the biregular Eisenstein series

$$\mathcal{E}(z,w) = \sum_{M:\mathcal{T}_p\backslash\Gamma_p} (1\|_{1,1}M)(z,w)$$

$$= \sum_{M:\mathcal{T}_p\backslash\Gamma_p} J_h(cz+d)^*\overline{J_l(wc^*+d^*)}, \quad (z,w) \in H_2^+(\mathbb{R}^n). \quad (4.10)$$

To verify that these series are non-trivial transcendental functions, consider

$$\lim_{x_n, y_n \to \infty} \mathcal{E}(\mathbf{e}_n x_n, \mathbf{e}_n y_n)$$

$$= \sum_{M: T_p \backslash \Gamma_p, c_M \neq 0} \underbrace{\lim_{x_n, y_n \to \infty} J_1(c\mathbf{e}_n x_n + d)^* \overline{J_1(\mathbf{e}_n y_n c^* + d^*)}}_{=0}$$

$$+ \sum_{M: T_p \backslash \Gamma_p, c_M = 0} J_1(d)^* \overline{J_1(d^*)}$$

$$= 2 \sum_{A \subseteq \{1, \ldots, p\}} e_A \overline{e_A} = 2^{p+1},$$

where we put $z = \sum_{j=1}^n x_j \mathbf{e}_j$ and $w = \sum_{j=1}^n y_j \mathbf{e}_j$. Also their restriction to the diagonal $\mathcal{E}(z, z)$ is thus a non-constant C^∞-automorphic form with respect to Γ_p. Once again, the series $G_\mathbf{m}$ can be used to generate classes of non-trivial examples for hypercomplex modular forms transforming as (4.6)—in this setting, even for obtaining families of non-trivial monogenic modular forms for the full modular group Γ_{n-1}—namely by applying (4.7) on

$$\tilde{f}(z, w) = G_\mathbf{m}(z + \mathbf{e}_n) G_\mathbf{m}(w + \mathbf{e}_n) \tag{4.11}$$

which is bounded on $H_2^+(\mathbb{R}^n)$. To show the non-triviality we consider again

$$\lim_{x_n, y_n \to \infty} \sum_{M: T_p \backslash \Gamma_p} J_1(c\mathbf{e}_n x_n + d)^* G_\mathbf{m}(M < \mathbf{e}_n x_n > + \mathbf{e}_n)$$

$$\times G_\mathbf{m}(M < \mathbf{e}_n y_n > + \mathbf{e}_n) \overline{J_1(\mathbf{e}_n y_n c^* + d^*)}$$

$$= (-8) \sum_{A \subseteq \{1, \ldots, p\}} e_A^* \|\zeta(\mathbf{m})\|^2 \overline{e_A}^*.$$

As mentioned above, there must be at least some indices $\mathbf{m} \in \mathbb{N}_0^{n-1}$ of odd length, for which $\zeta(\mathbf{m})$ and hence $\|\zeta(\mathbf{m})\|^2$ is different from zero. For these indices we have then that the limit considered is different from zero, proving the non-triviality of the associated series

$$\sum_{M: T_p \backslash \Gamma_p} (\tilde{f}\|_{h,l} M)(z, w)$$

and in particular that it is non-constant on the diagonal (z, z).

In this paper we proceed now to treat also biregular and polybiregular monogenic modular forms f that satisfy additionally

$$\lim_{x_n \to \infty} f(z, w) = \lim_{y_n \to \infty} f(z, w) = 0. \tag{4.12}$$

To this end it is suggestive to use the following variant of a biregular exponential function on $H_2^+(\mathbb{R}^n)$:

Definition 3. *For* $\mathbf{p}, \mathbf{q} \in \mathbb{R}^{n-1}\setminus\{0\}$ *and* $(z, w) \in H_2^+(\mathbb{R}^n)$ *define*

$$E_{\mathbf{pq}}(w, z) := (i\mathbf{e}_n + \frac{\mathbf{p}}{\|\mathbf{p}\|})e^{-x_n\|\mathbf{p}\|}e^{i<\mathbf{x},\mathbf{p}>}e^{i<\mathbf{y},\mathbf{q}>}e^{-y_n\|\mathbf{q}\|}(i\mathbf{e}_n + \frac{\mathbf{q}}{\|\mathbf{q}\|}).$$

This function is left monogenic in z and right monogenic in w on the whole domain $H_2^+(\mathbb{R}^n)$. Notice that left and right monogenicity is provided by the constant factors standing left and right from the exponential expressions. It is crucial that

$$\lim_{x_n \to \infty} E_{\mathbf{pq}}(z, w) = \lim_{y_n \to \infty} E_{\mathbf{pq}}(z, w) = 0$$

for any parameter $\mathbf{p}$ and $\mathbf{q}$ from $\mathbb{R}^{n-1}\setminus\{0\}$. In order to construct hypercomplex modular cusp forms that are left h-monogenic in z and right l-monogenic in w, the following construction is thus plausible:

$$\mathcal{P}_{h,l}(z, w; \mathbf{p}, \mathbf{q}) = \sum_{M:T_p\setminus\Gamma_p} (E_{\mathbf{p},\mathbf{q}}\|_{h,l}M)(z, w). \qquad (4.13)$$

The function $E_{\mathbf{p},\mathbf{q}}$ is bounded on $H_2^+(\mathbb{R}^n)$; thus the convergence abscissa of the series is $p < 2n - (h + l) - 1$. However, until now we have not managed yet to find an argument to show that these series do not vanish. Note that by construction

$$\lim_{x_n \to \infty} \mathcal{P}_{h,l}(z, w; \mathbf{p}, \mathbf{q}) = \lim_{y_n \to \infty} \mathcal{P}_{h,l}(z, w; \mathbf{p}, \mathbf{q}) = 0.$$

At the moment it is not clear how to extend to Clifford analysis the non-triviality arguments that are applied for cusp forms in the classical theory, as for instance in the one complex variable case.

Further possible candidates for modular forms with (4.12) that are left h-monogenic in z and right l-monogenic in w might be obtained by inserting for $\tilde{f}$ in the expression (4.7) the function

$$\tilde{f}(z, w) := \epsilon_{\mathbf{m}}(z + \mathbf{e}_n)\epsilon_{\mathbf{r}}(w + \mathbf{e}_n)$$

where $|\mathbf{r}|$ and $|\mathbf{m}|$ are assumed to be greater than or equal to 1, since

$$\lim_{x_n \to \infty} \epsilon_{\mathbf{m}}(z + \mathbf{e}_n) = 0$$

whenever $|\mathbf{m}| \geq 1$.

Finally, we also want to give some examples of non-trivial hypercomplex monogenic and polymonogenic modular forms on the Cartesian product of two unit balls in this paper.

Let

$$g = \begin{pmatrix} \frac{1}{\sqrt{2}} & -\frac{1}{\sqrt{2}}\mathbf{e}_n \\ -\frac{1}{\sqrt{2}}\mathbf{e}_n & \frac{1}{\sqrt{2}} \end{pmatrix}.$$

Then the associated Cayley transformation

$$(z, w) \to (g < z >, g < w >)$$

maps $H_2^+(\mathbb{R}^n)$ conformally onto

$$B^2(0,1) := B(0,1) \times B(0,1) = \{(z,w) \in \mathbb{R}^n \times \mathbb{R}^n; \|z\| < 1 \text{ and } \|w\| < 1\}.$$

In order to construct monogenic and polymonogenic automorphic forms on $B^2(0,1)$ it is suggestive to make a construction of the form

$$\sum_{M:gG'g^{-1}\backslash gGg^{-1}} (\tilde{f}\|_{h,l}M)(z,w) \qquad (4.14)$$

where $G' \leq G \leq \Gamma_{n-1}$ and where $\tilde{f}$ is a left h-monogenic function in z and l-monogenic in w that is bounded on $B^2(0,1)$ and *totally* invariant under $gG'g^{-1}$. Here, we observe that the transfer to the unit ball is not as direct, as one might logically think at the very first: Note that if f is monogenic in z, then in general $f(M < z >)$ will only remain monogenic, when M is a translation matrix.

Putting $G' = \mathcal{T}_{n-1}$ (which is natural), then not every monogenic $\tilde{f}$ has the property that $\tilde{f}((gG'g^{-1}) < z >)$ remains monogenic. The same holds also for l-monogenic functions. The simplest non-trivial (poly-) biregular examples that we can construct to the whole modular group $G := g\Gamma_p g^{-1}$ on $B^2(0,1)$ and for which we can avoid this problem, are the following (poly-) biregular Eisenstein series on the Cartesian product of the two unit balls.

We introduce:

Definition 4. *Let $p < 2n - (h+l) - 1$ where $h + l \equiv 0 \pmod{2}$ and let*

$$G_p := g\Gamma_p g^{-1}, \quad G'_p := g\mathcal{T}_p g^{-1}$$

where g stands for the Cayley transformation. Then, the (poly-) biregular Eisenstein series on the Cartesian product of the two unit balls $B^2(0,1)$ are defined by

$$\mathcal{H}(z,w) := \sum_{M:G'_p\backslash G_p} (1\|_{h,l}M)(z,w). \qquad (4.15)$$

The function $\tilde{f}(z,w) \equiv 1$ is of course left h-monogenic and right l-monogenic and totally invariant under G'_p. Thus, the limit function is left h-monogenic and right l-monogenic. The series has the same convergence abscissa as its analogue on the half-space, and its automorphy behavior

$$\mathcal{H}(z,w) = (\mathcal{H}\|_{h,l}M)(z,w) \quad \text{for all} \quad M \in G_p$$

can again be shown similarly as we did for the previous examples. To show that the series does not vanish identically on the diagonal whenever $h + l \equiv 0 \pmod{2}$, evaluate the function at the point $(0,0)$, which of course is a (poly-) regular point as it is inside the unit balls, and all the singularities are concentrated on the boundary of $B^2(0,1)$. The series is therefore normally convergent in any compact set around $(0,0)$ inside $B^2(0,1)$. Thus,

$$\mathcal{H}(0,0) = \sum_{M:G'_p\backslash G_p} J_h(d)^* \overline{J_l(d^*)} = \sum_{M:G'_p\backslash G_p} \|d\|^2 \neq 0 \qquad (4.16)$$

and the non-triviality of the series is proven.

REFERENCES

[1] O. Blumenthal, Über Modulfunktionen von mehreren Veränderlichen, *Math. Ann.* **56** (1903), 509–548 and **58** (1904), 497–527.

[2] D. Constales and R.S. Krausshar, Representation formulas for the general derivatives of the fundamental solution to the Cauchy–Riemann operator in Clifford Analysis and Applications, *Z. Anal. Anw.* **21**, No. 3 (2002), 579–597.

[3] R. Delanghe, F. Sommen and V. Souček, *Clifford Algebra and Spinor Valued Functions*, Kluwer, Dordrecht-Boston-London, 1992.

[4] A. Dixon, On the Newtonian potential, *Quarterly Journal of Mathematics* **35** (1904), 283–296.

[5] J. Elstrodt, F. Grunewald and J. Mennicke, Eisenstein series on three-dimensional hyperbolic space and imaginary quadratic number fields. *J. Reine Angew. Math.* **360** (1985), 160–213.

[6] J. Elstrodt, F. Grunewald and J. Mennicke, Kloosterman sums for Clifford algebras and a lower bound for the positive eigenvalues of the Laplacian for congruence subgroups acting on hyperbolic spaces, *Invent. Math.* **101**, No. 3 (1990), 641–668.

[7] E. Freitag and C. F. Hermann, Some modular varieties of low dimension, *Adv. Math.* **152**, No. 2 (2000), 203–287.

[8] R. Fueter, Über automorphe Funktionen in Bezug auf Gruppen, die in der Ebene uneigentlich diskontinuierlich sind, *Crelle's Journal* **137** (1927), 66–77.

[9] R. Fueter, Über vierfachperiodische Funktionen, *Monatshefte Math. Phys.* **48** (1939), 161–169.

[10] R. Fueter, *Functions of a Hyper Complex Variable*, Lecture notes written and supplemented by E. Bareiss, Math. Inst. Univ. Zürich, Fall Semester 1948/49.

[11] F. Gürsey and H. Tze, Complex and quaternionic analyticity in chiral and gauge theories I, *Annals of Physics* **128** (1980), 29–130.

[12] T. Hempfling and R.S. Krausshar, Order theory for isolated points of monogenic functions, *Archiv der Mathematik* **80** (2003), 406–423.

[13] R.S. Krausshar, *Eisenstein Series in Clifford Analysis*, Ph.D. Thesis RWTH Aachen, Aachener Beiträge zur Mathematik **28**, Wissenschaftsverlag Mainz, Aachen, 2000.

[14] R.S. Krausshar, Monogenic multiperiodic functions in Clifford analysis, *Complex Variables* **46**, No. 4 (2001), 337–368.

[15] R.S. Krausshar, On a new type of Eisenstein series in Clifford analysis, *Z. Anal. Anw.* **20**, No. 4 (2001), 1007–1029.

[16] R.S. Krausshar, Automorphic forms in Clifford Analysis, *Complex Variables* **47**, No. 5 (2002), 417–440.

[17] R.S. Krausshar, Eisenstein series in complexified Clifford analysis, *Computational Methods and Function Theory* **2**, No. 1 (2002), 29–65.

[18] R.S. Krausshar, Monogenic modular forms in two and several real and complex vector variables, *Computational Methods and Function Theory* **2**, No. 2 (2002).

[19] A. Krieg, *Modular Forms on Half-Spaces of Quaternions*. Springer Verlag, Berlin-Heidelberg, 1985.

[20] A. Krieg, Eisenstein series on real, complex and quaternionic half-spaces. *Pac. J. Math.* **133**, No. 2 (1988), 315–354.

[21] A. Krieg, Eisenstein series on the four–dimensional hyperbolic space, *J. Number Theory* **30** (1988), 177–197.

[22] G. Laville and I. Ramadanoff, Holomorphic Cliffordian functions, *Adv. Appl. Clifford Algebras* **8**, No. 2 (1998), 323–340.

[23] G. Laville and I. Ramadanoff, Elliptic Cliffordian functions, *Complex Variables* **45**, No. 4 (2001), 297–318.

[24] H. Maass, Zur Theorie der automorphen Funktionen von n Veränderlichen, *Math. Ann.* **117** (1940), 538–578.

[25] H. Maass, Automorphe Funktionen von mehreren Veränderlichen und Dirichletsche Reihen, *Abh. Math. Sem. Univ. Hamb.* **16** (1949), 53–104.

[26] O. Richter, Theta functions with harmonic coefficients over number fields, *J. Number Theory*, Vol. **95**, No. 1 (2002), 101–121.

[27] O. Richter and H. Skogman, Jacobi theta functions over number fields, preprint, Feb. 2002.

[28] J. Ryan, Clifford analysis with generalized elliptic and quasi elliptic functions, *Appl. Anal.* **13** (1982), 151–171.

[29] J. Ryan, Intertwining operators for iterated Dirac operators over Minkowski-type spaces, *J. Math. Anal. Appl.* **177**, No. 1, (1993), 1–23.

[30] M. Sce, Sulle serie di potenzi nei moduli quadratici, *Lincei Rend. Sci. Fis. Mat. et Nat.* **23** (1957), 220–225.

[31] C.L. Siegel, *Topics in Complex Function Theory Vol. III*, Wiley, New York-Chichester, 1988.

Rolf Sören Krausshar
Vakgroep Wiskundige Analyse
Universiteit Gent
Galglaan 2
B-9000 Gent (Belgium)
e-mail: krauss@cage.rug.ac.be

Submitted: June 21, 2002; Revised: November 7, 2002.

6

Harmonic Analysis for General First Order Differential Operators in Lipschitz Domains

Emilio Marmolejo-Olea and Marius Mitrea

ABSTRACT We develop a function theory associated with non-elliptic, variable coefficient operators of Dirac type on Lipschitz domains. Boundary behavior, global regularity, integral representation formulas, are studied by means of tools originating in PDE and harmonic analysis.

Keywords: Cauchy type integral operators, Lipschitz domains, Hardy spaces

6.1   Introduction

In this paper we discuss some basic aspects of the function theory associated with a general first-order differential operator D, such as: *Pompeiu* type integral representation formulas, *Cauchy* type integral operators, and *Hardy* spaces (with emphasis on boundary behavior and global regularity), in a subdomain Ω of a manifold M. For maximum applicability, it is important to allow *minimal smoothness assumptions* on $\partial\Omega$, *variable coefficients* and *vector character* for D, as well as operators with *non-invertible* (principal) symbols.

While our primary source of inspiration in pursuing the program sketched above has been the function theory of the classical Dirac operator associated with $C\ell(\mathbb{R}^m)$—the standard Clifford algebraic structure in $\mathbb{R}^m$—our goal here is to explore the extent to which such a theory can be essentially worked out by *real variable methods*, in the absence of the rich algebraic-geometric structural properties of $C\ell(\mathbb{R}^m)$.

Thus, loosely speaking, the aim is to do Clifford analysis without (extensive use of) Clifford algebras, whenever possible. On the other hand, results which are fundamentally dependent on the underlying algebraic structure are recognized as such.

This point of view is rather useful in many applications, particularly to PDE problems. Specific problems arising in mathematical physics have been treated in [7, 14, 15, 20].

AMS Subject Classification: 35F15, 34L40, 15A66, 32W05, 31C12, 32A30, 58C35.

The layout of the paper is as follows. In §2 we collect some basic definitions, and discuss an important integration by parts identity. In §3 we introduce the class of Dirac type operators—via a *nonstandard* definition—and present some relevant examples. Section 4 is devoted to discussing the role of unique continuation. Cauchy type operators are introduced in §5; here we also prove a general Pompeiu type integral identity. Some special cases have been previously treated in [3] and [24].

The interior regularity and boundary behavior of the aforementioned Cauchy integral operators make the object of §6. Here we develop the main tools needed in §7, where Hardy spaces in Lipschitz domains are studied. Our results in this part of the paper extend those of [6, 19]. In the next three sections, §8-10, we specialize some of our previous results to specific cases of particular interest: in §8 we look at Clifford separately monogenic (or multi-monogenic) functions, while in §9 we consider a similar theory in the context of octonions. In §10 we work out the details of Pompeiu's formula for operators such as $D = d + d^*$, and its complex counterpart, $D = \partial + \partial^*$ (rather in matrix form). Among other things, these are shown to contain the classical Bochner–Martinelli–Koppelman formulas as particular cases. In §11, we list several open problems closely related to the theory outlined in this paper. Finally, there is also an appendix, containing a self-contained proof of a Hartogs type theorem in the Clifford algebra context.

Acknowledgments. The work of M. M. has been supported in part by an NSF grant and a University of Missouri Research Board grant. The authors would like to thank Rafał Abłamowicz and John Ryan for their efforts to make the Cookeville meeting a success. The research described here was initiated when E. M.-O. visited M. M. at the University of Missouri-Columbia in 1999–2000. Both authors would also like to thank Dorina Mitrea for her role in bringing them together.

6.2   Background material

Let M be a smooth, compact, boundaryless, orientable, Riemannian manifold. Depending on the context, we may allow M to be either a real or a complex manifold and occasionally we may take M to be the entire (real or complex) Euclidean space. We denote by dV the volume element and by d the usual exterior differential operator on M. Other standard notation will be adopted without any special mention. Next, consider $\mathcal{E}, \mathcal{F} \to M$ two smooth, Hermitian vector bundles and let the linear map

$$D : C^\infty(M, \mathcal{E}) \longrightarrow C^\infty(M, \mathcal{F}), \qquad (2.1)$$

be a *differential operator* of order $k \geq 1$. According to a well-known characterization of J. Petree, this is equivalent to the property that D is *local*, i.e., D does not 'increase' the support. In the sequel, the superscripts 'star' and 't' are used to denote (complex) adjunction and (real) transposition, respectively. Also, 'bar' stands for complex conjugation. Define $\bar{D}u := \overline{D\bar{u}}$, so that $D^* = (\bar{D})^t = \overline{(D^t)}$.

Finally, $\langle \cdot, \cdot \rangle$ will denote the *bilinear* (rather than sesqui-linear) pointwise pairing in the fibers of a Hermitian vector bundle (each time, its specific meaning should be clear from the context). In particular, $|u|^2 = \langle u, \bar{u} \rangle$.

The *principal symbol* of D in (2.1) is the map

$$\sigma(D; \cdot) : T^*M \setminus 0 \longrightarrow \mathrm{Hom}(\mathcal{E}, \mathcal{F}) \tag{2.2}$$

defined as

$$\left(\sigma(D; \xi) e \right)_\alpha = \sum_{|\gamma|=k} a_\gamma^{\alpha\beta} (i\xi)^\gamma e_\beta, \tag{2.3}$$

if $(Du)_\alpha = \sum_{|\gamma|=k} a_\gamma^{\alpha\beta} \partial^\gamma u_\beta +$ lower order terms, in local coordinates.

Examples of first-order differential operators and their symbols:

- If 'wedge' $\wedge$ stands for the usual *exterior product* of differential forms on M, then $\sigma(d; \xi) = i\xi \wedge \cdot$ for any $\xi \in T^*M \equiv \Lambda^1 TM$.

- If the 'upside-down wedge' $\vee$ denotes the *interior product* of forms (i.e., the adjoint of wedge), then $\sigma(d^*; \xi) = -i\xi \vee \cdot, \forall \xi \in \Lambda^1 TM$.

- Denote by $C\ell(M)$ the exterior power bundle on M, equipped with the Clifford algebra structure induced by the Riemannian metric on M. Then, if $D = d + d^*$, it follows that $\sigma(D; \xi) = i\xi \wedge \cdot - i\xi \vee \cdot =: i\xi \cdot$, the Clifford algebra multiplication with $i\xi$ in $C\ell(M)$.

- More generally, let $\mathcal{E} \to M$ be a Hermitian vector bundle with the property that each fiber $\mathcal{E}_x$ is a Hermitian $C\ell(M)_x$-module. Assume that $\mathcal{E}$ is equipped with a connection $\nabla : C^\infty(M, \mathcal{E}) \to C^\infty(M, T^*M \otimes \mathcal{E})$, and denote by $m : C^\infty(M, T^*M \otimes \mathcal{E}) \to C^\infty(M, \mathcal{E})$, $m(\xi \otimes e) = \xi \cdot e$, the Clifford multiplication map. Finally, introducing $D := m \circ \nabla$, we have $\sigma(D; \xi) = i\xi \cdot$, the Clifford algebra multiplication with $i\xi$ in $\mathcal{E}_x$, for each $\xi \in T_x^*M$. See [25] for more details.

- In the case when $M = \mathbb{R}^m$, the flat, Euclidean space, and with $e_1, e_2, \ldots,$ e_m denoting the standard orthonormal basis in $\mathbb{R}^m$, the above construction yields $D := \sum e_\alpha \partial_\alpha$, the canonical Dirac operator in $\mathbb{R}^m$ (see [2]). Once again, $\sigma(D; \xi) = i\xi \cdot$, the Clifford algebra multiplication with $i\xi = i \sum \xi_\alpha e_\alpha$.

- Assume now that M is a *complex* manifold and consider the decomposition $\Lambda^1 TM = \Lambda^{1,0} TM \oplus \Lambda^{0,1} TM$. For a generic 1-form ξ we write $\xi = \xi^{1,0} + \xi^{0,1}$. Accordingly, the exterior differential operator d splits as $d = \partial + \bar{\partial}$. Then $\sigma(\partial; \xi) = i \, \xi^{1,0} \wedge \cdot$ and $\sigma(\bar{\partial}; \xi) = i \, \xi^{0,1} \wedge \cdot$.

Lemma 1. *For any differential operator* $D : \mathcal{E} \to \mathcal{F}$ *of order* k, *we have* $\sigma(D^t; \xi) = (-1)^k \sigma(D; \xi)^t$ *and* $\sigma(\bar{D}; \xi) = (-1)^k \overline{\sigma(D; \xi)}$. *In particular,*

$$\sigma(D^*; \xi) = \sigma(D; \xi)^*.$$

Finally, $\sigma(D_1 D_2; \xi) = \sigma(D_1; \xi)\sigma(D_2; \xi)$.

An integration by parts formula for *first-order* differential operators, very useful in the sequel, is recorded below. Before stating it, recall that a *Lipschitz* subdomain of M is an open set Ω with the property that, in appropriate local coordinates, $\partial\Omega$ is given by graphs of (Euclidean) Lipschitz functions. We denote by ds the surface element on $\partial\Omega$ (inherited from the metric on M), and by $\nu \in T^*M$ the outward unit conormal to Ω defined ds-a.e. on $\partial\Omega$.

Proposition 1. *Let Ω be a Lipschitz subdomain of M. Then, for any $u \in C^1(\bar{\Omega}, \mathcal{E})$ and $v \in C^1(\bar{\Omega}, \mathcal{F})$, we have*

$$\int_\Omega \langle Du, v \rangle \, d\mathcal{V} = \int_\Omega \langle u, D^t v \rangle \, d\mathcal{V} - \int_{\partial\Omega} \langle i\sigma(D;\nu)u, v \rangle \, ds. \qquad (2.4)$$

See also [25]. In closing, we would like to point out that (2.4) is the departure point in setting up areolar type derivatives, in the spirit of D. Pompeiu's original *dérivée aréolaire*, as well as Morera type theorems for the operator D.

6.3   Dirac type operators

The *Laplace operator* associated with a first-order differential operator as in (2.1), in short the *D-Laplacian*, is the self-adjoint, nonpositive definite, second order differential operator $-D^*D : \mathcal{E} \to \mathcal{E}$. Call D of *Dirac type* if its associated Laplacian has the property that

$$\sigma(D^*D;\xi) \text{ is scalar, real and invertible, } \forall \xi \in T^*M \setminus 0. \qquad (3.1)$$

The reader should be aware that this definition is non-standard (the ones commonly used in the literature are more restrictive). In particular, we do *not* require D to be *elliptic* (i.e., have an *invertible* symbol); this latter case has been considered in [20].

It is clear from definitions that our class of Dirac type operators is stable under conjugation and transposition and, hence, under adjunction also. A typical Dirac type operator is contained in the fourth example in §2; more examples are provided below.

6.3.1   Examples of Dirac type operators

- Consider the operator $D := d + d^*$. Upon recalling that $d^2 = 0$ and that $\Delta = -dd^* - d^*d$ is the *Hodge–Laplacian*, it follows that $D = D^*$ and $-D^2 = \Delta$. Hence, D is of Dirac type. A variant is to take $D := (d, d^*)$ and the same conclusion holds.

- A discussion similar in spirit applies to the case when $D := \partial + \partial^*$. This time, $\partial^2 = (\partial^*)^2 = 0$ and the *complex Laplacian*

$$\square := -\partial\partial^* - \partial^*\partial \qquad (3.2)$$

has a real, scalar, invertible principal symbol; see Theorem 3.16, p. 154 in Kodaira's book [12]. Since $D^* = D$ and $-D^2 = \Box$, we may conclude that D is of Dirac type. In particular, so is $\bar{D} = \bar{\partial} + \bar{\partial}^*$. Another related Dirac operator is $D := (\partial, \partial^*)$.

- Assume that

$$D := \begin{pmatrix} d & d^* \\ d^* & d \end{pmatrix}. \tag{3.3}$$

Then $-D^*D = -DD^* = \Delta I_{2\times 2}$, so that D is of Dirac type.

- Consider next the case when M is a complex manifold and set

$$D := \begin{pmatrix} \partial & \partial^* \\ \partial^* & \partial \end{pmatrix}. \tag{3.4}$$

It follows that $-D^*D = -DD^* = \Box I_{2\times 2}$, so that D is of Dirac type. In particular, so are

$$\bar{D} = \begin{pmatrix} \bar{\partial} & \bar{\partial}^* \\ \bar{\partial}^* & \bar{\partial} \end{pmatrix}, \quad D^* = \begin{pmatrix} \partial^* & \partial \\ \partial & \partial^* \end{pmatrix} \quad \text{and} \quad \bar{D}^* = \begin{pmatrix} \bar{\partial}^* & \bar{\partial} \\ \bar{\partial} & \bar{\partial}^* \end{pmatrix}.$$

- When $M = \mathbb{R}^m$ and $D := \sum \mathbf{e}_\alpha \partial_\alpha$, then $D^* = D$ and $-D^2 = \Delta$, the flat-space Laplacian. That D is self-adjoint is a consequence of the identity $\langle a \cdot f, g \rangle = \langle f, a^c \cdot g \rangle$, where the superscript c indicates standard conjugation in the Clifford algebra $C\ell(\mathbb{R}^m)$ (see [2]). Thus, D is of Dirac type.

- In the case when $M = \mathbb{R}^{m+1}$ and D is the Cauchy–Riemann operator

$$D := \partial_0 + \sum_{\alpha=1}^{m} \mathbf{e}_\alpha \partial_\alpha, \tag{3.5}$$

we have $D^* = -\partial_0 + \sum \mathbf{e}_\alpha \partial_\alpha$. Nonetheless, ultimately, $-D^*D = \Delta$, the Laplacian in $M = \mathbb{R}^{m+1}$, so D in (3.5) is of Dirac type.

- Let $M = \mathbb{R}^m$ and take $D := \sum_{\alpha=1}^{m} \mathbf{e}_\alpha \partial_\alpha + k\mathbf{e}_{m+1}$, where $k \in \mathbb{R}$. Then $D^* = \sum_{\alpha=1}^{m} \mathbf{e}_\alpha \partial_\alpha - k\mathbf{e}_{m+1}$ so $-D^*D = \Delta + k^2$, the Helmholtz operator in $\mathbb{R}^m$. Thus, D is a Dirac type. A similar conclusion holds for $D := ik + \sum_{\alpha=1}^{m} \mathbf{e}_\alpha \partial_\alpha$ with $k \in \mathbb{R}$.

- Analogous constructions (as in the last two examples above) can be carried out employing the embedding $\mathbb{R}^4 \hookrightarrow \mathbb{H}$, the skew field of quaternions. That is, if i, j, k are the standard anticommuting imaginary units in $\mathbb{H}$, consider $D := \partial_{x_0} + i\partial_{x_1} + j\partial_{x_2} + k\partial_{x_3}$, so that $D^* = -\partial_{x_0} + i\partial_{x_1} + j\partial_{x_2} + k\partial_{x_3}$. It follows that $-D^*D = -DD^* = \Delta$, so D is of Dirac type.

- Analogous considerations apply to the case of the Dirac operator associated with *octonions*; see §9 for more details.

- A simple, yet important, example of a Dirac type operator whose symbol is *not* invertible is $D :=$ grad, on scalar functions. Then $D^* = -$div, so $-D^*D = \Delta$, the Laplace–Beltrami operator on M.

- Suppose that several manifolds M_j and vector bundles $\mathcal{F}_j \to M_j$, $j = 1, \ldots, n$, have been given along with a vector space $\mathcal{A}$ (identified with a trivial vector bundle over each M_j). It is assumed that on each manifold M_j a Dirac type operator D_j (mapping $\mathcal{A}$-valued functions into sections in $\mathcal{F}_j$) has been fixed. Consider then the product manifold $\mathbb{M} := M_1 \times \cdots \times M_n$ together with bundle $\mathbb{F} := \oplus \mathcal{F}_j$ and, finally, the operator $\mathbb{D} := (D_j)_j$. That is, for each $u : \mathbb{M} \to \mathcal{A}$, we set $\mathbb{D}u = (D_ju)_j$, where D_ju means that D_j acts with respect to the variables on M_j.

It can then be checked that $\mathbb{D}^*(u_j)_j = \sum D_j^* u_j$ so that, ultimately,

$$-\mathbb{D}^*\mathbb{D} = \sum(-D_j^*D_j).$$

It follows that $\mathbb{D}$ is of Dirac type. It should be pointed out that, for $n > 1$, the operator $\mathbb{D}$ is *not* elliptic (in the sense that its symbol is not invertible).

This last example contains, in particular, the Dirac operator relevant in the case of *separately monogenic* (or multi-monogenic) functions. More specifically, consider the situation when $M_1 = M_2 = \cdots = M_n = \mathbb{R}^m$, $\mathcal{A} = C\ell(\mathbb{R}^m)$, and $D_j :=$ the standard Dirac operator with respect to the j-th copy of $\mathbb{R}^m$. Set

$$\mathbb{D} := (D_j)_j. \tag{3.6}$$

Then $u : \mathbb{R}^m \times \cdots \times \mathbb{R}^m \to C\ell(\mathbb{R}^m)$ is separately monogenic, i.e., $D_ju = 0$ for each j, if and only if $\mathbb{D}u = 0$. When each D_j is as in (3.5) and $m = 1$, then, of course, we are talking about holomorphic functions of several complex variables.

6.4   Unique continuation property

Suppose we are interested in inverting a first-order differential operator D as in (2.1) between appropriate Sobolev spaces. To this end, it is convenient to consider first the case of its associated Laplacian, $-D^*D$. Formally, $D^{-1} = -(-D^*D)^{-1}D^*$, so it suffices to invert the operator $-D^*D$. Since

$$\sigma(-D^*D; \xi) = -\sigma(D; \xi)^*\sigma(D; \xi),$$

it follows that $-D^*D$ is *strongly elliptic*, i.e., the (symmetric part of its) principal symbol is negative definite, if and only if

$$\sigma(D; \xi) \text{ is injective } \forall \xi \in T^*M \setminus 0. \tag{4.1}$$

Assuming that (4.1) holds, it follows from the classical theory that $-D^*D$ has index zero and a finite–dimensional kernel. In particular, the existence of an inverse $(-D^*D)^{-1}$ hinges on whether

$$\mathrm{Ker}(-D^*D) = \mathrm{Ker}\,D = 0. \tag{4.2}$$

Many times in practice, however, the above condition fails to hold (although there are remarkable exceptions). For example, if $D = d + d^*$, then $-D^*D = \Delta$ and the triviality of $\mathrm{Ker}\,\Delta$ ultimately depends on certain topological characteristics of the manifold M.

An alternative is to require the *weaker* condition that

$$Du = 0 \text{ on } M, \ \ u \equiv 0 \text{ on an open subset of } M \Rightarrow u \equiv 0 \text{ in } M. \tag{4.3}$$

In the sequel, the operator D is said to have the *unique continuation property* (abbreviated $D \in$ UCP henceforth) if (4.3) holds. The deep result of N. Aronszjan [1] asserts that all elliptic, second-order differential operators with a real, scalar principal part have the unique continuation property. This immediately implies that all Dirac type operators have the unique continuation property. The case of (rough) lower–order perturbations has been considered more recently in [8].

Consider now a fixed, scalar, real, nonnegative, smooth potential V on M, and introduce the formally self-adjoint, nonpositive, second-order differential operator

$$L := -D^*D - V, \ \ \ L : \mathcal{E} \longrightarrow \mathcal{E}. \tag{4.4}$$

Thus, L is a zero-order perturbation of the D-Laplacian. To discuss conditions under which L^{-1} exists, recall that for a Lipschitz domain Ω, the (zero-trace) Sobolev space $H_0^{1,2}(\Omega, \mathcal{E})$ is the closure of $C_o^\infty(\Omega, \mathcal{E})$ in the norm

$$\| \cdot \|_{L^2} + \| \nabla \cdot \|_{L^2}.$$

Proposition 2. *Let $D : \mathcal{E} \to \mathcal{F}$ be a first-order differential operator for which (4.1) holds and such that $D \in$ UCP. Then for each Lipschitz domain $\Omega \subseteq M$ (possibly the whole manifold M),*

$$u \in H_0^{1,2}(\Omega, \mathcal{E}) \ \& \ Lu = 0 \text{ in } \Omega \iff u = 0 \text{ in } \Omega. \tag{4.5}$$

In particular, L^{-1} exists if $V > 0$ on some open subset of M.

Proof. To justify (4.5), if $Lu = 0$ in Ω for some $u \in H_0^{1,2}(\Omega, \mathcal{E})$, then

$$0 = \int_\Omega \langle Lu, \bar{u} \rangle \, d\mathcal{V} = -\int_\Omega \left(|Du|^2 + V|u|^2 \right) d\mathcal{V}. \tag{4.6}$$

Hence, $Du = 0$ in Ω and $u = 0$ in $\mathrm{supp}\,V \cap \Omega$. In particular, with tilde denoting the extension by zero outside Ω, we see that $\tilde{u} \in H^{1,2}(M, \mathcal{E})$ and $D\tilde{u} = 0$ on M. Also, $\tilde{u}$ vanishes in some open subset of M. Indeed, this is clear when Ω is a proper subset of M, whereas when Ω is the whole manifold, the desired conclusion follows from the fact that $u = 0$ in supp V. At any rate, since $D \in$ UCP, these considerations imply $\tilde{u} \equiv 0$ in M and, ultimately, $u \equiv 0$ in Ω, proving (4.5).

That Ker $L = \{0\}$ on M also proves that L^{-1} exists globally.   □

6.5   Cauchy type operators

Assuming that the conditions (4.1) and (4.3) hold, let the distribution

$$E \in \mathcal{D}'(M \times M, \mathcal{E} \otimes \mathcal{E})$$

be the *Schwartz kernel* of L^{-1}. Then

$$L_x E(x, y) = L_y^t E(x, y) = \delta_x(y), \text{ the Dirac distribution on } M. \qquad (5.1)$$

The self-adjointness of L then translates into $E(x, y)^* = E(x, y)$. Set

$$\Gamma(x, y) := i\sigma(D^t; \nu_y)\bar{D}_y E(x, y), \quad \Gamma \in \mathcal{D}'(M \times M, \mathcal{E} \otimes \mathcal{E}). \qquad (5.2)$$

Loosely speaking, $E(x, y)$ and $\Gamma(x, y)$ play the roles of the Newtonian kernel and the Cauchy kernel, respectively. It is then natural to introduce the Cauchy type integral operator

$$\mathcal{C}f(x) := \int_{\partial\Omega} \langle \Gamma(x, y), f(y) \rangle \, ds_y, \quad x \notin \partial\Omega, \qquad (5.3)$$

where $f : \partial\Omega \to \mathcal{E}$ is an arbitrary section. Another way of understanding the operator $\mathcal{C}$, is to observe that the first–order differential operator

$$\partial_\nu := i\sigma(D^*; \nu)D \qquad (5.4)$$

is the *conormal derivative* corresponding to the factorization: top part of $L = -D^* D$. Consequently, $\partial_{\bar{\nu}} := i\sigma(D^t; \nu)\bar{D}$ is the complex conjugate of ∂_ν. Thus, on account of (5.2), $\mathcal{C}$ can be thought of as the natural *double layer* potential operator associated with the D-Laplacian. Let us also introduce the single layer potential operator

$$\mathcal{S}f(x) := \int_{\partial\Omega} \langle E(x, y), f(y) \rangle \, ds_y, \quad x \in M \setminus \partial\Omega. \qquad (5.5)$$

Our main result in this section is a generalization of the classical Pompeiu formula (from complex analysis, to the current setting).

Theorem 1. *Suppose that (4.1) and (4.3) hold for* D. *Then, for each* $x \in \Omega$ *and* $u \in C^2(\bar{\Omega}, \mathcal{E})$ *we have*

$$u(x) = \mathcal{C}(u|_{\partial\Omega})(x) - \mathcal{S}(Vu|_{\partial\Omega})(x) - \int_\Omega \langle \bar{D}_y E(x, y), (Du)(y) \rangle \, d\mathcal{V}_y$$

$$= \mathcal{C}(u|_{\partial\Omega}) - \mathcal{S}(\partial_\nu u|_{\partial\Omega}) - \int_\Omega \langle E(x, y), (Lu)(y) \rangle \, d\mathcal{V}_y. \qquad (5.6)$$

In particular,

$$V = 0 \quad on \quad \bar{\Omega} \quad and \quad Du = 0 \quad in \quad \Omega \Rightarrow u = \mathcal{C}(u|_{\partial\Omega}) \; on \; \Omega, \qquad (5.7)$$

$$Lu = 0 \quad in \quad \Omega \Rightarrow u = \mathcal{C}(u|_{\partial\Omega}) - \mathcal{S}(\partial_\nu u|_{\partial\Omega}) \quad on \quad \Omega. \qquad (5.8)$$

Proof. Since $L^t = -D^t \bar{D} - V$, both identities in (5.6) follow starting with

$$u(x) = \int_\Omega \langle [-D^t_y \bar{D}_y - V(y)] E(x, y), u(y) \rangle \, dV_y,$$

then invoking (2.4) twice. □

Note that a more 'standard' form of Pompeiu's formula is obtained from the first line of (5.6) by requiring that $V = 0$ on $\bar{\Omega}$. Also, (5.7) plays the role of *Cauchy's reproducing formula* in this setting, whereas (5.8) is *Green's representation formula* for null-solutions of L (in terms of double and single layer potentials). It should be mentioned that the conditions on the boundary behavior of u can be considerably relaxed. E.g., a proper control of the nontangential maximal function of u will do; see §6.

An intriguing aspect is that while the reproducing formula (5.7) holds for any null solution of D (which is reasonably well behaved near the boundary), in general it is *not* true that $D\mathcal{C} = 0$. It is therefore natural to try to find necessary conditions for this latter condition to hold.

Proposition 3. *Under the additional assumption that*

$$-DL^{-1}D^* = I \quad in \quad \mathcal{F}, \qquad (5.9)$$

the Cauchy operator (5.3) has the property that $D\mathcal{C} = 0$ in Ω.

Proof. At the level of Schwartz kernels, (5.9) entails $-D_x \bar{D}_y E(x, y) = \delta_x(y)$, the Dirac distribution with mass at x. Then, the desired property is an immediate consequence of this identity and the definition (5.2). □

Remark 1. *It is worth pointing out that (5.9) can only hold when D is elliptic and has a trivial kernel (globally, in L^2). When $\mathrm{Ker}\, D = 0$, the potential V can be chosen to be zero on M, without affecting the existence of L^{-1}. If, in addition, $\mathcal{E} = \mathcal{F}$ and $D^*D = DD^*$, then (5.9) automatically holds. This holds for most of the operators in §3 when M is, e.g., the entire space (the exceptions are the operators in the last two examples).*

6.6   Global regularity and boundary behavior

For a fixed, Lipschitz domain $\Omega \subset M$, let $\cdot|_{\partial\Omega}$ denote the nontangential boundary trace operator. That is,

$$u\big|_{\partial\Omega}(x) := \lim_{y \in \gamma(x)} u(y), \quad x \in \partial\Omega, \tag{6.1}$$

where $\gamma(x) \subseteq \Omega$ is a suitable nontangential approach region; see [17]. Also, let $\mathcal{N}$ stand for the nontangential maximal operator, i.e.,

$$\mathcal{N}u(x) := \sup\{|u(y)|;\ y \in \gamma(x)\}, \quad x \in \partial\Omega. \tag{6.2}$$

We shall also work with the boundary version of (5.3) and (5.5) given, respectively, by

$$Cf(x) := \text{p.v.} \int_{\partial\Omega} \langle \Gamma(x,y), f(y) \rangle \, ds_y, \quad x \in \partial\Omega, \tag{6.3}$$

$$Sf(x) := \int_{\partial\Omega} \langle E(x,y), f(y) \rangle \, ds_y, \quad x \in \partial\Omega. \tag{6.4}$$

In (6.3), 'p.v.' indicates that the integral is taken in the Cauchy *principal value* sense, i.e., by removing small geodesic balls and passing to the limit. Finally, we set $\Omega_+ := \Omega$ and $\Omega_- := M \setminus \overline{\Omega}$.

The next two theorems collect some of the basic properties of the operators (5.3)–(5.5) and their boundary versions (6.3)–(6.4). To state our first result, denote by $L_1^p(\partial\Omega, \mathcal{E})$ the Sobolev space of sections $f \in L^p(\partial\Omega, \mathcal{E})$ with $|\nabla_{\tan}f| \in L^p(\partial\Omega)$ and set $L_{-1}^p(\partial\Omega, \mathcal{E}) := (L_1^q(\partial\Omega, \mathcal{E}))^*$ if $\frac{1}{p} + \frac{1}{q} = 1$.

Theorem 2. *Let Ω be a Lipschitz subdomain of M and assume that V is an arbitrary, smooth, nonnegative, non-identically zero, scalar potential. Also, assume that the first-order differential operator D as in (2.1) satisfies (4.1) and (4.3). Then, for each $1 < p < \infty$, the following are true:*

(1) $\|\mathcal{N}(\nabla^s Cf)\|_{L^p(\partial\Omega)} \leq \kappa\|f\|_{L_s^p(\partial\Omega,\mathcal{E})},$ *for $s \in \{0,1\}$;*

(2) C *is bounded on $L^p(\partial\Omega, \mathcal{E})$ and on $L_1^p(\partial\Omega, \mathcal{E})$;*

(3) $Cf|_{\partial\Omega_\pm} = (\pm\frac{1}{2}I + C)f$ *for any $f \in L^p(\partial\Omega, \mathcal{E})$;*

(4) $\partial_\nu Sf|_{\partial\Omega_\pm} = (\mp\frac{1}{2}I + C^*)f$ *and $Sf|_{\partial\Omega_\pm} = Sf$ for any $f \in L^p(\partial\Omega, \mathcal{E})$;*

(5) $C \circ S = S \circ C^*$ *on $\partial\Omega$, and $L(Cf) = 0$, $L(Sf) = 0$ in $\Omega_\pm$;*

(6) $\partial_\nu C : L_1^p(\partial\Omega, \mathcal{E}) \to L^p(\partial\Omega, \mathcal{E})$ *is bounded, and $\partial_\nu Cf|_{\partial\Omega_+} = \partial_\nu Cf|_{\partial\Omega_-}$ for any $f \in L_1^p(\partial\Omega, \mathcal{E})$;*

(7) *The adjoint of the operator above is $\partial_\nu C : L^q(\partial\Omega, \mathcal{E}) \to L_{-1}^q(\partial\Omega, \mathcal{E})$, where $\frac{1}{p} + \frac{1}{q} = 1$;*

(8) $S : L^p(\partial\Omega, \mathcal{E}) \to L_1^p(\partial\Omega, \mathcal{E})$ *is bounded and its adjoint is the operator* $S : L_{-1}^q(\partial\Omega, \mathcal{E}) \to L^q(\partial\Omega, \mathcal{E})$, *where* $\frac{1}{p} + \frac{1}{q} = 1$;

(9) $\|\mathcal{N}(\nabla^s \mathcal{S}f)\|_{L^p(\partial\Omega)} \leq \kappa \|f\|_{L_{s-1}^p(\partial\Omega,\mathcal{E})}$ *for* $s \in \{0, 1\}$;

(10) $\| \, |\nabla^{1+s}\mathcal{C}f| \, \mathrm{dist}(\cdot, \partial\Omega)^{\frac{1}{2}} \|_{L^2(\Omega)} \leq \kappa \|f\|_{L_s^2(\partial\Omega,\mathcal{E})}$ *for* $s \in \{0, 1\}$.

Next we consider the action of $\mathcal{C}$ and C on other types of *Sobolev* and *Besov* spaces, denoted by L_s^p and $B_s^{p,q}$, respectively. These are considered either on Ω or $\partial\Omega$; see [9] for detailed definitions.

Theorem 3. *With the same hypotheses as above, the following operators are bounded on the indicated spaces:*

(1) C *on* $L_s^p(\partial\Omega, \mathcal{E})$ *for each* $1 < p < \infty$ *and* $0 \leq s \leq 1$;

(2) C *on* $B_s^{p,q}(\partial\Omega, \mathcal{E})$ *for each* $1 \leq p, q \leq \infty$ *and* $0 < s < 1$;

(3) $S : L_{-s}^p(\partial\Omega, \mathcal{E}) \to L_{1-s}^p(\partial\Omega, \mathcal{E})$ *for* $1 < p < \infty$ *and* $0 \leq s \leq 1$;

(4) $S : B_{-s}^{p,q}(\partial\Omega, \mathcal{E}) \to B_{1-s}^{p,q}(\partial\Omega, \mathcal{E})$ *for* $1 \leq p, q \leq \infty$ *and* $0 < s < 1$;

(5) $\partial_\nu C : L_s^p(\partial\Omega, \mathcal{E}) \to L_{s-1}^p(\partial\Omega, \mathcal{E})$ *for* $1 < p < \infty$ *and* $0 \leq s \leq 1$;

(6) $\partial_\nu C : B_s^{p,q}(\partial\Omega, \mathcal{E}) \to B_{s-1}^{p,q}(\partial\Omega, \mathcal{E})$ *for* $1 \leq p, q \leq \infty$ *and* $0 < s < 1$;

(7) $\mathcal{S} : B_{-s}^{p,q}(\partial\Omega, \mathcal{E}) \to B_{1-s+1/p}^{p,q}(\Omega, \mathcal{E})$ *for* $1 \leq p, q \leq \infty$, *and* $0 < s < 1$;

(8) $\mathcal{C} : L_s^p(\partial\Omega, \mathcal{E}) \to L_{s+1/p}^p(\Omega, \mathcal{E}) \cap B_{s+1/p}^{p,p}(\Omega, \mathcal{E})$ *for* $1 < p < \infty$ *and* $0 < s < 1$;

(9) $\mathcal{C} : B_s^{p,q}(\partial\Omega, \mathcal{E}) \to B_{s+1/p}^{p,q}(\Omega, \mathcal{E})$ *for* $1 \leq p, q \leq \infty$, *and* $0 < s < 1$;

(10) $\mathcal{C} : L_s^p(\partial\Omega, \mathcal{E}) \to B_{s+1/p}^{p,p^\#}(\Omega, \mathcal{E})$ *if* $1 < p < \infty$, *with* $p^\# := \max\{p, 2\}$, *and* $s \in \{0, 1\}$;

(11) $\mathcal{S} : L_{-s}^p(\partial\Omega, \mathcal{E}) \to L_{1-s+1/p}^p(\Omega, \mathcal{E}) \cap B_{1-s+1/p}^{p,p}(\Omega, \mathcal{E})$ *for* $1 < p < \infty$ *and* $0 < s < 1$;

(12) $\mathcal{S} : L_{s-1}^p(\partial\Omega, \mathcal{E}) \to B_{s+1/p}^{p,p^\#}(\Omega, \mathcal{E})$ *for* $1 < p < \infty$ *and* $s \in \{0, 1\}$.

The proof of Theorems 2–3 can be carried out using the techniques developed in [17] and [20]. These build on the work of many authors, including [4, 9, 23], (see also the references in [11]). For the case of constant coefficient operators see [19].

As pointed out in the statement of Theorem 1, null-solutions of D are 'reproduced' by $\mathcal{C}$. Next, we consider the *converse* issue (see also [13]), i.e., whether null-solutions of L which are reproduced by the Cauchy operator $\mathcal{C}$ are in fact null-solutions of D.

Theorem 4. *Retain the same hypotheses as in Theorem 2 and assume that $V \equiv 0$ near $\bar{\Omega}$. If $u \in C^1(\Omega, \mathcal{E})$ is such that $\mathcal{N}(u) \in L^2(\partial\Omega)$, then*

$$\mathcal{N}(Du) \in L^2(\partial\Omega) \ \& \ u = \mathcal{C}(u|_{\partial\Omega}) \text{ in } \Omega \iff Du = 0 \text{ in } \Omega. \quad (6.5)$$

Proof. The right-to-left implication is a consequence of Cauchy's reproducing formula (5.7). As for the opposite implication, first notice that $u = \mathcal{C}(u|_{\partial\Omega}) \implies Lu = 0$ in Ω. With this at hand, (2.4) gives

$$\int_\Omega |Du|^2 \, d\mathcal{V} = \int_\Omega \langle Du, \bar{D}\bar{u} \rangle \, d\mathcal{V} = \int_{\partial\Omega} \langle \partial_\nu u, \bar{u} \rangle \, ds, \quad (6.6)$$

since $V = 0$ near $\bar{\Omega}$. Thus, in order to conclude that $Du = 0$ in Ω, it suffices to show that $\partial_\nu u = 0$ on $\partial\Omega$.

With this goal in mind, note that our Green type identity (5.8), in concert with the assumption $u = \mathcal{C}(u|_{\partial\Omega})$, forces $\mathcal{S}(\partial_\nu u) = 0$ in Ω. Taking the boundary trace entails $S(\partial_\nu u) = 0$ on $\partial\Omega$. Assuming for a moment that S is injective on $L^2(\partial\Omega, \mathcal{E})$, we may finally conclude that $\partial_\nu u = 0$, as intended. Thus, the proof is finished, modulo showing that

$$S : L^2(\partial\Omega, \mathcal{E}) \longrightarrow L^2(\partial\Omega, \mathcal{E}) \quad (6.7)$$

is one-to-one.

To see this last claim, let $f \in L^2(\partial\Omega, \mathcal{E})$ be such that $Sf = 0$ on $\partial\Omega$, and set $u := \mathcal{S}f$ in $\Omega_\pm$. It follows that $u \in H_0^{1,2}(\Omega_\pm, \mathcal{E})$ and $Lu = 0$ in $\Omega_\pm$. Thus, $u = 0$ in $\Omega_\pm$, by virtue of Proposition 2. Let $\theta \in T^*M$ be transversal to $\partial\Omega$. Then,

$$0 = (\nabla_\theta u)|_{\partial\Omega_+} - (\nabla_\theta u)|_{\partial\Omega_-} = -i\langle \theta, \nu \rangle \sigma(L^{-1}; \nu)f \quad (6.8)$$

pointwise a.e. on $\partial\Omega$. Since $\langle \theta, \nu \rangle \neq 0$ on $\partial\Omega$ and since $\sigma(L^{-1}; \nu)$ is an isomorphism, the identity (6.8) forces $f = 0$. This justifies the claim made about (6.7) and finishes the proof of the theorem. $\square$

6.7 Hardy spaces in Lipschitz domains

For a first-order differential operator D as in (2.1) and for $0 < p \leq \infty$, define the *Hardy space* associated with D in a domain $\Omega \subset M$ as

$$\mathcal{H}^p(\Omega, D) := \{u : \Omega \to \mathcal{E}; \ Du = 0 \text{ in } \Omega, \text{ and } \mathcal{N}u \in L^p(\partial\Omega)\}. \quad (7.1)$$

The modern theory of Hardy spaces of holomorphic functions in Lipschitz subdomains of the complex plane originates in [4, 10]. Here we continue this line of work by proving the following.

Theorem 5. *Let Ω be a Lipschitz subdomain of M and assume that the first-order differential operator D satisfies (4.1), (4.3). Then, for each $u \in \mathcal{H}^p(\Omega, D)$, $1 < p < \infty$, the following hold.*

(1) u has a non-tangential limit $u|_{\partial\Omega}$ a.e. and $u|_{\partial\Omega} \in L^p(\partial\Omega, \mathcal{E})$. Furthermore, $\|u|_{\partial\Omega}\|_{L^p(\partial\Omega)} \approx \|\mathcal{N}u\|_{L^p(\partial\Omega)}$.

(2) $u|_{\partial\Omega} \in L_1^p(\partial\Omega, \mathcal{E})$ if and only if $\mathcal{N}(\nabla u) \in L^p(\partial\Omega)$. Moreover, there holds $\|u|_{\partial\Omega}\|_{L_1^p(\partial\Omega, \mathcal{E})} \approx \|\mathcal{N}(\nabla u)\|_{L^p(\partial\Omega)}$.

(3) $u|_{\partial\Omega} \in B_s^{p,q}(\partial\Omega, \mathcal{E})$ if and only if $u \in B_{s+1/p}^{p,q}(\Omega, \mathcal{E})$, with equivalence of norms.

(4) $u|_{\partial\Omega} \in L^2(\partial\Omega, \mathcal{E})$ if and only if $u \in L_{1/2}^2(\Omega, \mathcal{E})$ if and only if

$$|\nabla u|\, \mathrm{dist}(\cdot, \partial\Omega)^{\frac{1}{2}} \in L^2(\Omega);$$

(5) $u|_{\partial\Omega} \in L_1^2(\partial\Omega, \mathcal{E})$ if and only if $u \in L_{3/2}^2(\Omega, \mathcal{E})$ if and only if

$$|\nabla^2 u|\, \mathrm{dist}(\cdot, \partial\Omega)^{\frac{1}{2}} \in L^2(\Omega);$$

(6) Under the additional assumption (5.9), the Plemelj–Calderón decomposition $L^p(\partial\Omega, \mathcal{E}) = \mathcal{H}^p(\Omega_+, D)|_{\partial\Omega} \oplus \mathcal{H}^p(\Omega_-, D)|_{\partial\Omega}$ holds; here the direct sum is both algebraic and topological.

In the case when $M = \mathbb{C}$ and $D = \frac{\partial}{\partial\bar{z}}$, (4)–(5) go back to Kenig's area-function estimate in [10]; see also [4]. Regularity results in the same spirit for $M = \mathbb{C}^m$ and $D = \bar{\partial}$ have also recently been considered in [16].

Proof of Theorem 5. The idea is to choose $V \equiv 0$ near $\bar{\Omega}$, and then use the fact that any $u \in \mathcal{H}^p(\Omega, D)$ is the Cauchy integral extension of its boundary trace, i.e., $u = \mathcal{C}(u|_{\partial\Omega})$ in Ω. At this stage, we may rely on Theorems 2–3 in order to prove the claims (1)–(5). As for (6), the fact that $f = \mathcal{C}f|_{\partial\Omega_+} + \mathcal{C}f|_{\partial\Omega_-}$ for any $f \in L^p(\partial\Omega, \mathcal{E})$ shows that $\mathcal{H}^p(\Omega_\pm, D)|_{\partial\Omega}$ generate $L^p(\partial\Omega, \mathcal{E})$. To see that the sum is direct, assume that $u_\pm \in \mathcal{H}^p(\Omega_\pm, D)$ are such that $u_+|_{\partial\Omega} = u_-|_{\partial\Omega}$. Then, for every $x \in \Omega$, $u_+(x) = \mathcal{C}(u_+|_{\partial\Omega})(x) = \mathcal{C}(u_-|_{\partial\Omega})(x) = 0$, integrating by parts. Thus, $u_\pm \equiv 0$, and the desired conclusion follows. $\qquad\square$

For $(m-1)/m < p \leq 1$, similar properties hold provided one employs (local) *atomic* Hardy spaces $\mathfrak{h}_{at}^p(\partial\Omega)$ (see [11] for definitions) in place of L^p. We now proceed to discuss a compensated compactness type result, extending work from [18].

Proposition 4. *Let $0 < p, q \leq \infty$ and $(m-1)/m < r \leq \infty$ be such that $\frac{1}{p} + \frac{1}{q} = \frac{1}{r}$. Then*

$$\langle u|_{\partial\Omega}, \sigma(D^t; \nu)v|_{\partial\Omega}\rangle \in \mathfrak{h}_{at}^r(\partial\Omega), \tag{7.2}$$

for every $u \in \mathcal{H}^p(\Omega, D)$ and $v \in \mathcal{H}^q(\Omega, D^t)$.

Proof. This can be proved by mimicking the approach in [26]. There, the context is that of (generalized) Cauchy–Riemann systems in the upperhalf–space but since the approach utilizes only nontangential maximal function estimates and cancelations based on integrations by parts, it can be extended to the present setting.   □

6.8   Clifford analysis for functions of several variables

In this section we consider in more detail the last example in §3, corresponding to Clifford separately monogenic functions. For a fixed integer n, let $(\mathbb{R}^m)^n$ be embedded in $\mathcal{A} := \oplus C\ell(\mathbb{R}^m)$, the sum of n copies of $C\ell(\mathbb{R}^m)$. That is, if $x \in (\mathbb{R}^m)^n = \mathbb{R}^{mn}$ we write $x = (x_1, \ldots, x_n)$ with $x_j = \sum_{\alpha=1}^n x_j^\alpha \mathbf{e}_\alpha \in C\ell(\mathbb{R}^m)$ for $1 \le j \le n$. In this setting, the Dirac operator corresponding to the j-th copy of $C\ell(\mathbb{R}^m)$ is

$$D_j = \sum_{\alpha=1}^m \mathbf{e}_\alpha \frac{\partial}{\partial x_j^\alpha}, \qquad (8.1)$$

while the global Dirac operator reads $\mathbb{D} := (D_j)_j$. It follows that $\mathbb{D}$ is real (i.e., $\bar{\mathbb{D}} = \mathbb{D}$) and $-\mathbb{D}^*\mathbb{D} = \sum_j \Delta_j = \Delta$. Here Δ_j is the Laplacian in the j-th factor of the Cartesian product $(\mathbb{R}^m)^n$, for each $1 \le j \le n$, and Δ is the Laplacian in the whole space $\mathbb{R}^{mn}$. Call a $C\ell(\mathbb{R}^m)$-valued function u defined in a region Ω of $\mathbb{R}^{mn}$ *separately monogenic* if $D_j u = 0$ in Ω for each $j = 1, \ldots, n$. Note that this amounts to the requirement that $\mathbb{D}u = 0$; in particular, a separately monogenic function is harmonic.

Recall the standard fundamental solution for the Laplacian in $(\mathbb{R}^m)^n$, i.e.,

$$\mathbb{E}(x) = \begin{cases} -\frac{1}{2\pi}\log|x|, & \text{if } mn = 2; \\ \mathbb{E}(x) = -\frac{1}{(mn-2)\omega_{mn}} \frac{1}{|x|^{mn-2}}, & \text{assuming } mn \ge 3; \end{cases}$$

hereafter, ω_{mn} stands for the area of the unit sphere in $\mathbb{R}^{mn}$.

Starting from this, we set

$$E(x, y) := \sum_I \mathbb{E}(x - y)\mathbf{e}_I \otimes \mathbf{e}_I,$$

where $\mathbf{e}_I := \mathbf{e}_{i_1} \cdots \mathbf{e}_{i_\ell}$ if $I = (i_1, \ldots, i_\ell)$, and the sum is performed over strictly increasing multi–indices. This corresponds precisely to the setup discussed at the beginning of §5, adapted to the current context.

Going further, fix a bounded Lipschitz domain $\Omega \subset \mathbb{R}^{mn}$ and denote by

$$\nu = (\nu_j)_j : \partial\Omega \to (\mathbb{R}^m)^n \hookrightarrow \mathcal{A}$$

the outward unit normal to $\partial\Omega$ (with $\nu_j = \sum \nu_j^\alpha \mathbf{e}_\alpha$). In particular,

$$\partial_\nu = \sum \nu_j \cdot D_j,$$

where each product is to be understood in the Clifford algebra sense. One can then easily check that, with $\langle \cdot, \cdot \rangle$ denoting the natural (bilinear) pairing in $C\ell(\mathbb{R}^m)$, we have $\langle \partial_{\nu_y} E(x,y), f(y) \rangle = \Gamma(x,y) \cdot f(y)$ where

$$\Gamma(x,y) := \sum_{j=1}^{n} D_{y_j} \mathbb{E}(x,y) \cdot \nu_j(y) = -\frac{1}{\omega_{mn}} \sum_{j=1}^{n} \frac{x_j - y_j}{|x-y|^{mn}} \cdot \nu_j(y), \qquad (8.2)$$

can be thought of as the natural Cauchy kernel in this setting. Then, in complete analogy to (5.3), the corresponding Cauchy integral operator is defined by

$$\mathcal{C}f(x) = \int_{\partial\Omega} \Gamma(x,y) \cdot f(y)\, ds_y, \quad x \in \mathbb{R}^{mn} \setminus \partial\Omega, \qquad (8.3)$$

for $f : \partial\Omega \to C\ell(\mathbb{R}^m)$. Let us point out that while $\Gamma(x,y)$ is harmonic in x, it is not separately monogenic when $n > 1$. Thus, the Cauchy operator is separately monogenic if and only if $n = 1$. The fact that $\mathcal{C}$ reproduces separately monogenic functions (see Theorem 1) can then be viewed as a version of the classical Bochner–Martinelli formula in $\mathbb{C}^m$; we leave the easy details of this verification to the interested reader. At this stage, having completed the parallel with our earlier, general theory, we may conclude that

$$\textit{all main results in §6–§7 hold in the current setting.} \qquad (8.4)$$

Other, more specific results, are dealt with separately below.

A distinguished feature of the setting we are currently considering is the fact that a Hartogs type theorem is valid in this context. Specifically, we have the following.

Theorem 6. *Let Ω be an open set in $\mathbb{R}^{mn}$, $n > 1$, and let K be a compact subset of Ω with $\Omega \setminus K$ connected. Then each separately monogenic function $F : \Omega \setminus K \to C\ell(\mathbb{R}^m)$ extends to a separately monogenic function in Ω.*

The proof is presented in Appendix A. This result has many remarkable corollaries. Here we only want to single out a few.

Corollary 1. *Assume that $n > 1$ and fix $\Omega \subset \mathbb{R}^{mn}$ a bounded Lipschitz domain with a connected complement. Then*

$$\begin{array}{c} \textit{u is separately monogenic in } \mathbb{R}^{mn} \setminus \bar{\Omega} \\ \textit{and u is bounded near infinity} \end{array} \iff u = 0. \qquad (8.5)$$

Proof. This is a direct consequence of Theorem 6 and Liouville's theorem (for harmonic functions). $\qquad\qquad \square$

We now discuss conditions guaranteeing that a function f defined on $\partial\Omega$ extends to a separately monogenic function F in Ω.

Corollary 2. *Let $\Omega \subset \mathbb{R}^{mn}$, $n > 1$, be a bounded Lipschitz domain with a connected complement. Then for each $f \in L_1^2(\partial\Omega, C\ell(\mathbb{R}^m))$, the following are equivalent:*

(i) *There exists a (unique) separately monogenic function F in Ω such that $\mathcal{N}(F) \in L^2(\partial\Omega)$ and for which $F|_{\partial\Omega} = f$;*

(ii) *Same as above, except that also $\mathcal{N}(\nabla F) \in L^2(\partial\Omega)$;*

(iii) $Cf = \frac{1}{2}f$;

(iv) $C^2 f = \frac{1}{4}f$.

Proof. That (i) and (ii) are equivalent follows from (2) in Theorem 5. The implication $(i) \Rightarrow (iii)$ is a direct consequence of (5.7) and the point (3) in Theorem 2. Also, clearly, $(iii) \Rightarrow (iv)$. There remains to show that $(iv) \Rightarrow (i)$. To this end, let $F_\pm := Cf$ in $\Omega_\pm$, so $\mathcal{N}(\nabla F_\pm) \in L^2(\partial\Omega)$. Next, so we claim,

$$F_\pm = \pm C(F_\pm|_{\partial\Omega}) \quad \text{in} \quad \Omega_\pm. \tag{8.6}$$

Indeed, by uniqueness in the Dirichlet problem for harmonic functions in Dahlberg's sense (see [11]), it suffices to show that both sides in (8.6) have identical traces on $\partial\Omega$. In turn, this is a straightforward consequence of definitions, the point (3) in Theorem 2, and our working assumption, (iv).

Granted (8.6), it follows now from Theorem 4 that F_+, F_- are separately monogenic in Ω_+ and Ω_-, respectively. In particular, Corollary 1 forces $F_- = 0$ in Ω_-. Consequently, $f = F_+|_{\partial\Omega} - F_-|_{\partial\Omega} = F_+|_{\partial\Omega}$, so (i) is satisfied by choosing $F := F_+$. $\square$

It is important to point out that the above corollaries above *fail* for $n = 1$. The main difference, in this latter case, is that (iv) in Corollary 2 holds for *any* $f \in L^2(\partial\Omega, C\ell(\mathbb{R}^m))$; see [19].

In closing, let us mention that several other variants are possible. For instance, a completely analogous theory, indeed more akin to the classical theory of *several complex variables*, can be developed in connection with the Cauchy–Riemann operator (3.5).

6.9   Octonionic analysis

Let us first discuss a general algebraic construction. Assume that $\mathcal{A} = (\mathcal{A}, +, \cdot)$ is a real, unitary, associative algebra endowed with a linear involution $c \mapsto a^c$. The latter is called a *conjugation* if $(ab)^c = b^c a^c$ for every $a, b \in \mathcal{A}$. In particular, if $1_\mathcal{A}$ is the multiplicative unit in $\mathcal{A}$, then $1_\mathcal{A}^c = 1_\mathcal{A}$. For later reference, define the *trace* of $a \in \mathcal{A}$ as $\text{Tr}_\mathcal{A}(a) := \frac{1}{2}(a + a^c)$. Next, consider the functor $\mathcal{A} \mapsto K(\mathcal{A}) = (\mathcal{A} \times \mathcal{A}, +, \cdot)$ where the addition in $K(\mathcal{A})$ is componentwise while the multiplication and conjugation are, respectively,

$$(a, b) \cdot (\alpha, \beta) := (a\alpha - \beta^c b, \, \beta a + b\alpha^c), \quad (a, b)^c := (a^c, -b). \tag{9.1}$$

The following claims are straightforward (see also [5]).

Proposition 5. *Assume that the algebra $\mathcal{A}$ is as above. Then:*

(1) $K(\mathcal{A})$ is a real, unitary algebra, equipped with a conjugation.

(2) $(1_\mathcal{A}, 0)$ is the multiplicative unit in $K(\mathcal{A})$. Also, if $(e_j)_{j \in J}$ are imaginary units which anticommute in $\mathcal{A}$, then $(0, 1_\mathcal{A})$, $(e_j, 0)$, $(0, e_j)$, $j \in J$, are imaginary units which anticommute in $K(\mathcal{A})$.

(3) If $\mathcal{A}$ is commutative, then $K(\mathcal{A})$ is associative.

(4) If $\mathcal{A}$ is associative and if any element has a scalar trace, or, $\mathrm{Tr}_\mathcal{A}(a) \in \mathbb{R}$ for any $a \in \mathcal{A}$, then $K(\mathcal{A})$ is an alternative algebra, i.e.,

$$[x, y, z] := (xy)z - x(yz) \qquad (9.2)$$

is trilinear alternate, and any element in $K(\mathcal{A})$ has a scalar trace.

(5) If $\mathrm{Tr}_\mathcal{A}(ab) = \mathrm{Tr}_\mathcal{A}(ba)$ for any $a, b \in \mathcal{A}$, then we have

$$\mathrm{Tr}_{K(\mathcal{A})}(xy) = \mathrm{Tr}_{K(\mathcal{A})}(yx)$$

for any $x, y \in K(\mathcal{A})$.

(6) If $\mathcal{A}$ is associative and $\mathrm{Tr}_\mathcal{A}(ab) = \mathrm{Tr}_\mathcal{A}(ba)$ for any $a, b \in \mathcal{A}$, then

$$\mathrm{Tr}_{K(\mathcal{A})}((xy)z) = \mathrm{Tr}_{K(\mathcal{A})}(y(zx))$$

or, alternatively, $\mathrm{Tr}_{K(\mathcal{A})}[x, y, z] = 0$ for any $x, y, z \in K(\mathcal{A})$.

(7) If $\mathcal{A}$ is a normed algebra and $aa^c = |a|^2$, then $K(\mathcal{A})$ becomes a normed algebra with $|x|^2 = xx^c = x^c x = |a|^2 + |b|^2$ for every $x = (a, b) \in K(\mathcal{A})$. In particular, the multiplicative inverse of any element $x \in K(\mathcal{A}) \setminus 0$ is $x^{-1} = x^c / |x|^2$.

(8) Assume that $\mathcal{A}$ is a normed algebra such that $aa^c = |a|^2$, and so that any element in $\mathcal{A}$ has a scalar trace. Then $K(\mathcal{A})$ turns into a real Hilbert space with respect to the pairing

$$\langle x, y \rangle := \mathrm{Tr}_{K(\mathcal{A})}(xy^c) = \tfrac{1}{2}(xy^c + yx^c). \qquad (9.3)$$

(9) $K(\mathbb{R}) = \mathbb{C}$, the complex numbers, $K(\mathbb{C}) = \mathbb{H}$, the quaternions, and $K(\mathbb{H}) = \mathbb{O}$, the Cayley algebra of octonions.

Specializing these to the case of the Cayley algebra of octonions gives:

Proposition 6. *The following are true:*

(1) If i, j, k are the standard anticommuting imaginary units in $\mathbb{H}$, then $\xi_1 := (i, 0)$, $\xi_2 := (j, 0)$, $\xi_3 := (k, 0)$, $\xi_4 := (0, 1)$, $\xi_5 := (0, i)$, $\xi_6 := (0, j)$, $\xi_7 := (0, k)$ are anticommuting imaginary units which, along with $\xi_0 := (1, 0)$, the multiplicative unit in $\mathbb{O}$, form a basis for $\mathbb{O}$. Accordingly, one can embed $\mathbb{R}^8 \hookrightarrow \mathbb{O}$ by identifying each point $x = (x_\alpha)_\alpha \in \mathbb{R}^8$ with the Cayley number $\sum x_\alpha \xi_\alpha \in \mathbb{O}$.

(2) Equipped with the pairing (9.3), the Cayley algebra $\mathbb{O}$ becomes a real, eight-dimensional Hilbert space, with $\{\xi_\alpha\}_\alpha$ an orthonormal basis.

(3) If $x = \sum x_\alpha \xi_\alpha \in \mathbb{O}$, with $x_\alpha \in \mathbb{R}$, then $x^c = x_0 - x_1\xi_1 - \cdots - x_7\xi_7$. In particular, $\mathrm{Tr}_{\mathbb{O}}(x) = x_0$.

(4) $x(xy) = x^2 y$, $(xy)y = xy^2$ and $(xy)y^{-1} = x = y^{-1}(yx)$.

The first–order differential operator we intend to study in this section is

$$D := \sum_{\alpha=0}^{7} \xi_\alpha \frac{\partial}{\partial x_\alpha}, \qquad (9.4)$$

which acts on $\mathbb{O}$-valued functions (defined in $\mathbb{R}^8$) in a natural fashion. Set $D^c := 2\xi_0\partial_{x_0} - D$. An incisive observation is recorded below.

Proposition 7. *The adjoint of the operator (9.4) with respect to the pairing (9.3) is $-\partial_0 + \xi_1\partial_1 + \cdots + \xi_7\partial_7$. That is, $D^* = -D^c$. In particular, $-D^*D = \Delta$, the Laplacian in $\mathbb{R}^8$, so D is of Dirac type.*

Proof. The crucial observation is that

$$\langle x \cdot y, z \rangle = \langle y, x^c \cdot z \rangle, \qquad \forall\, x, y, z \in \mathbb{O}. \qquad (9.5)$$

Indeed, by (9.3), the above is equivalent to $\mathrm{Tr}_{\mathbb{O}}((xy)z^c) = \mathrm{Tr}_{\mathbb{O}}(y(z^c x))$ which holds true, by virtue of (6) in Proposition 5. With this at hand, it is then straightforward to finish the proof of the proposition. $\qquad\square$

Once the Dirac character of D in (9.4) has been established, *all our general results developed in §4–§7 will work when specialized to the present context as well.*

Thus, one can talk about analysis in several Cayley variables in $(\mathbb{R}^\xi)^n$; i.e., the function theory for the differential operator $\mathbb{D} := (D_j)_j$, where D_j is the octonionic Dirac operator (9.4) in the j-th factor in $(\mathbb{R}^8)^n$, for $1 \leq j \leq n$. Here we only want to point out that, in this latter context, the natural Cauchy operator in a Lipschitz domain $\Omega \subset (\mathbb{R}^8)^n$, $n \geq 1$, is

$$\mathcal{C}f(x) = \frac{1}{\omega_{8n}} \int_{\partial\Omega} \sum_{j=1}^{n} \frac{\overline{x_j - y_j}}{|x - y|^{8n}} \cdot (\nu_j(y) \cdot f(y))\, ds_y, \qquad x \notin \partial\Omega. \qquad (9.6)$$

Above, $x = (x_1, \ldots, x_n) \in (\mathbb{R}^8)^n$, and $\nu : \partial\Omega \to (\mathbb{R}^8)^n$, $\nu = (\nu_j)_j$, is the outward unit normal to $\partial\Omega$. That (9.6) is a particular case of (5.3) can be seen much as we have done for (8.3) in the Clifford analysis context.

6.10   Integral representation formulas

Here we work out Pompeiu's integral representation formula in some important special situations. First, let D be as in (3.3) and set $u := (f, 0)$, with f an arbitrary differential form. In this context, the first line in (5.6) gives

$$f(x) = i \int_{\partial\Omega} \left\{ \langle \sigma(d^*; \nu_y) d_y E(x,y), f(y) \rangle + \langle \sigma(d; \nu_y) d_y^* E(x,y), f(y) \rangle \right\} ds_y$$

$$- \int_{\Omega} \langle d_y E(x,y), df(y) \rangle \, dV_y + \int_{\Omega} \langle d_y^* E(x,y), d^* f(y) \rangle \, dV_y, \quad (10.1)$$

for each $x \in \Omega$, where $E(x,y)$ is the Schwartz kernel of $(\Delta - V)^{-1}$. If we integrate by parts in the last solid integrand, the identity (10.1) becomes

$$f(x) = i \int_{\partial\Omega} \langle \sigma(d^*; \nu_y) d_y E(x,y), f(y) \rangle \, ds_y \qquad (10.2)$$

$$- \int_{\Omega} \left\{ \langle d_y E(x,y), df(y) \rangle + \langle d_y d_y^* E(x,y), f(y) \rangle \right\} dV_y, \quad x \in \Omega.$$

Next, the identity $d^*(\Delta - V) = (\Delta - V)d^* - dV \vee \cdot$ translates, at the level of Schwartz kernels, into $d_y^* E(x,y) = d_x E(x,y) + R(x,y)$ where the (residual) double form $R(x,y)$ is $\mathcal{O}(|x - y|^{-(m-4)})$. In fact, $R \equiv 0$ when V is a constant. Accordingly, (10.2) further becomes

$$f(x) = \int_{\partial\Omega} \langle d_y E(x,y), (\nu \wedge f)(y) \rangle \, ds_y - \int_{\Omega} \langle d_y E(x,y), df(y) \rangle \, dV_y$$

$$- d_x \left[\int_{\Omega} \langle d_y E(x,y), f(y) \rangle \, dV_y \right] + \text{a residual term.} \qquad (10.3)$$

The residual term (a weakly singular integral operator acting on f) disappears when V is constant. E.g., when certain Betti numbers vanish for the manifold M (so that Δ^{-1} exists), or when $M = \mathbb{R}^m$, we can take $V = 0$.

The same program, carried out above, works equally well in the case when M is a complex manifold, and D is the complex conjugate of the operator (3.4). In this setting, the analogue of (10.3) reads

$$f(x) = \int_{\partial\Omega} \langle \partial_y E(x,y), (\nu^{1,0} \wedge f)(y) \rangle \, ds_y - \int_{\Omega} \langle \partial_y E(x,y), (\bar\partial f)(y) \rangle \, dV_y$$

$$- \bar\partial_x \left[\int_{\Omega} \langle \partial_y E(x,y), f(y) \rangle \, dV_y \right] + \text{a residual term.} \qquad (10.4)$$

Much as before, the residual term (in general, a weakly singular integral operator) is not present when V is constant.

This is a version of the Bochner–Martinelli–Koppelman formula (see [21] for the case when M is $\mathbb{C}^m$). Martinelli's original version was proved in 1938 for the ball in $\mathbb{C}^n$, then extended to general domains in $\mathbb{C}^n$ by Martinelli and Bochner in 1943 for holomorphic functions, and by Koppelman in 1967 when $\bar{\partial}f \neq 0$. Our contribution is to observe that a family of integral representation formulas, containing—in particular—all these versions, follow, more or less directly, from Pompeiu's formula in Theorem 1.

We also want to point out that, for the operator (3.4) in the flat, Euclidean space, some interesting calculations have been carried out in [22].

6.11   Some open problems

Here we collect a few open problems relevant to the topics presented so far.

(1) In the context of Theorem 4, does

$$\mathcal{N}(u) \in L^2(\partial\Omega) \ \& \ u = \mathcal{C}(u|_{\partial\Omega}) \text{ in } \Omega \implies Du = 0 \text{ in } \Omega, \qquad (11.1)$$

hold in place of (6.5)? Note that, this time, no *a priori* assumptions are made on Du, so the implication (11.1) is a stronger statement than (6.5).

(2) In the context of §8, if $f : \partial\Omega \to C\ell(\mathbb{R}^m)$ has a separately monogenic extension in a neighborhood of $\partial\Omega$, does it follow that $\mathcal{C}f$ is separately monogenic in Ω?

(3) Is there a natural concept of *tangential Cauchy–Riemann–Clifford equations*, akin to the situation in several complex variables (as in, e.g., [21, pages 159–166]? If so, can this be used to characterize (Clifford algebravalued) boundary functions which admit separately monogenic extensions inside of the domain? Note that Corollary 2 contains some criteria in this regard which involve boundary *integral* operators.

(4) Null-solutions F of a Dirac operator D (in the flat, Euclidean setting) have the property that $|F|^p$ is subharmonic for some $p < 1$ (a property discovered by E. Stein and G. Weiss; see the discussion in [6]). The critical (subharmonicity) exponent of D is the infimum of all such indices p.

For example, in the case of the octonionic Dirac operator (9.4), we claim that the critical exponent is $\frac{6}{7}$. Indeed, if $F(x) = x^c/|x|^8$, $x \in \mathbb{R}^8 \setminus 0$, then $\Delta|F(x)|^p = 7p(7p - 6)|x|^{7p-2}$ which is ≥ 0 only if $p \geq \frac{6}{7}$. This proves that the critical exponent of the octonionic Dirac operator is $\geq \frac{6}{7}$. On the other hand, the inequality

$$\sup_{0 \leq \alpha \leq 7} |\partial_\alpha F|^2 \leq \tfrac{7}{8} |\nabla F|^2 \qquad (11.2)$$

is valid for any $\mathbb{O}$-valued function F with $DF = 0$ (this can be proved by a direct analysis of the system $DF = 0$). Thus, according to the discussion on p. 105 in [6], the critical exponent is $\leq \frac{6}{7}$. The claim is proved.

The question here is to *explicitly compute the critical exponents corresponding to the constant coefficient Dirac operators in the examples discussed in §3.* It is important to consider natural *sub-bundles* as well.

(5) The Hardy space $\mathcal{H}^p(\Omega, D)$, introduced in (7.1) and equipped with the norm $u \mapsto \|\mathcal{N}u\|_{L^p(\partial\Omega)}$, is *Banach* when $1 \leq p \leq \infty$, but only *quasi-Banach* for $0 < p < 1$. Thus, the natural question here is to *identify the Banach envelope of $\mathcal{H}^p(\Omega, D)$ for $0 < p < 1$, i.e., the 'smallest' Banach spaces containing it.* A key technical aspect is proving the implications

$$u \in B^{1,1}_{2-s}(\Omega, \mathcal{E}), \ Lu = 0 \implies \partial_\nu u \in B^{1,1}_{-s}(\partial\Omega, \mathcal{E}), \tag{11.3}$$

$$u \in B^{1,1}_{1-s}(\Omega, \mathcal{E}), \ Du = 0 \implies \sigma(D;\nu)u \in B^{1,1}_{-s}(\partial\Omega, \mathcal{E}). \tag{11.4}$$

It is possible to show that the above hold true when $B^{1,1}_{-s}(\partial\Omega, \mathcal{E})$ is replaced by $(B^{\infty,\infty}_s(\partial\Omega, \mathcal{E}))^*$; the latter is, however, a strictly larger space.

A   Appendix

Here we provide the proof of Theorem 6. Accordingly, we retain the notation and terminology introduced in §8, and introduce

$$\Phi(z) := \frac{1}{\omega_m} \frac{z}{|z|^m}, \quad z \in \mathbb{R}^m \setminus \{0\}, \tag{A.1}$$

i.e., the *opposite* of the usual fundamental solution for the standard Dirac operator in $\mathbb{R}^m$. Recall that, for each $1 \leq j \leq n$, Δ_j is the Laplacian in the j-th copy of $\mathbb{R}^m$ in the Cartesian product $(\mathbb{R}^m)^n$. The departure point is the treatment of the inhomogeneous system for the operator $\mathbb{D} = (D_j)_j$.

Proposition 8. *Let $f_j \in C_o^\infty(\mathbb{R}^{mn})$ be n (where $n > 1$) given $C\ell(\mathbb{R}^m)$-valued functions. Then the following are equivalent:*

(1) The system

$$D_j g = f_j, \quad 1 \leq j \leq n, \tag{A.2}$$

has a solution $g \in C_o^\infty(\mathbb{R}^m)$;

(2) For each $j = 2, \ldots, n$, there holds

$$f_j(x_1, \ldots, x_n) = D_j\left[\int_{\mathbb{R}^m} \Phi(y)f_1(x_1 + y, x_2, \ldots, x_n)\, dy\right]. \tag{A.3}$$

(3) For each $1 \leq j, k \leq n$, there holds

$$\Delta_j f_k = -D_k D_j f_j. \tag{A.4}$$

In the case when one of the above conditions holds true, a solution g of (A.2) can be found with the additional property that

$$g \equiv 0 \text{ in the unbounded component of } \mathbb{R}^{mn} \setminus \bigcup_j \operatorname{supp} f_j. \qquad (A.5)$$

Proof. To justify the implication $(1) \Longrightarrow (2)$, we note that since $f_1 = D_1 g$,

$$\int_{\mathbb{R}^m} \Phi(y) f_1(x_1 + y, x_2, \ldots, x_n) \, dy = \int_{\mathbb{R}^m} \Phi(y)(D_1 g)(x_1 + y, x_2, \ldots, x_n) \, dy$$

$$= g(x_1, \ldots, x_n), \qquad (A.6)$$

via an integration by parts. Thus, the right side of (A.3) is

$$D_j g(x_1, \ldots, x_n) = f_j(x_1, \ldots, x_n)$$

as desired. As for $(2) \Longrightarrow (1)$, define

$$g(x_1, \ldots, x_n) := \int_{\mathbb{R}^m} \Phi(y) f_1(x_1 + y, \ldots, x_n) \, dy, \qquad (A.7)$$

so that $g \in C^\infty(\mathbb{R}^{mn})$. Also, by virtue of (A.3) we have $D_j g = f_j$ for $2 \leq j \leq n$. As for $j = 1$, differentiating under the integral sign and then invoking Pompeiu's formula, it follows that $D_1 g = f_1$, as wanted.

Finally, the fact that $D_j g = f_j$, $1 \leq j \leq n$, entails that g is harmonic near infinity. The fact that g vanishes in an unbounded open set (seen from (A.7)) forces g to have compact support, by unique continuation. In fact, a moment's reflection shows that (A.5) holds.

Now, $(A.2) \Rightarrow D_k D_j f_j = D_k D_j D_j g = D_k(-\Delta_j)g = -\Delta_j D_k g = -\Delta_j f_k$. Hence, $(1) \Rightarrow (3)$. Conversely, if (3) holds, let g be as in (A.7). Then

$$\Delta_j g(x_1, \ldots, x_n) = \int_{\mathbb{R}^m} \Phi(y) \Delta_j f_1(x_1 + y, \ldots, x_n) \, dy,$$

$$= -\int_{\mathbb{R}^m} \Phi(y)(D_1 D_j f_1)(x_1 + y, \ldots, x_n) \, dy,$$

$$= -D_j f_j(x_1, \ldots, x_n), \qquad (A.8)$$

integrating by parts in the last step. Thus, $\Delta_j g = -D_j f_j$ for each j. With this at hand we then compute

$$\Delta_k[D_j g - f_j] = D_j \Delta_k g - \Delta_k f_j = -D_j D_k f_k - \Delta_k f_j = 0,$$

by assumption. Summing up in k shows that $D_j g - f_j$ is harmonic in $\mathbb{R}^{mn}$ for each j. Since this is also compactly supported, it follows that $D_j g = f_j$ in $\mathbb{R}^{mn}$ for each j. This justifies the implication $(3) \Longrightarrow (1)$ and concludes the proof of Proposition 8. $\qquad \square$

We now turn to the

Proof of Theorem 6. Let $u \in C^\infty(\Omega \setminus K)$ be a $C\ell(\mathbb{R}^m)$-valued function such that $D_j u = 0$ for $1 \leq j \leq n$. As in the classical complex case, for some real-valued $\varphi \in C_o^\infty(\Omega)$, identically 1 in a neighborhood of K, introduce

$$v := (1 - \varphi)u \text{ in } \Omega \setminus K, \text{ and } v \equiv 0 \text{ on } K. \tag{A.9}$$

In particular, $v \in C^\infty(\Omega)$ and $\operatorname{supp}(D_j v) \subset \{0 < \varphi < 1\}$ for each j. Set $f_j := D_j v$, extended by zero outside Ω, so that $f_j \in C^\infty(\Omega)$, $1 \leq j \leq n$.

Assume for a moment that we can find $g \in C_o^\infty(\mathbb{R}^{mn})$ so that $D_j g = f_j$ and $g \equiv 0$ in the unbounded component of $\mathbb{R}^{mn} \setminus \{0 < \varphi < 1\}$. Then, $U := u - g|_\Omega$ is separately monogenic in Ω, and $U = u$ near $\partial\Omega$. In particular, $U = u$ in $\Omega \setminus K$ by unique continuation for harmonic functions.

In turn, by (3) in Proposition 8, the existence of g as above hinges on whether (A.4) holds for $1 \leq j, k \leq n$. Since $f_j := D_j v$ in our case, this is trivial to check. Thus g exists, and the proof is finished. $\qquad\square$

REFERENCES

[1] N. Aronszajn, A unique continuation theorem for solutions of elliptic differential equations or inequalities of second order, *Journ. de Math.* **36** (1957), 235–249.

[2] F. Brackx, R. Delanghe and F. Sommen, *Clifford Analysis*, Research Notes in Mathematics **76**, Pitman, Boston, MA, 1982.

[3] D. Calderbank, *Geometric Aspects of Spinor and Twistor Analysis*, Ph.D. thesis, Univ. Warwick, Warwick, 1995.

[4] R. Coifman, A. McIntosh, and Y. Meyer, L'intégrale de Cauchy definit un opérateur borné sur L^2 pour les courbes Lipschitziennes, *Annals of Math.* **116** (1982), 361–388.

[5] P. Dentoni and M. Sce, Funzioni regolari nell'algebra di Cayley, *Rend. Sem. Mat. Univ. Padova* **50** (1973), 251–267.

[6] J. Gilbert and M. A. M. Murray, *Clifford Algebras and Dirac Operators in Harmonic Analysis*, Cambridge Studies in Advanced Mathematics, 1991.

[7] K. Gürlebeck and W. Spröẞig, *Quaternionic Analysis and Elliptic Boundary Value Problems*, Birkhäuser Verlag, Basel, 1990.

[8] D. Jerison, Carleman inequalities for the Dirac and Laplace operators and unique continuation, *Adv. in Math.* **62** (1986), 118–134.

[9] D. Jerison and C. Kenig, The inhomogeneous Dirichlet problem in Lipschitz domains, *J. Funct. Anal.* **130** (1995), 161–219.

[10] C. Kenig, Weighted H_p spaces on Lipschitz domains, *Amer. J. Math.* **102** (1980), 129–163.

[11] C. Kenig, *Harmonic Analysis Techniques for Second Order Elliptic Boundary Value Problems*, CBMS Series in Mathematics **83**, AMS, 1994.

[12] K. Kodaira, *Complex Manifolds and Deformation of Complex Structures*, Springer-Verlag, New York, 1986.

[13] A. M. Kytmanov, *The Bochner–Martinelli Integral and Its Applications*, Birkhäuser-Verlag, Basel, Boston, Berlin, 1995.

[14] A. McIntosh and M. Mitrea, Clifford algebras and Maxwell's equations in Lipschitz domains, *Math. Methods Appl. Sci.* **22** (1999), 1599–1620.

[15] A. McIntosh, D. Mitrea, and M. Mitrea, Rellich type estimates for one-sided monogenic functions in Lipschitz domains and applications, *Analytical and Numerical Methods in Quaternionic and Clifford Algebras*, K. Gürlebeck and W. Sprössig eds., 1996, pp. 135–143.

[16] J. Michel and M.-C. Shaw, The $\bar\partial$-Neumann operator on Lipschitz pseudoconvex domains with plurisubharmonic defining functions, *Duke Math. J.* **108** (2001), 421–447.

[17] D. Mitrea, M. Mitrea and M. Taylor, Layer potentials, the Hodge Laplacian, and global boundary problems in nonsmooth Riemannian manifolds, *Mem. Amer. Math. Soc.* **150**, 2001.

[18] I. Mitrea and M. Mitrea, Monogenic Hardy spaces on Lipschitz domains and compensated compactness, *Complex Variables. Theory Appl.* **35** (1998), 225–282.

[19] M. Mitrea, *Clifford Wavelets, Singular Integrals, and Hardy Spaces*, Lecture Notes in Mathematics **1575**, Springer-Verlag, Berlin, Heidelberg, 1994.

[20] M. Mitrea, Generalized Dirac operators on nonsmooth manifolds and Maxwell's equations, *J. Fourier Anal. Appl.* **7** (2001), 207–256.

[21] M. Range, *Holomorphic Functions and Integral Representations in Several Complex Variables*, Springer-Verlag, New York, 1986.

[22] R. Rocha-Chávez, M. Shapiro and F. Sommen, *Integral theorems for functions and differential forms in $\mathbb{C}^m$*, Chapman & Hall/CRC, 2002.

[23] J. Ryan, Ed., *Clifford Algebras in Analysis and Related Topics*, Studies in Advanced Mathematics, CRC Press, Boca Raton, FL, 1996.

[24] J. Ryan, Clifford analysis on spheres and hyperbolae, *Math. Methods Appl. Sci.* **20** (1997), 1617–1624.

[25] M. Taylor, *Partial Differential Equations*, Springer-Verlag, 1996.

[26] J. M. Wilson, A simple proof of atomic decompositions for $H^p(\mathbb{R}^n)$, $0 < p \le 1$, *Studia Math.* **74** (1982), 25–33.

Emilio Marmolejo-Olea
Instituto de Matemáticas, Unidad Cuernavaca
Universidad Nacional Autónoma de México (UNAM)
A.P. 273-3 ADMON 3
Cuernavaca, Mor., 62251, México
E-mail: emilio@matcuer.unam.mx

Marius Mitrea
Department of Mathematics
University of Missouri-Columbia
Columbia, MO 65211, USA
E-mail: marius@math.missouri.edu

Submitted: August 15, 2002.

7

Paley–Wiener Theorems and Shannon Sampling in the Clifford Analysis Setting

Tao Qian

ABSTRACT This paper is concerned with the classical Paley–Wiener theorems in one and several complex variables, the generalization to Euclidean spaces in the Clifford analysis setting and their proofs. We prove a new Shannon sampling theorem in the Clifford analysis setting.

Keywords: Shannon sampling, Paley–Wiener theorem, interpolation, sinc function, monogenic function

7.1 Classical Paley–Wiener theorem and its generalizations

In the complex plane $\mathbb{C}$ the classical Paley–Wiener theorem reads

Theorem 1 (Paley–Wiener). *Let $f : \mathbb{C} \to \mathbb{C}$ be entire and $f|_{\mathbb{R}} \in L^2(\mathbb{R})$. Let σ be any positive number, then (i) there exists $C > 0$ such that $|f(z)| \leq Ce^{\sigma|z|}$ if and only if $\mathrm{supp}(\widehat{f|_{\mathbb{R}}}) \subset [-\sigma, \sigma]$; and, (ii) if one of the two conditions in (i) is fulfilled, then*

$$f(z) = \frac{1}{2\pi} \int_{\mathbb{R}} e^{\mathrm{i}\xi z} (\widehat{f|_{\mathbb{R}}})(\xi) \, d\xi.$$

An entire function is said to be in the class $PW(\sigma)$ if it satisfies one of the two conditions in the conclusion (i) of the theorem. The formula in (ii) can be used to obtain functions in PW classes from square integrable functions with compact support.

The theorem has generalizations to several complex variables (see, for instance, [2, 11]). Let $K \subset \mathbb{R}^n$ be convex, closed, symmetric, and $0 \in K$. Define

$$K^* = \{y \in \mathbb{R}^n \mid x \cdot y \leq 1, \ \forall x \in K\},$$
$$\|z\|^* = \sup_{y \in K} |z \cdot y| = \sup_{y \in K} |z_1 y_1 + \cdots + z_n y_n|,$$

AMS Subject Classification: 65M60, 65N30.

The work was supported by research grant of the University of Macau No. RG055/01-02S/QT/FST.

and the function class

$$\epsilon(K^*) =$$
$$\{F : \mathbb{C}^n \to \mathbb{C} \mid F \text{ is entire and satisfies } |F(z)| \leq A_\epsilon e^{(1+\epsilon)\|z\|^*}, \ \forall \epsilon > 0\}.$$

For example, if $K = \overline{B(0,\sigma)}$, where $B(0,\sigma) = \{x \in \mathbb{R}^n \mid |x| < \sigma\}$, then $\|z\|^* = \sigma |z|$.

Theorem 2 (The $\mathbb{C}^n$ version). *Let $F : \mathbb{C}^n \to \mathbb{C}$ be entire, $F|_{\mathbb{R}^n} \in L^2(\mathbb{R}^n)$. Then (i) $F(z)$ belongs to $\epsilon(K^*)$ if and only if $\mathrm{supp}(F|_{\mathbb{R}^n})\check{} \subset K$; and (ii) if one of the two conditions in (i) is fulfilled, then*

$$F(z) = \frac{1}{(2\pi)^n} \int_{\mathbb{R}^n} e^{i\xi \cdot z} (F|_{\mathbb{R}^n})\check{}(\xi) \, d\xi.$$

The proof of Theorem 2 is reduced to Theorem 1 (see [11]). The usual proof of Theorem 1 relies on a Phragmén–Lindelöf type result:

Lemma 1. *Let S be the region in $\mathbb{C}$ bounded by two rays meeting at the origin at an angle π/α. Suppose f is analytic on $\overline{S}$ and satisfies $|f(z)| \leq Ae^{|z|^\beta}$ for some β, $0 \leq \beta < \alpha$, where $z \in S$. Then, the condition $|f(z)| \leq M$ on the two boundary rays implies $|f(z)| \leq M$ for all $z \in S$.*

While Theorem 2 is a nice generalization to $\mathbb{C}^n$, it is not a generalization to $\mathbb{R}^n$. In the latter the right substitution of complex analyticity would be Clifford monogenicity. We use the usual formulation.

Denote

$$\mathbb{R}^n =$$
$$\{\underline{x} = x_1 \mathbf{e}_1 + \cdots + x_n \mathbf{e}_n \mid x_i \in \mathbb{R}, \mathbf{e}_i^2 = -1, \mathbf{e}_i \mathbf{e}_j = -\mathbf{e}_j \mathbf{e}_i, i = 1, \ldots, n, i \neq j\}$$

and

$$\mathbb{R}_1^n = \{x = x_0 + \underline{x} \mid x_0 \in \mathbb{R}, \ \underline{x} \in \mathbb{R}^n\}.$$

The definitions of left- and right-monogenicity are made as usual through the Cauchy–Riemann operator. If a function is both left- and right-monogenic, then it is said to be monogenic. A function left- or right-monogenic or monogenic in the whole $\mathbb{R}_1^n$ is said to be left- or right-entire, or entire, respectively. Now we have enough preparation to state (see [4])

Theorem 3 (The $\mathbb{R}^n$ version). *Let f be a function from $\mathbb{R}_1^n$ to $\mathbb{C}^{(n)}$, where $\mathbb{C}^{(n)}$ is the complex Clifford algebra generated by $\mathbf{e}_1, \cdots, \mathbf{e}_n$. Suppose f is left-entire and $f|_{\mathbb{R}^n} \in L^2(\mathbb{R}^n)$. Then:*

(i) $|f(x)| \leq Ce^{\sigma|x|}$ if and only if $\mathrm{supp}(f|_{\mathbb{R}^n})\check{} \subset \overline{B(0,\sigma)}$.

(ii) If one of the two conditions of (i) is fulfilled, then

$$f(x) = \frac{1}{(2\pi)^n} \int_{\mathbb{R}^n} e(x,\underline{\xi})(f|_{\mathbb{R}^n})\check{(\underline{\xi})}\, d\underline{\xi},$$

where

$$e(x,\underline{\xi}) = e^+(x,\underline{\xi}) + e^-(x,\underline{\xi}), \quad e^\pm(x,\underline{\xi}) = e^{i<\underline{x},\underline{\xi}>}e^{\mp x_0|\underline{\xi}|}\chi_\pm(\underline{\xi}),$$

and

$$\chi_\pm(\underline{\xi}) = \tfrac{1}{2}(1 + i\frac{\underline{\xi}}{|\underline{\xi}|}).$$

The theorem reduces to Theorem 1 when $n = 1$ and $e_1 = -i$. The proof of Theorem 1 by using the Phragmén–Lindelöf type result, however, does not seem to be directly adaptable to the Clifford context. It is an open question whether there exist Phragmén–Lindelöf type results in the Clifford context, where, for instance, the angle in the complex plane case is replaced by a cone with vertex at the origin. Here the Clifford setting encounters again the problem that the product of two monogenic functions is no longer monogenic. Indeed, in the complex case the proof of Lemma 1 uses the auxiliary analytic function $g(z) = f(z)e^{-\epsilon z^\gamma}$, $\beta < \gamma < \alpha$, $\epsilon > 0$ (see Lemma 1).

The striking feature of Theorem 3 is the use of the generalization of the exponential function, $e(x,\underline{\xi})$, that was first found in Alan McIntosh's lecture notes and subsequently in [6] and [7]. In [1] and the related early work similar generalizations may be found. McIntosh's form, however, has found rich applications to harmonic analysis in Euclidean spaces (see [3, 4, 7, 9]).

It would be interesting to look a bit more at the generalization. In the complex plane one could easily generalize e^{iuv}, $u, v \in \mathbb{R}$, to e^{izw}, where $z, w \in \mathbb{C}$, and usage of the latter is often crucial. In the Clifford setting, when generalizing $e^{i<\underline{x},\underline{\xi}>}$, $\underline{x} \in \mathbb{R}^n$ is extended to $x \in \mathbb{R}^n_1$ and $\underline{\xi} \in \mathbb{R}^n$ to $\zeta \in \mathbb{C}^n$, and the function becomes $e(x,\zeta)$ (see [6] and [7]). This is particularly convenient when studying Fourier theory. In this setting, the Fourier transform of a function on an n-dimensional manifold in $\mathbb{R}^n_1$ is a function of several complex variables.

In addition to usage of the generalized exponential function, the other important feature of the proof of Theorem 3 is that, in spite of the unavailability of Phragmén–Lindelöf type results, some explicit computations in one complex variable may be closely followed in the higher-dimensional cases with the Clifford analysis setting.

For example, below we compare a proof of Theorem 1 (see [13]) and the corresponding one for Theorem 3 with the Clifford analysis setting (see [4]). We will only discuss the hard part: Assuming the exponential-type bounds of the entire function whose restriction to $\mathbb{R}$ or $\mathbb{R}^n$ belongs to L^2 in the two contexts, respectively, we prove that the Fourier transform of the restricted function has compact support contained in the ball centred at the origin with radius θ as specified in Theorem 1 or 3.

Proof of Theorem 1 (see [13]):

(i) Expand $f(z)$ into its Taylor series whose coefficients can be explicitly computed with bounds derived from that of the entire function:

$$f(z) = \sum_{k=0}^{\infty} \frac{a_k}{k!} z^k, \quad a_k = \frac{k!}{2\pi i} \int_{\partial B(0,R)} \frac{f(\zeta)}{\zeta^{k+1}} d\zeta,$$

$$\lim_{N \to \infty} \sup_{k > N} |a_k| \le \sigma.$$

(ii) Consider the Hardy space decomposition of the L^2 function $(f|_{\mathbb{R}})\check{}$:

$$(f|_{\mathbb{R}})\check{} = (f|_{\mathbb{R}})\check{}_+ + (f|_{\mathbb{R}})\check{}_-, \quad (f|_{\mathbb{R}})\check{}_\pm(z) = \frac{1}{2\pi} \int_{\mathbb{R}} e^{itz} \chi_{(0,\infty)}(\pm t) \check{f}(t)\, dt.$$

For $|z| > \sigma$, using the series expansion of f in (i) we can expand $(f|_{\mathbb{R}})\check{}$ into

$$(f|_{\mathbb{R}})\check{}_\pm(z) = \pm \sum_{k=0}^{\infty} \frac{a_k}{(iz)^{k+1}}, \quad \pm \mathrm{Im}\, z > 0, \quad (\text{resp.})$$

(iii) For $|t| > \sigma$, the boundary value properties of the Hardy $H^2(\mathbb{R}_\pm)$ spaces imply that

$$(f|_{\mathbb{R}})\check{}(t) = \lim_{s \to 0+} (f|_{\mathbb{R}})\check{}_+(t + is) + \lim_{s \to 0+} (f|_{\mathbb{R}})\check{}_-(t - is).$$

(The relationship can also be obtained by using the Poisson type convolution integral expressions of $(f|_{\mathbb{R}})\check{}_\pm$.) From the explicit expressions of $(f|_{\mathbb{R}})\check{}_\pm$ obtained in (ii) we conclude $\mathrm{supp}(f|_{\mathbb{R}})\check{} \subset [-\sigma, \sigma]$.

$\square$

Proof of Theorem 3:

(i) In the context f also has a Taylor expansion

$$f(x) = \sum_{k=0}^{\infty} f_k(x) \quad (f_k \text{ are } k - \text{ solid harmonics})$$

$$= \sum_{k=0}^{\infty} \frac{1}{\omega_n} \int_{\partial B(0,r)} P^{(k)}(y^{-1}x) E(y) n(y) f(y)\, d\sigma(y),$$

where ω_n is the surface area of the n-dimensional unit sphere in $\mathbb{R}_1^n$, $P^{(k-1)}(x) = E(x)P^{(-k)}(x^{-1})$, where $E = \bar{x}/|x|^{n+1}$ is the Cauchy kernel in $\mathbb{R}_1^n$, $P^{(-k)}(x) = \Delta^{\frac{n-1}{2}}(x_0 + \underline{x})^{-k}$, $k = 1, 2, \ldots$, and $n(x) = \frac{x}{|x|}$ is the unit normal at x on the sphere $\partial B(0,r)$, and $d\sigma$ denotes the surface area measure on the sphere (see [10]).

In each general entry of the expansion, except for the $n = 1$ case, the variables x and y and the constant part (compare with (i) of the one complex variable case) cannot be separated, and thus there is no estimate for "the constants".

(ii) Define

$$(f|_{\mathbb{R}^n})_{\pm}(x) = \frac{1}{(2\pi)^n} \int_{\mathbb{R}^n} e^{i<\underline{x},\underline{\xi}>} e^{\mp x_0|\underline{\xi}|} \chi_{\pm}(\underline{\xi}) \hat{f}|_{\mathbb{R}^n}(\underline{\xi}) \, d\underline{\xi}, \text{ for } \pm x_0 > 0.$$

Using the estimates

$$|P^{(k)}(y^{-1}x)| \leq C_n k^n (|x|^k/|y|^k)$$

and

$$|P^{(k)}(y^{-1}\underline{D})E(-x)| \leq C_n k^n (|x|^k/|y|^k)$$

(see [10]), and the bounds of f, it may be shown that both the functions can be monogenically extended to $|x| > \sigma$ with the expressions

$$(f|_{\mathbb{R}^n})_{\pm}(x) =$$

$$\mp c_0 \sum_{k=0}^{\infty} \frac{i^k}{\omega_n} \int_{\partial B(0,r)} P^{(k)}(y^{-1}\underline{D})E(-x)E(y)n(y)f(y) \, d\sigma(y),$$

respectively, where

$$c_0 = 2^{n-1} \pi^{\frac{n-1}{2}} \Gamma(\tfrac{1}{2}n + \tfrac{1}{2}), \quad \underline{D} = \partial_1 e_1 + \cdots + \partial_n e_n.$$

(iii) It may be shown, through explicit computation with the Cauchy integral or the related results in [7] or [8], that

$$(f|_{\mathbb{R}^n})(\underline{x}) = \lim_{x_0 \to 0+} (f|_{\mathbb{R}^n})_{+}(x_0 + \underline{x}) + \lim_{x_0 \to 0+} (f|_{\mathbb{R}^n})_{-}(-x_0 + \underline{x}).$$

From this and the last expressions of $(f|_{\mathbb{R}^n})_{\pm}(x)$ in (ii) we obtain the desired relation $\operatorname{supp}(\widehat{f|_{\mathbb{R}^n}}) \subset \overline{B(0,\sigma)}$.

$\square$

The above comparison suggests that certain explicit computation using Laurent series, Fourier multipliers and convolution integrals in one-complex variable may be closely followed: the availability of Laurent series, Fourier multipliers and convolution integrals of Clifford monogenic functions makes things work.

7.2 Applications to interpolation using monogenic sinc function

On the real line the sinc function is defined by

$$\text{sinc}(x) = \frac{1}{2\pi} \int_{-\pi}^{\pi} e^{ixt}\, dt = \frac{\sin \pi x}{\pi x}.$$

It has a holomorphic extension to an entire function in $\mathbb{C}$:

$$\text{sinc}(z) = \frac{1}{2\pi} \int_{-\pi}^{\pi} e^{izt}\, dt = \frac{\sin \pi z}{\pi z}.$$

There has been a comprehensive study of the sinc function with applications to a number of aspects of numerical analysis, including Shannon sampling (exact interpolation) and quadrature (see [5] and [12]).

The function theory can be extended to the n-complex variables case in the $\mathbb{C}^n$ setting without essential difficulty. The extension is related to Theorem 2. The Clifford monogenic sinc function is another extension involving non-trivial estimates.

Define

$$\text{sinc}(x) = \frac{1}{(2\pi)^n} \int_{[-\pi,\pi]^n} e(x,\underline{\xi})\, d\underline{\xi}.$$

It is easy to see that

$$\text{sinc}(\underline{x}) = \prod_{i=1}^{n} \text{sinc}(x_i) = \prod_{i=1}^{n} \frac{\sin \pi x_i}{\pi x_i},$$

coincident with one in the $\mathbb{C}^n$ setting. Introduce the Paley–Wiener classes $W(\pi/h), h > 0$:

$$W(\pi/h) = \{f : \mathbb{R}_1^n \to \mathbb{C}^{(n)} \mid f \text{ is left entire, and } |f(x)| \le Ce^{\frac{\pi}{h}|x|}\}.$$

Theorem 3 implies that if $f \in W(\pi/h)$, then

$$f(x) = \frac{1}{(2\pi)^n} \int_{[-\pi/h,\pi/h]^n} e(x,\underline{\xi})(f|_{\mathbb{R}^n})\widehat{\ }(\underline{\xi})\, d\underline{\xi}.$$

First we consider the case $x = \underline{x}$. Expanding the function $e^{i<\underline{x},\underline{\xi}>}$ into its uniformly convergent Fourier series in the cube $[-\pi/h, \pi/h]^n$, we have

$$e^{i<\underline{x},\underline{\xi}>} = \sum_{\underline{k}\in\mathbb{Z}^n} e^{i<h\underline{k},\underline{\xi}>} \text{sinc}\left(\frac{x - h\underline{k}}{h}\right),$$

where the terms $\text{sinc}\left(\frac{x - h\underline{k}}{h}\right)$ are the Fourier coefficients (see [3]). Substituting into the right-hand side of the last integral yields

$$f(x) = \frac{1}{(2\pi)^n} \int_{[-\pi/h, \pi/h]^n} \left[\sum_{\underline{k} \in \mathbb{Z}^n} e^{i<h\underline{k}, \underline{\xi}>} \text{sinc}\left(\frac{x - h\underline{k}}{h}\right)\right] (f|_{\hat{\mathbb{R}}^n})(\underline{\xi}) \, d\underline{\xi}$$

$$= \sum_{\underline{k} \in \mathbb{Z}^n} \text{sinc}\left(\frac{x - h\underline{k}}{h}\right) \left[\frac{1}{(2\pi)^n} \int_{[-\pi/h, \pi/h]^n} e^{i<h\underline{k}, \underline{\xi}>} (f|_{\mathbb{R}^n})\hat{}(\underline{\xi}) \, d\underline{\xi}\right]$$

$$= \sum_{\underline{k} \in \mathbb{Z}^n} \text{sinc}\left(\frac{x - h\underline{k}}{h}\right) f(h\underline{k}).$$

The exchange of the summation and the integration is justified by the uniform convergence of the Fourier series of $e^{i<\underline{x}, \underline{\xi}>}$.

This is the exact interpolation in $\mathbb{R}^n$. To extend it to $\mathbb{R}_1^n$ one needs the non-trivial estimate

$$|\text{sinc}(x)| \leq C_n(x_0) \frac{e^{\sqrt{n}\pi|x_0|}}{\prod_{i=1}^n (1 + |x_i|)}, \tag{2.1}$$

where $C_n(x_0)$ is polynomial in x_0 with positive coefficients depending only of n. We have (See [3] for the detailed proof, or, see the proof of Theorem 5 below.)

Theorem 4 (Shannon sampling, exact interpolation). *A function f belongs to $W(\pi/h)$ if and only if*

$$f(x) = \sum_{\underline{k} \in \mathbb{Z}^n} \text{sinc}\left(\frac{x - h\underline{k}}{h}\right) f(h\underline{k}).$$

Theorem 4 has the following variation.

Theorem 5. *Let $h, T > 0$, f be left-Clifford entire whose restriction to $\mathbb{R}^n$ belongs to $L^2(\mathbb{R}^n)$, and $\text{supp}\,\hat{f} \subset [-T, T]^n$. Then: (i) When $h > \pi/T$, the sampling of f at the points $h\underline{k}$ does not recover f, even restricted to $\mathbb{R}^n$; (ii) If $h < \min\{\pi/T, 1\}$, for any complex-valued function ϕ such that $\hat{\phi}$ is infinitely differentiable with compact support in $[-\pi/h, \pi/h]^n$ being equal to 1 in $[-T, T]^n$, there exists exact interpolation*

$$f(x) = \sum_{\underline{k} \in \mathbb{Z}^n} \phi\left(\frac{x - h\underline{k}}{h}\right) f(h\underline{k}),$$

where

$$f(h\underline{k}) = \frac{1}{h^n} \int_{\mathbb{R}^n} f(\underline{x}) \overline{\phi}\left(\frac{x - h\underline{k}}{h}\right) d\underline{x}.$$

Proof. First prove (i). Let $0 < \epsilon < T - \pi/h$ and θ be a function not identical to zero such that $\hat{\theta}$ is infinitely differentiable and supported in $[-\epsilon, \epsilon]^n$. Let

$$f_0(\underline{x}) = \left(\prod_{i=1}^{n} (e^{-i\pi x_i/h} - e^{i\pi x_i/h}) \right) \theta(\underline{x}) = \left(\prod_{i=1}^{n} (-2i\sin(\pi h^{-1} x_i)) \right) \theta(\underline{x}).$$

It is easy to verify that $\mathrm{supp}(\hat{f}_0) \subset [-T, T]^n$. This function can be monogenically extended to $\mathbb{R}_1^n$ using the formula in Theorem 3. But $f_0(h\underline{k}) = 0$ for all $\underline{k}$. This implies that the sampling at the points $h\underline{k}$ does not recover f_0.

Now we prove (ii). First, expand the smooth function $e^{i\langle \underline{x}, \underline{\xi}\rangle} \hat{\phi}(\underline{\xi})$ (in variable $\underline{\xi}$) in the cube $[-\pi/h, \pi/h]^n$ into its uniformly convergent Fourier series:

$$e^{i\langle \underline{x}, \underline{\xi}\rangle} \hat{\phi}(\underline{\xi}) = \sum_{\underline{k} \in \mathbb{Z}^n} e^{i\langle h\underline{k}, \underline{\xi}\rangle} \phi\left(\frac{x - h\underline{k}}{h} \right),$$

where $\phi\left(\frac{x - h\underline{k}}{h} \right)$ are the scalar-valued Fourier coefficients given by

$$\frac{1}{(2\pi)^n} \int_{[-\pi/h, \pi/h]^n} e^{-i\langle h\underline{k}, \underline{\xi}\rangle} e^{i\langle \underline{x}, \underline{\xi}\rangle} \hat{\phi}(\underline{\xi}) d\underline{\xi} = \phi\left(\frac{x - h\underline{k}}{h} \right).$$

Since $\hat{\phi}$ is identically 1 on $[-T, T]^n$, when restricted to $[-T, T]^n$ the above expansion becomes

$$e^{i\langle \underline{x}, \underline{\xi}\rangle} = \sum_{\underline{k} \in \mathbb{Z}^n} e^{i\langle h\underline{k}, \underline{\xi}\rangle} \phi\left(\frac{x - h\underline{k}}{h} \right).$$

Since $\hat{f}$ is supported in $[-T, T]^n$, the argument preceding the statement of Theorem 4 gives

$$f(\underline{x}) = \sum_{\underline{k} \in \mathbb{Z}^n} \phi\left(\frac{x - h\underline{k}}{h} \right) f(h\underline{k}). \tag{2.2}$$

Now we consider the left-monogenic extension of ϕ to $\mathbb{R}_1^n$ using the formula in Theorem 3. Taking into account that $\hat{\phi} = \hat{\phi}\chi_{[-\pi/h, \pi/h]^n}$, we have

$$\phi(x) = \frac{1}{(2\pi)^n} \int_{\mathbb{R}^n} e(x, \underline{\xi}) \hat{\phi}(\underline{\xi}) \chi_{[-\pi/h, \pi/h]^n}(\underline{\xi}) d\underline{\xi}$$

$$= \frac{1}{(2\pi)^n} \int_{\mathbb{R}^n} e^{i\langle \underline{x}, \underline{\xi}\rangle} \hat{\phi}(\underline{\xi}) [(e^{-x_0|\underline{\xi}|}\chi_+(\underline{\xi}) + e^{x_0|\underline{\xi}|}\chi_-(\underline{\xi}))\chi_{[-\pi/h, \pi/h]^n}(\underline{\xi})] d\underline{\xi}$$

$$= \left(\phi(\underline{\cdot}) * \frac{1}{h^n} \mathrm{sinc}\left(\frac{x_0 + (\cdot)}{h} \right) \right)(\underline{x}).$$

The role of the convolution operator using ϕ as kernel is to smooth the test function. A detailed estimate will conclude the same estimate for $\phi(x)$:

$$|\phi(x)| \leq C_{n,h}(x_o) \frac{e^{\sqrt{n}\pi|x_0|}}{\prod_{i=1}^{n}(1+|x_i|)},\tag{2.3}$$

where $C_{n,h}(x_0)$ is a polynomial in x_0 with positive coefficients depending on n and h. Now consider the left-monogenic extension of both sides of (2.2). The left-hand side becomes the function $f(x)$ itself. The right-hand side is uniformly convergent in a neighbourhood of x. To see this, we use the Cauchy–Schwarz inequality and obtain

$$\left(\sum_{\underline{k}\in\mathbb{Z}^n}\left|\phi\left(\frac{x-h\underline{k}}{h}\right)f(h\underline{k})\right|\right)^2 \leq \left(\sum_{\underline{k}\in\mathbb{Z}^n}\left|\phi(\frac{x-h\underline{k}}{h})\right|^2\right)\left(\sum_{\underline{k}\in\mathbb{Z}^n}|f(h\underline{k})|^2\right).$$

Because $\hat{f}$ has compact support and $\hat{f}\in L^2$, the Plancherel theorem on compact intervals implies that the second series on the right-hand side is convergent. To estimate the first series, first using the estimate (2.3) and then decomposing the obtained series into a product of n convergent series, we obtain the uniform convergence of the first series in a neighborhood of x. The Morera theorem in the Clifford setting then implies that the function

$$\sum_{\underline{k}\in\mathbb{Z}^n}\phi\left(\frac{x-h\underline{k}}{h}\right)f(h\underline{k})$$

is left-monogenic. This function and the function $f(x)$ are both left-entire, and are the same since they agree when restricted to $\mathbb{R}^n$. We therefore obtain

$$f(x) = \sum_{\underline{k}\in\mathbb{Z}^n}\phi\left(\frac{x-h\underline{k}}{h}\right)f(h\underline{k}).$$

Using Parseval's identity to the integral, since $\mathrm{supp}\,\hat{f}\subset[-T,T]^n$, we have

$$\int_{\mathbb{R}^n}f(\underline{x})\overline{\phi}\left(\frac{x-h\underline{k}}{h}\right)d\underline{x} = h^n\int_{[-T,T]^n}\hat{f}(\underline{\xi})e^{i<\underline{\xi},h\underline{k}>}\overline{\hat{\phi}}(h\underline{\xi})\,d\underline{\xi}.$$

From $h<1$ and $\hat{\phi}\chi_{[-T,T]^n}(\underline{\xi})=\chi_{[-T,T]^n}(\underline{\xi})$, we have

$$\overline{\hat{\phi}}(h\underline{\xi})\chi_{[-T,T]^n}(\underline{\xi}) = \chi_{[-T,T]^n}(\underline{\xi}).$$

Therefore,

$$h^n\int_{[-T,T]^n}\hat{f}(\underline{\xi})e^{i<\underline{\xi},h\underline{k}>}\overline{\hat{\phi}}(h\underline{\xi})\,d\underline{\xi} = h^n\int_{\mathbb{R}^n}\hat{f}(\underline{\xi})e^{i<\underline{\xi},h\underline{k}>}\,d\underline{\xi} = h^n f(h\underline{k}).$$

References

[1] F. Brackx, R. Delanghe and F. Sommen, *Clifford Analysis*, Research Notes in Mathematics **76**, Pitman, Boston/London/Melbourne, 1982.

[2] I.M. Gel'fand and G.E. Shilov, *Generalized Functions, Volume 2, Spaces of Fundamental and Generalized Functions*, Academic Press, New York and London. 1968.

[3] K.I. Kou and T. Qian, Shannon sampling with the Clifford analysis setting, preprint.

[4] K.I. Kou and T. Qian, The Paley–Wiener Theorem in $\mathbb{R}^n$ with the Clifford Analysis Setting, *J. Func. Anal.* **189** (2002), 227–241.

[5] J. Lund and K.L. Bowers, *Sinc Methods for Quadrature and Differential Equations*, SIAM, Philadelphia, 1992.

[6] A. McIntosh, Clifford algebras, Fourier theory, singular integrals, and harmonic functions on Lipschitz domains, *Clifford Algebras in Analysis and Related Topics*, Studies in Advanced Mathematics, edited by John Ryan, CRC PRESS, Boca Raton, New York, London, Tokyo, 1996, pp. 33–88.

[7] C. Li, A. McIntosh and T. Qian, Clifford algebras, Fourier transform and singular integrals on Lipschitz surfaces, *Rev. Mat. Iberoamericana* **10** (1994), 665–721.

[8] M. Mitrea, *Clifford Wavelets, Singular Integrals, and Hardy Spaces*, Lecture Notes in Mathematics **1575**, Springer-Verlag, New York/Berlin, 1994.

[9] J. Peetre and T. Qian, Möbius covariance of iterated Dirac operators, *J. Austral. Math. Soc. (Series A)* **56** (1994), 403–414.

[10] T. Qian, On starlike Lipschitz surfaces in $\mathbb{R}^n$, *J. of Func. Anal.* **183** (2001), 370–412.

[11] E. Stein and G. Weiss, *Introduction to Fourier analysis on Euclidean spaces*, Princeton University Press, Princeton, NJ, 1987.

[12] F. Stenger, *Numerical Methods Based on Sinc and Analytic Functions*, Springer Series in Computational Mathematics **20**, New York, 1993.

[13] A. Timan, *Theory of Approximation of Functions of a Real Variable*, Fizmatgiz, Moscow. 1960; English translation *Internat. Ser. Monogr. Pure Appl. Math.* **34**, Macmillan, New York, 1963, 582–592.

Tao Qian
Faculty of Science and Technology
University of Macau
P.O. Box 3001
Macao (via Hong Kong)
E-mail: fsttq@umac.mo

Submitted: March 1, 2003.

8

Bergman Projection in Clifford Analysis

Guangbin Ren and Helmuth R. Malonek

ABSTRACT We study weighted Bergman projections in the monogenic Bergman spaces of the real unit ball $\mathbb{B}$ in $\mathbb{R}^n$. We extend results of Forelli–Rudin, Coifman–Rochberg, and Djrbashian to Clifford analysis. The main result is as follows:
Let P_α be the orthogonal projection from the Hilbert space $L^2(\mathbb{B}, C\ell_{0,n}, dV_\alpha)$ onto the subspace of monogenic functions $A^2(\mathbb{B}, C\ell_{0,n}, dV_\alpha)$. If $p(\alpha + 1) > \beta + 1$ with $1 \leq p < \infty$ and $\alpha, \beta > -1$, then the operator $P_\alpha : L^p(\mathbb{B}, C\ell_{0,n}, dV_\beta) \to A^p(\mathbb{B}, C\ell_{0,n}, dV_\beta)$ is bounded.

Keywords: Bergman kernel, Bergman projection, monogenic Bergman spaces

8.1   Introduction

In the function space theory, Bergman projection plays an important role since it produces integral representations for functions in consideration. The fact that the Bergman projection in L^p spaces is bounded has been considered by different authors in different settings, for example, Forelli and Rudin [11] for holomorphic functions in the complex unit ball, Coifman and Rochberg [4] for harmonic functions in the real unit ball and Djrbashian [9] for the functions associated to the Riesz system in the real unit ball.

The purpose of this paper is to study the monogenic Bergman space in the real unit ball, which consists of all Clifford algebra-valued functions annihilated by the Dirac operator. We shall give the explicit computation of the Bergman kernel for the monogenic Bergman space in closed form. For the weighted cases we obtain the exact estimates of weighted monogenic Bergman kernels. As a result, we can prove that the weighted Bergman projections in the corresponding L^p space of the real unit ball are bounded. This yields integral representations for monogenic Bergman functions. We refer to [2, 3, 5–7, 15, 17] for the investigation of Bergman kernels in Clifford analysis.

To motivate our consideration in the setting of Clifford analysis, in the rest of this section we recall in two parts the background for holomorphic functions in the unit disc as well as for the Riesz system in the real unit ball. We will show that

AMS Subject Classification: 30G35, 32A36, 35C10, 42B35.

it is natural to extend the results for Riesz systems to the general Clifford analysis case.

(I) Holomorphic Bergman spaces

Let $\mathbb{U}$ denote the open unit disc in $\mathbb{C}$, $d\nu$ the normalized Lebesgue measure in $\mathbb{U}$, and

$$d\nu_\alpha(z) = C_\alpha(1 - |z|^2)^\alpha \, d\nu(z)$$

with $\alpha > -1$ and $\nu_\alpha(\mathbb{U}) = 1$. The *Bergman space*

$$A^2(\mathbb{U}, \, d\nu_\alpha) \equiv L^2(\mathbb{U}, \, d\nu_\alpha) \cap H(\mathbb{U})$$

is a Hilbert space, where $H(\mathbb{U})$ is the set of all holomorphic functions in $\mathbb{U}$. For any $f \in A^2(\mathbb{U}, \, d\nu_\alpha)$, since $|f|^2$ is subharmonic, we have

$$|f(z)|^2 \leq \frac{1}{r^2} \int_{U(z,r)} |f(w)|^2 \, d\nu(w) \leq C(r) \int_{\mathbb{U}} |f(w)|^2 \, d\nu_\alpha(w) \, ,$$

where $U(z, r)$ is the disc in $\mathbb{U}$ with center z and radius r. Therefore, the point evaluation functional

$$\Phi_z : A^2(\mathbb{U}, \, d\nu_\alpha) \longrightarrow \mathbb{C}$$

defined by $\Phi_z(f) = f(z)$ is continuous. The Riesz representation theorem shows that

$$\Phi_z f = (f, k_z) = \int_{\mathbb{U}} f(w)\overline{k_z(w)} \, d\nu_\alpha(w)$$

for some $k_z \in A^2(\mathbb{U}, \, d\nu_\alpha)$. Denote $K_\alpha(z, w) \equiv \overline{k_z(w)}$, then

$$f(z) = \int_{\mathbb{U}} K_\alpha(z, w) f(w) \, d\nu_\alpha(w) \, , \quad z \in \mathbb{U},$$

for any $f \in A^2(\mathbb{U}, \, d\nu_\alpha)$. The kernel K_α is the so-called *Bergman kernel*. It is well known that

$$K_\alpha(z, w) = (1 - z\overline{w})^{-(2+\alpha)} \, , \quad z, w \in \mathbb{U}.$$

Associated with the Bergman kernel is the *Bergman projection*:

$$P_\alpha f(z) = \int_{\mathbb{U}} K_\alpha(z, w) f(w) \, d\nu_\alpha(w).$$

According to the Hilbert space theory,

$$P_\alpha : L^2(\mathbb{U}, \, d\nu_\alpha) \to A^2(\mathbb{U}, \, d\nu_\alpha)$$

is an orthogonal projection. Furthermore we have the following well-known result (see [4] and [11] for the generalization to higher dimensions):

Theorem 1. *Suppose* $1 \leq p < \infty$ *and* $-1 < \alpha, \beta < \infty$. *Then*

$$P_\alpha : L^p(\mathbb{U}, d\nu_\beta) \longrightarrow A^p(\mathbb{U}, d\nu_\beta)$$

is a continuous projection if and only if $p(\alpha + 1) > \beta + 1$.

Remark 1. One of the advantages of the theory of Bergman spaces over that of Hardy spaces is the abundance of holomorphic projections. But contrary to the case of

$$P_\alpha : L^1(\mathbb{U}, d\nu) \longrightarrow A^1(\mathbb{U}, d\nu)$$

for $\alpha > 0$, there exists no bounded projection

$$P : L^1(\partial\mathbb{U}) \longrightarrow H^1(\mathbb{U}),$$

where $H^1(\mathbb{U})$ denotes the Hardy space in $\mathbb{U}$.

(II) Riesz system

Let $\mathbb{B}$ be the open unit ball in $\mathbb{R}^n$ with boundary, dV the normalized Lebesgue measure in $\mathbb{B}$, and

$$dV_\alpha(x) = C_\alpha(1 - |x|^2)^\alpha \, dV(x)$$

with $\alpha > -1$ and $V_\alpha(\mathbb{B}) = 1$. Following [9], a mapping $F = (f_1, \cdots, f_n)$ is called a *Riesz system* if it satisfies the equation

$$\sum_{j=1}^n \frac{\partial f_j}{\partial x_k} = 0, \quad \text{with} \quad \frac{\partial f_j}{\partial x_k} = \frac{\partial f_k}{\partial x_j}, \quad 1 \leq j, k \leq n.$$

It is proved by Stein and Weiss [16] that F is a Riesz system if and only if $F = \nabla H$ for some harmonic function H.

We call $F \in L^p(\mathbb{B}, \mathbb{R}^n, dV_\alpha)$ if $f_j \in L^p(\mathbb{B}, dV_\alpha)$ for all j. The Bergman space $A^p(\mathbb{B}, \mathbb{R}^n, dV_\alpha)$ consists of all Riesz systems F in $L^p(\mathbb{B}, \mathbb{R}^n, dV_\alpha)$. For the point evaluation functional we have

$$|F(x)| \leq C(x)\|F\|_{A^2(\mathbb{B}, \mathbb{R}^n, dV_\alpha)}.$$

The construction of the Bergman kernel for a Riesz system is somewhat complicated. Let $H_k(\mathbb{R}^n)$ be the set of all homogeneous harmonic polynomials of degree k, and let

$$H_k(\partial\mathbb{B}) = \{f\,|_{\partial\mathbb{B}} : f \in H_k(\mathbb{R}^n)\}$$

be the set of spherical harmonic functions of degree k.

Denote by $\{Y_j^k\}_{1 \leq j \leq d_k}$ the real orthogonal basis of $H_k(\partial\mathbb{B})$, where

$$d_k = \dim H_k(\partial\mathbb{B}) = \binom{n+k-2}{n-2} + \binom{n+k-3}{n-2}.$$

Write

$$\varphi_{k,j}(x) = \sqrt{\frac{\Gamma(2k+n-1+\alpha)}{\Gamma(2k+n-1)\Gamma(\alpha+1)}} \nabla \left(|x|^k Y_j^k(x')\right), \quad x = |x|x'.$$

Then the weighted Bergman kernel is given by

$$K_\alpha(x,y) = \sum_{k=1}^\infty \sum_{j=1}^{d_k} \varphi_{kj}^*(x)\varphi_{kj}(y),$$

where the asterisk denotes the transposed transformation.

Then it follows by the Riesz representation theorem

$$F(x) = \int_{\mathbb{B}} K_\alpha(x,y)F(y)\,dV_\alpha(y), \quad F \in A^2(\mathbb{B}, \mathbb{R}^n, dV_\alpha).$$

We can consider the Bergman operator

$$P_\alpha F(x) = \int_{\mathbb{B}} K_\alpha(x,y)F(y)\,dV_\alpha(y).$$

The following result is due to Djrbashian [9] with the restriction $\alpha \in \mathbb{N}$. It can be extended to the general case $\alpha > -1$ by our approach in Section 8.4.

Theorem 2. *Let* $1 \le p < \infty$, $-1 < \alpha, \beta < \infty$. $\alpha \in \mathbb{N}$. *Then*

$$P_\alpha : L^p(\mathbb{B}, \mathbb{R}^n, dV_\beta) \longrightarrow A^p(\mathbb{B}, \mathbb{R}^n, dV_\beta)$$

is a continuous projection provided $p(\alpha + 1) > \beta + 1$.

Remark 2. In the language of Clifford analysis, F is a Riesz system if and only if the corresponding Clifford algebra-valued function f is monogenic and a pure 1-vector (i.e., $f : \mathbb{B} \to \mathbb{R}^n \subset C\ell_{0,n}$ and $Df = 0$).

8.2 Preliminaries

Let $C\ell_{0,n}$ be the *Clifford algebra* generated by $\mathbb{R}^n$. Fix an orthogonal basis $\{e_1, \cdots, e_n\}$ of $\mathbb{R}^n$ with the product rules

$$e_k e_j + e_j e_k = -2\delta_{kj}.$$

Then the Clifford algebra $C\ell_{0,n}$ considered as a real vector space has a basis

$$\{e_A : A \in \mathcal{PN}\},$$

where

$$\mathcal{PN} = \{(j_1, \ldots, j_k) : 1 \le j_1 < \cdots < j_k \le n, \quad 0 \le k \le n\}, \tag{2.1}$$

$$e_A = \mathbf{e}_{(j_1 \cdots j_k)} = \mathbf{e}_{j_1} \cdots \mathbf{e}_{j_k}, \quad e_\phi = e_0 = 1. \tag{2.2}$$

A real Hilbert space structure in $C\ell_{0,n}$ is given by the inner product

$$\langle a, b \rangle = \sum_{A \in PN} a_A b_A, \quad a = \sum_{A \in PN} a_A e_A, \quad b = \sum_{A \in PN} b_A e_A.$$

We embed $\mathbb{R}^n$ into $C\ell_{0,n}$ by identify $(x_1, \cdots, x_n)$ with $x_1 e_1 + \cdots + x_n e_n$. The *Clifford conjugation* as an involution of $C\ell_{0,n}$ is given by

$$\bar{1} = 1, \quad \bar{e}_j = -e_j, \quad e_A \bar{e}_A = \bar{e}_A e_A = 1.$$

We consider Clifford-valued functions of the form

$$f(x) = \sum_{A \in PN} f_A(x) e_A$$

with $f_A(x)$ being scalar–valued functions. The (left) Dirac operator is defined by

$$D = \sum_{j=1}^{n} e_j \frac{\partial}{\partial x_j}$$

with the action

$$Df = \sum_j e_j \frac{\partial}{\partial x_j} \left(\sum_A f_A e_A \right) = \sum_{j,A} \frac{\partial f_A}{\partial x_j} e_j e_A.$$

We call a function f (left) monogenic in $\mathbb{B}$ if $Df = 0$ in $\mathbb{B}$. f is called a pure 1-vector if

$$f = f_1 e_1 + \cdots + f_n e_n,$$

and it is easy to verify that $Df = 0$ is equivalent to the fact that the map $F :=$ $(f_1, \ldots, f_n)$ is a Riesz system in the sense of Section 8.1.

A polynomial which is homogeneous of degree k and monogenic is called an inner monogenic of degree k. The module of the collection of all inner monogenics of degree k is denoted by $\mathcal{M}_{+,k}$ (see [3] and [7]).

Let dV be the normalized Lebesgue measure on $\mathbb{B}$. For any $\alpha > -1$, we consider the normalized weighted measure on $\mathbb{B}$:

$$dV_\alpha(x) = C_\alpha (1 - |x|^2)^\alpha dV(x), \quad x \in \mathbb{B}.$$

A function f in $\mathbb{B}$ belongs to $L^p(\mathbb{B}, C\ell_{0,n}, dV_\alpha)$, if each of its components $f_A \in L^p(\mathbb{B}, dV_\alpha)$; or equivalently

$$\|f\|_{p,\alpha}^p = \int_{\mathbb{B}} |f(y)|^p \, dV_\alpha(x) < \infty,$$

where $|f(y)|^2 = \langle f(y), f(y) \rangle$.

The weighted monogenic Bergman space $A^p(\mathbb{B}, C\ell_{0,n}, dV_\alpha)$ is the subspace of all monogenic functions in $L^p(\mathbb{B}, C\ell_{0,n}, dV_\alpha)$. It is known that $A^2(\mathbb{B}, C\ell_{0,n}, dV_\alpha)$ is a Hilbert space with respect to the inner product

$$(f, g)_0 = \left[\int_{\mathbb{B}} \overline{f(x)} g(x) \, dV_\alpha(x)\right]_0$$

where the index 0 indicates the scalar part (0-vector part) of the corresponding expression. Furthermore, it is well known that $A^2(\mathbb{B}, C\ell_{0,n}, dV_\alpha)$ is a Hilbert module with reproducing kernel (see [7, 8]).

8.3   The closed form of the Bergman kernel

The Clifford algebra-valued Bergman kernel in series form is well known (see for instance [3] and [17]). More precisely, for any $f \in A^2(B, C\ell_{0,n}, dV)$ and $x \in \mathbb{B}$,

$$f(x) = \int_B \overline{K(x, y)} f(y) \, dV(y)$$

where $K(x, y)$ is the Bergman kernel given by

$$K(x, y) = 2 \sum_{k=0}^{\infty} (k + \tfrac{1}{2}n)|x|^k |y|^k \left[C_k^{\frac{n}{2}}(\langle x', y'\rangle) + y'x' C_{k-1}^{\frac{n}{2}}(\langle x', y'\rangle)\right]. \quad (3.1)$$

Here $C_k^{n/2}(r)$ is the Gegenbauer polynomial of degree k (see [3]).

We now present the Bergman kernel in a closed form:

Theorem 3. *For any $x \in \mathbb{B}$ denote $x = |x|x'$ with $x' \in \partial \mathbb{B}$. The Bergman kernel in the unit ball is given by*

$$K(x, y) = \frac{n(1 + yx)(1 - |x|^2 |y|^2) + 2yx|x||y - x'|^2}{||x|y - x'|^{n+2}}, \quad x, y \in \mathbb{B}. \quad (3.2)$$

Proof. Recall the generating function of the Gegenbauer polynomials

$$\frac{1}{(1 - 2rt + t^2)^\lambda} = \sum_{k=0}^{\infty} C_k^\lambda(r) t^k. \quad (3.3)$$

Applying the operator $\lambda + t\frac{d}{dt}$ on both sides of (3.3), we have

$$\frac{\lambda(1 - t^2)}{(1 - 2rt + t^2)^{\lambda+1}} = \sum_{k=0}^{\infty} (k + \lambda) C_k^\lambda(r) t^k. \quad (3.4)$$

Now taking $t = |x||y|$ and $r = \langle x', y'\rangle$, then we obtain

$$1 - 2rt + t^2 = ||x|y - x'|^2.$$

We can now rewrite (3.3) and (3.4) as follows:

$$\frac{1}{\||x|y - x'|^{2\lambda}} = \sum_{k=0}^{\infty} C_k^\lambda(\langle x', y'\rangle)|x|^k|y|^k, \tag{3.5}$$

$$\frac{\lambda(1 - |x|^2|y|^2)}{\||x|y - x'|^{2(\lambda+1)}} = \sum_{k=0}^{\infty}(k + \lambda)C_k^\lambda(\langle x', y'\rangle)|x|^k|y|^k. \tag{3.6}$$

Sum the above identities together and replace k by $k - 1$ to yield

$$\frac{\lambda(1 - |x|^2|y|^2) + \||x|y - x'|^2}{\||x|y - x'|^{2(\lambda+1)}}$$

$$= \sum_{k=0}^{\infty}(k + \lambda)C_{k-1}^\lambda(\langle x', y'\rangle)|x|^{k-1}|y|^{k-1}. \tag{3.7}$$

Now we set $\lambda = \frac{1}{2}n$ and calculate $2((3.6)) + yx(3.7))$. Since

$$yx = y'x'|x||y|,$$

we see that the calculation for the right side gives $K(x, y)$, as desired. $\square$

Corollary 1. *For any $x \in \mathbb{B}$,*

$$K(x, x) = \frac{n + (n - 2)|x|^2}{(1 - |x|^2)^n}, \tag{3.8}$$

$$K(x, -x) = \frac{n(1 - |x|^2) + 2|x|^2}{(1 + |x|^2)^n}. \tag{3.9}$$

The corollary shows that the Clifford algebra-valued Bergman kernel always remains positive along the diagonal $x = y$ or the line $x = -y$.

Proof of Corollary 1. If $y = x$, then the denominator in (3.2) becomes

$$\||x|x - x'|^{n+2} = (1 - |x|^2)^{n+2}.$$

Since $x^2 = -|x|^2$ for any $x \in \mathbb{B}$, the numerator in (3.2) with $y = x$ equals

$$n(1 - |x|^2)(1 - |x|^4) - |x|^2(1 - |x|^2)^2 = (1 - |x|^2)^2(n + (n - 2)|x|^2).$$

This proves (3.8); the second identity follows similarly. $\square$

Remark 3. We remark that in the classical case the Bergman kernel along the diagonal is important in the construction of intrinsic invariant metrics; see [13].

8.4   The estimate for the weighted Bergman kernel

Now we consider the weighted Bergman kernel.

Lemma 1. *Let $\alpha > -1$ and denote $x = |x|x'$ with $x' \in \partial\mathbb{B}$ for any $x \in \mathbb{B}$. The weighted Bergman kernel $K_\alpha(x, y)$ for any $x, y \in \mathbb{B}$ is given by*

$$2\sum_{k=0}^{\infty} \frac{\Gamma(\frac{n}{2} + k + \alpha + 1)}{\Gamma(\frac{n}{2} + k)\Gamma(\alpha + 1)} |x|^k |y|^k \left(C_k^{\frac{n}{2}}(\langle x', y'\rangle) + y'x' C_{k-1}^{\frac{n}{2}}(\langle x', y'\rangle) \right). \quad (4.1)$$

This can be checked as in [3] for the non-weighted case. In fact, notice that each function f in $A^2(B, C\ell_{0,n}, dV)$ can be decomposed into a series of inner monogenic functions

$$f(y) = \sum_{l=0}^{\infty} |y|^l P_l f(y')$$

and the function

$$P_k(x', y') = C_k^{\frac{n}{2}}(\langle x', y'\rangle) + y'x' C_{k-1}^{\frac{n}{2}}(\langle x', y'\rangle)$$

is the reproducing kernel for $\mathcal{M}_{+,k}$. Applying the orthogonality on $\partial\mathbb{B}$ of spherical monogenics one gets

$$\int_\mathbb{B} \overline{K_\alpha(x, y)} f(y) \, dV_\alpha(y)$$

$$= 2\sum_{k=0}^{\infty} |x|^k \int_0^1 \frac{\Gamma(\frac{1}{2}n + k + \alpha + 1)}{\Gamma(\frac{1}{2}n + k)\Gamma(\alpha + 1)} r^{2k+n-1}(1 - r^2)^\alpha \, dr$$

$$\times \int_{\partial\mathbb{B}} \left(C_k^{\frac{n}{2}}(\langle x', y'\rangle) + y'x' C_{k-1}^{\frac{n}{2}}(\langle x', y'\rangle) \right) P_k f(y') \, d\sigma(y')$$

$$= \sum_{k=0}^{\infty} |x|^k P_k f(x) = f(x).$$

In view of (4.1), one can not hope to get the closed form for the weighted Bergman kernel in the general case. But from the closed form of non-weighted Bergman kernel in (3.2), we have the simple estimate that

$$|K(x, y)| \leq C||x|y - x'|^{-n} \quad (4.2)$$

for any $x, y \in \mathbb{B}$, due to the fact that, for any $x, y \in \mathbb{B}$,

$$|1 + yx| = |-\frac{x^2}{|x|^2} + yx| = |x||\frac{x}{|x|^2} - y| = ||x|y - x'| \quad (4.3)$$

and

$$|1 - |x|^2|y|^2| \leq 2(1 - |x||y|) \leq 2||x|y - x'|.$$

We can show that a similar relation also holds for the weighted case, with the estimate of the non-weighted case (4.2) as starting point. This result turns out to be crucial in applications.

Theorem 4. *Let $\alpha > -1$. Then*

$$|K_\alpha(x,y)| \leq C||x|y - x'|^{-(n+\alpha)} \tag{4.4}$$

for any $x, y \in \mathbb{B}$.

To prove this theorem we need some lemmas.

Lemma 2. *For any $m \in \mathbb{N}$, there exists a constant $C = C(m) > 0$ such that, for any $x, y \in \mathbb{B}$ with $r = |x|$,*

$$\left|\left(r\frac{d}{dr}\right)^m K(x,y)\right| \leq \frac{C}{||x|y - x'|^{n+m}}.$$

Proof. From (3.2), if we denote

$$A_1 = 1 + yx, \quad A_2 = 1 - |x|^2|y|^2, \quad A_3 = ||x|y - x'|,$$

and the numerator in (3.2) by B, then $B = nA_1A_2 + 2yxA_3^2$ and

$$K(x,y) = \frac{B}{||x|y - x'|^{n+2}}.$$

It follows that

$$\frac{d}{dr}K(x,y) = -(n+2)\frac{B}{||x|y - x'|^{n+3}}\frac{dA_3}{dt}$$

$$+ \frac{n\frac{dA_1}{dt}A_2 + nA_1\frac{dA_2}{dt} + 2yxA_3\frac{dA_3}{dt} + 2\frac{d(yx)}{dt}A_3^2}{||x|y - x'|^{n+2}}.$$

Notice that from (4.2) and (4.3) we have

$$|A_1| = |A_3| = ||x|y - x'|, \quad \text{and} \quad |A_2| \leq 2||x|y - x'|,$$

so that $|B| \leq C||x|y - x'|^2$. Since $\frac{dA_i}{dt} \leq C$ for $i = 1, 2, 3$, we finally get

$$\frac{d}{dr}K(x,y) \leq C\frac{1}{||x|y - x'|^{n+1}}.$$

The general case follows by induction. $\qquad\qquad\qquad\qquad\qquad\qquad\square$

Lemma 3. *Let $0 < \delta < \lambda$. Then, for any points x and y in $\mathbb{B}$,*

$$\int_0^1 \frac{(1-t)^{\delta-1}}{|t|y|x - y'|^\lambda}\,dt \leq \frac{8^\lambda}{\delta(\lambda-\delta)}\frac{1}{||y|x - y'|^{\lambda-\delta}}.$$

Proof. Note that for any $t \in [0, 1]$ and $x, y \in \mathbb{B}$,

$$||y|x - y'| \leq 2|t|y|x - y'|. \tag{4.5}$$

Indeed, from the triangle inequality we have

$$|t|y|x - y'| \geq 1 - t,$$
$$|t|y|x - y'| \geq ||y|x - y'| - (1 - t),\tag{4.6}$$

so that summing up yields (4.5).

If $||y|x - y'| \geq 1$, then $|t|y|x - y'| \geq \frac{1}{2}$ from (4.5). Combining this with the inequality $||y|x - y'| \leq 2$, we get

$$\int_0^1 \frac{(1 - t)^{\delta - 1}}{|t|y|x - y'|^\lambda} \, dt \leq \frac{2^\lambda}{\delta} \leq \frac{2^{2\lambda - \delta}}{\delta} \frac{1}{||y|x - y'|^{\lambda - \delta}}.$$

Now assume $||y|x - y'| < 1$ and denote $r = 1 - ||y|x - y'|$, then $0 < r < 1$ and $1 - r = ||y|x - y'|$. From (4.5) and (4.6) we have

$$1 - rt = 1 - t + t||y|x - y'| \leq 3|t|y|x - y'|.$$

It leads to

$$\int_0^1 \frac{(1 - t)^{\delta - 1}}{|t|y|x - y'|^\lambda} \, dt \leq 3^\lambda \int_0^1 \frac{(1 - t)^{\delta - 1}}{(1 - tr)^\lambda} \, dt$$

$$\leq 3^\lambda \frac{\lambda}{\delta(\lambda - \delta)} \frac{1}{(1 - r)^{\lambda - \delta}}$$

$$\leq C \frac{1}{||y|x - y'|^{\lambda - \delta}}.$$

This completes the proof.   □

With the above lemmas, we are now able to prove Theorem 4. The proof is divided into two cases.

Proof of Theorem 4. Case (i) $\alpha = m \in \mathbb{N}$.

For convenience, we can omit the constant $2/\Gamma(\alpha + 1)$ in (4.1) and still denote the remaining function by $K_\alpha(x, y)$. It is easy to verify the induced formula

$$K_{m+1}(x, y) = \left(r \frac{d}{dr} + (m + 1 + \tfrac{1}{2}n) \right) K_m(x, y)).$$

Notice that $K_0(x, y) = K(x, y)$, so we have

$$K_m(x, y) = Q_m(r \frac{d}{dr}) K(x, y),$$

where $Q_m(t)$ is a polynomial of degree m. Therefore Lemma 2 implies the desired result.

Case (ii) $\alpha \notin \mathbb{N}$.

Denote $m = [\alpha] + 1$, then $m \in \mathbb{N}$ or $m = 0$. By the series expansion of the weighted Bergman kernel, it is easy to see that

$$K_\alpha(x, y) = \frac{2}{\Gamma(m - \alpha)} \int_0^1 t^{n+2\alpha+1}(1 - t^2)^{m-\alpha-1} K_m(tx, ty)\, dt\,.$$

The desired result follows from Lemma 3 and the estimate on K_m in the case (i).

$\square$

8.5   The bounded Bergman projection

Now we consider the Bergman operator:

$$P_\alpha f(x) = \int_\mathbb{B} \overline{K_\alpha(x, y)} f(y)\, dV_\alpha(y)\,.$$

Our main result is the following theorem.

Main Theorem. *Suppose* $1 \le p < \infty$, $-1 < \alpha, \beta < \infty$. *Then*

$$P_\alpha : L^p(\mathbb{B}, C\ell_{0,n}, dV_\beta) \longrightarrow A^p(\mathbb{B}, C\ell_{0,n}, dV_\beta)$$

is a continuous projection provided $p(\alpha + 1) > \beta + 1$.

To prove this theorem we need some preparation. In fact, we need some known properties of hypergeometric functions:

1. Bateman's integral formula [10, page 78]

$$F(a, b; c + \mu; s) = \frac{\Gamma(c + \mu)}{\Gamma(c)\Gamma(\mu)} \int_0^1 t^{c-1}(1 - t)^{\mu-1} F(a, b; c; ts)\, dt \quad (5.1)$$

with $c, \mu > 0$ and $s \in (-1, 1)$.

2. For any integer m ([14, page 69])

$$F(-m, a + m; c; 1) = \frac{(-1)^m(1 + a - c)_m}{(c)_m}\,. \quad (5.2)$$

Lemma 4. *Let* $t > 1$, $\lambda \in \mathbb{R}$ *and* $r \in (-1, 1)$. *Then*

$$\int_{-1}^1 \frac{(1 - u^2)^{\frac{t-3}{2}}}{(1 - 2ru + r^2)^\lambda}\, du = \frac{\Gamma(\frac{1}{2}t - \frac{1}{2})\Gamma(\frac{1}{2})}{\Gamma(\frac{1}{2}t)} F(\lambda, \lambda + 1 - \tfrac{1}{2}t; \tfrac{1}{2}t; r^2)\,. \quad (5.3)$$

Proof. Let $C_m^\lambda(u)$ be the Gegenbauer polynomials, defined by the generating function

$$(1 - 2ru + r^2)^{-\lambda} = \sum_{m=0}^{\infty} C_m^\lambda(u) r^m, \tag{5.4}$$

where

$$C_{2m}^\lambda(u) = (-1)^m \frac{(\lambda)_m}{m!} F(-m, m + \lambda; \tfrac{1}{2}; u^2),$$

$$C_{2m+1}^\lambda(u) = (-1)^m \frac{(\lambda)_m}{m!} 2u F(-m, m + \lambda + 1; \tfrac{3}{2}; u^2). \tag{5.5}$$

To calculate the integral in (5.3), we apply (5.4) and (5.5). Then we deduce that it is only needed to evaluate the integral

$$\int_{-1}^{1} (1 - u^2)^{\frac{t-3}{2}} F(-m, m + \lambda; \tfrac{1}{2}; u^2) \, du,$$

or rather, an integral over the interval $(0, 1)$ by the simple change of variables $t = u^2$. For this integral, we first use Bateman's integral formula (5.1) with $s = 1$ and apply (5.2), so that it can be represented by Pochhammer symbols. The calculation of the integral in (5.3) then leads to a series which by definition is the desired hypergeometric function.   □

Lemma 5. *Let* $\alpha > -1$ *and* $\beta \in \mathbb{R}$. *Then for any* $x \in \mathbb{B}$,

$$\int_{\mathbb{B}} \frac{(1 - |y|^2)^\alpha}{||x|y - x'|^{n+\alpha+\beta}} \, dy \approx \begin{cases} (1 - |x|^2)^{-\beta}, & \beta > 0, \\ \log \dfrac{1}{1 - |x|^2}, & \beta = 0, \\ 1, & \beta < 0. \end{cases}$$

The notion $a(x) \approx b(x)$ means that the ratio $a(x)/b(x)$ has a positive finite limit as $|x| \to 1$.

Proof of Lemma 5. Denote the above integral by $J_{\alpha,\beta}(x)$. Using Stirling's formula we can show that

$$J_{\alpha,\beta}(x) = \frac{\Gamma(\tfrac{1}{2}n + 1)\Gamma(\alpha + 1)}{\Gamma(\alpha + \tfrac{1}{2}n + 1)} F(\tfrac{1}{2}(n+\alpha+\beta), \tfrac{1}{2}(2+\alpha+\beta); \alpha + \tfrac{1}{2}n + \tfrac{-}{}; |x|^2).$$

Indeed, for any continuous function f of one variable and any $\eta \in \partial \mathbb{B}$, we have the formula (see [1, page 216])

$$\int_{\partial \mathbb{B}} f(< \zeta, \eta >) \, d\sigma(\zeta) = \frac{\Gamma(\tfrac{1}{2}n)}{\Gamma(\tfrac{1}{2}n - \tfrac{1}{2})\Gamma(\tfrac{1}{2})} \int_{-1}^{1} (1 - u^2)^{\frac{n-3}{2}} f(u) \, du,$$

where the symbol $< \zeta, \eta >$ stands for the inner product in $\mathbb{R}^n$. Taking

$$f(u) = (1 - 2ru + r^2)^{-\frac{n+\alpha+\beta}{2}}$$

for fixed $r \in (0, 1)$ and combining with Lemma 4 we have

$$\int_{\partial \mathbb{B}} (1 - 2r < \zeta, \eta > + r^2)^{-\frac{n+\alpha+\beta}{2}} \, d\sigma(\zeta)$$

$$= \frac{\Gamma(\frac{1}{2}n)}{\Gamma(\frac{1}{2}n - \frac{1}{2})\Gamma(\frac{1}{2})} \int_{-1}^{1} \frac{(1 - u^2)^{\frac{n-3}{2}}}{(1 - 2ru + r^2)^{\frac{n+\alpha+\beta}{2}}} \, du$$

$$= F(\tfrac{1}{2}(n + \alpha + \beta), \tfrac{1}{2}(2 + \alpha + \beta); \tfrac{1}{2}n; r^2).$$

Consequently, from the polar coordinates formula we get that $J_{\alpha,\beta}(x)$ equals

$$n \int_0^1 r^{n-1}(1 - r^2)^\alpha \int_S (1 - 2r|x| < x', \eta > + r^2|x|^2)^{-\frac{n+\alpha+\beta}{2}} \, d\sigma(\zeta)$$

$$= C \int_0^1 r^{n-1}(1 - r^2)^\alpha F(\tfrac{1}{2}(n + \alpha + \beta), \tfrac{1}{2}(2 + \alpha + \beta); \tfrac{1}{2}n; r^2|x|^2) \, dr.$$

The assertion now follows from Bateman's integral formula (5.1).   □

Now we come to the proof of the Main Theorem.

Proof of Main Theorem. We consider the integral operator

$$T_\alpha f(x) = \int_{\mathbb{B}} Q_\alpha(x, y) f(y) \, dV_\alpha(y),$$

where $Q_\alpha(x, y) = ||x|y - x'|^{-n+\alpha}$. By Theorem 4,

$$|K_\alpha(x, y)| \le C Q_\alpha(x, y)$$

for any $x, y \in \mathbb{B}$. Thus it is sufficient to show that the operator T is bounded on $L^p(\mathbb{B}, dV_\beta)$ for any $1 \le p < \infty$ and $p(\alpha + 1) > \beta + 1$.

The case of $p = 1$ is a direct consequence of Lemma 5 and Fubini's theorem. When $1 < p < \infty$, we rewrite the operator T as

$$T_\alpha f(x) = \int_{\mathbb{B}} Q_{\alpha,\beta}(x, y) f(y) \, dV_\beta(y),$$

where

$$Q_{\alpha,\beta}(x, y) = \frac{(1 - |x|^2)^{\alpha - \beta}}{||x|y - x'|^{n+\alpha}}.$$

To prove that the operator T_α is bounded in $L^p(\mathbb{B}, dV_\beta)$, we appeal to Schur's test [12]. Take

$$h(x) = (1 - |x|)^{-\frac{1+\beta}{pq}}$$

with $\frac{1}{p} + \frac{1}{q} = 1$. It is enough to verify that

$$\int_{\mathbb{B}} Q_{\alpha,\beta}(x, y) h(y)^q \, dV_\beta(y) \le C h(x)^q, \quad x \in B,$$

and

$$\int_{\mathbb{B}} Q_{\alpha,\beta}(x,y)h(x)^p \, dV_\beta(x) \le Ch(y)^p \,, \quad y \in B \,,$$

which follows from Lemma 5.   □

As a direct consequence of our Main Theorem, we have the integral formula for monogenic Bergman functions.

Theorem 5. *If* $f \in A^p(\mathbb{B}, C\ell_{0,n}, dV_\alpha)$ *with* $1 \le p < \infty$ *and* $\alpha > -1$, *then there holds the integral representation:*

$$f(x) = \int_{\mathbb{B}} \overline{K_\alpha(x,y)} f(y) \, dV_\alpha(y) \,, \quad x \in \mathbb{B} \,. \tag{5.6}$$

Acknowledgment

The work of the first author was partially supported by the NNSF of China (No. 10001030), and a Post-doctoral Fellowship of the University of Aveiro. UI&D "Matemática e Aplicações".

REFERENCES

[1] S. Axler, P. Bourdon, and W. Ramey, *Harmonic Function Theory*, Graduate Texts in Math. **137**, Springer-Verlag, New York, 1992.

[2] F. Brackx, R. Delanghe, and F. Sommen, *Clifford Analysis*, Research Notes in Mathematics 76. Pitman, Boston, MA, 1982.

[3] F. Brackx, F. Sommen, and N. Van Acker, Reproducing Bergman kernels in Clifford analysis, *Complex Variables Theory Appl.* **24** (1994), no. 3–4, 191–204.

[4] Coifman R. R. and Rochberg R., Representation theorems for holomorphic and harmonic functions in L^p, Representation theorems for Hardy spaces, *Astérisque* **77**, Soc. Math. France, Paris, 1980, 11–66.

[5] D. Constales and R. S. Krausshar, Szegö and polymonogenic Bergman kernels for half-space and strip domains, and single-periodic functions in Clifford analysis, *Complex Var. Theory Appl.* **47** (2002), no. 4, 349–360.

[6] R. Delanghe, On Hilbert modules with reproducing kernel, in *Function Theoretic Methods for Partial Differential Equations* (Proc. Internat. Sympos., Darmstadt, 1976), Lecture Notes in Math. **561**, Springer, Berlin, 1976, pp. 158–170.

[7] R. Delanghe and F. Brackx, Hypercomplex function theory and Hilbert modules with reproducing kernel, *Proc. London Math. Soc.* (3) 37 (1978), no. 3, 545–576.

[8] R. Delanghe, F. Sommen F., and V. Souček, *Clifford algebra and spinor-valued functions. A function theory for the Dirac operator*, Mathematics and its Applications **53**, Kluwer Academic Publishers Group, Dordrecht, 1992.

[9] Djrbashian A., Integral representations for Riesz systems in the unit ball and some applications, *Proc. Amer. Math. Soc.* **117** (1993), 395–403.

[10] Erdélyi A. et al., *Higher Transcendental Functions I*, McGraw-Hill, New York, 1953.

[11] F. Forelli and W. Rudin, Projections on spaces of holomorphic functions in balls, *Indiana Univ. Math. J.* **24** (1974/75), 593–602.

[12] H. Hedenmalm, B. Korenblum, and K. H. Zhu, *Theory of Bergman Spaces*, Graduate Texts in Mathematics **199**, Springer-Verlag, New York, 2000.

[13] S. G. Krantz, *Function Theory of Several Complex Variables*, AMS Chelsea Publishing, Providence, RI, 2001.

[14] E.D. Rainville, *Special Functions*, Chelsea Publishing Company, Bronx, New York, 1971.

[15] M. V. Shapiro and N. L. Vasilevski, On the Bergman kernel function in the Clifford analysis, in *Clifford algebras and their applications in mathematical physics*, Eds. F. Brackx, R. Delanghe, and H. Serras, *Fund. Theories Phys.*, **55**, Kluwer Acad. Publ., Dordrecht, 1993, pp. 183–192.

[16] E. M. Stein and G. Weiss, Generalization of the Cauchy–Riemann equations and representation of the rotation group, *Amer. J. Math.* **90**, 1968, 163–196.

[17] Z. Xu,, A function theory for the operator $D - \lambda$, *Complex Variables* **16** (1991), 27–42.

[18] K. Zhu, *Operator Theory in Function Spaces*, Monographs and Textbooks in Pure and Applied Mathematics, **139**. Marcel Dekker, Inc., New York, 1990.

Guangbin Ren
Department of Mathematics
University of Science and Technology of China,
Hefei, Anhui 230026, P. R. China
E-mail: rengb@ustc.edu.cn

Helmuth R. Malonek
Departamento de Matemática
Universidade de Aveiro
P-3810-193 Aveiro, Portugal
E-mail: hrmalon@mat.ua.pt

Submitted: August 15, 2002.

9

Quaternionic Calculus for a Class of Initial Boundary Value Problems

Wolfgang Sprössig

ABSTRACT We study Galpern–Sobolev equations with the help of a quaternionic operator calculus. Previous work is extended to the case of a variable dispersive term. We approximate the time derivative by forward finite differences. Solving the resulting stationary problems by means of a quaternionic calculus, we obtain representation formulae.

Keywords: Galpern–Sobolev equation, quaternionic analysis, operator calculus, pseudoparabolic differential equations

9.1   Introduction

Partial differential equations in mathematical physics describe time-dependent physical processes. In cooperation with several people from the Clifford analysis group such as K. Gürlebeck [7, 8], S. Bernstein [2, 3], U. Kähler [5] and M. Shapiro [1, 10] we have developed a quaternionic operator calculus for the solution of stationary e.g., time-independent boundary value problems.

In this case it was not necessary to leave real Clifford algebras. Most of the boundary value problems have been treated with the help of the algebraic structure of real quaternions. In order to include time in our conception we have to clear essential difficulties. It seems to be clear that we have to use complex Clifford algebras or at least the algebra of complex quaternions. So former results by J. Ryan [14], S. Bernstein [2], V. Kravchenko, M. Shapiro [13] and F. Sommen [4] could be interesting for our problems. In order to use as much as possible of our operator calculus we are going to focus our attention on a semi-discretization method. An alternative approach was proposed by G. Kaiser [11], who uses complex distances. In this formulation our basic operators D, T and F in our calculus change dramatically. We guess it will be very difficult to keep many of the useful topological properties.

Now we develop a method of time-discretization and have to prove approxima-

---

AMS Subject Classification: 30G35, 35K15.

tion, stability and convergence.

In [9] we presented first applications of the time-discretization method for the model of the *Galpern–Sobolev equation* (cf. [12, 16]). This important equation contains dissipative, dispersive and non-linear terms, and can be formulated as follows:

Let $G \subset \mathbb{R}^n$ be a bounded domain with a sufficiently smooth boundary Γ and $[0, T]$ a time interval. Further, let $f : G \times [0, T] \to \mathbb{R}^n$ and $u : G \times [0, T] \to \mathbb{R}^n$. Then we consider the differential equation

$$\partial_t(u - \eta \, \triangle \, u) - \nu \, \triangle \, u = f(x, t, u, \nabla u) \quad \text{in} \ \ G \times (0, T]$$

with the boundary condition $(Au)(x, t) = g(x, t)$ on Γ (A is a boundary operator) and the initial condition $u(x, 0) = u_0$. Here ν and η are non-negative constants.

Within the $(k + 1)$-th time interval we obtained the following representation formula for $u = u(x, t)$:

Theorem 1 ([9]). *Let $u_k = u(x, k\tau)$ and*

$$f_k(x) = \tau^{-1} \int\limits_{k\tau}^{(k+1)\tau} f(x, t) \, dt.$$

Then we have

$$u_{k+1} = (\alpha \, R_{i\beta} S_\eta)^{k+1} u_0 + \sum_{l=0}^{k} (\alpha \, R_{i\beta} S_\eta)^l (\alpha \tau \, R_{i\beta} f_{k-l} + \mathcal{H}_{k-l})$$

where

$$S_\eta = I - \eta \triangle, \quad R_{i\beta} = T_{-i\beta} Q_{i\beta} T_{i\beta}, \quad Q_{i\beta} = I - P_{i\beta},$$
$$\mathcal{H}_k = F_{-i\beta} g_k + T_{-i\beta} P_{i\beta} D_{-i\beta} H_k, \quad g_k = g(x, k\tau)$$

are the Dirichlet data on the boundary at time $k\tau$ and H_k its smooth extension to G. Furthermore, τ is the mesh width of the time-discretization $\eta + \nu\tau = \alpha^{-1}$ and $\beta = \sqrt{\alpha}$. The definition of the operators $F_{-i\beta}, T_{\pm i\beta}$ and $D_{-i\beta}$ is given in the next section. The operators $Q_{\pm i\beta}, P_{\pm i\beta}$ are defined as

$$Q_{i\beta} := I - P_{i\beta} \quad \text{and} \quad P_{i\beta} := F_{i\beta}(\operatorname{tr}_\Gamma T_{-i\beta} F_{i\beta})^{-1} T_{-i\mu},$$

where the last one can be seen as a modification of the Bergman projection.

Corollary 1. *The extension H_k is not unique.*

Let H_k^1 and H_k^2 be different smooth extensions. We consider the difference $H_k^1 - H_k^2$. Because of $D_{-i\beta}(H_k^1 - H_k^2) \in \operatorname{im} Q_{i\beta}$ we have

$$T_{-i\beta} P_{i\beta} D_{-i\beta}(H_k^1 - H_k^2) = 0.$$

Remark 1. This proposition shows that uniqueness is not needed here. Truncation error and stability estimates can be proved. We refer to our paper [9].

In [9] it was assumed that the coefficients of the dissipative term as well as the dispersive term has to be constant. In continuation of our considerations we will now calculate with a nonconstant dispersive term of the form $\alpha^{-1} D \alpha D u$ where

$$D = \sum_{i=1}^{3} e_i \partial_i, \quad DD = -\triangle \text{ and } \alpha = \alpha(x, u) : G \times \mathbb{H} \to \mathbb{R}^+. \text{ For } \alpha = \text{const. we}$$

again obtain the previous case.

9.2   Formulation of the problem

It is sufficient to consider the following initial boundary value problem $(\eta = 0)$:

Let $G \subset \mathbb{R}^3$ be a bounded domain with sufficient smooth boundary Γ. There are given the quaternionic–valued functions:

$$f \in C^1((0, T], L_2(G)), g \in C^1((0, T], W_2^{3/2}(\Gamma)), h \in W_2^2(G)$$

as well as the real-valued function $\alpha \in C^\infty(G)$, $\alpha \neq 0$. We look for the solution $u \in C^1([0, T], W_2^2(G))$ of the problem:

$$\partial_t u + \alpha^{-1} D \alpha D u = f \quad in \quad (0, T] \times G, \tag{2.1}$$

$$u = g \quad on \quad (0, T] \times \Gamma, \tag{2.2}$$

$$u = h \quad on \quad \{0\} \times G. \tag{2.3}$$

For this reason we carry out the following time discretization for the problem (2.1)–(2.3). We have to replace the time derivative $\partial_t u$ at $k\tau$ by

$$\frac{u(x, (k+1)\tau) - u(x, k\tau)}{\tau}, \quad (k = 0, \ldots, N - 1)$$

where $\tau = T/N$ is the mesh width of the discretization. Abbreviate

$$u(x, k\tau) = u_k(x), \quad (k = 0, \ldots, N).$$

We obtain the discretized equation

$$\frac{u_{k+1}}{\tau} + \alpha^{-1} D \alpha D u_{k+1} = f_k + \frac{u_k}{\tau}, \quad (k = 0, \ldots, N - 1).$$

The function $f_k = f_k(x)$ describes the mean value of f over the interval $[k\tau, (k+1)\tau]$:

$$f_k(x) := \frac{1}{\tau} \int_{k\tau}^{(k+1)\tau} f(x, t) \, dt.$$

Setting now $\mu := \tau^{-1/2}$ we obtain a *Yukawa-type equation*

$$(\alpha^{-1} D \, \alpha \, D + \mu^2) u_{k+1} = f_k + \mu^2 u_k, \quad (k = 0, 1, \ldots, N - 1).$$

In order to realize a factorization we now need complex quaternions. Note that $i e_j = e_j i$ $(j = 1, 2, 3)$. We have

$$\alpha^{-1}(D - i\mu)\alpha(D + i\mu)u_{k+1} = f_k - i\mu D\alpha u_{k+1} + \mu^2 u_k. \tag{2.4}$$

The last identity follows from the validity of the generalized Leibniz rule:

$$\alpha^{-1} D(\alpha i \mu u) = \mu i D u + (\alpha^{-1} i \mu D \alpha) u.$$

9.3   Quaternionic operators

Next we have to consider an appropriate operator which is right-inverse to our operators $D \pm i\mu$. This kind of operator has been well studied (cf. [7, 8] and [17]). For this reason we have to introduce the kernel function:

$$e_\mu(x) := -\left(\frac{i\mu}{2\pi}\right)^{\frac{3}{2}} [f_{i\mu}(|x|)w - g_{i\mu}(|x|)],$$

where

$$f_{i\mu}(t) := t^{-1/2} K_{3/2}(i\mu t), \quad g_{i\mu}(t) := t^{-1/2} K_{1/2}(i\mu t).$$

Clearly, $w = x/|x| \in S^2$ (unit sphere in $\mathbb{R}^3$) and $K(z)$ $(z \in \mathbb{C})$ denotes *Macdonald's function*. In this way $e_\mu(x)$ is a fundamental solution of the operator $D + i\mu$. Analogously $e_{-\mu}(x)$ is a fundamental solution of the operator $D - i\mu$. Now we are able to construct the right-inverses $T_{\pm i\mu}$ of the operators $D \pm i\mu$. We get for $u \in C(G)$,

$$(T_{\pm i\mu} u)(x) := \int_G e_{\pm\mu}(y - x)u(y)\,dy \quad (x \in G),$$

which is called a *modified Teodorescu transform*. We have

$$\left((D \pm i\mu)T_{\pm i\mu} u\right)(x) = \begin{cases} 0, & x \in \mathbb{R}^3 \setminus \overline{G}, \\ u(x), & x \in G. \end{cases}$$

Unfortunately, the Teodorescu transform is not also the left-inverse. So for $u \in C^1(G) \cap C(\overline{G})$ the so-called Borel–Pompeiu formula

$$(T_{\pm i\mu}(D \pm i\mu)u)(x) = u(x) - (F_{\pm i\mu} u)(x) \quad (x \in G) \tag{3.1}$$

is valid. Here

$$(F_{\pm i\mu} u)(x) := \int_\Gamma e_{\pm\mu}(x - y)n(y)u(y)\,d\Gamma_y,$$

where n is the unit vector of the outer-normal at the point y of Γ. The operator $F_{\pm i\mu}$ is called a *modified Cauchy–Bizadse operator*. It follows that:

$$((D \pm i\mu)F_{\pm i\mu}u)(x) = 0 \qquad (x \notin \Gamma).$$

The Borel–Pompeiu formula can be extended by continuity to functions $u \in W_2^1(G)$ and their traces in $W_2^{1/2}(\Gamma)$.

Applying the modified Teodorescu transforms $T_{\pm i\mu}$ to equation (4) from the left we obtain

$$u_{k+1} = T_{-i\mu}\alpha^{-1}T_{i\mu}\alpha f_k - i\mu T_{-i\mu}\alpha^{-1}T_{i\mu}(D\alpha)u_{k+1}$$
$$+ \mu^2 T_{-i\mu}\alpha^{-1}T_{i\mu}\alpha u_k + T_{-i\mu}\alpha^{-1}\phi_+ + \phi_- \quad (3.2)$$

where $\phi_\pm \in \ker(D \pm i\mu)$.

9.4   On a modified Bergman projection

Now we have to determine the functions $\phi_\pm$. Hence, we restrict both sides onto the boundary Γ of our domain G and apply from the left the Cauchy–Bizadse operator $F_{-i\mu}$. We already know that $F_{\pm i\mu}T_{\pm i\mu} = 0$. Because of $u_{k+1}|_\Gamma = g_{k+1}$ and $F_{-i\mu}u_{k+1}|_\Gamma = F_{-i\mu}g_{k+1} = \phi_-$, we get

$$\text{tr}_\Gamma \left[T_{-i\mu}\alpha^{-1}T_{i\mu}\alpha f_k + \mu^2 T_{-i\mu}\,\alpha^{-1}T_{i\mu}\alpha u_k \right.$$
$$\left. - T_{-i\mu}\alpha^{-1}T_{i\mu}(D\alpha)u_{k+1} + T_{-i\mu}\alpha^{-1}\phi_+ \right] = Q_{\Gamma,-i\mu}g_{k+1}.$$

In previous papers we proved that the operator

$$\text{tr}_\Gamma T_{\pm i\mu}\alpha^{-1}F_{\mp i\mu} : \text{im } P_{\Gamma,i\mu} \cap W_2^{k-1/2}(\Gamma) \to \text{im } Q_{\Gamma,-i\mu} \cap W_2^{k+1/2}(\Gamma)$$

is an isomorphism, where $P_{\Gamma,i\mu}$ and $Q_{\Gamma,-i\mu}$ are the *generalized Plemelj projections* on the corresponding Hardy spaces. As for $F_{i\mu}\phi_+ = \phi_+$ we have

$$\text{tr}_\Gamma \phi_+ = (\text{tr}_\Gamma T_{-i\mu}\alpha^{-1}F_{i\mu})^{-1}[-T_{-i\mu}\alpha^{-1}T_{i\mu}\alpha f_k - \mu^2 T_{-i\mu}\alpha^{-1}T_{i\mu}\alpha u_k$$
$$+ i\mu(T_{-i\mu}\alpha^{-1}T_{i\mu}D\alpha)u_{k+1}] + (\text{tr}_\Gamma T_{-i\mu}\alpha^{-1}F_{i\mu})^{-1}Q_{\Gamma.i\mu}g_{k+1}.$$

Hence

$$\phi_+ = F_{i\mu}(\text{tr}_\Gamma T_{-i\mu}\alpha^{-1}F_{i\mu})^{-1}\left[-T_{-i\mu}\alpha^{-1}T_{i\mu}(\alpha f_k + \mu^2\alpha u_k) \right.$$
$$\left. + i\mu(T_{-i\mu}\alpha^{-1}T_{i\mu}D\alpha)u_{k+1} \right] + F_{i\mu}(\text{tr}_\Gamma T_{-i\mu}\alpha^{-1}F_{i\mu})^{-1}Q_{\Gamma,-i\mu}g_{k+1}.$$

Replacing now ϕ_+ in (5) by the last expression we find

$$
\begin{aligned}
u_{k+1} = {} & T_{-i\mu}\alpha^{-1}T_{i\mu}\alpha f_k - i\mu T_{-i\mu}\alpha^{-1}T_{i\mu}(D\alpha)u_{k+1} \\
& + \mu^2 T_{-i\mu}\alpha^{-1}T_{i\mu}\alpha u_k \\
& + T_{-i\mu}\alpha^{-1}F_{i\mu}(\mathrm{tr}_\Gamma\, T_{-i\mu}\alpha^{-1}F_{i\mu})^{-1} \\
& \times \left[-T_{-i\mu}\alpha^{-1}T_{i\mu}(\alpha f_k + \mu^2\alpha u_k) + i\mu T_{-i\mu}\alpha^{-1}T_{i\mu}(D\alpha)u_{k+1}\right] \\
& + T_{-i\mu}\alpha^{-1}F_{i\mu}(\mathrm{tr}_\Gamma\, T_{-i\mu}\alpha^{-1}F_{i\mu})^{-1}Q_{\Gamma,-i\mu}g_{k+1} + F_{-i\mu}g_{k+1}\,.
\end{aligned}
$$

This leads to

$$
\begin{aligned}
u_{k+1} = {} & T_{-i\mu}Q_{i\mu}\alpha^{-1}T_{i\mu}(\alpha f_k + \mu^2\alpha u_k) - i\mu T_{-i\mu}Q_{i\mu}\alpha^{-1}T_{i\mu}(D\alpha)u_{k+1} \\
& + T_{-i\mu}\alpha^{-1}F_{i\mu}(\mathrm{tr}_\Gamma\, T_{-i\mu}\alpha^{-1}F_{i\mu})^{-1}Q_{\Gamma,-i\mu}g_{k+1} + F_{-iu}g_{k+1}\,,
\end{aligned}
$$

where $Q_{i\mu} = I - \mathcal{P}_{i\mu}$ and

$$
\mathcal{P}_{i\mu} = \alpha^{-1}F_{i\mu}(\mathrm{tr}_\Gamma\, T_{-i\mu}\alpha^{-1}F_{i\mu})^{-1}T_{-i\mu}
$$

is a *modified Bergman projection* onto the subspace

$$
\alpha^{-1}\ker(D+i\mu)\cap L_2(G).
$$

In order to get an explicit formula for u_{k+1} we have to study the operator

$$
L_\mu := i\mu T_{-i\mu}Q_{i\mu}\alpha^{-1}T_{i\mu}(D\alpha).
$$

Our goal is to show that for $\tau \to 0$ the norm $\|L_\mu\|_{L[L_2]} \to 0$.

The operator $Q_{i\mu}$ is a projection onto the space

$$
(D-i\mu)\,\overset{\circ}{W}{}^1_2(G)\cap L_2(G).
$$

The L_2-norm $\|Q_{i\mu}\|_{L(L_2)}$ is 1. This is not completely obvious, because we have here to take the complex-quaternionic Hilbert space $L_2 := L_{2,\mathbb{C}IH}$. The scalar product in this Hilbert space is given by

$$
\ll u, v \gg := \int_G \overline{u(x)}^{\,\mathbb{C}IH} v(x)\, dx = \langle u^1, v^1\rangle + \langle u^2, v^2\rangle + \left[\langle u^1, v^2\rangle - \langle u^2, v^1\rangle\right]
$$

and $\langle\cdot,\cdot\rangle$ takes into account the scalar product in the real quaternionic valued Hilbert space. Notice that

$$
|u|^2_{L_2,\mathbb{C}IH} = \mathrm{Re}_{IH} \ll u, u \gg = \mathrm{Re}_{\mathbb{C}} \ll u, u \gg = \mathrm{Re}_{\mathbb{C}IH} \ll u, u \gg.
$$

Now we get

$$
\|L_\mu u\|_{L_2} \le \tau^{-1/2}\|\alpha^{-1}\|_C\|D\alpha\|_C\|T_{-i\mu}\|_{L(L_2)}\|T_{i\mu}\|_{L(L_2)}\|u\|_{L_2}.
$$

Using our result $\|T_{\pm i\mu}\|_{L(L_2)} \le C\sqrt{\tau}$, which is published in [1], we obtain the estimate

$$\|L_\mu u\|_{L_2} \le C(\alpha)\sqrt{\tau}\|u\|_{L_2}$$

and see that for $\tau \to 0$, also $L_\mu u \to 0$. Therefore, for sufficiently small τ the operator $I - L_\mu$ is invertible in $L(L_2)$. Setting now $R_\mu = (I - L_\mu)^{-1}$,

$$R_\mu^{-1} u_{k+1} = T_{-i\mu} Q_{i\mu} \alpha^{-1} T_{i\mu}(\alpha f_k + \mu^2 \alpha u_k)$$
$$+ T_{-i\mu} \alpha^{-1} F_{i\mu}(\operatorname{tr}_\Gamma T_{-i\mu} \alpha^{-1} F_{i\mu})^{-1} Q_{\Gamma,-i\mu} g_{k+1} + F_{-i\mu} g_{k+1}$$

with $k = 0, \ldots, N-1$.

As $Q_{i\mu} w = (D - i\mu)\tilde{w}$ with $\tilde{w} \in \overset{\circ}{W}{}^1_2(G)$ we obtain

$$\operatorname{tr}_\Gamma T_{-i\mu} Q_{i\mu} w = \operatorname{tr}_\Gamma(\tilde{w} - F_{-i\mu}\tilde{w}) = 0.$$

In this way we have

$$\operatorname{tr}_\Gamma R_\mu^{-1} u_{k+1}$$
$$= \operatorname{tr}_\Gamma u_{k+1}$$
$$= (\operatorname{tr}_\Gamma T_{-i\mu} \alpha^{-1} F_{i\mu})(\operatorname{tr}_\Gamma T_{-i\mu} \alpha^{-1} F_{i\mu})^{-1} Q_{\Gamma,-i\mu} g_{k+1} + P_{\Gamma,-i\mu} g_{k+1}$$
$$= Q_{\Gamma,-i\mu} g_{k+1} + P_{\Gamma,-i\mu} g_{k+1} = g_{k+1}.$$

$P_{\Gamma,-i\mu}, Q_{\Gamma,-i\mu}$ are the corresponding Plemelj projections, which are associated to the Cauchy–Bizadse operator $F_{-i\mu}$. We can show that in each time layer the boundary condition is fulfilled.

Let us substitute the modified Bergman projection

$$\alpha^{-1} F_{i\mu}(\operatorname{tr}_\Gamma T_{-i\mu} \alpha^{-1} F_{i\mu})^{-1} T_{-i\mu} =: \mathcal{P}_{i\mu}.$$

Using Borel–Pompeiu's formula we come to the representation

$$R_\mu^{-1} u_{k+1} = T_{-i\mu} Q_{i\mu} \alpha^{-1} T_{i\mu}(\alpha f_k + \mu^2 \alpha u_k)$$
$$+ T_{-i\mu} \mathcal{P}_{i\mu} D_{-i\mu} H_{k+1} + F_{-i\mu} g_{k+1}.$$

Again H_{k+1} denotes a smooth (non-unique) extension of $g_{k+1} = g(x, k\tau)$ to G. Furthermore, let

$$\mathcal{H}_k := T_{-i\mu} \mathcal{P}_{i\mu} D_{-i\mu} H_{k+1} + F_{-i\mu} g_{k+1}.$$

This leads to

$$R_\mu^{-1} u_{k+1} = \left[T_{-i\mu} Q_{i\mu} \alpha^{-1} T_{i\mu} \alpha f_k + \mathcal{H}_k \right] + \mu^2 T_{-i\mu} Q_{i\mu} \alpha^{-1} T_{i\mu} \alpha u_k.$$

We obtain successively with $T_{-i\mu} Q_{i\mu} \alpha^{-1} T_{i\mu} =: S_{i\mu}$ and $u_0 := h$,

$$R_\mu^{-1} u_{k+1} = S_{i\mu} \alpha f_k + \mathcal{H}_k + \mu^2 S_{i\mu} R_\mu S_{i\mu} \alpha f_{k-1} + \mu^2 S_{i\mu} R_\mu \mathcal{H}_{k-1}$$
$$+ \mu^4 S_{i\mu} R_\mu S_{i\mu} R_\mu R_\mu^{-1} \alpha u_{k-1}$$
$$= S_{i\mu} \alpha f_k + \mu^2 S_{i\mu} R_\mu S_{i\mu} \alpha f_{k-1} + \mathcal{H}_k + \mu^2 S_{i\mu} R_\mu \mathcal{H}_{k-1}$$
$$+ (\mu^2 S_{i\mu} R_\mu)^2 R_\mu^{-1} \alpha u_{k-1},$$

and finally get the following representation

Theorem 2. *We have*

$$R_\mu^{-1} u_{k+1} = (\mu^2 S_{i\mu} R_\mu)^{k+1} R_\mu^{-1} h + \sum_{l=0}^{k} (\mu^2 S_{i\mu} R_\mu)^l (\mathcal{H}_{k-l} + S_{i\mu} \alpha f_{k-l}) . \quad (4.1)$$

9.5 Approximation property

Next we consider the approximation property of the finite difference operator (cf. [15]). We abbreviate:

$$Au(x,t) := \partial_t u(x,t) - \alpha^{-1} D \, \alpha \, D \, u(x,t)$$

and

$$A_\tau u(x, t+\tau) := 1/\tau [u(x, t+\tau) - u(x,t)] + \alpha^{-1} D \, \alpha \, D \, u(x, t+\tau) .$$

As usual we write $u_k := u(x, k\tau)$ and set $t = k\tau$; then it follows that

$$|A_\tau u_k - A u_k| \leq 1/\tau \left| (u_{k+1} - u_k - \tau \partial_t u_k) + \alpha^{-1} D \, \alpha \, D(u_{k+1} - u_k) \right| .$$

Hence

$$\begin{aligned}
|A_\tau u_k - A u_k| &\leq \left| \tfrac{1}{2}\tau (\partial_{tt} u)(x, t+\theta\tau) + (f_{k+1} - f_k) - \partial_t (u_{k+1} - u_k) \right| \\
&= \left| \tfrac{1}{2}\tau (\partial_{tt} u)(x, t+\theta\tau) + (f_{k+1} - f_k) - \tau \partial_{tt} u(x, t+\theta'\tau) \right| \\
&= \tau \left| \tfrac{1}{2}(\partial_{tt} u)(x, t+\theta\tau) - (\partial_{tt} u)(x, t+\theta'\tau) \right. \\
&\qquad \left. + \partial_t f(x, t+\theta''\tau) \right| \leq \tau \, C_k(u,f) .
\end{aligned}$$

If the solution u is sufficiently smooth, then we have estimated the truncation error.

9.6 Stability

Further, we need a stability estimation for our method. For this reason we consider:

$$\begin{aligned}
u_{k+1} = R_\mu T_{-i\mu} Q_{i\mu} \alpha^{-1} T_{i\mu} (\alpha f_k + \mu^2 \alpha u_k) \\
+ R_\mu T_{-i\mu} \mathcal{P}_{i\mu} D_{-i\mu} H_{k+1} + R_\mu F_{-i\mu} g_{k+1}
\end{aligned}$$

for $k = 0, \dots, N-1$.

Note that $\mathcal{H}_{k+1} := T_{-i\mu} \mathcal{P}_{i\mu} D_{+i\mu} H_{k+1} + F_{-i\mu} g_{k+1}$ solves the boundary value problem

$$(D + i\mu)\alpha(D - i\mu)v_k = 0,$$

$$v_k = g_{k+1}.$$

Indeed, we have

$$(D + i\mu)\alpha(D - i\mu)(T_{-i\mu}\mathcal{P}_{i\mu}D_{i\mu}H_{k+1} + F_{-i\mu}g_{k+1})$$
$$= (D + i\mu)F_{i\mu}(\mathrm{tr}_\Gamma \, T_{-i\mu}\alpha^{-1}F_{i\mu})^{-1}T_{-i\mu}D_{i\mu}H_{k+1} = 0 \, .$$

Using the Plemelj–Sokhotzki formulae we get

$$\mathrm{tr}_\Gamma \, T_{-i\mu}(I - Q_{i\mu}D_{-i\mu}H_{k+1} + P_{\Gamma,-i\mu}g_{k+1}$$
$$= \mathrm{tr}_\Gamma \, T_{-i\mu}D_{-i\mu}H_{k+1} - \mathrm{tr}_\Gamma \, T_{-i\mu}Q_{i\mu}D_{-i\mu}H_{k+1} + P_{\Gamma,-i\mu}g_{k+1}$$
$$= Q_{\Gamma,-i\mu}g_{k+1} + P_{\Gamma,-i\mu}g_{k+1} - \mathrm{tr}_\Gamma \, T_{-i\mu}Q_{i\mu}D_{-i\mu}H_{k+1} = g_{k+1} \, .$$

The last identity follows from the proposition that

$$\mathrm{tr}_\Gamma \, T_{-i\mu}w = 0 \quad \text{if and only if} \quad w \in \mathrm{im} \, Q_{i\mu} \, .$$

We know that the operator $T_{\pm i\mu}$ is well defined and for instance continuous from $L_p(G)$ to $W_p^1(G)(p > 1)$. This result can be found in [10]. So the operator $T_{-i\mu}Q_{i\mu}\alpha^{-1}T_{i\mu}\alpha$ is smoothing in the Sobolev scale. Moreover, the following estimate is valid:

$$\|R_\mu T_{-i\mu}Q_{i\mu}\alpha^{-1}T_{i\mu}\mu^2 u\|_{L_2}$$
$$\leq \|R_\mu\|_{L(L_2)}\|T_{i\mu}\|_{L(L_2)}\|T_{-i\mu}\|_{L(L_2)}\|\alpha\|_C\|\alpha^{-1}\|_C\|u\|_{L_2}$$
$$\leq \text{const.} \, (1 - \|L_\mu\|_{L(L_2)})^{-1}\|\alpha\|_C\|\alpha^{-1}\|_C\tau \, 1/\tau\|u\|_{L_2}$$
$$\leq \text{const.} \, \|\alpha\|_C\|\alpha^{-1}\|_C\|u\|_{L_2} \, .$$

We have proved that the operator $R_\mu T_{-i\mu}Q_{i\mu}\alpha^{-1}T_{i\mu}\alpha$ is uniformly bounded independently of the mesh width τ. By the way, it holds more exactly that

$$\|T_{\pm i\mu}\|_{L(L_2)} \leq n\sigma_n \frac{2^{\frac{1}{2}n-2}}{2\pi} \left(\Gamma(\tfrac{1}{2}n) + \frac{\Gamma(\tfrac{1}{2}n + 1)\Gamma(\tfrac{1}{2}n + 2)}{\pi^{\frac{1}{2}}\Gamma(\tfrac{1}{2}n + \tfrac{3}{2})} \right) \sqrt{\tau} \, .$$

(cf. [1]).

9.7   Generalizations

Assume now that the right-hand side depends on the unknown function u too, i.e., $f = f(x, t, u)$. Further, let f satisfy the Lipschitz condition:

$$|f(x, t, u_1) - f(x, t, u_2)| \leq L(x, t)|u_1 - u_2| \, .$$

Our problem (2.1)–(2.3) transforms after replacing the time-derivative by the forward finite difference and setting

$$f_k = f_k(x, u_k) := \frac{1}{\tau} \int\limits_{k\tau}^{(k+1)\tau} f(x, t, u_k) \, dt$$

with $\mu = \tau^{\frac{1}{2}}$ into a Yukawa-type equation

$$(\alpha^{-1} D \alpha D + \mu^2) u_{k+1} = f_k(u_k) + \mu^2 u_k, \quad (k = 0, 1, \ldots, N-1).$$

We obtain in a similar way

$$R_\mu^{-1} u_{k+1} = T_{-i\mu} Q_{i\mu} \alpha^{-1} T_{i\mu} \alpha f_k(u_k) + \mathcal{H}_{k+1} + \mu^2 T_{-i\mu} Q_{i\mu} \alpha^{-1} T_{i\mu} \alpha u_k.$$

The approximation property is satisfied. Indeed, we have

$$|f_{k+1} - f_k| \le L |u_{k+1} - u_k| \le \tau L |\partial_t u(x, t + \theta\tau)|.$$

In order to ensure the stability it is needed to assume $f(0) = 0$. Then we get from the Lipschitz condition that $|f(u)| \le L|u|$ and so

$$\| R_\mu T_{-i\mu} Q_{i\mu} \alpha^{-1} T_{i\mu} \alpha \left[f_k(u_k) + \mu^2 u_k \right] \|_{L_2}$$
$$\le \text{const.} \|\alpha\|_C \|\alpha^{-1}\| (L + \mu^2) \| u_k \|_{L_2},$$

which means the boundedness of the non-linear operator

$$R_\mu T_{-i\mu} Q_{i\mu} \alpha^{-1} T_{i\mu} \left[f_k(\cdot) + \mu^2 \right].$$

Theorem 3. *The iteration procedure*

$$u_{k+1}^{(j+1)} = R_\mu \mathcal{H}_{k+1} + R_\mu T_{-i\mu} Q_{i\mu} \alpha^{-1} T_{i\mu} \alpha f_{k+1}(u_{k+1}^{(j)}), \quad (j = 0, 1 \ldots.),$$
$$u_{k+1}^{(0)} = R_\mu \mathcal{H}_{k+1}$$

provides a unique solution of our semi-discretized problem

$$\alpha^{-1}(D - i\mu)\alpha(D + i\mu) u_{k+1} = f_{k+1} - i\mu \alpha^{-1}(D\alpha) u_{k+1} + \mu^2 u_k \quad in \quad G,$$
$$u_{k+1} = g_{k+1} \quad on \quad \Gamma,$$
$$u_0 = h.$$

The sequence $\left(u_{k+1}^{(j)} \right)$ converges in $W_2^1(G)$ in a neighborhood $B \subset W_2^1(G)$ of $R_\mu \mathcal{H}_{k+1}$ for a sufficiently small τ.

Proof. The proof of convergence is a consequence of the estimation

$$\| T_{\pm i\mu} \|_{L_2} \le C \sqrt{\tau}$$

and above outlined calculations. $\qquad\square$

REFERENCES

[1] Bahmann H., Gürlebeck K., Shapiro M. and Sprössig W., On a modified Teodorescu transform, *Integral Transforms and Special Functions*, Vol. **12**, Number 3 (2001), 213–226.

[2] Bernstein S., Operator calculus for elliptic boundary value problems in unbounded domains, *Zeitschrift f. Analysis u. Anwend.* **10**, 4 (1991), 447–460.

[3] Bernstein S., Fundamental solutions of Dirac type operators, Banach Center Publications, *Banach Center Symposium: Generalizations of Complex Analysis, May 30– July 1, 1994, Warsaw*, **37** (1996), pp. 159–172.

[4] Delanghe R., Sommen F. and Souček V., Residues in Clifford analysis. In: Begehr H. and Jeffrey A. (eds.), *Partial Differential Equations with Complex Analysis*, Pitman Res. Notes in Math. Ser. **262** (1992), pp. 61–92.

[5] Gürlebeck K., Kähler U., Ryan J. and Sprössig W., Clifford analysis over unbounded domains, *Advances in Applied Mathematics* **19** (1997), 216–239.

[6] Gürlebeck K., Hypercomplex factorization of the Helmholtz equation, *Z. Anal. Anwend.* **5** (1986), 125–131.

[7] Gürlebeck K. and Sprössig W., *Quaternionic Analysis and Boundary Value Problems*, Birkhäuser Verlag, Basel, 1990.

[8] Gürlebeck K. and Sprössig W., *Quaternionic and Clifford Calculus for Physicists and Engineers*, John Wiley, Chichester, 1997.

[9] Gürlebeck K. and Sprössig W., Representation theory for classes of initial value problems with quaternionic analysis, *Math. Meth. Appl. Sci.* **25** (2002), 1371–1382.

[10] Gürlebeck K., Shapiro M. and Sprössig W., On a Teodorescu transform for a class of metaharmonic functions, *J. Natural Geometry* Vol. **21** (2002), 17–38.

[11] Kaiser, G., Complex-distance potential theory and hyperbolic equations. In: Ryan J., Sprössig W. *Clifford Analysis*, Progress in Physics, Birkhäuser, Boston, Basel, 2000.

[12] Karch G., Asymptotic behaviour of solutions to some pseudoparabolic equations, *Math. Meth. Appl. Sci.* **20** (1997), 271–289.

[13] Kravchenko V. and Shapiro M., *Integral Representations for Spatial Models of Mathematical Physics*. Pitman Research Notes in Math. Series **351**, 1996.

[14] Ryan J., Complexified Clifford analysis. *Complex Variables Theory and Appl.* **1** (1982), 119–149.

[15] Samarskij A.A., *The Theory of Difference Methods*, Moscow, Nauka, 1997 (in Russian).

[16] Showalter R.E., Partial differential equations of Sobolev–Galpern type, *Pac. J. Math.* **31** (1969), 787–793.

[17] Sprössig S., On decomposition of the Clifford valued Hilbert space and their applications to boundary value problems, *Advances in Applied Clifford Algebras* Vol. **5**, No. 2 (1995), 167–185.

Wolfgang Sprössig
Freiberg University of Mining and Technology
Agricolastr. 1
09596 Freiberg (Sachsen) Germany
E-mail: sproessig@math.tu-freiberg.de

Submitted: June 2002.

PART II. GEOMETRY

10

A Nahm Transform for Instantons over ALE Spaces

Claudio Bartocci and Marcos Jardim

ABSTRACT We define a Nahm transform for instantons over hyperkähler ALE 4-manifolds, and explore some of its basic properties.

Keywords: Nahm transform, instantons, ALE spaces

10.1   Introduction

Since the complete classification of instantons on $\mathbb{R}^4$ was achieved by Atiyah, Drinfeld, Hitchin and Manin in the late 1970s, mathematicians and physicists have turned their attention to G-invariant instantons on $\mathbb{R}^4$, where G is typically a subgroup of orientation preserving isometries.

The case of G being a subgroup of translations is particularly interesting for physics. For example, monopoles and calorons can be obtained in this way. The main tool used to obtain existence results and study the moduli spaces of translation invariant instantons on $\mathbb{R}^4$ is the *Nahm transform*. Roughly speaking, let Λ be a subgroup of translations of $\mathbb{R}^4$, and define the *dual* subgroup:

$$\Lambda^* = \{\xi \in (\mathbb{R}^4)^* \mid \xi(z) \in \mathbb{Z}, \ \forall z \in \Lambda\}.$$

Then the Nahm transform is a 1-1 correspondence between Λ-invariant instantons on $\mathbb{R}^4$ and Λ^*-invariant instantons on $(\mathbb{R}^4)^*$. A more detailed exposition can be found in [15].

The case when G is a finite subgroup of $\mathrm{SU}(2)$ is also remarkable. Kronheimer and Nakajima described the existence of instantons on minimal resolutions of $\mathbb{R}^4/G$ [12] using a clever generalization of the ADHM construction. The moduli spaces of instantons over these so-called *ALE spaces* were then studied by Nakajima [13] and eventually lead to profound new results in the representation theory of affine Lie algebras [14].

In this research announcement, we will describe a generalization of the Nahm transform for instantons on minimal resolutions of $\mathbb{R}^4/G$, where G is a finite subgroup of $\mathrm{SU}(2)$. It is important to note that our construction implies that Nahm

---

AMS Subject Classification: 53C07, 65R10.

transforms are not only defined for *flat* Riemannian manifolds. Indeed, the constructions of [2, 3] and the one below indicates that Nahm transforms can be defined on *hyperkähler* manifolds.

To conclude this Introduction, let us briefly outline the contents of this paper. We start by reviewing the definition of hyperkähler ALE 4-manifolds and some basic properties of their moduli spaces of instantons. The Nahm transform is then defined in Section 10.3. Under certain circumstances, one can expect this transform to be invertible, and we announce an invertibility result in Section 10.4. Finally, Section 10.5 is dedicated to some differential geometric aspects of the transform.

10.2   Moduli spaces of instantons over ALE spaces

Let Γ be a finite subgroup of $\mathrm{SU}(2)$. The minimal resolution X of the quotient $\mathbb{C}^2/\Gamma$ has trivial canonical bundle, so it is a 2-dimensional complex manifold with a holomorphic symplectic form. It can be proved that X is a hyperkähler 4-manifold, that is it carries three almost complex structures (I, J, K) and a metric g, such that (I, J, K) are parallel with respect to the Levi-Civita connection and satisfy quaternionic relations $IJ = -JI = K$. Furthermore, the metric g is *asymptotically locally Euclidean* (ALE) in the following sense. Some open neighborhood V of infinity in X has a finite covering $\tilde{V}$ diffeomorphic to $\mathbb{R}^4 \backslash \overline{E(0, R)}$, for some $R > 0$, and in the induced coordinates x_i the metric g is required to satisfy the relation

$$g_{ij}(x) = \delta_{ij} + a_{ij}$$

with $|\partial^p a_{ij}(x)| = O(|x|^{(-4-p)})$, $p \geq 0$. Notice that these coordinates induce a diffeomorphism $V \simeq (R, \infty) \times S^3/\Gamma$. A large class of such manifolds, known in the physical literature under the name of gravitational (multi-) instantons, was first discovered by Gibbons and Hawking [8]; in [10, 11] Kronheimer proved that every complete hyperkähler ALE 4-manifold is diffeomorphic to a minimal resolution of $\mathbb{C}^2/\Gamma$.

Every ALE space (that is, complete hyperkähler ALE 4-manifold) admits a topological one-point compactification $\overline{X} = X \cup \{\infty\}$, which carries a natural orbifold structure. The hyperkähler metric g on X is conformally anti-self-dual; since this condition is conformally invariant, $\overline{X}$ is endowed with a conformally anti-self-dual orbifold metric $\bar{g}$. Consider the subset $\tilde{U} = \tilde{V} \cup \{\infty\}$; the singularity of $\overline{X}$ at ∞ is a finite quotient singularity modeled on $\tilde{U}/\Gamma = U = V \cup \{\infty\}$, where Γ is identified with a subgroup of $\mathrm{SO}(4)$.

The moduli spaces of instantons over ALE spaces have been studied in detail by Nakajima [13], and by Kronheimer and Nakajima [12]. Let $E \rightarrow X$ be a complex (smooth) vector bundle of rank n and trivial determinant (i.e., an $\mathrm{SU}(n)$ vector bundle). In order to define a suitable notion of connections that are "framed at infinity", we fix a group homomorphism $\rho : \Gamma \rightarrow \mathrm{SU}(n)$; this homomorphism will be identified with a flat $\mathrm{SU}(n)$ connection over S^3/Γ. By taking coordinates

x_i on $V \subset X$ as before, we can extend the function $r(p) = |x(p)|$ to a positive r function on all of X. Given a connection A_0, a weighted Sobolev norm $\| \cdot \|_{l,2,\delta}$ on the space of k-forms $\Omega^k(E)$ is defined as follows:

$$\|\alpha\|_{l,2,\delta} = \sum_{j=0}^{l} \|r^{j-(\delta+2)} \nabla_{A_0}^{(j)} \alpha\|_{L^2} \qquad (2.1)$$

for an integer $l \geq 0$ and $\delta \in \mathbb{R}$. We denote the completion of the space $\Omega^k(E)$ in this norm by $W_\delta^{l,2}(E \otimes \Lambda^k T^* X)$.

Now fix $l > 2$. We say that a connection A on E is *asymptotic* to ρ in $W_{-2}^{l,2}$ if there is a gauge such that $A = A_0 + \alpha$, where the restriction of (E, A_0) to

$$\{t\} \times S^3/\Gamma \subset (R, \infty) \times S^3/\Gamma \simeq V$$

is the flat bundle with connection ρ, for all $t > R$ and $\|\alpha\|_{l,2,-2} < \infty$. We denote by $\mathcal{A}_X^l(\rho)$ the space of such connections.

The space of anti-self-dual connections on E asymptotic to ρ and having topological charge k is described as follows:

$$\mathcal{A}_{X,asd}^l(E, \rho, k) =$$

$$\left\{ A \in \mathcal{A}_X^l(\rho) \mid A \text{ is anti-self-dual and } \frac{1}{8\pi^2} \int_X \|F_A\|^2 = k \right\}.$$

The corresponding moduli space is given by the quotient

$$\mathcal{M}_X(E, \rho, k) = \mathcal{A}_{X,asd}^l(E, \rho, k)/\mathcal{G}_0^{l+1}$$

where $\mathcal{G}_0^{l+1}$ is the gauge group of automorphisms of E converging to the identity, that is:

$$\mathcal{G}_0^{l+1} = \left\{ s \in W_{-1}^{l+1,2}(\text{End}(E)) \mid \|s - \mathbf{1}_E\|_{l+1,2,-1} < \infty \right\}.$$

We notice that, by virtue of Uhlenbeck's removable singularities theorem, *anti-self-dual connections A over X with finite action can be identified with anti-self-dual connections over $\overline{X}$.*

The following two results by Nakajima [13] are fundamental for this paper:

Theorem 1. *Each non-empty, non-compact 4-dimensional component of the moduli space $\mathcal{M}_X(E, \rho, k)$ is a complete hyperkähler ALE space.*

In other words, every such a component of the moduli space $\mathcal{M}_X(E, \rho, k)$ is diffeomorphic to a minimal resolution of $\mathbb{C}^2/\hat{\Gamma}$ for some discrete subgroup $\hat{\Gamma} \subset \text{SU}(2)$, which might be, in general, distinct from Γ.

Theorem 2. *Let E be a rank 2 complex vector bundle over an ALE space X. Let $\rho : \Gamma \hookrightarrow \text{SU}(2)$ be the inclusion map, with $|\Gamma|$ denoting the order of Γ. Then $\hat{X} = \mathcal{M}_X(E, \rho, \frac{|\Gamma|-1}{|\Gamma|})$ is diffeomorphic to X. Furthermore, if Γ is cyclic, then $\hat{X}$ is isomorphic to X as a hyperkähler manifold.*

In fact, the second part of Theorem 2 is also true when Γ is not cyclic, as it was pointed out in [12, page 302, example 3].

10.3   The transform

Let Y be a non-compact 4-dimensional component of the moduli space $\mathcal{M}_X(E, \rho, k)$ described in the previous section. According to Theorem 1, Y is also an ALE space. We will define a transform mapping instantons over X into instantons over Y, along the same guidelines as in the case of the Nahm transform for instantons on 4-tori [6, 7] and doubly-periodic instantons [9] or the Fourier–Mukai transform on K3 surfaces [2].

First, notice that any point $\xi \in Y$ may be thought of as an anti-self-dual connection on the bundle $E \to X$. Now let $F \to X$ be an $SU(r)$ vector bundle, and let $\sigma : \Gamma \to SU(r)$ be a group homomorphism. We take an anti-self-dual connection A asymptotic to σ and define the connection A_ξ so that

$$\nabla_{A_\xi} = \nabla_A \otimes 1_E + 1_F \otimes \nabla_\xi$$

on the bundle $F \otimes E$, for any $\xi \in Y$. Notice that A_ξ is also anti-self-dual, since A and ξ are.

We consider the family of Dirac operators $(l \geq 1)$

$$D_{A_\xi}^+ : W_{-1}^{l,2}(F \otimes E \otimes S^+) \to W_{-2}^{l-1,2}(F \otimes E \otimes S^-)$$

and

$$D_{A_\xi}^- : W_{-2}^{l,2}(F \otimes E \otimes S^-) \to W_{-3}^{l-1,2}(F \otimes E \otimes S^+)$$

obtained by coupling the connection A_ξ with the standard Dirac operators defined on the spinor bundles $S^\pm$ over X. It is proved in [12] that the operators $D_{A_\xi}^\pm$ are Fredholm and that $\operatorname{Ker} D_{A_\xi}^+ = 0$, $\forall \xi \in Y$. It is important to note that no further conditions on the connection A are necessary.

Thus, one can define the complex vector bundle $\hat{F} = -\operatorname{Ind}(D_{A_\xi}^+)$. Moreover, we regard $\hat{F}$ as a subbundle of the Hilbert space bundle $\hat{H}$, whose fibers are

$$L^2(F \otimes E \otimes S^-) \hookrightarrow W_{-2}^{l,2}(F \otimes E \otimes S^-).$$

We denote by $\varpi : \hat{H} \to \hat{F}$ the orthonormal projection with respect to the L^2-norm of harmonic spinors, and by $\iota : \hat{F} \to \hat{H}$ the natural inclusion. More precisely,

$$\varpi(\xi) = 1_{\hat{H}} - D_{A_\xi}^+ G_{A_\xi} D_{A_\xi}^- \tag{3.1}$$

where

$$G_{A_\xi} = (D_{A_\xi}^- D_{A_\xi}^+)^{-1} : W_{-3}^{l-1,2}(F \otimes E \otimes S^+) \to W_{-1}^{l+1,2}(F \otimes E \otimes S^+)$$

is the Green's operator.

We define the connection $\hat{A}$ on $\hat{F}$ by projection:

$$\nabla_{\hat{A}} s = \varpi \circ \underline{d} \circ \iota(s) \quad \text{for all sections } s \text{ of } \hat{F}, \tag{3.2}$$

where $\underline{d}$ is the trivial covariant derivative on $\hat{H}$. Clearly, $\hat{A}$ is unitary.

A choice of complex structure on X induces a choice of complex structure on Y. Since the transformed connection has type $(1,1)$ with respect to all such complex structures, we obtain from [3, 7] the following result:

Lemma 3. *The connection $\hat{A}$ is anti-self-dual.*

The pair $(\hat{F}, \hat{A})$ is called the *Nahm transform* of (F, A).

The next step is to show that the connection $\hat{A}$ has finite action. By Uhlenbeck's removable singularities theorem, we can identify the anti-self-dual connections on X with those on $\overline{X}$.

Let $[\xi_i]$ be a sequence in Y. By Uhlenbeck's compactness theorem, $[\xi_i]$ has a subsequence converging to an anti-self-dual connection Ξ on $\overline{X} \setminus \{$finitely many points$\}$. A result of Nakajima ([13], Theorem 5.2) tells us that there are two possibilities:

- The curvature concentration occurs at a point other than $\infty \in \overline{X}$, so a connection on the 4-sphere bubbles off from there. But that is not possible (for the dimension of the moduli space would be reduced by at least eight). So the limiting connection Ξ is defined on the whole X, and it yields a point in Y.

- The curvature concentration happens at ∞, and there exists a sequence $[\xi_i]$ with a subsequence converging to an unique anti-self-dual connection ξ_∞ away from ∞. Thus $\Xi = \xi_\infty$ is defined on the whole X, and it yields the point at infinity of $\overline{Y}$.

Therefore, any point $\xi \in \overline{Y} = Y \cup \{\xi_\infty\}$ can be thought as an anti-self-dual connection on X. In this way, we can construct a family of Dirac operators $D_{A_\xi}^{\pm}$ parameterized by $\xi \in \overline{Y}$. One still has $\operatorname{Ker} D_{A_\xi}^{+} = 0$ [11]. Hence, we get an index bundle $-\operatorname{Ind}(D_{A_\xi}^{+})$ on $\overline{Y}$ and a connection B (defined via the projection (3.1)) on it, whose restriction to Y clearly coincides with the pair $(\hat{F}, \hat{A})$. In other words, $\hat{A}$ can be extended to a connection in $\overline{Y}$; thus by Uhlenbeck's removable singularities theorem, we have proved the following result.

Lemma 4. *The connection $\hat{A}$ has finite action.*

We define:

$$\hat{c} = \frac{1}{8\pi^2} \int_Y \|F_{\hat{A}}\|^2 \tag{3.3}$$

which is the charge of the transformed instanton connection.

Lemma 5. *If A and B are two gauge equivalent connections on a vector bundle $F \to X$, then $\hat{A}$ and $\hat{B}$ are gauge equivalent connections on the transformed bundle $\hat{F} \to Y$.*

Proof. Since A and B are gauge equivalent, there is a bundle automorphism h : $F \to F$ such that $\nabla_B = h^{-1}\nabla_A h$. Take $g = h \otimes 1_E \in \mathrm{Aut}(F \otimes E)$, so that $\nabla_{B_\xi} = g^{-1}\nabla_{A_\xi} g$, hence $D_{B_\xi}^- = g^{-1}D_{A_\xi}^- g$, for all $\xi \in Y$. Thus if $\{\Psi_i\}$ is a basis for $\ker D_{A_\xi}^-$, then $\{\Psi_i' = g^{-1}\Psi\}$ is a basis for $\ker D_{B_\xi}^-$. So g can also be regarded as an automorphism of the transformed bundle $\hat{F}$. It is then easy to see that

$$\nabla_{\hat{B}} = \varpi_B \circ \underline{d} \circ \iota_B = (g^{-1}\varpi_B g) \circ \underline{d} \circ (g^{-1}\iota_B g) = g^{-1}\nabla_{\hat{A}} g$$

since $\underline{d}g^{-1} = 0$, for $g = h \otimes 1_E$ does not depend on ξ. $\square$

Therefore, the Nahm transform gives a well-defined map from moduli spaces of instantons over X to moduli spaces of instantons Y :

$$\mathcal{N} : \mathcal{M}_X(F, \sigma, c) \longrightarrow \mathcal{M}_Y(\hat{F}, \hat{\sigma}, \hat{c}) \tag{3.4}$$

where $\hat{\sigma} : \hat{\Gamma} \to \mathrm{SU}(2)$ is the representation describing the asymptotic behaviour of $\hat{A}$.

Explicit formula for $\hat{A}$.

The Nahm transformed connection $\hat{A}$ was defined above in a rather coordinate-free manner. For many calculations, it is important to have a more explicit description.

So let $\{\Psi_i = \Psi_i(x; \xi)\}_{i=1}^{\hat{r}}$ be linearly independent solutions of the Dirac equation $D_{A_\xi}^- \Psi_i = 0$, where $\hat{r} = \mathrm{rank}(\hat{F}) = L^2\text{-index}(D_{A_\xi}^-)$. We can assume that $\langle \Psi_i, \Psi_j \rangle = \delta_{ij}$, where $\langle \cdot, \cdot \rangle$ denotes the L^2 inner product on $\hat{H}$. Clearly, $\{\Psi_i\}_{i=1}^{\hat{r}}$ forms a local orthonormal frame for $\hat{F}$. In this choice of trivialization, the components of the connection matrix $\hat{A}$ can be written as

$$\hat{A}_{ij} = \langle \Psi_i, \nabla_{\hat{A}} \Psi_j \rangle = \langle \Psi_i, \underline{d}\Psi_j \rangle . \tag{3.5}$$

In this trivialization, using (3.1) and (3.2), the curvature can be expressed as follows:

$$(F_{\hat{A}})_{ij} = \langle \Psi_i, \nabla_{\hat{A}}\nabla_{\hat{A}}\Psi_j \rangle = \langle \Psi_i, \underline{d}\varpi\underline{d}\Psi_j \rangle$$
$$= \langle \Psi_i, \underline{d}D_{A_\xi}^+ G_{A_\xi} D_{A_\xi}^- \underline{d}\Psi_j \rangle = -\langle D_{A_\xi}^- \underline{d}\Psi_i, G_{A_\xi} D_{A_\xi}^- \underline{d}\Psi_j \rangle.$$

We define $\Omega = [D_{A_\xi}^-, \underline{d}]$. Note that is an algebraic operator acting as:

$$\Omega : \Gamma(X \times Y, \pi_1^*(F \otimes E \otimes S^-)) \longrightarrow \Gamma(X \times Y, \pi_1^*(F \otimes E \otimes S^+) \otimes \pi_2^*(T^*Y))$$

where π_1 and π_2 are the projection of $X \times Y$ onto the first and second factors. Clearly, if $\Psi \in \ker D_{A_\xi}^-$, then $\Omega(\Psi) = D_{A_\xi}^- \underline{d}\Psi$. We then have

$$(F_{\hat{A}})_{ij} = -\langle \Omega \bullet \Psi_i, G_{A_\xi} (\Omega \bullet \Psi_j) \rangle$$

where • denotes the Clifford multiplication.

To conclude this section, we would like to emphasize that *the Nahm transformed instanton connection has no singularities!* This fact contrasts with what happens with the usual Nahm transform for instantons over $T^n \times \mathbb{R}^{4-n}$, where $n = 1, 2, 3$, the other known examples of a Nahm transform of instantons over non-compact manifolds. We refer to [5] for a detailed exposition of the $n = 2$ case (full description of the two other cases have not been given yet).

Indeed, recall that the Nahm transform associates instantons on $T^n \times \mathbb{R}^{4-n}$ to (possibly singular) solutions of the dimensionally reduced anti-self-duality equations on $\hat{T}^n$, the torus dual to T^n. In this case, the instanton's asymptotic behaviour is translated into the singularities' data.

In the present situation, however, no singularities appear in the transformed connection, and we expect that the asymptotic behaviour of A will be converted into the asymptotic behaviour of $\hat{A}$. More precisely, we conjecture that $\hat{\sigma}$ depends only on σ and ρ.

10.4   Invertibility

All the examples of Nahm transform described in the literature have the important property of being *invertible*. However, in the general context of Theorem 1, it seems unreasonable to hope that such property will still hold. Furthermore, the Nahm transform is *not unique*, in the sense that there exist Nahm maps (3.4) relating instantons over X to instantons over each 4-dimensional component of the moduli space of instantons on X.

More precisely, let X and Y be two ALE spaces. We denote $X \prec Y$ whenever Y is diffeomorphic to a 4-dimensional component of the moduli space of instantons on X.

Definition 1. *The sequence $(X_1, X_2, \ldots, X_p)$ of non-diffeomorphic ALE spaces is said to be an ALE p-cycle if:*

$$X_1 \prec X_2 \prec \cdots \prec X_p \prec X_1.$$

In particular, X and Y are said to be dual ALE spaces if (X, Y) is a 2-cycle.

Theorem 2 guarantees the existence of 1-cycles, since any ALE space has a 4-dimensional component of its moduli space of instantons diffeomorphic to itself. One can apply the construction described in Section 10.3 in both directions, transforming instantons over X into instantons over $\hat{X}$ and vice-versa. The goal of the present section is to show that these transforms are the inverse of one another.

Theorem 6. *With the notation of Theorem 2, and under the same hypotheses, the Nahm map*

$$\mathcal{N} : \mathcal{M}_X(F, \sigma, c) \longrightarrow \mathcal{M}_{\hat{X}}(\hat{F}, \hat{\sigma}, \hat{c})$$

is invertible.

The result above reinforces the analogy between the familiar Nahm transform of translation invariant instantons and the transform defined in this paper. However, the Nahm transform for instantons over ALE spaces is potentially a lot richer. Let $(X_1, X_2, \ldots, X_p)$ be an ALE p-cycle. Then there is a chain of Nahm maps,

$$\mathcal{M}_{X_1} \xrightarrow{\mathcal{N}_1} \mathcal{M}_{X_2} \xrightarrow{\mathcal{N}_2} \cdots \xrightarrow{\mathcal{N}_{p-1}} \mathcal{M}_{X_p} \xrightarrow{\mathcal{N}_p} \mathcal{M}_{X_1},$$

where $\mathcal{M}_{X_j}$ denotes a certain component of the moduli space of instantons over X_j. Generalizing Theorem 6, we conjecture that the composition of Nahm maps $\mathcal{N}_p \circ \mathcal{N}_{p-1} \circ \cdots \circ \mathcal{N}_1$ yields the identity map $\mathcal{M}_{X_1} \to \mathcal{M}_{X_1}$. In particular, if X, Y are dual ALE spaces, then the Nahm maps,

$$\mathcal{M}_{X_1} \xrightarrow{\mathcal{N}_1} \mathcal{M}_{X_2} \quad \text{and} \quad \mathcal{M}_{X_2} \xrightarrow{\mathcal{N}_2} \mathcal{M}_{X_1},$$

are the inverse of one another (for more details, see [4]).

Outline of the proof of Theorem 6. Let us now give a brief outline of the proof of Theorem 6; the full argument will appear elsewhere [4]. Start by choosing a complex structure J on X, so that the anti-self-dual connection A induces a holomorphic structure on F; let us denote this holomorphic bundle by $\mathcal{F}$.

The choice of a complex structure on X induces a complex structure $\hat{J}$ on $\hat{X}$. Moreover, the transformed connection $\hat{A}$ induces a holomorphic structure on the transformed bundle $\hat{F}$, and the associated holomorphic bundle $\hat{\mathcal{F}}$ can then be described as follows:

$$\hat{\mathcal{F}} = R^1 \hat{\pi}_* (\pi^* \mathcal{F} \otimes \Upsilon) \tag{4.1}$$

where $\Upsilon \to X \times \hat{X}$ is the Atiyah–Singer universal bundle [1], with the holomorphic structure induced by its universal connection. The maps π and $\hat{\pi}$ are the natural projections of $X \times \hat{X}$ onto the first and second factors, respectively.

Now we can apply the Nahm transform to $(\hat{F}, \hat{A})$, obtaining a new bundle $\check{F}$ and a new instanton connection $\check{A}$ over X. The associated holomorphic bundle $\check{\mathcal{F}}$ can be characterized as

$$\check{\mathcal{F}} = R^1 \pi_* (\hat{\pi}^* \hat{\mathcal{F}} \otimes \Upsilon^\vee). \tag{4.2}$$

One can show that $\check{\mathcal{F}} \simeq \mathcal{F}$. Since this isomorphism holds for every choice of complex structure on X, we conclude that A and $\check{A}$ must be gauge equivalent. □

Finally, to check that the Nahm map described in the theorem above is not just the identity, one can use index theory to see that $\text{rank}(F) \neq \text{rank}(\hat{F}) = L^2$-$\text{index}(D_A^-)$.

10.5   Geometric properties

So far, we have only discussed the set-theoretical properties of the Nahm map (3.4). We will now study some of its differential geometric properties.

The first step is to compute the derivative of $\mathcal{N}$. Let $[A]$ denote a point in the moduli space $\mathcal{M}_X = \mathcal{M}_X(F, \sigma, c)$. Recall that the tangent space at $[A]$ is given by:

$$T_{[A]}\mathcal{M}_X = \{\alpha \in W^{l,2}_{-2}(F \otimes T^*X) \mid d^+_A \alpha = d^*_A \alpha = 0\}.$$

In other words, the 1-form α can be regarded as an infinitesimal change on the instanton A. The first condition $d^+_A \alpha = 0$ implies that $A(t) = A + t \cdot \alpha$ is anti-self-dual up to first order on the parameter t, while the second condition $d^*_A \alpha = 0$ implies that $A(t)$ is not gauge equivalent to A, up to first order on t.

Now let us understand how the harmonic spinors $\{\Psi_i\}$ change under an infinitesimal change on the instanton connection. Clearly, $A(t)_\xi = A_\xi + t \cdot \alpha$. Let $\Psi_i(t) = \Psi_i + t \cdot \psi_i$, where $\psi \in W^{l,2}_{-2}(F \otimes E \otimes S^-)$. Solving the Dirac equation $D^-_{A(t)_\xi} \Psi_i(t) = 0$ up to first order, we conclude that $D^-_{A_\xi} \psi_i = \alpha \bullet \Psi_i$, where $\bullet$ denotes the Clifford multiplication. Therefore, we have that

$$\psi_i = -D^+_{A_\xi} G_{A_\xi}(\alpha \bullet \Psi_i) \tag{5.1}$$

To see how an infinitesimal change on A affects the transformed connection, we proceed as in (3.5):

$$\widehat{A(t)}_{ij} = \langle \Psi_i(t), \underline{d}\Psi_j(t) \rangle$$
$$= \langle \Psi_i, \underline{d}\Psi_j \rangle + t \cdot (\langle \psi_i, \underline{d}\Psi_j \rangle + \langle \Psi_i, \underline{d}\psi_j \rangle) + t^2 \cdot \langle \psi_i, \underline{d}\psi_j \rangle.$$

Therefore, up to first order, we have that $\widehat{A(t)} = \hat{A} + t \cdot \hat{\alpha}$, where the matrix coefficients of $\hat{\alpha}$ are given by:

$$\hat{\alpha}_{ij} = \langle \psi_i, \underline{d}\Psi_j \rangle + \langle \Psi_i, \underline{d}\psi_j \rangle$$
$$= -\langle D^+_{A_\xi} G_{A_\xi}(\alpha \bullet \Psi_i), \underline{d}\Psi_j \rangle - \langle \Psi_i, \underline{d}D^+_{A_\xi} G_{A_\xi}(\alpha \bullet \Psi_j) \rangle$$
$$= -\langle G_{A_\xi}(\alpha \bullet \Psi_i), \Omega \bullet \Psi_j \rangle + \langle \Omega \bullet \Psi_i, G_{A_\xi}(\alpha \bullet \Psi_j) \rangle \tag{5.2}$$

and recall that $\Omega = [D^-_{A_\xi}, \underline{d}]$. Noting that the Green's operator G_{A_ξ} is self-adjoint, we can rewrite (5.2) as follows:

$$\hat{\alpha}_{ij} = \langle \Psi_i, \Omega^\dagger \bullet G_{A_\xi}(\alpha \bullet \Psi_j) \rangle - \langle \Psi_i, \overline{\alpha} \bullet G_{A_\xi}(\Omega \bullet \Psi_j) \rangle \tag{5.3}$$

where $\Omega^\dagger = [D^+_{A_\xi}, \underline{d}]$ and $\overline{\alpha}$ denotes complex conjugation only of the vector bundle part of α as a section of $E \otimes T^*X$ [1]. Alternatively, in a coordinate-free manner, we have:

$$\hat{\alpha}(s) = \varpi \left(\Omega^\dagger \bullet G_{A_\xi}(\alpha \bullet s) - \overline{\alpha} \bullet G_{A_\xi}(\Omega \bullet s) \right) \tag{5.4}$$

for $s \in \Gamma(\hat{F})$.

---

[1]Locally, α looks like $f(x_i)dx_i$ where x_i are local coordinates on X and $f(x_i)$ assumes values on the fibers of E. Then $\overline{\alpha}$ looks like $\overline{f(x_i)}dx_i$.

Now let M_1 and M_2 be hyperkähler manifolds, with complex structures given by (I_p, J_p, K_p), $p = 1, 2$. Recall that a differentiable map $f : M_1 \to M_2$ is said to be *hypercomplex* if the derivative $D_x f : T_x M_1 \to T_{f(x)} M_2$ commutes with all three complex structures, i.e., $D_x f \circ I_1 = I_2 \circ D_x f$, etc. In other words, f is holomorphic regardless of the choice of complex structures on M_1 and M_2.

Lemma 7. *If $X \prec Y$, then the Nahm maps $\mathcal{N} : \mathcal{M}_X \to \mathcal{M}_Y$ are hypercomplex.*

Proof. We must first check that $\mathcal{N}$ is differentiable, that is $d_{\hat{A}}^+ \hat{\alpha} = 0$ and $d_{\hat{A}}^* \hat{\alpha} = 0$.

Indeed, since $d_A^+ \alpha = 0$, we know that $F_{A(t)}^+ = 0$ up to first order on t. Then also $F_{\widehat{A(t)}}^+ = 0$ up to first order on t, which implies that $d_{\hat{A}}^+ \hat{\alpha} = 0$. The Coulomb gauge condition can be checked directly by applying the method of [6, Proposition 3.1].

Finally, it is also easy to see from (5.3) that the transformed deformation $\hat{\alpha}$ does not depend on a choice of complex structure on X. Thus $D_{[A]} \mathcal{N}$ commutes with all complex structures. $\square$

The Lemma can be strengthened when we know that the Nahm maps $\mathcal{N}$ are invertible, since both $\mathcal{N}$ and $\mathcal{N}^{-1}$ are differentiable. So from Theorem 6, we obtain:

Theorem 8. *Let X be an ALE space and let $\hat{X}$ be as in Theorem 2. Then the Nahm maps*

$$\mathcal{N} : \mathcal{M}_X(F, \sigma, c) \longrightarrow \mathcal{M}_{\hat{X}}(\hat{F}, \hat{\sigma}, \hat{c})$$

are hypercomplex diffeomorphisms.

Moreover, in view of the properties of the usual Nahm transform for translation invariant instantons, we also conjecture that the Nahm maps are hyperkähler isometries whenever the Nahm transform is invertible.

10.6   Conclusion

In this paper, we have investigated the basic properties of a Nahm transform for instantons over ALE spaces. Work is in progress [4] to address the following issues:

- compute the topological invariants (rank and charge) of the transformed instanton;

- show that the Nahm map in Theorem 10.4 is an isometry;

- describe ALE cycles and study the invertibility of the Nahm map under more general hypotheses.

Furthermore, taking into account the previous examples of Nahm transforms available in the literature, it seems clear that a purely algebraic geometric description of the transform here presented can be given along the same lines as the

Fourier–Mukai transform for sheaves on Abelian or K3 surfaces. Furthermore, such description should also yield an equivalence between the derived categories of coherent sheaves on X and Y, whenever $X \prec Y \prec X$. We also hope to address these issues in [4].

Finally, on a more speculative level, we expect an interpretation of the Nahm transform in terms of quiver varieties, possibly leading to interesting new results in representation theory in connection with Nakajima's work [14].

Acknowledgments.

M.J. would like to thank the conference organizers for their financial support. C.B. acknowledges the financial support of the Ministero dell'Università and the University of Genova through the national research project "Geometria dei sistemi integrabili".

REFERENCES

[1] M. Atiyah and I. Singer, Dirac operators coupled to vector potentials. *Proc. Natl. Acad. Sci. USA* **81** (1984), 2597–2600.

[2] C. Bartocci, U. Bruzzo and D. Hernández Ruipérez, A Fourier–Mukai transform for stable bundles on K3 surfaces. *J. Reine. Angew. Math.* **486** (1997), 1–16.

[3] C. Bartocci, U. Bruzzo and D. Hernández Ruipérez, A hyperkähler Fourier transform. *Differential Geom. Appl.* **8** (1998), 239–249.

[4] C. Bartocci and M. Jardim, Nahm-Mukai transform for ALE instantons. In preparation.

[5] O. Biquard and M. Jardim, Asymptotic behaviour and the moduli space of doubly-periodic instantons. *J. Eur. Math. Soc.* **3** (2001), 335–375.

[6] P. Braam and P. van Baal, Nahm's transformations for instantons. *Commun. Math. Phys.* **122** (1989), 267–280.

[7] S.K. Donaldson and P.B. Kronheimer, *The Geometry of Four-Manifolds*. Oxford University Press, New York 1990.

[8] G.W. Gibbons and S. Hawking, Gravitational multi-instantons. *Phys. Lett. B* **78** (1978), 430–432.

[9] M. Jardim, Nahm transform and spectral curves for doubly-periodic instantons. *Commun. Math. Phys.* **225** (2002), 639–668.

[10] P.B. Kronheimer, The construction of ALE spaces as hyperkähler quotients. *J. Diff. Geom.* **29** (1989), 665–683.

[11] P.B. Kronheimer, A Torelli-type theorem for gravitational instantons. *J. Diff. Geom.* **29** (1989), 685–697.

[12] P.B. Kronheimer, and H. Nakajima, Yang-Mills instantons and ALE gravitational instantons. *Math. Ann.* **288** (1990), 263–307.

[13] H. Nakajima, Moduli spaces of anti-self-dual connections on ALE gravitational instantons. *Invent. Math.* **102** (1990), 267–303.

[14] H. Nakajima, Instantons on ALE spaces, quiver varieties, and Kac–Moody algebras. *Duke Math. J.* **76** (1994), 365–416.

[15] H. Nakajima, Monopoles and Nahm's equations. In: *Einstein Metrics and Yang–Mills Connections*, T. Mabuchi and S. Mukai, eds., Lecture Notes in Pure and Appl. Math. **145**, Dekker, New York, 1993, pp. 193–211.

Claudio Bartocci
Università degli Studi di Genova
Dipartimento di Matematica
Via Dodecaneso 35
16146 Genova, ITALY
E-mail: bartocci@dima.unige.it

Marcos Jardim
University of Massachusetts at Amherst
Department of Mathematics and Statistics
Amherst, MA 01003-9305, USA
E-mail: jardim@math.umass.edu

Submitted: August 1, 2002; Revised: June 23, 2003.

11

Hyper-Hermitian Manifolds and Connections with Skew-Symmetric Torsion

Gueo Grantcharov

ABSTRACT The aim of the present paper is to review general results and some constructions of the hyper-Kähler geometry with torsion. This is the geometry of a special type of hyper-Hermitian metrics on a hypercomplex manifold related to some questions in theoretical physics. In particular, we show that there is a local existence of such metrics based on an HKT-potential theory, a moment map and reduction theory, as well as a global non-existence property.

Keywords: hypercomplex manifolds, HKT structures

11.1   Introduction

It is known that the internal space for $N = 2$ supersymmetric one-dimensional sigma model is a Kähler manifold [36] and the internal space for $N = 4$ supersymmetric one-dimensional sigma model is a hyper-Kähler manifold [4, 18]. This means that there exists a torsion-free connection with holonomy in $\mathrm{U}(n)$ or $\mathrm{Sp}(n)$, respectively, on the internal space.

It has also been known for a fairly long time that when the Wess–Zumino term is present in the sigma model, the internal space has linear connections with holonomy in $\mathrm{U}(n)$ or $\mathrm{Sp}(n)$ depending on the number of supersymmetries. However, the connection has torsion and the torsion tensor is totally skew-symmetric [12, 15, 20]. The geometry of a connection with totally skew-symmetric torsion and holonomy in $\mathrm{U}(n)$ is referred to as KT-geometry by physicists. When the holonomy is in $\mathrm{Sp}(n)$, the geometry is referred to HKT-geometry.

If one ignores the metric and the connection of an HKT-geometry, the remaining object on the manifold is a hypercomplex structure. The subject of hypercomplex manifolds has been studied by many people since the publication of [2] and [33]. A considerable amount of information is known. It has a twistor cor-

AMS Subject Classification: 53C15, 53C25.
Partially supported by NSF Grant DMS-0209306 and EDGE, Research Training Network HPRN-CT-2000-00101.

respondence [29, 33]. There are homogeneous [23] as well as inhomogeneous examples [3, 30]. There is a reduction construction modeled on symplectic reduction and hyper-Kähler reduction [22]. However, all these works focus on the hypercomplex structure and the associated Obata connection which is a torsion-free connection preserving the hypercomplex structure.

On the other hand, Hermitian connections on almost Hermitian manifolds are studied rather thoroughly by Gauduchon [14]. He considered a subset of Hermitian connections determined by the form of their torsion tensor, called canonical connections.

Guided by the physicists' work and based on the results on hypercomplex manifolds in [16], we reviewed and further developed the theory of HKT-geometry. After [16] many papers on HKT-geometry appeared in the mathematics literature e.g., [6–8, 10, 31, 32]. We also note that many of the results in [16] were re-interpretations of previous results by the physicists, especially those in [19, 20], and [28, 34]. The aim of the present paper is to review the general results and constructions in HKT geometry at the moment.

In Section 11.2, we review the basic definition of HKT-geometry along the line of classical Hermitian geometry developed by Gauduchon [14]. Based on Joyce's construction of homogeneous hypercomplex manifolds [23], we review the construction of an invariant HKT-geometry on some compact semi-simple Lie groups [28].

In Section 11.2.4, we find that a hyper-Hermitian manifold admits HKT-connection if and only if for each complex structure, there is a holomorphic $(0, 2)$-form. This characterization easily implies that some hyper-Hermitian manifolds are HKT: for example this is always true in the four-dimensional case.

In Section 11.3 we study the potential theory for HKT-geometry which is based on the results in Section 11.2.4. Similarly to Kähler geometry, the potentials allow one to prove that any hypercomplex manifold locally admits an HKT metric. In particular, we show that hyper-Kähler potentials generate many HKT-potentials. The results in this section and Section 11.2.4 allow us to construct a large family of inhomogeneous HKT-structures on compact manifolds including $S^1 \times S^{4n+3}$. A novelty in the present paper is the observation that these spaces do not admit any strong HKT metric.

In Sections 11.4 and 11.5 we study the reduction theory for HKT structures. It is based on the reduction of symplectic manifolds. The symplectic or Marsden–Weinstein reduction has been extended to hyper-Kähler manifolds in [18]. More generally it has been shown that if M is a hypercomplex manifold admitting a tri-holomorphic group action, then the reduced space $M//G$ is also hypercomplex [22]. The details of this construction will be summarized in Section 11.4.2. Here it is worth mentioning that in the context of hypercomplex reductions, moment maps do not arise naturally because in the generic case there are no symplectic forms which are preserved by the group action. Instead it is assumed that one can find such functions on M which have the required properties.

In the next section, we assume the existence of a G-moment map on M and study the geometry on the reduced space N. We show that the reduction of a KT-space is a KT-space and the reduction of an HKT-space is again an HKT-space. The result on KT-space in Section 11.4.1 is not surprising because a Hermitian structure can easily be found on a reduced space and every Hermitian structure has a unique KT-connection. The existence of an HKT-connection on the reduction of an HKT-space is less trivial.

In Section 11.5, we provide a construction of a moment map on strong KT-manifolds and strong HKT-manifolds modulo some restrictions. It again resembles that of a symplectic moment map. In the absence of symplectic forms however, this is a non-trivial result as one usually generates moment maps through the Kähler form. In Section 11.6 we discuss when a potential function on HKT-space may descend to a potential function on the reduced HKT-space.

The local existence of HKT structure on hypercomplex manifolds as well as the various examples and constructions naturally lead to the following question: *Does any hypercomplex manifold admit an HKT-metric?* In the last section we provide an example of a hypercomplex nilmanifold, which does not admit a compatible HKT metric.

Acknowledgments: This paper consists mainly of results obtained in previous publications. I would like to take this opportunity to thank my collaborators G. Papadopoulos and A. Fino for their help and the discussions we have had, and especially Y.S. Poon who initiated our work on HKT-problems.

11.2 HKT-geometry of a hypercomplex manifold

11.2.1 KT geometry of a complex manifold

Let M be a smooth manifold with a Riemannian metric g. An integrable complex structure J is an endomorphism of the tangent bundle TM with $J^2 = -Id$ such that

$$N(X, Y) = [X, Y] + J[JX, Y] + J[X, JY] - [JX, JY] = 0.$$

M is called a Hermitian manifold if $g(JX, JY) = g(X, Y)$. The Kähler form F is a type (1,1)-form defined by $F(X, Y) = g(JX, Y)$.

A linear connection ∇ on M is Hermitian if it preserves the metric g and the complex structure J, i.e., $\nabla g = 0$ and $\nabla J = 0$. Since the connection preserves the metric, it is uniquely determined by its torsion tensor T. We shall also consider the (3,0)-tensor

$$c(X, Y, Z) = g(X, T(Y, Z)). \tag{2.1}$$

Physicists find that the presence of the Wess–Zumino term in $N = 2$ supersymmetry yields a Hermitian connection whose torsion c is totally skew-symmetric. In other words, c is a 3-form. Such a connection turns out to be another distinguished Hermitian connection [1, 14]. The geometry of such a connection is called by physicists a KT-connection. Among some mathematicians, this connection is called the Bismut connection. According to Gauduchon [14], on any Hermitian manifold there exists a unique Hermitian connection whose torsion tensor c is a 3-form. Moreover, the torsion form can be expressed in terms of the complex structure and the Kähler form. Recall the following definitions and conventions [5, Equations 2.8 and 2.15–2.17]. For any n-form ω, let

$$(J\omega)(X_1, \ldots, X_n) = (-1)^n \omega(JX_1, \ldots, JX_n) \quad \text{and}$$
$$d^c \omega = (-1)^n JdJ\omega, \tag{2.2}$$

and moreover

$$\partial = \tfrac{1}{2}(d + id^c) = \tfrac{1}{2}(d + (-1)^n iJdJ),$$
$$\bar{\partial} = \tfrac{1}{2}(d - id^c) = \tfrac{1}{2}(d - (-1)^n iJdJ). \tag{2.3}$$

By [14], the torsion 3-form of the Bismut connection is

$$c(X, Y, Z) = -d^c F(X, Y, Z). \tag{2.4}$$

11.2.2 Hyper-hermitian connections and HKT-geometry

By definition three complex structures I_1, I_2 and I_3 on M form a hypercomplex structure if

$$I_1^2 = I_2^2 = I_3^2 = -1, \quad \text{and} \quad I_1 I_2 = I_3 = -I_2 I_1. \tag{2.5}$$

A triple of such complex structures is equivalent to the existence of a 2-sphere worth of integrable complex structures:

$$\mathcal{I} = \{a_1 I_1 + a_2 I_2 + a_3 I_3 : a_1^2 + a_2^2 + a_3^2 = 1\}. \tag{2.6}$$

When g is a Riemannian metric on the manifold M such that it is Hermitian with respect to every complex structure in the hypercomplex structure, $(M, \mathcal{I}, g)$ is called a hyper-Hermitian manifold. Note that g is hyper-Hermitian if and only if

$$g(X, Y) = g(I_1 X, I_1 Y) = g(I_2 X, I_2 Y) = g(I_3 X, I_3 Y). \tag{2.7}$$

By a standard result any hypercomplex manifold admits hyper-Hermitian metric. On a hyper-Hermitian manifold, there are two natural torsion-free connections, namely the Levi-Civita connection and the Obata connection. However, in general, the Levi-Civita connection does not preserve the hypercomplex structure and the Obata connection does not preserve the metric. We are interested in the following types of connections.

Definition 1. *A linear connection* ∇ *on a hyper-Hermitian manifold* $(M, \mathcal{I}, g)$ *is hyper-Hermitian if*

$$\nabla g = 0, \quad \text{and} \quad \nabla I_1 = \nabla I_2 = \nabla I_3 = 0. \tag{2.8}$$

Definition 2. *A linear connection ∇ on a hyper-Hermitian manifold $(M, \mathcal{I}, g)$ is HKT if its torsion tensor c is totally skew-symmetric.*

In the physics literature an HKT-connection is also called a hyper-Kähler connection. The geometry of this connection is also referred to as HKT-geometry. Note that an HKT-connection is also the Bismut connection for each complex structure in the given hypercomplex structure. For the complex structures $\{I_1, I_2, I_3\}$, we consider their corresponding Kähler forms $\{F_1, F_2, F_3\}$ and the complex operators $\{d_1, d_2, d_3\}$ where $d_i = d_i^c$. Due to Gauduchon's characterization of the Bismut connection, we have

Proposition 1. *A hyper-Hermitian manifold $(M, \mathcal{I}, g)$ admits an HKT-connection if and only if $d_1 F_1 = d_2 F_2 = d_3 F_3$. If it exists, it is unique.*

In view of the uniqueness, we say that $(M, \mathcal{I}, g)$ is an HKT-structure if it admits a hyper-Kähler connection. If the hyper-Kähler connection is also torsion-free, then the HKT-structure is a hyper-Kähler structure.

In physics literature the notion of strong HKT-structure also appears.

Definition 3. *An HKT-structure on a manifold is called strong if the corresponding torsion 3-form c is closed.*

11.2.3 First examples—compact Lie groups

As shown by Joyce [23], there is a family of homogeneous hypercomplex structures associated to any compact semi-simple Lie group. In this section, we briefly review this construction and demonstrate, as Opfermann and Papadopoulos did [28], the existence of left-invariant HKT-connections. The construction works also in the more general setting of homogeneous spaces.

Let G be a compact semi-simple Lie group and U a maximal torus. Let $\mathfrak{g}$ and $\mathfrak{u}$ be their algebras. Choose a system of ordered roots with respect to $\mathfrak{u}_{\mathbb{C}}$. Let α_1 be a maximal positive root, and $\mathfrak{h}_1$ the dual space of α_1. Let ∂_1 be the $\mathfrak{sp}(1)$-subalgebra of $\mathfrak{g}$ such that its complexification is isomorphic to $\mathfrak{h}_1 \oplus \mathfrak{g}_{\alpha_1} \oplus \mathfrak{g}_{-\alpha_1}$, where $\mathfrak{g}_{\alpha_1}$ and $\mathfrak{g}_{-\alpha_1}$ are the root spaces for α_1 and $-\alpha_1$ respectively. Let $\mathfrak{b}_1$ be the centralizer of ∂_1. Then there is a vector subspace $\mathfrak{f}_1$ composed of root spaces such that $\mathfrak{g} = \mathfrak{b}_1 \oplus \partial_1 \oplus \mathfrak{f}_1$. If $\mathfrak{b}_1$ is not Abelian, Joyce applies this decomposition to it. By inductively searching for $\mathfrak{sp}(1)$ subalgebras, he finds the following [23, Lemma 4.1].

Lemma 1. *The Lie algebra $\mathfrak{g}$ of a compact Lie group G decomposes as*

$$\mathfrak{g} = \mathfrak{b} \oplus_{j=1}^{n} \partial_j \oplus_{j=1}^{n} \mathfrak{f}_j, \tag{2.9}$$

with the following properties. (1) *$\mathfrak{b}$ is Abelian and ∂_j is isomorphic to $\mathfrak{sp}(1)$.* (2) *$\mathfrak{b} \oplus_{j=1}^{n} \partial_j$ contains $\mathfrak{u}$.* (3) *Set $\mathfrak{b}_0 = \mathfrak{g}$, $\mathfrak{b}_n = \mathfrak{b}$ and $\mathfrak{b}_k = \mathfrak{b} \oplus_{j=k+1}^{n} \partial_j \oplus_{j=k+1}^{n} \mathfrak{f}_j$. Then $[\mathfrak{b}_k, \partial_j] = 0$ for $k \geq j$.* (4) *$[\partial_l, \mathfrak{f}_l] \subset \mathfrak{f}_l$.* (5) *The adjoint representation of ∂_l on $\mathfrak{f}_l$ is reducible to a direct sum of the irreducible 2-dimensional representations of $\mathfrak{sp}(1)$.*

When the group G is semi-simple, the Killing–Cartan form is a negative definite inner product on the vector space $\mathfrak{g}$. One can also check that [16].

Lemma 2. *The Joyce decomposition of a compact semi-simple Lie algebra is an orthogonal decomposition with respect to the Killing–Cartan form.*

Let G be a compact semi-simple Lie group with rank r. Then

$$(2n - r)\mathfrak{u}(1) \oplus \mathfrak{g} \cong \mathbb{R}^n \oplus_{j=1}^n \partial_j \oplus_{j=1}^n \mathfrak{f}_j. \tag{2.10}$$

At the tangent space of the identity element of $T^{2n-r} \times G$, i.e., the Lie algebra $(2n - r)\mathfrak{u}(1) \oplus \mathfrak{g}$, a hypercomplex structure $\{I_1, I_2, I_3\}$ is defined as follows. Let $\{E_1, \ldots, E_n\}$ be a basis for $\mathbb{R}^n$. Choose isomorphisms ϕ_j from $\mathfrak{sp}(1)$, the real vector space of imaginary quaternions, to ∂_j. It gives a real linear identification of the quaternions $\mathbb{H}$ with $\langle E_j \rangle \oplus \partial_j$. If H_j, X_j and Y_j form a basis for ∂_j such that $[H_j, X_j] = 2Y_j$ and $[H_j, Y_j] = -2X_j$, then

$$I_1 E_j = H_j, I_2 E_j = X_j, I_3 E_j = Y_j. \tag{2.11}$$

Define the action of I_a on $\mathfrak{f}_j$ by $I_a(v) = [v, \phi_j(\iota_a)]$ where $\iota_1 = i, \iota_2 = j, \iota_3 = k$. The complex structures $\{I_1, I_2, I_3\}$ at the other points of the group $T^{2n-r} \times G$ are obtained by left translations. These complex structures are integrable and form a hypercomplex structure [23]. To define also a metric we use:

Lemma 3. [16] *When G is a compact semi-simple Lie group with rank r, there exists a negative definite bilinear form $\hat{B}$ on the decomposition*

$$(2n - r)\mathfrak{u}(1) \oplus \mathfrak{g} \cong \mathbb{R}^n \oplus_{j=1}^n \partial_j \oplus_{j=1}^n \mathfrak{f}_j$$

such that (1) *its restriction to $\mathfrak{g}$ is the Killing–Cartan form,* (2) *it is hyper-Hermitian with respect to the hypercomplex structure, and* (3) *the above decomposition is orthogonal.*

Let g be the left-translation of the extended Killing–Cartan form $-\hat{B}$. It is a bi-invariant metric on the manifold $T^{2n-r} \times G$. The Levi-Civita connection D is the bi-invariant connection. Let ∇ be the left-invariant connection defined by having all left-invariant vector fields being parallel. When X and Y are left-invariant vector fields,

$$D_X Y = \tfrac{1}{2}[X, Y] \quad \text{and} \quad \nabla_X Y = 0.$$

Since the hypercomplex structure and the hyper-Hermitian metric are left-invariant, the left-invariant connection is hyper-Hermitian. The torsion tensor for the left-invariant connection is $T(X, Y) = -[X, Y]$. The (3,0)-torsion tensor is

$$c(X, Y, Z) = -\hat{B}([X, Y], Z).$$

It is well known that c is a totally skew-symmetric and closed 3-form. Therefore, the left-invariant connection is a strong HKT-structure on the group manifold $T^{2n-r} \times G$. Theorem 4 in the next section shows that some HKT manifolds do not admit a strong HKT structure.

11.2.4 Characterization of HKT-structures

In this section, we characterize HKT-structures in terms of the existence of a holomorphic object with respect to any complex structure in the hypercomplex structure. The results seem to indicate that the holomorphic characterization developed here will serve all the purposes that one wants the twistor theory of HKT-geometry as developed in [19] to serve. In [16] we proved:

Proposition 2. *Let $(M, \mathcal{I}, g)$ be a hyper-Hermitian manifold and F_a be the Kähler form for (I_a, g). Then $(M, \mathcal{I}, g)$ is an HKT-structure if and only if $\partial_1(F_2 + iF_3) = 0$; or equivalently $\overline{\partial}_1(F_2 - iF_3) = 0$.*

This has an immediate consequence which has been proven also in [13, Section 2.2].

Corollary 1. *Any 4-dimensional hyper-Hermitian manifold is HKT.*

On any hypercomplex manifold $(M, \mathcal{I})$, if $F_2 - iF_3$ is a 2-form such that $-F_2(I_2X, Y) = g(X, Y)$ is positive definite and is a non-holomorphic (0,2)-form with respect to I_1, then $(M, g, \mathcal{I})$ is a hyper-Hermitian manifold but it is not an HKT-structure. For example, a conformal change of an HKT-structure by a generic function gives a hyper-Hermitian structure which is not an HKT-structure so long as the dimension of the underlying manifold is at least eight. Similarly to Proposition 2 we have:

Theorem 1. *[16] Let $(M, \mathcal{I})$ be a hypercomplex manifold and $F_2 - iF_3$ be a (0,2)-form with respect to I_1 such that $\overline{\partial}_1(F_2 - iF_3) = 0$ or equivalently $\partial_1(F_2 + iF_3) = 0$ and $-F_2(I_2X, Y) = g(X, Y)$ is a positive definite symmetric bilinear form. Then $(M, \mathcal{I}, g)$ is an HKT-structure.*

Remark 1. M. Verbitsky [35], developing further the properties of HKT manifolds due to existence of a form as in the above theorem, proved that the Dolbeault (0,p)-cohomology $H^{0,p}(M, K^{\frac{1}{2}})$ of the "half" of the canonical sheaf with respect to one of the structures in $\mathcal{I}$ carries a Lefschetz-type $sl(2)$-action. This is similar to the $sl(2)$-action on the Dolbeault cohomology of a Kähler manifold.

11.3 Potential theory

Theorem 1 shows that the form $F_2 + iF_3$ is a ∂_1-closed (2,0)-form on an HKT-manifold. It is natural to consider a differential form β_1 as a potential 1-form for $F_2 + iF_3$ if $\partial_1\beta_1 = F_2 + iF_3$. A priori, the 1-form β_1 depends on the choice of the complex structure I_1. The potential 1-form for $F_3 + iF_1$, if it exists, depends on I_2, and so on. In this section, we seek a function that generates all Kähler forms.

11.3.1 Potential functions

A function μ is a potential function for a hyper-Kähler manifold $(M, \mathcal{I}, g)$ if the Kähler forms F_a are equal to $dd_a\mu$. Since $d_a = (-1)^n I_a dI_a$ on n-forms, $d_a\mu = I_a d\mu$. Therefore,

$$d_1 d_2 \mu = d_1 I_2 d\mu = -I_1 dI_1 I_2 d\mu$$
$$= -I_1 dI_3 d\mu = -I_1 dd_3 \mu = -I_1 \Omega_3 = \Omega_3 = dd_3 \mu.$$

Now we generalize this concept to HKT-manifolds.

Definition 4. [16] *Let $(M, \mathcal{I}, g)$ be an HKT-structure with Kähler forms F_1, F_2 and F_3. A possibly locally defined function μ is a potential function for the HKT-structure if*

$$F_1 = \tfrac{1}{2}(dd_1 + d_2 d_3)\mu, \quad F_2 = \tfrac{1}{2}(dd_2 + d_3 d_1)\mu, \quad F_3 = \tfrac{1}{2}(dd_3 + d_1 d_2)\mu. \quad (3.1)$$

Due to the identities $dd_a + d_a d = 0$ and $d_a d_b + d_b d_a = 0$, μ is a potential function if and only if

$$F_{\vec{a}} = \tfrac{1}{2}(dd_{\vec{a}} + d_{\vec{b}} d_{\vec{c}})\mu,$$

when $\vec{a} = \vec{b} \times \vec{c}$ and $F_{\vec{a}}$ is the Kähler form for the complex structure $I_{\vec{a}} = a_1 I_1 + a_2 I_2 + a_3 I_3$. Moreover, the torsion 3-form is equal to $-\tfrac{1}{4} d_1 d_2 d_3 \mu$. Furthermore, since $\partial_a = \tfrac{1}{2}(d + id_a)$ and $\overline{\partial}_a = \tfrac{1}{2}(d - id_a)$,

$$F_2 + iF_3 = \tfrac{1}{2}(dd_2 + idd_3 + id_1 d_2 - d_1 d_3)\mu = 2\partial_1 I_2 \overline{\partial}_1 \mu. \quad (3.2)$$

Conversely, if a function μ satisfies the above identity, it satisfies the last two identities in (3.1). Since the metric is hyper-Hermitian, for any vectors X and Y, $F_1(X, Y) = F_2(I_3 X, Y)$. Through the integrability of the complex structures I_1, I_2, I_3, the quaternion identities (2.5) and the last two identities in (3.1), one derives the first identity in (3.1). Therefore, we have the following theorem which justifies our definition for potential functions.

Theorem 2. [16] *Let $(M, \mathcal{I}, g)$ be an HKT-structure with Kähler form F_1, F_2 and F_3. A possibly locally defined function μ is a potential function for the HKT-structure if*

$$F_2 + iF_3 = 2\partial_1 I_2 \overline{\partial}_1 \mu. \quad (3.3)$$

In this context, an HKT-structure is hyper-Kähler if and only if the potential function satisfies the identities

$$dd_1 \mu = d_2 d_3 \mu, \quad dd_2 \mu = d_3 d_2 \mu, \quad dd_3 \mu = d_1 d_2 \mu. \quad (3.4)$$

Corollary 2. [16] *Every hypercomplex manifold locally admits an HKT-metric.*

Remark 2. As in the Kähler case, compact manifolds do not admit globally defined HKT potential. To verify, let f be a potential function and g be the corresponding induced metric. Define the complex Laplacian of f with respect to g:

$$\overline{\partial}^* \overline{\partial} f = \triangle^c f = g(dd_1 f, F_1).$$

Then $0 \leq 2g(F_1, F_1) = g(dd_1 f + d_2 d_3 f, F_1) = 2\triangle^c f$, because

$$g(d_2 d_3 f, F_1) = g(-I_2 dd_1 f, F_1) = -g(dd_1 f, I_2 F_1) = g(dd_1 f, F_1) = \triangle^c f.$$

Now the remark follows from the standard arguments involving the maximum principle for second–order elliptic differential equation just as in the Kähler case since $\triangle^c f$ does not have zero-order terms.

Remark 3. If we introduce the following quaternionic operators acting on the left on quaternion-valued forms: $\partial^H = d + id_1 + jd_2 + kd_3$, and $\overline{\partial}^H = d - id_1 - jd_2 - kd_3$, then a real-valued function μ is an HKT-potential if $\partial^H \overline{\partial}^H \mu = -2iF_1 - 2jF_2 - 2kF_3$.

If we identify $\mathbb{H}^n$ with $\mathbb{C}^{2n}$, we deduce as in Corollary 2 that any pluri-subharmonic function in the domain of $\mathbb{C}^{2n}$ is an HKT-potential. The converse however is wrong. For example the function $|z|^2$ is an HKT potential in $\mathbb{C}^{2n}$ but is not strictly pluri-subharmonic.

Remark 4. Given an HKT-metric g with Kähler forms F_1, F_2 and F_3, for any real-valued function μ we consider

$$\hat{F}_2 + i\hat{F}_3 = F_2 + iF_3 + \partial_1 I_2 \overline{\partial}_1 \mu.$$

According to Theorem 1 and other results in this section, whenever the form $\hat{g}(X, Y) := -\hat{F}_2(I_2 X, Y)$ is positive definite, we obtain a new HKT-metric with respect to the old hypercomplex structure.

11.3.2   Inhomogeneous HKT-structures on $S^1 \times S^{4n-3}$

On the complex vector space $(\mathbb{C}^n \oplus \mathbb{C}^n) \backslash \{0\}$, let (z_α, w_α), $1 \leq \alpha \leq n$, be its coordinates. We define a hypercomplex structure to contain this complex structure as follows.

$$I_1 dz_\alpha = -idz_\alpha, \quad I_1 dw_\alpha = -idw_\alpha, \quad I_1 d\overline{z}_\alpha = id\overline{z}_\alpha, \quad I_1 d\overline{w}_\alpha = id\overline{w}_\alpha.$$
$$I_2 dz_\alpha = d\overline{w}_\alpha, \quad I_2 dw_\alpha = -d\overline{z}_\alpha, \quad I_2 d\overline{z}_\alpha = dw_\alpha, \quad I_2 d\overline{w}_\alpha = -dz_\alpha.$$
$$I_3 dz_\alpha = id\overline{w}_\alpha, \quad I_3 dw_\alpha = -id\overline{z}_\alpha, \quad I_3 d\overline{z}_\alpha = -idw_\alpha, \quad I_3 d\overline{w}_\alpha = idz_\alpha.$$

The function $\mu = \frac{1}{2}(|z|^2 + |w|^2)$ is the hyper-Kähler potential for the standard Euclidean metric:

$$g = \frac{1}{2}(dz_\alpha \otimes d\overline{z}_\alpha + d\overline{z}_\alpha \otimes dz_\alpha + dw_\alpha \otimes d\overline{w}_\alpha + d\overline{w}_\alpha \otimes dw_\alpha). \tag{3.5}$$

Since $|\nabla \mu|^2 = 2\mu$, the function $f(\mu) = \ln \mu$ is the HKT-potential for an HKT-metric $\hat{g}$ on $\mathbb{C}^{2n} \backslash \{0\}$. Next for any real number r, with $0 < r < 1$, and $\theta_1, \ldots, \theta_n$ modulo 2π, we consider the integer group $\langle r \rangle$ generated by the following action on $(\mathbb{C}^n \oplus \mathbb{C}^n) \backslash \{0\}$.

$$(z_\alpha, w_\alpha) \mapsto (re^{i\theta_\alpha} z_\alpha, re^{-i\theta_\alpha} w_\alpha). \tag{3.6}$$

One can check that the group $\langle r \rangle$ is a group of hypercomplex transformations. As observed in [30], the quotient space of $(\mathbb{C}^n \oplus \mathbb{C}^n)\backslash\{0\}$ with respect to $\langle r \rangle$ is the manifold $S^1 \times S^{4n-1} = S^1 \times \mathrm{Sp}(n)/\mathrm{Sp}(n-1)$. Since the group $\langle r \rangle$ is also a group of isometries with respect to the HKT-metric $\hat{g}$ determined by $f(\mu) = \ln \mu$, the HKT-structure descends from $(\mathbb{C}^n \oplus \mathbb{C}^n)\backslash\{0\}$ to an HKT-structure on $S^1 \times S^{4n-1}$. Since the hypercomplex structures on $S^1 \times S^{4n-1}$ are parameterized by $(r, \theta_1, \ldots, \theta_n)$ and a generic hypercomplex structure in this family is inhomogeneous [30], we obtain a family of inhomogeneous HKT-structures on the manifold $S^1 \times S^{4n-1}$.

Theorem 3. [16] *Every hypercomplex deformation of the homogeneous hypercomplex structure on $S^1 \times S^{4n-1}$ admits an HKT-metric.*

Furthermore, $\hat{F}_2 + i\hat{F}_3 = 2\partial_1 I_2 \overline{\partial}_1 \mu$ descends to $S^1 \times S^{4n-1}$. However, the function μ does not descend to $S^1 \times S^{4n-1}$. Therefore, this (2,0)-form has a potential form $I_2 \overline{\partial}_1 \mu$ but not a globally defined potential function.

Remark 5. The HKT-potential function was also independently introduced in [26] in connection with the study of a (supersymmetric) quantum mechanical system. It also plays a role in the description of HKT-structure on the moduli spaces of very special black holes. In some of the models the potential is proportional to the norm of a conformally-Killing vector field of special kind. This is further investigated from a mathematical perspective in [31, 32]. In [26] it is shown that any HKT-metric on an open set of $\mathbb{H}^n$ arises from a potential. This leads to the question: *Does any HKT-metric locally arise from a potential?* We do not yet know the answer to this question.

Theorem 4. *The manifold $S^1 \times S^{4n-1}$ with its invariant hypercomplex structure does not admit any strong HKT structure for $n > 1$.*

Proof. Any strong HKT structure should have $dd^c F_a = 0$. But this is impossible due to Theorem 8.2 of [9]. $\qquad\qquad\square$

11.4   Reduction

There is a procedure of reduction which parallels the symplectic reduction for other types of manifolds with additional structure. Examples are the contact and Sasakian reduction. We concentrate here on the reduction of complex and hypercomplex manifolds. Assuming the existence of "moment maps", we examine the geometry on the reduced space in the next two sections.

11.4.1   KT or Hermitian reduction

Before we explain HKT reduction, it is instructive to consider first the reduction of KT manifolds, i.e., Hermitian manifolds equipped with a hermitian connection whose torsion is a 3-form.

Let M be a KT manifold and let G be a compact group of complex isometries on M. Denote the algebra of holomorphic vector fields by $\mathfrak{g}$. Next introduce a G-equivariant map $\nu : M \to \mathfrak{g}$ satisfying the transversality condition, i.e., $Id\nu(X) \neq 0$ for all $X \in \mathfrak{g}$. We remark that a map ν is equivariant if $\nu(g \cdot x) = Ad\,g^*(\nu(x))$.

Definition 5. *A map ν is called a G-moment map if and only if (i) it is equivariant and (ii) it satisfies the transversality condition.*

We remark that for simply connected Kähler manifolds the moment map can be constructed using the invariance of Kähler form and complex structure and it satisfies the transversality property. However additional conditions are required in order for the moment map to be equivariant.

Next, given a point $\zeta \in \mathfrak{g}$, denote the level set $\nu^{-1}(\zeta)$ by P. Since the map ν is G-equivariant, level sets are invariant if the group G is Abelian or if the point ς is invariant. Assuming that the level set P is invariant, and the action of G on P is free, then the quotient space $N = P/G$ is a smooth manifold. Let $\pi : P \to N$ be the quotient map.

It can be shown that in fact $N = P/G$ is a complex manifold. This construction can be done as follows. For each point m in the space P, its tangent space is

$$T_m P = \{t \in T_m M : d\nu(t) = 0\}.$$

Consider the vector subspace

$$\mathcal{U}_m = \{t \in T_m P : Id\nu(t) = 0\}.$$

Due to the transversality condition, this space is transversal to the vectors generated by elements in $\mathfrak{g}$. In addition, this space is a vector subspace of $T_m P$ with co-dimension $\dim \mathfrak{g}$, hence it is a vector subspace of $T_m M$ with co-dimension $2 \dim \mathfrak{g}$. The same condition implies that, as a subbundle of $TM_{|P}$, $\mathcal{U}$ is closed under I. Moreover there is a G-invariant splitting

$$TP = \mathcal{U} \oplus \mathcal{V} \tag{4.1}$$

where $\mathcal{V}$ is the tangent space to the orbits of G and coincides with the bundle of kernels of $d\pi$. We use the terms "horizontal" and "vertical" for $\mathcal{U}$ and $\mathcal{V}$.

As the projection π is an isomorphism on $\mathcal{U}$, for any tangent vector $\hat{A}$ at $\pi(m)$, there exists a unique element A^u in $\mathcal{U}_m$ such that $d\pi(A^u) = \hat{A}$. We call A^u the horizontal lift of $\hat{A}$. The complex structure on N is defined by

$$I\hat{A} = d\pi(IA^u), \quad \text{i.e.,} \quad (I\hat{A})^u = IA^u. \tag{4.2}$$

Theorem 5. [17] *Let $(M, \mathcal{I}, g)$ be a KT-manifold. Suppose that G is a compact group of complex isometries admitting a G-moment map ν. Then the reduced space $N = M//G$ inherits a complex and KT structure.*

11.4.2   HKT reduction

We shall begin with a description of the hypercomplex reduction as developed by Joyce [22]. Let G be a compact group of hypercomplex automorphisms of M. Denote the algebra of hyper-holomorphic vector fields by $\mathfrak{g}$. Suppose that $\nu = (\nu_1, \nu_2, \nu_3) : M \longrightarrow \mathbb{R}^3 \otimes \mathfrak{g}$ is a G-equivariant map satisfying the the Cauchy–Riemann condition, $I_1 d\nu_1 = I_2 d\nu_2 = I_3 d\nu_3$, and the transversality condition, $I_a d\nu_a(X) \neq 0$ for all $X \in \mathfrak{g}$. In analogy with a similar definition given in the previous section, any map satisfying these conditions is called a G-moment map. Given a point $\zeta = (\zeta_1, \zeta_2, \zeta_3)$ in $\mathbb{R}^3 \otimes \mathfrak{g}$, denote the level set $\nu^{-1}(\zeta)$ by P. Assuming that the level set P is invariant, and the action of G on P is free, then the quotient space $N = P/G$ is a smooth manifold.

Joyce proved that the quotient space $N = P/G$ inherits a natural hypercomplex structure [22]. His construction runs as follows. For each point m in the space P, its tangent space is

$$T_m P = \{t \in T_m M : d\nu_1(t) = d\nu_2(t) = d\nu_3(t) = 0\}.$$

Consider the vector subspace

$$\mathcal{U}_m = \{t \in T_m P : I_1 d\nu_1(t) = I_2 d\nu_2(t) = I_3 d\nu_3(t) = 0\}. \qquad (4.3)$$

Due to the transversality condition, this space is transversal to the vectors generated by elements in $\mathfrak{g}$. Due to the Cauchy–Riemann condition, this space is a vector subspace of $T_m P$ with co-dimension $\dim \mathfrak{g}$, hence it is a vector subspace of $T_m M$ with co-dimension $4 \dim \mathfrak{g}$.

The same condition implies that, as a subbundle of $TM_{|P}$, $\mathcal{U}$ is closed under I_a. Moreover there is a G-invariant splitting

$$TP = \mathcal{U} \oplus \mathcal{V} \qquad (4.4)$$

where $\mathcal{V}$ is the tangent space to the orbits of G and coincides with the bundle of kernels of $d\pi$. Again, we use the terms "horizontal" and "vertical" for $\mathcal{U}$ and $\mathcal{V}$, although the two spaces are not necessarily orthogonal. Following the techniques and notation of the last section, a hypercomplex structure on N is defined by

$$I_a \hat{A} = d\pi(I_a A^u), \quad \text{i.e.,} \quad (I_a A)^u = I_a A^u. \qquad (4.5)$$

Theorem 6. [17] *Let $(M, \mathcal{I}, g)$ be an HKT-manifold. Suppose that G is a compact group of hypercomplex isometries admitting a G-moment map ν. Then hypercomplex reduced space $N = M//G$ inherits an HKT structure.*

11.5   Moment maps for strong KT and HKT-spaces

As we have seen, the construction of new HKT manifolds using the HKT reduction necessitates the existence of a G-moment map satisfying the requirements of Theorem 6. This moment map is not specified within the theory, as it is the case for the hyper-Kähler reduction, but rather its existence is an additional assumption for the construction. However as we shall see in the special case of reduction for strong KT (and HKT) manifolds, under certain assumptions such a moment map arises naturally. The local construction of a moment map for KT and HKT geometries presented below parallels the construction of an action for two-dimensional (2,0)- and (4,0)-supersymmetric gauged sigma models with Wess–Zumino term in [21], respectively. Again, we focus on a reduction theory for strong KT-structures first. The reduction theory for strong HKT-structures follows.

11.5.1   Existence of moment maps for strong KT and HKT structures

When the torsion 3-form c is closed, any vector field preserving the metric and the complex structure preserves the 3-form c as well. In particular locally $i_X c = du_X$ for some 1-form u_X. Assuming that this condition holds globally, and that the manifold satisfies the $\partial\bar{\partial}$-lemma, then we can "integrate" further u_X to obtain a function as is shown in [17]. This will produce a map ν_X.

In the case when the group G is Abelian, the issue of equivariance is absent and hence the map ν so constructed is the moment map. The general conditions for invariance are given in [17] and are expressed in terms of some cohomology groups $H^1_{\delta}(\mathcal{O} \otimes \mathfrak{g})$ of a complex which is a resolution of the sheaf $\mathcal{O} \otimes \mathfrak{g}$. This sheaf is the tensor product of the Lie algebra $\mathfrak{g}$ of G and the sheaf of the germs of holomorphic functions $\mathcal{O}$. Moreover the condition $i_a c = 0$ in $H^2(M, \mathbb{R})$ is described in terms of equivariant cohomology of M. Here we present a simplified version for torus actions on compact manifolds:

Theorem 7. [17] *Let M be a compact strong KT manifold with an action of a torus T via holomorphic isometries and M satisfies the $\partial\bar{\partial}$-lemma. If $i_X c$ is a trivial class in $H^2(M, \mathbb{R})$ for the vector field X generated by the action, then there exists a (canonical) moment map.*

Similarly, we formulate a particular case of strong HKT reduction which follows :

Theorem 8. [17] *Let M be a compact strong HKT manifold satisfying the $\partial\bar{\partial}$-lemma and G be a torus acting on M and leaving invariant the HKT structure. If the torsion 3-form c admits an extension as an equivariantly closed form in terms of equivariant cohomoogy, then $M//G$ is an HKT manifold.*

11.5.2   Potential functions and reduction

Recall that if $(M, \mathcal{I}, g)$ is an HKT manifold with Kähler forms ω_a, an *HKT potential* is a function μ such that

$$2\omega_1 = dd_1\mu + d_2 d_3\mu,\ 2\omega_2 = dd_2\mu + d_3 d_1\mu,\ 2\omega_3 = dd_3\mu + d_1 d_2\mu.$$

In this section, we follow the methods in [24] to find a potential function on the reduced space.

Theorem 9. [17] *Let* $(M, \mathcal{I}, g)$ *be an HKT manifold with HKT potential function* ρ. *Suppose that* G *is a compact group of hypercomplex isometries leaving* μ *invariant with moment map* $\nu = (\nu_1, \nu_2, \nu_3)$ *such that the tangent vectors to the orbits of* G *in* $\nu^{-1}(0)$ *are in the* $\ker(d_a\mu)$, *for* $a = 1, 2, 3$. *Then the function* ρ *induces an HKT potential function on the reduced space* $N = M//G$.

Remark 6. In the case of when the torsion vanishes, the condition in the above theorem is equivalent to the one proposed by Kobak and Swann [24]. In both cases the crucial point is to ensure $i^* I_a d\mu = I_a i^* d\mu$. In both cases $d\mu(X^v) = 0$ since μ is invariant.

It is known that $SU(3)$ admits an invariant hypercomplex structure, constructed by Joyce. Moreover Pedersen and Poon [30] considered the deformation of this structure and succeeded to represent any "small" deformation as a hypercomplex reduced space of the space $S^1 \times S^{11}$ under an appropriate S^1 action. As it is shown in [16] and [28], the space $S^1 \times S^{11}$ is HKT and one can check that the S^1-actions considered in [30, Section 6.3] are HKT-isometries. Now according to the theorem of section 2.2 we have:

Theorem 10. [17] *Any small deformation of the invariant hypercomplex structure on* $SU(3)$ *admits an HKT structure.*

In view of Theorem 10 and Theorem 4, we may ask: *Is any small hypercomplex deformation of HKT-structure again HKT?* The answer to this question is negative. We refer to [11] for the example.

11.6   Hypercomplex manifold admitting no HKT metric

After we have proved the local existence of an HKT metric and provided a rich supply of global examples, a natural question arises (see [16]): *Does every hypercomplex manifold admit an HKT metric?* We give an example providing a negative answer to this question.

A hypercomplex structure on a Lie algebra $\mathfrak{g}$ is a triple of complex structures $\{J_i\}_{i=1,2,3}$ satisfying the quaternion relations $J_i^2 = -I$, $a = 1, 2, 3$, $J_1 J_2 = -J_2 J_1 = J_3$ and the integrability conditions $N_a(X, Y) = 0$ for $X, Y \in \mathfrak{g}$.

The hypercomplex structure will be called Abelian if

$$[J_a X, J_a Y] = [X, Y], \quad X, Y \in \mathfrak{g}.$$

Let g be an r product compatible with the hypercomplex structure.

By [6] $(\{J_i\}_{i=1,2,3}, g)$ is an HKT structure on $\mathfrak{g}$ if and only if g satisfies the extra condition

$$g([J_1 X, J_1 Y], Z) + g([J_1 Y, J_1 Z], X) + g([J_1 Z, J_1 X], Y)$$
$$= g([J_2 X, J_2 Y], Z) + g([J_2 Y, J_2 Z], X) + g([J_2 Z, J_2 X], Y)$$
$$= g([J_3 X, J_3 Y], Z) + g([J_3 Y, J_3 Z], X) + g([J_3 Z, J_3 X], Y).$$

When the hypercomplex structure is Abelian the previous condition is always satisfied for any inner product g compatible with the hypercomplex structure, hence it gives rise to an HKT-structure on G.

We say that a Lie group G has an invariant HKT structure ($\{J_i\}_{i=1,2,3}, g$) if the hypercomplex structure $\{J_i\}_{i=1,2,3}$ and the metric g arise from corresponding left-invariant tensors.

The groups we want to consider belong to the class of the so-called 2-step nilpotent Lie groups. They are characterized by the condition that their Lie algebra $\mathfrak{g}$ satisfies

$$[[\mathfrak{g}, \mathfrak{g}], \mathfrak{g}] = 0.$$

In [6] it is shown that, if G is a 2-step nilpotent Lie group carrying an invariant hypercomplex structure and an invariant HKT structure, then the hypercomplex structure is Abelian. Then one has the following:

Lemma 4. *A 2-step nilpotent Lie group G with a non-abelian hypercomplex structure admits no invariant HKT metric compatible with such a hypercomplex structure.*

Consider the Lie algebra $\mathfrak{n}_3$ of a nilpotent simply connected Lie group N_3, defined by its structure equations

$$de^i = 0, \quad i = 1, \ldots, 5, \qquad de^6 = e^1 \wedge e^2 + e^3 \wedge e^4,$$
$$de^7 = e^1 \wedge e^3 - e^2 \wedge e^4, \qquad de^8 = e^1 \wedge e^4 + e^2 \wedge e^3.$$

It carries a non-abelian hypercomplex structure given by

$$J_1(e_1) = e_2, \quad J_1(e_3) = e_4, \quad J_1(e_5) = e_6, \quad J_1(e_7) = e_8,$$
$$J_2(e_1) = e_3, \quad J_2(e_2) = -e_4, \quad J_1(e_5) = e_7, \quad J_1(e_6) = -e_8.$$

If G is a simply-connected nilpotent Lie group and, if the structure equations of its Lie algebra are rational, then there exists a discrete subgroup Γ of G for which $M = \Gamma \backslash G$ is compact [25]. Any invariant HKT structure on G will pass to an HKT structure on M.

By [27, Lemma 6.2] if a Lie group G admits a discrete subgroup Γ with compact quotient, then G is unimodular, i.e., its left-invariant Haar measure is also right-invariant. Then, any simply-connected nilpotent Lie group G with rational structure equations admits a bi-invariant volume form $d\mu$. Then by "symmetrization" one obtains:

Theorem 11. [11] *Any compact quotient* $M = \Gamma \backslash G$ *of a 2-step nilpotent Lie group* G *with a non-abelian left-invariant hypercomplex structure* $\{J_i\}_{i=1,2,3}$ *admits no HKT metric compatible with such a hypercomplex structure (for example* N_3 *above).*

REFERENCES

[1] J.-M. Bismut. A local index theorem for non-Kähler manifolds, *Math. Ann.* **284** (1989), 681–699.

[2] C. Boyer. A note on hyperhermitian four-manifolds, *Proc. Amer. Math. Soc.* **102** (1988), 157–164.

[3] C. Boyer, K. Galicki, B. Mann. Hypercomplex structures on Stiefel manifolds, *Ann. Global Anal. Geom.* **14** (1996), 81–105.

[4] T.L. Curtright, D. Z. Freedman. Nonlinear σ-models with extended supersymmetry in four dimensions, *Phys. Lett.* **90B** (1980), p. 71.

[5] A. Besse. *Einstein Manifolds*, Ergebnisse der Mathematik und ihrer Grenzgebiete, 3. Folge **10**, Springer-Verlag, New York, 1987.

[6] I. Dotti, A. Fino. Hyperkähler torsion structures invariant by nilpotent Lie groups, *Classical Quantum Gravity* **19** (2002), 551–562.

[7] I. Dotti, A. Fino. Abelian hypercomplex 8-dimensional nil manifolds, *Ann. Glob. Anal. and Geom.* **18** (2000), 47–59.

[8] I. Dotti, A. Fino. Hypercomplex 8-dimensional nilpotent Lie groups, *J. Pure Appl. Alg.* Vol. **184** (2003), 41–57.

[9] N. Egidi. Special metrics on compact complex manifolds, *Diff. Geom. Appl.* **14** (2001), 217–234.

[10] B. Feix, H. Pedersen. Hyper-Kähler structures with torsion on nilpotent Lie groups, IMADA-preprint SDU (2002).

[11] A. Fino, G. Grantcharov. On some properties of the manifolds with skew-symmetric torsion and holonomy SU(n) and Sp(n), preprint math.DG/0302358.

[12] S. J. Gates, C. M. Hull, M. Roček. Twisted multiplets and new supersymmetric non-linear sigma models, *Nucl. Phys.* **B248** (1984), 157–186.

[13] P. Gauduchon, K.P. Tod. Hyper-Hermitian metrics with symmetry, *Journ. Geom. Phys.* **25** (1998), 291–304.

[14] P. Gauduchon. Hermitian connections and Dirac operators, *Bollettino U.M.I.* **11B** (1997), 257–288.

[15] G.W. Gibbons, G. Papadopoulos, K.S. Stelle. HKT and OKT geometries on soliton black hole moduli spaces, *Nucl. Phys.* **B508** (1997), 623–658; hep-th/9706207.

[16] G. Grantcharov, Y.S. Poon. Geometry of Hyper-Kähler connections with torsion, *Comm. Math. Phys.* **213** (2000), 19–37.

[17] G. Grantcharov, G. Papadopoulos, Y.S.Poon. Reduction of HKT structures. *J. Math. Phys.* **43** (2002), 3766–3783.

[18] N.J. Hitchin, A. Karlhede, U. Lindström, M. Roček. Hyper-Kähler metrics and supersymmetry, *Commun. Math. Phys.* **108** (1987), 535–589.

[19] P.S. Howe, G. Papadopoulos. Twistor spaces for HKT manifolds, *Phys. Lett.* **B379** (1996), 80–86; hep-th/9602108.

[20] P.S. Howe, G. Papadopoulos. Holonomy groups and W-symmetries, *Commun. Math. Phys.* **151** (1993), 467–480.

[21] C.M. Hull, G. Papadopoulos, B. Spence. Gauge symmetries for (p,q) supersymmetric sigma models, *Nucl. Phys.* **B363** (1991), 593–621.

[22] D. Joyce. The hypercomplex quotient and quaternionic quotient, *Math. Ann.* **290** (1991), 323–340.

[23] D. Joyce. Compact hypercomplex and quaternionic manifolds, *J. Diff. Geom.* **35** (1992), 743–761.

[24] P. Kobak, A. Swann. Hyper Kähler potentials via finite-dimensional quotients, *Geom. Dedicata* **88** (2001), 1–19.

[25] A.I. Malcev. On a class of homogeneous spaces, reprinted in *Amer. Math. Soc. Translations*, Series 1, **9** (1962), 276–307.

[26] J. Michelson, A. Strominger. The geometry of (super) conformal quantum mechanics, *Comm. Math. Phys.* **213** (2000), 1–17.

[27] J. Milnor. Curvature of left invariant metrics on Lie groups, *Adv. Math.* **21** (1976), 293–329.

[28] A. Opfermann, G. Papadopoulos. Homogeneous HKT and QKT manifolds, preprint; math-ph/9807026.

[29] H. Pedersen, Y.S. Poon. Deformations of hypercomplex structures, *J. reine angew. Math.* **499** (1998), 81–99.

[30] H. Pedersen, Y.S. Poon. Inhomogeneous hypercomplex structures on homogeneous manifolds, *J. reine angew. Math.* **516** (1999), 159–181.

[31] Y.S. Poon, A.F. Swann. Potential functions on HKT spaces, *Class. Quant. Gravity* **18** (2001), 4711–4714.

[32] Y.S. Poon, A.F. Swann. Superconformal symmetry and hyper-Kähler manifolds with torsion, preprint, math. DG/0111276.

[33] S.M. Salamon. Differential geometry of quaternionic manifolds, *Ann. scient. Éc. Norm. Sup. 4e* **19** (1986), 31–55.

[34] Ph. Spindel, A. Sevrin, W. Troost, A. Van Proeyen. Extended supersymmetric σ-models on group manifolds, *Nucl. Phys.* **B308** (1988), 662–698.

[35] M. Verbitsky. Hyper-Kähler manifolds with torsion, supersymmetry and Hodge theory, *Asian J. Math.* **6** (2002), 679–712.

[36] B. Zumino. Supersymmetry and Kähler manifolds, *Phys. Lett.* **87B** (1979), 203–206.

Gueo Grantcharov
Department of Mathematics
Florida International University
Miami FL 33199, USA
On leave from the University of Sofia
E-mail: grantchg@fiu.edu

Submitted: January 2, 2002.

12

Casimir Elements and Bochner Identities on Riemannian Manifolds

Yasushi Homma

ABSTRACT We show that the principal symbols of first order geometric differential operators on Riemannian manifolds are controlled by the enveloping algebra and higher Casimir elements of $\mathfrak{so}(n)$. Then we give all the Bochner identities for the operators explicitly.

Keywords: gradients, Bochner identities, SO(n)-modules, Casimir elements

12.1 Introduction

Let M be an n-dimensional oriented Riemannian manifold with metric g, and $\mathbf{SO}(M)$ be the principal SO(n) bundle of the oriented orthonormal frames on M. Given an irreducible unitary representation (π_ρ, V_ρ) of SO(n) with highest weight ρ, we have an associated vector bundle $\mathbf{S}_\rho = \mathbf{SO}(M) \times_\rho V_\rho$ on M. Then we construct first-order geometric differential operators called *gradients* on M as follows. Let ∇ be the covariant derivative on $\mathbf{S}_\rho$ induced by the Levi-Civita connection,

$$\Gamma(M, \mathbf{S}_\rho) \xrightarrow{\nabla} \Gamma(M, \mathbf{S}_\rho \otimes T^*(M)).$$

Since the tensor bundle $\mathbf{S}_\rho \otimes T^*(M)$ is also an associated vector bundle, we can decompose it into irreducible bundles with respect to SO(n),

$$\mathbf{S}_\rho \otimes T^*(M) = \sum_\lambda \mathbf{S}_\lambda.$$

Under this decomposition, the covariant derivative ∇ splits into the sum of differential operators, $\nabla = \sum_\lambda D_\lambda^\rho$. Thus, we have a first-order differential operator D_λ^ρ given by

$$D_\lambda^\rho : \Gamma(M, \mathbf{S}_\rho) \xrightarrow{\nabla} \Gamma(M, \mathbf{S}_\rho \otimes T^*(M)) \xrightarrow{\Pi_\lambda^\rho} \Gamma(M, \mathbf{S}_\lambda),$$

where Π_λ^ρ is the orthogonal projection from $\mathbf{S}_\rho \otimes T^*(M)$ onto $\mathbf{S}_\lambda$.

AMS Subject Classification: 53B20, 58J60, 17B35.

Many geometric first order differential operators can be realized as gradients. For example, the exterior derivative, its adjoint operator, and the conformal Killing operator are gradients on Riemannian manifolds. For the spin case, the Dirac operator, the twistor operator, and the Rarita–Schwinger operator are gradients on spin manifolds.

A significant feature common to the gradients is conformal covariance. For a smooth function σ on M, we change the Riemannian metric g to $g' = e^{2\sigma}g$ conformally. Then we have a principal bundle isomorphism

$$\mathbf{SO}(M) \ni p = \{\mathbf{e}_i\}_{i=1}^n \mapsto p' = \{\mathbf{e}'_i = e^{-\sigma}\mathbf{e}_i\}_{i=1}^n \in \mathbf{SO}'(M),$$

and a bundle isometry

$$\psi_\rho : \mathbf{S}_\rho = \mathbf{SO}(M) \times_\rho V_\rho \ni [p, \phi] \mapsto [p', \phi] \in \mathbf{SO}'(M) \times_\rho V_\rho = \mathbf{S}'_\rho.$$

Since the gradients are defined through the Riemannian metric, the gradients change under conformal deformation as follows (see [7]):

$$D'^\rho_\lambda = (e^{-(m(\lambda)+1)\sigma}\psi_\lambda) \circ D^\rho_\lambda \circ (e^{-m(\lambda)\sigma}\psi_\rho)^{-1} : \Gamma(M, \mathbf{S}'_\rho) \to \Gamma(M, \mathbf{S}'_\lambda).$$

Here, $m(\lambda)$ is a constant called *the conformal weight* depending on ρ and λ.

From recent research on gradients, we know various properties and applications of them: Bochner identities [1], vanishing theorems, eigenvalue estimates [3, 4], ellipticities [1], refined Kato inequalities [2, 6], spherical harmonics [5, 9], and gradients on Kähler manifolds [10]. Through these papers, we see the importance of conformal weights to study gradients.

Let us see how conformal weights are used in the author's paper [10], where the gradients on Kähler manifolds called *Kählerian gradients* are discussed. A result in [10] is to give Bochner identities or Bochner–Weitzenböck formulas for Kählerian gradients, and its method is as follows:

(i) We relate the principal symbols of Kählerian gradients with the enveloping algebra of $\mathfrak{u}(m)$ by using conformal weights. Here, m is the complex dimension of the underlying manifold. (ii) We find relations among the enveloping algebra and Casimir elements which correspond to the Bochner identities. (iii) These observations allow us to give the Bochner identities.

We can expect that this method is valid for gradients on Riemannian manifolds. Indeed, in this paper, we discuss the principal symbols of the gradients and give Bochner identities on Riemannian manifolds.

In [1], using spectral resolution on the standard sphere, Branson proved that, if there are N gradients $\{D^\rho_{\lambda_i}\}_{i=1}^N$ on the associated bundle $\mathbf{S}_\rho$, then there exist $[\frac{1}{2}N]$ independent Bochner identities. More precisely, the space of $(b_{\lambda_1}, \cdots, b_{\lambda_N}) \in \mathbb{R}^N$ such that

$$\sum_{1 \leq i \leq N} b_{\lambda_i}(D^\rho_{\lambda_i})^* D^\rho_{\lambda_i} = \text{(curvature action)} \tag{1.1}$$

has dimension $[\frac{1}{2}N]$ in $\mathbb{R}^N$, and the vector $(b_{\lambda_1}, \ldots, b_{\lambda_N})$ satisfies the following linear simultaneous equations:

$$\sum_{1 \leq i \leq N} b_{\lambda_i} \tilde{c}_{\rho \lambda_i} (m(\lambda_i) - \tfrac{1}{2}n + \tfrac{1}{2})^{2j} = 0, \quad j = 0, \ldots, [\tfrac{1}{2}(N+1)] - 1,$$

where $\tilde{c}_{\rho \lambda_i}$ is a constant depending on ρ and λ_i. Note that, in the same paper [1], Branson also answered the ellipticity problem. Thus, the Bochner identities on Riemannian manifolds are already known. But, our method is quite different from Branson's, and relevant to the paper [6] by Calderbank, Gauduchon and Herzlich. We deal with the principal symbols directly and find that their structure is controlled by the enveloping algebra and Casimir elements of $\mathfrak{so}(n)$. This observation leads us to explicit formulas of the Bochner identities for the gradients.

This paper is organized as follows. In section 12.2, we study the enveloping algebra and Casimir elements of $\mathfrak{so}(n)$, and give relations among them corresponding to Bochner identities. In section 12.3, we define the principal symbols of gradients called *the Clifford homomorphisms*. We relate them with the enveloping algebra by using conformal weights. As a corollary, we calculate the eigenvalues of Casimir elements. In the last section, we give all the Bochner identities for the gradients explicitly. In other words, we give explicit formulas of the vector $(b_{\lambda_1}, \ldots, b_{\lambda_N})$ and curvature action in (1.1).

Acknowledgement. The author is partially supported by the Grant-in-Aid for Scientific Research (No. 13740120) from the Ministry of Education, Culture, Sports, Science and Technology.

12.2   Enveloping algebra and Casimir elements

Let $\mathbb{R}^n$ be the n-dimensional Euclidean space with the standard basis $\{e_i\}_{i=1}^n$, and $\mathfrak{so}(n)$ be the Lie algebra of the spin group $\mathrm{Spin}(n)$ or the special orthogonal group $\mathrm{SO}(n)$. It is known that there is a natural isomorphism between $\bigwedge^2(\mathbb{R}^n)$ and $\mathfrak{so}(n)$ by associating $e_i \wedge e_j$ with a skew-symmetric endomorphism

$$(e_i \wedge e_j)(v) = \langle e_i, v \rangle e_j - \langle e_j, v \rangle e_i \quad \text{for any } v \in \mathbb{R}^n.$$

The basis of $\mathfrak{so}(n)$ consists of $\{e_i \wedge e_j\}_{i<j}$. For $i, j = 1, \ldots, n$, we put $e_{ij} := e_i \wedge e_j$ and have relations

$$e_{ij} = -e_{ji},$$
$$[e_{kl}, e_{ij}] = \delta_{ki} e_{lj} + \delta_{kj} e_{il} - \delta_{il} e_{kj} - \delta_{lj} e_{ik}$$

in $\mathfrak{so}(n)$.

We denote the complexification of $\mathfrak{so}(n)$ by $\mathfrak{so}(n, \mathbb{C})$ and its enveloping algebra by $U(\mathfrak{so}(n, \mathbb{C}))$. The center $\mathfrak{Z}$ of the enveloping algebra is characterized as the invariant subalgebra in $U(\mathfrak{so}(n, \mathbb{C}))$ under adjoint action of $\mathrm{SO}(n)$ or $\mathfrak{so}(n)$, that is,

$$\mathfrak{Z} = U(\mathfrak{so}(n, \mathbb{C}))^{\mathrm{SO}(n)}$$

(see [12]). We call elements in $\mathfrak{Z}$ *Casimir elements*. It follows from Schur's lemma that a Casimir element is a multiple of the identity on any irreducible $\mathfrak{so}(n)$-module.

We shall define some elements in $U(\mathfrak{so}(n, \mathbb{C}))$ whose traces are Casimir elements. For any non-negative integer q, we define an element $\mathbf{e}_{ij}^q$ with degree q in $U(\mathfrak{so}(n, \mathbb{C}))$ by

$$\mathbf{e}_{ij}^q := \begin{cases} \sum_{1 \le i_1, \dots, i_{q-1} \le n} \mathbf{e}_{ii_1} \mathbf{e}_{i_1 i_2} \cdots \mathbf{e}_{i_{q-1} j} & q \ge 1, \\ \delta_{ij} & q = 0. \end{cases} \tag{2.1}$$

This $\mathbf{e}_{ij}^q$ behaves like $\mathbf{e}_{ij}$ under adjoint action of $\mathfrak{so}(n)$.

Lemma 1. *The elements $\{\mathbf{e}_{ij}^q | q \in \mathbb{Z}_{\ge 0}, i, j = 1, \dots, n\}$ satisfy that*

$$[\mathbf{e}_{kl}, \mathbf{e}_{ij}^q] = \delta_{ki} \mathbf{e}_{lj}^q + \delta_{kj} \mathbf{e}_{il}^q - \delta_{il} \mathbf{e}_{kj}^q - \delta_{lj} \mathbf{e}_{ik}^q, \tag{2.2}$$

$$\sum_{1 \le k \le n} \mathbf{e}_{ik}^p \mathbf{e}_{kj}^q = \mathbf{e}_{ij}^{p+q}. \tag{2.3}$$

Proof. The second equation follows from the definition (2.1). To prove the first equation, we consider the adjoint action of $\mathfrak{so}(n)$ on $\mathbf{e}_{ij}^q$. Then we have

$$[\mathbf{e}_{kl}, \mathbf{e}_{ij}^q]$$
$$= \sum_{i_1, \dots, i_{q-1}} [\mathbf{e}_{kl}, \mathbf{e}_{ii_1}] \mathbf{e}_{i_1 i_2} \cdots \mathbf{e}_{i_{q-1} j} + \cdots + \sum_{i_1, \dots, i_{q-1}} \mathbf{e}_{ii_1} \mathbf{e}_{i_1 i_2} \cdots [\mathbf{e}_{kl}, \mathbf{e}_{i_{q-1} j}]$$
$$= \delta_{ki} \mathbf{e}_{lj}^q + \delta_{kj} \mathbf{e}_{il}^q - \delta_{il} \mathbf{e}_{kj}^q - \delta_{lj} \mathbf{e}_{ik}^q.$$

Thus, we have proved the lemma. $\qquad\square$

It follows from (2.2) that the trace $\sum_i \mathbf{e}_{ii}^q$ is invariant under adjoint action of $\mathfrak{so}(n)$.

Corollary 1. *The trace of $\mathbf{e}_{ij}^q$ is in the center $\mathfrak{Z}$ of $U(\mathfrak{so}(n, \mathbb{C}))$. Thus, we have the Casimir element $c_q := \sum_{i=1}^n \mathbf{e}_{ii}^q$ with degree q for $q = 0, 1, \dots$.*

Remark 1. We remark that c_0 is equal to n, c_1 is equal to 0, and $c_2 = \sum_{ij} \mathbf{e}_{ij} \mathbf{e}_{ji}$ is the usual Casimir element.

In the case of $n = 2m$, we have another Casimir element pf given by

$$pf := \sum_{\sigma \in \mathfrak{S}_{2m}} \operatorname{sign}(\sigma) \mathbf{e}_{\sigma(1)\sigma(2)} \mathbf{e}_{\sigma(3)\sigma(4)} \cdots \mathbf{e}_{\sigma(2m-1)\sigma(2m)}. \tag{2.4}$$

We call this Casimir element *the Pfaffian element*. From [11] or [12], we know the following facts about these Casimir elements:

1. In the case of $n = 2m$, $\{c_2, c_4, \dots, c_{2m-2}, pf\}$ generate the center $\mathfrak{Z}$ algebraically.

2. In the case of $n = 2m + 1$, $\{c_2, c_4, \ldots, c_{2m}\}$ generate the center $\mathfrak{Z}$ algebraically.

Now, the purpose in this section is to give algebraic identities for $\{\mathbf{e}_{ij}^q\}_q$, which will correspond to the Bochner identities. We can show by an induction that $\mathbf{e}_{ij}^q$ is a linear combination of $\{\mathbf{e}_{ji}^p\}_{p=0}^q$ whose coefficients are Casimir elements,

$$\mathbf{e}_{ij}^q = \sum_{p=0}^q a_{q,p} \mathbf{e}_{ji}^p, \quad a_{q,p} \in \mathfrak{Z}.$$

Here, we can obtain a recursion formula for $\{a_{q,p}\}_{q \geq p \geq 0}$. Since it is complicated to solve the recursion formula, we have to translate $\mathbf{e}_{ij}^q$ to another element $\hat{\mathbf{e}}_{ij}^q$. We define a translated element $\hat{\mathbf{e}}_{ij}$ by

$$\hat{\mathbf{e}}_{ij} := \mathbf{e}_{ij} + \tfrac{1}{2}(n-1)\delta_{ij},$$

and define $\hat{\mathbf{e}}_{ij}^q$ with degree q by

$$\hat{\mathbf{e}}_{ij}^q := \begin{cases} \sum_{1 \leq i_1, \ldots, i_{q-1} \leq n} \hat{\mathbf{e}}_{ii_1} \hat{\mathbf{e}}_{i_1 i_2} \cdots \hat{\mathbf{e}}_{i_{q-1} j}, & q \geq 1, \\ \delta_{ij}, & q = 0. \end{cases}$$

Note that $\hat{\mathbf{e}}_{ij}^q$ is related to $\mathbf{e}_{ij}^q$ with the equation

$$\hat{\mathbf{e}}_{ij}^q = \sum_{p=0}^q \binom{q}{p} \left(\frac{n-1}{2}\right)^{q-p} \mathbf{e}_{ij}^p.$$

We can easily show the following relations for $\{\hat{\mathbf{e}}_{ij}^q\}_q$ in the same way as Lemma 1.

Lemma 2. *The translated elements* $\{\hat{\mathbf{e}}_{ij}^q | q \in \mathbb{Z}_{\geq 0}, \ i, j = 1, \ldots, n\}$ *satisfy that*

$$[\hat{\mathbf{e}}_{kl}, \hat{\mathbf{e}}_{ij}^q] = \delta_{ki}\hat{\mathbf{e}}_{lj}^q + \delta_{kj}\hat{\mathbf{e}}_{il}^q - \delta_{il}\hat{\mathbf{e}}_{kj}^q - \delta_{lj}\hat{\mathbf{e}}_{ik}^q, \tag{2.5}$$

$$\sum_k \hat{\mathbf{e}}_{ik}^p \hat{\mathbf{e}}_{kj}^q = \hat{\mathbf{e}}_{ij}^{p+q}, \tag{2.6}$$

$$\hat{\mathbf{e}}_{ij} = -\hat{\mathbf{e}}_{ji} + (n-1)\delta_{ij}. \tag{2.7}$$

By using this lemma, we have algebraic identities for $\hat{\mathbf{e}}_{ij}^q$ and $\hat{c}_q := \sum_i \hat{\mathbf{e}}_{ii}^q$.

Theorem 1. *The translated element* $\hat{\mathbf{e}}_{ij}^q$ *is a linear combination of* $\{\hat{\mathbf{e}}_{ji}^p\}_{p=0}^q$ *whose coefficients are Casimir elements,*

$$\hat{\mathbf{e}}_{ij}^q = (-1)^q \hat{\mathbf{e}}_{ji}^q - \tfrac{1}{2}(1 - (-1)^q)\hat{\mathbf{e}}_{ji}^{q-1} + \sum_{p=0}^{q-1} (-1)^p \hat{c}_{q-1-p} \hat{\mathbf{e}}_{ji}^p. \tag{2.8}$$

Here, $\hat{c}_q := \sum_i \hat{e}_{ii}^q$ is in the center $\mathfrak{Z}$. Thus, we have

$$\hat{e}_{ij}^{2q} = \hat{e}_{ji}^{2q} + \sum_{p=0}^{2q-1} (-1)^p \hat{c}_{2q-1-p} \hat{e}_{ji}^p, \tag{2.9}$$

$$\hat{e}_{ij}^{2q+1} = -\hat{e}_{ji}^{2q+1} - \hat{e}_{ji}^{2q} + \sum_{p=0}^{2q} (-1)^p \hat{c}_{2q-p} \hat{e}_{ji}^p. \tag{2.10}$$

Proof. From (2.5)–(2.7), we have

$$\begin{aligned}
\hat{e}_{ij}^{q+1} &= \sum_k [\hat{e}_{ik}, \hat{e}_{kj}^q] + \sum_k \hat{e}_{kj}^q \hat{e}_{ik} \\
&= \sum_k (\delta_{ik} \hat{e}_{kj}^q + \delta_{ij} \hat{e}_{kk}^q - \delta_{kk} \hat{e}_{ij}^q - \delta_{kj} \hat{e}_{ki}^q) \\
&\qquad\qquad\qquad + \sum_k \hat{e}_{kj}^q (-\hat{e}_{ki} + (n-1)\delta_{ki}) \\
&= \delta_{ji} \hat{c}_q - \hat{e}_{ji}^q - \sum_{k=1}^n \hat{e}_{kj}^q \hat{e}_{ki}. \tag{2.11}
\end{aligned}$$

Setting $\hat{e}_{ij}^q = \sum_{p=0}^q \hat{a}_{q,p} \hat{e}_{ji}^p$, we shall produce a recursion formula for $\{\hat{a}_{q,p}\}_{q,p}$, where each $\hat{a}_{q,p}$ is in the center $\mathfrak{Z}$. It follows from (2.11) that

$$\begin{aligned}
\hat{e}_{ij}^{q+1} &= \delta_{ji} \hat{c}_q - \hat{e}_{ji}^q - \sum_k \hat{e}_{kj}^q \hat{e}_{ki} \\
&= \delta_{ji} \hat{c}_q - \hat{e}_{ji}^q - \sum_k \sum_p \hat{a}_{q,p} \hat{e}_{jk}^p \hat{e}_{ki} \\
&= \delta_{ji} \hat{c}_q - \hat{e}_{ji}^q - \sum_{p=0}^q \hat{a}_{q,p} \hat{e}_{ji}^{p+1} \\
&= -\hat{a}_{q,q} \hat{e}_{ji}^{q+1} + (-\hat{a}_{q,q-1} - 1)\hat{e}_{ji}^q - \sum_{p=0}^{q-2} \hat{a}_{q,p} \hat{e}_{ji}^{p+1} + \hat{c}_q \delta_{ji} \\
&= \sum_{p=0}^{q+1} \hat{a}_{q+1,p} \hat{e}_{ji}^p.
\end{aligned}$$

Then we have a recursion formula for $\{\hat{a}_{q,p}\}_{q \geq p \geq 0}$,

$$\hat{a}_{q+1,p} = \begin{cases}
-\hat{a}_{q,q}, & p = q+1, \\
-\hat{a}_{q,q-1} - 1, & p = q, \\
-\hat{a}_{q,p-1}, & 1 \leq p \leq q-1, \\
\hat{c}_q, & p = 0.
\end{cases}$$

We can easily solve it and have

$$
\hat{a}_{q+1,p} = \begin{cases} (-1)^{q+1}, & p = q+1, \\ (-1)^q(n-1) - \frac{1-(-1)^q}{2} = (-1)^q \hat{c}_0 - \frac{1-(-1)^{q+1}}{2}, & p = q, \\ (-1)^p \hat{c}_{q-p}, & 0 \le p \le q-1. \end{cases}
$$

Thus, we have proved the theorem. □

Take the trace on the equation (2.8), and we have identities for the Casimir elements $\{\hat{c}_q\}_q$. The trace of (2.9) is a trivial identity, while the trace of (2.10) gives the following corollary.

Corollary 2. *The Casimir elements* $\{\hat{c}_0, \hat{c}_1, \hat{c}_2, \dots\}$ *satisfy*

$$
2\hat{c}_{2q+1} = -\hat{c}_{2q} + \sum_{p=0}^{2q} (-1)^p \hat{c}_{2q-p} \hat{c}_p \tag{2.12}
$$

for $q = 0, 1, \dots$. *In particular, the Casimir element* $\hat{c}_{2q+1}$ *with odd degree can be represented as a polynomial of* $\{\hat{c}_0, \hat{c}_2, \dots, \hat{c}_{2q}\}$.

12.3   The Clifford homomorphisms: The principal symbols of gradients

In this section, we shall discuss the Clifford homomorphisms introduced by the author as a generalization of the Clifford multiplication in [8] and [10].

Let (π_ρ, V_ρ) be an irreducible unitary representation of the Lie algebra $\mathfrak{so}(n)$ with highest weight ρ. The highest weight ρ is a vector $\rho = (\rho^1, \dots, \rho^m)$ in $\mathbb{Z}^m \cup (\mathbb{Z} + \frac{1}{2})^m$ satisfying the dominant condition (see [12]):

$$
\rho^1 \ge \rho^2 \ge \cdots \ge \rho^{m-1} \ge |\rho^m|, \qquad n = 2m,
$$
$$
\rho^1 \ge \rho^2 \ge \cdots \ge \rho^m \ge 0, \qquad n = 2m+1.
$$

Conversely, every dominant integral or half-integral weight is the highest weight of an irreducible unitary representation. We denote the standard basis of $\mathbb{Z}^m$ by $\{\mu_i\}_{i=1}^m$,

$$
\mu_i = (\underbrace{0, \dots, 0}_{i-1}, 1, \underbrace{0, \dots, 0}_{m-i}).
$$

With this notation, (π_{μ_1}, V_{μ_1}) is the natural representation of $\mathfrak{so}(n)$ on $\mathbb{R}^n \otimes \mathbb{C}$.

Now, we consider the tensor representation $(\pi_\rho \otimes \pi_{\mu_1}, V_\rho \otimes_{\mathbb{C}} \mathbb{R}^n)$ and its irreducible decomposition

$$
V_\rho \otimes \mathbb{R}^n = \sum_\lambda V_\lambda,
$$

where we adopt the inner product on V_λ induced by the tensor inner product on $V_\rho \otimes \mathbb{R}^n$. The highest weights of irreducible components occur with multiplicity 1, and are characterized as follows (see [7]):

1. When $n = 2m$ or when $n = 2m + 1$ and $\rho^m = 0$, $\lambda = \rho \pm \mu_i$ for $i = 1, \ldots, m$ such that λ is dominant.

2. When $n = 2m + 1$ and $\rho^m > 0$, $\lambda = \rho$ or $\lambda = \rho \pm \mu_i$ for $i = 1, \ldots, m$ such that λ is dominant.

The Clifford homomorphism is defined through the orthogonal projection from $V_\rho \otimes \mathbb{R}^n$ onto each irreducible component.

Definition 1. *For ξ in $\mathbb{R}^n$, we define a linear mapping $p_\lambda^\rho(\xi)$ from V_ρ to V_λ by*

$$\mathbb{R}^n \times V_\rho \ni (\xi, \phi) \mapsto p_\lambda^\rho(\xi)\phi := \Pi_\lambda^\rho(\phi \otimes \xi) \in V_\lambda,$$

where Π_λ^ρ is the orthogonal projection from $V_\rho \otimes \mathbb{R}^n$ onto irreducible component V_λ. We denote by $p_\lambda^\rho(\xi)^*$ the adjoint of $p_\lambda^\rho(\xi)$ with respect to the inner products on V_ρ and V_λ. We call these linear mappings $p_\lambda^\rho(\xi)$ and $p_\lambda^\rho(\xi)^*$ the Clifford homomorphisms associated to ρ and λ.

Example 1. Let $(\pi_{\Delta^\pm}, V_{\Delta^\pm})$ be the spinor representation with highest weight $\Delta^\pm = (\frac{1}{2}, \ldots, \frac{1}{2}, \pm\frac{1}{2})$. There are two components $V_{\Delta^\mp}$ and $V_{T^\pm}$ in $V_{\Delta^\pm} \otimes \mathbb{R}^n$, where $T^\pm$ is $(\frac{3}{2}, \frac{1}{2}, \ldots, \frac{1}{2}, \pm\frac{1}{2})$. The Clifford homomorphism $p_{\Delta^\mp}^{\Delta^\pm}(u) : V_{\Delta^\pm} \to V_{\Delta^\mp}$ is the usual Clifford multiplication on spinor space up to a normalization.

The next proposition says that the Clifford homomorphisms are related to the enveloping algebra.

Proposition 1 ([6, 8]). *Let $\{\mathbf{e}_i\}_{i=1}^n$ be the standard orthonormal basis of $\mathbb{R}^n$. The Clifford homomorphisms $\{p_\lambda^\rho\}_\lambda$ satisfy*

$$\sum_\lambda (-m(\lambda))^q p_\lambda^\rho(\mathbf{e}_i)^* p_\lambda^\rho(\mathbf{e}_j) = \pi_\rho(\mathbf{e}_{ij}^q) \tag{3.1}$$

for $q = 0, 1, \ldots$, and $i, j = 1, \ldots, n$. The constant $m(\lambda)$ is called the conformal weight associated to ρ and λ defined by

$$m(\lambda) = \tfrac{1}{2}(n - \|\delta + \lambda\|^2 + \|\delta + \rho\|^2 - 1),$$

where δ is half the sum of the positive roots of $\mathfrak{so}(n)$, and $\| \cdot \|$ is the standard norm on the space of weights. In particular, we have

$$\sum_\lambda (-m(\lambda))^q \sum_i p_\lambda^\rho(\mathbf{e}_i)^* p_\lambda^\rho(\mathbf{e}_i) = \pi_\rho(c_q).$$

Remark 2. We can easily show that the conformal weights associated to irreducible components differ from each other except the following case (see [6]). When $n = 2m$, $\rho^{m-1} > 0$, and $\rho^m = 0$, we have irreducible components $V_{\rho+\mu_m}$ and $V_{\rho-\mu_m}$ in $V_\rho \otimes \mathbb{R}^n$ whose conformal weights coincide, $m(\rho + \mu_m) = m(\rho - \mu_m)$. We call this case *the exceptional case*.

Proof. First, we shall prove (3.1) for $q = 0$. For ϕ and ψ in V_ρ, we have

$$\langle \phi \otimes \mathbf{e}_i, \psi \otimes \mathbf{e}_j \rangle = \langle \phi, \psi \rangle \langle \mathbf{e}_i, \mathbf{e}_j \rangle = \delta_{ij} \langle \phi, \psi \rangle.$$

On the other hand, we show from the definition of the Clifford homomorphisms that

$$\langle \phi \otimes \mathbf{e}_i, \psi \otimes \mathbf{e}_j \rangle = \sum_\lambda \langle p_\lambda^\rho(\mathbf{e}_i)\phi, p_\lambda^\rho(\mathbf{e}_j)\psi \rangle = \Big\langle \sum_\lambda p_\lambda^\rho(\mathbf{e}_j)^* p_\lambda^\rho(\mathbf{e}_i)\phi, \psi \Big\rangle.$$

Since the above two equations are valid for any ϕ and ψ, we have

$$\sum_\lambda p_\lambda^\rho(\mathbf{e}_j)^* p_\lambda^\rho(\mathbf{e}_i) = \delta_{ji}. \tag{3.2}$$

In particular, for ξ and η in $\mathbb{R}^n$,

$$\sum_\lambda p_\lambda^\rho(\xi)^* p_\lambda^\rho(\eta) = \langle \xi, \eta \rangle.$$

Next, to prove (3.1) for $q \geq 1$, we need an operator C on $V_\rho \otimes \mathbb{R}^n$ defined by

$$C := \pi_\rho \otimes \pi_{\mu_1}(c_2) - \pi_\rho(c_2) \otimes \mathrm{id} - \mathrm{id} \otimes \pi_{\mu_1}(c_2).$$

Here, c_2 is the usual Casimir element $c_2 = \sum_{ij} \mathbf{e}_{ij}\mathbf{e}_{ji}$, which is

$$2(\|\delta + \rho\|^2 - \|\delta\|^2)\mathrm{id}$$

on irreducible $\mathfrak{so}(n)$-module V_ρ (see [7]). Then the operator C is $-4m(\lambda)\mathrm{id}$ on irreducible component V_λ of $V_\rho \otimes \mathbb{R}^n$. On the other hand, we can show that the operator C is realized by

$$C = 2 \sum_{ij} \pi_\rho(\mathbf{e}_{ij}) \otimes \pi_{\mu_1}(\mathbf{e}_{ji}).$$

So, for $\phi \otimes \mathbf{e}_i$ in $V_\rho \otimes \mathbb{R}^n$, we have

$$C(\phi \otimes \mathbf{e}_i) = 2 \sum_{kl} \pi_\rho(\mathbf{e}_{kl})\phi \otimes \pi_{\mu_1}(\mathbf{e}_{lk})\mathbf{e}_i = 2 \sum_{kl} \pi_\rho(\mathbf{e}_{kl})\phi \otimes (\delta_{il}\mathbf{e}_k - \delta_{ki}\mathbf{e}_l)$$

$$= 4 \sum_k \pi_\rho(\mathbf{e}_{ki})\phi \otimes \mathbf{e}_k = 4 \sum_\lambda \sum_k p_\lambda^\rho(\mathbf{e}_k)\pi_\rho(\mathbf{e}_{ki})\phi.$$

As a result, we have $\sum_k p_\lambda^\rho(\mathbf{e}_k)\pi_\rho(\mathbf{e}_{ki}) = -m(\lambda)p_\lambda^\rho(\mathbf{e}_i)$ for any λ. This equation and (3.2) induce (3.1) for $q \geq 1$. $\square$

If we translate $m(\lambda)$ to $\hat{m}(\lambda) := m(\lambda) - \frac{1}{2}n + \frac{1}{2}$, then we can similarly show that

$$\sum_\lambda (-\hat{m}(\lambda))^q p_\lambda^\rho(\mathbf{e}_i)^* p_\lambda^\rho(\mathbf{e}_j) = \pi_\rho(\hat{\mathbf{e}}_{ij}^q). \tag{3.3}$$

Since we have known the relation (2.8) for $\hat{\mathbf{e}}_{ij}^q$, we obtain algebraic relations for the Clifford homomorphisms.

Theorem 2. *Let $\hat{m}(\lambda) = m(\lambda) - \frac{1}{2}n + \frac{1}{2}$ be the translated conformal weight associated to ρ and λ. Then there are the following algebraic relations for the Clifford homomorphisms: for ξ and η in $\mathbb{R}^n$,*

$$\sum_\lambda \hat{m}(\lambda)^{2q} p^\rho_\lambda(\xi)^* p^\rho_\lambda(\eta)$$

$$= \sum_\lambda \left\{ \hat{m}(\lambda)^{2q} + \sum_{p=0}^{2q-1} \hat{m}(\lambda)^p \pi_\rho(\hat{c}_{2q-1-p}) \right\} p^\rho_\lambda(\eta)^* p^\rho_\lambda(\xi), \quad (3.4)$$

$$\sum_\lambda \hat{m}(\lambda)^{2q+1} p^\rho_\lambda(\xi)^* p^\rho_\lambda(\eta)$$

$$= \sum_\lambda \left\{ -\hat{m}(\lambda)^{2q+1} + \hat{m}(\lambda)^{2q} - \sum_{p=0}^{2q} \hat{m}(\lambda)^p \pi_\rho(\hat{c}_{2q-p}) \right\} p^\rho_\lambda(\eta)^* p^\rho_\lambda(\xi). \quad (3.5)$$

and

$$\sum_\lambda p^\rho_\lambda(\xi)^* p^\rho_\lambda(\eta) = \langle \xi, \eta \rangle. \quad (3.6)$$

Example 2. On spinor space, these relations give the usual Clifford relation $\xi \cdot \eta \cdot + \eta \cdot \xi \cdot = -2\langle \xi, \eta \rangle$.

For $n = 2m$, we get another identity through the Pfaffian element pf in (2.4). Considering the action pf on $V_\rho \otimes \mathbb{R}^n$, we can relate the Pfaffian element pf to the Clifford homomorphisms $\{p^\rho_\lambda\}_\lambda$ as follows.

Proposition 2 ([8]). *The Clifford homomorphisms $\{p^\rho_\lambda\}_\lambda$ satisfy that*

$$\sum_\lambda \pi_\lambda(pf) p^\rho_\lambda(e_j)^* p^\rho_\lambda(e_i) = \delta_{ij} \pi_\rho(pf)$$

$$+ 2m(1 - \delta_{ij}) \text{sign} \left(\begin{smallmatrix} 1 & 2 & 3 & \cdots & 2m \\ i & j & 1 & \cdots & 2m \end{smallmatrix} \right)$$

$$\times \sum_{\sigma \in \mathfrak{S}^{i,j}_{2m}} \text{sign}(\sigma) \pi_\rho(e_{\sigma(1)\sigma(2)}) \cdots \pi_\rho(e_{\sigma(2m-1)\sigma(2m)}),$$

where $\mathfrak{S}^{i,j}_{2m}$ is the permutation group of $\{1, \ldots, 2m\} \setminus \{i, j\}$. In particular, we have

$$\sum_\lambda \pi_\lambda(pf)(p^\rho_\lambda(\xi)^* p^\rho_\lambda(\eta) + p^\rho_\lambda(\eta)^* p^\rho_\lambda(\xi)) = 2\pi_\rho(pf)\langle \xi, \eta \rangle \quad (3.7)$$

for ξ and η in $\mathbb{R}^n$.

In the rest of this section, we calculate the constant $\pi_\rho(\hat{c}_q)$, that is, the eigenvalues of $\hat{c}_q$. Here, our method is based on [6] or [11]. First, we need the following lemma.

Lemma 3. *The orthogonal projection $\Pi^\rho_\lambda : V_\rho \otimes \mathbb{R}^n \to V_\lambda \subset V_\rho \otimes \mathbb{R}^n$ is realized as follows:*

$$\Pi^\rho_\lambda(\phi \otimes u) = \sum_i p^\rho_\lambda(e_i)^* p^\rho_\lambda(u)\phi \otimes e_i. \quad (3.8)$$

Proof. We can easily show that the Clifford homomorphisms are compatible with the action of $SO(n)$,

$$p_\lambda^\rho(\pi_{\mu_1}(g)u) = \pi_\lambda(g)p_\lambda^\rho(u)\pi_\rho(g^{-1})$$

for g in $SO(n)$ and u in $\mathbb{R}^n$. It follows that the following mapping is an $\mathfrak{so}(n)$-equivariant injection,

$$\Phi_\lambda : V_\lambda \ni \psi \mapsto \sum_i p_\lambda^\rho(\mathbf{e}_i)^* \psi \otimes \mathbf{e}_i \in V_\rho \otimes \mathbb{R}^n.$$

Then we decompose $\phi \otimes u$,

$$\phi \otimes u = \sum_i \langle \mathbf{e}_i, u \rangle \phi \otimes \mathbf{e}_i$$

$$= \sum_i \sum_\lambda p_\lambda^\rho(\mathbf{e}_i)^* p_\lambda^\rho(u)\phi \otimes \mathbf{e}_i = \sum_\lambda \sum_i p_\lambda^\rho(\mathbf{e}_i)^* p_\lambda^\rho(u)\phi \otimes \mathbf{e}_i,$$

and have the projection formula (3.8). □

Proposition 3. *We set* $d(\rho) := \dim V_\rho$. *The Clifford homomorphism* p_λ^ρ *satisfies*

$$\sum_i p_\lambda^\rho(\mathbf{e}_i)^* p_\lambda^\rho(\mathbf{e}_i) = d(\lambda)/d(\rho).$$

Hence, eigenvalues of c_q *and* $\hat{c}_q$ *on irreducible* $\mathfrak{so}(n)$-*module* V_ρ *are*

$$\pi_\rho(c_q) = \frac{1}{d(\rho)} \sum_\lambda (-m(\lambda))^q d(\lambda), \quad \pi_\rho(\hat{c}_q) = \frac{1}{d(\rho)} \sum_\lambda (-\hat{m}(\lambda))^q d(\lambda).$$

Moreover, we have a relation for eigenvalues of the Pfaffian element,

$$n\pi_\rho(pf) = \frac{1}{d(\rho)} \sum_\lambda \pi_\lambda(pf)d(\lambda).$$

Proof. Let $\{\phi_s\}_{s=1}^{\dim V_\rho}$ be an orthonormal basis of V_ρ. Taking the trace of Π_λ^ρ, we have

$$d(\lambda) = \sum_{s,i} (\Pi_\lambda^\rho(\phi_s \otimes \mathbf{e}_i), \phi_s \otimes \mathbf{e}_i)$$

$$= \sum \left(\sum_j p_\lambda^\rho(\mathbf{e}_j)^* p_\lambda^\rho(\mathbf{e}_i)(\phi_s) \otimes \mathbf{e}_j, \phi_s \otimes \mathbf{e}_i \right)$$

$$= \sum \left(\sum_j p_\lambda^\rho(\mathbf{e}_j)^* p_\lambda^\rho(\mathbf{e}_i)(\phi_s), \phi_s \right) \delta_{ij} = \sum_s (\phi_s, \phi_s) \sum_i p_\lambda^\rho(\mathbf{e}_i)^* p_\lambda^\rho(\mathbf{e}_i)$$

$$= d(\rho) \sum_i p_\lambda^\rho(\mathbf{e}_i)^* p_\lambda^\rho(\mathbf{e}_i).$$

Remark 3. In [11] or [12], we know that the eigenvalue of the Pfaffian element on irreducible $\mathfrak{so}(n)$-module V_ρ is

$$\pi_\rho(pf) = (\sqrt{-1})^m 2^m m!(\rho^1 + m - 1)(\rho^2 + m - 2)\cdots(\rho^{m-1} + 1)\rho^m. \quad (3.9)$$

12.4   The Bochner identities for the gradients

In this section, we give explicit formulas of all the Bochner identities for gradients on Riemannian or spin manifolds. We consider only the gradients on Riemannian manifolds. Since the discussions in this section are valid for the gradients on spin manifolds, the spin case is left to the readers. Let M be an oriented Riemannian manifold, and $\mathbf{SO}(M)$ be the principal $SO(n)$ bundle of the oriented orthonormal frames on M. For an irreducible unitary representation (π_ρ, V_ρ) of $SO(n)$, we have an associated vector bundle $\mathbf{S}_\rho := \mathbf{SO}(M) \times_\rho V_\rho$. The gradients are first-order differential operators realized as irreducible components of the covariant derivative ∇ as follows:

$$D_\lambda^\rho : \Gamma(M, \mathbf{S}_\rho) \xrightarrow{\nabla} \Gamma(M, \mathbf{S}_\rho \otimes T^*(M)) \xrightarrow{\Pi_\lambda^\rho} \Gamma(M, \mathbf{S}_\lambda).$$

Here, the covariant derivative ∇ is the one induced by the Levi-Civita connection and Π_λ^ρ is the orthogonal projection defined fiberwise. We call this first-order differential operator D_λ^ρ the *gradient associated to ρ and λ*.

We have already known the principal symbol of the gradient D_λ^ρ, that is, the Clifford homomorphism associated to ρ and λ. So there is a formula for the gradient D_λ^ρ like the Dirac operator,

$$D_\lambda^\rho = \sum_i p_\lambda^\rho(\mathbf{e}_i) \nabla_{\mathbf{e}_i} : \Gamma(M, \mathbf{S}_\rho) \to \Gamma(M, \mathbf{S}_\lambda),$$

where $\{\mathbf{e}_i\}_{i=1}^n$ is a local orthonormal frame. Furthermore, we have a formula of the formal adjoint operator of D_λ^ρ,

$$(D_\lambda^\rho)^* = \sum_i -p_\lambda^\rho(\mathbf{e}_i)^* \nabla_{\mathbf{e}_i} : \Gamma(M, \mathbf{S}_\lambda) \to \Gamma(M, \mathbf{S}_\rho).$$

The algebraic identities (3.4)–(3.6) allow us to have the Bochner identities for the gradients $\{D_\lambda^\rho\}_\lambda$. But, because of non-commutativity of ∇, we need the following curvature endomorphisms on $\mathbf{S}_\rho$.

Definition 2. *Let $R_\rho(X, Y)$ be the curvature of ∇ on $\mathbf{S}_\rho$ for vector fields X and Y, and $\{\mathbf{e}_i\}_{i=1}^n$ be a local orthonormal frame of M. We define the curvature endomorphism $\hat{R}_\rho^q$ for $q = 1, 2, \ldots,$ by*

$$\hat{R}_\rho^q := \sum_{i,j} \pi_\rho(\hat{\mathbf{e}}_{ij}^q) R_\rho(\mathbf{e}_i, \mathbf{e}_j) \in \Gamma(M, \mathrm{End}(\mathbf{S}_\rho)).$$

Now, we are in a position to obtain the Bochner identities.

Theorem 3 (The Bochner identities for the gradients). *Let $\{D_\lambda^\rho\}_\lambda$ be the gradients on $\Gamma(M, \mathbf{S}_\rho)$, and $\{(D_\lambda^\rho)^*\}_\lambda$ be their formal adjoints. There exist the following identities for the gradients:*

$$\sum_\lambda \left\{ \sum_{p=0}^{2q-1} \pi_\rho(\hat{c}_{2q-1-p}) \hat{m}(\lambda)^p \right\} (D_\lambda^\rho)^* D_\lambda^\rho = \hat{R}_\rho^{2q} \tag{4.1}$$

for $q = 1, 2, \ldots,$ and

$$\sum_\lambda (D_\lambda^\rho)^* D_\lambda^\rho = \nabla^* \nabla. \tag{4.2}$$

Here, $\hat{R}_\rho^q$ is the curvature endomorphism and $\nabla^* \nabla$ is the connection Laplacian on $\mathbf{S}_\rho$. In the exceptional case such that $n = 2m$, $\rho^{m-1} > 0$, and $\rho^m = 0$, there exists another identity

$$2((D_{\rho+\mu_m}^\rho)^* D_{\rho+\mu_m}^\rho - (D_{\rho-\mu_m}^\rho)^* D_{\rho-\mu_m}^\rho)$$
$$= -\sum_{i,j}(p_{\rho+\mu_m}^\rho(\mathbf{e}_i)^* p_{\rho+\mu_m}^\rho(\mathbf{e}_j) - p_{\rho-\mu_m}^\rho(\mathbf{e}_i)^* p_{\rho-\mu_m}^\rho(\mathbf{e}_j)) R_\rho(\mathbf{e}_i, \mathbf{e}_j). \tag{4.3}$$

Proof. We define a second-order differential operator $\nabla_{X,Y}^2$ on $\Gamma(M, \mathbf{S}_\rho)$ for vector fields X and Y by $\nabla_{X,Y}^2 := \nabla_X \nabla_Y - \nabla_{\nabla_X Y}$. It follows from (3.3) that

$$\sum_\lambda (-\hat{m}(\lambda))^q (D_\lambda^\rho)^* D_\lambda^\rho = -\sum_{ij} \pi_\rho(\hat{\mathbf{e}}_{ij}^q) \nabla_{\mathbf{e}_i, \mathbf{e}_j}^2.$$

In particular, when q is zero, we have (4.2).

To give the identities (4.1), we use the algebraic relation (2.9) for the enveloping algebra. Then,

$$\hat{R}_\rho^{2q} = \sum_{i,j} \pi_\rho(\hat{\mathbf{e}}_{ij}^{2q})(\nabla_{\mathbf{e}_i, \mathbf{e}_j}^2 - \nabla_{\mathbf{e}_j, \mathbf{e}_i}^2)$$

$$= -\sum_\lambda \hat{m}(\lambda)^{2q}(D_\lambda^\rho)^* D_\lambda^\rho - \sum_{i,j} \pi_\rho\Big(\hat{\mathbf{e}}_{ji}^{2q} + \sum_p (-1)^p \hat{c}_{2q-1-p} \hat{\mathbf{e}}_{ji}^p\Big)\nabla_{\mathbf{e}_j, \mathbf{e}_i}^2$$

$$= \sum_\lambda \Big\{ \sum_{p=0}^{2q-1} \pi_\rho(\hat{c}_{2q-1-p})\hat{m}(\lambda)^p \Big\}(D_\lambda^\rho)^* D_\lambda^\rho.$$

In the exceptional case, we show from (3.9) that the eigenvalue $\pi_\lambda(pf)$ on V_λ is zero except $\lambda = \rho \pm \mu_m$. Furthermore, we know $\pi_\rho(pf) = 0$ and $\pi_{\rho+\mu_m}(pf) = -\pi_{\rho-\mu_m}(pf)$. So, the relation (3.7) for the exceptional case becomes

$$(p_{\rho+\mu_m}^\rho(\mathbf{e}_j)^* p_{\rho+\mu_m}^\rho(\mathbf{e}_i) + p_{\rho+\mu_m}^\rho(\mathbf{e}_i)^* p_{\rho+\mu_m}^\rho(\mathbf{e}_j))$$
$$= (p_{\rho-\mu_m}^\rho(\mathbf{e}_j)^* p_{\rho-\mu_m}^\rho(\mathbf{e}_i) + p_{\rho-\mu_m}^\rho(\mathbf{e}_i)^* p_{\rho-\mu_m}^\rho(\mathbf{e}_j)).$$

This equation induces the identity (4.3).   □

Remark 4. The algebraic relation (2.10) induces other identities

$$\hat{R}_\rho^{2q+1} = \sum_\lambda \Big\{ 2\hat{m}(\lambda)^{2q+1} - \hat{m}(\lambda)^{2q} + \sum_{p=0}^{2q} \pi_\rho(\hat{c}_{2q-p})\hat{m}(\lambda)^p \Big\}(D_\lambda^\rho)^* D_\lambda^\rho$$

for $q = 0, 1, \ldots$. But, from the discussion below, these identities are linearly dependent on (4.1).

We shall discuss linear independence of our Bochner identities (4.1). We assume that there are N irreducible components in $V_\rho \otimes \mathbb{R}^n$, and hence, N gradients $\{D^\rho_{\lambda_i}\}_{i=1}^N$ on $\Gamma(M, \mathbf{S}_\rho)$. Our aim is to prove that the identities (4.1) for $q = 1, \ldots, [\frac{1}{2}N]$ are independent. In other words, if we define the vector $v(q)$ of the coefficients in (4.1) by

$$v(q) := \left(\sum_{p=0}^{2q-1} \pi_\rho(\hat{c}_{2q-1-p})\hat{m}(\lambda_1)^p, \ldots, \sum_{p=0}^{2q-1} \pi_\rho(\hat{c}_{2q-1-p})\hat{m}(\lambda_N)^p \right) \in \mathbb{R}^N,$$

then we would prove that $v(1), v(2), \ldots, v([\frac{1}{2}N])$ are linearly independent in $\mathbb{R}^N$.

For $q = 1, 2, \ldots$, we decompose $(v(1), v(2), \ldots, v(q))$ into the product of a $q \times 2q$ matrix $C(q)$ and a $2q \times N$ matrix $M(q)$ given by

$$C(q) := \begin{pmatrix} \pi_\rho(\hat{c}_1) & \pi_\rho(\hat{c}_0) & 0 & 0 & \cdots & 0 & 0 \\ \pi_\rho(\hat{c}_3) & \pi_\rho(\hat{c}_2) & \pi_\rho(\hat{c}_1) & \pi_\rho(\hat{c}_0) & \cdots & 0 & 0 \\ \cdots & \cdots & \cdots & \cdots & \cdots & \cdots & \cdots \\ \pi_\rho(\hat{c}_{2q-1}) & \pi_\rho(\hat{c}_{2q-2}) & \cdots & \cdots & \cdots & \pi_\rho(\hat{c}_1) & \pi_\rho(\hat{c}_0) \end{pmatrix},$$

$$M(q) := \begin{pmatrix} 1 & 1 & 1 & \cdots & 1 \\ \hat{m}(\lambda_1) & \hat{m}(\lambda_2) & \hat{m}(\lambda_3) & \cdots & \hat{m}(\lambda_N) \\ \hat{m}(\lambda_1)^2 & \hat{m}(\lambda_2)^2 & \hat{m}(\lambda_3)^2 & \cdots & \hat{m}(\lambda_N)^2 \\ \cdots & \cdots & \cdots & \cdots & \cdots \\ \hat{m}(\lambda_1)^{2q-1} & \hat{m}(\lambda_2)^{2q-1} & \hat{m}(\lambda_3)^{2q-1} & \cdots & \hat{m}(\lambda_N)^{2q-1} \end{pmatrix}.$$

Since the conformal weights are different from each other, the rank of the matrix $(v(1), v(2), \ldots, v([\frac{1}{2}N])) = C([\frac{1}{2}N])M([\frac{1}{2}N])$ is $[\frac{1}{2}N]$ except the exceptional case. For the exceptional case, the rank of $C([\frac{1}{2}N])M([\frac{1}{2}N])$ is $[\frac{1}{2}N] - 1$. But, there is another identity (4.3) independent of (4.1).

Thus, if there are N gradients $\{D^\rho_{\lambda_i}\}_{i=1}^N$, then we have at least $[\frac{1}{2}N]$ independent Bochner identities. In [1], by using spectral resolution on the standard sphere S^n, Branson proved that the number of independent identities is just $[\frac{1}{2}N]$. Hence we conclude that

Corollary 3. *The identities* (4.1) *and* (4.3) *give all the Bochner identities for the gradients.*

REFERENCES

[1] T. Branson, Stein-Weiss operators and ellipticity, *J. Funct. Anal.* **151** (1997), 334–383.

[2] T. Branson, Kato constants in Riemannian geometry, *Math. Res. Lett.* **7** (2000), 245–261.

[3] T. Branson and O. Hijazi, Vanishing theorems and eigenvalue estimates in Riemannian spin geometry, *Internat. J. Math.* **8** (1997), 921–934.

[4] T. Branson and O. Hijazi, Improved forms of some vanishing theorems in Riemannian spin geometry, *Internat. J. Math.* **11** (2000), 291–304.

[5] J. Bureš, The higher spin Dirac operators, in *Differential Geometry and Applications*, Masaryk Univ., Brno, 1999, pp. 319–334.

[6] D. Calderbank, P. Gauduchon, M. Herzlich, Refined Kato inequalities and conformal weights in Riemannian geometry, *J. Funct. Anal.* **173** (2000), 214–255.

[7] H. D. Fegan, Conformally invariant first order differential operators, *Quart. J. Math. Oxford* **27** (1976), 371–378.

[8] Y. Homma, Clifford homomorphisms and higher spin Dirac operators, preprint, arXiv.math.DG/0007052.

[9] Y. Homma, Spherical harmonic polynomials for higher bundles, Int. Conf. on Clifford Analysis, Its Appl. and Related Topics. Beijing, *Adv. Appl. Clifford Algebras* **11** (S2) (2001), 117–126.

[10] Y. Homma, The Bochner identities for the Kählerian gradients, preprint, arXiv.math.DG/0207031.

[11] S. Okubo, Casimir invariants and vector operators in simple and classical Lie algebras, *J. Math. Phys.* **18** (1977), 2382–2394.

[12] D. P. Želobenko, *Compact Lie Groups and Their Representations*, Trans. Math. Monographs, Vol. **40** A.M.S. 1973.

Yasushi Homma
Department of Mathematical Sciences
Waseda University
3-4-1 Ohkubo, Shinjuku-ku, Tokyo, 169-8555
E-mail: homma@gm.math.waseda.ac.jp

Submitted: August 25, 2002.

13

Eigenvalues of Dirac and Rarita–Schwinger Operators

Doojin Hong

ABSTRACT Let $M = S^1 \times S^{n-1}$ with metric Lorentzian or Riemannian and nontrivial spin structure on S^1, Riemannian metric and standard spin structure on S^{n-1}, and n even. We give explicit formulas for the eigenvalues of Dirac and Rarita–Schwinger operators on M.

Keywords: Dirac, Rarita–Schwinger, spinor, twistor

13.1 Introduction

Stein and Weiss introduced generalized gradients in [10]. Fegan showed in [8] that each gradient is conformally covariant. When realized on the sphere S^n, these gradients are automatically intertwining for principal series representations of the conformal group $SO_0(n + 1, 1)$ or its cover $\mathrm{Spin}_0(n + 1, 1)$. In 1997, Branson computed the spectrum of D^*D on the standard sphere S^n for each gradient in [3]. For operators having the same source and target bundles like the Dirac and Rarita–Schwinger operators, this also gives the spectrum of D as shown in [5]. One use of explicit spectra, among other things, would be in application to the Polyakov formula in even dimensions (see [2]) for the quotient of functional determinants of operators such as the squares of the Dirac, Rarita–Schwinger, and higher spin operators. Some general theory asserts that the precise form of these Polyakov formulas depends only on some precise constants that appear in the spectral asymptotics of the operator in question. Potentially, the more test manifolds one has explicit spectra on, the more such precise constants one can "pick off".

In this paper, we analyze spinors and twistors on $S^1 \times S^{n-1}$ to use the spectral data on the sphere and apply Bochner–Weitzenböck formulas to get spectra of the squares of the Dirac and Rarita–Schwinger operators.

AMS Subject Classification: 53C27, 53C30, 53C50.

13.2 Spinors and twistors on M

Let $M = S^1 \times S^{n-1}$, n even, and metric $\begin{cases} \text{Riemannian} & \text{on } S^{n-1} \\ \eta = \pm 1 & \text{on } S^1. \end{cases}$

A spinor on M can be viewed as a pair of time-dependent spinors on S^{n-1}, i.e., $\begin{pmatrix} \varphi \\ \psi \end{pmatrix}$, where φ and ψ are t-dependent spinors on S^{n-1}.

According to the decomposition above, the fundamental tensor-spinor on M takes the form

$$\gamma^i = \begin{pmatrix} \alpha^i & 0 \\ 0 & -\alpha^i \end{pmatrix} \quad \text{and} \quad \gamma^0 = \begin{pmatrix} 0 & 1 \\ -\eta & 0 \end{pmatrix},$$

where α is the fundamental tensor-spinor on S^{n-1}, i.e., a smooth section of the bundle $TS^{n-1} \otimes \text{End}(\Sigma)$ with $\alpha^i \alpha^j + \alpha^j \alpha^i = -2g^{ij}$ and $\nabla \alpha \equiv 0$ (see [7] for details).

It is easy to check the following Clifford relation (see, e.g., [9])

$$\gamma^\alpha \gamma^\beta + \gamma^\beta \gamma^\alpha = -2g^{\alpha\beta}$$

and

$$\nabla_\alpha \gamma^\beta \equiv 0.$$

A twistor is a spinor-one-form killed by the interior Clifford multiplication. So, on M, it is

$$\Phi = dt \wedge \begin{pmatrix} \varphi \\ \psi \end{pmatrix} + \begin{pmatrix} \varphi_i \\ \psi_i \end{pmatrix},$$

where $\begin{pmatrix} \varphi \\ \psi \end{pmatrix}$ is a pair of t-dependent spinors and $\begin{pmatrix} \varphi_i \\ \psi_i \end{pmatrix}$ is a pair of t-dependent spinor-one-forms on S^{n-1} with the interior condition. Write $\Phi = \begin{pmatrix} \varphi_0 & \varphi_i \\ \psi_0 & \psi_i \end{pmatrix}$.

Then the interior condition is $\gamma^\alpha \begin{pmatrix} \varphi_\alpha \\ \psi_\alpha \end{pmatrix} = 0$. Let us now consider the chirality operators

$$\chi = \gamma^0 \gamma^1 \cdots \gamma^{n-1} \quad \text{on } M \quad \text{and} \quad c = \alpha^1 \cdots \alpha^{n-1} \quad \text{on } S^{n-1}.$$

Then, using the Clifford relation, we get $\chi = \pm\sqrt{\eta}\, c$ and this allows us to express spinors and twistors as follows. For $\begin{pmatrix} \varphi \\ \psi \end{pmatrix}$, spinor on M,

$$\begin{pmatrix} \varphi \\ \psi \end{pmatrix} = \begin{pmatrix} \mp\frac{\psi}{\sqrt{\eta}} \\ \psi \end{pmatrix} = \begin{pmatrix} \epsilon\frac{\psi}{\sqrt{\eta}} \\ \psi \end{pmatrix}, \quad \epsilon = \begin{cases} -1 & \Rightarrow \text{Chirality I spinor} \\ +1 & \Rightarrow \text{Chirality II spinor.} \end{cases}$$

For $\Phi = \begin{pmatrix} \varphi_0 & \varphi_i \\ \psi_0 & \psi_i \end{pmatrix}$, twistor on M,

$$\Phi = \begin{pmatrix} \frac{\epsilon}{\sqrt{\eta}}\psi_0 & \frac{\epsilon}{\sqrt{\eta}}\psi_i \\ \psi_0 & \psi_i \end{pmatrix} = \begin{pmatrix} -\eta\alpha^i\psi_i & \frac{\epsilon}{\sqrt{\eta}}\psi_i \\ -\frac{\epsilon}{\sqrt{\eta}}\alpha^i\psi_i & \psi_i \end{pmatrix}, \text{ since } \gamma^\alpha \Phi_\alpha = 0.$$

In particular, twistors on M are determined by t-dependent spinor-one-forms on S^{n-1}.

13.3   Decomposition of spinor-one-forms on S^{n-1}

Let θ_j be a spinor-one-form on S^{n-1}. Then, it can be written as

$$\theta_j = \alpha_j\left(-\frac{1}{n-1}\alpha^i\theta_i\right) + \left(\theta_j + \frac{1}{n-1}\alpha_j\alpha^i\theta_i\right) = \alpha_j\theta + \pi_j,$$

where θ is a spinor and π_j is a twistor on S^{n-1}. To decompose the twistor part further, we consider Hodge decomposition of twistors on S^{n-1} (see, e.g., [11]).

$$\Gamma(\Sigma) \xrightarrow{T} \Gamma(Tw) \xrightarrow{T^2 = G_Y} \Gamma(Tw^2).$$

Let $D = \alpha^k \nabla_k$ be the Dirac operator on S^{n-1}. Then the twistor operator T and the gradient $T^2 = G_Y$ are defined by

$$(T\varphi)_i = \nabla_i\varphi + \frac{1}{n-1}\alpha_i D\varphi,$$

$$(T^2\psi)_{ij} = \frac{1}{2}\{\nabla_i\psi_j - \nabla_j\psi_i\} + \frac{1}{2(n-3)}\{\alpha_i D\psi_j - \alpha_j D\psi_i\}$$
$$- \frac{1}{2(n-2)(n-3)}\{\alpha_i\alpha_j - \alpha_j\alpha_i\}\nabla^k\psi_k,$$

and Tw^2 is the space of spinor-two-forms η_{ij} with $\alpha^i\eta_{ij} = 0$ as in [6]. Note that T^2 is not a square of T. Note also that T^2T is an action of the Weyl conformal tensor by Lemmas 2.1 and 2.2 in [4]. Thus, it is 0, since S^{n-1} is conformally flat.

On leading symbol level,

$$\Sigma_x \xrightarrow{\sigma_1(T)(\xi)} Tw_x \xrightarrow{\sigma_1(T^2)(\xi)} Tw_x^2,$$

we know that $\mathcal{R}(\sigma_1(T)(\xi)) = \mathcal{N}(\sigma_1(T^2)(\xi))$. So it is an elliptic complex. By Hodge's theorem, Tw splits up as follows.

$$Tw = \mathcal{R}(T) \oplus \mathcal{R}(T^{2^*}) \oplus \text{Harmonic}. \qquad (3.1)$$

Now we look at the harmonic condition, $T^*\varphi = 0$ and $T^2\varphi = 0$. Theorem 8.3 in [5] tells us that $\mathcal{N}(T^*) \cap \mathcal{N}(T^2) = \{0\}$ on S^{n-1}. So, there is no harmonic piece in the decomposition (3.1) and for $\Psi \in \Gamma(Tw)$ on S^{n-1}, we have

$$(\Psi)_i = (T\pi + \operatorname{div}\eta)_i \quad \text{for some} \quad \pi \in \Gamma(\Sigma),\ \eta \in \Gamma(Tw^2)$$

$$= \nabla_i\pi + \frac{1}{n-1}\alpha_i D\pi - \nabla^j\eta_{ji} \quad (D \text{ is the Dirac on } S^{n-1})$$

$$= T_i\pi - \nabla^j\eta_{ji}.$$

Therefore, twistors on M can be decomposed as follows.

$$\Psi = \begin{pmatrix} -\eta\alpha^i\psi_i & \frac{\epsilon}{\sqrt{\eta}}\psi_i \\ -\frac{\epsilon}{\sqrt{\eta}}\alpha^i\psi_i & \psi_i \end{pmatrix}$$

$$= \begin{pmatrix} -\eta\alpha^i\alpha_i\theta & \frac{\epsilon}{\sqrt{\eta}}\alpha_i\theta \\ -\frac{\epsilon}{\sqrt{\eta}}\alpha^i\alpha_i\theta & \alpha_i\theta \end{pmatrix} + \begin{pmatrix} -\eta\alpha^iT_i\tau & \frac{\epsilon}{\sqrt{\eta}}T_i\tau \\ -\frac{\epsilon}{\sqrt{\eta}}\alpha^iT_i\tau & T_i\tau \end{pmatrix}$$

$$+ \begin{pmatrix} -\eta\alpha^i(-\nabla^j\eta_{ji}) & \frac{\epsilon}{\sqrt{\eta}}(-\nabla^j\eta_{ji}) \\ -\frac{\epsilon}{\sqrt{\eta}}\alpha^i(-\nabla^j\eta_{ji}) & -\nabla^j\eta_{ji} \end{pmatrix}$$

$$= \begin{pmatrix} (n-1)\eta\theta & \frac{\epsilon}{\sqrt{\eta}}\alpha_i\theta \\ (n-1)\frac{\epsilon}{\sqrt{\eta}}\theta & \alpha_i\theta \end{pmatrix} + \begin{pmatrix} 0 & \frac{\epsilon}{\sqrt{\eta}}T_i\tau \\ 0 & T_i\tau \end{pmatrix}$$

$$+ \begin{pmatrix} 0 & \frac{\epsilon}{\sqrt{\eta}}(-\nabla^j\eta_{ji}) \\ 0 & -\nabla^j\eta_{ji} \end{pmatrix}$$

$$= \langle\theta\rangle + \{\tau\} + [B]. \tag{3.2}$$

The zeros appearing in the last two matrices are due to the fact that $T_i\tau$ and $-\nabla^j\eta_{ji}$ are killed by Clifford multiplication.

13.4   Eigenvalues of the Dirac operator

Let $\slashed{\nabla} = \gamma^\alpha\nabla_\alpha$ be the Dirac operator on M. We have

$$\slashed{\nabla}\begin{pmatrix} \frac{\epsilon}{\sqrt{\eta}}\psi \\ \psi \end{pmatrix} = (\gamma^0\nabla_0 + \gamma^i\nabla_i)\begin{pmatrix} \frac{\epsilon}{\sqrt{\eta}}\psi \\ \psi \end{pmatrix}$$

$$= \begin{pmatrix} 0 & 1 \\ -\eta & 0 \end{pmatrix}\begin{pmatrix} \frac{\epsilon}{\sqrt{\eta}}\dot\psi \\ \dot\psi \end{pmatrix} + \begin{pmatrix} \frac{\epsilon}{\sqrt{\eta}}D\psi \\ -D\psi \end{pmatrix}$$

$$= \begin{pmatrix} \dot\psi + \frac{\epsilon}{\sqrt{\eta}}D\psi \\ -\sqrt{\eta}\epsilon\psi - D\psi \end{pmatrix},$$

where $D = \alpha^i \nabla_i$ is the Dirac operator on S^{n-1}. Note that eigenspinors on S^1 take the form e^{ift} with $f \in \frac{1}{2} + \mathbb{Z}$, since we imposed nontrivial spin structure on S^1. And, for the eigenvalues on the sphere S^{n-1}, we have explicit formula $K = \pm(\frac{1}{2}(n-1)+j)$, where $j = 0,1,2,\ldots$ as in Example 10.3 in [5]. From now on, f and K will always have the above meaning throughout this paper. By an ϵ-spinor on $e^{ift}E_j$ over M, we mean a t-dependent ϵ-spinor $\begin{pmatrix} \epsilon \frac{\psi}{\sqrt{\eta}} \\ \psi \end{pmatrix}$ on S^{n-1} with $\dot{\psi} = if\psi$ and $D\psi = K\psi$. Thus, we have

$$\nabla \begin{pmatrix} \frac{\epsilon}{\sqrt{\eta}}\psi \\ \psi \end{pmatrix} = \begin{pmatrix} if\psi + \frac{\epsilon}{\sqrt{\eta}}K\psi \\ -\sqrt{\eta}\epsilon if\psi - K\psi \end{pmatrix}.$$

However, it does not give eigenvalues, since ∇ changes chirality. So we look at the square of the Dirac operator.

$$\nabla^2 \begin{pmatrix} \varphi \\ \psi \end{pmatrix} = (\gamma^0 \nabla_0 + \gamma^i \nabla_i)(\gamma^0 \nabla_0 + \gamma^i \nabla_i) \begin{pmatrix} \varphi \\ \psi \end{pmatrix}$$

$$= (\gamma^0 \gamma^0 \partial_t^2 + (\gamma^0 \gamma^i + \gamma^i \gamma^0)\nabla_0 \nabla_i + D^2) \begin{pmatrix} \varphi \\ \psi \end{pmatrix}$$

$$= (g^{00}\gamma_0 \gamma^0 \partial_t^2 - 2g^{0i}\nabla_0 \nabla_i + D^2) \begin{pmatrix} \varphi \\ \psi \end{pmatrix}$$

$$= (-\eta \partial_t^2 + D^2) \begin{pmatrix} \varphi \\ \psi \end{pmatrix},$$

since M is a product space of S^1 and S^{n-1}. So, for an ϵ-spinor on $e^{ift}E_j$,

$$\nabla^2 \begin{pmatrix} \frac{\epsilon}{\sqrt{\eta}}\psi \\ \psi \end{pmatrix} = (\eta f^2 + K^2) \begin{pmatrix} \frac{\epsilon}{\sqrt{\eta}}\psi \\ \psi \end{pmatrix}.$$

Therefore, eigenvalues of the Dirac operator on M are $\pm\sqrt{\eta f^2 + K^2}$, where $f \in \frac{1}{2} + \mathbb{Z}$ and $K = \pm(\frac{1}{2}(n-1)+j)$ with $j \in \mathbb{N}$.

13.5   Rarita–Schwinger operator

The covariant derivative ∇ takes sections of twistor bundle T into those of $T^*M \otimes T$. By the classical selection rule, we have

$$T^*M \otimes T \cong \Sigma \oplus T \oplus Y \oplus Z,$$

where Σ is the spinor bundle, $T = T_w$ is the twistor bundle, $Y = T_w^2$ is the bundle of spinor-two-forms annihilated by Clifford multiplication, and Z is the bundle of trace-free symmetric spinor-two-tensors annihilated by Clifford multiplication. Let G_Σ (resp. G_T, G_Y, G_Z) be the operator ∇ followed by the projection into Σ (resp. T, Y, Z). Then $\mathcal{S}$, the Rarita–Schwinger operator on twistor bundle over M given by

$$\varphi \longmapsto \gamma^\lambda \nabla_\lambda \varphi_\alpha - \frac{2}{n}\gamma_\alpha \nabla^\lambda \varphi_\lambda,$$

is $G_T^* G_T$ up to a constant multiple (see Section 4 in [6]). And, two Bochner–Weitzenböck formulas (see Theorem 4 in [6]) on a twistor bundle are

$$\frac{(n-3)(n-2)}{2n}G_\Sigma^* G_\Sigma - \frac{(n+2)(n-3)}{2n}G_T^* G_T + G_Y^* G_Y$$
$$= \frac{1}{8}C\diamond + \frac{n-3}{2(n-2)}b\cdot - \frac{(n-2)(n-3)}{8n(n-1)}R \quad (5.1)$$

and

$$-\frac{(n-1)(n+2)}{2n}G_\Sigma^* G_\Sigma - \frac{(n+1)(n-2)}{2n}G_T^* G_T + G_Z^* G_Z$$
$$= \frac{3}{8}C\diamond - \frac{n+1}{2(n-2)}b\cdot - \frac{(n+2)(n+1)}{8n(n-1)}R, \quad (5.2)$$

where $C\diamond$ is the Weyl conformal tensor action, $b\cdot$ is the trace-free Ricci tensor action, and R is the scalar curvature. Note also that since S^{n-1} is conformally flat, $C\diamond = 0$ on S^{n-1}, so it is also 0 on M. Thus, those two formulas (5.1) and (5.2) above with the fact

$$\nabla^* \nabla = G_\Sigma^* G_\Sigma + G_T^* G_T + G_Y^* G_Y + G_Z^* G_Z$$

give

$$G_T^* G_T = \frac{2n}{(n-2)^2(n+2)}b\cdot + \frac{n^2-n+4}{4(n-2)(n-1)(n+2)}R$$
$$- \frac{4(n-1)}{(n-2)(n+2)}G_\Sigma^* G_\Sigma + \frac{n}{(n-2)(n+2)}\nabla^* \nabla. \quad (5.3)$$

We want to compute $b\cdot$, R, $G_\Sigma^* G_\Sigma$, and $\nabla^* \nabla$ on the section space of the twistor bundle which is $\Gamma(Tw) = \langle\theta\rangle \oplus \{\tau\} \oplus [B]$ by (3.2). The scalar curvature R on $M = S^1 \times S^{n-1}$ is simply

$$R = R_1 + R_2 = R_2 = (n-1)(n-2). \quad (5.4)$$

To get the b action formula, let us consider a trace-free Ricci (Einstein) tensor

$$b_{\alpha\beta} = r_{\alpha\beta} - \frac{R}{n}g_{\alpha\beta} \quad \text{on } M.$$

Note first that a Ricci tensor on M looks like

$$r = \begin{pmatrix} 0 & 0 \\ 0 & r_{ij} = (n-2)\delta_{ij} \end{pmatrix},$$

since
$$r_{ij} = g^{hk}R_{kihj} = g^{hk}(g_{hk}g_{ij} - g_{kj}g_{ih}).$$

So the trace-free Ricci takes the following form:

$$b = \begin{pmatrix} -\dfrac{\eta(n-1)(n-2)}{n} & 0 \\[2mm] 0 & \dfrac{n-2}{n}\delta_{ij} \end{pmatrix}.$$

Since a b action on $\varphi \in \Gamma(Tw)$ over M is given by

$$(b \cdot \varphi)_j := \frac{1}{n}\left[(n-2)b_{j\lambda}\varphi^\lambda - b_{\alpha\lambda}\gamma^\alpha\gamma_j\varphi^\lambda\right],$$

we have, for $\varphi = \begin{pmatrix} \lambda_j \\ \mu_j \end{pmatrix} \in \Gamma(Tw)$,

$$(b \cdot \varphi)_j = \frac{n-2}{n}\begin{pmatrix} \lambda_j \\ \mu_j \end{pmatrix} + \frac{n-2}{n}\alpha_j\begin{pmatrix} \alpha^i\lambda_i \\ \alpha^i\mu_i \end{pmatrix}.$$

Some direct computations now lead us to the following formula.

$$b \cdot \begin{pmatrix} \langle\theta\rangle \\ \{\tau\} \\ [B] \end{pmatrix} = \begin{pmatrix} -\dfrac{(n-2)^2}{n} & 0 & 0 \\[2mm] 0 & \dfrac{n-2}{n} & 0 \\[2mm] 0 & 0 & \dfrac{n-2}{n} \end{pmatrix}\begin{pmatrix} \langle\theta\rangle \\ \{\tau\} \\ [B] \end{pmatrix}. \qquad (5.5)$$

Before we proceed to $G_\Sigma^* G_\Sigma$, let us see the explicit relationship between $G_\Sigma^* G_\Sigma$ and $T\delta$.

$$\begin{aligned}
(T\delta\varphi)_i &= \left(T(-\nabla^\alpha\varphi_\alpha)\right)_i \\
&= \nabla_i(-\nabla^\alpha\varphi_\alpha) + \frac{1}{n}\gamma_i\gamma^k\nabla_k(-\nabla^\alpha\varphi_\alpha) \\
&= \tfrac{1}{2}(\gamma_k\gamma_i + \frac{n-2}{n}\gamma_i\gamma_k)\nabla^k\nabla^\alpha\varphi_\alpha \\
&= (n-1)G_\Sigma^* G_\Sigma.
\end{aligned}$$

So we may compute $T\delta$ instead of $G_\Sigma^* G_\Sigma$. By a little bit long but straightforward calculations, on $e^{ift}E_j$, we have a matrix expression for $T\delta$ as follows with respect to $\{\langle\theta\rangle, \{\tau\}, [B]\}$.

$$\begin{pmatrix} \dfrac{n-1}{n}f^2\eta + \dfrac{K^2}{n(n-1)} & (n-2)(\dfrac{K^2}{n-1} - \dfrac{n-1}{4})(-\dfrac{K}{n(n-1)} + \dfrac{1}{n}if\epsilon\sqrt{\eta}) & 0 \\[2mm] -(n-1)if\sqrt{\eta}\epsilon - K & (n-2)(\dfrac{K^2}{n-1} - \dfrac{n-1}{4}) & 0 \\[2mm] 0 & 0 & 0 \end{pmatrix}.$$

$$(5.6)$$

Finally, we want to know what $\nabla^*\nabla$ does to $\langle\theta\rangle$, $\{\tau\}$, and $[B]$ parts on $e^{ift}E_j$. On $\langle\theta\rangle$,

$$\nabla^*\nabla\langle\theta\rangle = (-\nabla^0\nabla_0 + \triangledown^*\triangledown)\langle\theta\rangle = \left(\eta f^2 + K^2 - \tfrac{1}{4}(n-1)(n-2)\right)\langle\theta\rangle$$

by Theorem 1.1 in [1], where the smaller $\triangledown^*\triangledown$ is the Bochner Laplacian on S^{n-1}. On $\{\tau\}$,

$$\nabla^*\nabla\{\tau\} = \eta f^2\{\tau\} + \triangledown^*\triangledown T\begin{pmatrix}\frac{\epsilon}{\sqrt{\eta}}\tau \\ \tau\end{pmatrix}.$$

But

$$\triangledown^*\triangledown T = T(\triangledown^*\triangledown - 2J) = T\left(\triangledown^*\triangledown - (n-1)\right),$$

since S^{n-1} is a space of constant curvature and

$$J := \frac{R}{2(m-1)}$$

where m is the dimension of the manifold. Since our manifold is S^{n-1} and $R = (n-1)(n-2)$, J equals $\tfrac{1}{2}(n-1)$. So,

$$\nabla^*\nabla\{\tau\} = \left(\eta f^2 + K^2 - \tfrac{1}{4}R_2 - (n-1)\right)\{\tau\}$$
$$= \left(\eta f^2 + K^2 - \tfrac{1}{4}(n-1)(n+2)\right)\{\tau\},$$

where R_2 is the scalar curvature on S^{n-1}. On $[B]$,

$$\nabla^*\nabla[B] = \eta f^2[B] + \triangledown^*\triangledown\begin{pmatrix}-\frac{\epsilon}{\sqrt{\eta}}\nabla^i\eta_{ij} \\ -\nabla^i\eta_{ij}\end{pmatrix}.$$

Notice that $-\nabla^i\eta_{ij}$ are of Spin(n)-types $(\tfrac{3}{2}+j, \boxed{\tfrac{3}{2}}, \tfrac{1}{2}, \ldots, \tfrac{1}{2}, \pm\tfrac{1}{2})$.

Applying Theorem 1.1 in [1], we know that $\triangledown^*\triangledown$ acts as multiplication by $\left(K^2 - \tfrac{1}{4}n^2 + \tfrac{3}{4}n - \tfrac{3}{2}\right)$ on Spin(n)-types $(\tfrac{3}{2}+j, \boxed{\tfrac{3}{2}}, \tfrac{1}{2}, \ldots, \tfrac{1}{2}, \pm\tfrac{1}{2})$. Thus,

$$\nabla^*\nabla[B] = \left(\eta f^2 + K^2 - \tfrac{1}{4}n^2 + \tfrac{3}{4}n - \tfrac{3}{2}\right)[B].$$

Therefore, $\nabla^*\nabla$ is

$$\left(\eta f^2 + K^2 - \tfrac{1}{4}(n-1)(n-2)\right)E_{11} + \left(\eta f^2 + K^2 - \tfrac{1}{4}(n-1)(n+2)\right)E_{22}$$
$$+ \left(\eta f^2 + K^2 - \tfrac{1}{4}n^2 + \tfrac{3}{4}n - \tfrac{3}{2}\right)E_{33}, \quad (5.7)$$

with respect to $\{\langle\theta\rangle, \{\tau\}, [B]\}$, where E_{ij} is the 3×3 matrix whose (i,j)-th entry is 1 and 0 elsewhere.

Now, putting all the information (5.4), (5.5), (5.6), and (5.7) into (5.3), we get $G_T^*G_T$ as follows.

- On (1,1),

$$\frac{n-2}{n(n+2)}\eta f^2 + \frac{n^2+n+2}{n(n-1)(n+2)}K^2 - \frac{1}{n+2}.$$

- On (1,2),

$$\left[-\frac{4i\sqrt{\eta}\epsilon K^2}{n(n-1)(n+2)} + \frac{(n-1)i\sqrt{\eta}\epsilon}{n(n+2)}\right]f + \frac{4K^3}{n(n-1)^2(n+2)} - \frac{K}{n(n+2)}.$$

- On (2,1),

$$\frac{4\left[i(n-1)f\epsilon\sqrt{\eta}+K\right]}{(n-2)(n+2)}.$$

- On (2,2),

$$\frac{n\eta f^2}{(n-2)(n+2)} + \frac{(n^2-5n+8)K^2}{(n-2)(n-1)(n+2)} - \frac{1}{n+2}.$$

- On (3,3),

$$\frac{(\eta f^2 + K^2)n}{(n-2)(n+2)}.$$

- On (1,3), (2,3), (3,1), and (3,2), it is 0.

As a consequence, we have the following result for Lorentzian M (i.e., $\eta = -1$).

Theorem. *Let S be the Rarita–Schwinger operator on $S^1 \times S^{n-1}$ with $n \geq 4$ even and the metric standard Lorentzian (i.e., $g_M = -g_{S^1} + g_{S^{n-1}}$). Then $\mathcal{N}(S) \neq \{0\}$.*

Proof. Note first that $S^2 = \dfrac{(n-2)(n+2)}{n}G_T^* G_T$. By considering the characteristic polynomial of the matrix of $G_T^* G_T$, we know that $G_T^* G_T$ has 0 eigenvalue

$$\Leftrightarrow -\frac{n}{(n-2)(n+2)^3}(f^4 - 2f^2K^2 - 2f^2 + K^4 - 2K^2 + 1)(f^2 - K^2) = 0$$

$$\Leftrightarrow (f+K)(f-K)(f+1+K)(f+1-K)(f-1+K)(f-1-K) = 0.$$

Since $f \in \frac{1}{2} + \mathbb{Z}$, $K = \pm(\frac{1}{2}(n-1)+j)$ and n is even, both f and K are half integers and this completes the proof. $\qquad\square$

Remark. In the matrix for $T\delta$, the first 2×2 entries are nonzero. So the matrix gives information about products of eigenvalues rather than eigenvalues themselves. This is due to the fact that the Spin(2) × Spin(n) isotypic summands over the twistor bundle on M occur with multiplicity 2. That is one reason why the Dirac square case is simpler. In fact, the multiplicity goes up with the spin.

REFERENCES

[1] T. Branson, Harmonic analysis in vector bundles associated to the rotation and spin groups, *J. Funct. Anal.* **106**, (1992), 314–328.

[2] T. Branson, Sharp inequalities, the functional determinant, and the complementary series, *Trans. Amer. Math. Soc.* **347**, (1995), 3671–3742.

[3] T. Branson, Stein–Weiss operators and ellipticity, *J. Funct. Anal.* **151**, (1997), 334–383.

[4] T. Branson, Second order conformal covariants, *Proc. Amer. Math. Soc.* **126**, (1998), 1031–1042.

[5] T. Branson, Spectra of self-gradients on spheres, *J. Lie Theory.* **9**, (1999), 491–506.

[6] T. Branson and O. Hizaji, Bochner–Weitzenböck formulas associated with the Rarita–Schwinger operator. arXiv:hep-th/0110014v1, 1 Oct 2001.

[7] T. Branson, Clifford bundles and Clifford algebras, in *Lectures on Clifford Geometric Algebras and Applications*, R. Abłamowicz and G. Sobczyk, Eds., Birkhäuser, Boston, 2003, Lecture 6, pp. 163–196.

[8] H. Fegan, Conformally invariant first order differential operators, *Quart. J. Math. (Oxford)* **27**, 1976, 371–378.

[9] H. Lawson and M. Michelsohn, *Spin Geometry*, Princeton University Press. Princeton, New Jersey, 1989.

[10] E. Stein and G. Weiss, Generalization of the Cauchy-Riemann equations and representations of the rotation group, *Amer. J. Math.* **90**, (1968), 163–196.

[11] F. Warner, *Foundations of Differentiable Manifolds and Lie Groups*, Scott, Foresman and Company, Glenview, Illinois, 1971.

Doojin Hong
Department of Mathematics
University of Iowa
Iowa City, IA, 52242
E-mail: dohong@math.uiowa.edu

Submitted: August 31, 2002.

14

Differential Forms Canonically Associated to Even-Dimensional Compact Conformal Manifolds

William J. Ugalde

ABSTRACT On a 6-dimensional, conformal, oriented and compact manifold M without boundary we compute a whole family of differential forms $\Omega_6(f, h)$ of order 6 with f, $h \in C^\infty(M)$. Each of these forms will be symmetric on f and h, conformally invariant, and such that $\int_M f_0 \Omega_6(f_1, f_2)$ defines a Hochschild 2-cocycle over the algebra $C^\infty(M)$. In the particular 6-dimensional conformally flat case, we compute a unique form satisfying $\mathrm{Wres}(f_0[F, f][F, h]) = \int_M f_0 \Omega_6(f, h)$ for the Fredholm module $(\mathcal{H}, F)$ associated by A. Connes [6] to the manifold M, and the Wodzicki residue Wres.

Keywords: conformal geometry, Wodzicki residue, Fredholm module

14.1 Introduction

For a compact, oriented manifold M of even dimension $n = 2l$, endowed with a conformal structure, there is a canonically associated Fredholm module $(\mathcal{H}, F)$ [6]. $\mathcal{H}$ is the Hilbert space of square integrable forms of middle dimension $L^2(M, \Lambda_{\mathbb{C}}^{n/2} T^* M)$ in which functions on M act as multiplication operators. F is the pseudodifferential operator of order 0 acting in $\mathcal{H}$ obtained from the orthogonal projection P on the image of d by the relation $F = 2P - 1$. From the Hodge decomposition theorem [16] it is easy to see that (i) F preserves the finite–dimensional space of harmonic forms H^l, and, (ii) F restricted to the $\mathcal{H} \ominus H^l$ is given by

$$F = \frac{d\delta - \delta d}{d\delta + \delta d}, \tag{1.1}$$

in terms of a Riemannian metric compatible with the conformal structure of M. Both $\mathcal{H}$ and F are independent of the metric in the conformal class [7, Section IV.4.γ].

AMS Subject Classification: 53A30, 35S05, 58J40, 58B34, 46L87.

By considering a Riemann surface Σ, and by considering instead of dX its quantized version $[F, X]$, Connes (Chapter 4 in [7]) quantized the Polyakov action as the Dixmier trace of the operator $\eta_{ij} dX^i dX^j$,

$$\frac{1}{2\pi} \int_\Sigma \eta_{ij} \, dX^i \wedge \star dX^j = -\frac{1}{2} \operatorname{trace}_\omega \left(\eta_{ij} [F, X^i][F, X^j] \right). \tag{1.2}$$

Because of the fact that the Wodzicki residue extends uniquely the Dixmier trace as a trace on the algebra of pseudodifferential operators [17], this quantized Polyakov action has sense in the general even-dimensional case. Because of the Connes' trace theorem (see Theorem 7.18 in [9]), the Dixmier trace and the Wodzicki residue of an elliptic pseudodifferential operator of order $-n$ in an n-dimensional manifold are proportional by a factor of $n(2\pi)^n$. In the 2-dimensional case the factor is $8\pi^2$ and so the quantized Polyakov action (1.2) can be written as

$$-16\pi^2 I = \operatorname{Wres}\left(\eta_{ij} [F, X^i][F, X^j] \right) \tag{1.3}$$

which determines, by using the general formula for the total symbol of the product of two pseudodifferential operators, an n-dimensional differential form Ω_n. Note that we decided to write the constant on the left of (1.3) to simplify the typing. This differential form is **symmetric, conformally invariant** and **uniquely** determined, for every $f_0, f, h \in C^\infty(M)$, by the relation:

$$\operatorname{Wres}\left(f_0 [F, f][F, h] \right) = \int_M f_0 \Omega_n(f, h). \tag{1.4}$$

$\Omega_n(f, h)$ is given as the Wodzicki 1-density (which we denote wres following [9]) of the product of commutators $[F, f][F, h]$:

$$\Omega_n(f, h) = \operatorname{wres}([F, f][F, h]))$$

$$= \left\{ \sum \frac{A_{\alpha', \alpha'', \beta, \delta}}{\alpha'! \alpha''! \beta! \delta!} \, D_x^\beta(f) D_x^{\alpha'' + \delta}(h) \right\} d^n x, \tag{1.5}$$

with the sum taken over $|\alpha'| + |\alpha''| + |\beta| + |\delta| + j + k = n$, $|\beta| \geq 1$, $|\delta| \geq 1$, and

$$A_{\alpha', \alpha'', \beta, \delta} = \int_{|\xi| = 1} \operatorname{trace} \left\{ \partial_\xi^{\alpha' + \alpha'' + \beta} \left(\sigma_{-j}(F) \right) \partial_\xi^\delta \left(D_x^{\alpha'} \left(\sigma_{-k}(F) \right) \right) \right\} d^{n-1}\xi$$

where $\sigma_{-j}(x, \xi)$ is the component of order $-j$ in the total symbol of P, $|\xi| = 1$ means the Euclidean norm of the coordinate vector $(\xi_1, \ldots, \xi_n)$ in $\mathbb{R}^n$, and $d^{n-1}\xi$ is the normalized volume on $\{ |\xi| = 1 \}$.

In the 4-dimensional case Connes [6] showed that the Paneitz operator P_4, analogue of the scalar Laplacian in 4-dimensional conformal geometry, can be derived from the Wodzicki residue by dropping some information. That is to say, by setting $f_0 = 1$ and integrating by parts he has obtained

$$\int_M \Omega_4(f, h) = \int_M f P_4(h) \, dv \tag{1.6}$$

producing P_4 by the arbitrariness of f and h.

Relation (1.6) is of importance in the study of conformally invariant differential operators generalizing the Yamabe operator. There are (see [11]) invariant operators (*GJMS operators*) on scalar densities with principal parts Δ^k, unless the dimension is even and $2k > n$. The n-th order operator is called a *critical* GJMS operator. When one calculates the Polyakov action for the quotient of functional determinants of a conformally covariant operator D at conformally related metrics, the operator P_n appears, see for example [3] and [5]. That is why the study of the differential form Ω_n is of considerable importance in the case $n \geq 4$. We propose an approach using automated symbolic computation in (1.5).

In this paper, we present partial results from [14], see also [15]. In section 2 we introduce a general construction which associates to any pseudodifferential operator S of order 0 acting on sections of a bundle B on a compact manifold without boundary M, a differential form $\Omega_{n,S}(f, h)$ of order n acting on $C^\infty(M) \times C^\infty(M)$. This $\Omega_{n,S}(f, h)$ is uniquely given by the relation

$$\mathrm{Wres}(f_0[S, f_1][S, f_2]) = \int_M f_0 \Omega_{n,S}(f_1, f_2) \, \mathrm{dvol}$$

for every $f_i \in C^\infty(M)$, with Wres the Wodzicki residue. The first result is given in Lemma 2 where we give an explicit expression for $\Omega_{n,S}$ in terms of the total symbol of S. In the particular case of a compact, conformal, oriented, even-dimensional manifold, with $S = F$ the operator given by (1.1), the differential form $\Omega_{n,F}$ is furthermore, symmetric on f and h and conformally invariant (Theorem 2). The rest of the paper focuses on computing $\Omega_{n,F} = \Omega_n$ in this case, in particular for $n = 6$. In section 3, we give an explicit expression for $\Omega_n(f_1, f_2)$ in the flat case. This expression (see Proposition 2) is given in terms of the Taylor expansion of the function $\mathrm{trace}(\sigma_L^F(\xi)\sigma_L^F(\eta))$ (see (3.1)). In section 4, we present $\Omega_6(f_1, f_2)$, given by Theorem 2, in the 6-dimensional conformally flat case. The last result is presented as Theorem (4) in section 6, where we compute a whole family $\Omega_6(f, h)$ of differential forms of order 6 associated to a conformal, oriented, compact manifold without boundary. Each of these forms will be symmetric on f, and h, conformally invariant and such that $\int_M f_0 \Omega_6(f_1, f_2)$ defines a Hochschild 2-cocycle over the algebra $C^\infty(M)$. To compute the unique form Ω_6 satisfying the relation $\mathrm{Wres}(f_0[F, f_1][F, f_2]) = \int_M f_0 \Omega_6(f_1, f_2)$ more information is needed in the 6-dimensional case.

My special thanks to T. Branson for his constant support and guidance. I am also in debt to the referee of this paper for valuable suggestions.

14.1.1   *Notation and conventions*

A conformal manifold is an equivalence class of Riemannian manifolds where two metrics g and $\hat{g}$ are said to be equivalent if one is a positive scalar multiple of the other; for this work, it is convenient to write $\hat{g} = e^{2\eta}g$ for some $\eta \in C^\infty(M)$.

The Laplace–Beltrami operator on k-forms is defined as $\Delta = d\delta + \delta d$ where we assume the sign convention $\Delta = -\frac{\partial}{\partial x} - \frac{\partial}{\partial y}$ on $\mathbb{R}^2$. The contraction of a k-form η

with a vector field X is defined by

$$\iota_\eta(X)(X_1,\ldots,X_{k-1}) := \eta(X,X_1,\ldots,X_{k-1}).$$

The contraction of η with a 1-form ξ_X which is determined by the vector field X is given by $\iota_\eta(\xi_X) = \iota_\eta(X)$. The exterior multiplication by a k-form η will be denoted by $\varepsilon_\eta : \xi \mapsto \eta \wedge \xi$.

In this work, the Riemann curvature tensor will be represented with the letter R, the Ricci tensor will be represented by $\mathrm{Rc}_{ij} = R^k{}_{ikj}$, and the scalar curvature by $\mathrm{Sc} = \mathrm{Rc}^i{}_i$. The conformal change equation for the Ricci tensor,

$$\eta_{;ij} = -V_{ij} - \eta_{;i}\,\eta_{;j} + \tfrac{1}{2}\eta_{;k}\,\eta_{;}{}^k\,g_{ij}, \tag{1.7}$$

allows us to replace, in the case $g = e^{2\eta}g_{\text{flat}}$, the second derivatives on η with terms of the Ricci tensor. V represents a normalized translation of the Ricci tensor, useful in conformal geometry, given in terms of the normalized scalar curvature J by

$$V = \frac{\mathrm{Rc} - Jg}{n-2} \quad \text{with} \quad J = \frac{\mathrm{Sc}}{2(n-1)}.$$

In (1.7), the indices after the semicolon represents covariant derivatives, $\eta_{;ij} = \nabla_j\nabla_i\eta$. In terms of V, the relation between the Weyl tensor and the Riemann tensor is given by

$$W^i{}_{jkl} = R^i{}_{jkl} + V_{jk}\delta^i{}_l - V_{jl}\delta^i{}_k + V^i{}_l g_{jk} - V^i{}_k g_{jl}$$

where δ represents the Kronecker's delta tensor.

If needed, we will "raise" and "lower" indices without explicit mention following for example, $g_{mi}R^i{}_{jkl} = R_{mjkl}$.

When working with the total symbol of a pseudodifferential operator P, we will denote its leading symbol by σ_L^P, or $\sigma_L(P)$ in case P has a long expression. If the operator P is of order k, then its total symbol (in some given local coordinates) will be represented as

$$\sigma(P) = \sigma_k^P + \sigma_{k-1}^P + \sigma_{k-2}^P + \cdots,$$

where $\sigma_L^P = \sigma_k^P$. It is important to note that the different σ_j^P for $j < k$ are defined only in local charts and are not diffeomorphism invariant [12]. However, Wodzicki [17] has shown that the term σ_{-n}^P enjoys a very special significance. For a pseudodifferential operator P, acting on sections of a bundle B over a manifold M, there is a 1-density on M expressed in local coordinates by

$$\mathrm{wres}(P) = \int_{|\xi|=1} \left\{ \mathrm{trace}(\sigma_{-n}^P(x,\xi))\, d^{n-1}\xi \right\} d^n x. \tag{1.8}$$

This *Wodzicki residue density* is independent of the local representation. Here we are using the same notation as in [9], where an elementary proof of this matter can be found. The Wodzicki residue, $\mathrm{Wres}(P)$, is then computed [6] by choosing any local coordinates x^j on M and any local basis of sections s_k for B. P is represented in terms of the chosen basis s_k as a matrix P^l_k of scalar pseudodifferential operators: $P(f^k \alpha_k) = (P^i_k f^k) \alpha_i$. The residue $\mathrm{Wres}(P)$ is given by

$$\mathrm{Wres}(P^k_k) = \int_M \left\{ \int_{|\xi|=1} \mathrm{trace}(\sigma_{-n}(x, \xi)) \, d^{n-1}\xi \right\} d^n x$$

where $\sigma_{-n}(x, \xi)$ is the component of order $-n$ in the total symbol of P, $|\xi| = 1$ means the Euclidean norm of the coordinate vector $(\xi_1, \cdots, \xi_n)$ in $\mathbb{R}^n$, and $d^{n-1}\xi$ is the normalized volume on $\{ |\xi| = 1 \}$. $\mathrm{Wres}(P)$ is independent of the choice of the local coordinates on M, the local basis (s_k) of B, and defines a trace (see [17]).

To study the conformal invariance, there is no need to study the whole conformal deformation. It is enough to study the conformal deformation up to order one in η as follows. If we set a metric g in M and consider another metric $\hat{g}$ conformally related to g by the relation $\hat{g} = e^{2z\eta}g$ where $\eta \in C^\infty(M)$ and z a constant, then the conformal variation of each expression is a polynomial in z whose coefficients are expressions in the metric and the conformal factor η (actually, this is an abuse of the language since the conformal factor is $e^{2z\eta}$). In this way, the conformal deformation up to order one in η is given by $\frac{d}{dz}\big|_{z=0}$. As stated in [10], if a natural tensor or a differential operator is invariant up to order 1 in η, i.e., if its conformal deformation up to order 1 is equal to zero, then by integration it follows that it is fully invariant, for details see [2].

14.2   Existence of Ω_n

In this section, we associate to any pseudodifferential operator S of order 0 acting on sections of a bundle B on an arbitrary compact manifold without boundary M, a bilinear differential form $\Omega_{n,S}$ acting on $C^\infty(M) \times C^\infty(M)$. Most of the properties of $\Omega_{n,S}$ are related with the properties of the Wodzicki residue. The total symbol up to order $-n$, of the pseudodifferential operator of order -2 given by the product $P = f_0[S, f_1][S, f_2]$ with each $f_i \in C^\infty(M)$, is represented as a sum of $r \times r$ matrices of the form $\sigma^P_{-2} + \sigma^P_{-3} + \cdots + \sigma^P_{-n}$, with r the rank of B. We aim to study

$$\mathrm{Wres}(P) = \int_M \left\{ \int_{|\xi|=1} \mathrm{trace}(\sigma^P_{-n}(x, \xi)) \, d^{n-1}\xi \right\} d^n x.$$

In general, the total symbol of the product of two pseudodifferential operators P_1 and P_2 is given by

$$\sigma(P_1 P_2) = \sum \frac{1}{\alpha!} \partial^\alpha_\xi(\sigma^{P_1}) D^\alpha_x(\sigma^{P_2}) \tag{2.1}$$

where $\alpha = (\alpha_1, \ldots, \alpha_n)$ is a multi-index, $\alpha! = \alpha_1!\alpha_2! \cdots \alpha_n!$ and $D_x^\alpha = (-i)^{|\alpha|}\partial_x^\alpha$. Using this formula it is possible to deduce an expression for $\sigma_{-n}([S, f_1][S, f_2])$ finding first $\sigma([S, f])$.

Lemma 1. $[S, f]$ *is a pseudodifferential operator of order* -1 *with total symbol* $\sigma([S, f]) = \sum_{k \geq 1} \sigma_{-k}([S, f])$ *where*

$$\sigma_{-k}([S, f]) = \sum_{|\beta|=1}^{k} \frac{1}{\beta!} D_x^\beta(f) \partial_\xi^\beta (\sigma_{-(k-|\beta|)}^S).$$

Lemma 2. *With the sum taken over* $|\alpha'| + |\alpha''| + |\beta| + |\delta| + j + k = n$, $|\beta| \geq 1$, *and* $|\delta| \geq 1$,

$$\sigma_{-n}([S, f_1][S, f_2]) =$$
$$\sum \frac{1}{\alpha'!\alpha''!\beta!\delta!} D_x^\beta(f_1) D_x^{\alpha''+\delta}(f_2) \partial_\xi^{\alpha'+\alpha''+\beta}(\sigma_{-j}^S) \partial_\xi^\delta (D_x^{\alpha'}(\sigma_{-k}^S)). \quad (2.2)$$

As a consequence

$$\text{wres}([S, f_1][S, f_2]) = \left\{ \int_{|\xi|=1} \text{trace}\left\{ \sum \frac{1}{\alpha'!\alpha''!\beta!\delta!} D_x^\beta(f_1) D_x^{\alpha''+\delta}(f_2) \right. \right.$$
$$\left. \left. \times \partial_\xi^{\alpha'+\alpha''+\beta}(\sigma_{-j}^S) \partial_\xi^\delta (D_x^{\alpha'}(\sigma_{-k}^S)) \right\} d^{n-1}\xi \right\} d^n x. \quad (2.3)$$

Definition 1. *For every* f_1 *and* f_2 *in* $C^\infty(M)$ *we define*

$$\Omega_{n,S}(f_1, f_2) := \text{wres}([S, f_1][S, f_2]). \quad (2.4)$$

Theorem 1. *For any pseudodifferential operator* S *of order* 0 *acting on sections of a bundle* B *on a manifold* M, *there is a unique* n-*differential form* $\Omega_{n,S}$ *such that*

$$\text{Wres}(f_0[S, f_1][S, f_2]) = \int_M f_0 \Omega_{n,S}(f_1, f_2)$$

for all $f_i \in C^\infty(M)$.

We restrict ourselves to an even-dimensional, compact, oriented, conformal manifold without boundary M, and (B, S) given by the canonical Fredholm module $(\mathcal{H}, F)$ associated to M by A. Connes [6].

In this particular case, $F = (d\delta - \delta d)(d\delta + \delta d)^{-1}$ is the pseudodifferential operator of order 0 acting on the component $\text{Im}(d + \delta)$ of $\mathcal{H} = L^2(M, \Lambda_{\mathbb{C}}^l T^* M)$. We can conclude that the differential form $\Omega_{n,F}$ is symmetric and conformally invariant.

Theorem 2. *In the particular case in which* M *is an even-dimensional compact conformal manifold without boundary and* $(\mathcal{H}, F)$ *is the Fredholm module associated to* M *by A. Connes* [6], *there is a unique, symmetric, and conformally* *invariant n-differential form* $\Omega_n = \Omega_{n,F}$ *such that*

$$\text{Wres}(f_0[F, f_1][F, f_2]) = \int_M f_0 \Omega_n(f_1, f_2)$$

for all $f_i \in C^\infty(M)$.

All the proofs for the lemmas and theorems in this paper, as well as the detail for the computations presented here can be read in [15].

14.3 Ω_n in the flat case

Proposition 1. *The leading symbol of F is given by*

$$\sigma_L^F(x, \xi) = \sigma_L^F(\xi) = |\xi|^{-2}(\varepsilon_\xi \iota_\xi - \iota_\xi \varepsilon_\xi)$$

for all $(x, \xi) \in T^*M$, $\xi \neq 0$. *In the particular case of a flat metric, we also have* $\sigma_{-k}^F = 0$ *for all* $k \geq 1$.

Using this result, and the explicit expression for Ω_n (2.4), it is possible to give a formula for Ω_n in the flat case using the Taylor expansion of the function

$$\text{trace}(\sigma_L^F(\xi)\sigma_L^F(\eta)) = a_{n,m} \frac{\langle \xi, \eta \rangle^2}{|\xi|^2 |\eta|^2} + b_{n,m}. \tag{3.1}$$

Here $\xi, \eta \in T_x^*M \setminus \{0\}$, $\sigma_L^F(\xi)\sigma_L^F(\eta)$ is acting on $n/2$-forms, and the values of the constants are given by

$$\binom{n}{m} - a_{n,m} = b_{n,m} = \binom{n-2}{m-2} + \binom{n-2}{m} - 2\binom{n-2}{m-1}.$$

Because $\sigma_{-k}^F(x, \xi) = 0$ for all $k > 0$ in the flat case, (2.4) reduces to

$$\Omega_{n\,\text{flat}}(f, h) =$$

$$\left\{ \int_{|\xi|=1} \text{trace}\left(\sum \frac{1}{\alpha!\beta!\delta!} (D_x^\beta f)(D_x^{\alpha+\delta} h)\partial_\xi^{\alpha+\beta}(\sigma_L^F)\partial_\xi^\delta(\sigma_L^F) \right) d^{n-1}\xi \right\} d^n x,$$

with the sum taken over $|\alpha| + |\beta| + |\delta| = n$, $1 \leq |\beta|$, $1 \leq |\delta|$.

We denote by $T_n'\psi(\xi, \eta, u, v)$ the term of order n in the Taylor expansion of

$$\psi(\xi, \eta) = \text{trace}(\sigma_L^F(\xi)\sigma_L^F(\eta))$$

minus the terms with only powers of u or only powers of v. That is to say,

$$T_n'\psi(\xi, \eta, u, v) = \sum_{|\beta|+|\delta|=n, |\beta| \geq 1, |\delta| \geq 1} \frac{u^\beta v^\delta}{\beta! \delta!} \text{trace}\left(\partial_\xi^\beta(\sigma_L^F(\xi))\partial_\eta^\delta(\sigma_L^F(\eta))\right).$$

Proposition 2.

$$\Omega_{n\,\text{flat}}(f,h) = \left(\sum A_{a,b}(D_x^a f)(D_x^b h)\right) d^n x,$$

where

$$\sum A_{a,b} u^a v^b = \int_{|\xi|=1} \left(T'_n \psi(\xi,\xi,u+v,v) - T'_n \psi(\xi,\xi,v,v)\right) d^{n-1}\xi.$$

14.4   The 6-dimensional conformally flat case

In the 6-dimensional case, symmetry and conformal invariance are not enough to fully describe Ω_6 as we will find terms like

$$\left\{A\, f_{;ih}{}_{;}{}^i W_{jklm} W^{jklm} + B\, f_{;ih}{}_{;j} W^i{}_{klm} W^{jklm}\right\} d^6 x$$

which are symmetric on f and h, and conformally invariant.

Another important property that we will exploit, is the fact that, by definition, $\mathrm{Wres}(f_0[F,f_1][F,f_2])$ is a Hochschild 2-cocycle over the algebra $C^\infty(M)$. If we define $\tau(f_0,f,h)$ as

$$\tau(f_0,f,h) := \int_M f_0\left\{A\, f_{;ih}{}_{;}{}^i W_{jklm} W^{jklm} + B\, f_{;ih}{}_{;j} W^i{}_{klm} W^{jklm}\right\} d^6 x,$$

then we obtain a Hochschild 2-cocycle for any value of A and B, that is to say [9]

$$\begin{aligned}
0 &= (b\tau)(f_0,f_1,f_2,f_3)\\
&= \tau(f_0 f_1, f_2, f_3) - \tau(f_0, f_1 f_2, f_3) + \tau(f_0, f_1, f_2 f_3) - \tau(f_3 f_0, f_1, f_2).
\end{aligned}$$

In the 4-dimensional case, this property was used merely to make sure the constants found had the right values. In the 6-dimensional case, as we shall see, this property will play a more important role in the non-conformally flat case; even so, a full description of Ω_6 escapes these properties, requiring some more information to be used.

In the 6-dimensional flat case, using Lemma 2 to compute Ω_6 we have found

$$\begin{aligned}
\Omega_{6\,\text{flat}}(f,h) &= Q_6(df,dh)\\
&= \Big\{12(f_{;ih}{}_{;}{}^i{}_j{}^j{}_k{}^k + f_{;ij}{}^j{}_k{}^k h_{;}{}^i) + 24\,(f_{;ij}h_{;}{}^{ij}{}_k{}^k + f_{;ijk}{}^k h_{;}{}^{ij})\\
&\quad + 6\,(f_{;i}{}^i h_{;j}{}^j{}_k{}^k + f_{;i}{}^i{}_j{}^j h_{;k}{}^k) + 24\,f_{;ij}{}^j h_{;}{}^i{}_k{}^k + 16\,f_{;ijk} h_{;}{}^{ijk}\Big\} d^6 x\\
&= \Big\{12\,\Delta^2(\langle df, dh\rangle) - 6\,\Delta(\Delta f \Delta h) - 12\,\langle \nabla(\Delta f), \nabla(\Delta h)\rangle\\
&\quad + 24\,\Delta(\langle \nabla df, \nabla dh\rangle) + 16\,\langle \nabla^2 df, \nabla^2 dh\rangle\Big\} d^6 x, \tag{4.1}
\end{aligned}$$

where each summand in the last expression is explicitly symmetric on f and h.

Studying Ω_n in the flat metric gives information about the conformally flat case, in particular, by growing the flat expression for Ω_n we can find its expression in the conformally flat case. To do that, we consider a metric $\hat{g}$ conformally related to the flat metric g by the relation $\hat{g} = e^{2\eta}g$ with η a smooth function on M. Each time we express a component of (4.1) in the conformally related metric, terms containing derivatives on η will show. Using the conformal change equation for the Ricci tensor (1.7) until we reduce all the higher derivatives on η to derivatives of order 1, the Ricci tensor, and the scalar curvature related to g, we obtain the expression for Ω_n in the conformally flat case. The computations could be a little tedious, so we did them using `Ricci.m` [13] to obtain:

$$\Omega_{6\,\text{conf. flat}}(f, h) = \Omega_{6\,\text{flat}}(f, h)$$

$$+ \left\{ \left(-72(f_{;ij}{}^j h_{;}{}^i + f_{;i} h_{;}{}^i{}_j{}^j) - 24 f_{;i}{}^i h_{;j}{}^j - 96 f_{;ij} h_{;}{}^{ij} \right) J \right.$$

$$+ 96 f_{;i} h_{;}{}^i J^2$$

$$+ 24(f_{;i}{}^i h_{;j} J_{;}{}^j + f_{;i} h_{;j}{}^j J_{;}{}^i) - 24(f_{;ij} h_{;}{}^i J_{;}{}^j + f_{;i} h_{;}{}^i{}_j J_{;}{}^j)$$

$$- 24 f_{;i} h_{;}{}^i J_{;j}{}^j + 64 f_{;i} h_{;j} J V^{ij} - 32(f_{;ij}{}^j h_{;k} V^{ik} + f_{;i} h_{;jk}{}^k V^{ij})$$

$$+ 64(f_{;ijk} h_{;}{}^i V^{jk} + f_{;i} h_{;}{}^i{}_{jk} V^{jk}) + 96(f_{;ij} h_{;k}{}^k V^{ij} + f_{;i}{}^i h_{;jk} V^{jk})$$

$$- 192 f_{;ij} h_{;}{}^j{}_k V^{jk} - 64 f_{;i} h_{;}{}^i V_{jk} V^{jk} + 128 f_{;i} h_{;j} V^i{}_k V^{jk} \left. \right\} d^6 x. \quad (4.2)$$

An interesting introduction to automated symbolic computation can be found in [1].

Theorem 3. *In the 6-dimensional conformally flat case, the expression for Ω_6 given by Theorem 2 as a sum of explicitly symmetric components on f and h, is given by*

$$\Omega_{6\text{conf. flat}}(f, h) = \left\{ 12\,\Delta^2(\langle df, dh \rangle) - 6\Delta(\Delta f \Delta h) - 12\langle \nabla \Delta f, \nabla \Delta h \rangle \right.$$

$$+ 24\,\Delta(\langle \nabla df, \nabla dh \rangle) + 16\,\langle \nabla^2 df, \nabla^2 dh \rangle + 72\,\Delta \langle df, dh \rangle J$$

$$- 24\Delta(f)\Delta(h)J + 48\,\langle \nabla df, \nabla dh \rangle J + 96\,\langle df, dh \rangle J^2$$

$$+ 24\,\langle df, dh \rangle \Delta(J) - 24\langle \Delta(f)dh + \Delta(h)df, dJ \rangle$$

$$- 24\,\langle d(\langle df, dh \rangle), dJ \rangle - 96\langle \Delta(h)\nabla df + \Delta(f)\nabla dh, V \rangle$$

$$+ 32\,\langle \nabla(\Delta(f) \otimes dh) + \nabla(\Delta(h) \otimes df), V \rangle$$

$$+ 64\,\langle \nabla^2(\langle df, dh \rangle), V \rangle - 64\,\langle df, dh \rangle \langle V, V \rangle$$

$$- 128\,\text{trace}((df \otimes dh)V^2) + 64\,\text{trace}((\text{Hess } f)(\text{Hess } h)V) \left. \right\} d^6 x. \quad (4.3)$$

In the last two terms, both factors are considered as $(1, 1)$ tensors (one contravariant and one covariant).

Actually, the difference between the two expressions (4.2) and (4.3) is given by the term

$$(4.3) - (4.2) = 96\, f_{;ij}h_{;kl}W^{iljk} - 32(f_{;ij}h_{;k}W^{ijk}{}_{l}{}_{;}{}^{l} + f_{;i}h_{;jk}W^{ijk}{}_{l}{}_{;}{}^{l})$$

which vanishes in the conformally flat case.

Leaving for an instant the conformally flat case, in the general conformally curved case, the conformal variation of $\Omega_6(f,h)$, up to order 1 in η is given by

$$-32\Big\{\eta_{;i}f_{;j}h_{;k}W^{ijk}{}_{l}{}_{;}{}^{l} + \eta_{;i}f_{;j}h_{;k}W^{ikj}{}_{l}{}_{;}{}^{l}$$
$$+ \eta_{;i}f_{;jk}h_{;l}W^{ijkl} - \eta_{;i}f_{;j}h_{;kl}W^{ikjl}\Big\}\,d^6x, \quad (4.4)$$

which vanishes in the conformally flat case, meaning that our expression is conformally invariant inside the conformally flat class of metrics on M. In the general conformally curved case, this variation will be useful in finding the extra terms we are missing, that is to say, those terms that vanish in the conformally flat case.

If we define using (4.2) the trilinear form on $C^\infty(M)$,

$$\tau(f_0, f_1, f_2) := \int_M f_0 \Omega_{6,\text{conf. flat}}(f_1, f_2),$$

then

$$(b\tau)(f_0, f_1, f_2, f_3)$$
$$= \int_M f_0\Big(-96\,(f_{1\,;j}f_{2\,;i}f_{3\,;k}W^{ijk}{}_{l}{}_{;}{}^{l} + f_{1\,;j}f_{2\,;i}f_{3\,;k}W^{ikj}{}_{l}{}_{;}{}^{l})$$
$$+ 128\,(f_{1\,;jk}f_{2\,;i}f_{3\,;l}W^{ijkl} + f_{1\,;j}f_{2\,;i}f_{3\,;kl}W^{ikjl})\Big)\,d^6x \quad (4.5)$$

which vanishes in the conformally flat case, meaning that τ is a Hochschild 2-cocycle, in the conformally flat case.

14.5   A filtration by degree

To simplify the notation, and because of the factor $d^n x$ in the definition of Ω_n, we will write $\Omega_n(f,h) = \omega_n(f,h)\,d^n x$ where,

$$\omega_n(f,h) = \sum \frac{1}{\alpha'!\alpha''!\beta!\delta!} D_x^\beta(f) D_x^{\alpha''+\delta}(h) \times$$
$$\int_{|\xi|=1} \text{trace}\big(\partial_\xi^{\alpha'+\alpha''+\beta}(\sigma_{-j}^F)\partial_\xi^\delta(D_x^{\alpha'}(\sigma_{-k}^F))\big)\,d^{n-1}\xi.$$

The expression $\omega_n(f,h)$ is a sum of homogeneous polynomials in the ingredients $\nabla^\alpha df$, $\nabla^\beta dh$, and $\nabla^\gamma R$ for multi-indices α, β, and γ, in the following sense.

Each monomial must satisfy the "homogeneity condition" given by the rule (see [2]):

$$\text{twice the appearances of } R + \text{ number of covariant derivatives} = n,$$

where for covariant derivatives we count all of the derivatives on R, f, and h, and any occurrence of W, Rc, V, Sc, or J is counted as an occurrence of R. By closing under addition, we denote by $\mathcal{P}_n$ the space of these polynomials.

This same idea is used in [4] to study leading terms in the heat invariants produced by the Laplacian of de Rham and other complexes. We borrow from there the idea of *filtration by degree*. For a homogeneous polynomial Q in $\mathcal{P}_n$, we denote by k_R its degree in R and by k_∇ its degree in ∇. In this way, $2k_R + k_\nabla = n$. Because $|\beta| \geq 1$, and $|\delta| \geq 1$ in Lemma 2, we have $k_\nabla \geq 2$ and hence $2k_R \leq n - 2$.

We say that Q is in $\mathcal{P}_{n,l}$ if Q can be written as a sum of monomials with $k_R \geq l$, or equivalently, $k_\nabla \leq n - 2l$. We have

$$\mathcal{P}_n = \mathcal{P}_{n,0} \supseteq \mathcal{P}_{n,1} \supseteq \mathcal{P}_{n,2} \supseteq \cdots \supseteq \mathcal{P}_{n,\frac{n-2}{2}},$$

and $\mathcal{P}_{n,l} = 0$ for $l > (n-2)/2$. There is an important observation to make. An expression which a priori appears to be in, say $\mathcal{P}_{6,1}$, may actually be in a subspace of it, like $\mathcal{P}_{6,2}$. For example,

$$\underbrace{f_{;i}h_{;jkl}W^{ijlk}}_{\in \mathcal{P}_{6,1}} = \underbrace{f_{;i}h_{;j}V_{kl}W^{ikjl} + f_{;i}h_{;j}W^{i}{}_{klm}W^{j}{}_{klm}}_{\in \mathcal{P}_{6,2}},$$

by reordering covariant derivatives and making use of the symmetries of the Weyl tensor. Because of this filtration, we use a fixed convention on how the indices should be placed when representing each expression in its index notation. For example, $f_{;ijk}h_{;}{}^{ijk}$ will be preferred over $f_{;ikj}h_{;}{}^{ijk}$. Also $f_{;ij}{}^{j}{}_{k}{}^{k}$ will be preferred over $f_{;i}{}^{i}{}_{jk}{}^{k}$ or any other variation. Once we have defined the filtration on $\mathcal{P}_n$, and accepted our notational convention, we can state the following proposition.

Proposition 3. *There exists a universal bilinear form* $Q_n(df, dh)$ *in* $\mathcal{P}_{n,0} \smallsetminus \mathcal{P}_{n,1}$ *and a form* $Q_{R,n}(df, dh)$ *in* $\mathcal{P}_{n,1}$ *such that*

$$\Omega_n(f, h) = Q_n(df, dh) + Q_{R,n}(df, dh).$$

In the particular case of the flat metric, $\Omega_n(f, h) = Q_n(df, dh)$ *since the curvature vanishes.*

In the particular case $n = 4$, k_R can be 0 or 1, hence Ω_4 can be written as

$$\Omega_4(f, h) = \underbrace{Q_4(df, dh)}_{\in \mathcal{P}_{4,0}} + \underbrace{Q_{R,4}(df, dh)}_{\in \mathcal{P}_{4,1}}$$

where $Q_{R,4}(df, dh)$ is a trilinear form on R, df, and dh.

In the 6-dimensional case, $k_R \in \{0, 1, 2\}$, thus

$$\Omega_6(f, h) = Q_6(df, dh)$$
$$+ Q_{R,6}^{(1,0)}(df, dh) + Q_{R,6}^{(1,1)}(df, dh) + Q_{R,6}^{(1,2)}(df, dh)$$
$$+ Q_{R,6}^{(2,0)}(df, dh), \quad (5.1)$$

where

- $Q_{R,6}^{(1,0)}(df, dh) \in \mathcal{P}_{6,1} \smallsetminus \mathcal{P}_{6,2}$, without covariant derivatives on R,

- $Q_{R,6}^{(1,1)}(df, dh) \in \mathcal{P}_{6,1} \smallsetminus \mathcal{P}_{6,2}$, with a single covariant derivative on R,

- $Q_{R,6}^{(1,2)}(df, dh) \in \mathcal{P}_{6,1} \smallsetminus \mathcal{P}_{6,2}$, with two covariant derivatives on R,

- $Q_{R,6}^{(2,0)}(df, dh) \in \mathcal{P}_{6,2}$, without covariant derivatives on R.

From the previous expressions, it is evident that there exists a sub-filtration inside each $\mathcal{P}_{n,l}$ for $l \geq 1$. Such a filtration is a lot more complicated to describe in higher dimension because of the presence of terms like $\nabla^a R \nabla^c R \cdots$.

14.6   The 6-dimensional non-conformally flat case

We do not restrict ourselves anymore to the conformally flat case. Now we are going to find those terms we need to add to $\Omega_{6,\text{conf. flat}}$ in order to get the expression for Ω_6 in the general conformally curved case.

The first set of terms to be added will complete the expression for $Q_{R,6}^{(1,0)}$. In this case, there is just one possibility to consider, that is

$$f_{;ij} h_{;kl} W^{ikjl}.$$

Any other possibility is ruled out by the relation

$$f_{;i} h_{;jkl} W^{ijkl} = f_{;i} h_{;j} V_{kl} W^{ikjl} + f_{;i} h_{;j} W_i{}^{klm} W^{jlkm}$$

which express the right-hand side, an element of $\mathcal{P}_{6,1}$, as the sum of two elements of $\mathcal{P}_{6,2}$.

The second set of terms completes the expression for $Q_{R,6}^{(1,1)}$; it is given by the following symmetric term on f and h:

$$f_{;ij} h_{;k} W^i{}_l{}^{jk}{}_;{}^l + h_{;ij} f_{;k} W^i{}_l{}^{jk}{}_;{}^l.$$

The only term to complete the expression for $Q_{R,6}^{(1,2)}$ is

$$f_{;i} h_{;j} W^i{}_k{}^j{}_l{}_;{}^{kl},$$

symmetric on f and h.

For $Q_{R,6}^{(2,0)}$ we consider at this time, just one term

$$f_{;i}h_{;j}V_{kl}W^{ikjl}.$$

It happens that the other possible terms

$$f_{;i}h_{;}{}^{i}W_{jklm}W^{jklm}\,d^6x \quad \text{and} \quad f_{;i}h_{;j}W^{i}{}_{klm}W^{jklm}\,d^6x \tag{6.1}$$

are conformally invariant.

Up to this point, what we must add to $\Omega_{6,\text{conf. flat}}$ is a linear combination of the form

$$\left\{ A f_{;ij}h_{;kl}W^{ikjl} + B\left(f_{;ij}h_{;k}W^{i}{}_{l}{}^{jk}{}_{;}{}^{l} + h_{;ij}f_{;k}W^{i}{}_{l}{}^{jk}{}_{;}{}^{l}\right) \right.$$
$$\left. + C f_{;i}h_{;j}W^{i}{}_{k}{}^{j}{}_{l;}{}^{kl} + D f_{;i}h_{;j}V_{kl}W^{ikjl}\right\} d^6x. \tag{6.2}$$

Its conformal variation up to order 1 in η is given by

$$\left\{ (B+2C)\left(\eta_{;i}f_{;j}h_{;k}W^{ijk}{}_{l;}{}^{l} + \eta_{;i}f_{;j}h_{;k}W^{ikj}{}_{l;}{}^{l}\right) \right.$$
$$+ (3B-2A)\left(\eta_{;i}f_{;jk}h_{;l}W^{ijkl} + \eta_{;i}f_{;j}h_{;kl}W^{ilkj}\right)$$
$$\left. + (D-3C)\left(\eta_{;ij}f_{;k}h_{;l}W^{ikjl}\right)\right\} d^6x. \tag{6.3}$$

By comparing it with the conformal variation of $\Omega_{6,\text{conf. flat}}(f,h)$ (4.4), we deduce the conditions $B + 2C = -32 = 3B - 2A$, and $D - 3C = 0$, which means, conformal invariance and symmetry is not enough to find the right values for all the constants. So far, the term to be added to $\Omega_6(f,h)$ is given by

$$\left\{ -(32+3C) f_{;ij}h_{;kl}W^{ikjl} \right.$$
$$- (32+2C)\left(f_{;ij}h_{;k}W^{i}{}_{l}{}^{jk}{}_{;}{}^{l} + h_{;ij}f_{;k}W^{i}{}_{l}{}^{jk}{}_{;}{}^{l}\right.$$
$$+ C f_{;i}h_{;j}W^{i}{}_{k}{}^{j}{}_{l;}{}^{kl} + 3C f_{;i}h_{;j}V_{kl}W^{ikjl}$$
$$\left. + E f_{;i}h_{;}{}^{i}W_{jklm}W^{jklm} + G f_{;i}h_{;j}W^{i}{}_{klm}W^{jklm}\right\} d^6x \tag{6.4}$$

where the last two terms come from (6.1).

Using the Hochschild 2-cocycle property

Proposition 4. *If we define the trilinear form on $C^\infty(M)$,*

$$\tau'(f_0, f_1, f_2)$$
$$:= \int_M f_0 \left\{ C f_{1;i}f_{2;j}W^{i}{}_{k}{}^{j}{}_{l;}{}^{kl} + D f_{1;i}f_{2;j}V_{kl}W^{ikjl} \right.$$
$$\left. + E f_{1;i}f_{2;}{}^{i}W_{jklm}W^{jklm} + G f_{1;i}f_{2;j}W^{i}{}_{klm}W^{jklm}\right\} d^6x,$$

then $(b\tau')(f_0, f_1, f_2, f_3) = 0$ *for any* $f_i \in C^\infty(M)$, *meaning that we obtain a Hochschild 2-cocycle on the algebra* $C^\infty(M)$ *for any value of the constants* $C, D, E,$ *and* G.

On the other hand, if we define

$$\tau''(f_0, f_1, f_2) := \int_M f_0 \Big\{ A\, f_{1\,;\,ij} f_{2\,;\,kl} W^{ikjl}$$
$$+ B\, (f_{1\,;\,ij} f_{2\,;\,k} W^i{}_l{}^{jk}{}^{;\,l} + f_{2\,;\,ij} f_{1\,;\,k} W^i{}_l{}^{jk}{}^{;\,l}) \Big\} d^6 x,$$

then

$$(b\tau'')(f_0, f_1, f_2, f_3)$$
$$= \int_M f_0 \Big(3\, B\, (f_{1\,;\,j} f_{2\,;\,i} f_{3\,;\,k} W^{ijk}{}_l{}^{;\,l} + f_{1\,;\,j} f_{2\,;\,i} f_{3\,;\,k} W^{ikj}{}_l{}^{;\,l})$$
$$- 2\, A\, (f_{1\,;\,jk} f_{2\,;\,i} f_{3\,;\,l} W^{ijkl} + f_{1\,;\,j} f_{2\,;\,i} f_{3\,;\,kl} W^{ikjl}) \Big) d^6 x. \quad (6.5)$$

To have that $\int_M f_0 \Omega_6(f_1, f_2)$ is a Hochschild 2-cocycle on $C^\infty(M)$, we need $(4.5) + (6.5) = 0$, for any $f_i \in C^\infty(M)$. Thus $3B = 96$ and $2A = 128$. Because $B + 2C = -32 = 3B - 2A$, we must have $C = -32$ and hence using (4.2), (6.1), and (6.4) we conclude

$$\Omega_6(f, h) = \Omega_{6\,\mathrm{conf.\,flat}}(f, h)$$
$$+ \Big\{ 64\, f_{;\,ij} h_{;\,kl} W^{ikjl} + 32\, (f_{;\,ij} h_{;\,k} W^{ijk}{}_l{}^{;\,l} + h_{;\,ij} f_{;\,k} W^{ijk}{}_l{}^{;\,l})$$
$$- 32\, f_{;\,i} h_{;\,j} W^i{}_k{}^j{}_l{}^{;\,kl} - 96\, f_{;\,i} h_{;\,j} V_{kl} W^{ikjl}$$
$$+ E\, f_{;\,i} h_{;}{}^i W_{jklm} W^{jklm} + G\, f_{;\,i} h_{;\,j} W^i{}_{klm} W^{jklm} \Big\} d^6 x. \quad (6.6)$$

where the last two terms are the needed ones to fully complete the expression for $Q_{R,6}^{(2,0)}$ as in (6.1).

Theorem 4. *The expression (6.6) gives a family of 6-dimensional differential forms associated to* M. *Each of these differential forms is symmetric on* f *and* h, *conformally invariant, and defines a Hochschild 2-cocycle on the algebra* $C^\infty(M)$ *by the relation* $\tau(f_0, f_1, f_2) = \int_M f_0 \Omega_6(f_1, f_2)$.

In the 6-dimensional conformally curved case, more information is needed to find the unique one satisfying the relation

$$\mathrm{Wres}(f_0 [F, f_1][F, f_2]) = \int_M f_0 \Omega_6(f_1, f_2).$$

REFERENCES

[1] T. P. Branson, Automated symbolic computations in spin geometry, In *Clifford Analysis and its Applications*, F. Brackx, J.S.R. Chisholm, and V. Souček, eds., pp. 27–38. NATO Science Series II, Vol. 25, Kluwer Academic Publishers, Dordrecht, 2001.

[2] T. P. Branson, Differential operators canonically associated to a conformal structure, *Math. Scand.* **57** (1985), 293–345.

[3] T. Branson, An Anomaly associated with 4-dimensional quantum gravity, *Commun. Math. Phys.* **178** (1996), 301–309.

[4] T.P. Branson, P.B. Gilkey, and Bent Ørsted, Leading terms in the heat invariants for the Laplacians of the de Rham, signature, and spin complexes, *Math. Scand.* **66** (1990), 307–319.

[5] T. Branson and B. Ørsted, Explicit functional determinants in four dimensions, *Proc. Amer. Math. Soc.* **113** (1991), 669–682.

[6] A. Connes, Quantized calculus and applications, *XI-th International Congress of Mathematical Physics* (Paris, 1994), pp. 15–36, Internat. Press, Cambridge, MA, 1995.

[7] A. Connes, *Noncommutative Geometry*, Academic Press, London and San Diego, 1994.

[8] A. R. Gover, L. J. Peterson, Conformally invariant powers of the Laplacian, Q-curvature and tractor calculus, math-ph/0201030.

[9] J. M. Gracia-Bondía, J. C. Várilly and H. Figueroa, *Elements of Noncommutative Geometry*, Birkhäuser Advanced Texts, Birkhäuser, Boston, 2001.

[10] C. R. Graham, Conformally invariant powers of the Laplacian, II: Nonexistence, *J. London Math. Soc.* (2) **46** (1992), 566–576.

[11] R. Graham, R. Jenne, L. Mason, and G. Sparling, Conformally invariant powers of the Laplacian, I: Existence, *J. London Math. Soc.* (2) 46 (1992), 557–565.

[12] H. Kumano-go, *Pseudo-Differential Operators*, The MIT Press, Cambridge, Massachusetts, 1981.

[13] J. M. Lee, A Mathematica package for doing tensor calculations in differential geometry, http://www.math.washington.edu/~lee/Ricci/.

[14] W. J. Ugalde, *Differential forms canonically associated to even-dimensional compact conformal manifolds*, Thesis, axXiv:math.DG/0211240 .

[15] W. J. Ugalde, Differential forms and the Wodzicki residue, axXiv:math/DG0211361.

[16] F. W. Warner, *Foundations of Differential Manifolds and Lie Groups*, Springer-Verlag, New York, 1983.

[17] M. Wodzicki, *Noncommutative Residue, Part I. Fundamentals, K-Theory, arithmetic and geometry*, (Moscow 1984–86), pp. 320–399, Lecture Notes in Mathematics, 1289, Springer, Berlin, 1987.

William J. Ugalde[1]
Department of Mathematics
14 MacLean Hall
The University of Iowa
Iowa City, Iowa, 52242
E-mail: wugalde@math.uiowa.edu

Submitted: August 17, 2002; Revised: March 03, 2003.

[1]Current address: Department of Mathematics, Purdue University, 150 N. University Street, West Lafayette, IN 47907-2067, E-mail: ugalde@math.purdue.edu.

15

The Interface of Noncommutative Geometry and Physics

Joseph C. Várilly

ABSTRACT As a mathematical theory, noncommutative geometry (NCG) is by now well established. From the beginning, its progress has been crucially influenced by quantum physics: we briefly review this development in recent years.

The standard model of fundamental interactions, with its central role for the Dirac operator, led to several formulations culminating in the concept of a real spectral triple. String theory then came into contact with NCG, leading to an emphasis on Moyal-like algebras and formulations of quantum field theory on noncommutative spaces. Hopf algebras have yielded an unexpected link between the noncommutative geometry of foliations and perturbative quantum field theory.

The quest for a suitable foundation of quantum gravity continues to promote fruitful ideas, among them the spectral action principle and the search for a better understanding of "noncommutative spaces".

Keywords: standard model, noncommutative field theory, Hopf algebra

15.1 Introduction

About 20 years ago, the mathematical theory nowadays known as noncommutative geometry (NCG) began taking shape. A landmark paper of Connes (1980) ushered in a differential geometric treatment of the noncommutative torus [1] (further developed and classified by Rieffel [2]), which remains the paradigm of a noncommutative space. Its differential calculus was put in a more general framework at the Oberwolfach meeting in September–October 1981, where Connes unveiled a "homology of currents for operator algebras" [3], which soon became known as cyclic cohomology [4]. This was developed in detail in his "Noncommutative Differential Geometry" [5], in preprint form around Christmas 1982; the related periodic cyclic cohomology is a precise generalization, in algebraic language, of the de Rham homology of smooth manifolds.

The same algebraic approach, applied to the theory of foliations [6], led Connes to emphasize the notion of Fredholm module, which is a cornerstone of his work with Karoubi on canonical quantization [7]. A key observation here is that anoma-

---

AMS Subject Classification: 58B34, 81T10.

lous commutators form a cyclic 1-cocycle [8], so that in the noncommutative approach to quantum field theory, the Schwinger terms are built in.

Noncommutative geometry, then, is an operator-algebraic reformulation of the foundations of geometry, extending to noncommutative spaces. It allows consideration of "singular spaces" erasing the distinction between the continuous and the discrete. On the mathematical side, current topics of interest include index theory and groupoids, mathematical quantization, the Baum–Connes conjectures on the K-theory of group algebras, locally compact quantum groups, second quantization in the framework of spectral triples, and the Riemann hypothesis. Our focus here, however, is on its interface with physics.

15.2   NCG and the Standard Model

We say interface because one should not speak of the "application" of NCG to physics, but rather of mutual intercourse. Indeed, the first use of noncommutative geometry in physics did not attempt to derive the laws of physics from some NCG construct, but simply and humbly, to learn from the mainstream physical theories —concretely, the standard model (SM) of fundamental interactions— what the (noncommutative) geometry of the world could be.

The crucial concepts of the SM are those of gauge fields and of chiral fermions: they correspond to two basic notions of NCG, namely connections and Dirac operators. Indeed, the algebraic definition of linear connection is imported verbatim into NCG. Chiral fermions, for their part, are acted on by Dirac and Dirac–Weyl operators.

Dirac operators are a source of NCG: any complex spinor bundle on a smooth manifold $S \to M$ gives rise to a generalized Dirac operator D on the spinor space $L^2(M, S)$, whose sign operator $F = D|D|^{-1}$ determines a Fredholm module; its K-homology class $[F] \in K_\bullet(M)$ depends only on the underlying spin^c-structure [9, 10]. Since the spin^c structure determines the orientation of the manifold, this fundamental class —sometimes called a K-orientation [11]— is a finer invariant than the usual fundamental class in homology.

The approach to the SM by Connes and Lott [12] used a noncommutative algebra to describe the electroweak sector, plus a companion algebra to incorporate colour symmetries (see [13] and [14] for reviews of this preliminary approach). Later on [15], a better understanding of the role of the charge conjugation allowed this pair of algebras to be replaced by a *single* algebra acting bilaterally.

The gauge potentials appearing in the SM may be collected into a single package of differential forms:

$$\mathbb{A}' = i(B, W, A),$$

where

$$B = -\tfrac{i}{2}g_1 \mathbf{B}_\mu \, dx^\mu, \quad W = -\tfrac{i}{2}g_2 \, \boldsymbol{\tau} \cdot \mathbf{W}_\mu \, dx^\mu \quad \text{and} \quad A = -\tfrac{i}{2}g_3 \, \boldsymbol{\lambda} \cdot \mathbf{A}_\mu \, dx^\mu,$$

with $\mathbf{B}$, $\mathbf{W}$ and $\mathbf{A}$ denoting respectively the hypercharge, weak isospin and colour gauge potentials; W is to be regarded as a quaternion-valued 1-form. Thus, $\mathbb{A}'$ is an element of $\Lambda^1(M) \otimes \mathcal{A}_F$, where the noncommutative algebra $\mathcal{A}_F := \mathbb{C} \oplus \mathbb{H} \oplus M_3(\mathbb{C})$, that we have called the "Eigenschaften algebra" [16], plays the crucial role.

We next collect all chiral fermion fields into a multiplet Ψ and denote by J the charge conjugation; then the fermion kinetic term is rewritten as follows:

$$I(\Psi, \mathbb{A}', J) = \langle \Psi \mid (i\partial\!\!\!/ + \mathbb{A}' + J\mathbb{A}'J^\dagger)\Psi \rangle.$$

To incorporate the Yukawa part of the SM Lagrangian, let ϕ be a Higgs doublet with vacuum expectation value $v/\sqrt{2}$, normalized by setting $\Phi := \sqrt{2}\,\phi/v$. We need both

$$\Phi = \begin{pmatrix} \Phi_1 \\ \Phi_2 \end{pmatrix} \quad \text{and} \quad \widetilde{\Phi} := \begin{pmatrix} -\bar{\Phi}_2 \\ \bar{\Phi}_1 \end{pmatrix}.$$

The Higgs may be properly regarded as a quaternion-valued field; by introducing $q_\Phi = \begin{pmatrix} \Phi_1 & \bar{\Phi}_2 \\ -\Phi_2 & \bar{\Phi}_1 \end{pmatrix}$, where $\langle q_\Phi \rangle = 1$, we may write, schematically for a right-left splitting of the fermion multiplets:

$$\mathbb{A}'' = \begin{pmatrix} & M^\dagger(q_\Phi - 1) \\ (q_\Phi - 1)^\dagger M & \end{pmatrix},$$

where M denotes the mass matrix for quarks (including the Cabibbo–Kobayashi–Maskawa parameters) and leptons. Denoting by $\mathcal{D}_F$ the Yukawa operator which relates the left- and right-handed chiral sectors in the space of internal degrees of freedom, the Yukawa terms for both particles and antiparticles (for the first generation) can now be written as

$$\begin{aligned} I(\Psi, \mathbb{A}'', J) &:= \langle \Psi \mid (\mathcal{D}_F + \mathbb{A}'' + J\mathbb{A}''J^\dagger)\Psi \rangle \\ &= \bar{q}_L \Phi\, m_d\, d_R + \bar{q}_L \widetilde{\Phi}\, m_u\, u_R + q_R \bar{\Phi}\, \bar{m}_d\, \bar{d}_L + q_R \bar{\widetilde{\Phi}}\, \bar{m}_u\, \bar{u}_L \\ &\quad + \bar{\ell}_L \Phi\, m_e\, e_R + \bar{\ell}_L \widetilde{\Phi}\, m_\nu\, \nu_R + \ell_R \bar{\Phi}\, \bar{m}_e\, \bar{e}_L + \ell_R \bar{\widetilde{\Phi}}\, \bar{m}_\nu\, \bar{\nu}_L + \text{h.c.} \end{aligned}$$

Altogether, we get a Dirac–Yukawa operator $\mathcal{D} = i\partial\!\!\!/ \oplus \mathcal{D}_F$. With $\mathbb{A} := \mathbb{A}' \oplus \mathbb{A}''$, the *whole* fermionic sector of the SM is recast as

$$I(\Psi, \mathbb{A}, J) = \langle \Psi \mid (\mathcal{D} + \mathbb{A} + J\mathbb{A}J^\dagger)\Psi \rangle.$$

The upshot is that the ordinary gauge fields and the Higgs are combined as entries of a *generalized gauge potential*. The Yukawa terms come from the minimal coupling recipe applied to the gauge field in the internal space. The Dirac–Yukawa operator is seen to contain in NCG all the relevant information pertaining to the SM.

This Connes–Lott reconstruction of the SM gave rise to two "predictions". (At that time, the top quark had not been seen, and the best estimates for its mass ranged around 130 GeV.) The NCG model sort of explains why the masses of the top quark, the W and Z particles and the Higgs particle should be of the same order, and gave right away

$$m_{\text{top}} \geq \sqrt{3}\, m_W \approx 139\,\text{GeV}.$$

With a bit of renormalization group running [17], it fell right on the mark. On the other hand, the "prediction" for the Higgs mass from Connes' NCG has remained stuck around 200 GeV, while the current phenomenological prejudice is that it should be much lower.

A major limitation of the Connes–Lott approach is that the fermion mass matrix must be taken as an input. A different though less ambitious proposal, put forward about the same time, was the Mainz–Marseille scheme, based on organizing the (W, B) forms and the Higgs field components as a 3×3 matrix in the Lie superalgebra $\mathfrak{su}(2|1)$. The known families of quarks and leptons can then be fitted into (reducible but indecomposable) $\mathfrak{su}(2|1)$ representations, and some relations among the quark masses and CKM mixing parameters emerge [18]; this analysis applies likewise to lepton masses and neutrino mixing.

This "bottom-up" interaction between physics and NCG yielded an important dividend. The clarification of the role of J as a "Tomita conjugation" [19] fostered the emergence of the concept of a *real spectral triple* —the word "real" being taken in the sense of Atiyah's "Real K-theory" [20]— which led to a construction of noncommutative spin manifolds [21]. This construction is explained in our book [22]. Thus, we now know how to put fermion fields on a noncommutative manifold.

A *spectral triple* $(\mathcal{A}, \mathcal{H}, D)$ consists of a (unital) *algebra* $\mathcal{A}$ represented on a Hilbert space $\mathcal{H}$, plus a selfadjoint operator D on $\mathcal{H}$, such that $[D, a]$ is bounded for all $a \in \mathcal{A}$ and D^{-1} is compact. It is *even* if there is a grading operator χ (or "γ_5") on H with respect to which $\mathcal{A}$ is even and D is odd. It is *real* if there is an antiunitary operator J on H such that $J^2 = \pm 1$, $JD = \pm DJ$ and $J\chi = \pm\chi J$ (even case); the signs depend on a certain dimension mod 8. From these data, by imposing a few extra conditions, spin manifolds can be reconstructed [23].

15.3   The spectral action principle

The early Connes–Lott models did not take account of gravity. To remedy that, Connes and Chamseddine [24] proposed a universal formula for an action associated with a noncommutative spin geometry, modeled by a real spectral triple $(\mathcal{A}, \mathcal{H}, D, J)$. The action $S(D) = B_\phi[D] + \langle \Psi \mid D\Psi \rangle$ is based on the spectrum of the Dirac operator and is a geometric invariant. Automorphisms of the algebra $\mathcal{A}$ combine ordinary diffeomorphisms with internal symmetries which alter the metric by $D \mapsto D + A + JAJ^\dagger$.

The bosonic part of the action functional is $B_\phi[D] = \text{Tr}\,\phi(D^2)$, where ϕ is an "arbitrary" positive function (a regularized cutoff) of D. Chamseddine and

Connes argue that B_ϕ has an asymptotic expansion

$$B_\phi[D/\Lambda] \sim \sum_{n=0}^{\infty} f_n \Lambda^{4-2n} a_n(D^2) \quad \text{as } \Lambda \to \infty,$$

where the a_n are the coefficients of the heat kernel expansion for D^2 and

$$f_0 = \int_0^\infty x\phi(x)\,dx, \quad f_1 = \int_0^\infty \phi(x)\,dx, \quad f_2 = \phi(0), \quad f_3 = -\phi'(0),$$

and so on: this is in fact a Cesàro asymptotic development [25]. On computing this expansion for the Dirac–Yukawa operator of the standard model, they found all terms in the bosonic part of the SM action, plus unavoidable gravity couplings. That is to say, the spectral action for the standard model unifies with gravity at a very high energy scale.

Recently, Wulkenhaar [26] has conjectured that on θ-deformed spacetime, the spectral action may have the necessary additional symmetries to renormalize gauge theories. In this regard, Langmann [27] has managed to prove that the effective action of fermions coupled to a Yang–Mills field contains the usual Yang–Mills bosonic action. To check the conjecture, one first needs to extend the spectral action to the context of noncompact NC manifolds.

By "noncompact noncommutative spin geometry" we understand a real spectral triple $(\mathcal{A}, \mathcal{H}, D, J)$ where $\mathcal{A}$ is a *nonunital* algebra, where $[D, a]$ is bounded and $a|D|^{-1}$ is compact for all $a \in \mathcal{A}$. Geometries of this type are discussed in [28, 29, 44]: the analytic toolbox of NC spin geometries [22] extends to the noncompact case if suitable multiplier algebras are employed.

15.4   Noncommutative field theory

The next phase of the dialogue between NCG and the physics of fundamental interactions was characterized by a "top-down" approach. An important precursor is the 1947 paper by Snyder, "Quantized space-time" [30], where it was first suggested that coordinates x^μ may be noncommuting operators; the six commutators are of the form $[x^\mu, x^\nu] = (ia^2/\hbar)\, L^{\mu\nu}$ where a is a basic unit of length and the $L^{\mu\nu}$ are generators of the Lorentz group; throughout, Lorentz covariance is maintained. Then as now, noncommuting coordinates were used to describe spacetime in the hope of improving the renormalizability of QFT and of coming to terms with the nonlocality of physics at the Planck scale.

In a similar vein, Doplicher, Fredenhagen and Roberts [31] have considered a model with commutation relations

$$[x^\mu, x^\nu] = i\, Q^{\mu\nu},$$

where the $Q^{\mu\nu}$ are the components of a tensor, but commute among themselves and with each x^μ. Thus in their formalism, Lorentz invariance is also explicitly kept.

String theorists have recently revived this top-down approach. In their most popular model, the commutation relations are simply of the form

$$[x^\mu, x^\nu] = i\,\theta^{\mu\nu}, \tag{15.4.1}$$

where the $\theta^{\mu\nu}$ are c-numbers, breaking Lorentz invariance. As anticipated by Sheikh-Jabbari [32] and plausibly argued by Seiberg and Witten [33], open strings with allowed endpoints on 2D-branes in a B-field background act as electric dipoles of the Abelian gauge field of the brane; the endpoints live on the non-commutative space determined by (15.4.1), as pointed out by [34].

Even before that, Connes, Douglas and Schwarz [35] had shown that com-pactification of M-theory, in the context of dimensionally reduced gauge theory actions, leads to spaces with embedded noncommutative tori. Soon after, Dou-glas and Hull saw that gauge theories on noncommutative spaces arise naturally from string theory [36]. See also [37] for the relation between noncommutative geometry and strings.

An important feature of [33] is the "Seiberg–Witten map" in gauge theory, which relates gauge fields and gauge variations in a noncommutative theory with their commutative counterparts. In the NC theory, multiplication is replaced by the Moyal product $\star_\theta$ with parameter $\theta = [\theta^{\mu\nu}]$; in order to preserve gauge equiv-alence (whenever A and A' are equivalent gauge fields, so should be the NC gauge fields $\widehat{A}$ and $\widehat{A}'$), Seiberg and Witten found θ-dependent formulas for the latter. As explained by Jackiw and Pi [38] (see also [39]), these formulas correspond to an infinitesimal 1-cocycle for a projective representation of the underlying gauge group in the Moyal algebra.

The Moyal product which appears here is nonperturbatively defined, for non-degenerate skewsymmetric θ, as

$$f \star_\theta g(u) := (\pi\theta)^{-4} \int_{\mathbb{R}^4 \times \mathbb{R}^4} f(u+s)g(u+t)\, e^{2is\theta^{-1}t}\, d^4s\, d^4t,$$

and this gives rise to the commutation relations (15.4.1). Those are just the com-mutation relations of quantum mechanics, when $\hbar$ replaces θ! The precise relation of this integral formula to the asymptotic development usually put forward as the Moyal product was spelled out some time ago in [40]. This product is the basis of the Weyl–Wigner–Moyal or phase-space approach to quantum mechanics [41], which already had a long history when (a version of) the Moyal product was re-discovered by string theorists. It should be said that many of the recent papers which purport to use this product in string theory or NC field theory are rather careless; some are unaware of the mathematical properties of the Moyal product, which are outlined, for instance, in our [42, 43] or in [44].

It is worth pointing out that noncommutative field theory can be developed independently of its string theory motivation, and indeed preexisted the Seiberg–Witten paper. Quantum field theory has an algebraic core which is independent of the nature of spacetime. From the representation theory of the infinite dimen-sional orthogonal group (or an appropriate subgroup), with the input of a one-particle space, one can derive all Fock space quantities of interest: nothing really changes if the "matter field" evolves on a noncommutative space. That is to say, one can apply the canonical quantization machinery to a noncommutative kind of one-particle space [45]. The long-standing hope, that giving up *locality* in the interaction of fields would be rewarded with a better ultraviolet behaviour, was now amenable to rigorous scrutiny, and it is not borne out. QFT on noncommutative manifolds also requires renormalization. This, in some sense the first result of NCFT, was proved in general by Gracia-Bondía and myself in [45], using a cohomological argument internal to noncommutative geometry.

Of course, one can prove the same in the context of a *particular* NCG model, by writing down the integral corresponding to a Feynman diagram, and finding it to be divergent. That had been shown previously by Filk [46], for the scalar Lagrangian theory associated to the Moyal product algebra. Filk made the point that the momentum integrals for planar Feynman graphs are identical to those in the commutative theory, and the contributions from nonplanar graphs cannot cancel them. The same basic point had been made much earlier in [47], with regard to the continuum limit of a reduced model of large N field theory.

FIGURE 15.1. Planar and nonplanar tadpole diagrams in NC ϕ^4 theory

The distinction between planar and nonplanar Feynman diagrams is an essential feature of NC field theory. Consider, for instance, the theory given by the action functional

$$S = \int d^4x \left(\frac{1}{2} \frac{\partial \phi}{\partial x^\mu} \frac{\partial \phi}{\partial x_\mu} + \frac{1}{2} m^2 \phi^2 + \frac{g}{4!} \phi \star_\theta \phi \star_\theta \phi \star_\theta \phi \right).$$

The Feynman rules yield the same propagators as in the commutative theory, but the vertices get in momentum space an extra factor proportional to

$$\exp\left(-\frac{i}{2} \sum_{k<l} p_{k\mu} \theta^{\mu\nu} p_{l\nu} \right),$$

where $p_1, \dots, p_r$ are the momenta incoming on the vertex, in cyclic order. Planar diagrams get overall phase factors depending only on external momenta; for nonplanar diagrams, the phase factors may also depend on loop variables, and the corresponding integrals may become convergent. For the tadpole diagrams of Figure 15.1, we get amplitudes of the form

$$\Gamma_{\mathrm{pl}}(p) \propto \int \frac{d^4k}{k^2 + m^2}, \qquad \Gamma_{\mathrm{npl}}(p) \propto \int \frac{d^4k}{k^2 + m^2} e^{-ip\theta k},$$

and the second integral is finite for $p \neq 0$.

However, nonplanar diagrams may become divergent again for particular values of the momenta (try $p = 0$ in the previous example). For complicated diagrams with subdivergencies, this dependence of the amplitude behaviour on p is troublesome, because such diagrams may unexpectedly become divergent again. This is the notorious UV/IR mixing [48], which tends to spoil renormalizability. For Moyal NC Yang–Mills theory, this happens already at the 2-loop level.

It was pointed out by Gomis and Mehen [49] that whenever there is time-like noncommutativity $\theta^{0i} \neq 0$, one encounters a violation of unitarity of the S-matrix. However, Bahns et al. [50] have argued that, if the above Lagrangian approach is replaced by a Hamiltonian approach to NC field theory, the apparent failure of unitarity disappears. The change in viewpoint concerns only the nonplanar diagrams. This should not really be surprising: even for ordinary Yang–Mills theories, the full equivalence of both approaches has never been established [51]. Actually, the indications seem to be that they are *not* equivalent in noncommutative field theory [52, 53].

The literature on NC field theories is already very large, and of uneven quality; we cannot really do it justice here. For recent extensive reviews, see [54] and [55].

Finally, it is now possible to combine these NC field theories with the Connes–Lott approach, by taking the tensor product $\mathcal{A}_{NC} \otimes \mathcal{A}_F$ of a noncommutative spacetime algebra and the Eigenschaften algebra $\mathcal{A}_F$. This has been done by Morita [56], Chaichian et al [57] and the München group [58], in various ways (and with different outcomes). However, there is some doubt as to whether these models are anomalous [59], and therefore nonrenormalizable [60].

15.5   Noncommutative spaces

In an eventual noncommutative approach to quantum gravity, one must be able to sum over families of noncommutative spaces. This, together with the need for good examples, has inspired a search for noncommutative manifolds. Connes [61] has suggested that "NC spheres" may be obtained from two homological conditions: (1) the Chern character form vanishes in all intermediate degrees; and (2) the metric may vary while keeping the volume form fixed. In even dimensions $n = 2m$, this comes down to setting

$$\mathrm{ch}_k(e) \equiv (-1)^k \frac{(2k)!}{k!} \mathrm{tr}((e - \tfrac{1}{2})\,(de\,de)^k) = 0 \qquad (15.5.1a)$$

for $k = 0, 1, \ldots, m - 1$, and

$$\pi_D(\mathrm{ch}_m(e)) = \chi, \qquad (15.5.1b)$$

where $e = e^2 = e^* \in M_{2^m}(\mathcal{A})$ is an orthogonal projector, χ is the chiral grading operator on $\mathcal{H}$, and $\pi_D(a_0\,da_1 \ldots da_n) := a_0\,[D, a_1] \ldots [D, a_n]$; the domain of π_D is the universal graded differential algebra over $\mathcal{A}$, which may be regarded as the space of chains for Hochschild homology. Briefly, one finds that (15.5.1a) makes $\mathrm{ch}_m(e)$ a Hochschild cycle, and the (15.5.1b) says that this cycle gives the desired volume form [23]. Now (15.5.1) becomes a system of equations which impose severe restrictions on the algebra $\mathcal{A}$ to which the matrix elements of e belong.

For $n = 2$, there is only the commutative solution [61], $\mathcal{A} = C^\infty(\mathbb{S}^2)$. For $n = 4$, Connes and Landi [62] found "θ-twisted 4-spheres" $\mathbb{S}^4_\theta$ with embedded copies of the NC 2-torus $\mathbb{T}^2_\theta$. Later, Connes and Dubois-Violette [63] showed that there is a 3-parameter family of NC 3-spheres, including a θ-twisted subfamily $\mathbb{S}^3_\theta$.

These θ-twisted spheres can all be described as quantum homogeneous spaces [63, 64]: in fact, we can construct M_θ by twisting whenever $M = G/H$ is a quotient of compact Lie groups with rank $H \geq 2$. The noncommutative algebra $C^\infty(M_\theta)$ is simply $C^\infty(M)$ equipped with a periodic version of the Moyal product [65], and the symmetry group G is correspondingly deformed to a quantum group [64]. If M is spin, then so is M_θ; it carries an NC spin geometry obtained by isospectral deformation from that of M [62]. Noncommutative twistors can also be obtained in this manner. $SU_q(2)$-invariant homogeneous spaces are developed in [66] and [67].

One motivation for constructing such examples of noncommutative spaces is to come back to quantum gravity by (a) allowing for metric fluctuations with fixed volume; and eventually (b) relaxing the Hochschild condition to incorporate "virtual" NC manifolds, whereby the condition (15.5.1b) would appear as the signal of a "true" manifold [68].

15.6 The Connes–Kreimer Hopf algebras

Bogoliubov's renormalization scheme in dimensional regularization can be summarized as follows. Let Γ be a one-particle irreducible (1PI) graph (i.e., a connected graph which cannot be disconnected by removing a single line), with amplitude $f(\Gamma)$; if Γ is primitive (i.e., has no subdivergencies), set

$$C(\Gamma) := -T(f(\Gamma)), \quad \text{and then} \quad R(\Gamma) := f(\Gamma) + C(\Gamma),$$

where $C(\Gamma)$ is the *counterterm*, $R(\Gamma)$ is the desired finite value, and T projects on the pole part: in other words, for primitive graphs, one simply removes the pole part. We recursively define Bogoliubov's $\overline{R}$-operation by setting

$$\overline{R}(\Gamma) = f(\Gamma) + \sum_{\emptyset \subsetneq \gamma \subsetneq \Gamma} C(\gamma) \, f(\Gamma/\gamma),$$

where $C(\gamma_1 \ldots \gamma_r) := C(\gamma_1) \ldots C(\gamma_r)$, whenever $\gamma = \gamma_1 \ldots \gamma_r$ is a disjoint union of several pieces. Then we remove the pole part of the previous expression:

$C(\Gamma) := -T(\overline{R}(\Gamma))$ and $R(\Gamma) := \overline{R}(\Gamma) + C(\Gamma)$. Overall,

$$C(\Gamma) := -T\left[f(\Gamma) + \sum_{\emptyset \subsetneq \gamma \subsetneq \Gamma} C(\gamma) f(\Gamma/\gamma) \right], \qquad (15.6.1a)$$

$$R(\Gamma) := f(\Gamma) + C(\Gamma) + \sum_{\emptyset \subsetneq \gamma \subsetneq \Gamma} C(\gamma) f(\Gamma/\gamma). \qquad (15.6.1b)$$

Now let Φ stand for any particular QFT. There is an associated Hopf algebra H_Φ [69, 70] which is, first of all, a commutative algebra generated by the 1PI graphs Γ of Φ. The product is given by the disjoint union of graphs. The counit ε is defined on generators by $\varepsilon(\Gamma) := 0$ unless Γ is empty, and $\varepsilon(\emptyset) := 1$; and the unit map η is determined by $\eta(1) := \emptyset$. The *coproduct* Δ is given by

$$\Delta\Gamma := \sum_{\emptyset \subseteq \gamma \subseteq \Gamma} \gamma \otimes \Gamma/\gamma,$$

where the sum ranges over all subgraphs γ which are divergent and proper (i.e., removing one internal line cannot make more connected components); γ itself need not be connected. The terms for $\gamma = \emptyset$ and $\gamma = \Gamma$ in the sum are $\Gamma \otimes 1 + 1 \otimes \Gamma$. The notation Γ/γ denotes the (connected, 1PI) graph obtained from Γ by replacing each component of γ by a single vertex. One checks that Δ is coassociative [69], so H_Φ is a bialgebra.

Here are some coproducts[1] for $\Phi = \varphi_4^4$, taken from [71]:

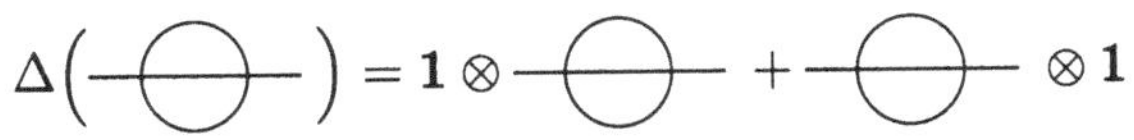

FIGURE 15.2. The "setting sun": a primitive diagram

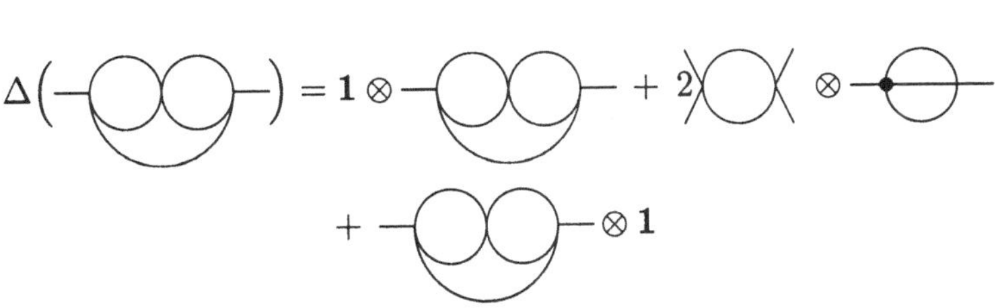

FIGURE 15.3. The "double ice cream in a cup" (depth = 2)

---

[1] Note: the setting sun diagram is primitive in position space, and is usually not considered primitive when working in momentum space. Of course, the associated amplitude must not depend on this description; and it does not [72]. This piece of wisdom seems less well known than it should be.

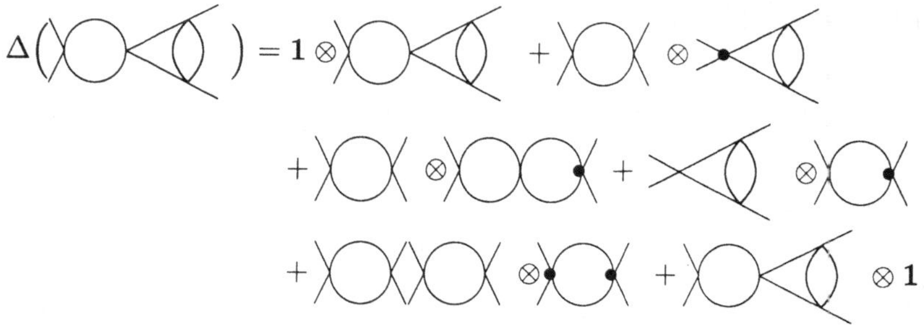

FIGURE 15.4. The "rag-doll" (depth = 3)

A grading, which ensures that H_Φ is a Hopf algebra, is provided by *depth* [60]: a graph Γ has depth k (or is "k-primitive") if

$$P^{\otimes(k+1)}(\Delta^k \Gamma) = 0 \quad \text{and} \quad P^{\otimes k}(\Delta^{k-1}\Gamma) \neq 0$$

where P is the projection $\eta\varepsilon - \text{id}$. Depth measures the maximal length of the inclusion chains of subgraphs appearing in the Bogoliubov recursion. In dimensional regularization, a graph of depth l is expected to display a pole of order l. The antipode S can now be defined as the inverse of $\text{id} = \eta\varepsilon - P$ for the convolution; if Γ_l is a graph of depth l, one finds

$$S(\Gamma_l) := \sum_{k=1}^{l} P^{*k} \Gamma_l = -\Gamma_l + \sum_{\emptyset \subsetneq \gamma \subsetneq \Gamma_l} S(\gamma)\, \Gamma_l/\gamma. \qquad (15.6.2)$$

As it stands, the Hopf algebra H_Φ corresponds to a formal manipulation of graphs. These formulas can be matched to expressions for numerical values, as follows. The Feynman rules for the unrenormalized theory prescribe an algebra homomorphism

$$f : H_\Phi \to \mathcal{A}$$

into some commutative algebra $\mathcal{A}$; that is, f is linear and $f(\Gamma_1\Gamma_2) = f(\Gamma_1)\, f(\Gamma_2)$. In dimensional regularization, $\mathcal{A}$ is an algebra of Laurent series in a complex parameter ε, and $\mathcal{A}$ is the direct sum of two *subalgebras*:

$$\mathcal{A} = \mathcal{A}_+ \oplus \mathcal{A}_-,$$

where $\mathcal{A}_+$ is the holomorphic subalgebra of Taylor series and $\mathcal{A}_-$ is the subalgebra of polynomials in $1/\varepsilon$ without constant term. The projection $T\colon \mathcal{A} \to \mathcal{A}_-$, with $\ker T = \mathcal{A}_+$, picks out the pole part, in a minimal subtraction scheme. Now T is not a homomorphism, but the property that both its kernel and image are subalgebras is reflected in a "multiplicativity constraint":

$$T(ab) + T(a)\, T(b) = T(T(a)\, b) + T(a\, T(b)).$$

The equation (15.6.1a) means that "the antipode delivers the counterterm": one replaces S in the calculation (15.6.2) by C to obtain the right-hand side, before projection with T. From the definition of the coproduct in H_Φ, (15.6.1b), which extracts the finite value, is a *convolution* in $\mathrm{Hom}(H_\Phi, \mathcal{A})$, namely, $R = C * f$. To show that R is multiplicative, it is enough to verify that the counterterm map C is multiplicative: convolution of homomorphisms is a homomorphism because $\mathcal{A}$ is commutative. The multiplicativity of C follows from the constraint on T, as shown by Connes and Kreimer in [69]. See also [73] and [74] in regard to these convolution formulas.

The previous discussion is logically independent of NCG, but there is an important historical link. The Hopf algebra approach to renormalization theory arose in parallel with the Connes–Moscovici noncommutative theory of foliations. In that theory, a foliation is described by a noncommutative algebra of functions twisted by local diffeomorphisms, $\mathcal{A} = C_c^\infty(F) \rtimes \Gamma$; horizontal and vertical vector fields on the frame bundle $F \to M$ are represented on $\mathcal{A}$ by the action of a certain Hopf algebra H_{CM} which provides a way to compute a local index formula in NCG [75]. One can map H_{CM} into an extension of the *Hopf algebra of rooted trees*, a precursor of the Connes–Kreimer graphical Hopf algebras which is described in detail in [76] and [22]. On extending the Hopf algebra H_Φ of graphs by incorporating operations of insertion of subgraphs, one obtains a noncommutative Hopf algebra of the H_{CM} type, which gives a supplementary handle on the combinatorial structure of H_Φ [77].

15.7   Outlook

Noncommutative geometry has had, for many years now, a mutually rewarding conversation with quantum physics. The underlying motif of this conversation can be said to be the belief that quantum field theory encodes the true geometry of the world, and that the mathematical task is to elucidate this geometrical structure. The payback to physics takes the form of new tools and methods; and the work is far from over. For the biggest challenge, that of understanding quantum gravity, there is a long way yet to travel.

Just as the effort to understand the gauge symmetries of the standard model led, in due time, to the introduction of real structures for spectral triples and from there to a noncommutative understanding of spin geometries, we may likewise expect that the NC approach to gravity will help to clarify our still imperfect understanding of the nature of noncommutative manifolds. The story continues ...

Acknowledgements

I am grateful to the organizers of the 6th Conference on Clifford Algebras and their Applications in Mathematical Physics for inviting me to speak about noncommutative geometry and its interface with physics. Support from the organizers and the Vicerrectoría de Investigación of the Universidad de Costa Rica is acknowledged. Much of what is written here emerged from lengthy discussions and past collaborations with José M. Gracia-Bondía. Finally, I owe a debt of gratitude to Alain Connes, whose continuing creation of noncommutative geometry is an inspiration to us all.

REFERENCES

[1] A. Connes, C^*-algèbres et géométrie différentielle, *C. R. Acad. Sci. Paris* **290** (1980), 599–604.

[2] M.A. Rieffel, C^*-algebras associated with irrational rotations, *Pac. J. Math.* **93** (1981), 415–429.

[3] A. Connes, Spectral sequence and homology of currents for operator algebras, *Tagungsbericht* 42/81, Mathematisches Forschungszentrum Oberwolfach, 1981.

[4] A. Connes, Cohomologie cyclique et foncteurs Ext^n, *C. R. Acad. Sci. Paris* **296** (1983), 953–958.

[5] A. Connes, Noncommutative differential geometry, *Publ. Math. IHES* **39** (1985), 257–360.

[6] A. Connes, Cyclic cohomology and the transverse fundamental class of a foliation, in *Geometric Methods in Operator Algebras*, Eds. H. Araki and E. G. Effros; Pitman Research Notes in Math. **123** (1986), pp. 52–144.

[7] A. Connes and M. Karoubi, Caractère multiplicatif d'un module de Fredholm, *K-Theory* **2** (1988), 431–463.

[8] H. Araki, Schwinger terms and cyclic cohomology, in *Quantum Theories and Geometry*, Eds. M. Cahen and M. Flato, Kluwer, Dordrecht, 1988; pp. 1–22.

[9] P. Baum and R. G. Douglas, Index theory, bordism and K-homology, in *Operator Algebras and K-Theory*, Eds. R. G. Douglas and C. Schochet; Contemp. Math. **10** (1982), pp. 1–31.

[10] N. Higson and J. Roe, *Analytic K-Homology*, Oxford University Press, Oxford, 2000.

[11] M. F. Atiyah, R. Bott and A. Shapiro, Clifford modules, *Topology* **3** (1964), 3–38.

[12] A. Connes and J. Lott, Particle models and noncommutative geometry, *Nucl. Phys. B (Proc. Suppl.)* **18** (1990), 29–47.

[13] J. C. Várilly and J. M. Gracia-Bondía, Connes' noncommutative differential geometry and the standard model, *J. Geom. Phys.* **12** (1993), 223–301.

[14] D. Kastler and T. Schücker, A detailed account of Alain Connes' version of the standard model in noncommutative differential geometry. IV, *Rev. Math. Phys.* **8** (1996), 205–228.

[15] A. Connes, Noncommutative geometry and reality, *J. Math. Phys.* **36** (1995), 6194–6231.

[16] C. P. Martín, J. M. Gracia-Bondía and J. C. Várilly, The standard model as a noncommutative geometry: the low energy regime, *Phys. Reports* **294** (1998), 363–406.

[17] E. Álvarez, J. M. Gracia-Bondía and C. P. Martín, Parameter constraints in a noncommutative geometry model do not survive standard quantum corrections, *Phys. Lett.* **B306** (1993), 55–58.

[18] F. Scheck, The standard model within noncommutative geometry: A comparison of models, Talk at the Ninth Max Born Symposium, Karpacz, Poland, September 1996; hep-th/9701073, Mainz, 1997.

[19] M. Takesaki, *Tomita's Theory of Modular Hilbert Algebras*, Springer, Berlin, 1970.

[20] M. F. Atiyah, K-theory and reality, *Quart. J. Math.* **17** (1966), 367–386.

[21] A. Connes, La notion de variété et les axiomes de la géométrie, Cours au Collège de France, Paris, January – March 1996.

[22] J. M. Gracia-Bondía, J. C. Várilly and H. Figueroa, *Elements of Noncommutative Geometry*, Birkhäuser, Boston, 2001.

[23] A. Connes, Gravity coupled with matter and foundation of noncommutative geometry, *Commun. Math. Phys.* **182** (1996), 155–176.

[24] A. H. Chamseddine and A. Connes, The spectral action principle, *Commun. Math. Phys.* **186** (1997), 731–750.

[25] R. Estrada, J. M. Gracia-Bondía and J. C. Várilly, On summability of distributions and spectral geometry, *Commun. Math. Phys.* **191** (1998), 219–248.

[26] R. Wulkenhaar, Nonrenormalizability of θ-expanded noncommutative QED, *J. High Energy Phys.* **0203** (2002), 024.

[27] E. Langmann, Generalized Yang–Mills actions from Dirac operator determinants, *J. Math. Phys.* **42** (2001), 5238–5256.

[28] A. Rennie, Poincaré duality and spin^c structures for complete noncommutative manifolds, math-ph/0107013, Adelaide, 2001.

[29] A. Rennie, Smoothness and locality for nonunital spectral triples, *K-Theory* **28** (2003), 127–165.

[30] H. S. Snyder, Quantized space-time, *Phys. Rev.* **71** (1947), 38–41.

[31] S. Doplicher, K. Fredenhagen and J. E. Roberts, The quantum structure of spacetime at the Planck scale and quantum fields, *Commun. Math. Phys.* **172** (1995), 187–220.

[32] M. M. Sheikh-Jabbari, Open strings in a B-field background as electric dipoles, *Phys. Lett.* **B455** (1999), 129–134.

[33] N. Seiberg and E. Witten, String theory and noncommutative geometry, *J. High Energy Phys.* **9** (1999), 032.

[34] V. Schomerus, D-branes and deformation quantization, *J. High Energy Phys.* **9906** (1999), 030.

[35] A. Connes, M. R. Douglas and A. Schwartz, Noncommutative geometry and Matrix theory: compactification on tori, *J. High Energy Phys.* **9802** (1998), 003.

[36] M. R. Douglas and C. M. Hull, D-branes and the noncommutative torus, *J. High Energy Phys.* **9802** (1998), 008.

[37] G. Landi, F. Lizzi and R. J. Szabo, String geometry and the noncommutative torus, *Commun. Math. Phys.* **206** (1999), 603–637.

[38] R. Jackiw and S.-Y. Pi, Noncommutative 1-cocycle in the Seiberg–Witten map, *Phys. Lett. B* **534** (2002), 181–184.

[39] B. Jurčo, P. Schupp and J. Wess, Noncommutative line bundle and Morita equivalence, *Lett. Math. Phys.* **61** (2002), 171–186.

[40] R. Estrada, J. M. Gracia-Bondía and J. C. Várilly, On asymptotic expansions of twisted products, *J. Math. Phys.* **30** (1989), 2789–2796.

[41] J. E. Moyal, Quantum mechanics as a statistical theory, *Proc. Cambridge Philos. Soc.* **45** (1949), 99–124.

[42] J. M. Gracia-Bondía and J. C. Várilly, Algebras of distributions suitable for phase-space quantum mechanics. I, *J. Math. Phys.* **29** (1988), 869–879.

[43] J. C. Várilly and J. M. Gracia-Bondía, Algebras of distributions suitable for phase-space quantum mechanics. II. Topologies on the Moyal algebra, *J. Math. Phys.* **29** (1988), 880–887.

[44] J. M. Gracia-Bondía, F. Lizzi, G. Marmo and P. Vitale, Infinitely many star-products to play with, *J. High Energy Phys.* **0204** (2002), 026.

[45] J. C. Várilly and J. M. Gracia-Bondía, On the ultraviolet behaviour of quantum fields over noncommutative manifolds, *Int. J. Mod. Phys.* **A14** (1999), 1305–1323.

[46] T. Filk, Divergences in a field theory on quantum space, *Phys. Lett.* **B376** (1996), 53–58.

[47] A. González-Arroyo and C. P. Korthals-Altes, Reduced model for large N continuum field theories, *Phys. Lett.* **B131** (1983), 396–398.

[48] S. Minwalla, M. V. Raamsdonk and N. Seiberg, Noncommutative perturbative dynamics, *J. High Energy Phys.* **0002** (2000), 020.

[49] J. Gomis and T. Mehen, Space-time noncommutative field theories and unitarity, *Nucl. Phys.* **B591** (2000), 265–270.

[50] D. Bahns, S. Doplicher, K. Fredenhagen and G. Piacitelli, On the unitarity problem in space/time noncommutative theories, *Phys. Lett.* **B533** (2002), 178–181.

[51] H. Cheng, How to quantize Yang–Mills theory, in *Chen Ning Yang: A Great Physicist of the Twentieth Century*, Eds. C. S. Liu and S.-T. Yau, International Press, Cambridge, MA, 1995; pp. 49–57.

[52] C. Rim and J. H. Yee, Unitarity in space-time noncommutative field theories, hep-th/0205193, Chonbuk, Korea, 2002.

[53] Y. Liao and K. Sibold, Time-ordered perturbation theory on noncommutative space-time I: basic rules, *Eur. Phys. J.* **C25** (2002), 469–477; II: unitarity, *Eur. Phys. J.* **C25** (2002), 479–486.

[54] M. R. Douglas and N. A. Nekrasov, Noncommutative field theory, *Rev. Mod. Phys.* **73** (2002), 977–1029.

[55] R. J. Szabo, Quantum field theory on noncommutative spaces, *Physics Reports* **378** (2003), 207–299.

[56] K. Morita, Connes' gauge theory on noncommutative spacetimes, hep-th/0011080, Nagoya, 2000.

[57] M. Chaichian, P. Prešnajder, M. M. Sheikh-Jabbari and A. Tureanu, Noncommutative Standard Model: model building, *Eur. Phys. J.* **C29** (2003), 413–432.

[58] X. Calmet, B. Jurčo, P. Schupp, J. Wess and M. Wohlgenannt, The standard model on noncommutative spacetime, *Eur. Phys. J.* **C23** (2002), 363–376.

[59] J. M. Gracia-Bondía and C. P. Martín, Chiral gauge anomalies on noncommutative $\mathbb{R}^4$, *Phys. Lett.* **B479** (2000), 321–328.

[60] J. M. Gracia-Bondía, Noncommutative geometry and fundamental interactions: the first ten years, *Ann. Phys. (Leipzig)* **11** (2002), 479–495.

[61] A. Connes, A short survey of noncommutative geometry, *J. Math. Phys.* **41** (2000), 3832–3866.

[62] A. Connes and G. Landi, Noncommutative manifolds, the instanton algebra and isospectral deformations, *Commun. Math. Phys.* **221** (2001), 141–159.

[63] A. Connes and M. Dubois-Violette, Noncommutative finite-dimensional manifolds. I. Spherical manifolds and related examples, *Commun. Math. Phys.* **230** (2002), 539–579.

[64] J. C. Várilly, Quantum symmetry groups of noncommutative spheres, *Commun. Math. Phys.* **221** (2001), 511–523.

[65] M. A. Rieffel, *Deformation Quantization for Actions of* $\mathbb{R}^d$, Memoirs of the AMS **506**, Providence, RI, 1993.

[66] P. S. Chakraborty and A. Pal, Equivariant spectral triples on the quantum $SU(2)$ group, *K-Theory* **28** (2003), 107–126.

[67] A. Connes, Cyclic cohomology, quantum group symmetries and the local index formula for $SU_q(2)$, math.QA/0209142, IHES, 2002.

[68] A. Connes, Talk at the *Third Meeting on Nichtkommutative Geometrie*, Oberwolfach, March 2002.

[69] A. Connes and D. Kreimer, Renormalization in quantum field theory and the Riemann–Hilbert problem I: the Hopf algebra structure of graphs and the main theorem, *Commun. Math. Phys.* **210** (2000), 249–273.

[70] A. Connes and D. Kreimer, Renormalization in quantum field theory and the Riemann–Hilbert problem II: the β-function, diffeomorphisms and the renormalization group, *Commun. Math. Phys.* **216** (2001), 215–241.

[71] J. M. Gracia-Bondía and S. Lazzarini, Connes–Kreimer–Epstein–Glaser renormalization, hep-th/0006106, Marseille and Mainz, 2000.

[72] W. Zimmermann, Remark on equivalent formulations for Bogoliubov's method of renormalization, in *Renormalization Theory*, G. Velo and A. S. Wightman, eds., NATO ASI Series C **23** (D. Reidel, Dordrecht, 1976).

[73] J. C. Várilly, Hopf algebras in noncommutative geometry, in *Geometrical and Topological Methods in Quantum Field Theory*, Eds. A. Cardona, H. Ocampo and S. Paycha, World Scientific, Singapore, 2003; hep-th/0109077.

[74] F. Girelli, P. Martinetti and T. Krajewski, The Hopf algebra of Connes and Kreimer and wave function renormalization, *Mod. Phys. Lett.* **A16** (2001), 299–303.

[75] A. Connes and H. Moscovici, Hopf algebras, cyclic cohomology and the transverse index theorem, *Commun. Math. Phys.* **198** (1998), 198–246.

[76] A. Connes and D. Kreimer, Hopf algebras, renormalization and noncommutative geometry, *Commun. Math. Phys.* **199** (1998), 203–242.

[77] A. Connes and D. Kreimer, Insertion and elimination: the doubly infinite Lie algebra of Feynman graphs, *Ann. Henri Poincaré* **3** (2002), 411–433.

Joseph C. Várilly
Departamento de Matemática
Universidad de Costa Rica
2060 San José, Costa Rica
(Regular Associate of the Abdus Salam ICTP)
E-mail: varilly@cariari.ucr.ac.cr

Submitted: September 11, 2002.

PART III. MATHEMATICAL STRUCTURES

16

The Method of Virtual Variables and Representations of Lie Superalgebras

Andrea Brini, Francesco Regonati, and Antonio Teolis

ABSTRACT We provide a brief account of Capelli's method of virtual variables and of its relations with representations of general linear Lie superalgebras. More specifically, we study letterplace superalgebras regarded as bimodules under the action of superpolarization operators and exhibit complete decomposition theorems for these bimodules as well as for the operator algebras acting on them.

Keywords: Capelli operators, biproducts, symmetrized bitableaux, Gordan–Capelli series, Young–Capelli symmetrizers

16.1   Introduction

The purpose of the present work is to provide a systematic and synthetic account of some aspects of the so-called Capelli's method of virtual variables and of its deep relations with the representation theory of general linear Lie superalgebras. The basic mathematical structure we consider is the supersymmetric letterplace algebra $Super[\mathcal{L}|\mathcal{P}]$ over a pair of finite signed, or $\mathbb{Z}_2-$graded, alphabets $\mathcal{L}$ and $\mathcal{P}$. In spite of their quite elementary definition, letterplace algebras admit a wide variety of semantics and interpretations both in mathematics and physics. From our point of view, letterplace algebras are quite significant when regarded as bimodules over a pair of general linear Lie superalgebras $pl(\mathcal{L})$ and $pl(\mathcal{P})$ that are naturally associated to the letter alphabet $\mathcal{L}$ and the place alphabet $\mathcal{P}$, respectively.

The first main problem we consider is that of finding a way to describe in a simple and combinatorial way the actions of these general linear Lie algebras on letterplace algebras. The basic idea, borrowed from an old idea of Alfredo Capelli, is that of embedding the algebra $Super[\mathcal{L}|\mathcal{P}]$, called the "proper letterplace algebra", into a larger algebra $Super[\mathcal{X}|\mathcal{Y}]$, called the "virtual letterplace algebra", built over a pair of signed sets $\mathcal{X}$ and $\mathcal{Y}$ obtained from $\mathcal{L}$ and $\mathcal{P}$ by adjoining new symbols of both signatures called "virtual symbols". We exploit this point of

AMS Subject Classification: 05E15, 17B, 05E10.

view by describing in detail how some relevant and quite complicated operators from the universal enveloping algebras $U(pl(\mathcal{L}))$ and $U(pl(\mathcal{P}))$ may be replaced by what we call "virtual operators"; the main point (see Section 3.4) is that the actions of virtual operators on the proper letterplace algebra are the same as the actions of operators from the enveloping algebras $U(pl(\mathcal{L}))$ and $U(pl(\mathcal{P}))$. The description and the study of virtual operators and their actions are, in general, much more manageable than those of their non-virtual companions. As an example of this approach, in Sections 3.5–3.7 we exhibit virtual forms of generalized Capelli operators as products of virtual polarizations, as well as a few lines proof of the classical Capelli identities.

In the second part of the paper, we extend the previous ideas to objects associated to pairs of superstandard Young tableaux. Specifically, thanks to the virtual method, it is possible to define a special set of elements of the proper letterplace superalgebra, called "symmetrized bitableaux", and a special set of generators of the operator algebra generated by the action of the Lie superalgebra $pl(\mathcal{L})$ on the proper letterplace algebra $Super[\mathcal{L}|\mathcal{P}]$, called "Young–Capelli symmetrizers". The crucial fact is that the ordered set of standard symmetrized bitableaux is a $\mathbb{K}$−linear basis of the letterplace algebra $Super[\mathcal{L}|\mathcal{P}]$ (called the "Gordan–Capelli series") and that the ordered set of Young–Capelli symmetrizers acts in a non-degenerate triangular way on this basis. This result yields concrete complete decompositions of the proper letterplace algebra as a bimodule over the pair of Lie superalgebras $pl(\mathcal{L})$ and $pl(\mathcal{P})$, concrete complete decompositions of the operator algebras generated by the actions of $pl(\mathcal{L})$ and $pl(\mathcal{P})$, and a double commutator theorem for the representations of $U(pl(\mathcal{L}))$ and $U(pl(\mathcal{P}))$.

By way of application, in the final section, we consider the remarkable Berele–Regev representation theory of the symmetric group S_n over $\mathbb{Z}_2$−graded tensor spaces $T^n[V_0 \oplus V_1]$ built on $\mathbb{Z}_2$−graded finite dimensional vector spaces. We first recognize that the Berele–Regev action of S_n on $T^n[V_0 \oplus V_1]$ admits a quite natural description in terms of letterplace superalgebras and, furthermore, that the action of the symmetric group induces an operator algebra generated by a special subset of the set of Young–Capelli symmetrizers. Hence, we obtain the main results of this theory (double commutator theorem, decomposition theorems, etc.) as special cases of some of the preceding results on letterplace superalgebras.

16.2   Preliminaries

16.2.1   Letterplace superalgebras

Let $\mathcal{X} = \mathcal{X}^- \cup \mathcal{X}^+$ and $\mathcal{Y} = \mathcal{Y}^- \cup \mathcal{Y}^+$ be signed (i.e., $\mathbb{Z}_2$-graded) alphabets ($\mathcal{X}^-, \mathcal{X}^+, \mathcal{Y}^-, \mathcal{Y}^+$ numerable sets), called the *letter alphabet* and the *place alphabet*, respectively.

The *letterplace alphabet* $[\mathcal{X}|\mathcal{Y}] = \{(x|y);\ x \in \mathcal{X},\ y \in \mathcal{Y}\}$ inherits a signature (i.e., $\mathbb{Z}_2$-gradation) by setting $|(x|y)| = |x| + |y| \in \mathbb{Z}_2$. In the following, $\mathbb{K}$ will denote a field, $\mathrm{char}(\mathbb{K}) = 0$.

The *letterplace superalgebra* $Super[\mathcal{X}|\mathcal{Y}]$ (over $\mathbb{K}$) is the quotient algebra of the free associative $\mathbb{K}$-algebra with 1 generated by the letterplace alphabet $[\mathcal{X}|\mathcal{Y}]$ modulo the bilateral ideal generated by the elements of the form:

$$(x|y)(z|t) - (-1)^{(|x|+|y|)(|z|+|t|)}(z|t)(x|y), \quad x, z \in \mathcal{X}, \ y, t \in \mathcal{Y}.$$

Remark 1. We point out that

- $Super[\mathcal{X}|\mathcal{Y}]$ is a commutative superalgebra (i.e., it is $\mathbb{Z}_2$-graded and super-symmetric);

- $Super[\mathcal{X}|\mathcal{Y}]$ is an $\mathbb{N}$-graded algebra

$$Super[\mathcal{X}|\mathcal{Y}] = \bigoplus_{n \in \mathbb{N}} Super_n[\mathcal{X}|\mathcal{Y}]$$

$$Super_n[\mathcal{X}|\mathcal{Y}] = \langle (x_{i_1}|y_{i_1})(x_{i_2}|y_{i_2}) \cdots (x_{i_n}|y_{i_n}); \ x_{i_h} \in \mathcal{X}, \ y_{j_k} \in \mathcal{Y} \rangle_{\mathbb{K}};$$

- the $\mathbb{Z}_2$-gradation and the $\mathbb{N}$-gradation of $Super[\mathcal{X}|\mathcal{Y}]$ are coherent, that is

$$(Super[\mathcal{X}|\mathcal{Y}])^i = \bigoplus_{n \in \mathbb{N}} (Super_n[\mathcal{X}|\mathcal{Y}])^i, \quad i \in \mathbb{Z}_2.$$

16.2.2 Letter-polarizations and place-polarizations

Let $x', x \in \mathcal{X}$. The *superpolarization of the letter x to the letter x'* is defined to be the $\mathbb{K}$-linear operator

$$\mathcal{D}_{x',x} : Super[\mathcal{X}|\mathcal{Y}] \to Super[\mathcal{X}|\mathcal{Y}]$$

such that

- $\mathcal{D}_{x',x}(z|t) = \delta_{x,z}(x'|t)$, for every $(z|t) \in [\mathcal{X}|\mathcal{Y}]$;

- $\mathcal{D}_{x',x}$ is a left superderivation of $\mathbb{Z}_2$-grade $|x'| + |x|$, that is

$$\mathcal{D}_{x',x}(MM') = \mathcal{D}_{x',x}(M)M' + (-1)^{(|x'|+|x|)|M|} M \mathcal{D}_{x',x}(M'),$$

for all monomials $M, M' \in Super[\mathcal{X}|\mathcal{Y}]$.

Let $y', y \in \mathcal{Y}$. The *superpolarization of the place y to the place y'* is defined to be (mirror notation) the $\mathbb{K}$-linear operator

$$Super[\mathcal{X}|\mathcal{Y}] \leftarrow Super[\mathcal{X}|\mathcal{Y}] : {}_{y,y'}\mathcal{D}$$

such that

- $(z|t) \ {}_{y,y'}\mathcal{D} = \delta_{t,y}(z|y')$, for every $(z|t) \in [\mathcal{X}|\mathcal{Y}]$;

- $_{y,y'}\mathcal{D}$ is a right superderivation of $\mathbb{Z}_2$-grade $|y| + |y'|$, that is

$$(-1)^{(|y|+|y'|)|M'|}(M)\,_{y,y'}\mathcal{D}M' + M(M')\,_{y,y'}\mathcal{D} = (MM')\,_{y,y'}\mathcal{D},$$

for all monomials $M, M' \in Super[\mathcal{X}|\mathcal{Y}]$.

Remark 2. We have

$$\mathcal{D}_{x'x} = \sum_{y\in\mathcal{Y}}(x'|y)\frac{\partial}{\partial(x|y)},$$

where $\frac{\partial}{\partial(x|y)}$ denotes the left superderivation, homogeneous of $\mathbb{Z}_2$-grade $|x|+|y|$, defined by setting $\frac{\partial}{\partial(x|y)}(z|t) = \delta_{xz}\delta_{yt}$. Note that the action of the preceding "infinite" sum on any element of $Super[\mathcal{X}|\mathcal{Y}]$ produces all but a finite number of summands equal to zero.

Obviously, a parallel remark holds for place polarizations.

16.2.3   The commutation property

Every letter-polarization operator commutes with every place-polarization operator. In symbols, for every $x', x \in \mathcal{X}$ and $y, y' \in \mathcal{Y}$

$$(\mathcal{D}_{x',x}(M))_{y,y'}\mathcal{D} = \mathcal{D}_{x',x}((M)_{y,y'}\mathcal{D})$$

for every $M \in Super[\mathcal{X}|\mathcal{Y}]$.

16.3   The method of virtual variables

16.3.1   Proper letterplace superalgebras

Let $\mathcal{L} = \mathcal{L}^+ \cup \mathcal{L}^- \subset \mathcal{X}$ and $\mathcal{P} = \mathcal{P}^+ \cup \mathcal{P}^- \subset \mathcal{Y}$ be finite subsets of the "universal" letter and place alphabets $\mathcal{X}$ and $\mathcal{Y}$, respectively.

The elements $x \in \mathcal{L}$ are called *proper* letters of the *proper* letter alphabet $\mathcal{L}$. The elements $x \in \mathcal{X} \setminus \mathcal{L}$ are called *virtual* letters of the "universal" letter alphabet $\mathcal{X}$.

The elements $y \in \mathcal{P}$ are called *proper* places of the *proper* place alphabet $\mathcal{P}$. The elements $y \in \mathcal{Y} \setminus \mathcal{P}$ are called *virtual* places of the "universal" place alphabet $\mathcal{Y}$.

The signed subset

$$[\mathcal{L}|\mathcal{P}] = \{(x|y);\ x \in \mathcal{L},\ y \in \mathcal{P}\} \subset [\mathcal{X}|\mathcal{Y}]$$

is called a *proper letterplace alphabet*.

The $\mathbb{Z}_2$-graded subalgebra $Super[\mathcal{L}|\mathcal{P}] \subset Super[\mathcal{X}|\mathcal{Y}]$ generated by the set of proper letterplace variables $[\mathcal{L}|\mathcal{P}]$ is called the *proper letterplace superalgebra* generated by the proper alphabets $\mathcal{L}$ and $\mathcal{P}$.

16.3.2   Tensor product interpretation of $Super[\mathcal{L}|\mathcal{P}]$

Set

$$V = \langle \mathcal{L} \rangle_{\mathbb{K}} = \langle \mathcal{L}^+ \rangle_{\mathbb{K}} \oplus \langle \mathcal{L}^- \rangle_{\mathbb{K}} = V_0 \oplus V_1,$$
$$W = \langle \mathcal{P} \rangle_{\mathbb{K}} = \langle \mathcal{P}^+ \rangle_{\mathbb{K}} \oplus \langle \mathcal{P}^- \rangle_{\mathbb{K}} = W_0 \oplus W_1.$$

It follows that, canonically,

$$Super[\mathcal{L}|\mathcal{P}] \cong Super[V \otimes_{\mathbb{K}} W]$$
$$= Sym[(V_0 \otimes_{\mathbb{K}} W_0) \oplus (V_1 \otimes_{\mathbb{K}} W_1)] \otimes_{\mathbb{K}} \bigwedge[(V_0 \otimes_{\mathbb{K}} W_1) \oplus (V_1 \otimes_{\mathbb{K}} W_0)].$$

16.3.3   Superpolarizations and representations of general linear Lie superalgebras

Consider the general linear Lie superalgebras

$$pl(V) = pl(\mathcal{L}), \quad pl(W) = pl(\mathcal{P})$$

and their standard bases

$$\{E_{x'x}; \; x, x' \in \mathcal{L}\}, \quad \{E_{yy'}; \; y, y' \in \mathcal{P}\}.$$

The mappings

$$E_{x'x} \mapsto \mathcal{D}_{x',x}, \; x, x' \in \mathcal{L}, \quad E_{yy'} \mapsto {}_{y',y}\mathcal{D}, \; y, y' \in \mathcal{P}$$

induce Lie superalgebra actions of $pl(\mathcal{L})$ and $pl(\mathcal{P})$ on any $\mathbb{N}$-homogeneous component $Super_n[\mathcal{L}|\mathcal{P}]$ of the proper letterplace algebra. Furthermore, by the commutation property,

$$pl(\mathcal{L}) \cdot Super_n[\mathcal{L}|\mathcal{P}] \cdot pl(\mathcal{P})$$

is a bimodule over the general linear Lie superalgebras $pl(\mathcal{L})$ and $pl(\mathcal{P})$.

16.3.4   The metatheoretic significance of Capelli's idea of virtual variables

Let $\mathcal{L} \subset \mathcal{X}$, $\mathcal{P} \subset \mathcal{Y}$ be proper alphabets, and recall that a virtual letter is an element $\alpha \in \mathcal{X} \setminus \mathcal{L}$. Consider an operator of the form

$$\mathcal{D}_{x_{i_1} \alpha_{i_1}} \cdots \mathcal{D}_{x_{i_n} \alpha_{i_n}} \mathcal{D}_{\alpha_{i_1} x_{j_1}} \cdots \mathcal{D}_{\alpha_{i_n} x_{j_n}} \tag{3.1}$$

with $x_{i_1}, \ldots, x_{i_n}, x_{j_1}, \ldots, x_{j_n} \in \mathcal{L}$ proper letters and $\alpha_{i_1}, \ldots, \alpha_{i_n}$ virtual letters, not necessarily distinct; in plain words, an operator of type (3.1) is an operator that creates some virtual letters (with some prescribed multiplicities) times an operator that annihilates the same virtual letters (with the same prescribed multiplicities). Clearly, the proper letterplace superalgebra $Super[\mathcal{L}|\mathcal{P}]$ is left invariant under the action of an operator of type (3.1).

Example 1. *Let* $x, y, z \in \mathcal{L}$ *and* $\alpha_1, \alpha_2 \in \mathcal{X} \setminus \mathcal{L}$. *Then*

$$\mathcal{D}_{x\alpha_1} \mathcal{D}_{y\alpha_2} \mathcal{D}_{y\alpha_2} \mathcal{D}_{\alpha_2 z} \mathcal{D}_{\alpha_2 x} \mathcal{D}_{\alpha_1 z}$$

is an operator that causes one creation/annihilation of α_1 *and two creations/annihilations of* α_2.

Theorem 1. *The action of an operator of type (3.1) on the proper letterplace algebra* $Super[\mathcal{L}|\mathcal{P}]$ *is the same as the action of a "polynomial" operator in the proper polarizations* $\mathcal{D}_{x_{i_h} x_{j_k}}$, x_{i_h}, $x_{j_k} \in \mathcal{L}$, *that is, an operator that does not involve virtual variables.*

Roughly speaking, an operator of type (3.1) is of $pl(\mathcal{L})$-representation theoretic meaning, and, in general, is much more manageable than its "non-virtual companion".

In the following, we will use the symbol $\cong$ to mean that two operators are the same when restricted to the proper letterplace algebra $Super[\mathcal{L}|P]$.

Example 2. *Let* $x, y \in \mathcal{L}^-$, *with* $x \neq y$, *and* $\alpha \in \mathcal{X} \setminus \mathcal{L}$, *with* $|\alpha| = 0$. *Then*

$$
\begin{aligned}
\mathcal{D}_{y\alpha} & \mathcal{D}_{x\alpha} \mathcal{D}_{\alpha x} \mathcal{D}_{\alpha y} \\
&= -\mathcal{D}_{y\alpha} \mathcal{D}_{\alpha x} \mathcal{D}_{xa} \mathcal{D}_{\alpha y} + \mathcal{D}_{y\alpha} \mathcal{D}_{xx} \mathcal{D}_{\alpha y} + \mathcal{D}_{y\alpha} \mathcal{D}_{\alpha\alpha} \mathcal{D}_{\alpha y} \\
&= +\mathcal{D}_{y\alpha} \mathcal{D}_{\alpha x} \mathcal{D}_{\alpha y} \mathcal{D}_{x\alpha} - \mathcal{D}_{y\alpha} \mathcal{D}_{\alpha x} \mathcal{D}_{xy} - \mathcal{D}_{xx} \mathcal{D}_{\alpha y} \mathcal{D}_{y\alpha} \\
&\quad + \mathcal{D}_{xx} \mathcal{D}_{yy} - \mathcal{D}_{xx} \mathcal{D}_{\alpha\alpha} + \mathcal{D}_{y\alpha} \mathcal{D}_{\alpha y} \mathcal{D}_{\alpha\alpha} + \mathcal{D}_{y\alpha} \mathcal{D}_{\alpha y} \\
&\cong -\mathcal{D}_{y\alpha} \mathcal{D}_{\alpha x} \mathcal{D}_{xy} + \mathcal{D}_{xx} \mathcal{D}_{yy} + \mathcal{D}_{y\alpha} \mathcal{D}_{\alpha y} \\
&\cong \cdots \\
&\cong -\mathcal{D}_{yx} \mathcal{D}_{xy} + \mathcal{D}_{xx} \mathcal{D}_{yy} + \mathcal{D}_{yy}.
\end{aligned}
$$

16.3.5   Virtual forms of Capelli operators

An operator of the form

$$\mathcal{D}_{y_1\alpha} \mathcal{D}_{y_2\alpha} \cdots \mathcal{D}_{y_n\alpha} \mathcal{D}_{\alpha z_1} \mathcal{D}_{\alpha z_2} \cdots \mathcal{D}_{\alpha z_n}, \tag{3.2}$$

where $y_1, \ldots, y_n, z_1, \ldots, z_n \in \mathcal{L}$, not necessarily distinct, $\alpha \in \mathcal{X} \setminus \mathcal{L}$ a virtual letter, is called a (virtual) *Capelli operator*.

As we shall see, Capelli operators are supersymmetrization operators in disguise.

16.3.6   Devirtualization of Capelli operators and Laplace expansion type identities

If we consider the restricted action of a Capelli operator from the proper superalgebra $Super[\mathcal{L}|\mathcal{P}]$ to itself, we get the following Laplace expansion type identities.

Theorem 2. *Let $\alpha \in \mathcal{X} \setminus \mathcal{L}$ be a virtual letter, $|\alpha| = 0$; then*

$$\mathcal{D}_{y_1 \alpha} \mathcal{D}_{y_2 \alpha} \cdots \mathcal{D}_{y_n \alpha} \mathcal{D}_{\alpha z_1} \mathcal{D}_{\alpha z_2} \cdots \mathcal{D}_{\alpha z_n} \cong$$

$$\sum_{i=1}^{n} \pm (\mathcal{D}_{y_i z_1} - (-1)^{|y_i||z_1|}(n-1)\delta_{y_i z_1} I) \, \mathcal{D}_{y_1 \alpha} \cdots \widehat{\mathcal{D}_{y_i \alpha}} \cdots \mathcal{D}_{y_n \alpha} \mathcal{D}_{\alpha z_2} \cdots \mathcal{D}_{\alpha z_n}$$

where $\pm$ is the sign associated to the pair of words $y_1 \cdots y_i \cdots y_n z_1 \cdots z_n$ and $y_i z_1 y_1 \cdots \widehat{y_i} \cdots y_n z_2 \cdots z_n$; that is

$$(-1)^{|z_1|(|y_1|+\cdots+|\widehat{y_i}|+\cdots+|y_n|)+|y_i|(|y_1|+\cdots+|y_{i-1}|)}.$$

A similar result holds in the case $|\alpha| = 1$.

Remark 3. By iterating the preceding identity, one can eliminate the virtual variable α and, therefore, obtain a devirtualization of the operator. The crucial point of the virtual method is that, in the study of the actions, the virtual form is much more preferable than a devirtualized form.

Example 3.

1. *Let x, y with $x \neq y$ be proper letters, α be a virtual letter, and $|\alpha| = |x| = |y| = 0$; then*
$$\mathcal{D}_{x\alpha}^n \mathcal{D}_{\alpha y}^n \cong n! \mathcal{D}_{xy}^n.$$

2. *Let x be a proper letter, α be a virtual letter, and $|\alpha| = |x| = 0$; then*

$$\mathcal{D}_{x\alpha}^n \mathcal{D}_{\alpha x}^n \cong n! \mathcal{D}_{xx}(\mathcal{D}_{xx} - I) \cdots (\mathcal{D}_{xx} - (n-1)I) = (n!)^2 \binom{\mathcal{D}_{xx}}{n}.$$

3. *Let $y_{i_1}, \ldots, y_{i_n}, x_{j_1}, \ldots, x_{j_n}$ be two n-tuples of proper letters of the same signature, say $|y_{i_h}| = |x_{j_k}| = 0$, for every $h, k = 1, \ldots, n$. Assume that the two n-tuples above have no letters in common. Then*

$$\mathcal{D}_{y_{i_1}\alpha} \cdots \mathcal{D}_{y_{i_n}\alpha} \mathcal{D}_{\alpha x_{j_n}} \cdots \mathcal{D}_{\alpha x_{j_1}} \cong \begin{cases} \mathrm{per}[\mathcal{D}_{y_{i_h} x_{j_k}}]_{h,k}, & |\alpha| = 0 \\ \det[\mathcal{D}_{y_{i_h} x_{j_k}}]_{h,k}, & |\alpha| = 1. \end{cases}$$

4. *Let $\mathcal{L} = \mathcal{L}^- = \{x_1, \ldots x_m\}$ be a linearly ordered set of (distinct) negative proper letters; then, we get the classical Capelli operator:*

$$\mathcal{D}_{x_m \alpha} \cdots \mathcal{D}_{x_1 \alpha} \mathcal{D}_{\alpha x_1} \cdots \mathcal{D}_{\alpha x_m} \cong$$

$$\det \begin{bmatrix} \mathcal{D}_{x_1 x_1} + (m-1)I & \mathcal{D}_{x_1 x_2} & \cdots & \mathcal{D}_{x_1 x_m} \\ \mathcal{D}_{x_2 x_1} & \mathcal{D}_{x_2 x_2} + (m-2)I & \cdots & \mathcal{D}_{x_2 x_m} \\ \vdots & \vdots & \ddots & \vdots \\ \mathcal{D}_{x_m x_1} & \mathcal{D}_{x_m x_2} & \cdots & \mathcal{D}_{x_m x_m} \end{bmatrix} = H_m,$$

where the "determinant" is expanded according to the Weyl convention [11].

16.3.7 The classical general and special Capelli identities

Let $\mathcal{L} = \{x_1, x_2, \ldots, x_m\} = \mathcal{L}^-$ and $\mathcal{P} = \{1, 2, \ldots, n\} = \mathcal{P}^-$, with $m \geq n$. In this case, we have

$$Super_n[\mathcal{L}|\mathcal{P}] \cong \mathbb{K}[(x_i|j)]_{i=1,\ldots,m; j=1,\ldots,n},$$

the ordinary polynomial algebra in the variables $(x_i|j) = x_{ij}$.

Remark 4. Note that, for any $\alpha \in \mathcal{X}^+ \setminus \mathcal{L}$,

$$\mathcal{D}_{x_{i_n}\alpha} \cdots \mathcal{D}_{x_{i_2}\alpha} \mathcal{D}_{x_{i_1}\alpha}((\alpha|1)(\alpha|2)\cdots(\alpha|n))$$
$$= \det[x_{i_h k}]_{h,k=1,2,\ldots,n} = [x_{i_1}, x_{i_2}, \ldots, x_{i_n}],$$

which is the usual bracket.

Theorem 3. *Under the above assumptions,*

$$H_m = \begin{cases} 0, & m > n \\ [x_1, x_2, \ldots, x_n]\Omega_n, & m = n \end{cases}$$

where Ω_n denotes the Cayley Ω-process, that is,

$$\Omega_n = \det \begin{bmatrix} \frac{\partial}{\partial x_{11}} & \frac{\partial}{\partial x_{12}} & \cdots & \frac{\partial}{\partial x_{1n}} \\ \frac{\partial}{\partial x_{21}} & \frac{\partial}{\partial x_{22}} & \cdots & \frac{\partial}{\partial x_{2n}} \\ \vdots & \vdots & \ddots & \vdots \\ \frac{\partial}{\partial x_{n1}} & \frac{\partial}{\partial x_{n2}} & \cdots & \frac{\partial}{\partial x_{nn}} \end{bmatrix}.$$

Proof. By Section 2.2, the virtualization part $\mathcal{D}_{\alpha x_1} \mathcal{D}_{\alpha x_2} \cdots \mathcal{D}_{\alpha x_m}$ can be expanded as follows:

$$\sum_{1 \leq h_1 \leq n} (\alpha|h_1)\frac{\partial}{\partial(x_1|h_1)} \sum_{1 \leq h_2 \leq n} (\alpha|h_2)\frac{\partial}{\partial(x_2|h_2)} \cdots \sum_{1 \leq h_m \leq n} (\alpha|h_m)\frac{\partial}{\partial(x_m|h_m)}$$

$$= \sum_{1 \leq h_1, h_2, \ldots, h_m \leq n} (\alpha|h_1)(\alpha|h_2)\cdots(\alpha|h_m)\frac{\partial^m}{\partial(x_1|h_1)\partial(x_2|h_2)\cdots\partial(x_m|h_m)}.$$

Since the variables $(\alpha|h_i)$ are skew-commutative, all the terms of the preceding sum equal 0 whenever $m > n$. On the other hand, if $n = m$ the sum can be rewritten as:

$$\sum_{\sigma=(h_1, h_2, \ldots, h_n) \in S_n} (-1)^{\sigma}(\alpha|1)(\alpha|2)\cdots(\alpha|n)\frac{\partial^n}{\partial(x_1|h_1)\partial(x_2|h_2)\cdots\partial(x_n|h_n)}$$

$$= (\alpha|1)(\alpha|2)\cdots(\alpha|n)\sum_{\sigma=(h_1, h_2, \ldots, h_n) \in S_n} (-1)^{\sigma}\frac{\partial^n}{\partial(x_1|h_1)\partial(x_2|h_2)\cdots\partial(x_n|h_n)}.$$

Since

$$\mathcal{D}_{x_n\alpha}\cdots\mathcal{D}_{x_2\alpha}\mathcal{D}_{x_1\alpha}((\alpha|1)(\alpha|2)\cdots(\alpha|n)) = [x_1, x_2, \ldots, x_n],$$

the assertion follows. $\qquad\qquad\qquad\qquad\qquad\qquad\qquad\qquad\qquad\qquad\Box$

16.3.8   Capelli operators and supersymmetries in $Super[\mathcal{L}|\mathcal{P}]$

Let $x_1, \ldots, x_n \in \mathcal{L}$ and $y_1, \ldots, y_n \in \mathcal{P}$ be proper letters and places, respectively. Let $\alpha \in \mathcal{X} \setminus \mathcal{L}$ be a virtual letter. By applying the Capelli operator to the generic monomial $(x_1|y_1)(x_2|y_2)\cdots(x_n|y_n) \in Super_n[\mathcal{L}|\mathcal{P}]$, we get

$$\mathcal{D}_{x_1\alpha}\mathcal{D}_{x_2\alpha}\cdots\mathcal{D}_{x_n\alpha}\mathcal{D}_{\alpha x_n}\cdots\mathcal{D}_{\alpha x_2}\mathcal{D}_{\alpha x_1}(x_1|y_1)(x_2|y_2)\cdots(x_n|y_n)$$
$$= \pm k\mathcal{D}_{x_1\alpha}\mathcal{D}_{x_2\alpha}\cdots\mathcal{D}_{x_n\alpha}(\alpha|y_1)(\alpha|y_2)\cdots(\alpha|y_n)$$

which is

- supersymmetric in the x's and the y's if $|\alpha| = 0$;

- "dual" supersymmetric in the x's and the y's if $|\alpha| = 1$.

16.3.9   Biproducts as basic symmetrized elements in $Super[\mathcal{L}|\mathcal{P}]$

The argument of the previous subsection leads naturally to a virtual definition of the basic supersymmetric and of the basic dually supersymmetric objects in $Super_n[\mathcal{L}|\mathcal{P}]$ that are both associated to pairs of sequences of the same length in $\mathcal{L}$ and $\mathcal{P}$. These objects, here presented in their three different virtual forms, are called *biproducts* and **-biproducts*, respectively; in particular, the biproducts coincide, in characteristic 0, with the Grosshans–Rota–Stein biproducts [9].

Let $x_1, \ldots, x_n \in \mathcal{L}$ and $y_1, \ldots, y_n \in \mathcal{P}$ be proper letters and places, respectively. Let $\alpha \in \mathcal{X} \setminus \mathcal{L}$ be a virtual letter and $\beta \in \mathcal{Y} \setminus \mathcal{P}$ be a virtual place, with $|\alpha| = |\beta|$.

The element of $Super_n[\mathcal{L}|\mathcal{P}]$,

$$\mathcal{D}_{x_1\alpha}\mathcal{D}_{x_2\alpha}\cdots\mathcal{D}_{x_n\alpha}(\alpha|y_1)(\alpha|y_2)\cdots(\alpha|y_n)$$
$$= \mathcal{D}_{x_1\alpha}\mathcal{D}_{x_2\alpha}\cdots\mathcal{D}_{x_n\alpha}(\frac{(\alpha|\beta)^n}{n!})_{\beta y_1}\mathcal{D}_{\beta y_2}\mathcal{D}\cdots_{\beta y_n}\mathcal{D}$$
$$= (x_1|\beta)(x_2|\beta)\cdots(x_n|\beta)_{\beta y_1}\mathcal{D}_{\beta y_2}\mathcal{D}\cdots_{\beta y_n}\mathcal{D}$$

is called, for $|\alpha| = |\beta| = 0$, the *biproduct* of $x_1\cdots x_n$ and $y_1\cdots y_n$, denoted by

$$(x_1 x_2 \cdots x_n | y_1 y_2 \cdots y_n), \qquad\qquad\qquad (3.3)$$

and, for $|\alpha| = |\beta| = 1$, the **-biproduct* of $x_1\cdots x_n$ and $y_1\cdots y_n$, denoted by

$$(x_1 x_2 \cdots x_n | y_1 y_2 \cdots y_n)^*. \qquad\qquad\qquad (3.4)$$

Remark 5. Laplace expansions of biproducts and *–biproducts correspond to the Leibniz rule for superderivations.

Example 4.

1. *Let* $x_1, \ldots, x_n \in \mathcal{L}^+ = \mathcal{L}$ *and* $y_1, \ldots, y_n \in \mathcal{P}^+ = \mathcal{P}$ *be proper letters and places, respectively; then* $Super_n[\mathcal{L}|\mathcal{P}] = Sym_n[\mathcal{L}|\mathcal{P}]$. *We have*

$$(x_1 x_2 \cdots x_n | y_1 y_2 \cdots y_n) = \mathrm{per} \left((x_i | y_j) \right)_{i,j=1,\ldots,n},$$

$$(x_1 x_2 \cdots x_n | y_1 y_2 \cdots y_n)^* = (-1)^{\binom{n}{2}} \det \left((x_i | y_j) \right)_{i,j=1,\ldots,n}$$

2. *Let* $x_1, x_2 \in \mathcal{L}^+ = \mathcal{L}$, $y_1, y_2 \in \mathcal{P}^- = \mathcal{P}$; *then* $Super[\mathcal{L}^+|\mathcal{P}^-] = \bigwedge [\mathcal{L}|\mathcal{P}]$. *We have*

$$(x_1 x_2 | y_1 y_2) = \mathcal{D}_{x_1 \alpha} \mathcal{D}_{x_2 \alpha} \left((\alpha|y_1)(\alpha|y_2) \right)$$
$$= (x_2|y_1)(x_1|y_2) + (x_1|y_1)(x_2|y_2);$$

on the other hand

$$(x_1 x_2 | y_1 y_2) = \left((x_1|\beta)(x_2|\beta) \right) {}_{\beta y_1} \mathcal{D} {}_{\beta y_2} \mathcal{D}$$
$$= -(x_1|y_2)(x_2|y_1) + (x_1|y_1)(x_2|y_2).$$

Note that $(x_1 x_2 | y_1 y_2)$ *is symmetric in the* x's *and skew-symmetric in the* y's.

16.3.10   Enter representations

We will study the n-th homogeneous component $Super_n[\mathcal{L}|\mathcal{P}]$ of the proper letterplace superalgebra as a bimodule

$$pl(\mathcal{L}) \cdot Super_n[\mathcal{L}|\mathcal{P}] \cdot pl(\mathcal{P})$$

over the general linear Lie superalgebras $pl(\mathcal{L})$ and $pl(\mathcal{P})$, or, equivalently, as a bimodule over the universal enveloping algebras $\mathcal{U}(pl(\mathcal{L}))$ and $\mathcal{U}(pl(\mathcal{P}))$. In plain words, we will study the bimodule

$$\mathcal{B}_n \cdot Super_n[\mathcal{L}|\mathcal{P}] \cdot_n \mathcal{B}$$

where

$$\mathcal{B}_n, \, _n\mathcal{B} \subset \mathrm{End}(Super_n[\mathcal{L}|\mathcal{P}])$$

are the subalgebras generated by the proper letter and place polarization operators, respectively, and we will exhibit concrete complete decomposition theorems for these modules and algebras.

The focal point is the definition, through the method of virtual variables, of the "symmetrized bitableaux" and of the "Young–Capelli symmetrizers".

16.4   A glimpse on basic combinatorics of tableaux

16.4.1   Young tableaux

A *Young tableau* over an alphabet $\mathcal{A}$ is a sequence $S = (w_1, w_2, \ldots, w_p)$ of words $w_i = a_{i1}a_{i2}\cdots a_{i\lambda_i}$, $a_{ij} \in \mathcal{A}$, whose lengths form a weakly decreasing sequence, i.e., a partition $\lambda = (\lambda_1 \geq \lambda_2 \geq \cdots \geq \lambda_p) = sh(S)$, called the *shape* of S. The word juxtaposition of the words w_i,

$$w = w_1 w_2 \cdots w_p = w(S),$$

is called the *word* of S. If n is the length of w, then λ is a partition of n, in symbols $\lambda \vdash n$.

We will frequently represent tableaux in the array notation:

$$
S = (abb,\ bae,\ c) = \begin{array}{ccc} a & b & b \\ b & a & e \\ c & & \end{array}.
$$

The set of all the tableaux over $\mathcal{A}$ is denoted by $Tab(\mathcal{A})$.

16.4.2   Co-Deruyts and Deruyts tableaux

A tableau C is said to be of *co-Deruyts* type whenever any two symbols in the same row of C are equal, while any two symbols in the same column of C are distinct. For example:

$$
C = \begin{array}{ccccc} a & a & a & a & a \\ b & b & b & & \\ c & & & & \end{array}.
$$

A tableau D is said to be of *Deruyts* type whenever any two symbols in the same column of D are equal, while any two symbols in the same row of D are distinct. For example:

$$
D = \begin{array}{ccccc} a & b & c & d & e \\ a & b & c & & \\ a & & & & \end{array}.
$$

16.4.3   Standard Young tableaux

Assume now that the alphabet

$$\mathcal{A} = \mathcal{A}^+ \cup \mathcal{A}^-$$

is signed and linearly ordered.

A Young tableaux S over $\mathcal{A}$ is called *(super) standard* when each row of S is non-decreasing, with no negative repeated symbols and each column of S is non-decreasing, with no positive repeated symbols. For example:

$$S = \begin{matrix} a & a & c & c & d \\ b & c & d & & \\ b & & & & \end{matrix} \qquad \text{for } a, c \in \mathcal{A}^+, \, b, d \in \mathcal{A}^-.$$

The set of all the standard tableaux over $\mathcal{A}$ is denoted by $Stab(\mathcal{A})$.

16.4.4   The Berele–Regev hook property

Assume now that the linearly ordered signed alphabet $\mathcal{A}$ is finite, with $|\mathcal{A}^+| = r$ and $|\mathcal{A}^-| = s$. The hook set of $\mathcal{A}$ is

$$H(\mathcal{A}) = \{(\lambda_1 \geq \lambda_2 \geq \cdots); \quad \lambda_{r+1} < s + 1\}.$$

Proposition 1. [1, 9] *There is some standard tableau of shape λ over $\mathcal{A}$ iff $\lambda \in H(\mathcal{A})$. Furthermore, the number $f_\lambda(\mathcal{A})$ of standard tableaux of any given shape λ over $\mathcal{A}$ is independent from the linear order defined over $\mathcal{A}$.*

16.5   The actions of the algebras of proper polarizations on $Super_n[\mathcal{L}|\mathcal{P}]$

16.5.1   Bitableau monomials and bitableau polarizations

In this subsection, we introduce some definitions and notations that will be useful in the sequel.

Let $T \in Tab(\mathcal{X})$, $U \in Tab(\mathcal{Y})$, with $sh(T) = sh(U) \vdash n$. The bitableau monomial $\langle T|U \rangle$ is defined as follows:

$$\langle T|U \rangle = (t_1|u_1)(t_2|u_2) \cdots (t_n|u_n) \in Super_n[\mathcal{X}|\mathcal{Y}],$$

where $t_1 \cdots t_n = w(T)$ and $u_1 \cdots u_n = w(U)$.

Let $S', S \in Tab(\mathcal{X})$, with $sh(S') = sh(S) \vdash n$. The letter bitableau polarization of the tableau S to the tableau S' is defined to be the $\mathbb{K}$-linear operator

$$\mathcal{D}_{S'S} = \mathcal{D}_{x'_1 x_1} \mathcal{D}_{x'_2 x_2} \cdots \mathcal{D}_{x'_n x_n} : Super_n[\mathcal{X}|\mathcal{Y}] \to Super_n[\mathcal{X}|\mathcal{Y}],$$

where $x'_1 \cdots x'_n = w(S')$ and $x_1 \cdots x_n = w(S)$.

Let $V, V' \in Tab(\mathcal{Y})$, with $sh(V) = sh(V') \vdash n$. The place bitableau polarization of the tableau V to the tableau V' is defined to be the $\mathbb{K}$-linear operator

$$Super_n[\mathcal{X}|\mathcal{Y}] \leftarrow Super_n[\mathcal{X}|\mathcal{Y}] : {}_{y_1 y'_1}\mathcal{D} \, {}_{y_2 y'_2}\mathcal{D} \cdots {}_{y_n y'_n}\mathcal{D} = {}_{VV'}\mathcal{D},$$

where $y_1 \cdots y_n = w(V)$ and $y'_1 \cdots y'_n = w(V')$.

16.5.2   Symmetrized bitableaux

For every $\lambda \vdash n$ and every $T \in Tab(\mathcal{L})$, $U \in Tab(\mathcal{P})$, with $sh(T) = \lambda = sh(U)$, we define the *symmetrized bitableau*

$$\left(T \middle\| \boxed{U}\right) \in Super_n[\mathcal{L}|\mathcal{P}]$$

by setting

$$\left(T \middle\| \boxed{U}\right) = \mathcal{D}_{TC^+} \langle C^+ | D^- \rangle {}_{D^-U} \mathcal{D}$$

where $C^+ \in Tab(\mathcal{X}^+ \setminus \mathcal{L})$ is any virtual tableau of co-Deruyts type, with $sh(C^+) = \lambda$ and $D^- \in Tab(\mathcal{Y}^- \setminus \mathcal{P})$ is any virtual tableau of Deruyts type, with $sh(D^-) = \lambda$.

Example 5.

$$\left(\begin{array}{c} abc \\ de \end{array} \middle\| \boxed{\begin{array}{c} uvw \\ xy \end{array}} \right) =$$

$$\mathcal{D}_{a\alpha}\mathcal{D}_{b\alpha}\mathcal{D}_{c\alpha}\mathcal{D}_{d\beta}\mathcal{D}_{e\beta} \cdot \left\langle \begin{array}{c} \alpha\alpha\alpha \\ \beta\beta \end{array} \middle| \begin{array}{c} \gamma\delta\epsilon \\ \gamma\delta \end{array} \right\rangle \cdot {}_{\gamma u}\mathcal{D}\ {}_{\delta v}\mathcal{D}\ {}_{\epsilon w}\mathcal{D}\ {}_{\gamma x}\mathcal{D}\ {}_{\delta y}\mathcal{D} =$$

$$\mathcal{D}_{a\alpha}\mathcal{D}_{b\alpha}\mathcal{D}_{c\alpha}\mathcal{D}_{d\beta}\mathcal{D}_{e\beta} \cdot (\alpha|\gamma)(\alpha|\delta)(\alpha|\epsilon)(\beta|\gamma)(\beta|\delta) \cdot {}_{\gamma u}\mathcal{D}\ {}_{\delta v}\mathcal{D}\ {}_{\epsilon w}\mathcal{D}\ {}_{\gamma x}\mathcal{D}\ {}_{\delta y}\mathcal{D}.$$

The symmetrized bitableau $\left(T \middle\| \boxed{U}\right)$ is supersymmetric in the rows of T, i.e., any transposition of two letters t' and t in the same row of T gives rise to a sign change

$$(-1)^{|t'||t|},$$

and dual supersymmetric in the columns of U, i.e., any transposition of two places u' and u in the same column of U gives rise to a sign change

$$(-1)^{(|u'|+1)(|u|+1)}.$$

A symmetrized bitableau is called *standard* when both its letter tableau and its place tableau are standard.

Theorem 4 (Gordan–Capelli Series [3]). *The set of standard symmetrized bitableaux* $\left(T \middle\| \boxed{U}\right)$ *whose shapes are partitions of n is a $\mathbb{K}$-linear basis of* $Super_n[\mathcal{L}|\mathcal{P}]$.

This assertion may be regarded as a char$(\mathbb{K}) = 0$ strong form of the super-straightening formula of Grosshans–Rota–Stein [9].

16.5.3   Young–Capelli symmetrizers

For every $\lambda \vdash n$ and every $S', S \in Tab(\mathcal{L})$, with $sh(S') = \lambda = sh(S)$, we define the *Young–Capelli symmetrizer*

$$\gamma_n(S', S) : Super_n[\mathcal{L}|\mathcal{P}] \to Super_n[\mathcal{L}|\mathcal{P}]$$

by setting

$$\gamma_n(S', S) = \mathcal{D}_{S'C^+} \mathcal{D}_{C^+D^-} \mathcal{D}_{D^-S} \in \mathcal{B}_n,$$

where $C^+ \in Tab(\mathcal{X}^+ \setminus \mathcal{L})$ is any virtual tableau of co-Deruyts type, with $sh(C^+) = \lambda$ and $D^- \in Tab(\mathcal{X}^- \setminus \mathcal{L})$ is any virtual tableau of Deruyts type, with $sh(D^-) = \lambda$.

A Young–Capelli symmetrizer is called *standard* when both its tableaux are standard.

16.5.4 Factorization properties

Using the notation of the previous subsection, we have the following

Proposition 2. [4] *For any standard tableau T, $sh(T) = \lambda$, with no letter in common with the tableaux C^+ and D^-, the Young–Capelli symmetrizer $\gamma_n(S', S)$ can be factorized as follows:*

$$\gamma_n(S', S) = c_T \mathcal{D}_{S'C^+} \mathcal{D}_{C^+T} \mathcal{D}_{TD^-} \mathcal{D}_{D^-S}, \quad c_T \in \mathbb{K}^*.$$

Note that, if the tableau T has no letter in common with the tableaux S' and S, then the operator $\mathcal{D}_{S'C^+} \mathcal{D}_{C^+T}$ can be expressed as the product of the Capelli operators relative to pairs of corresponding rows of S' and T; analogously the operator $\mathcal{D}_{TD^-} \mathcal{D}_{D^-S}$ can be expressed as the product of the Capelli operators relative to pairs of corresponding columns of T and S.

16.5.5 Triangularity

We define a linear order on $Tab_n(\mathcal{L})$ in the following way:

$$S < T \quad \text{iff} \quad \begin{cases} sh(S) < sh(T), & \text{or} \\ sh(S) = sh(T), & w(S) > w(T) \end{cases},$$

where the shapes (n-partitions) and the words are ordered lexicographically.

Theorem 5. [4] *The action of standard Young–Capelli symmetrizers on standard symmetrized bitableaux is given by*

$$\gamma_n(S', S)\left(T\boxed{U}\right) = \begin{cases} \theta_{S,T}\left(S'\boxed{U}\right), & sh(S) = sh(T) \\ 0, & \text{otherwise} \end{cases}$$

where the matrix $(\theta_{S,T})$ is lower triangular with nonzero integral diagonal entries.

The matrix

$$(\rho_{S,T}) = (\theta_{S,T})^{-1}$$

is called the **Rutherford matrix**.

The theorem admits a rather elementary proof based on the virtual presentation of Young–Capelli symmetrizers and of symmetrized bitableaux

$$\gamma_n(S',S)(T\boxed{U}) = \mathcal{D}_{S'C_1^+}\mathcal{D}_{C_1^+D_1^-}\mathcal{D}_{D_1^-S}\mathcal{D}_{TC_2^+}\langle C_2^+|D_2^-\rangle_{D_2^-U}\mathcal{D},$$

combined with a refined version of a combinatorial theorem of Gale and Ryser on interpolating matrices [4, 9].

16.5.6   Orthonormal generators

For every $S', S \in Stab(\mathcal{L})$, with $sh(S') = sh(S) \vdash n$, we define the *orthonormal generator*

$$Y_n(S',S) : Super_n[\mathcal{L}|\mathcal{P}] \to Super_n[\mathcal{L}|\mathcal{P}]$$

by setting

$$Y_n(S',S) = \sum_{T \in Stab(\mathcal{L})} \rho_{ST}\gamma_n(S',T).$$

Theorem 6. *The action of the orthonormal generators on the standard symmetrized bitableaux is given by*

$$Y_n(S',S)(T\boxed{U}) = \delta_{S,T}(S'\boxed{U}).$$

Corollary 1. *The orthonormal generators $Y_n(S',S)$ form a $\mathbb{K}$-linear basis of the algebra $\mathcal{B}_n$.*

16.5.7   Complete decomposition theorems

Theorem 7. *We have the following complete decomposition of $Super_n[\mathcal{L}|\mathcal{P}]$ as a module with respect to the action of the general linear Lie superalgebra $pl(\mathcal{L})$:*

$$Super_n[\mathcal{L}|\mathcal{P}] = \bigoplus_{\substack{\lambda \in H(\mathcal{L}) \cap H(\mathcal{P}) \\ \lambda \vdash n}} \bigoplus_{\substack{T \in Stab(\mathcal{P}) \\ sh(T)=\lambda}} \langle (S\boxed{T}), \ S \in Stab(\mathcal{L})\rangle,$$

where the outer sum indicates the isotypic decomposition of the semisimple module, and the inner sum describes a complete decomposition of each isotypic component into irreducible submodules.

Remark 6. From the straightening formula of Grosshans–Rota–Stein, it follows that the set

$$\{(S\boxed{T}), \ S \in Stab(\mathcal{L})\}$$

is a $\mathbb{K}$-linear basis of the irreducible subspace

$$\langle (S\boxed{T}), \ S \in Tab(\mathcal{L})\rangle_{\mathbb{K}}.$$

Theorem 8. *We have the following complete decomposition for the operator algebra* $\mathcal{B}_n$ *generated by the letter polarization operators acting on* $Super_n[\mathcal{L}|\mathcal{P}]$:

$$\mathcal{B}_n = \bigoplus_{\substack{\lambda \in H(\mathcal{L}) \cap H(\mathcal{P}) \\ \lambda \vdash n}} \bigoplus_{\substack{S \in Stab(\mathcal{L}) \\ sh(S) = \lambda}} \langle Y_n(S', S), \ S' \in Stab(\mathcal{L}) \rangle,$$

where the outer sum indicates the isotypic components of the semisimple algebra, and the inner sum describes a complete decomposition of each simple subalgebra into minimal left ideals.

Corollary 2 (Structure Theorem).

$$\mathcal{B}_n \cong \bigoplus_{\substack{\lambda \in H(\mathcal{L}) \cap H(\mathcal{P}) \\ \lambda \vdash n}} M_{f_\lambda(\mathcal{L})}(\mathbb{K}),$$

where $M_{f_\lambda(\mathcal{L})}(\mathbb{K})$ *is the full matrix algebra of square matrices of order* $f_\lambda(\mathcal{L})$.

16.5.8   The Double Centralizer Theorem

A completely parallel theory holds for the operator algebra $_n\mathcal{B}$ generated by the proper place polarization operators acting on $Super_n[\mathcal{L}|\mathcal{P}]$. As a consequence, we get:

Theorem 9. *The algebras*

$$_n\mathcal{B}, \ \mathcal{B}_n \subset \mathrm{End}_{\mathbb{K}}(Super_n[\mathcal{L}|\mathcal{P}])$$

are the centralizer of each other.

16.6   The Berele–Regev theory

From now on let $\mathcal{L} = \{1, 2, \ldots, n\} = \mathcal{L}^+$ be an alphabet of n positive letters, and $\mathcal{P} = \mathcal{P}^+ \cup \mathcal{P}^-$ an arbitrary finite alphabet of places.

A monomial of $Super_n[\mathcal{L}|\mathcal{P}]$ is said to be letter-*multilinear* whenever it contains each letter of $\mathcal{L}$ exactly once; a Young tableau T over $\mathcal{L}$ is said to be *multilinear* whenever each letter of $\mathcal{L}$ appears exactly once in T; obviously, if T is multilinear over $\mathcal{L}$, then $sh(T) \vdash n$.

We consider the bimodule

$$\underline{\mathcal{B}_n} \cdot \underline{Super_n[\mathcal{L}|\mathcal{P}]} \cdot_n \mathcal{B},$$

where

- $\underline{Super_n[\mathcal{L}|\mathcal{P}]}$ is the subspace of $Super_n[\mathcal{L}|\mathcal{P}]$ generated by the letter-multilinear monomials or, equivalently, by the letter-multilinear symmetrized bitableaux;

- $\underline{\mathcal{B}}_n$ is the subalgebra of the operator algebra $\mathcal{B}_n$ linearly generated by the multilinear orthonormal generators $Y_n(S', S)$. Note that

$$Super_n[\underline{\mathcal{L}}|\mathcal{P}]$$

is invariant under the action of $\underline{\mathcal{B}}_n$; hence, with a little abuse of notation, we will indicate by the same symbol its homomorphic image in

$$\mathrm{End}_{\mathbb{K}}(Super_n[\underline{\mathcal{L}}|\mathcal{P}]);$$

- $_n\mathcal{B}$ is the operator algebra generated by the proper place polarizations, or, equivalently, the operator algebra induced by the action of the general linear Lie superalgebra $pl(\mathcal{P})$.

By the decomposition theorems, the operator algebras $\underline{\mathcal{B}}_n$ and $_n\mathcal{B}$ are each the centralizer of the other in the endomorphism algebra of $Super_n[\underline{\mathcal{L}}|\mathcal{P}]$.

Remark 7. Let

$$W = \langle \mathcal{P} \rangle_{\mathbb{K}} = \langle \mathcal{P}^+ \rangle_{\mathbb{K}} \oplus \langle \mathcal{P}^- \rangle_{\mathbb{K}} = W_0 \oplus W_1,$$

then

$$Super_n[\underline{\mathcal{L}}|\mathcal{P}] \cong T^n[W_0 \oplus W_1]$$

via the linear isomorphism

$$(1|y_{i_1}) \cdots (n|y_{i_n}) \mapsto y_{i_1} \otimes \cdots \otimes y_{i_n}.$$

Therefore, any construction and any result we will present below may be interpreted on the space of the n-tensors over the $\mathbb{Z}_2$-graded vector space $W = W_0 \oplus W_1$.

Note that a group action of the symmetric group S_n over $Super_n[\underline{\mathcal{L}}|\mathcal{P}]$ is consistently defined by setting

$$\sigma \cdot (i_1|y_{j_1})(i_2|y_{j_2}) \cdots (i_n|y_{j_n}) = (\sigma(i_1)|y_{j_1})(\sigma(i_2)|y_{j_2}) \cdots (\sigma(i_n)|y_{j_n}).$$

This action defines a representation

$$\rho : \mathbb{K}[S_n] \to \mathrm{End}_{\mathbb{K}}[Super_n[\underline{\mathcal{L}}|\mathcal{P}]] \cong \mathrm{End}_{\mathbb{K}}[T^n[W_0 \oplus W_1]],$$

called a Berele–Regev $\mathbb{Z}_2$-graded representation of S_n.

Proposition 3. $\underline{\mathcal{B}}_n = \rho(\mathbb{K}[S_n])$.

Proof. To begin, we show that each operator associated to a permutation belongs to the algebra $\underline{\mathcal{B}}_n$. Indeed, by the remark of Section 6.7, for any permutation

$\sigma \in S_n$, and for any multilinear standard symmetrized bitableau $(T \boxed{} \boxed{U})$, we have

$$\sigma \cdot (T \boxed{} \boxed{U}) = (\sigma \cdot T \boxed{} \boxed{U}) = \sum_{S'} c_{TS'}^{\sigma} (S' \boxed{} \boxed{U})$$

$$= \Big(\sum_{S',S} c_{SS'}^{\sigma} \, Y_n(S', S) \Big) (T \boxed{} \boxed{U}),$$

where S' and S range over all the letter–multilinear standard letter tableaux.

On the other hand any multilinear Young–Capelli symmetrizer belongs to the subalgebra $\rho(\mathbb{K}[S_n])$. Indeed, by the factorization proposition of Section 5.4, the action of $\gamma_n(S', S)$ over $Super_n[\mathcal{L}|\mathcal{P}]$ is the same, up to a scalar factor, of the action of an operator of the form

$$\mathcal{D}_{S'C} \mathcal{D}_{CT} \mathcal{D}_{TD} \mathcal{D}_{DS}$$

where T denotes any multilinear standard tableau of the same shape as S' and S, filled with positive virtual letters $\alpha_1, \ldots, \alpha_n$ not appearing in the tableaux C and D. Furthermore, by the final remarks of the same subsection and the third example of Section 3.6, the action of this operator is the same, up to a scalar factor, of the action of an operator of the form

$$\sum_{\sigma} \mathcal{D}_{\sigma(1)\alpha_1} \mathcal{D}_{\sigma(2)\alpha_2} \cdots \mathcal{D}_{\sigma(n)\alpha_n} \sum_{\tau} (-1)^{\tau} \mathcal{D}_{\alpha_1 \tau(1)} \mathcal{D}_{\alpha_2 \tau(2)} \cdots \mathcal{D}_{\alpha_n \tau(n)}$$

$$= \sum_{\sigma, \tau} \pm \mathcal{D}_{\sigma(1)\alpha_1} \mathcal{D}_{\sigma(2)\alpha_2} \cdots \mathcal{D}_{\sigma(n)\alpha_n} \mathcal{D}_{\alpha_1 \tau(1)} \mathcal{D}_{\alpha_2 \tau(2)} \cdots \mathcal{D}_{\alpha_n \tau(n)},$$

where σ, τ range over suitable sets of permutations.

Finally, notice that each summand acts over $Super_n[\mathcal{L}|\mathcal{P}]$ in the same way as a permutation. $\square$

Remark 8. If $S' = S$, then the action of $\gamma_n(S, S)$ is the same as the action of the classical Young symmetrizer associated to the standard tableau S.

By the previous proposition we get immediately the following results.

Theorem 10. [1, 2]

1. $\rho(\mathbb{K}[S_n])$ *and* $_n\mathcal{B}$ *are the centralizers of each other in the endomorphism algebra of* $Super_n[\mathcal{L}|\mathcal{P}]$;

2. *A complete decomposition of* $Super_n[\mathcal{L}|\mathcal{P}]$ *with respect to the action of the symmetric group* S_n *is given by:*

$$T^n[W_0 \oplus W_1] = Super_n[\mathcal{L}|\mathcal{P}]$$

$$= \bigoplus_{\substack{\lambda \in H(\mathcal{P}) \\ \lambda \vdash n}} \bigoplus_{\substack{T \in Stab(\mathcal{P}) \\ sh(T) = \lambda}} \langle (S \boxed{} \boxed{T}), \; S \in Stab(\mathcal{L}), \; S \; multilinear \rangle;$$

3. *A complete decomposition of* $Super_n[\underline{\mathcal{L}}|\mathcal{P}]$ *with respect to the action of the general linear Lie superalgebra* $pl(\mathcal{P})$ *is given by:*

$$T^n[W_0 \oplus W_1] = Super_n[\underline{\mathcal{L}}|\mathcal{P}]$$

$$= \bigoplus_{\substack{\lambda \in H(\mathcal{P}) \\ \lambda \vdash n}} \quad \bigoplus_{\substack{S \in Stab(\mathcal{L}),\, S\ multilinear \\ sh(S)=\lambda}} \quad \langle (S\boxed{T}),\ T \in Stab(\mathcal{P}) \rangle.$$

REFERENCES

[1] A. Berele and A. Regev, Hook Young diagrams, combinatorics and representations of Lie superalgebras, *Bull. Am. Math. Soc.* **8** (1983), 337–339.

[2] A. Berele and A. Regev, Hook Young diagrams with applications to combinatorics and to representations of Lie superalgebras, *Adv. Math.* **64** (1987), 118–175.

[3] A. Brini, A. Palareti, and A. Teolis, Gordan–Capelli series in superalgebras, *Proc. Natl. Acad. Sci. USA* **85** (1988), 1330–1333.

[4] A. Brini and A. Teolis, Young–Capelli symmetrizers in superalgebras, *Proc. Natl. Acad. Sci. USA* **86** (1989), 775–778.

[5] A. Brini, R. Q. Huang and A. Teolis, The umbral symbolic method for supersymmetric tensors, *Adv. Math.* **96** (1992), 123–193.

[6] A. Brini, F. Regonati, and A. Teolis, Grassmann geometric calculus, invariant theory and superalgebras, in *Algebraic Combinatorics and Computer Science*, Eds. H. Crapo and D. Senato, Springer Italia, Milan, 2001, pp. 151–196.

[7] M. Clausen, Letter Place Algebras and a Characteristic-Free Approach to the Representation Theory of the General Linear and Symmetric Groups, I, *Adv. Math.* **33** (1979), 161–191.

[8] J. Désarménien, J. P. S. Kung, and G.-C. Rota, Invariant theory, Young bitableaux and combinatorics, *Adv. Math.* **27** (1978), 63–92.

[9] F. Grosshans, G.-C. Rota, and J. Stein, *Invariant theory and Superalgebras*, Am. Math. Soc., Providence, RI, 1987.

[10] M. Scheunert, *The Theory of Lie Superalgebras: An Introduction*, Lecture Notes in Mathematics, Vol. **716**, Springer Verlag, New York, 1979.

[11] H. Weyl, *The Classical Groups*, Princeton Univ. Press, Princeton, NY, 1946.

Andrea Brini
Dipartimento di Matematica
Universita di Bologna
40126 Bologna, Italy
E-mail: brini@dm.unibo.it

Francesco Regonati
Dipartimento di Matematica
Universita di Bologna
40126 Bologna, Italy
E-mail: regonati@dm.unibo.it

Antonio Teolis
Regione Emilia-Romagna
40100 Bologna, Italy
E-mail: ateolis@regione.emilia-romagna.it

Submitted: July 27, 2002; Revised: July 1, 2003.

17

Algebras Like Clifford Algebras

Michael Eastwood

ABSTRACT We compare various algebras associated with a representation of a semisimple Lie algebra. Their general construction is akin to that of the Clifford algebra (arising from the defining representation of the orthogonal algebra). These algebras arise from the symmetric product, skew product, or Cartan product of representations together with some additional data.

Keywords: Clifford algebra, Lie algebra, Cartan product

Introduction

This article elaborates on a point made near the end of a talk entitled 'Symmetries and Differential Invariants' presented at the Sixth Clifford Algebra Conference in Cookeville, Tennessee, in May 2002. An outline of this talk is in the appendix of this article and serves as motivation. Details are available separately [2]. I would like to thank Rafał Abłamowicz and John Ryan for their invitation to speak and kind hospitality in Cookeville.

The article is organised as follows. In section 17.1 are some familiar examples and one not-so-familiar, the symmetry algebra of the Laplacian in section 17.1.7. Since most of these examples are so well known, especially to participants in the Clifford Algebra Conferences, citations seemed superfluous. A general construction along these lines is introduced in section 17.2 and the examples of section 17.1 are repeated to make sure they are covered by this viewpoint. Each of these algebras is filtered and it should be possible to determine the corresponding graded algebra. This is discussed in section 17.3. There is a conjectured answer and some considerable evidence in favour of this conjecture.

17.1   Examples

All the following examples are associative algebras constructed in similar fashion, namely as quotients of the tensor algebra

$$\bigotimes V = \bigoplus_{s=0}^{\infty} \bigotimes{}^s V$$

for V a finite-dimensional real or complex vector space.

AMS Subject Classifications: 15A78, 15A66, 15A75, 16S99, 17B10, 17B35.
This research was supported by the Australian Research Council.

17.1.1 Grassmann

Also known as the exterior algebra, this is the quotient

$$\bigwedge V = \bigotimes V / (v \otimes w + w \otimes v)$$

where $(v \otimes w + w \otimes v)$ is the two-sided ideal generated by $v \otimes w + w \otimes v$ for $v, w \in V$. If we write

$$v \otimes w = v \odot w + v \wedge w = \underbrace{\tfrac{1}{2}(v \otimes w + w \otimes v)}_{\text{symmetric part}} + \underbrace{\tfrac{1}{2}(v \otimes w - w \otimes v)}_{\text{skew part}},$$

then, instead,

$$\bigwedge V = \bigotimes V / (v \odot w) = \bigotimes V / (v \otimes w - v \wedge w).$$

The Grassmann algebra is graded: $\bigwedge V = \bigoplus_{s=0}^{\infty} \bigwedge^s V$.

17.1.2 Clifford

Now suppose that V is real and is equipped with an inner product

$$V \otimes V \ni v \otimes w \mapsto \langle v, w \rangle \in \mathbb{R}.$$

The Clifford algebra of V is the quotient

$$C\ell(V) = \bigotimes V / (v \odot w + \langle v, w \rangle) = \bigotimes V / (v \otimes w - v \wedge w + \langle v, w \rangle). \quad (1.1)$$

We could also consider $\bigotimes V / (v \otimes w - v \wedge w + \lambda \langle v, w \rangle)$ for any $\lambda \in \mathbb{R}$ but, mapping $v \mapsto \sqrt{\lambda} v$, they are all isomorphic for $\lambda > 0$. They are similarly all isomorphic for $\lambda < 0$ and the Grassmann algebra if $\lambda = 0$. The Clifford algebra is not a graded algebra but is filtered by degree with $\bigwedge V$ as corresponding graded algebra. In fact, this lifts to a canonical isomorphism $C\ell(V) \cong \bigwedge V$ but only as vector spaces. If V is complex and

$$V \otimes V \ni v \otimes w \mapsto \langle v, w \rangle \in \mathbb{C}$$

is a non-degenerate symmetric bilinear form, then we can form the complex Clifford algebra $C\ell(V)$ in the same way (1.1).

17.1.3 Symmetric

For real or complex V we can consider the symmetric algebra

$$\bigodot V = \bigotimes V / (v \wedge w) = \bigotimes V / (v \otimes w - v \odot w).$$

It is graded: $\bigodot V = \bigoplus_{s=0}^{\infty} \bigodot^s V$.

17.1.4 Weyl

Also called the symplectic Clifford algebra, this is defined for a real or complex V equipped with a non–degenerate skew bilinear form $\langle\ ,\ \rangle$. It is the quotient $\bigotimes V/(v \wedge w + \langle v, w\rangle)$ and is isomorphic to the algebra of linear differential operators with polynomial coefficients on a vector space of half the dimension (with composition as the algebra operation). It is a filtered algebra with $\bigodot V$ as the corresponding graded algebra.

17.1.5 Enveloping

Now suppose V is a real or complex Lie algebra. Tradition dictates that we write it as $\mathfrak{g}$ and its typical elements as X and Y. It is equipped with a Lie bracket $[\ ,\ ] : \mathfrak{g} \otimes \mathfrak{g} \to \mathfrak{g}$. The universal enveloping algebra of $\mathfrak{g}$ is defined as the quotient

$$\mathfrak{U}(\mathfrak{g}) = \bigotimes \mathfrak{g}/(2X \wedge Y - [X, Y])$$

It is filtered by degree. As is well known, the corresponding graded algebra is $\bigodot \mathfrak{g}$, lifting to a canonical vector space isomorphism $\mathfrak{U}(\mathfrak{g}) \cong \bigodot \mathfrak{g}$.

17.1.6 Symmetric trace-free

Let us first rewrite the symmetric algebra using indices. An element of V will be written ϕ_a, an element of V^* as ψ^b, the dual pairing between V and V^* as $\phi_a \psi^a$ (with a repeated index), an element of $\bigotimes^2 V$ as θ_{ab}, the natural projection $\bigotimes^2 V \to \bigodot^2 V$ as $\theta_{ab} \mapsto \theta_{(ab)} = \frac{1}{2}(\theta_{ab} + \theta_{ba})$, and so on. This is Penrose's 'abstract index' notation: no particular bases are implied although, of course, it reflects the form taken by these various tensors if a basis were to be chosen. The symmetric algebra of V is now

$$\bigodot V = \bigotimes V/(\phi_a \psi_b - \phi_{(a} \psi_{b)}) = \bigotimes V/(\theta_{ab} - \theta_{(ab)}).$$

Suppose V is equipped with a non-degenerate symmetric form $g_{ab} \in \bigodot^2 V$. We may regard its inverse as an element $g^{ab} \in \bigodot^2 V^*$. We define the symmetric trace-free algebra of V as

$$\bigodot_\circ V = \bigotimes V/(\theta_{ab} - \theta_{(ab)} + \tfrac{1}{m} g^{cd}\theta_{cd}g_{ab}),$$

where $m = \dim V$. Equivalently, if we define $\odot : \bigotimes^2 V \to \bigotimes^2 V$ by

$$\theta_{ab} \longmapsto \theta_{(ab)} - \frac{1}{m} g^{cd}\theta_{cd}g_{ab},$$

then

$$\bigodot_\circ V = \bigotimes V/(v \otimes w - v \odot w).$$

We shall come back to this algebra shortly.

17.1.7 Laplace

This is the most complicated of the examples and, perhaps, should be avoided on first reading. It arises from symmetries of the Laplacian as outlined in the appendix to this article. As in section 17.1.6, suppose V is equipped with a non-degenerate symmetric form g_{ab}, writing g^{ab} for its inverse, and m for dim V. Suppose $m > 2$ and define a linear transformation

$$\odot : \bigotimes^2 \left(\wedge^2 V \right) \to \bigotimes^2 \left(\wedge^2 V \right) \tag{1.2}$$

by sending θ_{abcd}, skew in the indices ab and cd, to

$$R_{abcd} -$$
$$\frac{1}{m-2}(R_{ac}g_{bd} - R_{bc}g_{ad} + R_{bd}g_{ac} - R_{ad}g_{bc} - \frac{R}{m-1}(g_{ac}g_{bd} - g_{bc}g_{ad})),$$

where

$$R_{abcd} = \tfrac{1}{3}\theta_{abcd} + \tfrac{1}{3}\theta_{cdab} + \tfrac{1}{6}\theta_{acbd} - \tfrac{1}{6}\theta_{adbc} + \tfrac{1}{6}\theta_{bdac} - \tfrac{1}{6}\theta_{bcad},$$

$$R_{ac} = g^{bd}R_{abcd} \quad \text{and} \quad R = g^{ac}g^{bd}R_{abcd}.$$

Now $\wedge^2 V$ is also a Lie algebra under the Lie bracket

$$[\phi, \psi]_{ad} = \phi_{ab}g^{bc}\psi_{cd} - \psi_{ab}g^{bc}\phi_{cd}$$

and is itself equipped with a non–degenerate symmetric form

$$\langle \phi, \psi \rangle = \phi_{ab}g^{ac}g^{bd}\psi_{cd}.$$

Define the algebra $\mathcal{A}^\lambda$ as the quotient

$$\mathcal{A}^\lambda = \bigotimes \left(\wedge^2 V \right) / (\phi \otimes \psi - \phi \odot \psi - \tfrac{1}{2}[\phi, \psi] - \lambda\langle\phi, \psi\rangle).$$

17.2   The general construction

Let $\mathfrak{g}$ be a semisimple Lie algebra and W an irreducible representation of $\mathfrak{g}$. Then $W \otimes W$ decomposes into irreducibles amongst which there may be one or more copies of the trivial representation and also of W itself. In other words, there may be homomorphisms of $\mathfrak{g}$-modules:

$$\Phi : W \otimes W \to \mathbb{R} \text{ or } \mathbb{C} \quad \text{and} \quad \Psi : W \otimes W \to W.$$

If V and W are irreducible representations of $\mathfrak{g}$ with highest weights λ and μ, respectively, then in $V \otimes W$ there is the unique irreducible subspace with highest weight $\lambda + \mu$. This is called the *Cartan product* [1] of V and W. Let us denote this representation by $V \odot W$ and the natural projection $V \otimes W \to V \odot W \hookrightarrow V \otimes W$

by $v \otimes w \mapsto v \odot w$. When $V = W$, it factors through the symmetric product $v \odot w$. In fact, when $\mathfrak{g} = \mathfrak{sl}(n, \mathbb{R})$ and $W = \mathbb{R}^n$ is the defining representation, $v \odot w = v \odot w$. When $V = W$, there is also the exterior product $v \otimes w \mapsto v \wedge w$.

Having chosen homomorphisms Φ and Ψ (which may be zero), there are three natural algebras we can write down:

$$\text{I} \quad \otimes W/(v \otimes w - v \odot w - \Psi(v, w) - \Phi(v, w)),$$

$$\text{II} \quad \otimes W/(v \otimes w - v \odot w - \Psi(v, w) - \Phi(v, w)),$$

$$\text{III} \quad \otimes W/(v \otimes w - v \wedge w - \Psi(v, w) - \Phi(v, w)).$$

More generally, we can replace $\odot$, by any $\mathfrak{g}$-endomorphism of $W \otimes W$. All of the algebras considered in section 17.1 are of this form.

17.2.1 Grassmann

This is type III with $\mathfrak{g} = \mathfrak{sl}(n, \mathbb{R})$, W the standard representation on $\mathbb{R}^n$, $\Phi = 0$, and $\Psi = 0$. For the complex case, take $\mathfrak{g} = \mathfrak{sl}(n, \mathbb{C})$ acting on $\mathbb{C}^n$.

17.2.2 Clifford

This is type III with $\mathfrak{g} = \mathfrak{so}(n)$, W the standard representation on $\mathbb{R}^n$, $\Phi = -\langle \ , \ \rangle$, and $\Psi = 0$. The complex case is with $\mathfrak{g} = \mathfrak{so}(n, \mathbb{C})$.

17.2.3 Symmetric

This is type I or II, with $\mathfrak{g} = \mathfrak{sl}(n, \mathbb{R})$, W the standard representation on $\mathbb{R}^n$, $\Phi = 0$, and $\Psi = 0$.

17.2.4 Weyl

This is type I or II, with $\mathfrak{g} = \mathfrak{sp}(n, \mathbb{R})$, W the standard representation on $\mathbb{R}^{2n}$, $\Phi = -\langle \ , \ \rangle$, and $\Psi = 0$.

17.2.5 Enveloping

This is type II, with W the adjoint representation, $\Phi = 0$, and $\Psi = \frac{1}{2}[\ , \ ]$.

17.2.6 Symmetric trace-free

This is type I with $\mathfrak{g} = \mathfrak{so}(p, q)$, W the standard representation on $\mathbb{R}^{p+q}$, $\Phi = 0$, and $\Psi = 0$.

17.2.7 Laplace

This is type I with $\mathfrak{g} = \mathfrak{so}(p, q)$, W the adjoint representation, Φ a chosen multiple of the Killing form, and Ψ half the Lie bracket. Of course, there is something to check here, namely that the linear transformation (1.2) is the Cartan product. This is actually a well known case. The mapping $\theta_{abcd} \mapsto R_{abcd}$ is the tensor form of the projection onto the first irreducible in the decomposition

under $\mathfrak{sl}(m, \mathbb{R})$. The remaining formulae project onto the first irreducible in the decomposition

under $\mathfrak{so}(p, q)$, where $\circ$ denotes the trace-free part. This is the familiar decomposition of the Riemannian curvature tensor, the first irreducible being the Weyl tensor. Altogether,

and, up to scale, we see $\circledcirc$ as projection onto the first factor, Φ onto the third factor, and Ψ onto the fifth factor.

17.3 Results

Unfortunately, the general results are conjectural and specific results are examples of these conjectures or relations between them. Let U, V and W be irreducible representations of a semisimple Lie algebra $\mathfrak{g}$.

Conjecture 1. *Suppose $S \subset U \otimes V$ and $T \subset V \otimes W$ are irreducible. Then $(S \otimes W) \cap (U \otimes T) \subset U \otimes V \otimes W$ is irreducible.*

Conjecture 2. *For the Cartan product $\circledcirc$, we have*

$$((U \circledcirc V) \otimes W) \cap (U \otimes (V \circledcirc W)) = U \circledcirc V \circledcirc W. \tag{3.1}$$

Conjecture 3. *Suppose we decompose $U \otimes V$ and $V \otimes W$ into the Cartan product and the remaining irreducibles:*

$$U \otimes V = (U \circledcirc V) \oplus \mathcal{I}, \qquad V \otimes W = (V \circledcirc W) \oplus \mathcal{J}.$$

Then,

$$U \otimes V \otimes W = (U \odot V \odot W) \oplus ((\mathcal{I} \otimes W) + (U \otimes \mathcal{J})). \qquad (3.2)$$

Theorem 1. *Conjecture 1 implies Conjecture 2.*

Proof. By definition, $U \odot V$ and $V \odot W$ are irreducible so, if Conjecture 1 holds, then the left–hand side of (3.1) is irreducible. This left–hand side also contains a weight vector whose weight is the sum of the highest weights of U, V, and W, respectively. But the right–hand side of (3.1) is, by definition, characterised by these properties. ☐

Theorem 2. *For any particular $\mathfrak{g}$, U, V, and W, Conjectures 2 and 3 are equivalent.*

Proof. Evidently,

$$((U \odot V) \otimes W) \cap (U \otimes (V \odot W)) \supseteq U \odot V \odot W.$$

Suppose that Conjecture 3 holds and that

$$v \in ((U \odot V) \otimes W) \cap (U \otimes (V \odot W)).$$

According to (3.2) we may remove the component in $U \odot V \odot W$ and suppose that $v \in ((\mathcal{I} \otimes W) + (U \otimes \mathcal{J}))$. Then, in particular,

$$v \in ((U \odot V) \otimes W) \cap ((\mathcal{I} \otimes W) + (U \otimes \mathcal{J}))$$

whence $v \in U \otimes \mathcal{J}$. But $v \in (U \otimes (V \odot W))$ too, so $v = 0$.

Conversely, suppose that Conjecture 2 holds. Recall Weyl's unitary trick: let G denote the simply connected compact real form of $\mathfrak{g}$, regard U, V, and W as representations of G and endow each of them with G-invariant inner products. Denote by P and Q, orthogonal projection onto

$$(U \odot V) \otimes W \quad \text{and} \quad U \otimes (V \odot W),$$

respectively. Given (3.1), we are required to show (3.2), which now reads

$$U \otimes V \otimes W = (\operatorname{im} P \cap \operatorname{im} Q) \oplus (\ker P + \ker Q).$$

This fact concerns orthogonal projections. The composition QP preserves $(\operatorname{im} P \cap \operatorname{im} Q)^{\perp}$ and is strictly norm-decreasing there. Thus, $\operatorname{Id} - QP$ is invertible on this subspace. Therefore, we may write

$$v = (\operatorname{Id} - QP)(\operatorname{Id} - QP)^{-1}v = ((\operatorname{Id} - P) + (\operatorname{Id} - Q)P)(\operatorname{Id} - QP)^{-1}v$$

for $v \in (\operatorname{im} P \cap \operatorname{im} Q)^{\perp}$, an expression evidently in $\ker P + \ker Q$. ☐

The procedure just described to obtain (3.2) from (3.1) is actually quite feasible. The projections P and Q can usually be defined algebraically and the only problem is to invert QP on $(U \odot V \odot W)^\perp$. To do this, we can break $(U \odot V \odot W)^\perp$ into irreducibles and employ Schur's lemma. But, since there can be many such irreducibles, Conjecture 2 is much easier to verify than is Conjecture 3.

Let us now gather evidence for these conjectures and see how they apply to the algebras in section 17.1 and the general construction of section 17.2. The simplest case is $\mathfrak{sl}(m, \mathbb{C})$ acting on $U = V = W = \mathbb{C}^m$ in the standard manner. Then, elements of $U \otimes V \otimes W$ are tensors ϕ_{abc} with three indices. Let us consider Conjecture 1 in this case. There are only two possibilities for the irreducible $S \subset U \otimes V$, namely $S = \bigotimes^2 \mathbb{C}^m$ or $S = \bigwedge^2 \mathbb{C}^m$. Similarly, for $T \subset V \otimes W$. The conjecture is true and there are essentially three possibilities:

- If ϕ_{abc} is symmetric in ab and bc, then it is totally symmetric.

- If ϕ_{abc} is symmetric in ab and skew in bc, then it is zero.

- If ϕ_{abc} is skew in ab and bc, then it is totally skew.

The first of these is Conjecture 2. To deduce Conjecture 3, we follow the procedure in the proof of Theorem 2. Thus,

$$\phi_{abc} \xrightarrow{P} \tfrac{1}{2}(\phi_{abc} + \phi_{bac}), \qquad \phi_{abc} \xrightarrow{Q} (\phi_{abc} + \phi_{acb})$$

and we find

$$\begin{aligned}
\phi_{abc} &= \tfrac{1}{6}(\phi_{abc} + \phi_{bca} + \phi_{cab} + \phi_{bac} + \phi_{acb} + \phi_{cba}) \\
&\quad + (\mathrm{Id} - QP)\tfrac{1}{18}(17\phi_{abc} - 7\phi_{bca} - \phi_{cab} - \phi_{bac} - \phi_{acb} - 7\phi_{cba}) \\
&= \tfrac{1}{6}(\phi_{abc} + \phi_{bca} + \phi_{cab} + \phi_{bac} + \phi_{acb} + \phi_{cba}) \\
&\quad + \tfrac{1}{6}(3\phi_{abc} - \phi_{bca} + \phi_{cab} - 3\phi_{bac} + \phi_{acb} - \phi_{cba}) \\
&\quad + \tfrac{1}{6}(2\phi_{abc} - 2\phi_{cab} + 2\phi_{bac} - 2\phi_{acb}),
\end{aligned} \qquad (3.3)$$

which manifestly lies in

$$(\mathbb{C}^m \odot \mathbb{C}^m \odot \mathbb{C}^m) \oplus ((\mathbb{C}^m \wedge \mathbb{C}^m \otimes \mathbb{C}^m) + (\mathbb{C}^m \otimes \mathbb{C}^m \wedge \mathbb{C}^m)), \qquad (3.4)$$

as required. This result has an immediate consequence for the symmetric algebra of section 17.1.3 and 17.2.3. By construction, elements in

$$(\mathbb{C}^m \wedge \mathbb{C}^m \otimes \mathbb{C}^m) + (\mathbb{C}^m \otimes \mathbb{C}^m \wedge \mathbb{C}^m)$$

are precisely the elements of degree 3 in the two-sided ideal generated by $\mathbb{C}^m \wedge \mathbb{C}^m$. We deduce, from (3.4), that the degree 3 part of this graded algebra is $\odot^3 \mathbb{C}^m$. Of course, this continues to higher degree in view of

$$((\odot^{p+q} \mathbb{C}^m) \otimes (\odot^r \mathbb{C}^m)) \cap ((\odot^p \mathbb{C}^m) \otimes (\odot^{q+r} \mathbb{C}^m)) = \odot^{p+q+r} \mathbb{C}^m$$

as another case of Conjecture 2. Whilst this may seem like a ridiculously compli-cated way of deriving very well known facts about the symmetric algebra $\odot\, \mathbb{C}^m$, the point is that this reasoning extends to many other examples where the form of the graded algebra is far from obvious.

Another simple example concerns the symmetric trace-free algebra of sec-tion 17.1.6 and 17.2.6. Let $\mathfrak{g} = \mathfrak{so}(m)$ acting on $U = V = W = \mathbb{C}^m$. The first two conjectures are easily verified in this case. To see the validity of Conjec-ture 3 directly is already somewhat tricky. Of course, it follows from Theorem 2 but to follow the construction given in the proof is tedious. An easier route is firstly to use (3.3) to reduce to the case of ϕ_{abc} being symmetric. Then,

$$\phi_{abc} - \frac{1}{m+2}(g_{ab}\tau_c + g_{bc}\tau_a + g_{ca}\tau_b), \quad \text{where } \tau_a = g^{bc}\phi_{abc},$$

is symmetric trace-free whilst

$$g_{ab}\tau_c + g_{bc}\tau_a + g_{ca}\tau_b = 2g_{ab}\tau_c + \tau_a g_{bc} - (g_{ab}\tau_c - g_{ac}\tau_b)$$

manifestly has the form required in Conjecture 3. Thus, writing $\mathbb{C}^m = V$, we have shown that

$$V \otimes V \otimes V = (V \odot V \odot V) \oplus ((\mathcal{I} \otimes V) + (V \otimes \mathcal{I}))$$

where $\mathcal{I}$ is the linear span of elements of the form $v \otimes w - v \odot w$. Similar reasoning at higher degree shows that the degree s part of $\odot_\circ V$ is $\odot_\circ^s V$, as the notation suggests. Again, though this is well known, it does require proof when the algebra is defined as in section 17.1.6.

A very similar algebra may be defined as

$$\odot_\circ V = \bigotimes V / (v \otimes w - v \odot w - \langle v, w \rangle).$$

It is a type I example of the general construction. Since the ideal here is not homogeneous, the algebra is filtered rather than graded. The corresponding graded algebra is $\odot_\circ V$.

The Laplace algebra of section 17.1.7 and 17.2.7 is, perhaps, the next simplest case of the general construction. Already it is quite complicated. The presence of the skew form Ψ means that it is non-commutative. Comparison with the universal enveloping algebra of section 17.1.5 and 17.2.5 shows that it is a quotient of $\mathfrak{U}(\mathfrak{so}(p,q))$. The corresponding graded algebra is

$$\bigotimes (\wedge^2 V)/(\phi \otimes \psi - \phi \odot \psi)$$

and, were Conjecture 3 to hold, we would be able to identify it as

$$\odot (\wedge^2 V) = \bigoplus_{s=0}^{\infty} \odot^s (\wedge^2 V) = \bigoplus_{s=0}^{\infty} \underbrace{\boxed{\begin{array}{|c|c|c|c|c|c|c|} \hline & & & \cdots & & & \\ \hline & & & \cdots & & & \\ \hline \end{array}}}_{s}{}_\circ . \tag{3.5}$$

In fact, Conjecture 2, and hence Conjecture 3, does hold in this case. To see this, we need to know the symmetries that characterise $\odot^2(\wedge^2 V)$, the range of (1.2). These are the symmetries of the Weyl tensor:

$$C_{abcd} = -C_{abdc} = -C_{bacd} = C_{acbd} + C_{adcb} \quad \text{and} \quad g^{ac}C_{abcd} = 0. \quad (3.6)$$

Therefore, to show that

$$\odot^3(\wedge^2 V) = (\odot^2(\wedge^2 V) \otimes (\wedge^2 V)) \cap (\wedge^2 V) \otimes \odot^2(\wedge^2 V)).$$

as required by Conjecture 2, we must show that if ϕ_{abcdef} satisfies (3.6) in the indices $abcd$ and $cdef$, then ϕ_{abcdef} is totally trace-free and

$$\phi_{abcdef} + \phi_{cbedaf} + \phi_{ebadcf} = \phi_{cbadef} + \phi_{abedcf} + \phi_{ebcdaf}.$$

This may be verified directly. Similar direct reasoning shows

$$\odot^s(\wedge^2 V) = (\odot^{s-1}(\wedge^2 V) \otimes (\wedge^2 V)) \cap ((\wedge^2 V) \otimes \odot^{s-1}(\wedge^2 V)).$$

This is sufficient to establish (3.5).

More generally, we can construct algebras from any semisimple Lie algebra $\mathfrak{g}$ by

$$\mathcal{A}_{\mathfrak{g}}^\lambda = \bigotimes \mathfrak{g}/(\phi \otimes \psi - \phi \odot \psi - \tfrac{1}{2}[\phi,\psi] - \lambda\langle\phi,\psi\rangle)$$

where $\langle\ ,\ \rangle$ is the Killing form. The corresponding graded algebra is

$$\bigotimes \mathfrak{g}/(\phi \otimes \psi - \phi \odot \psi)$$

and, were Conjecture 3 to hold, we would be able to identify this as

$$\odot\mathfrak{g} = \bigoplus_{s=0}^{\infty} \odot^s\mathfrak{g}$$

with algebra operation the Cartan product.

A   Appendix

This is an outline of the talk 'Symmetry and Differential Invariants', as presented at the conference. The mathematical details may be found in [2]. The abstract for this talk was the following moral:

> On a homogeneous manifold, the full symmetry group may be employed to study its geometry and analysis. Invariant constructions in differential geometry often come from symmetry considerations on various "flat models," namely homogeneous versions of the geometry in question.

<u>Question</u>   Which linear differential operators preserve harmonic functions?
The answer in $\mathbb{R}^3$ is linear combinations of the following.

0th order $\qquad\qquad\qquad\qquad\qquad f \longmapsto \text{constant} \times f$,

1st order $\qquad P_1 \equiv \nabla_1 = \partial/\partial x_1$, $P_2 \equiv \nabla_2 = \partial/\partial x_2$, $P_3 \equiv \nabla_3 = \partial/\partial x_3$,

$$J_1 \equiv x_2 \nabla_3 - x_3 \nabla_2\,,\ J_2 \equiv x_3 \nabla_1 - x_1 \nabla_3\,,\ J_3 \equiv x_1 \nabla_2 - x_2 \nabla_1\,,$$

$$D \equiv x_1 \nabla_1 + x_2 \nabla_2 + x_3 \nabla_3 + \tfrac{1}{2}\,,$$

$$K_1 \equiv (x_1{}^2 - x_2{}^2 - x_3{}^2)\nabla_1 + 2x_1 x_2 \nabla_2 + 2x_1 x_3 \nabla_3 + x_1\,,$$

$$K_2 \equiv (x_2{}^2 - x_3{}^2 - x_1{}^2)\nabla_2 + 2x_2 x_3 \nabla_3 + 2x_2 x_1 \nabla_1 + x_2\,,$$

$$K_3 \equiv (x_3{}^2 - x_1{}^2 - x_2{}^2)\nabla_3 + 2x_3 x_1 \nabla_1 + 2x_3 x_2 \nabla_2 + x_3\,.$$

We have arranged the first order operators to close under commutation. Evidently,
they form a 10-dimensional Lie algebra. It is isomorphic to the conformal alge-
bra $\mathfrak{so}(4,1)$! Precisely, if we take the defining quadratic form of $\mathfrak{so}(4,1)$ to be
represented by the symmetric matrix

$$\begin{pmatrix} 0 & 0 & 0 & 0 & 1 \\ 0 & 1 & 0 & 0 & 0 \\ 0 & 0 & 1 & 0 & 0 \\ 0 & 0 & 0 & 1 & 0 \\ 1 & 0 & 0 & 0 & 0 \end{pmatrix},$$

then

$$\begin{pmatrix} 0 & 0 & 0 & 0 & 0 \\ -1 & 0 & 0 & 0 & 0 \\ 0 & 0 & 0 & 0 & 0 \\ 0 & 0 & 0 & 0 & 0 \\ 0 & 1 & 0 & 0 & 0 \end{pmatrix} \leftrightarrow P_1 \qquad \begin{pmatrix} 0 & 0 & 0 & 0 & 0 \\ 0 & 0 & 0 & 0 & 0 \\ -1 & 0 & 0 & 0 & 0 \\ 0 & 0 & 0 & 0 & 0 \\ 0 & 0 & 1 & 0 & 0 \end{pmatrix} \leftrightarrow P_2 \qquad \begin{pmatrix} 0 & 0 & 0 & 0 & 0 \\ 0 & 0 & 0 & 0 & 0 \\ 0 & 0 & 0 & 0 & 0 \\ -1 & 0 & 0 & 0 & 0 \\ 0 & 0 & 0 & 1 & 0 \end{pmatrix} \leftrightarrow P_3$$

$$\begin{pmatrix} 0 & 0 & 0 & 0 & 0 \\ 0 & 0 & 0 & 0 & 0 \\ 0 & 0 & 0 & 1 & 0 \\ 0 & 0 & -1 & 0 & 0 \\ 0 & 0 & 0 & 0 & 0 \end{pmatrix} \leftrightarrow J_1 \qquad \begin{pmatrix} 0 & 0 & 0 & 0 & 0 \\ 0 & 0 & 0 & -1 & 0 \\ 0 & 0 & 0 & 0 & 0 \\ 0 & 1 & 0 & 0 & 0 \\ 0 & 0 & 0 & 0 & 0 \end{pmatrix} \leftrightarrow J_2 \qquad \begin{pmatrix} 0 & 0 & 0 & 0 & 0 \\ 0 & 0 & 1 & 0 & 0 \\ 0 & -1 & 0 & 0 & 0 \\ 0 & 0 & 0 & 0 & 0 \\ 0 & 0 & 0 & 0 & 0 \end{pmatrix} \leftrightarrow J_3$$

$$\begin{pmatrix} 1 & 0 & 0 & 0 & 0 \\ 0 & 0 & 0 & 0 & 0 \\ 0 & 0 & 0 & 0 & 0 \\ 0 & 0 & 0 & 0 & 0 \\ 0 & 0 & 0 & 0 & -1 \end{pmatrix} \leftrightarrow D$$

$$\begin{pmatrix} 0 & 2 & 0 & 0 & 0 \\ 0 & 0 & 0 & 0 & -2 \\ 0 & 0 & 0 & 0 & 0 \\ 0 & 0 & 0 & 0 & 0 \\ 0 & 0 & 0 & 0 & 0 \end{pmatrix} \leftrightarrow K_1 \quad \begin{pmatrix} 0 & 0 & 2 & 0 & 0 \\ 0 & 0 & 0 & 0 & 0 \\ 0 & 0 & 0 & 0 & -2 \\ 0 & 0 & 0 & 0 & 0 \\ 0 & 0 & 0 & 0 & 0 \end{pmatrix} \leftrightarrow K_2 \quad \begin{pmatrix} 0 & 0 & 0 & 2 & 0 \\ 0 & 0 & 0 & 0 & 0 \\ 0 & 0 & 0 & 0 & 0 \\ 0 & 0 & 0 & 0 & -2 \\ 0 & 0 & 0 & 0 & 0 \end{pmatrix} \leftrightarrow K_3$$

At second order there is the Laplacian itself or indeed $f \mapsto g\Delta f$ for any smooth
function g. If we ignore these, then we find another 35-dimensional space of
second–order operators. They arise from the symmetric product $\odot^2 \mathfrak{so}(4,1)$ but

only as one irreducible part:

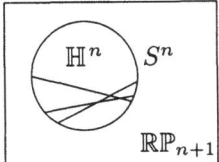

$$\bigodot{}^2 \mathfrak{so}(4,1) = \underbrace{\square \odot \square}_{55} = \underbrace{\boxed{}{}_{\circ}}_{35} \oplus \underbrace{\square\square{}_{\circ}}_{14} \oplus \underset{1}{\mathbb{R}} \oplus \left.\square\right\}{}_{5}$$

dimensions   55   35   14

To investigate third and higher order symmetries it is helpful to describe the Laplacian in a conformally invariant manner. This may be accomplished in two steps. The first is to use stereographic projection to compactify $\mathbb{R}^n$ as the sphere S^n. It is well known that stereographic projection is a conformal mapping. The second step is to write the sphere as the boundary of hyperbolic space:

$\mathbb{H}^n$ S^n $\mathbb{RP}_{n+1}$

Beltrami model of hyperbolic space

orbits of

$SO(n+1,1)$

Let us write $[x^A]$ for homogeneous coordinates on $\mathbb{RP}_{n+1}$ and $\widetilde{g}_{AB}$ for the quadratic form preserved by $SO(n+1,1)$. Write

$$r = \widetilde{g}_{AB} x^A x^B \quad \text{and} \quad \widetilde{\Delta} = \widetilde{g}_{AB} \frac{\partial^2}{\partial x^A \partial x^B}.$$

Now suppose that $f = f(x^A)$ is defined on the sphere $S^n = \{r = 0\}$ and is homogeneous of degree w. Pick an arbitrary extension $\widetilde{f}$ of f into the ambient $\mathbb{RP}_{n+1}$. The general such extension is $\widetilde{f} + rg$ for g homogeneous of degree $w - 2$. An easy computation gives

$$\widetilde{\Delta}(rg) = r\widetilde{\Delta}g + 2\underbrace{(n + 2w - 2)}\, g$$
$$= 0 \text{ if } w = 1 - \tfrac{1}{2}n.$$

It follows that

$$f \longmapsto \widetilde{f} \longmapsto \widetilde{\Delta}\widetilde{f} \longmapsto \widetilde{\Delta}\widetilde{f}|_{S^n} \equiv \Delta f \tag{A.1}$$

is invariant under the action of $SO(n+1,1)$ if $w = 1 - \tfrac{1}{2}n$. This argument, which is due to Hughston and Hurd, was generalised to the curved setting by Fefferman and Graham to obtain, under $g_{ab} \mapsto \widehat{g}_{ab} = \Omega^2 g_{ab}$,

$$\Omega^{-1-\frac{1}{2}n}\underbrace{\left(\Delta - \frac{n-2}{4(n-1)}R\right)}f = \left(\widehat{\Delta} - \frac{n-2}{4(n-1)}\widehat{R}\right)\Omega^{1-\frac{1}{2}n}f$$

Conformal Laplacian

by an ambient construction. They are early examples of the 'AdS/CFT correspondence'.

To return to symmetries of the Laplacian, we shall say that a linear differential operator $\mathcal{D}$ is a *symmetry* if and only if $\Delta\mathcal{D} = \delta\Delta$ for some linear differential operator δ. Clearly, such symmetries preserve harmonic functions. Any operator of the form $\mathcal{P}\Delta$ is a symmetry but of a rather trivial kind since it kills harmonic functions. Let us say that two symmetry operators are *equivalent* if and only if their difference is of the form $\mathcal{P}\Delta$ for some linear differential operator $\mathcal{P}$. The *symmetry algebra* $\mathcal{A}_n$ comprises the symmetries of Δ on $\mathbb{R}^n$ up to equivalence with algebra operation induced by composition.

Theorem 3. *If*

$$\mathcal{D} = \underbrace{V^{bc\cdots d}}_{symmetric\ trace\text{-}free} \nabla_b \nabla_c \cdots \nabla_d + \text{lower order terms} \tag{A.2}$$

is a symmetry of the Laplacian, then $V^{bc\cdots d}$ is a conformal Killing tensor:

$$\text{symmetric trace-free part of } (\nabla^a V^{bc\cdots d}) = 0. \tag{A.3}$$

Theorem 4. *If $V^{bc\cdots d}$ is a conformal Killing tensor on $\mathbb{R}^n$, then there is a canonically defined symmetry of the Laplacian $\mathcal{D}_V$ of the form* (A.2).

Of these, Theorem 3 is quite straightforward. Theorem 4 is more difficult and is best proved by an ambient construction outlined below. The direct formulae on $\mathbb{R}^n$ are quite complicated. When the conformal Killing tensor has three indices, for example,

$$\mathcal{D}_V f = V^{abc}\nabla_a\nabla_b\nabla_c + \frac{3(n+2)}{2(n+4)}(\nabla_a V^{abc})\nabla_b\nabla_c f$$

$$+ \frac{3(n+2)n}{4(n+4)(n+3)}(\nabla_a\nabla_b V^{abc})\nabla_c f + \frac{n(n-2)}{8(n+4)(n+3)}(\nabla_a\nabla_b\nabla_c V^{abc})f.$$

The conformal Killing equation (A.3) is conformally invariant. Its solutions on $\mathbb{R}^n$ therefore define a representation of $\mathfrak{so}(n+1,1)$.

Theorem 5. *The conformal Killing tensors on $\mathbb{R}^n$ of valence s may be identified with the totally trace-free tensors $V^{BC\cdots DQR\cdots S}$ on $\mathbb{R}^{n+2}$ that are symmetric in the s indices $BC\cdots D$ and $PQ\cdots R$ and such that symmetrising over any $s+1$ indices gives zero.*

This space is an irreducible representation of $\mathfrak{so}(n+1,1)$, and Theorems 3, 4, and 5 identify $\mathcal{A}_n$ as (3.5), but only as a vector space. With further effort we can identify $\mathcal{A}_n$ as an algebra, namely

$$\bigotimes(\mathfrak{so}(n+1,1))/(\phi \otimes \psi - \phi \odot \psi - \frac{1}{2}[\phi,\psi] - \frac{n-2}{4(n+1)}\langle\phi,\psi\rangle),$$

as already discussed in sections 17.1.7 and 17.2.7.

It remains to indicate the proof of Theorem 4. For $V^{BC\cdots DQR\cdots S}$ as in Theorem 5, consider the ambient differential operator

$$\mathfrak{D}_V = V^{BC\cdots DQR\cdots S} x_B x_C \cdots x_D \frac{\partial^s}{\partial x^Q \partial x^R \cdots \partial x^S},$$

where $x_B = x^A \widetilde{g}_{AB}$ and so on. The symmetries of $V^{BC\cdots DQR\cdots S}$ imply

$$\mathfrak{D}_V(rg) = r\mathfrak{D}_V g \quad \text{and} \quad \widetilde{\Delta}\mathfrak{D}_V = \mathfrak{D}_V \widetilde{\Delta}.$$

The first of these shows that, as with (A.1),

$$f \longmapsto \widetilde{f} \longmapsto \mathfrak{D}_V \widetilde{f} \longmapsto \mathfrak{D}_V \widetilde{f}|_{S^n} \equiv \mathcal{D}_V f$$

is well defined and the second shows that this is a symmetry.

REFERENCES

[1] E.B. Dynkin, The maximal subgroups of the classical groups, *Amer. Math. Soc. Transl. Series 2* vol. **6** (1957), 245–378.

[2] M.G. Eastwood, Higher symmetries of the Laplacian, hep-th/0206233.

Michael Eastwood
Department of Pure Mathematics
University of Adelaide
South Australia 5005
E-mail: meastwoo@maths.adelaide.edu.au

Submitted: August 6, 2002.

18

Grade Free Product Formulæ from Grassmann–Hopf Gebras

Bertfried Fauser

ABSTRACT In the traditional approaches to Clifford algebras, the Clifford product is evaluated by recursive application of the product of a one-vector (span of the generators) on homogeneous, i.e., sums of decomposable (Grassmann), multi-vectors and later extended by bilinearity. The Hestenesian 'dot' product, extending the one-vector scalar product, is even worse having exceptions for scalars and the need for applying grade operators at various times. Moreover, the multivector grade is not a generic Clifford algebra concept. The situation becomes even worse in geometric applications if a meet, join or contractions have to be calculated.

Starting from a naturally graded Grassmann–Hopf gebra, we derive general formulæ for the products: meet and join, comeet and cojoin, left/right contraction, left/right cocontraction, Clifford and co-Clifford products. All these product formulæ are valid for any grade and any inhomogeneous multivector factors in Clifford algebras of any bilinear form, including non-symmetric and degenerated forms. We derive the three well-known Chevalley formulæ as a specialization of our approach and will display co-Chevalley formulæ. The Rota–Stein cliffordization is shown to be the generalization of Chevalley deformation. Our product formulæ are based on invariant theory and are not tied to representations/matrices and are highly computationally effective. The method is applicable to symplectic Clifford algebras too.

Keywords: Grassmann–Hopf gebra, contraction, cocontraction, Chevalley deformation, Rota–Stein cliffordization, Clifford product, Clifford co-product, meet, join, comeet, cojoin, contractions, cocontractions, linear duality, categorial duality, Grassmann–Cayley algebra

18.1   Introduction

18.1.1   Preliminary note

Beside some rumor during the conference, we continue to use algebra, cogebra and Hopf gebra as technical terms. In our eyes these names fit into mathematical nomenclature having also a linguistic background. The most striking argument is, however, that it is misleading to call a cogebra a co*al*gebra making use of and

AMS Subject Classification: 16W30, 15A66.

pointing to the term *algebra*. By duality one sees that cogebras contain in principle the same amount of information as algebras. One could (should?) come up with a linear cogebra theory not making use of any algebraic structure or knowledge. It seems necessary to us to put the finger into the wound of the missed opportunity [8] to develop algebra and coalgebra on the same footing and beg for pardon to those who feel linguistically offended by our naming.

18.1.2   Synopsis

The present paper will gather *grade free* product formulæ for almost all algebra and cogebra products related to Grassmann–, Grassmann–Cayley, and Clifford algebras. This does not mean that we abandon the multivector structures of these algebras but that we come up with formulæ which are valid for general multivector polynomials, i.e., for general elements x from the algebra A or cogebra C. In present literature important product formulæ are given only on generators or homogeneous elements of certain grades, and have to be expanded by iteration and linearity to the general case. Among these the most important Clifford product has to be calculated this way!

In [17] we find formulæ (1.21a–c), (1.22a,c), (1.23a,b), (1.25b,c), etc., where even restrictions such as that the grade of one homogeneous algebra element has to be less than or equal to the grade of another such element, e.g., (1.23a) and (1.25b,c), have to be assumed. The situation even gets worse if dot and inner products are considered. It is not our aim to criticise but to overcome these deficiencies. During this course we will gain lots of insight into the (almost) perfectly dual structure of algebras and cogebras.

To reach our goal we will see that we *have to* employ algebra and cogebra structures. Furthermore we will take as our point of departure the well behaved Grassmann–Hopf algebra. First, we will show that the Grassmann–Cayley algebra is related to the Grassmann–Hopf algebra by dualizing the co-product. Then, by deformation, we will reach contractions and Clifford algebras. It will turn out that the Chevalley deformation having a grade restriction is a particular case of the general Rota–Stein *cliffordization* obeying no grade restriction.

Using categorial duality, we can write down immediately dualized versions of all algebraic well-known structures coming up with a self dual Grassmann–Cayley double algebra, cocontractions, Chevalley codeformation, and with Clifford cogebras etc.

Categorial duality employs the most powerful and beautiful symmetry. To reach our results, *cogebra structures* are indispensable. We strongly belief that only a fully dual treatment of (projective) geometry, linear co/al-gebra, invariant and deformation theory will prove powerful enough to overcome recent problems in mathematics and physics.

18.1.3   The grading

Since in various discussions, which took place during the ICCA 6, it became clear to me that the concept of grading and filtration seems not to be common ground, we will first settle this issue here.

A *graded* $\Bbbk$-*module* A (graded $\Bbbk$-vector space, or simply *linear space*) is a (finite) family of $\Bbbk$-modules $\{A_n\}$ where n runs through the non-negative integers. n is called *degree* or step, grade etc. The degree of an element a is denoted in various ways, $\partial a = |a| = \text{length}(a) = \deg(a) = \ldots$. Let A, B be graded $\Bbbk$-modules. A *graded morphism* f is a family of morphisms $\{f_i\}$ such that the $f_i : A_i \to B_i$ are morphisms of $\Bbbk$-modules. An element $a \in A$ is called *homogeneous* of degree r iff one has $a \in A_i$ and $a \notin A_j, i \neq j$. Any module can be trivially graded by declaring its degree to be zero.

A grading can be introduced also by the action of an Abelian group G such that the modules A_i are invariant subspaces of the group labeled by a representation index (character): $G \bullet A_i \subset A_i$, $\chi_j(A_i) = i\delta_{i,j}$.

A *filtration* is defined in an analogous way demanding of the weaker obstruction to modules, morphisms, etc., that they consist of or map into spaces of the same or lower degree

$$f_i : A_i \to \oplus_{j \leq i} B_j. \tag{1.1}$$

We will later note that products emerging from cliffordization will not be, in general, graded morphisms as they will only obey a filtration.

Example 1. *Consider a polynomial* $p(x) = \alpha_0 + \alpha_1 x + \alpha_2 x^2 + \cdots \in \Bbbk[[x]]$ *in one variable* x *over the ring (field)* $\Bbbk$. *The 1-dimensional spaces spanned by* x^i *are* $\Bbbk$-*modules* A_i. *Monomials* $\alpha\, x^q$ *are homogenous elements of degree* q. *Polynomials in several commuting complex variables* $\mathbb{C}[[z, w]]$ *can be graded by their* total *degree:* $p(z, w) = \alpha_0 + \alpha_{1,0}z + \alpha_{0,1}w + \alpha_{2,0}z^2 + \alpha_{1,1}zw + \alpha_{0,2}w^2 + \cdots$. *The* $\mathbb{C}$-*spaces of degree* q *have dimensions* $q + 1$. *Observe that one could introduce a finer grading by specifying a* multidegree *composed from the degrees in* z *and* w, *e.g., degree* $\partial(z^4 w^3) = \partial(z^4) + \partial(w^3) = 4 + 3 = 7$, *multidegree* $\partial(z^4 w^3) = (4, 3)$.

A binary *product* m (a binary multiplication) is a morphism from the space $B \simeq A \otimes A$ into the space A. If A is a graded space we can define a grading of $A \otimes A$ to be the sum of the grades of the homogeneous factors, i.e., $\partial(A_i \otimes A_j) = \partial A_i + \partial A_j$, turning B into a graded module. A product $m : B \to A$ is graded if it is a graded morphism. In other terms

$$m : A_i \otimes A_j \subset A_{i+j}. \tag{1.2}$$

Note that this definition of a product implies bilinearity but not associativity. We denote the product by $m(a \otimes b) = m(a, b)$, or in infix notation, or even by juxtaposition, as $a\, m\, b = ab$. Let α, β be ring elements. Then, we have right and left *distributive laws* (linearity)

$$m(\alpha\, a + \beta\, b, c) = \alpha\, m(a, c) + \beta\, m(b, c),$$
$$m(a, \alpha\, b + \beta\, c) = \alpha\, m(a, b) + \beta\, m(a, c). \tag{1.3}$$

If an associative product acts on two adjacent slots of a higher tensor space (two out of a larger number of arguments) it is easily proven to be multilinear. Since we mainly deal with associative products, from now on we assume m to be associative. In this case we can define $m(a \otimes \cdots \otimes b) = m(a, \ldots, b) = m(a, m(\cdots))$ where the order of binary multiplications is irrelevant.

Example 2. *Canonical examples of graded binary products are the tensor product, the Grassmann product, and the symmetric product. Let $A = a \ldots b$, $B = c \ldots d$, be two words in a tensor algebra $T(V)$ generated by the letters $a, b, \ldots \in V$ linearly over some ring $\Bbbk$. A grading in $T(V)$ is defined via the lengths of the words $length(A) = r$ and $length(B) = s$, while the tensor product is a concatenation. We find*

$$m(A \otimes B) = AB, \quad length(m(A \otimes B)) = length(A) + length(B). \quad (1.4)$$

This allows one to decompose the tensor algebra, viewed as a module, into a sum of disjoint submodules containing homogeneous elements $T(V) = \Bbbk \oplus V \oplus V^{\otimes^2} \oplus \cdots$. The elements of $V^{\otimes^r}$ need not be decomposable (products of generators) but may be sums of products of generators.

Let $e_1, e_2, \ldots$ be generators of a Grassmann algebra $V^\wedge = \Bbbk \oplus V \oplus V^{\wedge^2} + \ldots$. A grading in $V^\wedge$ can be defined by the number of generators in a monomial. If we define 0 to have any grade, we find that the Grassmann wedge product is a graded product

$$V^{\wedge^r} \wedge V^{\wedge^s} \subset V^{\wedge^{r+s}},$$

$$length(V^{\wedge^r} \wedge V^{\wedge^s}) = length(V^{\wedge^r}) + length(V^{\wedge^s}) = r + s. \quad (1.5)$$

A symmetric product in $\Bbbk[[a, b, \ldots]]$, defined as the usual point wise product of polynomials, is graded.

$$m(a^3 b^2 \otimes (c^2 d + d^3)) = a^3 b^2 c^2 d + a^3 b^2 d^3, \quad (1.6)$$

$$length(m(a^3 b^2 \otimes (c^2 d + d^3))) = length(a^3 b^2) + length(c^2 d + d^3) = 5 + 3.$$

It is possible to derive algebras from the tensor algebra by factoring out bilateral ideals. These ideals are generated by elements fulfilling some relations. In the case of the Grassmann and symmetric algebras they are, for $x, y \in V$,

$$\mathcal{I}_{Gr} = \text{gen}\{x \otimes y + y \otimes x\},$$
$$\mathcal{I}_{sym} = \text{gen}\{x \otimes y - y \otimes x\}. \quad (1.7)$$

Since these ideals are graded, the factored algebras are also graded by the $\mathbb{Z}$-grading inherited from the tensor algebra. Without going into detail of this construction, we find immediately that the ideal of a Clifford algebra generated as

$$\mathcal{I}_{Cl} = \text{gen}\{x \otimes y + y \otimes x - 2g(x, y)\text{Id}\}, \quad (1.8)$$

where $g(x, y)$ is the symmetric polar bilinear form of a quadratic form Q on V, is *no longer* $\mathbb{Z}$*-graded* since tensors of different degrees are identified.

A good example to this claim and a counter example to a widely accepted assumption that the Clifford algebra comes up with generic 'multivectors', i.e., a $\mathbb{Z}$-grading, are the quaternions.

Example 3. *Let* $1, i, j, k$ *be the standard basis of the quaternions, obeying the relations*

$$k = ij, \quad ij = -ji, \quad jk = -kj, \quad ki = -ik, \quad ii = jj = kk = ijk = -1. \quad (1.9)$$

We obtain a grading using the following length function. Assume that i, j *are generators and define*

$$length(1) = 0, \quad length(i) = 1, \quad length(j) = 1, \quad length(k) = 2. \quad (1.10)$$

However, the roles of i, j, k *are fully symmetric and we could have chosen* j, k *as generators so that* $i = jk$, *which would have lead us to a second, different,* $\mathbb{Z}$*-grading*

$$length(1) = 0, \quad length(j) = 1, \quad length(k) = 1, \quad length(i) = 2. \quad (1.11)$$

Hence, there is no unique such grading present in the quaternions. The argument above using tensor algebra and factorization shows that such a grading cannot uniquely be established in any Clifford algebra. Only the $\mathbb{Z}_2$*-grading or* parity grading *defined by the length function modulo 2 is generic.*

Adding a multivector structure to a Clifford algebra depends on additional choices, e.g., the choice of particular elements as generators. In fact one has to decide in which way a Grassmann algebra having multivectors is embedded in a Clifford algebra. We are consequently using such an identification in the present work and all gradings we refer to are derived from the grading of the tensor and Grassmann algebras.

18.1.4 Algebra and cogebra

We will informally introduce the notion of a cogebra by dualizing the algebra structure. In category theory one uses *commutative diagrams* (CD) for this purpose, however, we will also frequently use *tangles*, see discussion and references in [12]. The difference between both pictures is that they are dual in the sense that arrows and objects change their graphical representation. A product m can be seen as a morphism (arrow) acting on objects (source and target points) in a CD. In the tangle analog we represent morphisms by points and objects by lines (arrows, implicitly read downwards unless otherwise specified). Categorial duality is the operation which reverses all arrows or mirrors all tangles at a horizontal line. The therefrom generated dualized morphisms are named using the prefix '*co*', e.g., a product changes to a *co-product*. In graphical notation we get

$$
\begin{array}{c} A \otimes A \\ \Big\downarrow m \\ A \end{array} \;\cong\; \bigcup m \qquad \Leftarrow \text{duality} \Rightarrow \qquad \begin{array}{c} C \\ \Big\downarrow \Delta \\ C \otimes C \end{array} \;\cong\; \bigwedge_{\Delta} \quad . \tag{1.12}
$$

Tangles can be read like processes in physics, e.g., think of Feynman diagrams, or flow diagrams in computer science. Elements or spaces enter at the top, flow down and suffer at the vertices, representing morphisms, some action. A binary product combines two inputs into one output, while a binary co-product has one input and two outputs. Such a representation is called *graphical calculus*. Some details and references may be found in [12]. If one calculates with tangles, an equality is sometimes called a *move*. The co-product is a $1 \to 2$ map algebraically given as

$$
\Delta \;:\; C \to C \otimes C. \tag{1.13}
$$

The co-product is in general an indecomposable tensor. It is very convenient to introduce the Sweedler notation [26]

$$
\Delta(x) = \sum_r a_r \otimes b_r = \sum_{(x)} x_{(1)} \otimes x_{(2)} = x_{(1)} \otimes x_{(2)}
$$

$$
\Delta(x^i) = \sum_i \Delta_i^{jk}\, x^i = \sum_{(r)} a_{(r)}^j \otimes b_{(r)}^k \quad \text{(w.r.t. arbitrary basis).} \tag{1.14}
$$

The Δ_i^{jk} are called *section coefficients* and constitute a sort of *comultiplication table*. Associativity dualizes to coassociativity and its axiom reads as a commutative diagram or a tangle as follows:

$$
\begin{array}{ccc}
C & \xrightarrow{\;\Delta\;} & C \otimes C \\
\Delta \downarrow & & \downarrow \mathrm{Id} \otimes \Delta \\
C \otimes C & \xrightarrow{\;\Delta \otimes \mathrm{Id}\;} & C \otimes C \otimes C
\end{array}
\qquad
\bigwedge = \bigwedge
\tag{1.15}
$$

A co-product may have a *counit* ϵ which is defined once more by dualizing the axioms of the unit. We find $(\epsilon \otimes \mathrm{Id})\Delta = \mathrm{Id} = (\mathrm{Id} \otimes \epsilon)\Delta$ or, graphically,

$$
\begin{array}{ccc}
\Bbbk \otimes C & \xleftarrow{\;\epsilon \otimes \mathrm{Id}\;} C \otimes C \xrightarrow{\;\mathrm{Id} \otimes \epsilon\;} & C \otimes \Bbbk \\
& \underset{\approx}{\searrow} \;\; \Delta \uparrow \;\; \underset{\approx}{\swarrow} & \\
& C &
\end{array}
\qquad
\overset{\epsilon}{\bigwedge} = \Big| = \overset{}{\bigwedge}_{\epsilon}
\tag{1.16}
$$

The pair $\mathcal{A} = (A, m)$ is called an (associative, possibly unital) algebra and the dualized structure $\mathcal{C} = (C, \Delta)$ is called a (coassociative, possibly co-unital) *cogebra*.

18.1.5   Linear duality

Since we have already used categorial duality we need to introduce a technical term *linear duality* for the conventional duality. Any possibly graded finite–
dimensional $\Bbbk$-module A comes naturally, i.e., functorially, with a *linear dual*
$A^* \simeq \text{lin-hom}(A, \Bbbk)$. Elements ω of A^* are called linear forms. We will freely
use the notation

$$\omega(x) = \langle \omega \mid x \rangle = \text{eval}(\omega \otimes x). \tag{1.17}$$

Arrows are used to indicate the type of space. Downward oriented lines represent
the space A while upward oriented lines depict the dual space A^*. The action of
a linear dual on a space is an *evaluation map*, denoted as eval. Due to a symmetry
we can define the action the opposite way around also, thereby identifying A with
its double dual A^{**}. In terms of tangles we write:

$$\text{(1.18)}$$

eval   eval

18.1.6   Product co-product duality (by evaluation)

The evaluation map provides a natural (functorial) connection of products and co-
products on A and A^*. The action of a linear form ω on a product $m(a \otimes b)$
shall be rewritten as a sum of scalar products of actions of some tensor $\omega_{(1)} \otimes \omega_{(2)}$
on the argument $a \otimes b$ of m. In the tangle picture this means, for example, that
one pulls the product from two down-strands on the right to a single up-strand on
the left (see the first graphical equation in (1.19)). During this process, the product
tangle gets mirrored (rotated by π) and becomes a co-product tangle acting on
the dual space. The second equation in (1.19) dualizes a product on $A^* \otimes A^*$.

$$\text{(1.19)}$$

In terms of algebraic formulæ we can write this as

$$\text{eval}(m(\omega \otimes \omega') \otimes x) = (\text{eval} \otimes \text{eval})(\omega \otimes \omega' \otimes \Delta(x)) = \omega(x_{(2)})\omega'(x_{(1)}),$$
$$\text{eval}(\omega \otimes m(x \otimes y)) = (\text{eval} \otimes \text{eval})(\Delta(\omega) \otimes x \otimes y) = \omega_{(1)}(y)\omega_{(2)}(x). \tag{1.20}$$

Using the evaluation map any product induces a co-product on the dual space
and vice versa [20].

$$\Delta : A \to A \otimes A \quad \Leftrightarrow \quad m^* : A^* \otimes A^* \to A^*,$$
$$m : A \otimes A \to A \quad \Leftrightarrow \quad \Delta^* : A^* \to A^* \otimes A^*. \tag{1.21}$$

Working with a space A and its dual A^* we are still free to choose (i) a product and co-product on A *or* on A^*, (ii) a product m on A and m^* on A^*, i.e., Grassmann–Cayley case, or, (iii) a co-product Δ on A and Δ^* on A^*.

18.2   Grassmann–Hopf algebra

We will define the Grassmann–Hopf algebra using the notion of letters and words, i.e., choosing a basis. Of course one could reformulate the following results basis free also. However, it will become important that the structure is unique up to isomorphy only. The notion of the Grassmann–Hopf algebra is standard and may be found in [26], however we need to introduce some subtleties which will be used later on and are explained at length in [12]. The terms Hopf *al*gebra and Hopf gebra denote, in general, different structures that coincide in the Grassmann–Hopf case. For example, these terms are distinct for Clifford Hopf al/gebras.

A *Euclidean co-product* $\vee$ (meet of hyperplanes) on A^* of the Grassmann exterior product when acting on an element $x \in A$ is defined as a sum over all those tensors $x_{(1)i} \otimes x_{(2)i}$ which multiply back to the element x. Our terminology reflects our usage of the Euclidean dual isomorphism $\delta : V \to V^*$. Hence we consider the splits

$$I_x \equiv (x) := \Big\{ (a, b) \mid m(a \otimes b) = x \Big\},$$
$$I_{x(1)} \equiv x_{(1)} = a, \quad I_{x(2)} \equiv x_{(2)} = b, \tag{2.1}$$

and obtain

$$\Delta(x) := \sum_{I_x}^{|I_x|} I_{x(1)} \otimes I_{x(2)} = \sum_{(x)} x_{(1)} \otimes x_{(2)} = x_{(1)} \otimes x_{(2)},$$
$$m \circ \Delta(x) = \sum_{(x)} m(x_{(1)} \otimes x_{(2)}) = |I_x|\, x\,. \tag{2.2}$$

The Grassmann exterior algebra over a vector space $V^\wedge$ with a wedge product $\wedge$ can be obtained by factoring the tensor product modulo the antisymmetrization. A signed transposition needed for antisymmetrization is called Grassmann crossing or *graded switch*, and is defined as

$$\otimes \xrightarrow{\pi_{\hat\tau}} \wedge$$
$$\hat\tau(A \otimes B) = (-1)^{\partial A \partial B} B \otimes A \qquad \text{on homogeneous elements.} \tag{2.3}$$

We obtain

$$\Delta_\wedge(\mathrm{Id}) = \mathrm{Id} \otimes \mathrm{Id}$$
$$\Delta_\wedge(a) = a \otimes \mathrm{Id} + \mathrm{Id} \otimes a$$
$$\Delta_\wedge(a \wedge b) = a \wedge b \otimes \mathrm{Id} + a \otimes b - b \otimes a + \mathrm{Id} \otimes a \wedge b$$

$$\vdots$$

$$\Delta_\wedge(x) = x_{(1)} \otimes x_{(2)}. \tag{2.4}$$

The sign stemming from the permutations is included in Sweedler notation. To establish a basis in a Grassmann algebra, we need a term ordering on the elements $a, b, c \ldots \in V$ extended to $V^\wedge$ in order to decide if we should solve for ab or $ba = -ab$. The splits of a word $A = ab \ldots d$ into two blocks $B = a \ldots c$, $C = b \ldots d$ are such that in every block B, C the term ordering remains valid. In the Grassmann case we find that a word of length r obeys 2^r such splits. The Euclidean dualized wedge co-product is found to be: (i) co-unital with a counit $\epsilon : V^\wedge \to \Bbbk$, (ii) co-associative, (iii) (linear) dual to the exterior product (denoted as 'vee' $\vee$) on the dual space of linear forms $V^{*\wedge} : \Delta^* \equiv \vee$, (iv) can be *obtained in a combinatorial way* by a sum of all 'splits' of the exterior products into two blocks.

The pair $(V^\wedge, \wedge)$ is called a *Grassmann algebra* while the pair $(V^\wedge, \Delta_\wedge)$ is called a *Grassmann cogebra*. If the co-product is dualized from the vee $\vee$ we denote it as $\Delta_\vee$.

If the following compatibility laws are valid, and if we can prove that an *antipode* exists, then we can establish a *Grassmann–Hopf algebra*.

In Hopf *algebras* one demands as compatibility laws that the product and the unit are cogebra morphisms, and that the co-product and the counit are algebra morphisms:

$$\tag{2.5}$$

Finally, we give the axioms for the *antipode*, an anti-homomorphism, and a generalization of the inverse:

$$S(x_{(1)}) \wedge x_{(2)} = \epsilon(x)\,\mathrm{Id} = x_{(1)} \wedge S(x_{(2)}), \quad \forall x \in V^\wedge,$$

$$U = \mathrm{Id} \circ \epsilon = \mathrm{Id}_{\mathrm{conv}}. \tag{2.6}$$

A *Grassmann–Hopf algebra* is defined as the following septuple:

$$H^\wedge = (V^\wedge, \wedge, \mathrm{Id}, \Delta_\wedge, \epsilon, \hat{\tau}, S)$$

fulfilling the above axioms. A classification of convolution algebras obeying a product and a co-product can be found in [12]. There it was demonstrated that if a convolutive unit Id_{conv} and an antipode exits, then the product and co-product induce all other structure tensors in a Hopf gebra. This idea goes back to Z. Oziewicz [22, 24].

The rest of the paper is devoted to showing that almost all algebraic structures needed in geometry and physics can be *derived* in a plain and natural way from the common generic root of Grassmann–Hopf gebra. In this way we follow Oziewicz [21] from Grassmann to Grassmann–Cayley, Clifford, etc., adding at the same time the dual structures:

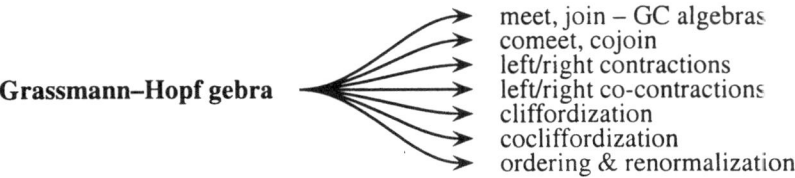

Grassmann–Hopf gebra

meet, join – GC algebras
comeet, cojoin
left/right contractions
left/right co-contractions
cliffordization
cocliffordization
ordering & renormalization

We will have no space to discuss the last point here, see [4, 5, 12].

18.3   Grassmann–Cayley double algebra

18.3.1   Integrals and the bracket

A *left (right) integral* is an element μ_L (μ_R) $\in A^*$, i.e., it is a comultivector of the unital cogebra A^* fulfilling:

$$
\qquad = \quad \mu_R \qquad\qquad\qquad = \quad \mu_L \qquad\qquad (3.1)
$$

$$
(\mathrm{Id} \otimes \mu_R)\Delta(x) = \mu_R(x)\mathrm{Id} \qquad (\mu_L \otimes \mathrm{Id})\Delta(x) = \mu_L(x)\mathrm{Id}\,. \qquad (3.2)
$$

Grassmann–Hopf gebras are bi-augmented, bi-connected (see [12, 20]) and possess a unique left/right integral μ. Integrals in general do *not* exist in Clifford Hopf gebras [12].

The *bracket* $[\ldots]$ of invariant theory is defined to be a multilinear alternating normalized map of s multivector arguments having total degree n, i.e., $\partial A_0 + \partial A_1 + \cdots + \partial A_s = n = \dim V$ and otherwise zero.

$$
[A_0, \ldots, A_s] \;:\; \otimes^s V^{s_i} \longrightarrow \mathbb{k}
$$
$$
[A_0, \ldots, A_s] \equiv (\mu \circ \wedge^s)(A_0 \otimes \ldots \otimes A_s)\,. \qquad (3.3)
$$

In fact, this is a determinantal map. The unique integral μ of a Grassmann–Hopf algebra turns out to be the projection onto the coefficient of the highest grade element. This allows one to define the bracket in Grassmann–Hopf algebraic terms:

$$[A_0, \ldots, A_s]_\mu \;\cong\; \begin{array}{c} A_0 \otimes \cdots \otimes A_s \\ \bigcup\nolimits_{\mu} \end{array} \tag{3.4}$$

In what follows it is important to realize, that the bracket is a sort of *cup-tangle* on n-strands or, equivalently, an $n \to 0$ *map*. While the evaluation map in (1.18) was a pairing of a space and a dual space, the bracket, using two arguments, constitutes a self pairing $[\cdot, \cdot] : V^\wedge \otimes V^\wedge \to \Bbbk$. In terms of tangles we can, however, easily transfer notions from one case to the other. This will be used in the next subsection.

18.3.2   Meet and join, linear logic

Let A be an *extensor*, i.e., a homogenous decomposable multivector which can be written as $A = a_0 \wedge \cdots \wedge a_r$. The linear space $\overline{A} = \text{span}\{a_0, \ldots, a_r\}$ is called the *support of* A. The *join* ($A \wedge B$) is defined as the *disjoint union* of the supports $\overline{A}$, $\overline{B}$, i.e., $\overline{A \wedge B} = \overline{A} \cap \overline{B}$, and zero, otherwise [3, 7]. In logical terms this is an *exclusive or (XOR)* on linear spaces.

Geometrically speaking, the join connects disjoint geometric elements. Two points are joined to span a line, a point and a line may span a plane, etc. It was already clear to Grassmann that one needed a second operation called a *meet* (his regressive product, a section) which allowed him to compute common subspaces thereby lowering the degree of the algebraic objects. We will show that this notion is natural to a Grassmann–Hopf algebra.

Historical Note. The *meet* or the $\vee$-product was introduced by H. Grassmann as 'eingewandtes Produkt' in [15] using what later became called the *rule of the common factor*. He weakened this concept and renamed the operation as *regressive product* in his second Ausdehnungslehre [A2, 1862] [14] using there a unary operation of 'Ergänzung'. This is the notion of an orthogonal complement and it is denoted by a vertical line $a \to |a$ such that $a \wedge |a = I$ where I is an element of maximal grade. In logical terms this operation is a *negation* on a linear space. The Ergänzung makes an explicit use of the total dimension of the underlying space V via the element I, as it is also well known in logic that negation is based on a maximal element in an orthomodular lattice. This Ergänzungs operation of taking the orthogonal complement *necessarily* needs, speaking in geometrical terms, a symmetric polarity which leads to a symmetric polar bilinear form! It is this place where the restriction enters. Hence we can address the Ergänzung as *linear NOT* in linear logic.

We refer to the following rule as *de Morgan law for linear spaces*. It can be found in Grassmann's A2, [14] and it was reinvented several times (see the second line below).

$$|(A \vee B) = (|A) \wedge (|B) \qquad\qquad [1862, A2]$$
$$A \vee B = I^{-1} \cdot ((I \cdot A) \wedge (I \cdot B)) \qquad [17] \text{ needs a 'dot' product} \qquad (3.5)$$

It should be remarked that the usage of a dot- or the scalar product is still more restrictive than the assumption of orthomodularity which fixes only a class of polarities having the same determinant.

A universal or *master formula* for the meet of r factors not using any symmetric polarity was given by Alfred Lotze in 1955 [18][1]. Lotze showed in a note added in proof of the above cited paper, that the meet product turns out to be an exterior product also. Moreover, Lotze showed that the 'double meet' (meet w.r.t. the meet) is again probably up to a sign the original wedge product. This is a remarkable and beautiful duality. Furthermore, it shows that we can safely reject Rota's idea to switch the notions of wedge and vee products to come up with a direct analogy to set theory since duality spoils a fixed relation. Finally, this duality shows that it is irrelevant what a point and a hyperplane are since these notions can be interchanged *provided* one interchanges also the meanings of the meet and the join. This is the celebrated duality of the projective geometry.

We are ready to define the meet now entirely in terms of the Grassmann–Hopf algebra as (the signs are due to a reordering of factors):

$$A \vee B := (a_1 \wedge \cdots \wedge a_r) \vee (b_1 \wedge \cdots \wedge b_s) =$$
$$[B_{(1)}, A]\, B_{(2)} = A_{(1)}\, [B, A_{(2)}] = \pm [A, B_{(1)}]\, B_{(2)} = \pm A_{(1)}\, [A_{(2)}, B]. \qquad (3.6)$$

The tangle definition of the meet reads:

$$(3.7)$$

This definition still needs the notion of a maximal grade to exist, but it works out properly for arbitrary, not necessarily symmetric, non-degenerate bilinear forms too. The meet is a sort of contraction w.r.t. the self pairing induced by the bracket.

18.3.3 Comeet and cojoin

Having the tangle definition, it is simply a matter of dualizing to come up with the notion of a cojoin and comeet. The *cojoin* turns out to be just the Grassmann

---

[1]We make use of the definition given by Doubilet, Rota & Stein [7] which is for two factors, but uses a more compact notation.

co-product $\Delta_\wedge$. The *co-meet* $\Delta_\vee$ is given by the categorial duality and involves the obvious notion of a *cointegral*.

$$\Delta_\vee \;:=\; \pm \left(\Delta_\wedge \quad \wedge \right) \;=\; \pm \left(\wedge \quad \Delta_\wedge \right) \tag{3.8}$$

The comeet is a co-product, i.e., a $1 \to 2$ map, hence it may be called cocontraction w.r.t the cobracket.

18.3.4 Grassmann–Cayley and fourfold algebra

The *Grassmann–Cayley algebra* is defined to be the di-algebra $GC(\vee, \wedge)$ having two associative unital binary products. The various duality relations allow us to identify the Grassmann–Cayley algebra with the Grassmann–Hopf algebra $H_\wedge$ or $H_\vee$ over $V^\wedge$ or $V^{*\vee}$, and to introduce a Grassmann–Cayley cogebra $GC(\Delta_\vee, \Delta_\wedge)$. In a CD these dualities read as:

$$
\begin{array}{ccc}
GC(\vee, \wedge) & \longleftrightarrow & GC(\Delta_\vee, \Delta_\wedge) \\
\updownarrow & & \updownarrow \\
H_\wedge(\wedge, \Delta_\vee) & \longleftrightarrow & H_\vee(\vee, \Delta_\wedge)
\end{array}
\tag{3.9}
$$

Note that in Grassmann–Hopf algebras the exterior product and the exterior coproduct are *independent*. This has some subtle consequences and was the motivation to use the wedge and the vee for the exterior products on $V^\wedge$ and $V^{*\vee}$, see [12]. This independence makes it useful to introduce a *fourfold algebra*:

$$H_\wedge \oplus H_\vee \simeq GC(\wedge, \Delta_\vee, \vee, \Delta_\wedge). \tag{3.10}$$

It would be interesting to investigate in what way this is a Grassmann–Cayley Hopf di-algebra. A reasonable assumption is to relate the wedge $\wedge$ and the $\vee$ vee products using an analogy of a co-(quasi) triangular structure (which might be trivial), see [5].

18.4 Bilinear forms and contractions

18.4.1 Scalar and coscalar products

A *scalar product* B on $V \otimes V$ is a map in the set lin-hom$(V \otimes V, \Bbbk)$ or, similarly on the dual space, $D \in$ lin-hom$(V^* \otimes V^*, \Bbbk)$. A *coscalar product* C is an element in the set lin-hom$(\Bbbk, V \otimes V)$ or lin-hom$(\Bbbk, V^* \otimes V^*)$.

$$
V \underset{D}{\overset{B}{\rightleftarrows}} V^*, \quad
V \otimes V \underset{C}{\overset{B}{\rightleftarrows}} \Bbbk \underset{E}{\overset{D}{\rightleftarrows}} V^* \otimes V^* \tag{4.1}
$$

Scalar products are $2 \to 0$ maps, i.e., they are cup-tangles while coscalar products are $0 \to 2$ maps, i.e., they are cap-tangles. However, on the linear spaces $V^\wedge$ and $V^{*\vee}$ we have to give a meaning to a scalar product $B^\wedge$, resp. $D^\vee$ in a *canonical way*. Later on we will investigate Clifford algebras where the scalar product is the polar bilinear form of a quadratic form on V and the algebra structure allows one to define a unique generalization.

If we demand that the scalar product be extended by an exponential map, then one can check that this is related to co-(quasi) triangular structures. Furthermore, one can show that only exponentially generated scalar products $B^\wedge$ on $V^\wedge \otimes V^\wedge$ come up with an associative algebraic structure during a deformation process [4, 5, 12].

The cup-tangles for scalar and coscalar products can be looked at in two ways, either as the scalar products or as the duality in lin-hom$(V^\wedge, V^{*\vee})$ resp. lin-hom$(V^{*\vee}, V^\wedge)$. This reads:

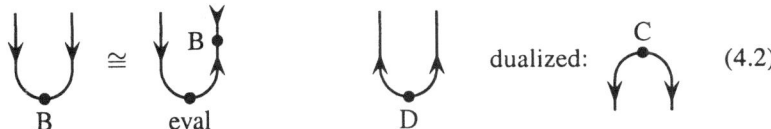

$$\qquad (4.2)$$

Hence we define the *canonically induced scalar product* $B^\wedge$, which fulfills the axioms of a co-(quasi) triangular structure, as:

$$B^\wedge = \exp_\wedge(B) = \epsilon \otimes \epsilon + B_{ij}\epsilon^i \otimes \epsilon^j + B_{[i_1 i_2],[j_1 j_2]}\epsilon^{i_1} \wedge \epsilon^{i_2} \otimes \epsilon^{j_1} \wedge \epsilon^{j_2} + \cdots$$

$$\qquad (4.3)$$

The coscalar product $C^{\triangle\vee}$ is obtained in the same way by categorial duality, i.e., by mirroring the tangle horizontally (rotating by π).

18.4.2 Contractions

Using the scalar product $B^\wedge$ as the cup-tangle, we can once more exploit the product co-product duality. This time all input spaces are of the same type and we get

$$\qquad (4.4)$$

This motivates our definition of a *right contraction* in terms of a tangle equation as

$$(4.5)$$

Moving the product from left to right in the product co-product duality (1.19) gives

$$(4.6)$$

which motivates our definition of a *left contraction* as:

$$(4.7)$$

These two definitions are valid for arbitrary inhomogeneous elements of any grade. While in textbooks one finds such definitions using the pairing, e.g., [16], there is no direct constructive rule for their evaluation. Since we can directly compute co-products, our formulæ

$$\lrcorner_{B^\wedge}(A \otimes B) = B^\wedge(A, B_{(1)}) \, B_{(2)},$$
$$\llcorner_{B^\wedge}(A \otimes B) = A_{(1)} \, B^\wedge(A_{(2)}, B)$$

$$(4.8)$$

are constructive and free of any grade, homogeneity or decomposability restrictions.

18.4.3   Chevalley formulæ for any grade

In [6] Chevalley introduced a recursive method to compute the contraction. From the properties of the pairing he derived the following well-known rules for the left contraction. Of course, analogous formulæ hold for the right contraction. Let $x, y \in V$ and $u, v, w \in V^\wedge$; then, the left contraction obeys

$$(i) \qquad \qquad x \lrcorner_B y = B(x, y) \, \mathrm{Id} = \epsilon(x \circ y) \, \mathrm{Id}$$
$$(ii) \qquad \qquad x \lrcorner_B (u \wedge v) = (x \lrcorner_B u) \wedge v + \hat{u} \wedge (x \lrcorner_B v)$$
$$(iii) \qquad \qquad u \lrcorner_B (v \lrcorner_B w) = (u \wedge v) \lrcorner_B w,$$

$$(4.9)$$

where $\hat{u} = (-1)^{\partial u} \, u$ is the grade involution which turns out to be the antipode of the Grassmann–Hopf algebra [12].

To show that the above tangle definition of the left contraction is a generalization of Chevalley deformation, we have to show that the three rules (4.9i–iii) follow from the tangle definition. But our main goal is to generalize the Chevalley relations to arbitrary inhomogeneous algebra elements of any grade.

Theorem 1. *The left contraction as defined in (4.7) for arbitrary algebra elements generalizes the Chevalley formulæ (4.9i–iii) and reduces to them for the one-vector specialization. The graded crossing $\hat{\tau}$ induces the grade involution (antipode) in the graded Leibnitz rule (4.9ii).*

Proof of 4.9(i). We compute the defining tangle of the left contraction on two grade one elements $a, b \in V$:

$$\lrcorner_B(a \otimes b) = (B^\wedge \otimes \mathrm{Id}) \left((\mathrm{Id} \otimes \Delta)(a \otimes b) \right)$$
$$= (B^\wedge \otimes \mathrm{Id})(a \otimes b \otimes \mathrm{Id} + a \otimes \mathrm{Id} \otimes b) = B(a, b)\mathrm{Id} \qquad (4.10)$$

recalling that $\Delta(b) = b \otimes \mathrm{Id} + \mathrm{Id} \otimes b$.

But the tangle definition is now valid for arbitrary elements

$$\lrcorner_B(u \otimes v) = (B^\wedge \otimes \mathrm{Id}) \left((\mathrm{Id} \otimes \Delta)(u \otimes v) \right) = \left(B^\wedge(u, v_{(1)}) \right) v_{(2)} \qquad (4.11)$$

where only those terms survive having $\partial u = \partial v_{(1)}$. $\qquad \square$

Proof of 4.9(iii). We compute using tangles the following equation:

$$(4.12)$$

We have used the definition of the contraction (4.7) in a), product co-product duality (1.19) in b), coassociativity (1.15) in c). This formula was already valid for any grade. $\qquad \square$

Proof of 4.9(ii). The most complicated case is relation (4.9ii). We compute firstly the tangle equation for the general case and prove that the restriction to a one-vector argument yields the well-known graded Leibnitz rule.

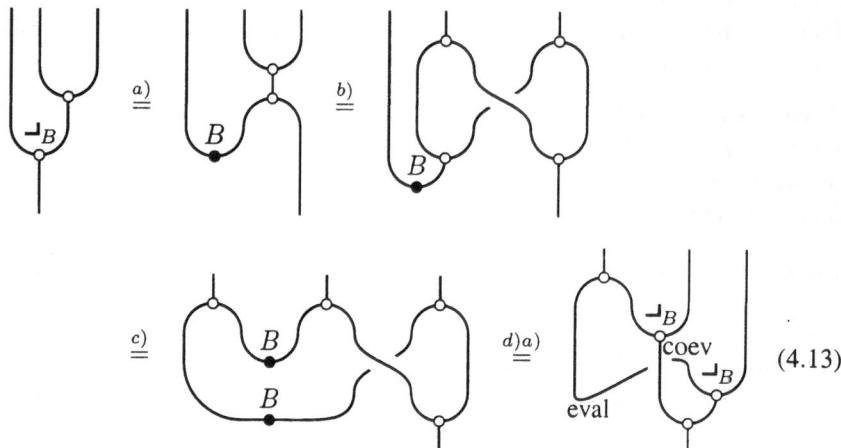

(4.13)

We have used the definition of the contraction (4.7) in a), the compatibility of algebra and cogebra structure (2.5) in b), product co-product duality (1.19) in c), and the following property (4.14) of the crossing in d):

$$B^{cs}\tau^{ab}_{sd} = \tau^{ca}_{ds}B^{sb}$$

(4.14)

In algebraic terms the tangle equation (4.13) reads:

$$w \lrcorner_B (u \wedge v) = (-1)^{(|w_{(1)}||w_{(2)} \lrcorner_B u|)} (w_{(2)} \lrcorner_B u) \wedge (w_{(1)} \lrcorner_B v).$$

(4.15)

To finish the proof we reduce the general formula to the case of a one-vector contraction, i.e., we let $w \to a \in V$,

$$a \lrcorner_B (u \wedge v) = (a \lrcorner_B u) \wedge v + \hat{u} \wedge (a \lrcorner_B v),$$

(4.16)

remembering the definition of the graded switch (2.3) and specializing also to a one-vector argument in the first tensor slot

$$\hat{\tau}(a \otimes u) = (-1)^{\partial a \partial u}(u \otimes a) = ((-1)^{\partial u}u) \otimes a = \hat{u} \otimes a,$$

(4.17)

we obtain the well-known graded Leibnitz rule (4.9ii).   □

Our calculation shows that the grade involution $\hat{u}$ originates in the graded switch $\hat{\tau}$. The crossing is thus related to the derivation property. This observation has tremendous impact on commutation relations. Let

$$a_i^\dagger \quad \Leftrightarrow \quad a_i \wedge \qquad \text{and} \qquad a_i \quad \Leftrightarrow \quad a_i \lrcorner_\delta$$

(4.18)

and note that equation (4.9ii) defines then the commutation relations of such creation and annihilation operations, i.e., a CAR algebra

$$a_i a_j^\dagger \mid \phi\rangle = \langle a_i \mid a_j^\dagger\rangle_\delta \mid \phi\rangle - a_j^\dagger a_i \mid \phi\rangle. \tag{4.19}$$

We have thus shown that all Chevalley deformation formulæ follow from the Grassmann–Hopf gebra generically. Note that our formulæ allow us to compute expressions having arbitrary grade or even being inhomogeneous. This will eventually be explored but see [9, 11].

18.4.4   Left/right cocontractions

Recalling that we have defined cap-tangles from co-scalar products, we can employ dualized product co-product duality to define left and right cocontractions. We write the tensor of the coscalar product as $C_{(1)}^\wedge \otimes C_{(2)}^\wedge$ and define the left cocontraction via:

$$\tag{4.20}$$

The right cocontraction follows from

$$\tag{4.21}$$

In terms of algebraic formulæ we find:

$$\Delta_{\lrcorner_c}(x) = C_{(1)}^\wedge \otimes (C_{(2)}^\wedge \wedge x) \quad \text{and} \quad \Delta_{\llcorner_c}(x) = (x \wedge C_{(1)}^\wedge) \otimes C_{(2)}^\wedge. \tag{4.22}$$

18.4.5   Co-Chevalley formulæ

Having an exterior co-product Δ and a cocontraction we can write down immediately the formulæ of co-Chevalley deformation. Let $x \in V$, $u \in V^\wedge$, we find

$(i)\ \Delta_{\lrcorner_C}(\mathrm{Id}) = C^{\wedge}_{(1)} \otimes C^{\wedge}_{(2)}$

$(i)'\ \Delta_{\lrcorner_C}(u) = C^{\wedge}_{(1)} \otimes C^{\wedge}_{(2)} \wedge u$

$(ii)\ (\mathrm{Id} \otimes \Delta)\Delta_{\lrcorner_C}(u) = C^{\wedge}_{(1)} \otimes \Delta(C^{\wedge}_{(2)} \wedge u)$

$$= (-1)^{\partial u_{(1)}\, \partial C^{\wedge}_{(22)}}\ C^{\wedge}_{(1)} \otimes C^{\wedge}_{(21)} \wedge u_{(1)} \otimes C^{\wedge}_{(22)} \wedge u_{(2)}$$

$(ii)'\ (\mathrm{Id} \otimes \Delta)\Delta_{\lrcorner_C}(x) = C^{\wedge}_{(1)} \otimes C^{\wedge}_{(21)} \otimes C^{\wedge}_{(22)} \wedge x$

$$+ (-1)^{\partial C^{\wedge}_{(22)}}\ C^{\wedge}_{(1)} \otimes C^{\wedge}_{(21)} \wedge x \otimes C^{\wedge}_{(22)}$$

$(iii)\ (\Delta \otimes \mathrm{Id})\Delta_{\lrcorner_C}(u) = C^{\wedge}_{(11)} \otimes C^{\wedge}_{(12)} \otimes C^{\wedge}_{(2)} \wedge u \qquad (4.23)$

where (i) is equivalent to the coscalar product, $(i)'$ is the general left cocontraction, (ii) is the general cocontraction on a co-product and $(ii)'$ is the corresponding co-Leibnitz rule for a one-vector argument, while (iii) dualizes the general formula (4.9iii). These formulæ are new according to our knowledge.

18.5   Deformation and cliffordization

18.5.1   Chevalley deformation, cliffordization

Composing the contraction and the exterior multiplication Chevalley [6] defined the Clifford product, denoted here as $\&c$, as an element γ_x of the endomorphism algebra $\mathrm{End}(V^{\wedge})$. Let $x \in V$ and $u \in V^{\wedge}$, he defined

$$\gamma_x \in \mathrm{End}(V^{\wedge}) \qquad\qquad \gamma : V \otimes V^{\wedge} \to V^{\wedge}$$
$$\gamma_x u := x \lrcorner_B u + x \wedge u = x\,\&c\,u. \qquad (5.1)$$

Having now general expressions for the contraction and the wedge product at hand, it is an easy task to write down a grade free Clifford product. This formula, i.e., the leftmost tangle in (5.2), was obtained by Rota and Stein [25] using Laplace Hopf algebras and has been called 'Rota-sausage' for obvious reasons by Oziewicz [13, 23]. This process is a very general deformation of an algebra and it is not limited to the Clifford case only. Rota and Stein coined the term *cliffordization* but it may also be addressed as a Drinfeld twist in certain circumstances. The cup-tangle in the deformation is a co-(quasi) triangular structure. However, *only* our approach makes it explicit that cliffordization is nothing but the generalized Chevalley deformation and that it is directly composed from left or right contraction and the exterior product. Examine only the middle and the right tangle in the following:

$$(5.2)$$

Note that this product is no longer graded since we find

$$\&c : V^{\wedge^r} \otimes V^{\wedge^s} \to V^{\wedge^{r+s}} \oplus \cdots \oplus V^{\wedge|r-s|} \tag{5.3}$$

but, instead, it obeys only a filtration. This filtration depends on the chosen generators, i.e., the basis. Since the proof that this deformation gives the Clifford product is given in [25] we give only a few examples.

Example 4. *Product of two one-vectors using the middle and the right tangle in (5.2) yields:*

$$a \,\&c\, b = a_{(1)} \wedge (a_{(2)} \,\lrcorner_B b) = a \wedge (Id \,\lrcorner_B b) + Id \wedge (a \,\lrcorner_B b)$$
$$= a \wedge b + a \,\lrcorner_B b = \gamma_a\, b$$
$$a \,\&c\, b = (a \,\llcorner_B b_{(1)}) \wedge b_{(2)} = (a \,\llcorner_B Id) \wedge b + (a \,\llcorner b) \wedge Id$$
$$= a \wedge b + b \,\llcorner_B a = \gamma_a\, b. \tag{5.4}$$

Example 5. *Product of two bivectors using just the middle tangle in (5.2) yields:*

$$(a \wedge b) \,\&c\, (x \wedge y) = (a \wedge b) \wedge (x \wedge y) + a \wedge (b \,\lrcorner_B ((x \wedge y)))$$
$$- b \wedge (a \,\lrcorner_B ((x \wedge y))) + Id \wedge ((a \wedge b) \,\lrcorner_B (x \wedge y))$$
$$= a \wedge (b \wedge (x \wedge y) + b \,\lrcorner_B (x \wedge y))$$
$$+ a \,\lrcorner_B (b \wedge (x \wedge y) + b \,\lrcorner_B (x \wedge y)) - (a \,\lrcorner_B b)(x \wedge y)$$
$$= \gamma_a(\gamma_b(x \wedge y)) - (a \,\lrcorner_B b)(x \wedge y)$$
$$= (\gamma_a \wedge \gamma_b)(x \wedge y) = \gamma_{a \wedge b}(x \wedge y). \tag{5.5}$$

One can prove that [12]:

- The Grassmann–Hopf gebra *unit Id* remains to be the unit, also denoted as Id, of the Clifford product if $B^\wedge$ is exponentially generated.

- The Clifford product is *associative* if and only if $B^\wedge$ is exponentially generated.

- The counit projects products onto the bilinear form

$$\epsilon(u \,\&c\, v) = B^\wedge(u, v) = \langle 0 \mid u \,\&c\, v \mid 0 \rangle.$$

This can be used as *vacuum expectation value* in quantum field theory.

- If $F^\wedge$ is exponentially generated from an antisymmetric bilinear form $F = -F^T$, then the deformed product $\&c = \dot{\wedge}$ is again an *exterior product*.

- The deformation w.r.t such an F encodes the Wick transformation of (fermionic) quantum field theory in Hopf algebraic terms [11].

- Exponentially generated bilinear forms fulfill the axioms of a *co-(quasi) triangular structure*.

- Clifford Hopf gebras are biaugmented but they are neither connected nor coconnected, see [12, 20] for definitions.

These properties result from considering an arbitrary, not exponentially generated, bilinear form $\mathcal{BF}$ on $V^\wedge \otimes V^\wedge$ in the cliffordization

$$a \circ b := \mathcal{BF}(a_{(2)}, b_{(1)}) a_{(1)} \wedge b_{(2)} \,. \tag{5.6}$$

Examining unit, associativity, etc. yields the above claims, see [12].

18.5.2   Co-cliffordization, co-Chevalley deformation

From the Rota sausage tangle (5.2), that is from the tangle definition of the Clifford product, we derive by categorial duality the cocliffordization and the *Clifford co-product* denoted as Δ_c.

$$\tag{5.7}$$

Obviously duality tells us that this co-product is derived from the co-Chevalley deformation also

$$\tag{5.8}$$

and, thus, it depends on the coscalar product in the same manner as the Clifford product depends on the scalar product.

18.5.3 Deformation from cochains

It was shown in [11] that the Wick transformation of normal–ordered operator products into (non–renormalized) time ordered operator products can be given by a cliffordization w.r.t. an antisymmetric scalar product F exponentially generalized to $F^\wedge$. This is important since renormalization can then be introduced using the Epstein–Glaser formalism. There is a hope that this can also be achieved by a product deformation [4, 5]. We will not go into this difficult case, but we will only try to show that the normal ordering transformation is topologically trivial. Therefore we show that the antisymmetric exponentially generated bilinear form $F^\wedge$ can be derived from a 1-cochain. For precise definitions see [19].

An r-*cochain* is defined to be a map $\mathsf{p} : \otimes^r V^\wedge \to \Bbbk$. A cochain may act in a convolution product, defined as $f \star g = m(f \otimes g)\Delta$, like an endomorphism $\mathcal{P} = \mathsf{p} \star \mathrm{Id} = \mathrm{Id} \star \mathsf{p}$ from $V^\wedge \xrightarrow{\mathcal{P}} V^\wedge$ where Δ is the Grassmann co-product and m is the product in $\Bbbk$. Let furthermore ∂ be a (group–like) co-boundary operator, so that $\mathcal{P} = \partial\mathsf{p}$ is a 2-*cocycle* which is a coboundary.

Tailoring a 1-cochain to obtain $F^\wedge$ as a particular 2-cocycle for Wick reordering leads to the following requirements for the special cochain p. Let $\mathsf{p}(\mathrm{Id}) = 1$, $\mathsf{p}(a) = 0$, $\forall a \in V$, $\mathsf{p}(a \wedge b) = \mathsf{p}_{ab} \in \Bbbk$ and expand the cochain via the Laplace like property $\mathsf{p}(u \wedge v \wedge w) = \pm\, \mathsf{p}(u_{(1)}, v)\mathsf{p}(u_{(2)}, w)$ to $V^\wedge$. Then define operators $\mathcal{P}$ and $\mathcal{P}^{-1}(x) : V^\wedge \to V^\wedge$,

$$\mathcal{P}(x) = \mathsf{p}(x_{(1)})x_{(2)}, \qquad \mathcal{P}^{-1}(x) = \mathsf{p}^{-1}(x_{(1)})x_{(2)} \tag{5.9}$$

which are assumed to be commutative under convolution, i.e.,

$$\mathcal{P} = \mathsf{p} \star \mathrm{Id} = \mathrm{Id} \star \mathsf{p}$$
$$\mathcal{P}^{-1} = \mathsf{p}^{-1} \star \mathrm{Id} = \mathrm{Id} \star \mathsf{p}^{-1}. \tag{5.10}$$

The circle product $\circ^{\mathsf{p}}$ is defined as product homomorphic to the exterior wedge product under $\mathcal{P}$.

$$\mathcal{P}(x \circ^{\mathsf{p}} y) = \mathcal{P}(x) \wedge \mathcal{P}(y), \qquad \mathcal{P}^{-1}(x \wedge y) = \mathcal{P}^{-1}(x) \circ^{\mathsf{p}} \mathcal{P}^{-1}(y). \tag{5.11}$$

The 2-cocycle derived from the cochain p is a bilinear form denoted as $\partial\mathsf{P}$. It is formally invertible and reads explicitly as

$$\partial\mathsf{P}(u, v) = \mathsf{p}(u_{(1)})\mathsf{p}(v_{(2)})\mathsf{p}^{-1}(u_{(2)} \wedge v_{(1)}),$$
$$\partial\mathsf{P}^{-1}(u, v) = \mathsf{p}^{-1}(u_{(1)})\mathsf{p}^{-1}(v_{(2)})\mathsf{p}(u_{(2)} \wedge v_{(1)}). \tag{5.12}$$

One needs to show that this bilinear form is (i) antisymmetric, (ii) exponentially generated, and, (iii) that the above homomorphism can be rewritten as a cliffordization w.r.t. this bilinear form. In terms of tangles one has to prove the '=' in the tangle equation

$$
\begin{array}{ccc}
\text{(tangle } \circ^P\text{)} & := & \text{(tangle with } p\ p \text{ and } p^{-1}) & = & \text{(owl tangle with } p\ p,\ p^{-1})
\end{array}
\qquad (5.13)
$$

where the rightmost tangle is called *owl tangle*. This was done in [12]. However, it is well known from deformation quantization that not every deformation can be written as a homomorphism of products. Furthermore, since the bilinear form ∂P is equivalent to a 2-coboundary, we see that both products, wedge and $\circ^P \equiv \dot{\wedge}$, are topologically equivalent. However, the related Hopf algebras are quite different [11]. While the Grassmann–Hopf algebra w.r.t. the wedge is biconnected, the algebra w.r.t. the dotted wedge $\dot{\wedge}$ is not even Hopf in general, but may be a braided Hopf algebra.

18.6   Outlook

For lack of space we will not give a summary but want to recall shortly the main idea and its further eventual development. Indeed the most striking feature of our approach is its complete duality between algebra and cogebra structures. Moreover we might have convinced the reader that cogebra structures are implicitly used e.g., in determinants, combinatorial identities, more explicitly in the Grassmann–Cayley di-algebra having two associative products, one being related to a co-product on the dual space, and, most strikingly, in the Clifford product and the very general procedure of cliffordization. It has become clear during the course of our work that the Grassmann–Hopf algebra is the core and the starting point for a systematic development of almost all algebraic structures and less known costructures. We might remark at this point that it is possible on a formal level to perform the same reasoning starting with the symmetric Hopf algebra and deforming it into Weyl or synonymously symplectic Clifford algebras. We have no time to show that cliffordization is also computationally very efficient but see [1, 2].

The most intriguing questions for further research are, among others, the following:

(i) Can a linear cogebra theory be developed including geometrical meaning without making recourse to the algebra side of the world?

(ii) Is there a set of axioms which directly characterizes Clifford Hopf algebras? There is a steady progress by Oziewicz in this direction, but there is no affirmative answer yet.

(iii) What is the link of this bigebraic mathematics to geometry and physics? We already know that the deformation has to do with quantization and the propagator of quantum field theory [9, 10], but this relation should be further investigated.

(iv) Since we deal with the alternating multivector fields, this structure is very close to string and M-theory; What is the concrete relation?

Lots more questions could be added, but we will insist in a final statement, probably of moral nature. Regarding the present development one *cannot go for the "algebra only" approach* any longer. We hope that this chapter will push forward this idea.

Acknowledgment: I would gratefully thank Prof. Heinz Dehnen and the organizers of the ICCA 6 for financial support.

REFERENCES

[1] Rafał Abłamowicz and Bertfried Fauser. Efficient algorithms for Clifford products using Hopf gebra methods. 2002. in preparation.

[2] Rafał Abłamowicz and Bertfried Fauser. Efficient algorithms for meet and join in Grassmann–Cayley algebras. 2002. in preparation.

[3] Marilena Barnabei, Andrea Brini, and Gian-Carlo Rota. On the exterior calculus of invariant theory. *J. Algebra* **96**, 120–160, 1985.

[4] Christian Brouder. A quantum field algebra. preprint, 2002. math-ph/0201033.

[5] Christian Brouder, Bertfried Fauser, Alessandra Frabetti, and Robert Oeckl. Let's twist again. 2002. preprint.

[6] Claude Chevalley. *The Algebraic Theory of Spinors and Clifford Algebras.* Springer-Verlag, Berlin, 1997. Collected Works Vol. 2, Pierre Cartier, Catherine Chevalley Eds.

[7] Peter Doubilet, Gian-Carlo Rota, and Joel Stein. On the foundations of combinatorial theory: IX Combinatorial methods in invariant theory. *Studies in Applied Mathematics* LIII(3), 185–216, September 1974.

[8] Freeman J. Dyson. Missed opportunities. *Bull. Amer. Math. Soc.* **78**, 635–652, 1974.

[9] Bertfried Fauser. On an easy transition from operator dynamics to generating functionals by Clifford algebras. *J. Math. Phys.* **39** (9), 4928–4947, 1998. hep-th/9710186.

[10] Bertfried Fauser. Clifford geometric parameterization of inequivalent vacua. *Mathematical Methods in the Applied Sciences* **24**, 885–912, 2001. hep-th/9710047v2.

[11] Bertfried Fauser. On the Hopf-algebraic origin of Wick normal–ordering. *J. Phys. A: Mathematical and General* **34**, 105–115, 2001. hep-th/0007032.

[12] Bertfried Fauser. A Treatise on Quantum Clifford Algebras. Konstanz, 2002. Habilitationsschrift, arXiv:math.QA/0202059.

[13] Bertfried Fauser and Rafał Abłamowicz. On the decomposition of Clifford algebras of arbitrary bilinear form. In Rafał Abłamowicz and Bertfried Fauser, editors, *Clifford Algebras and their Applications in Mathematical Physics*, pp. 341–366, Boston, 2000. Birkhäuser. math.QA/9911180.

[14] Hermann Grassmann. *Die Ausdehnungslehre; Vollständig und in strenger Form bearbeitet.* Verlag von Th. Chr. Fr. Enslin (Adolph Enslin), Berlin, 1862.

[15] Hermann Grassmann. *Die lineale Ausdehnungslehre; ein neuer Zweig der Mathematik.* Verlag Otto Wigand, Leipzig, 1878. (reprint of the original 1844 edition, with annotations of Grassmann).

[16] Werner Greub. *Linear Algebra.* Springer-Verlag, Berlin, 1976. Heidelberger Taschenbücher, 179.

[17] David Hestenes and Garret Sobczyk. *Clifford Algebra to Geometric Calculus.* Kluwer Academic Publishers, Dordrecht, 1992. 1st. ed 1984, Reprinted with corrections.

[18] Alfred Lotze. Über eine neue Begründung der regressiven Multiplikation extensiver Grössen in einem Hauptgebiet n-ter Stufe. *Jahresbericht der DMV* **57**, 102–110, 1955.

[19] Shahn Majid. *Foundations of Quantum Group Theory.* Cambridge University Press, Cambridge, 1995.

[20] John W. Milnor and John C. Moore. On the structure of Hopf algebras. *Annals of Mathematics* **81**, 211–264, 1965.

[21] Zbigniew Oziewicz. From Grassmann to Clifford. In J.S.R. Chisholm and A.K. Common, editors, *Clifford Algebras and their Applications in Mathematical Physics*, 245–256, Dordrecht, 1986. Kluwer. Canterbury, 1985.

[22] Zbigniew Oziewicz. Clifford Hopf gebra and biuniversal Hopf gebra. *Czechoslovak Journal of Physics* **47** (12), 1267–1274, 1997. q-alg/9709016.

[23] Zbigniew Oziewicz. Guest editor's note: Clifford algebras and their applications. *International Journal of Theoretical Physics* **40** (1), 1–13, 2001.

[24] Zbigniew Oziewicz and Jose de Jesu de Cruz Guzman. Unital and antipodal biconvolution and Hopf gebra. *Miscellanea Algebraicae*, 2001. in press.

[25] Gian-Carlo Rota and Joel A. Stein. Plethystic Hopf algebras. *Proc. Natl. Acad. Sci. USA* **91**, 13057–13061, December 1994.

[26] Moss E. Sweedler. *Hopf Algebras.* W. A. Benjamin, INC., New York, 1996.

Bertfried Fauser
Universität Konstanz
Fachbereich Physik, Fach M678
D-78457 Konstanz, Germany
E-mail: Bertfried.Fauser@uni-konstanz.de
URL: clifford.physik.uni-konstanz.de/~fauser/

Submitted: August 12, 2002.

19

The Clifford Algebra in the Theory of Algebras, Quadratic Forms, and Classical Groups

Alexander Hahn[1]

ABSTRACT This article is an expanded version of my plenary lecture for the conference. It was the aim of the lecture to introduce the participants of the conference—their diverse realms of expertise ranged from theoretical physics, to computer science, to pure mathematics—to the algebraic matters of the title above. The basic texts listed in the references—note that this listing is by no means complete—serve as illustration of the rich and persistent interest in these topics and provide a reader with an opportunity to explore them in detail.

Keywords: Clifford algebra, quadratic form, classical group, involutions, Clifford modules

19.1 Basic quadratic forms

We will set the stage in very general terms. Let R be a commutative (and associative) ring with 1. A *quadratic form* in n variables over R is a homogeneous polynomial of degree 2

$$p(X_1, \ldots, X_n) = \sum_{1 \leq i, j \leq n} a_{i,j} X_i X_j$$

with coefficients $a_{i,j}$ in R. Two quadratic forms are *equivalent* if there is an invertible linear substitution of variables that transforms the one into the other.

Example 1. *If $1 + 1 = 2$ is an invertible element in R, then $X_1^2 - X_2^2$ and $Y_1 Y_2$ are equivalent. What is the linear change of variables and why does 2 have to be invertible?*

[1] Many thanks go to Rafał Abłamowicz and John Ryan for their wonderful organizational effort in bringing together in stimulating conversation scholars from all disciplines of the multifaceted reality that is the Clifford algebra. I wish to dedicate this article to the memory of one of them: Many of us will miss the dynamic mathematical and personal presence of Pertti Lounesto.
AMS Subject Classification: 11E81, 11E88, 16D70, 20G15.

Let M be a free left R-module with basis $\{x_1, \ldots, x_n\}$ and let p be a quadratic form in n variables. Define

$$q : M \to R$$

by setting $q(r_1 x_1 + \cdots + r_n x_n) = p(r_1, \ldots, r_n)$. With the properties of this example in mind, we now turn to a more general concept of quadratic form.

Let M be any module over R. A quadratic form on M is a map

$$q : M \to R$$

such that $q(rx) = r^2 q(x)$ for all $r \in R$ and $x \in M$, and

$$b : M \times M \to R$$

given by $b(x, y) = q(x+y) - q(x) - q(y)$ is bilinear. The composite $M = (M, q)$ is a *quadratic module* over R. If M is finitely generated projective and q *nonsingular*—this means that the map $M \to M^*$ provided by b is an isomorphism— then M is a *quadratic space*. A quadratic module is *isotropic* if $q(x) = 0$ for some non-zero x in M.

Let S be a commutative ring containing R. The quadratic form

$$p(X_1, \ldots, X_n) = \sum_{1 \le i,j \le n} a_{i,j} X_i X_j$$

over R is also a quadratic form over S. In terms of quadratic modules this is captured by going from (M, q) to the tensor product $(M \otimes_R S, q_S)$.

Two quadratic modules (M, q) and (M', q') over R are *equivalent* or *isometric* if there is an invertible linear map $\varphi : M \to M'$ such that $q'(\varphi x) = q(x)$ for all x in M. If this is so, we will write $M \cong M'$. The notation

$$M \cong \sum_{1 \le i,j \le n} a_{i,j} X_i X_j$$

means that the quadratic module M is isometric to a quadratic module constructed from the quadratic form $\sum_{1 \le i,j \le n} a_{i,j} X_i X_j$ in the way described above. Check that if $M \cong X_1 X_2$, then M is a two-dimensional isotropic quadratic space. The following "diagonalization theorem" is a very basic result in the theory of quadratic forms.

Theorem 1. *If R is a field of characteristic not 2 and M is a finite dimensional quadratic module over R, then*

$$M \cong a_1 X_1^2 + a_2 X_2^2 + \cdots + a_n X_n^2,$$

where $n = \dim M$ and $a_i \in R$.

For an introduction to the basic concepts and facts discussed above, see [6, Chapter IV], [7, Chapter I], [8, Chapter IV], [9, Chapter 1], [14, Chapter 1], or [16, Chapter I].

19.2   Basic Clifford algebras

Let $M = (M, q)$ with associated bilinear form b be a quadratic module over R. A *Clifford algebra* of M is a pair $C(M) = (C(M), \gamma)$ such that

(a) $C(M)$ is an R-algebra and

(b) $\gamma : M \to C(M)$ is an R-module map satisfying $\gamma(x)^2 = q(x)1_C$ and $\gamma(x)\gamma(y) + \gamma(y)\gamma(x) = b(x, y)1_C$ for all x and y in M, and

(c) $(C(M), \gamma)$ is "minimal" (in the sense of a universal mapping property) with respect to (a) and (b).

The following very basic facts introduce concepts that will be relevant in the discussions that follow.

Theorem 2. *Any (M, q) has a unique Clifford algebra. If M is finitely generated projective, then γ is injective. There is a unique anti-automorphism on $C(M)$ taking γx to γx for all x. This anti-automorphism is an involution, as two successive applications of it give the identity map on $C(M)$.*

Theorem 3. *If $\{x_i\}_{i \in I}$, where I is an index set, spans M as R-module, then 1_C together with all the elements of the form*

$$\gamma x_{i_1} \gamma x_{i_2} \cdots \gamma x_{i_k}, \quad i_1 < \cdots < i_k$$

span $C(M)$ as R-module.

Taking the submodules of $C(M)$ spanned by 1_C and all the elements above with k even, and then in turn the submodule spanned by all the elements above with k odd, splits $C(M)$ into an even and odd part $C(M) = C_0(M) \oplus C_1(M)$, and provides $C(M)$ with a $\mathbb{Z}_2$-grading, in other words, a two-component grading.

Note in particular that $C_0(M)$ is a subalgebra of $C(M)$ and $C_1(M)$ is a $C_0(M)$-module.

Theorem 4. *If M is free with finite basis $\{x_1, \ldots, x_n\}$, then $C(M)$ is a free R-module with basis*

$$\{1_C, \gamma x_{i_1} \gamma x_{i_2} \cdots \gamma x_{i_k}\},$$

where $1 \le k \le n$ and $i_1 < \cdots < i_k$. So $C(M)$ is a free R-module of dimension 2^n. It follows that $C_0(M)$ and $C_1(M)$ are free R-modules as well and that both have dimension 2^{n-1}.

Example 2. *If $q = 0$, then $C(M)$ is an exterior algebra. If R is a field, $\dim M = 2$, and q is non-singular, then $C(M)$ is either the matrix algebra $\mathrm{Mat}_2(R)$ or a quaternion division algebra over R.*

The *discriminant algebra* $A(M)$ is the centralizer of $C_0(M)$ in $C(M)$. More precisely,

$$A(M) = \{c \in C(M) \mid cd = dc, \forall\, d \in C_0(M)\}.$$

The next theorem illustrates the important role that $A(M)$ plays in the structure theory of $C(M)$ and $C_0(M)$.

Theorem 5. *Let R be a field and M a quadratic space over R. Then*

(1) $A = A(M) \cong R[X]/(X^2 - aX - b)$ *with* $a^2 + 4b \neq 0$.

(2) *Suppose* $\dim_R M$ *is even. Then there is a division algebra D over R such that*
$$C(M) \cong \mathrm{Mat}_k(D)$$
with $\dim_R D$ *and* k *both powers of 2. So* $k = 2m$ *for some* m.

(2a) *Suppose* $X^2 - aX - b$ *has a root in R. Then* $A \cong R \oplus R$, *and*
$$C_0(M) \cong \mathrm{Mat}_m(D) \oplus \mathrm{Mat}_m(D).$$

(2b) *Suppose* $X^2 - aX - b$ *does not have a root in R, but does have a root in D. Then A is a subfield of D, the centralizer D_0 of the root in D is a central division algebra over A, and*
$$C_0(M) \cong \mathrm{Mat}_m(D_0).$$

(2c) *Suppose* $X^2 - aX - b$ *does not have a root in D. Then $D \otimes_R A$ is a central division algebra over A and*
$$C_0(M) \cong \mathrm{Mat}_m(D \otimes_R A).$$

(3) *Suppose* $\dim_R M$ *is odd. Then there is a division algebra D over R such that $C_0(M) \cong \mathrm{Mat}_k(D)$, with $\dim_R D$ and k both powers of 2. Statements analogous to those for $C_0(M)$ above hold for the algebra $C(M)$.*

Corollary 1. *If n is even, then $C(M)$ is a central simple algebra over R. If n is odd, then $C_0(M)$ is a central simple algebra over R.*

See [6, Chapter V], [9, Chapter 5], [14, Chapter 9], [16, Chapter IV], [18, Chapters 5, 9, and 11], and [22, Chapter 3] for the details.

19.3   CSAs with involution

We have seen that the existence of an involution and—over a field—the property of being central and simple are basic features of the Clifford algebra. Could it possibly be that any finite dimensional central simple algebra that comes equipped with an involution is isomorphic to a Clifford algebra? If the requirement "isomorphic" is replaced by the weaker "is similar to," that is to say, "is Brauer equivalent to," then the answer is yes!

Theorem A. *Let R be a field and A a finite-dimensional central simple algebra over R with an involution. Then there is a quadratic space M of even rank (and discriminant 1) such that A is similar to $C(M)$.*

At the core of the proof lies a deep theorem of Merkurjev asserting that the map from a certain "K-group" to the 2 part of the Brauer group is surjective.

If R is specialized to be a global field, then Theorem A can be sharpened. Recall that a *global field* is a finite extension field of either the rational numbers or of the quotient field of a polynomial ring in one variable over a finite field. A global field is an *algebraic number field* in the first case and an *algebraic function field* in the second.

Theorem B. *Let R be a global field and let A be a finite-dimensional central simple algebra over R with an involution. Then there is a quadratic space M of dimension 2 such that A is similar to $C(M)$.*

This follows from the fact that the only division algebras with involution over such an R are the quaternion division algebras (and R itself). This fact also provides the stronger isomorphism result for global fields.

Theorem C. *Let R be a global field and let A be a finite-dimensional central simple algebra over R with an involution. Then there is a quadratic space M of even dimension such that A is isomorphic to $C(M)$.*

The question arises as to whether these field theoretic results have ring theoretic analogues, once "finitely generated (as R-module), central separable" is inserted in place of "finite-dimensional central simple." In the situation of Theorem B, there is a theorem that says "yes," if R is an *arithmetic Dedekind domain*. There are a number of equivalent definitions of Dedekind domain. According to one of them, an integral domain R is Dedekind if every ideal of R is the product of prime ideals. A Dedekind domain is *arithmetic* if its field of fractions is a global field.

Theorem B'. *Let R be an arithmetic Dedekind domain and let A be any central separable algebra over R with an involution. Then there is a quadratic space M over R of rank 4 (and discriminant 1) such that A is similar to $C(M)$.*

Do there exist ring theoretic analogues of Theorems A and C? The article R. Parimala and R. Sridharan, Non-surjectivity of the Clifford invariant map, *Proc. Indian Acad. Sci.* **104** (1994), 49–56, shows that if R is the affine algebra of the smooth projective hyperelliptic curve over the 3-adic numbers defined by

$$y^2 = (t^2 - 3)(t^4 + t^3 + t^2 + t + 1).$$

then there are central separable algebras over R (finitely generated as R-modules) that are *not* similar to the Clifford algebra of any quadratic space. (The concepts just used can be found in R. Hartshorne, *Algebraic Geometry*, Springer-Verlag, Berlin Heidelberg New York, 1977). In particular, there appears to be no obvious analogue for Theorem A. My guess is that this is also true for Theorem C. I have given the matter no thought, but there ought to be versions of Theorems A–C that take the graded structure of the Clifford algebra into account. A graded Brauer group—see [9, Chapter 4] and [16, Chapter III (§6)]—is available for the purpose.

See [11, Chapters 7 and 8], [14, Chapter 2 (§14)] or [15, Sections 1.5 and 2.3] for introductions to the theorem of Merkurjev and [18, Chapters 11 and 14] for the classification results described above. Refer to [14, Chapter 8], [21, Chapter I], and [22, Chapter 6] for related concerns.

19.4   Quadratic forms over fields

We all know that a large part of the agenda of mathematics is the matter of classification. We just saw an example of this: Clifford algebras were identified as structures with a number of interesting properties, and the question was as to whether any structure with these properties is equivalent or similar to a Clifford algebra. This section discusses this question for quadratic forms with coefficients in a field. Here too, the Clifford algebra has a role to play.

Let F be a field of characteristic not 2 and let M (with q and b) be a quadratic space over F. Since $b(x, x) = 2q(x)$, b and q determine each other, so that the theories of quadratic and symmetric bilinear forms are the same. We call M *hyperbolic* if dim M is even and there is a basis $\{x_1, \ldots, x_{2k}\}$ of M such that there is the equality

$$(b(x_i, x_j)) = \begin{bmatrix} O & I \\ I & O \end{bmatrix}$$

of $2k \times 2k$ matrices, where O and I are the $k \times k$ zero and identity matrices respectively. Notice that hyperbolic spaces have lots of isotropy (indeed, they have maximal isotropy). Two quadratic spaces M and M' are *Witt equivalent* if there are hyperbolic spaces H and H' such that

$$M \perp H \cong M' \perp H'.$$

The equivalence classes form an additive monoid. This monoid can be expanded to a *Grothendieck group* (this is like going from the additive set of non-zero positive integers to the group $\mathbb{Z}$). With the tensor product of modules this additive group becomes the *Witt ring* $W(F)$ of F. The equivalence classes of even–dimensional spaces determine the *fundamental ideal* $I(F)$ of $W(F)$.

Theorem 6. *Let M and M' be quadratic spaces. Then $M \cong M'$ if and only if* dim $M =$ dim M' *and* $[M] = [M']$ *in* $W(F)$.

Consider the special cases $F = \mathbb{C}$ and $F = \mathbb{R}$. The diagonalization theorem of Section 19.1 plus the fact that everything in $\mathbb{C}$ is a square, tells us that the dimension alone suffices to characterize complex quadratic spaces. By similar considerations, any quadratic space M in the real case satisfies

$$M \cong X_1^2 + \cdots + X_s^2 - X_{s+1}^2 - \cdots - X_{s+r}^2,$$

where $s + r = \dim M$. The fact that s and r are uniquely determined by M (this is a theorem of Sylvester), provides the *signature* $s(M) = s - r \in \mathbb{Z}$ of M. Observe that the dimension together with the signature characterize real quadratic spaces. The assignment $\mathrm{sig}[M] \mapsto s(M)$ defines an isomorphism

$$\mathrm{sig} : W(\mathbb{R}) \to \mathbb{Z}.$$

Return to an arbitrary F. In addition to the surjective group homomorphism

$$\dim : W(F) \to \mathbb{Z}_2$$

given by dimension mod 2, there are the surjective group homomorphisms

$$\mathrm{disc} : I(F) \to Qu(F),$$

where $Qu(F)$ is a group of equivalence classes of quadratic algebras and $\mathrm{disc}[M] = [A(M)]$ with $A(M)$ the discriminant algebra of M, and

$$\mathrm{Cliff} : I^2(F) \to \mathrm{Br}_2(F),$$

where $\mathrm{Br}_2(F)$ are the elements of order 2 (and 1) in the Brauer group of F and $\mathrm{Cliff}[M] = [C(M)]$. The kernels of these three maps are, respectively, the fundamental ideal $I(F)$ and its powers $I^2(F)$ and $I^3(F)$. There are also graded analogues of these homomorphisms defined on all of $W(F)$.

If F is a local field (one that is equipped with a discrete and complete valuation with respect to which the field is locally compact), then dim, disc, and Cliff suffice to classify quadratic spaces. The same is true if F is an algebraic function field. However, if F is an algebraic number field, then one more invariant is needed.

Let M be a quadratic space over an algebraic number field F. Suppose that F has k real embeddings. For each of these, there is a signature via $M \otimes_F \mathbb{R}$. Putting all these signatures together, we get the *total signature*

$$\mathrm{sig} : W(F) \to \mathbb{Z}^k.$$

The fact is that dim, disc, Cliff, and sig classify quadratic spaces over F.

Example 3. *The quadratic spaces $M_1 \cong -X_1^2 + 3X_2^2 + 5X_3^2$ and $M_2 \cong X_1^2 + 7X_2^2 - 105X_3^2$ over $\mathbb{Q}$ are isometric, but the spaces $M_3 \cong X_1^2 - 6X_2^2 + 15X_3^2$ and $M_4 \cong 3X_1^2 - 10X_2^2 + 3X_3^2$ over $\mathbb{Q}$ are not isometric.*

All four quadratic spaces of the example are three–dimensional with signature 1. In view of the earlier discussion, they are all isometric over $\mathbb{R}$. Refer to [6, Chapter VI], [9, Chapters 2, 4, and 6] and [14, Chapters 2, 5, and 6] for the details and much more.

The suggestion that the three surjective homomorphisms described above might be the beginning of an infinite pattern of invariants that would classify quadratic forms over any field was made by Milnor in 1970. The proof of Milnor's conjecture became a holy grail of quadratic form theory. More precisely, the question was whether for any field F of characteristic not 2, there exist for each n, a suitable Galois cohomology group $H^n(F, \mathbb{Z}_2)$ and a surjective homomorphism

$$e_n : I^n(F) \to H^n(F, \mathbb{Z}_2)$$

with kernel $I^{n+1}(F)$. If the answer were yes, then these invariants would classify quadratic spaces over F as follows: For two quadratic spaces M_1 and M_2, let $[M_1] - [M_2] = [M] \in W(F)$. If $e_n[M] = 0$, then $[M]$ is in $I^{n+1}(F)$, so if all invariants of the difference $[M]$ are zero, then $[M] = 0$, and $[M_1]$ and $[M_2]$ are equal in $W(F)$.

Voevodsky announced a proof of Milnor's conjecture a few years ago and he provided complete details in the summer of 2001. For this impressive accomplishment, he was awarded the Fields Medal at the International Congress of Mathematicians in Beijing in 2002. For a glimpse of how Voevodsky did it and for the broader implications of what he did, refer to the surveys A. Pfister, On the Milnor Conjectures: History, Influence, and Applications, *Jahresbericht d. Dt. Math.-Verein.* **102** (2000), 15–41, and W. Scharlau, On the History of the Algebraic Theory of Quadratic Forms, *Contemporary Mathematics* **272** (2000), 229–259.

19.5   The spin group

We turn next to a connection between the Clifford algebra and an interesting slice of group theory. We continue to let F be a field of characteristic not 2 and M a quadratic space over F.

Recall that $\gamma : M \to C(M)$ is injective and that there is a unique involution $^{-}$ on $C(M)$ taking γx to γx for all x. Consider M to be a subset of $C(M)$ via γ, and define the group $\mathrm{Spin}(M)$ by

$$\mathrm{Spin}(M) = \{c \in C_0(M)^\times \mid cMc^{-1} = M, \, c\bar{c} = 1_C\},$$

where $C_0(M)^\times$ is the group of invertible elements of the ring $C_0(M)$. The isometries from M onto M constitute the orthogonal group $O(M)$ and $SO(M)$ is the subgroup of elements of determinant 1. For c in $\mathrm{Spin}(M)$, define

$$\pi c : M \to M$$

by $\pi c(x) = cxc^{-1}$. This provides a homomorphism

$$\pi : \mathrm{Spin}(M) \to SO(M).$$

By a theorem of Cartan and Dieudonné, any element σ in $O(M)$ is a product $\sigma = \tau_{y_1} \cdots \tau_{y_k}$ of hyperplane reflections τ_{y_i}. The assignment

$$\Theta(\sigma) = q(y_1) \cdots q(y_k)(F^\times)^2$$

defines the *spinor norm* homomorphism

$$\Theta : SO(M) \to F^\times/(F^\times)^2.$$

Theorem 7. *The sequence*

$$1 \to \{\pm 1_C\} \hookrightarrow \mathrm{Spin}(M) \xrightarrow{\pi} SO(M) \xrightarrow{\Theta} F^\times/(F^\times)^2$$

is exact. So π induces an isomorphism

$$\mathrm{Spin}(M)/\{\pm 1_C\} \cong O'(M)$$

where $O'(M) = \ker \Theta$.

Let $\Omega(M)$ be the commutator subgroup of $O(M)$. If dim $M \geq 5$, then

$$\Omega(M)/\mathrm{Center}\,\Omega(M)$$

is a simple group if F is a finite field, a local, or a global field, and for any field F if M is isotropic. It turns out that in all the situations just singled out, $\Omega(M) = O'(M)$. In particular,

$$\mathrm{Spin}(M)/\{\pm 1_C\} \cong \Omega(M),$$

Spin(M) is a central extension of $\Omega(M)$, and the quotient

$$SO(M)/\Omega(M)$$

is an elementary Abelian 2-group. Our discussion has connected the Clifford algebra to some interesting group theory for the orthogonal group: generators and relations and questions of length, analysis of central extensions, characterizations of the normal subgroups, as well as the theory of finite simple groups (there are also generalizations to some commutative rings).

The Spin groups have an impact on another issue that has received much attention over the years. Let M be a finitely generated projective module over a ring R. Suppose that M is equipped with a non-singular quadratic, alternating, or hermitian form, or the form that is identically zero. The group $I(M)$ of isometries of M is an orthogonal, symplectic, or unitary group, or a general linear group (in the case of the zero form). These groups, certain of their subgroups, and their quotients by scalar transformations are the *classical groups*.

Consider two such modules M_1 and M_2 over the rings R_1 and R_2 and let $I(M_1)$ and $I(M_2)$ be the respective groups of isometries. A *semi-isometry* α from M_1 to M_2 is an invertible map that is linear up to a ring isomorphism from R_1 to R_2 and that preserves the forms up to multiplication by an invertible ring element. Such a map induces a group isomorphism

$$\Phi_\alpha : I(M_1) \longrightarrow I(M_2)$$

in a natural way. This construction raises the following question. Pick two classical groups at "random" but suppose that they are isomorphic. Could it be that the isomorphism comes (via restriction and/or factoring scalars) from an isomorphism Φ_α as just defined. In particular, are the underlying modules semi-isometric? The short and generic answer is "yes."

Theorem 8. *Let G_1 and G_2 be two isomorphic classical groups over integral domains (possibly non-commutative). If the dimensions of the underlying spaces are large enough (say at least 6, and not both 8 in the case of two orthogonal groups), isotropic enough (say at least three independent orthogonal isotropic vectors), then the underlying modules are semi-isometric and the isomorphism is induced by a semi-isometry.*

For instance, a unitary group cannot be isomorphic to a linear, symplectic, or orthogonal group, and it can be isomorphic to another unitary group only if the underlying (hermitian) geometries are semi-isometric. The case of two 8-dimensional orthogonal groups needs to be excluded due to the existence of the exceptional triality isomorphisms.

What about the low–dimensional barrier in the theorem? There are exceptions that come from Spin groups! Suppose dim $M = 6$. If M is hyperbolic, then

$$\text{Spin}(M) \cong SL_4(F)$$

and if M is "almost" hyperbolic (there exist two independent orthogonal isotropic vectors rather than three), then

$$\text{Spin}(M) \cong SU_4(F).$$

If dim $M = 5$ and M is almost hyperbolic, then

$$\text{Spin}(M) \cong Sp_4(F).$$

By factoring centers we get isomorphic classical groups in these low–dimensional situations that have different underlying geometries. So these examples go "counter" to the theorem.

See [4, Chapters II and IV], [6, Chapters IV, V, IX (§95), and X (§101)], [15], [16, Chapter V], and [21, Chapter VI (§27)] for lots of information about the concerns of this section. The recent article A. Hahn, The Zassenhaus Decomposition for the Orthogonal Groups: Properties and Applications, *Documenta Mathematica* (2001), 165–181, accessible at

http://www.mathematik.uni-bielefeld.de/documenta/

also falls into this realm.

19.6   Quadratic forms over arithmetic domains

This section provides an overview of that part of the arithmetic theory of quadratic forms that attempts to come to grips with the classification of quadratic forms over arithmetic Dedekind domains. The Clifford algebra is connected to this enterprise via the spinor norm.

Let R be a Dedekind domain with global field of fractions F. We let $M = (M, q)$ and $M' = (M', q')$ be quadratic spaces over R and turn to the fundamental question: When is $(M, q) \cong (M', q')$? Suppose that $M \cong M'$. Since they are equivalent over R, they are equivalent after any extension of scalars. In particular,

$$(M \otimes_R F, q_F) \cong (M' \otimes_R F, q'_F).$$

This allows for the following reformulation of our question: Assume that both M and M' sit inside and span the same quadratic space V over F. Does there exist a $\sigma \in O(V)$ such that $\sigma M = M'$? Defining

$$\text{class } M = \{\sigma M \mid \sigma \in O(V)\},$$

the original question becomes: When is M' in the class of M?

Let $\mathfrak{p}$ be a non-trivial valuation on F. Let $F_\mathfrak{p}$ be the completion of F at $\mathfrak{p}$. There are three possibilities:

$$F_\mathfrak{p} \cong \mathbb{R}, \quad F_\mathfrak{p} \cong \mathbb{C}, \quad \text{or} \quad F_\mathfrak{p} \text{ is a local field (with } \mathfrak{p} \text{ discrete)}.$$

For $\mathfrak{p}$ discrete, let $R_\mathfrak{p}$ be the ring of integers in $F_\mathfrak{p}$. The fact is that there is a set S of discrete valuations such that

$$R = \bigcap_{\mathfrak{p} \in S} R_\mathfrak{p}.$$

Our Dedekind domain is a *Hasse domain* if every valuation of F—except for a finite number of them—is equivalent to one of the valuations in S. This is what we will now assume. If F is an algebraic number field and R its ring of algebraic integers, then R is a Hasse domain. The integers $\mathbb{Z}$ in $\mathbb{Q}$ and the Gaussian integers $\mathbb{Z}[i]$ in $\mathbb{Q}(i)$ are standard examples.

Consider for any $\mathfrak{p}$ the extensions of scalars, or "localizations,"

$$M_\mathfrak{p} = M \otimes_R R_\mathfrak{p}, \quad M'_\mathfrak{p} = M' \otimes_R R_\mathfrak{p}, \quad \text{and} \quad V_\mathfrak{p} = V \otimes_F F_\mathfrak{p}.$$

We say that the spaces M and M' are in the same *genus* if for any $\mathfrak{p} \in S$ there is a $\sigma_\mathfrak{p} \in O(V_\mathfrak{p})$ so that $\sigma_\mathfrak{p}(M_\mathfrak{p}) = M'_\mathfrak{p}$. This concept defines the set genus M. Notice that

$$\text{class } M \subseteq \text{genus } M.$$

There are invariants that let us decide whether M' is in genus M. But the question of when M' is in class M is much more subtle. There is an "intermediate" concept that splits it into two parts. We say that M' is in the *spinor genus* of M if there exist a $\sigma \in O(V)$ and for every $\mathfrak{p} \in S$ an element $\Sigma_\mathfrak{p} \in O'_\mathfrak{p}(V)$, such that

$$M'_\mathfrak{p} = \sigma_\mathfrak{p} \Sigma_\mathfrak{p} M_\mathfrak{p}.$$

The corresponding equivalence class gives rise to the set spingenus M. Notice that

$$\text{class } M \subseteq \text{spingenus } M \subseteq \text{genus } M.$$

The set spingenus M is partitioned into classes and genus M is partitioned into both classes and spinor genera.

The basic facts are these: the number of classes in any genus M is finite, and the number of spinor genera in any genus M is a power of 2. By varying M within V, any 2-power can be obtained. If the completion of V is isotropic for at least one $\mathfrak{p} \in S$—this is the so-called *indefinite* case—then spingenus M = class M for any M. Among the results that additional analysis provides is the following

Theorem 9. *Suppose M is a quadratic space over $\mathbb{Z}$ of rank n. Assume that $M \otimes_{\mathbb{Z}} \mathbb{R}$ is isotropic (this is the indefinite condition), $b(M, M) = \mathbb{Z}$, and $\mathrm{disc} M = \pm 1$. Then*

$$M \cong a_1 X_1^2 + a_2 X_2^2 + \cdots + a_n X_n^2$$

with all $a_i = \pm 1$.

Refer for these matters and much more to [1, Chapters IV, V, and VI], [6, Chapters VIII, IX, and X], [12], [17, Chapters 5, 6, and 7], and [20], as well as to the recent survey, J. Hsia, Arithmetic of Indefinite Quadratic Forms, *Contemp. Math.* Vol. **249** (1999), 1–15.

19.7   Generalized even Clifford algebras

There is a generalization of the Clifford algebra—more precisely of the even subalgebra—that plays an important role in the analysis of involutions on central simple algebras as well as the classification of algebraic groups over arbitrary fields.

Let F be a field, let A be a finite-dimensional central simple algebra over F, and let σ be an involution on A. The restriction of σ to F is an involution on F. If this involution is the identity on F, then σ is of the *first kind* and if not, σ is of the *second kind*. This section will concentrate exclusively on involutions of the first kind. A large class of examples of such involutions can be constructed as follows. Let V be a finite-dimensional vector space over F and consider the central simple algebra $\mathrm{End}_F(V)$. Take a non-singular bilinear from b on V (as given by an appropriate invertible matrix for instance) that is either symmetric or alternating (i.e., symmetric up to a $-$ sign). For any $t \in \mathrm{End}_F(V)$ there is a unique $t' \in \mathrm{End}_F(V)$ such that $b(x, ty) = b(t'x, y)$ for all $x, y \in V$. The assignment $t \mapsto t'$ defines an involution of the first kind on $\mathrm{End}_F(V)$. This is the *adjoint involution* corresponding to b. The fact is that every involution of the first kind on $\mathrm{End}_F(V)$ arises in this way.

Let us return to our general central simple algebra A with its involution σ of the first kind. It is a classical fact that there is a finite separable extension K of F such that

$$A \otimes_F K \cong \mathrm{Mat}_k(K).$$

Because $\sigma \otimes \mathrm{id}_K$ defines an involution of the first kind on $\mathrm{Mat}_k(K)$, it arises from a non-singular quadratic or alternating form. The involution σ is *orthogonal* or *symplectic* accordingly. Incidentally, the isomorphism above also supplies the *reduced trace* $\text{Trd}_A(a)$ for any $a \in A$ as the matrix trace of the image of $a \otimes 1$.

The definition of the generalized even Clifford algebra has as starting point the concept of a quadratic pair. Recall first that the dimension of A over F has the form m^2 for a positive integer m known as the *degree* of A over F. Consider the sets

$$\text{Sym}(A, \sigma) = \{a \in A \mid \sigma a = a\} \quad \text{and} \quad \text{Skew}(A, \sigma) = \{a \in A \mid \sigma a = -a\}$$

and observe that $\text{Sym}(A, \sigma) \supseteq F$. A *quadratic pair* (σ, f) on A consists of an involution σ on A and a linear map $f : \text{Sym}(A, \sigma) \to F$ such that

(i) $\text{Sym}(A, \sigma)$ has dimension $\frac{1}{2}m(m + 1)$ over F,

(ii) $\text{Trd}_A(\text{Skew}(A, \sigma)) = \{0\}$, and

(iii) f is F-linear and $f(a + \sigma a) = \text{Trd}_A(a)$ for all $a \in A$.

A quadratic space (V, q) over F gives rise to a quadratic pair as follows. Let b be the associated non-singular bilinear form of V. Let σ_b be the corresponding adjoint involution on the algebra $\text{End}_F(V)$. Fix $v \in V$ and note that the element in $\text{End}_F(V)$ defined by $x \mapsto vb(v, x)$ is fixed by σ_b. There is a unique linear map

$$f_q : \text{Sym}(\text{End}_F(V), \sigma_b) \longrightarrow F$$

such that $f_q(x \mapsto vb(v, x)) = q(v)$. It turns out that (σ_b, f_b) is a quadratic pair on the central simple algebra $\text{End}_F(V)$.

Return to an arbitrary quadratic pair (σ, f) on a central simple algebra A. If char $F \neq 2$, then the defining conditions imply that σ is orthogonal and that f is determined by $f(s) = \frac{1}{2}\text{Trd}_A(s)$. If char $F = 2$, then σ is symplectic and m is even. In this case, there are a number of quadratic pairs with the same σ. In any characteristic there exists an element $l \in A$ such that $f(s) = \text{Trd}_A(ls)$ for all $s \in \text{Sym}(A, \sigma)$.

Focus next on the F-vector space structure of A. Start with the tensor algebra $T(A)$ of A. Let I be the ideal of $T(A)$ generated by the elements of the form $s - f(s)1$ with $s \in \text{Sym}(A, \sigma)$. Consider the isomorphism

$$A \otimes_F A \xrightarrow{\varphi} \text{End}_F(A)$$

that satisfies $\varphi(a \otimes b)x = axb$ for all a, b, and x in A. Let J be the ideal of $T(A)$ generated by all $u - (\varphi u)1$ for all $u \in A \otimes_F A$ such that $(\varphi u)(x) = (\varphi u)(\sigma x)$ for all $x \in A$. The *Clifford algebra* $C(A, \sigma, f)$ is the F-algebra defined by

$$C(A, \sigma, f) = \frac{T(A)}{I + J}.$$

The involution on $T(A)$ defined by $a_1 \otimes \cdots \otimes a_r \mapsto \sigma a_r \otimes \cdots \otimes \sigma a_1$ induces an involution $\underline{\sigma}$ on $C(A, \sigma, f)$.

Let (V, q) be a quadratic space over F and consider the quadratic pair (σ_b, f_q) on $\operatorname{End}_F(V)$ that it defines. The corresponding Clifford algebra

$$C((\operatorname{End}_F(V), \sigma_b, f_q)$$

turns out to be isomorphic to the even Clifford algebra $C_0(V)$ as follows. The standard identification $V \otimes V \to \operatorname{End}_F(V)$ given by $(v, w) \mapsto (x \mapsto vb(w, x))$ defines an isomorphism $T(V \otimes V) \cong T(\operatorname{End}_F(V))$ which in turn induces an isomorphism

$$C_0(V) \to C((\operatorname{End}_F(V), \sigma_b, f_q)$$

that satisfies

$$vw \mapsto (x \mapsto vb(w, x)) + (I + J).$$

The involution $\underline{\sigma_b}$ corresponds to the restriction to $C_0(V)$ of the involution defined in Section 19.2.

To define the generalized Spin group, start with the submodule $T_+(A)$ of $T(A)$ given by

$$T_+(A) = \bigoplus_{k \geq 1} A^{\otimes k}.$$

The multiplication of $T(A)$ provides $T_+(A)$ with the structure of a right $T(A)$-module. Letting $T(A)$ act on $T_+(A)$ on the left by a permuted version $*$ of this multiplication, gives $T_+(A)$ the structure of a $T(A)$-bimodule. By going to the induced structure, the quotient

$$B(A, \sigma, f) = \frac{T_+(A)}{[I * T_+(A)] + [T_+(A) \cdot I]}$$

becomes a $C(A, \sigma, f)$-bimodule. In the example just discussed, $B(A, \sigma, f)$ corresponds to the $C_0(V)$-bimodule $V \otimes C_1(V)$ where both actions are given by the multiplication of $C(V)$. From $a \mapsto a \in T_+(A)$, we get an injective F-linear map $A \to B(A, \sigma, f)$. The image of A is written A^b. The Spin group of (A, σ, f) is defined by

$$\operatorname{Spin}(A, \sigma, f) = \{c \in C(A, \sigma, f)^\times \mid c^{-1} * A^b \cdot c \subseteq A^b \text{ and } \underline{\sigma} c = c^{-1}\}.$$

Theorem 10. *Suppose that G is a simply connected semisimple group of type D_n over F. If $n \neq 4$, then G is isomorphic—as group scheme over F—to $\operatorname{Spin}(A, \sigma, f)$ for some central simple F-algebra A of degree $2n$ with quadratic pair (σ, f).*

When $n = 4$ the groups $\operatorname{Spin}(A, \sigma, f)$ are still simply connected semisimple groups of type D_n, but the additional symmetry (triality) of the Dynkin diagram gives rise to more such groups. Refer to [21, Chapters I (§5), II (§8, 9), III (§13), and VI] for the complete exposition of the theory outlined above. The study of these generalized even Clifford algebras goes back to N. Jacobson, Clifford algebras for algebras with involution of type D, *J. Algebra*, **1** (1964), 288–300, and J. Tits, Formes quadratiques, groupes orthogonaux et algèbres de Clifford, *Inventiones Math.* **5** (1968), 19–41.

Two additional developments deserve attention. The first is related to the theme of Section 19.3.

Theorem 11. *Let F be a field of characteristic not 2. Suppose that the ideal $I(F)^3$ of the Witt ring of F is zero. Local fields and algebraic number fields with no real embeddings satisfy this condition. Let A be a central simple algebra over F. Then all symplectic involutions on A are isomorphic, and two orthogonal involutions on A are isomorphic if and only if they have isomorphic Clifford algebras.*

This theorem is proved in D. Lewis and J.-P. Tignol, Classification theorems for central simple algebras with involution, *Manuscripta Math.* **100** (1999), 259–276.

A different generalization of the Clifford algebra over a field F of characteristic not 2 is developed in K. Hannabuss, Bilinear forms, Clifford algebras, q-commutation relations, and quantum groups, *J. Algebra* **228** (2000), 227–256.

It is defined for any vector space V over F equipped with a non-degenerate bilinear form B. If B is symmetric, it specializes to the classical Clifford algebra of Section 19.2. Like the classical Clifford algebra, this generalized Clifford algebra is a quotient of the tensor algebra of V. This generalization is related to the quantized function algebras studied in Y. Manin, *Topics in non-commutative Geometry*, Princeton, 1991.

19.8   Clifford modules and vector fields on spheres

Clifford modules, especially over the real numbers $\mathbb{R}$, are a point of departure for some important interconnections and applications. One of the most spectacular of these is Adams's solution of the vector field problem for the sphere. We will briefly recall the essence of his achievement on the occasion of its 40th anniversary.

Let V be a quadratic space of dimension n over $\mathbb{R}$. Recall from Section 19.4 that

$$V \cong X_1^2 + X_2^2 + \cdots + X_s^2 - X_{s+1}^2 \cdots - X_{s+r}^2,$$

where $s + r = n$. Because the signature is well defined, s and r are uniquely determined by V. It has become custom to denote the Clifford algebra $C(V)$ by $C\ell_{r,s}$. An $\mathbb{R}$-*representation* of $C\ell_{r,s}$ is a homomorphism of $\mathbb{R}$-algebras

$$\rho : C\ell_{r,s} \to \mathrm{End}_{\mathbb{R}}(W),$$

where W is a finite-dimensional vector space over $\mathbb{R}$. It turns out that every representation of $C\ell_{r,s}$ is a direct sum of irreducible representations.

Theorem 12. *Let $w_{r,s}$ be the number of inequivalent irreducible $\mathbb{R}$-representations of $C\ell_{r,s}$. Then*

$$w_{r,s} = \begin{cases} 1, & \text{if } r - s \text{ is even;} \\ 2, & \text{if } r - s \text{ is odd.} \end{cases}$$

We now turn to vector fields on the sphere. Let N be a positive integer, consider $\mathbb{R}^{N+1}$, and let

$$S^N = \{x \in \mathbb{R}^{N+1} \mid \|x\|^2 = 1\}$$

be the N-sphere. A *tangent vector field on S^N* is a continuous function $v : S^N \to \mathbb{R}^{N+1}$ such that $v(x)$ is tangent to S^N at x for all x.

Example 4. *Let N be odd. Then $v : S^N \to \mathbb{R}^{N+1}$ defined by*

$$v(x) = (x_1, -x_0, x_3, -x_2, \ldots, x_{N-2}, x_{N-1})$$

is a tangent vector field on S^N that is nowhere 0. For N even, on the other hand, it turns out that every tangent vector field on S^N vanishes somewhere.

A set of tangent vector fields $v_1, \ldots, v_n$ on S^N is *linearly independent*, if $\{v_1(x), \ldots, v_n(x)\}$ is an independent set of vectors in $\mathbb{R}^{N+1}$ for every x in S^N.

The existence of linearly independent tangent vector fields on S^N is connected to the representation theory of Clifford algebras as follows. Turn to the special case $s = 0$ and $r = n$, and denote $C\ell_{r,s}$ by $C\ell_n$. Let $\rho : C\ell_n \to \mathrm{End}_{\mathbb{R}}(W)$ be an $\mathbb{R}$-representation. Defining $cw = (\rho c)w$ for $c \in C\ell_n$ and $w \in W$, makes W a module over the algebra $C\ell_n$. Conversely, any $C\ell_n$-module arises from such a representation. It is not very hard to prove the following

Theorem 13. *Suppose that $\mathbb{R}^{N+1}$ is a $C\ell_n$-module. Then there exist n linearly independent vector fields on S^N.*

Let m be a positive integer and write $m = 2^{4a+b}m_0$ with m_0 odd and $0 \le b \le 3$. The number $\rho(m) = 8a + 2^b$ is the *Radon–Hurwitz number* of m. Set $k = 4a + b$. One can check that

$$\rho(m) = \begin{cases} 2k + 1 & \text{if } k \equiv 0 \,(\mathrm{mod}\,4); \\ 2k & \text{if } k \equiv 1 \,(\mathrm{mod}\,4); \\ 2k & \text{if } k \equiv 2 \,(\mathrm{mod}\,4); \\ 2k + 2 & \text{if } k \equiv 3 \,(\mathrm{mod}\,4). \end{cases}$$

Theorem 14. *The integer $\rho(N+1) - 1$ is the largest integer n such that $\mathbb{R}^{N+1}$ is a $C\ell_n$-module.*

Check that this conclusion is consistent with the earlier example, and that

Theorem 15. *There exist $\rho(N+1) - 1$ linearly independent vector fields on S^N.*

is an obvious consequence of the two theorems above. The proofs of the results so far are largely routine. But now comes the important and very difficult question that the celebrated theorem of Adams answers: What is the largest number of linearly independent tangent vector fields that S^N can support?

Theorem 16. *The number $\rho(N + 1) - 1$ is in fact the largest number of linearly independent vector fields that can exist on S^N.*

The proof analyzes the group of stable fiber homotopy classes on real projective space and applies topological K-theory in the process. Refer to D. Husemoller, *Fibre Bundles,* third edition, Springer-Verlag, Berlin Heidelberg, New York, 1994, for the complete details, and to [22, Chapters 0 and 12] for background information and other connections.

Acknowledgments

I wish to thank Richard Elman, Max-Albert Knus, Carl Riehm, and especially Jean-Pierre Tignol and Adrian Wadsworth, for the insights, suggestions, and corrections that they provided.

REFERENCES

[1] B. Jones, *The Arithmetic Theory of Quadratic Forms,* Math. Assoc. of America, Buffalo, 1950.

[2] M. Eichler, *Quadratische Formen und Orthogonale Gruppen,* Springer-Verlag, Berlin, 1952.

[3] C. Chevalley, *The Algebraic Theory of Spinors,* Columbia University Press, New York, 1954.

[4] J. Dieudonné, *La Géométrie des Groupes Classiques,* Springer-Verlag, Berlin, 1955.

[5] E. Artin, *Geometric Algebra,* Wiley, New York, 1957.

[6] O. T. O'Meara, *Introduction to Quadratic Forms,* Springer-Verlag, Berlin Heidelberg New York, 1963. Reprinted in the Classics of Mathematics Series, Springer-Verlag, 2000.

[7] J. Milnor, D. Husemoller, *Symmetric Bilinear Forms,* Springer-Verlag, Berlin Heidelberg New York, 1973.

[8] J.-P. Serre, *A Course in Arithmetic,* Springer-Verlag, Berlin Heidelberg New York, 1973.

[9] T.Y. Lam, *The Algebraic Theory of Quadratic Forms,* Benjamin, Reading, MA, 1973.

[10] M. Kneser, *Quadratische Formen,* Mathematisches Institut der Universität, Göttingen, 1973–4.

[11] I. Reiner, *Maximal Orders,* Academic Press, New York London, 1975.

[12] J. Cassels, *Rational Quadratic Forms,* Academic Press, London New York San Francisco, 1978.

[13] R. Pierce, *Associative Algebras,* Springer-Verlag, Berlin Heidelberg New York, 1982.

[14] W. Scharlau, *Quadratic and Hermitian Forms*, Springer-Verlag, Berlin Heidelberg New York, 1985.

[15] A. Hahn, O.T. O'Meara, *The Classical Groups and K-Theory*, Springer-Verlag, Berlin Heidelberg New York, 1989.

[16] M.-A. Knus, *Quadratic and Hermitian Forms over Rings*, Springer-Verlag, Berlin Heidelberg New York, 1991.

[17] Y. Kitaoka, *Arithmetic of Quadratic Forms*, Cambridge University Press, Cambridge, 1993.

[18] A. Hahn, *Quadratic Algebras, Clifford Algebras, and Arithmetic Witt Groups*, Springer-Verlag, Berlin Heidelberg New York, 1994.

[19] I. Porteous, *Clifford algebras and the classical groups*, Cambridge University Press, Cambridge, 1995.

[20] J. Conway, *The sensual quadratic form*, Math. Assoc. of America, Providence, 1997.

[21] M.-A. Knus, A. Merkurjev, M. Rost, J.-P. Tignol, *The Book of Involutions*, American Math. Soc. Providence, 1998.

[22] D. Shapiro, *Composition of Quadratic Forms*, Walter de Gruyter, Berlin New York, 2000.

Alexander J. Hahn
Department of Mathematics
University of Notre Dame
Notre Dame, Indiana 46556
E-mail: hahn.1@nd.edu

Submitted: September 17, 2002.

20

Lipschitz's Methods of 1886 Applied to Symplectic Clifford Algebras

Jacques Helmstetter

ABSTRACT When Clifford algebras are studied in relation with exterior algebras, it is easy to undertake a parallel study of "symplectic Clifford algebras", also called "Weyl algebras": it suffices to replace exterior algebras with symmetric algebras, to remove all "twisting signs" from all places where they are present, and to intrude them into all places where they are absent. This enables one to imagine a symplectic counterpart of a theorem of Lipschitz about orthogonal transformations; unfortunately this counterpart needs an "enlargement" of the Weyl algebra, and leads to infinite sums and convergence problems. Some specific problems of the symplectic case, that result from this enlargement and that cannot be treated by purely algebraic means, are commented upon.

Keywords: Clifford algebras, Weyl algebras, Lipschitz groups, metaplectic groups

Symplectic Clifford algebras, formerly called Weyl algebras, have been rather neglected in the last conferences on Clifford algebras; therefore I am very grateful to Professor Abłamowicz because he has accepted my lecture proposal on this topic. I will begin this lecture at the lowest possible level, because I wish to make as many people as possible acquainted with these algebras, and convince them that such algebras really belong to the natural domain of cliffordian research.

20.1   Elementary description of Clifford algebras and Weyl algebras

Let φ be a nondegenerate symmetric bilinear form on a vector space M of dimension m over a field K which here is $\mathbb{R}$ or $\mathbb{C}$, and β a bilinear form $M \times M \to K$ such that, for all a and b in M,

$$\beta(a,b) + \beta(b,a) = \varphi(a,b) = \varphi(b,a). \tag{1.1}$$

AMS Subject Classification: 13F25, 15A66, 16S10, 32D15.

In the Clifford algebra of (M, φ) the equality

$$ab + ba = \varphi(a,b) = \varphi(b,a) \tag{1.2}$$

must be true for all a and b in M. As a vector space, this Clifford algebra is isomorphic to the exterior algebra $\bigwedge(M)$, which is already provided with an exterior multiplication satisfying this equality, whenever x and y are even or odd elements of $\bigwedge(M)$:

$$x \wedge y = (-1)^{\partial x \partial y} y \wedge x; \tag{1.3}$$

an even or odd element x has a degree ∂x in $\mathbb{Z}/2\mathbb{Z} = \{0,1\}$ indicating whether it belongs to the even subalgebra $\bigwedge_0(M)$ or to the odd subspace $\bigwedge_1(M)$. The bilinear form β allows us to carry the Clifford multiplication onto $\bigwedge(M)$ in such a way that the exterior and Clifford multiplications are related by these two equalities, for all $a, b_1, \ldots, b_k$ in M :

$$a\,(b_1 \wedge b_2 \wedge \cdots \wedge b_k)$$
$$= a \wedge b_1 \wedge b_2 \wedge \cdots \wedge b_k + \sum_{j=0}^{k}(-1)^{j-1}b_1 \wedge \cdots \beta(a,b_j) \cdots \wedge b_k; \tag{1.4}$$

$$(b_1 \wedge b_2 \wedge \cdots \wedge b_k)\,a$$
$$= b_1 \wedge b_2 \wedge \cdots \wedge b_k \wedge a + \sum_{j=0}^{k}(-1)^{k-j}b_1 \wedge \cdots \beta(b_j,a) \cdots \wedge b_k; \tag{1.5}$$

the notation $\bigwedge(M,\beta)$ means the vector space $\bigwedge(M)$ provided with these two multiplications. Among all bilinear forms β satisfying (1.1) there is a unique one that is symmetric: $\beta = \varphi/2$; it is called the canonical choice of β, and it allows us to write this useful equality (for all a and x as above):

$$ax + (-1)^{\partial x}xa = 2\,a \wedge x \quad \text{if} \quad \beta = \varphi/2. \tag{1.6}$$

At last, the Clifford algebra is provided with a reversion ρ that leaves all elements of M invariant, and such that, for all x and y in $\bigwedge(M,\beta)$

$$\rho(xy) = \rho(y)\rho(x); \tag{1.7}$$

the calculation of ρ is very simple for the canonical β :

$$\forall x \in \bigwedge^{k}(M), \quad \rho(x) = (-1)^{k(k-1)/2}\,x \quad \text{if} \quad \beta = \varphi/2. \tag{1.8}$$

In some of these eight formulas the sign $-$ appears whenever two odd factors have been reversed; this is obviously the effect of the twisting sign $(-1)^{\partial x \partial y}$ in (1.3), and similar twisting signs (taking into account $\partial a = 1$) also appear in (1.4), (1.5), (1.6). Notwithstanding that no such twisting signs appear in (1.1), (1.2) or (1.7); and in (1.8) there is no reversion of factors.

When the dimension m is even, all this can be repeated when φ is replaced by a symplectic form ψ (a nondegenerate skew symmetric bilinear form on M), provided that subsequent modifications are performed (as explained in [7]).

We must replace β by a bilinear form γ satisfying (1.9) beneath; there is still a canonical choice of γ, the only one that is skew symmetric, namely $\gamma = \psi/2$. And above all, the exterior algebra must be replaced by the infinite dimensional symmetric algebra $S(M)$. Here are the modifications that must be performed in our eight formulas; now $a, b, \ldots$ are still arbitrary elements of M, but x and y are even or odd elements of $S(M)$:

$$\gamma(a, b) - \gamma(b, a) = \psi(a, b) = -\psi(b, a) \tag{1.9}$$

$$ab - ba = \psi(a, b) = -\psi(b, a) \tag{1.10}$$

$$x \vee y = y \vee x \tag{1.11}$$

$$a\,(b_1 \vee b_2 \vee \cdots \vee b_k) = a \vee b_1 \vee b_2 \vee \cdots \vee b_k$$

$$+ \sum_{j=0}^{k} b_1 \vee \cdots \gamma(a, b_j) \cdots \vee b_k \tag{1.12}$$

$$(b_1 \vee b_2 \vee \cdots \vee b_k)\,a = b_1 \vee b_2 \vee \cdots \vee b_k \vee a$$

$$+ \sum_{j=0}^{k} b_1 \vee \cdots \gamma(b_j, a) \cdots \vee b_k \tag{1.13}$$

$$ax + xa = 2\,a \vee x \quad \text{if} \quad \gamma = \psi/2 \tag{1.14}$$

$$\rho(xy) = (-1)^{\partial x \partial y}\,\rho(y)\,\rho(x) \tag{1.15}$$

$$\forall x \in S^k(M), \quad \rho(x) = (-1)^{k(k-1)/2}\,x \quad \text{if} \quad \gamma = \psi/2 . \tag{1.16}$$

The modifications to be performed in these eight formulas, and also in a lot of other formulas that are not mentioned here, are very simple to explain: when the reversion of two odd factors is not compensated by a twisting sign, we must intrude a twisting sign in the symplectic counterpart of the concerned formula; but when there is already a twisting sign, we must suppress it. In particular the equality (1.15) means that now we meet a twisted reversion. The notation $S(M, \gamma)$ means the vector space $S(M)$ provided with its symmetric multiplication, and the other multiplication defined by (1.12) or (1.13), that lets $S(M)$ become an algebra isomorphic to the Weyl algebra of (M, ψ), also called its symplectic Clifford algebra.

Notwithstanding that Weyl algebras also require formulas without counterparts in Clifford algebras, because every nonzero $a \in M$ admits an infinite sequence of nonzero powers in $S(M)$. The following equality is valid for all choices of γ, provided that the powers of a and b are understood as Weyl powers; nevertheless, when $\gamma = \psi/2$, it is easy to deduce from (1.12) or (1.13) that the Weyl powers of any element a of M are equal to its symmetric powers, since $\gamma(a, a) = 0$. In this equality the summation runs on all (j, k) such that $0 \leq j \leq \inf(p, s)$ and

$0 \leq k \leq \inf(q,r)$:

$$\left(\frac{a^p}{p!} \vee \frac{b^q}{q!}\right)\left(\frac{a^r}{r!} \vee \frac{b^s}{s!}\right) =$$

$$\sum_{j,k} \frac{\gamma(a,b)^j}{j!} \frac{\gamma(b,a)^k}{k!} \left(\frac{a^{p-j}}{(p-j)!} \frac{a^{r-k}}{(r-k)!}\right) \vee \left(\frac{b^{q-k}}{(q-k)!} \frac{b^{s-j}}{(s-j)!}\right). \quad (1.17)$$

20.2   A theorem of Lipschitz (in a modernized version)

Lipschitz was the first to relate orthogonal transformations to inner automorphisms of Clifford algebras; this accounts for the name "Lipschitz group", often preferred to a "Clifford group". Before dealing with orthogonal transformations, let us consider the Lie algebra $\mathfrak{so}(M,\varphi)$ of the orthogonal group. When $\beta = \varphi/2$ (as it is assumed up to the end), it is well known that $\bigwedge^2(M)$ is a Lie algebra in $\bigwedge(M,\beta)$, and that we get an isomorphism of Lie algebras $\bigwedge^2(M) \to \mathfrak{so}(M,\varphi)$ if we map every bivector u to the infinitesimal orthogonal transformation F_u defined by $F_u(a) = ua - au$.

Whereas classical Lie theory suggests studying the Clifford exponentials $\exp(u)$ of the elements u of $\bigwedge^2(M)$, Lipschitz considered their exterior exponentials $\exp^\wedge(u)$ (which are finite sums), and proved this theorem (see [9] and [10], or even [15]), involving the infinitesimal orthogonal transformation F_u defined above.

Lipschitz Theorem. *Let u be any element of $\bigwedge^2(M)$, and let $x = \exp^\wedge(u)$, whence $\rho(x) = \exp^\wedge(-u)$; when x is invertible for the Clifford multiplication, the corresponding inner automorphism of $\bigwedge(M,\beta)$ leaves M invariant and induces an orthogonal transformation of M :*

$$x \, a \, x^{-1} = (\mathrm{id} - \tfrac{1}{2}F_u)^{-1}(\mathrm{id} + \tfrac{1}{2}F_u)(a) \, ;$$

besides, x is Clifford-invertible if and only if F_u has no eigenvalue equal to ± 2, because

$$x \, \rho(x) = \rho(x) \, x = \det(\mathrm{id} - \tfrac{1}{2}F_u) = \det(\mathrm{id} + \tfrac{1}{2}F_u) \, .$$

Because of a well–known theorem of Cayley, recalled just beneath, Lipschitz could claim that every orthogonal transformation of (M,φ) without eigenvalue equal to -1 was the restriction to M of some inner automorphism of the Clifford algebra. Then he extended this result to all orthogonal transformations g with determinant 1. The case of an orthogonal transformation with determinant -1 requires a correct treatment of parity gradings that was still unknown at Lipschitz's time. Besides, the equalities $g(a) = (-1)^{\partial x} x a x^{-1}$ and $g(a) = -a$ imply

$a \wedge x = 0$ because of (1.6), and this explains why eigenvalues -1 raise difficulties. If -1 is an eigenvalue of the orthogonal transformation g, and if $(b_1, \ldots, b_k)$ is a basis of $\ker(g + \mathrm{id})$, there is an associated x in the Lipschitz group that is equal to $b_1 \wedge \cdots \wedge b_k \wedge \exp^\wedge(u)$ for some $u \in \bigwedge^2(M)$; obviously this agrees with the equality $a \wedge x = 0$ whenever $g(a) = -a$.

The following theorem has a symplectic counterpart that is not stated here, since every reader can guess it without being helped.

Cayley Theorem. *The mapping $f \longmapsto (\mathrm{id} - f)^{-1}(\mathrm{id} + f)$ is a bijection from the set of all elements $f \in \mathfrak{so}(M, \varphi)$ without eigenvalue ± 1 onto the set of all elements $g \in \mathrm{SO}(M, \varphi)$ without eigenvalue -1.*

20.3   A dangerous illusion

Now let us consider $S(M, \gamma)$, where up to the end γ is assumed to be the canonical choice $\psi/2$. It is still true that $S^2(M)$ is a Lie subalgebra of $S(M, \gamma)$, and that we get an isomorphism $S^2(M) \to \mathfrak{sp}(M, \psi)$ if we map every $u \in S^2(M)$ to the infinitesimal symplectic transformation F_u still defined by $F_u(a) = ua - au$. This faithful analogy leads us to imagine that a symplectic Lipschitz group might also be derived from the Weyl algebra $S(M, \gamma)$; unfortunately this algebra contains no invertible element outside $K = S^0(M)$, neither for the symmetric multiplication nor for the Weyl multiplication. Consequently we must enlarge the space $S(M)$.

The easiest enlargement is the algebra $\bar{S}(M)$ that is the direct product of all spaces $S^k(M)$ (whereas $S(M)$ is their direct sum). Unfortunately the extension of the Weyl multiplication to $\bar{S}(M)$ involves infinite sums; indeed the Weyl product of two elements respectively in $S^p(M)$ and $S^q(M)$ may have a nonzero component in $S^{p+q-k}(M)$ for $k = 0, 2, 4, \ldots, \inf(2p, 2q)$; in general these infinite sums are not convergent. When dealing with the enlargement $\bar{S}(M)$, the main problem comes from the general divergence of these infinite sums. Even the exponential series $\exp(u)$ (with Weyl powers) are in general divergent in $\bar{S}(M)$. Unfortunately some people have overlooked this convergence problem, and have gotten seduced by this illusory theorem.

False Theorem (1975–1985). *For every symplectic transformation g of (M, ψ), there is an even element $x \in \bar{S}(M)$ (unique up to a factor in the group $K^\times$ of invertible scalars), that is invertible for the extended Weyl multiplication, and that allows us to write $g(a) = xax^{-1}$ for all $a \in M$.*

It must be explained that this illusory theorem is incompatible with elementary cliffordian algebra; the following two objections give evidence of this incompatibility.

First Objection. *Since the transformation $-\mathrm{id}$ is in the center of the symplectic group $\mathrm{Sp}(M, \psi)$, every element $x \in \bar{S}(M)$ lying above it (such that $xax^{-1} = -a$ for all $a \in M$) must be an eigenvector for all symplectic transformations; but all* common eigenvectors in $\bar{S}(M)$ of all symplectic transformations are in $K = S^0(M)$, and lie above id. Compare $\bar{S}(M)$ with $\bigwedge(M)$, in which all orthogonal transformations have two spaces of common eigenvectors, namely $K = \bigwedge^0(M)$ and $\bigwedge^m(M)$, if $m = \dim(M)$; the former lies over id, and the latter over $-$ id.

Second Objection. *If g is a symplectic transformation with eigenvalue -1, the equality $xax^{-1} = -a$ together with (1.14) implies $a \vee x = 0$; but this equality is impossible when a and x are nonzero, because the algebra $\bar{S}(M)$ contains no divisors of zero. Compare it with the analogous equality $a \wedge x = 0$ already mentioned above.*

After these two purely algebraic objections, a third one of topological nature must be presented. From a symplectic Lipschitz group lying above $\mathrm{Sp}(M, \varphi)$, it should be possible to extract a 2-sheet covering group over $\mathrm{Sp}(M, \varphi)$, quite analogous to the spinorial group which is a 2-sheet covering group over the orthogonal group. When $K = \mathbb{R}$, such a 2-sheet covering group of $\mathrm{Sp}(M, \varphi)$ exists, and is called the metaplectic group. But when $K = \mathbb{C}$, all symplectic groups are simply connected, and refuse to be covered by a metaplectic group. The mathematicians who have supported this illusory theorem have never explained why it failed when $K = \mathbb{C}$, whereas it supposedly succeeded when $K = \mathbb{R}$.

After having established its absolute falseness about 1979, I endeavored to get convergence theorems allowing construction of at least a local symplectic Lipschitz group, representing all symplectic transformations in some neighborhood of id_M. Since I had already discovered again Lipschitz's theorem, I considered symmetric exponentials $\exp^\vee(u)$ of elements $u \in S^2(M)$, instead of Weyl exponentials. As explained in [5] and [8], I managed to prove that, when u and v were in some neighborhood of 0 in $S^2(M)$, the Weyl product of $\exp^\vee(u)$ and $\exp^\vee(v)$ could be calculated by means of convergent infinite sums, and that all products involving such symmetric exponentials together with other factors in $S(M)$ were associative. This enabled me to reduce the difficult problems of convergence to algebraic problems: indeed if we replace the field K by the ring $K[[t]]$ of formal series, and if we replace γ by $t\gamma$, all concerned infinite sums become formally convergent, with sums that can be calculated. When later we replace t by 1, we know that the involved infinite sums converge when the concerned symplectic transformations are near enough to id_M; these infinite sums are analytical functions, and by analytical extension we can extend the results obtained over this neighborhood of id_M as long as these functions have no singularity; this means, as long as we have not yet reached any symplectic transformation with eigenvalue -1.

20.4   A symplectic counterpart of Lipschitz's theorem

In order to get the above mentioned "local group" lying above a neighborhood of id in $\mathrm{Sp}(M, \varphi)$, it is sensible to look for an adaptation of Lipschitz's theorem to this localized symplectic setting (see [5] and [8]).

Main Theorem. *Let u be an element of $S^2(M)$ in some neighborhood of 0, and let us set $x = \exp^\vee(u)$, whence $\rho(x) = \exp^\vee(-u)$; when x is invertible for the Weyl multiplication, it induces a symplectic transformation of (M, ψ) :*

$$x \, a \, x^{-1} \; = \; (\mathrm{id} - \tfrac{1}{2}F_u)^{-1}(\mathrm{id} + \tfrac{1}{2}F_u)(a);$$

besides, x is Weyl-invertible as long as all eigenvalues of F_u in $\mathbb{C}$ have a module < 2, because

$$x \, \rho(x) \; = \; \rho(x) \, x \; = \; \det(\mathrm{id} - \tfrac{1}{2}F_u)^{-1} \; = \; \det(\mathrm{id} + \tfrac{1}{2}F_u)^{-1}.$$

Before sketching a proof, I must point out the "inversion rule", according to which every determinant appearing in a formula concerning the Lipschitz group of (M, φ) must be replaced by the inverse of this determinant in the analogous formula concerning (M, ψ); as far as I know, this rule was first discovered by Sato, Miwa, Jimbo (see [13] and the Appendix in [14]), and independently (but slightly later) by myself (see [7]).

Proof of Main Theorem. Lipschitz's proof of the first assertion is still valid in the symplectic case. Indeed, according to (1.13), $\exp^\vee(u) \, a$ is the sum of $\exp^\vee(u) \vee a$ and another term $D_a(\exp^\vee(u))$ in which D_a represents a derivation of the algebra $\bar{S}(M)$; consequently $D_a(\exp^\vee(u)) = D_a(u) \vee \exp^\vee(u)$; easy calculations show that $F_u(a) = 2D_a(u)$; consequently

$$\exp^\vee(u)\, a \; = \; (a + \tfrac{1}{2}F_u(a)) \vee \exp^\vee(u) \; ;$$

similarly from (1.12) we deduce, for all $b \in M$,

$$b\exp^\vee(u) \; = \; (b - \tfrac{1}{2}F_u(b)) \vee \exp^\vee(u) \; ;$$

therefore the equality $bx = xa$ is valid whenever

$$b \; = \; (\mathrm{id} - \tfrac{1}{2}F_u)^{-1}(\mathrm{id} + \tfrac{1}{2}F_u)(a) \; ;$$

this proves the first assertion. When x is Weyl-invertible, by means of a classical argument first discovered by Lipschitz, we can prove that the symplectic transformations induced by x and $\rho(x)$ are reciprocal to each other; consequently the Weyl product of x and $\rho(x)$ belongs to $K^\times$. At this point I can no longer adapt the remainder of Lipschitz's proof, because he used tricky calculations with Pfaffians of skew symmetric matrices; therefore I continue with other methods.

We may assume that $K = \mathbb{C}$, and observe that almost all elements u of $S^2(M)$ can be written

$$u \; = \; \sum_{k=1}^{n} \lambda_k a_k \vee b_k \, ,$$

with all $\lambda_k \in K$ and $n = m/2$ in such a way that the n planes $Ka_k \oplus Kb_k$ are pairwise orthogonal for ψ; an argument of continuity allows us only to consider such elements u, and then a classical theorem about the Weyl algebra of an orthogonal direct sum allows us to reduce the problem to the case of a space M of dimension 2. Consequently we suppose that (a, b) is a basis of M, and that $u = \lambda a \vee b$. A direct application of (1.17) (with $p = q$ and $r = s$) shows that $x\,\rho(x)$, in other words $\exp^\vee(u)\exp^\vee(-u)$, is equal to

$$\sum_{p,r,j,k} \frac{p!\,r!\,(-1)^r\,\lambda^{p+r}\,\gamma(a,b)^j\,\gamma(b,a)^k\,a^{p+r-j-k}\vee b^{p+r-j-k}}{j!\,k!\,(p-j)!\,(p-k)!\,(r-j)!(r-k)!} \ ;$$

since we know that the result belongs to K, it suffices to consider the terms such that $p = r = j = k$:

$$\exp^\vee(u)\exp^\vee(-u) \;=\; \sum_k \lambda^{2k}\gamma(a,b)^{2k} \;=\; \frac{1}{1 - \lambda^2\gamma(a,b)^2}$$

if $|\lambda\gamma(a,b)| < 1$. A last easy calculation shows the eigenvalues of F_u :

$$F_u(a) \;=\; 2\lambda\,\gamma(b,a)\,a \quad \text{and} \quad F_u(b) \;=\; 2\lambda\,\gamma(a,b)\,b\,.$$

20.5   Towards a global symplectic Lipschitz group

Here M is a real vector space; but since beneath it is necessary to use complex functions $M^* \to \mathbb{C}$ defined on the dual space M^*, a complex Weyl algebra underlies the following considerations. The elements of $S(M)$ may be identified with polynomial functions on M^*, and the $\vee$-multiplication with the ordinary multiplication of functions on M^*. Every $u \in S^2(M)$ may be treated as a quadratic form on M^*, and thus $\exp^\vee(ru)$ becomes an analytic function on M^* for every $r \in \mathbb{C}$.

The local symplectic Lipschitz group only covers some neighborhood of id in the symplectic group $\mathrm{Sp}(M, \psi)$, but this neighborhood is large enough to contain $-\,\mathrm{id}$ in its frontier. This suggests using the symplectic counterpart of Lipschitz's theorem to construct a path from id to $-\,\mathrm{id}$ that is covered by the local symplectic Lipschitz group except at the end, and to see whether the elements of this local group over this path approach a limit when the path approaches $-\,\mathrm{id}$. This can be done in this way:

Let u be an element of $S^2(M)$ such that F_u is a bijective endomorphism of M, and for every scalar r let us set

$$x_r \;=\; \exp^\vee(ru) \quad \text{and} \quad g_r \;=\; (\mathrm{id} - \tfrac{1}{2}rF_u)^{-1}\,(\mathrm{id} + \tfrac{1}{2}rF_u)$$

provided that $\pm 2/r$ is not an eigenvalue of F_u; when $|r|$ becomes infinitely large, g_r approaches $-id$; consequently it suffices to let r vary on a path that starts at the point 0, avoids the values for which g_r is not defined, and allows $|r|$ to vary from 0 to infinity. Besides we may multiply the element x_r lying over g_r by any scalar μ_r that would prove to be helpful. When g_r approaches $-$ id, there is no limit for $\mu_r x_r$ in $\bar{S}(M)$, whatever μ_r we have chosen; but sometimes a limit appears when we consider x_r as a function on M^*, and multiply it by $\mu_r = |r|^{m/2}$. When a limit appears, up to a scalar factor it is the Dirac distribution δ at the origin of M^*.

This leads us to imagine a global symplectic Lipschitz group made of distributions on M^* (as in [6] and [8]); of course this is not an easy thing, as I explained at Canterbury in 1985 (see [1]). In particular, the equality (1.10) must be replaced by $ab - ba = i\,\psi(a,b)$, where $i = \sqrt{-1}$, and (1.12) and (1.13) must be modified accordingly. In this setting, the real Lie algebra $\mathfrak{sp}(M,\psi)$ is isomorphic to $i\,S^2(M)$; in other words, every element of $\mathfrak{sp}(M,\psi)$ is equal to F_{iu} for some unique $u \in S^2(M)$.

Let us verify that the three objections presented in Section 20.3 are no longer valid when δ lies over the symplectic transformation $-$ id.

First δ is invariant by all symplectic transformations. Secondly every $a \in M$ is a linear form on M^*, that is a continuous function vanishing at the origin of M^*, and consequently the product $a \vee \delta$ vanishes. Thirdly the emergence of δ as a limit (as explained above) results from the following assertion, in which r is a variable nonnegative *real* number, θ belongs to the interval $[-\pi/2, \pi/2]$ and $\chi(t)$ is a smooth function $\mathbb{R} \to \mathbb{C}$ rapidly vanishing at infinity:

$$\int_{\mathbb{R}} \sqrt{r}\exp\left(-\tfrac{1}{2}re^{i\theta}t^2\right)\chi(t)\,dt \to \sqrt{2\pi}\,e^{-i\theta/2}\chi(0) \quad \text{when} \quad r \to +\infty\;;$$

the condition imposed on θ allows $ire^{i\theta}$ to run through $\mathbb{R}$ but not through $\mathbb{C}$, and this explains why we cannot obtain symplectic Lipschitz groups over $\mathbb{C}$, and why we must put a factor i beside ψ to obtain symplectic Lipschitz groups over $\mathbb{R}$.

The above assertion allows one to find the limit of $r^{k/2}\exp^{\vee}(riu)$ when u is an element of $S^2(M)$ of rank k, and $r \to +\infty$ in $\mathbb{R}$; such a calculation is a particular case of the "theorem of the stationary phase". It is worth pointing out that there is a faithfully analogous situation in the exterior algebra $\bigwedge(M)$. Indeed, let u be an element of $\bigwedge^2(M)$ of rank k (necessarily even); when $r \to \infty$, then $r^{-k/2}\exp^{\wedge}(ru)$ has a nonzero limit: the exterior $(k/2)$-power of u divided by $(k/2)!$. The multiplication by $r^{-k/2}$ (instead of $r^{k/2}$) agrees with the "inversion rule" mentioned just after the Main Theorem.

20.6   Questions

Two questions came from the audience at the Cookeville conference. The first one mentioned the Moyal products of functions that had been imagined in quantum mechanics already a long time ago. Indeed, when $\gamma = \psi/2$ and when the elements of $S(M)$ are treated as functions on M^*, the Moyal product of these polynomial functions coincides with their Weyl product in $S(M, \gamma)$. Notwithstanding, when physicists actually use a Moyal product, the fundamental equality is not exactly (1.10), but rather $ab - ba = i\kappa\,\psi(a,b)$, where $i = \sqrt{-1}$ and κ is some physical constant involving Planck's constant. Remember that in Section 20.5 I also recommended the presence of the factor i, but for quite different reasons; once again independent mathematical and physical speculations have led to the same mathematical objects.

The second question mentioned Crumeyrolle's symplectic works. His first publication [2] about symplectic Clifford algebras appeared in 1975, three years after the common paper [11] of Nouazé and Revoy, in which Revoy presented probably the first *cliffordian* treatment of Weyl algebras (published again later in [12]). But Crumeyrolle managed to draw much more attention to this topic; in particular I became interested in this topic because of the mistrust raised among the symplectic mathematicians by his paper [3] of 1977. His own contribution to this topic contains at least these three ideas:

- The enlargement $S(M) \to \bar{S}(M)$ (see Section 20.3);

- The symplectic transformations associated with elements $\exp(\lambda a^2)$ (where $a \in M$ and $\lambda \in K$), which were interesting in 1975, but far behind the results soon later obtained by Sato, Miwa, Jimbo;

- A technical idea leading to the selection of some subspaces of $\bar{S}(M)$, which proved to be helpful in convergence problems when correctly used by other persons.

Any attempt to lengthen this short list is welcome, provided that it is based on a serious examination of his symplectic works. Unfortunately Crumeyrolle is also the inventor of the illusory theorem recalled in Section 20.3, and when in 1985 at Canterbury he acknowledged its falseness, he replaced it by another theorem (stated in [1] and [4]) which in my opinion was still worse. To be brief, I only mention that in 1986 the symplectic mathematicians that had examined his symplectic works, rejected all of them as being false beyond correction, and anyhow unusable; and they refused any longer to discuss them with him. I bear witness that it was impossible to discuss them with him; whenever he was driven back by grave reproaches, he tried to escape by making optimistic claims: Anyhow if there is an error, it suffices to correct it. But in my opinion the symplectic mathematicians misbehaved when they also refused to discuss the issues with persons that he had influenced. To promote quiet and helpful discussions after 20 years of unsuccessful discussions, I will say that any attempt to defend his symplectic works against so absolute a rejection is welcome, provided that the defender knows the reasons of the rejection, and tries to refute them with a serious argument.

REFERENCES

[1] J.R.S. Chisholm and A.K. Common, Eds., *Clifford Algebras and Their Applications in Mathematical Physics*, Proceedings of the Canterbury Workshop, D. Reidel Pub-

lishing Company, 1986, 517–529 (Crumeyrolle's contribution) and 559–564 (Helmstetter's contribution).

[2] A. Crumeyrolle, Algèbres de Clifford symplectiques et revêtements spinoriels du groupe symplectique, *C.R. Acad. Sc. Paris* **280**, 1975, 1689–1692.

[3] A. Crumeyrolle, Algèbre de Clifford symplectique et revêtements spinoriels du groupe symplectique, *J. Math. pures et appl.*, **56**, 1977, 205–230.

[4] A. Crumeyrolle, *Orthogonal and Symplectic Clifford Algebras*, Kluwer Academic Publishers, Math. and its appl. **57**, 1990, chapters 17, 18, 19, 22.

[5] J. Helmstetter, Produits intérieurs de séries formelles et algèbres de Clifford symplectiques, *C.R. Acad. Sc. Paris*, **293**, 1981, 63–66.

[6] J. Helmstetter, Fibré de Maslov, groupe métaplectique et ⋆-produits, *C.R. Acad. Sc. Paris*, **294**, 1982, 361–364.

[7] J. Helmstetter, Algèbres de Clifford et algèbres de Weyl, *Cahiers Mathématiques* **25**, Montpellier, 1982.

[8] J. Helmstetter, Algèbres de Weyl et ⋆-produit, *Cahiers Mathématiques* **34**, Montpellier, 1985.

[9] R. Lipschitz, Principes d'un calcul algébrique ..., *C.R. Acad. Sc. Paris* **91**, 1880, 619–621 and 660–664. Reprinted in Bull. Sci. Math. (2), **11**, 1887, 115–120.

[10] R. Lipschitz, *Untersuchungen über die Summen von Quadraten*, Max Cohen und Sohn, Bonn 1886 (147 pages). French translation of the first chapter in *J. Math. pures appl.* **2**, 1886, 373–439. French summary in *Bull. Sci. Math.* (2), **10**, 1886, 163–183.

[11] Y. Nouazé, Ph. Revoy, Sur les algèbres de Weyl généralisées, *Bull. Sci. Math.* (2), **96**, 1972, 27–47.

[12] Ph. Revoy, Autour des formes quadratiques, *Cahiers Mathématiques* **13**, Montpellier, 1978.

[13] M. Sato, T. Miwa, M. Jimbo, Holonomic quantum fields I, *Publ. R.I.M.S.*, Kyoto Univ., **14**, 1978, 223–267.

[14] M. Sato, T. Miwa, M. Jimbo, Holonomic quantum fields IV, *Publ. R.I.M.S.*, Kyoto Univ., **15**, 1979, 871–972.

[15] A. Weil, Correspondence, *Ann. of Math.* **69**, 1959, 247–251. Reprinted in his *Collected Papers*, Vol. II, 557–561, Springer Verlag, 1979.

Jacques Helmstetter
Université de Grenoble I
Institut Fourier, B. P. 74
38402 Saint-Martin d'Hères Cedex, France
E-mail: Jacques.Helmstetter@ujf-grenoble.fr

Submitted: June 29, 2002.

21

The Group of Classes of Involutions of Graded Central Simple Algebras

Jacques Helmstetter

ABSTRACT Some properties of Clifford algebras are actually common properties of all graded central simple algebras A provided with an involution ρ; with ρ is associated a "complex divided trace" (a complex number r such that $r^8 = 1$), and thus all such involutions are classified by a cyclic group of order 8. Complex divided traces are also involved in the Brauer–Wall group of the field $\mathbb{R}$, and they bring efficiency and enlightenment in the study of bilinear forms on graded A-modules.

Keywords: graded central simple algebras, involutions, Clifford algebras

The following classification is valid for every field K of characteristic $\neq 2$, but in order to avoid misunderstandings, I explain at once that, when f is an *involutive* endomorphism of a finite–dimensional vector space V, the notation $\mathrm{tr}(f)$ means no longer the trace of f, but the difference

$$\mathrm{tr}(f) = \dim(\ker(f - \mathrm{id}_V)) - \dim(\mathrm{im}(f - \mathrm{id}_V)).$$

Besides, this integer $\mathrm{tr}(f)$ will be silently treated as an element of the field $\mathbb{C}$ of complex numbers. Of course when K is a subfield of $\mathbb{C}$, then $\mathrm{tr}(f)$ is actually the trace of f.

21.1 Graded central simple algebras (or shortly G.C.S. algebras)

Here we only consider finite–dimensional K-algebras $A = A_0 \oplus A_1$ that are graded over the group $\mathbb{Z}/2\mathbb{Z} = \{0, 1\}$; every nonzero homogeneous (even or odd) $x \in A$ has a degree (or a parity) $\partial x \in \{0, 1\}$. For such an algebra we must distinguish two centers, the ordinary center $Z(A) = Z_0(A) \oplus Z_1(A)$ and the graded center $Z^g(A)$ that has the same even component $Z_0^g(A) = Z_0(A)$, but another odd component:

$$Z_1^g(A) = \{x \in A_1 \; ; \; \forall y \in A_0 \cup A_1, \; xy = (-1)^{\partial x \partial y} \, yx \}.$$

AMS Subject Classification: 11E81, 11E88, 13E10, 15A66, 17A35.

Such an algebra A is said to be *graded central* if $Z^g(A)$ is K (that is the natural image of K in A, since the unit element of A is also denoted as 1). It is said to be *graded simple* if every graded ideal J (such that $J = (J \cap A_0) \oplus (J \cap A_1)$) is equal to 0 or to A.

A graded algebra D is said to be a *graded division* algebra if every nonzero homogeneous element of D is invertible. If its grading is trivial (in other words, if $D_1 = 0$), then D is merely an ordinary division algebra, and with each pair (m, n) of nonnegative integers we associate a graded matrix algebra $\mathcal{M}(m, n; D)$ in which the even (resp. odd) elements are the square matrices of order $m + n$ looking like

$$\begin{pmatrix} a & 0 \\ 0 & d \end{pmatrix} \quad (\text{resp. } \begin{pmatrix} 0 & b \\ c & 0 \end{pmatrix}) ;$$

a and d are square submatrices of order respectively m and n, whereas b and c are rectangular submatrices. When the grading of D is not trivial, the matrix algebra $\mathcal{M}(m, D)$ inherits a nontrivial grading from D : a matrix is even (resp. odd) if all its entries are even (resp. odd). Any attempt to define an algebra $\mathcal{M}(m, n; D)$ (when D_1 contains invertible elements) would bring a graded algebra isomorphic to $\mathcal{M}(m + n, D)$.

A (finite–dimensional graded) algebra A is G.C.S. if and only if it is isomorphic to an algebra $\mathcal{M}(m, n; D)$ or $\mathcal{M}(m, D)$ with D a graded central division algebra (either trivially or nontrivially graded); thus A is classified by the isomorphy class of D, and the set of isomorphy classes of graded central division algebras constitutes the graded Brauer group, or Brauer–Wall group $BW(K)$.

For instance, the group $BW(\mathbb{R})$ of the field of real numbers is a cyclic group of order 8 ; two classes in $BW(\mathbb{R})$ are represented by trivially graded division rings, namely $\mathbb{R}$ itself and $\mathbb{H}$ (the division ring of real quaternions); the six other classes are represented by Clifford algebras according to Table 1, in which $C\ell(p, q)$ is the Clifford algebra of a real quadratic space of dimension $p + q$ containing positive (resp. negative) subspaces of dimension p (resp. q). The first row shows one graded central division algebra D in each one of the eight isomorphy classes; the second row (in which D_{ng} means "nongraded D") shows what D looks like when its grading is forgotten; and the third line shows what the even subalgebra D_0 looks like.

D	$\mathbb{R}$	$C\ell(1,0)$	$C\ell(2,0)$	$C\ell(3,0)$	$\mathbb{H}$	$C\ell(0,3)$	$C\ell(0,2)$	$C\ell(0,1)$
D_{ng}	$\mathbb{R}$	$\mathbb{R}^2$	$\mathcal{M}(2,\mathbb{R})$	$\mathcal{M}(2,\mathbb{C})$	$\mathbb{H}$	$\mathbb{H}^2$	$\mathbb{H}$	$\mathbb{C}$
D_0	$\mathbb{R}$	$\mathbb{R}$	$\mathbb{C}$	$\mathbb{H}$	$\mathbb{H}$	$\mathbb{H}$	$\mathbb{C}$	$\mathbb{R}$

TABLE 1.

21.2   Involutions (also called anti-involutions)

Here we only consider G.C.S. algebras A provided with an involution ρ, that is an involutive K-linear anti-automorphism such that $\rho(A_j) = A_j$ for $j = 0, 1$. The K-linearity of ρ means that it operates trivially on $K = Z^g(A)$. Some authors prefer the name "anti-involution", and I am still hesitating about the best name. It is well known that every Clifford algebra $C\ell(M, Q)$ (with Q a nondegenerate quadratic form on M) is a G.C.S. algebra provided with an involution which leaves invariant every element of M, and which is called the "reversion".

When A and B are G.C.S. algebras, their twisted tensor product $A \hat{\otimes} B$ is still G.C.S., and this property supplies the classifying set $BW(K)$ with its structure of a group. When A and B are provided with involutions ρ and σ respectively, then $A \hat{\otimes} B$ is provided with a unique involution $\rho \tilde{\otimes} \sigma$ extending ρ and σ; for all homogeneous x and y respectively in A and B,

$$\rho \tilde{\otimes} \sigma \, (x \otimes y) = (-1)^{\partial x \partial y} \, \rho(x) \otimes \sigma(y) \,.$$

The classification of these involutions requires that a class should be attributed to each one in such a way that the class of $\rho \tilde{\otimes} \sigma$ is always the product of the classes of ρ and σ. This can be achieved in this way: Let ρ_0 and ρ_1 be the restrictions of ρ to A_0 and A_1, and let $\mathrm{cp.dv.tr}(\rho)$ (complex divided trace of ρ) be the element of $\mathbb{C}$ defined by this formula (in which $i = \sqrt{-1}$) :

$$\mathrm{cp.dv.tr}(\rho) = \frac{\mathrm{tr}(\rho_0) + i\,\mathrm{tr}(\rho_1)}{\sqrt{\dim(A)}} \,;$$

it is easy to prove that $\mathrm{cp.dv.tr}(\rho\tilde{\otimes}\sigma)$ is the product of the complex divided traces of ρ and σ, but it is less trivial to prove that division by the square root of $\dim(A)$ makes $\mathrm{cp.dv.tr}(\rho)$ fall into the group R_8 of the eighth roots of 1 in $\mathbb{C}$. Besides, this proof also shows that the complex divided trace of the matrix transposition in $\mathcal{M}(m, n; K)$ is always 1. Thus we have classified the concerned involutions by means of a cyclic group of order 8, canonically isomorphic to R_8.

Some pieces of information can be immediately derived from the square of $r = \mathrm{cp.dv.tr}(\rho)$. When $r^2 = 1$ (resp. $r^2 = -1$), then the center $Z(A_0)$ of the even subalgebra has dimension 2 (whereas $Z(A) = K$), and ρ operates trivially (resp. nontrivially) on $Z(A_0)$. When $r^2 = i$ (resp. $r^2 = -i$), then the center $Z(A)$ has dimension 2 (whereas $Z(A_0) = K$), and ρ operates trivially (resp. nontrivially) on $Z(A)$. More refined pieces of information (depending on r itself) appear when we consider bilinear forms on A-modules (see Section 21.4).

There is a natural group morphism from $BW(K)$ onto $R_2 = \{1, -1\}$, which attributes the value 1 or -1 to the class of A according to whether $Z(A)$ is or is not equal to K; and there is also a natural morphism $R_8 \rightarrow R_2$ defined by $r \mapsto r^4$. The information given just above shows that $\mathrm{cp.dv.tr}(\rho)$ has the same image in R_2 as the Brauer–Wall class of A. The couples (A, ρ) are classified by the couples $(b, r) \in BW(K) \times R_8$ such that b and r have the same image in R_2, and such that b corresponds to G.C.S. algebras actually provided with involutions.

With every involution ρ is associated a conjugate involution defined by $x \mapsto (-1)^{\partial x}\rho(x)$; its complex divided trace is the inverse of that of ρ. For instance $K \times K$, with the grading determined by its two diagonals, is a graded central division algebra over K which admits two involutions: the identity mapping and the swap automorphism $(\lambda, \mu) \mapsto (\mu, \lambda)$; their complex divided traces are respectively $(1+i)/\sqrt{2}$ and its inverse $(1-i)/\sqrt{2}$.

Remark 1. When K is a field of characteristic 2, then $\mathrm{tr}(\rho) = \sqrt{\dim(A)}$ for any involution ρ of any G.C.S. algebra A, and we must classify ρ by means of

$$\mathrm{tw.dv.tr}(\rho) = (\mathrm{tr}(\rho_0) - \mathrm{tr}(\rho_1)) / \sqrt{\dim(A)} \in R_2 = \{1, -1\};$$

this twisted divided trace only reveals whether ρ operates trivially or non-trivially on $Z(A_0)$, which has always dimension 2 in this case. In my opinion it is not possible to refine this classification if we want these involutions to be classified by a group.

21.3   Involutions on real G.C.S. algebras

When K has characteristic 0, every G.C.S. algebra A is the direct sum of K and $A_1 + [A, A]$ (where $[A, A]$ is the subspace spanned by all $xy - yx$), whence a projection $T : A \to K$. When $K = \mathbb{R}$, an involution ρ of A is said to be positive if the quadratic form $x \mapsto T(x\rho(x))$ is positive. It is easy to prove that $\rho \,\tilde{\otimes}\, \sigma$ is positive whenever ρ and σ are positive. Besides, the graded central division algebras over $\mathbb{R}$ (see Table 1 above) admit positive involutions, and the matrix transposition in $\mathcal{M}(m, n; \mathbb{R})$ is also a positive involution; consequently every G.C.S. algebra A over $\mathbb{R}$ admits positive involutions. Lastly, all the positive involutions of A have the same complex divided trace (determined by the quadratic form $x \mapsto T(x^2)$). Thus we get an isomorphism $BW(\mathbb{R}) \to R_8$ if we map the class of A to the complex divided trace of its positive involutions; in other words, $BW(\mathbb{R})$ is canonically isomorphic to the group of classes of involutions.

The couples (A, ρ) over $\mathbb{R}$ are classified by the 32 couples $(r', r) \in R_8 \times R_8$ such that $r'^4 = r^4$; r and r' are respectively the complex divided trace of ρ and that of a positive involution of A. In particular, if we calculate (r', r) for the above defined Clifford algebra $C\ell(p, q)$, we get this result, that corroborates the already known classification of real Clifford algebras:

$$(r', r) = \left(\exp(\tfrac{1}{4}(p-q)i\pi),\ \exp(\tfrac{1}{4}(p+q)i\pi)\right);$$

since we have classified the couples (A, ρ) with a *group*, it suffices to verify this equality when (p, q) is equal to $(1, 0)$ or $(0, 1)$.

Almost all the content of this section was already revealed by Wall [5], but with very few explanations and without proofs. It is a pity that his paper has been so seldom mentioned; I became aware of its existence through the book of Porteous [3].

21.4   Bimodules

Let A be a G.C.S. algebra, and M a graded bimodule over A; thus there are two multiplications $A \times M \to M$ and $M \times A \to M$ satisfying the associativity condition $(xz)y = x(zy)$ and the grade condition $\partial(xzy) = \partial x + \partial z + \partial y$ for all homogeneous $x, y \in A$, and $z \in M$. The graded centralizer of A in M is the additive subgroup $Z^g(A, M)$ spanned by all homogeneous $z \in M$ such that $xz = (-1)^{\partial x \partial z} zx$ for all homogeneous $x \in A$; it is a vector space over $K = Z^g(A)$. The natural mapping

$$A \otimes Z^g(A, M) \to M$$

defined by $x \otimes z \mapsto xz$ is bijective; this is even one of the many characteristic properties of G.C.S. algebras, as pointed out by Bass [1] and Small [4].

21.4.1   First example

Let S be a finite–dimensional graded left module over A; for instance A may be a Clifford algebra, and S a graded spinor space. The algebra $\mathrm{End}(S)$ of K-linear endomorphisms of S inherits a parity grading from S, in such a way that $\partial f(s) = \partial f + \partial s$ for all homogeneous $f \in \mathrm{End}(S)$ and $s \in S$. It becomes a graded bimodule over A if we set $(xfy)(s) = xf(ys)$. Here $Z^g(A, \mathrm{End}(S))$ is usually noted $\mathrm{End}^g_A(S)$; it is a graded subalgebra of $\mathrm{End}(S)$, the homogeneous elements of which are characterized by the equality $f(xs) = (-1)^{\partial f \partial x} xf(s)$, valid for all homogeneous $x \in A$ and $s \in S$. Consequently we get an algebra morphism $A \hat{\otimes} \mathrm{End}^g_A(S) \to \mathrm{End}(S)$, and since A is G.C.S., it is an algebra isomorphism. It follows that $\mathrm{End}^g_A(S)$ is G.C.S. too; its class in $BW(K)$ is the inverse of that of A.

21.4.2   Second example

The space $\mathcal{B}(S)$ of bilinear forms $\varphi : S \times S \to K$ also inherits a parity grading from S, in such a way that $\varphi(s, t) = 0$ whenever $\partial \varphi \neq \partial s + \partial t$. When A is provided with an involution ρ, $\mathcal{B}(S)$ becomes a graded bimodule over A if we set $(x \varphi y)(s, t) = \varphi(ys, \rho(x)t)$. Thus the homogeneous elements of $Z^g(A, \mathcal{B}(S))$ are characterized by the equality

$$\varphi(xs, t) = (-1)^{\partial \varphi \partial x} \varphi(s, \rho(x)t)$$

valid for all $s, t \in S$, and all homogeneous $x \in A$. This implies that $Z^g(A, \mathcal{B}(S))$ is a right module over $\mathrm{End}^g_A(S)$ in a natural way: $(\varphi f)(s, t) = \varphi(f(s), t)$. Since the natural mapping $A \otimes Z^g(A, \mathcal{B}(S)) \to \mathcal{B}(S)$ is bijective, $Z^g(A, \mathcal{B}(S))$ has the same dimension as $\mathrm{End}^g_A(S)$.

Let τ be the involutive endomorphism of $\mathcal{B}(S)$ that maps every φ to $\tau(\varphi)$ defined by $(s, t) \mapsto \varphi(t, s)$; it is clear that τ leaves $Z^g_j(A, \mathcal{B}(S))$ invariant for $j =$

$0, 1$; consequently $Z^g(A, \mathcal{B}(S))$ is the direct sum of four subspaces $Z^g_{j,k}(A, \mathcal{B}(S))$ with $(j, k) \in (\mathbb{Z}/2\mathbb{Z})^2$; the index j indicates the parity, and k the symmetry property (symmetric if $k = 0$, skew symmetric if $k = 1$). By first examining the case of an irreducible graded left module S, it is possible to prove that in all cases at least one of the four subspaces $Z^g_{j,k}(A, \mathcal{B}(S))$ contains nondegenerate bilinear forms.

Let φ be a nondegenerate element in $Z^g_{j,k}(A, \mathcal{B}(S))$; thus $\partial\varphi = j$ and $\varphi(t, s) = (-1)^k \varphi(s, t)$; this φ induces an involution σ on $\mathrm{End}^g_A(S)$ such that

$$\varphi(f(s), t) = (-1)^{\partial\varphi \partial f} \varphi(s, \sigma(f)(t)),$$

and it is easy to prove that $\mathrm{cp.dv.tr}(\sigma)$ is the product of $\mathrm{cp.dv.tr}(\tau)$ by the factor $s \in R_4$ given in Table 2. In order to know $\mathrm{cp.dv.tr}(\sigma)$ and the dimensions of the

(j, k)	$(0, 0)$	$(0, 1)$	$(1, 0)$	$(1, 1)$
s	1	-1	$-i$	i

TABLE 2.

four subspaces $Z^g_{j,k}(A, \mathcal{B}(S))$, it suffices to know the complex divided trace of τ; this essential piece of information is given by this wonderful formula:

$$\mathrm{cp.dv.tr}(\rho)\, \mathrm{cp.dv.tr}(\tau) = 1.$$

Now a lot of other information becomes attainable. I only mention that the above φ can be "extended" to a bilinear mapping $\Phi : S \times S \to A$ by A-linearity on the left side:

$$\Phi(xs, t) = x\Phi(s, t) \quad \text{and} \quad \varphi = T \circ \Phi$$

where T is defined as in Section 21.3 (at least when K has characteristic 0). This implies

$$\Phi(s, yt) = \Phi(s, t)\, \rho(y) \quad \text{and} \quad \Phi(t, s) = (-1)^k \rho(\Phi(s, t)).$$

A similar "extension" $S \times S \to \mathrm{End}^g_A(S)$ is also possible, and involves σ instead of ρ. Even an "extension" $S \times S \to A \hat{\otimes} \mathrm{End}^g_A(S)$ is possible; it yields an isomorphism of bimodules $S \otimes S \to A \hat{\otimes} \mathrm{End}^g_A(S)$ which reveals the "bilinear covariants" of any group of homogeneous invertible elements of A operating in S; this can be used for the spinorial group when A is a Clifford algebra.

If we compare the above developments suggested by Bass's works, and the developments expounded in Porteous's book, we observe that many important objects are liable to be described according to both kinds of developments. But since the starting points are rather far from each other, it is not immediately apparent that the resulting descriptions are equivalent; a "dictionary" is necessary to yield the suitable "translations".

21.5   Remarks

(a) With every nongraded A-module S is associated a graded A-module $S \times S$, the even (resp. odd) component of which is $S \times 0$ (resp. $0 \times S$); if x is an even (resp. odd) element of A, and (s, t) any element of $S \times S$, by definition $x(s, t)$ is (xs, xt) (resp. (xt, xs)). Therefore the previous considerations are also helpful for nongraded modules.

(b) The usual spinor spaces in quantum mechanics are graded modules over the corresponding Clifford algebra A (that is $C\ell(1, 3)$ or $C\ell(3, 1)$), because they are also vector spaces over $\mathbb{C}$. Indeed in this case $Z(A_0)$ contains an element ω such that $\omega^2 = -1$ and $\omega x = (-1)^{\partial x} x \omega$ for all homogeneous $x \in A$; if A acts in a $\mathbb{C}$-space S, the mapping $s \mapsto i\omega s$ is an involutive $\mathbb{R}$-linear endomorphism of S, and its eigenspaces determine a parity grading of S that makes it become a graded A-module. Conversely any graded A-module S becomes a vector space over $\mathbb{C}$ if we set $is = (-1)^{\partial s} \omega s$ for all homogeneous $s \in S$; this multiplication by i commutes with the operations of all $x \in A$.

REFERENCES

[1] H. Bass, *Topics in Algebraic K-theory*, Tata Institute, Bombay, 1967.

[2] M-A. Knus, A. Merkurjev, M. Rost, J-P. Tignol, *The Book of Involutions*, American Math. Society, 1998.

[3] I. Porteous, *Clifford Algebras and the Classical Groups*, Cambridge University Press, 1995, 2000.

[4] Ch. Small, The Brauer–Wall group of a commutative ring, *Trans. Amer. Math. Soc.* **156**, 1971, 455–491.

[5] C.T.C. Wall, Graded algebras, anti-involutions, simple groups and symmetric spaces, *Bull. Amer. Math. Soc.* **74**, 1968, 198–202.

Jacques Helmstetter
Université de Grenoble I
Institut Fourier, B. P. 74
38402 Saint-Martin d'Hères Cedex, France
E-mail: Jacques.Helmstetter@ujf-grenoble.fr

Submitted: June 29, 2002.

22

A Binary Index Notation
for Clifford Algebras

Dennis W. Marks

ABSTRACT Hagmark and Lounesto's binary labeling of the generators of a Clifford algebra can be extended by ordering the basis elements in either ascending order, denoted by a post-scripted binary index, or descending order, denoted by a pre-scripted binary index. Reversion is then represented by swapping prescripts and postscripts, grade involution by changing the sign of the index, and conjugation by both swapping and changing the sign. Bit inversion of the binary index is a duality operation that does not suffer the handedness problems of the Clifford dual or of the Hodge dual. Generators whose binary indices are bit inverses of each other commute (anti-commute) if the product of their grades is even (odd). If the number of generators is even (odd), the commutators (anti-commutators) of basis vectors labeled with binary indices and of covectors labeled by bit inversion of binary indices yield generalizations of the Heisenberg commutation relations.

Keywords: anti-commutators, binary label, Clifford algebras, commutators, duality, grade

22.1 Binary index notation

22.1.1 Dirac matrices

Consider a set of orthonormal Dirac matrices γ_μ $(\mu = 0, 1, \ldots, n-1)$ that satisfy

$$\gamma_\mu \cdot \gamma_\nu = \tfrac{1}{2} \left(\gamma_\mu \gamma_\nu + \gamma_\nu \gamma_\mu \right) \doteq \eta_{\mu\nu} I, \tag{1.1}$$

where

$$\eta_{\mu\nu} = \begin{cases} +1 & \text{if } \mu = \nu = 0, 1, \ldots, p-1, \text{ or } 0, \ldots, \frac{n+s}{2} - 1; \\ -1 & \text{if } \mu = \nu = p, p+1, \ldots, p+q-1, \text{ or } \frac{n+s}{2}, \ldots, n-1; \\ 0 & \text{if } \mu \neq \nu. \end{cases} \tag{1.2}$$

The number of Dirac matrices is $n = p + q$; the signature of the metric $\eta_{\mu\nu}$ is $s = p - q$. Starting the count at 0, rather than at 1, simplifies certain formulae for the binary notation to be introduced below.

AMS Subject Classification: 15A66, 16W50, 03F15.

22.1.2 Prescripts and postscripts, subscripts and superscripts

Products of the γ_μ's generate elements e_m of a Clifford algebra $C\ell_{p,q}$. By repeated application of equation (1.1), the γ_μ's in the products can be arranged with the indices monotonically either increasing or decreasing. We introduce a convention that prescripted indices are decreasing and postscripted indices are increasing. We retain the convention that subscripted indices are covariant and superscripted indices are contravariant.

$$e_m = e_{m_o \ldots m_\mu \ldots m_{n-1}} = (\gamma_0)^{m_0} \cdots (\gamma_\mu)^{m_\mu} \cdots (\gamma_{n-1})^{m_{n-1}} \tag{1.3}$$

$$_m e = {}_{m_{n-1} \ldots m_\mu \ldots m_0} e = (\gamma_{n-1})^{m_{n-1}} \cdots (\gamma_\mu)^{m_\mu} \cdots (\gamma_0)^{m_0} \tag{1.4}$$

$$e^m = e^{m_o \ldots m_\mu \ldots m_{n-1}} = (\gamma^0)^{m_0} \cdots (\gamma^\mu)^{m_\mu} \cdots (\gamma^{n-1})^{m_{n-1}} \tag{1.5}$$

$$^m e = {}^{m_{n-1} \ldots m_\mu \ldots m_0} e = (\gamma^{n-1})^{m_{n-1}} \cdots (\gamma^\mu)^{m_\mu} \cdots (\gamma^0)^{m_0} \tag{1.6}$$

From equation (1.1), the reciprocal basis vectors are $\gamma^\lambda = (\gamma_\lambda)^{-1}$. Then

$$e_m = ({}^m e)^{-1}; \quad {}_m e = (e^m)^{-1}; \quad e^m = ({}_m e)^{-1}; \quad {}^m e = (e_m)^{-1}. \tag{1.7}$$

Taking the inverse turns subscripts into superscripts (and vice versa) and prescripts into postscripts (and vice versa).

The digits of a number are usually written from left to right with the least significant digit last; that is, in order of *descending* powers of the base. With this choice of order, the preferred binary labeling for Clifford elements is $_m e$. Whatever the order, since $\gamma_\mu^2 = \pm I$, each of the γ_μ is in the product e_m either once or not at all, so $m_\mu = 0$ or 1. This suggests treating m_μ as the bits of an binary number m :

$$m = \sum_{\mu=0}^{n-1} m_\mu 2^\mu \equiv m_\mu 2^\mu, \tag{1.8}$$

where we have extended the Einstein summation convention to exponents. The labels of the generating elements are positive integer powers of two: $e_1 = \gamma_0, e_2 = \gamma_1, e_4 = \gamma_2, e_8 = \gamma_3, \ldots$, or, in general,

$$e_{2^\mu} = \gamma_\mu. \tag{1.9}$$

This formula would not be as elegant if μ ran from 1 to n, instead of 0 to $n - 1$. For labels that are not powers of two, the Clifford element is the unique product of those generating elements whose labels sum to the label of the element; for example $_5 e = (_4 e)(_1 e) = \gamma_2 \gamma_0$.

Others ([2, 3, 5, 9], and [7, Chapter 21]) have utilized various binary notations for the elements of Clifford algebras, but with the bits labeled in ascending order from 1 to n. The binary labels introduced in this paper meet the need, emphasized recently in [4], for indices that show the behavior of Clifford elements under coordinate transformations (in this paper by the grade k_m calculated below for each m and by the position of the index as subscript or superscript) and under reversion (in this paper by the position of the index as prescript or postscript).

22.1.3   Identity and co-identity matrices

The number of bits in binary labels equals n, the number of generating elements. The $n \times n$ identity matrix is

$$I_n = e_{\underbrace{0...0}_{n \text{ zeroes}}} = (\gamma_0)^0 \cdots (\gamma_\mu)^0 \cdots (\gamma_{n-1})^0. \tag{1.10}$$

Similarly, we define the $n \times n$ co-identity matrix as

$$J_n = e_{2^n - 1} = e_{\underbrace{1...1}_{n \text{ ones}}} = (\gamma_0)^1 \cdots (\gamma_\mu)^1 \cdots (\gamma_{n-1})^1 \tag{1.11}$$

and likewise for $_nJ$, J^n, and ^nJ, where we use the same convention as before about prescripts and postscripts, and subscripts and superscripts. As with any e_m, taking the inverse of J_n turns subscripts into superscripts (and vice versa) and prescripts into postscripts (and vice versa), so

$$(J_n)^{-1} = ({}^nJ). \tag{1.12}$$

Note that

$$(J_n)^2 = (-1)^{\frac{n(n-1)}{2}} (I_n)^p (-I_n)^q = (-1)^{\frac{n^2-s}{2}} I_n. \tag{1.13}$$

J_n is the n-dimensional volume element. For space-time ($n = 4, s = 2$), $(J_4)^2 = -I_4$, so that J_4 plays the role of the imaginary i in quantum mechanics [6]. $J_4 = e_F = e_{15} = e_1 e_2 e_4 e_8 = \gamma_0 \gamma_1 \gamma_2 \gamma_3$ is often denoted γ_5.

22.1.4   Grade

The number of γ_μ's in an element e_m is its grade k_m. Elements of grade $0, 1, 2, \ldots, n - 1, n$ are scalars, vectors, bivectors, ..., covectors, and coscalars, respectively. The number of elements in the kth grade is the binomial factor $\binom{n}{k}$, so the kth grade and the $(n - k)^{th}$ grade have the same number of elements. Since m_μ equals 1 if γ_μ is present in a product and 0 if it is not, the sum of the bits of the binary label of e_m equals its grade k_m :

$$k_m = \sum_{\mu=0}^{n-1} m_\mu \equiv m_\mu 1^\mu. \tag{1.14}$$

22.1.5   (Anti-)automorphisms within grade k

Clifford algebras have two automorphisms and two anti-automorphisms within grade k_m [8]. The automorphism of a product of elements is the product of the automorphisms of the elements; the anti-automorphism of the product of elements is the product of the anti-automorphisms with the order reversed [7]. The automorphisms are the identity mapping and grade involution; the anti-automorphisms are reversion and conjugation. We extend the binary notation to denote the various (anti-)automorphisms. Grade involution is induced by changing the sign of each of the bases: $\gamma_\mu \mapsto -\gamma_\mu$. We denote its effect on each element of the Clifford algebra by changing the sign of the binary index: $e_m \mapsto e_{-m}$, where

$$e_{-m} = (-1)^{\sum_{\mu=0}^{n-1} m_\mu} e_m = (-1)^{k_m} e_m. \tag{1.15}$$

In particular,

$$e_{-0} = (-1)^{k_0} e_0 = (-1)^0 e_0 = e_0, \tag{1.16}$$

so that the label 0 is indeed the number 0, which is its own negative. Reversion reverses the order of the products. We have already introduced left and right indices to indicate the order of the products, so reversion swaps left and right indices: $e_m \mapsto {}_m e$ and ${}_m e \mapsto e_m$, where

$$_m e = (\gamma_{n-1})^{m_{n-1}} \cdots (\gamma_0)^{m_0} = (-1)^{\frac{k_m(k_m-1)}{2}} e_m. \tag{1.17}$$

Conjugation is the composition (in either order) of involution and reversion. We denote its effect on each element of the Clifford algebra by combining the notation for involution and reversion, that is, by changing the sign of the binary index and by switching left and right indices: $e_m \mapsto {}_{-m}e$, where

$$_{-m}e = (-1)^{k_m} (-1)^{\frac{k_m(k_m-1)}{2}} e_m = (-1)^{\frac{k_m(k_m+1)}{2}} e_m. \tag{1.18}$$

22.1.6   Dualities: isomorphisms between grades k and $n-k$

Duality operations generate isomorphisms between grades k and $n-k$. There are several different duals, including the Clifford dual, which we will write as e_{m*}, defined [7] as

$$e_{m*} = e_m J_n, \tag{1.19}$$

and the Hodge dual, which we will write as ${}_m e_*$, defined [7] as

$$_m e_* = ({}_m e) J_n = (-1)^{\frac{k_m(k_m-1)}{2}} e_m J_n = (-1)^{\frac{k_m(k_m-1)}{2}} e_{m*}. \tag{1.20}$$

One can create other duals by pre- or post-multiplying any of the automorphisms or anti-automorphisms of e_m by any of the variants of J_n. We extend the binary notation to these duals by using a subscripted (superscripted) asterisk on the left to indicate pre-multiplication by ${}_n J$ (${}^n J$) and a subscripted (superscripted) asterisk on the right to indicate post-multiplication by J_n (J^n). Taking the inverse again turns subscripts into superscripts (and vice versa), and prescripts into postscripts (and vice versa):

$$(e_m^*)^{-1} = (e_m J^n)^{-1} = (J^n)^{-1} (e_m)^{-1} = ({}_n J)({}^m e) = ({}_*^m e). \tag{1.21}$$

A dual distinct from the preceding duals is induced by bit inversion that maps $e_m \mapsto e_{\overline{m}}$, where $\overline{m}$ is the bit inverse of m :

$$e_{\overline{m}} = (\gamma_0)^{\overline{m}_0} \cdots (\gamma_\mu)^{\overline{m}_\mu} \cdots (\gamma_{n-1})^{\overline{m}_{n-1}}. \tag{1.22}$$

The bit inverse is related to the Clifford dual by

$$e_{\overline{m}} = (-1)^{\sum_{\mu=0}^{n-1} \mu m_\mu} e_*^m, \tag{1.23}$$

since, for the basis vector

$$\gamma_\mu = e_{2^\mu} = e_{\underset{0}{0}\ldots 0\underset{\mu}{1}0\ldots\underset{n-1}{0}}, \tag{1.24}$$

the (postscripted, superscripted) Clifford dual is the covector (grade $n-1$)

$$\gamma_\mu^* = \gamma_\mu e^{\underset{0}{1}\ldots 1\underset{\mu}{1}1\ldots\underset{n-1}{1}} = (-1)^\mu e^{\underset{0}{1}\ldots 1\underset{\mu}{0}1\ldots\underset{n-1}{1}} = (-1)^\mu e^{\overline{2^\mu}}. \tag{1.25}$$

Since $(\gamma_\mu)^{-1} = \gamma^\mu$, the form γ_μ^* is preferred over the customary $\gamma_{\mu*}$, which is $\gamma_\mu \gamma_5 = -\gamma_5 \gamma_\mu$ in four dimensions.

The bit inverse of the bit inverse is the original element: $e_{\overline{\overline{m}}} = e_m$. For the various duals involving J_n, the dual of the dual of e_m is $\pm e_m$, depending variously on dimension n, signature s, and grade k_m; for example,

$$(e_{m*})_* = (e_m J_n) J_n = e_m J_n^2 = (-1)^{\frac{n^2-s}{2}} e_m. \tag{1.26}$$

The duals involving J_n also suffer from dependency on the handedness of the base coordinate system, because J_n changes sign with the swapping of two of the bases. Bit inversion does not have these shortcomings.

22.1.7   Multiplication

The formula for the multiplication of two Clifford elements $_\ell e$ and $_m e$ is

$$(_\ell e)(_m e) = (-1)^{\sum_{\mu=1}^{n-1} m_\mu \sum_{\lambda=0}^{\mu-1} \ell_\lambda} (\gamma_{n-1})^{\ell_{n-1}+m_{n-1}} \cdots (\gamma_0)^{\ell_0+m_0}. \tag{1.27}$$

The exponents of the γ_μ in equation (1.27) can be expressed in terms of addition mod 2. Since ℓ_μ and m_μ are bits, either 0 or 1,

$$\ell_\mu \oplus m_\mu \equiv (\ell_\mu + m_\mu)_{mod2} = \ell_\mu + m_\mu - 2\ell_\mu m_\mu, \tag{1.28}$$

so

$$(\gamma_\mu)^{\ell_\mu+m_\mu} = (\gamma_\mu^2)^{\ell_\mu m_\mu}(\gamma_\mu)^{\ell_\mu \oplus m_\mu} = (\pm I)^{\ell_\mu m_\mu}(\gamma_\mu)^{\ell_\mu \oplus m_\mu}, \tag{1.29}$$

where the $+$ sign holds for $\mu = 0, \ldots, p-1$ and the $-$ sign holds for $\mu = p, \ldots, n-1$. Then equation (1.27) becomes

$$(_\ell e)(_m e) = (-1)^{\sum_{\mu=1}^{n-1} \sum_{\lambda=0}^{\mu-1} \ell_\lambda m_\mu} (-1)^{\sum_{\mu=p}^{n-1} \ell_\mu m_\mu} \prod_{\mu=n-1}^{0} (\gamma_\mu)^{\ell_\mu \oplus m_\mu}. \tag{1.30}$$

Equivalent formulae are given by [9] for $C\ell_{1,q}$, by [2] for $C\ell_{n,0}$, by [3] for $C\ell_{0,n}$, and by [5] and [7] for the general $C\ell_{p,q}$, all in ascending order rather than in the more natural descending order.

22.1.8   Commutators and anti-commutators

Reversing the order of the multiplicands in equation (1.30), we have

$$(_me)\,(_\ell e) = (-1)^{\sum_{\mu=1}^{n-1}\sum_{\lambda=0}^{\mu-1} m_\lambda \ell_\mu}\,(-1)^{\sum_{\mu=p}^{n-1} m_\mu \ell_\mu}\,\prod_{\mu=n-1}^{0}(\gamma_\mu)^{m_\mu\oplus\ell_\mu}. \tag{1.31}$$

Comparing equations (1.30 and 1.31), we see that all of the terms commute except for the first term on the right-hand side, which is either $+1$ or -1. Therefore each element $_\ell e$ either commutes or anti-commutes with every other element $_m e$. To ascertain which, we use the graded commutator

$$(a,b)_\epsilon = ab + \epsilon ba, \tag{1.32}$$

so $(a,b)_{-1} = ab - ba = [a,b]$ and $(a,b)_{+1} = ab + ba = \{a,b\}$. We write ℓ and m in binary notation with the bits arranged in descending order and then form a matrix of the bit products. Let

$$n_A = \sum_{\lambda=1}^{n-1}\sum_{\mu=0}^{\lambda-1}\ell_\lambda m_\mu \tag{1.33}$$

be the number of 1's above the diagonal of the product matrix,

$$n_B = \sum_{\mu=1}^{n-1}\sum_{\lambda=0}^{\mu-1}\ell_\lambda m_\mu \tag{1.34}$$

be the number of 1's below the diagonal, and

$$n_C = \sum_{\mu=0}^{n-1}\ell_\mu m_\mu \tag{1.35}$$

be the number of 1's on the diagonal (the number of γ_μ's that $_\ell e$ and $_m e$ have in common). Then equations (1.30) and (1.31) can be written as

$$(_\ell e)\,(_m e) = (-1)^{n_B}\,\prod_{\mu=n-1}^{0}\gamma_\mu^{\ell_\mu+m_\mu} \tag{1.36}$$

and

$$(_m e)\,(_\ell e) = (-1)^{n_A}\,\prod_{\mu=n-1}^{0}\gamma_\mu^{m_\mu+\ell_\mu}. \tag{1.37}$$

Element $_\ell e$ either commutes or anti-commutes with $_m e$ depending on whether the parity of n_A agrees or disagrees with the parity of n_B. Let

$$\Delta(\,_\ell e,\,_m e) = (n_A - n_B)_{\mathrm{mod}\,2} = (n_A + n_B)_{\mathrm{mod}\,2}. \tag{1.38}$$

In terms of the graded commutator,

$$(_\ell e, \ _m e)_\epsilon = 0 \text{ for } \epsilon = (-1)^{1-\Delta}. \tag{1.39}$$

Since $\ell_\lambda m_\mu$ is 1 rather than 0 only if both ℓ_λ and m_μ are 1,

$$n_A + n_B + n_C = k_\ell k_m. \tag{1.40}$$

Substituting equation (1.40) into equation (1.38), we get

$$\Delta (_\ell e, \ _m e) = (n_A + n_B)_{\mathrm{mod}\, 2} = (k_\ell k_m - n_C)_{\mathrm{mod}\, 2}. \tag{1.41}$$

Combining equations (1.36) and (1.37), we have

$$(_\ell e, \ _m e)_\epsilon = (-1)^{n_A} \left[(-1)^{\Delta(_\ell e, \ _m e)} + \epsilon \right] \prod_{\mu=n-1}^{0} \gamma_\mu^{\ell_\mu + m_\mu}. \tag{1.42}$$

If $_\ell e$ is the bit inverse $_{\overline{m}} e$ of the element $_m e$, then $_{\overline{m}} e$ and $_m e$ have no basis elements in common, so that $n_C = 0$; furthermore $\overline{m}_\mu + m_\mu = 1$ and $k_{\overline{m}} + k_m = n$; then equation (1.42) gives

$$(_{\overline{m}} e, \ _m e)_\epsilon = (-1)^{n_A} \left[(-1)^{(k_{\overline{m}} k_m)} + \epsilon \right] (_n J). \tag{1.43}$$

The bit inverse $_{\overline{m}} e$ either commutes or anti-commutes with element $_m e$:

for $k_{\overline{m}} k_m$ even, $(_{\overline{m}} e, \ _m e)_{-1} = 0 \Rightarrow \ _{\overline{m}} e$ and $_m e$ commute; $\tag{1.44}$

for $k_{\overline{m}} k_m$ odd, $(_{\overline{m}} e, \ _m e)_{+1} = 0 \Rightarrow \ _{\overline{m}} e$ and $_m e$ anti-commute. $\tag{1.45}$

If e_ℓ is a basis vector $\gamma_\lambda = e_{2\lambda}$ and e_m is a covector $\gamma_\mu^* = (-1)^\mu e^{\overline{2\mu}}$ [from equation (1.25)], then

$$(\gamma_\lambda, \gamma_\mu^*)_\epsilon = \gamma_\lambda \gamma_\mu J^n + \epsilon \gamma_\mu J^n \gamma_\lambda = [\gamma_\lambda \gamma_\mu + \epsilon \gamma_\mu (-1)^{n-1} \gamma_\lambda] J^n. \tag{1.46}$$

Consequently, for odd n,

$$(\gamma_\lambda, \gamma_\mu^*)_{+1} = (\gamma_\lambda \gamma_\mu + \gamma_\mu \gamma_\lambda) J^n = 2\eta_{\lambda\mu} J^n \tag{1.47}$$

and for even n

$$(\gamma_\lambda, \gamma_\mu^*)_{-1} = (\gamma_\lambda \gamma_\mu + \gamma_\mu \gamma_\lambda) J^n = 2\eta_{\lambda\mu} J^n. \tag{1.48}$$

These are generalizations of the Heisenberg commutation relation,

$$\left(x^\lambda, p_\mu \right)_{-1} = \left[x^\lambda, p_\mu \right] = -i\hbar \delta^\lambda{}_\mu. \tag{1.49}$$

Here $\gamma_\lambda = e_{2\lambda}$ is the basis vector for x^λ and $-\frac{1}{2}\hbar \gamma_\mu^* = -\frac{1}{2}\hbar(-1)^\mu e^{\overline{2\mu}}$ is its conjugate covector basis for p_μ. For even n, their commutator is $\hbar J^n$, the quantum of phase space in n dimensions. Complementarity between space-time and momentum-energy is achieved by bit inversion, which inter-converts between position representation and momentum representation. Treating momentum as a Clifford covector has the virtue of automatically enforcing the Heisenberg commutation relation as a consequence of the commutation and anti-commutation properties of the Clifford elements [1].

22.1.9   Acknowledgments

The author wishes to thank Valdosta State University for a Faculty Leave and Georgia Institute of Technology for the opportunity to be a Visiting Scholar. He is grateful to David Finkelstein for welcoming him into the Quantum Relativity Group and for his intellectual stimulation and encouragement. He appreciates helpful discussions with Larry Arbuckle, James Baugh, Bertfried Fauser, Andrei Galiautdinov, William Kallfelz, Jason Looper, Pertti Lounesto, Heinrich Saller, Mohsen Shiri-Garakani, and Tony Smith.

REFERENCES

[1] J. Baugh, D.R. Finkelstein, A. Galiautdinov, and H. Saller, Clifford algebra as quantum language, *J. Math. Phys.* **42** (2001), 1489–1505.

[2] R. Brauer and H. Weyl, Spinors in *n* dimensions, *Amer. J. Math.* **57** (1935), no. 2, 425–449.

[3] R. Delanghe and F. Brackx, Hypercomplex function theory and Hilbert modules with reproducing kernal, *Proc. London Math. Soc.* (3) **37** (1978), 545–576.

[4] A. Gsponer and J.-P. Hurni, Comment on formulating and generalizing Dirac's, Proca's, and Maxwell's equations with biquaterions or Clifford numbers, *Found. of Phys. Letters* **14** (2001), no. 1, 77–85.

[5] P-E. Hagmark and P. Lounesto, Walsh functions, Clifford algebras, and Cayley-Dickenson process, in *Clifford Algebras and their Applications in Mathematical Physics*, Eds. J. S. R. Chisholm and A. K. Common, D. Reidel, Dordrecht, 1986, pp. 531–540.

[6] D. Hestenes, Observables, operators, and complex numbers in the Dirac theory, *J. Math. Phys.* **16** (1975), 556–572.

[7] P. Lounesto, *Clifford Algebras and Spinors*, Cambridge University Press, Cambridge, 1997.

[8] I.R. Porteous, *Clifford Algebras and the Classical Groups*, Cambridge University Press, Cambridge, 1995.

[9] K.Th. Vahlen, Über höhere komplexe zahlen, *Schriften der phys.-ökon. Gesellschaft zu Königsberg* **38** (1897), 72–78.

Dennis W. Marks
Department of Physics, Astronomy, and Geosciences
Valdosta State University
Valdosta, Georgia 31698-0055
E-mail: dmarks@valdosta.edu

Submitted: May 7, 2002; Revised: June 11, 2002.

23

Transposition in Clifford Algebra: SU(3) from Reorientation Invariance

Bernd Schmeikal

ABSTRACT Recoding base elements in a spacetime algebra is an act of cognition. But at the same time this act refers to the process of nature. That is, the internal interactions with their standard symmetries reconstruct the orientation of spacetime. This can best be represented in the Clifford algebra $C\ell_{3,1}$ of the Minkowski spacetime. Recoding is carried out by the involutive automorphism of transposition. The set of transpositions of *erzeugende Einheiten* (primitive idempotents), as Hermann Weyl called them, generates a finite group: the reorientation group of the Clifford algebra. Invariance of physics laws with respect to recoding is not a mere matter of computing, but one of physics. One is able to derive multiplets of strong interacting matter from the recoding invariance of $C\ell_{3,1}$ alone. So the $SU(3)$ flavor symmetry essentially turns out to be a spacetime group. The original quark multiplet independently found by Gell-Mann and Zweig is reconstructed from Clifford algebraic eigenvalue equations of isospin, hypercharge, charge, baryon number and flavors treated as geometric operators. Proofs are given by constructing six possible commutative color spinor spaces $\mathbb{C}h_{\chi}$, or color tetrads, in the noncommutative geometry of the Clifford algebra $C\ell_{3,1}$. Calculations are carried out with CLIFFORD, Maple V package for Clifford algebra computations. Color spinor spaces are isomorphic with the quaternary ring $^4\mathbb{R} = \mathbb{R} \oplus \mathbb{R} \oplus \mathbb{R} \oplus \mathbb{R}$. Thus, the differential (Dirac) operator takes a very handsome form and equations of motion can be handled easily. Surprisingly, elements of $C\ell_{3,1}$ representing generators of $SU(3)$ bring forth (1) the well-known grade-preserving transformations of the Lorentz group together with (2) the *heterodimensional Lorentz transformations*, as Jose Vargas denoted them: *Lorentz transformations of inhomogeneous differential forms*. That is, trigonal tetrahedral rotations do not preserve the grade of a multivector but instead, they permute the base elements of the color tetrad having grades 0, 1, 2 and 3. In the present model the elements of each color space are exploited to reconstruct the flavor $SU(3)$ such that each single commutative space contains three flavors and one color. Clearly, the six color spaces do not commute, and color rotations act in the noncommutative geometry. To give you a picture: Euclidean space with its reorientation group, i.e., the

This work is an outgrowth of a paper presented at the 6th International Conference on Clifford Algebras and their Applications in Mathematical Physics, Cookeville, Tennessee, May 20–May 27, 2002.

AMS Subject Classification: 03G10, 15A66, 14L35, 17B45, 14L24.

again embed the root spaces of the 6 flavor su(3). Color su(3) is exact because it does not involve relativistic effects. Flavor does and is therefore inexact. It seems that $C\ell_{3,1}$ comprises enough structure for both the color- and the flavor-SU(3). In this very first approach both symmetries are reconstructed exact, whereas in reality the flavor SU(3) is only approximate. Here, the only way to make a difference between a color- and a flavor-rotation may be to distinguish between commutative and noncommutative geometry.

We conclude that an extended heterodimensional Lorentz invariance of the Minkowski spacetime and the SU(3) of strongly interacting quarks result from each other. This involves a nongrade preserving-degree of freedom of motion taking spatial lines to spacetime areas and areas to spacetime-volumes and back to lines and areas. Although the form of the Dirac equation is preserved, a thorough study of the involved nonlinearities of equations of motion is still outstanding.

Keywords: Clifford algebra, Lie algebra, idempotent lattice, color spinor space, recoding invariance, spinors, Dirac operator, standard model, Clifford normal real form, Cartan decomposition, inhomogeneous Lorentz transformation, heterodimensional transformation, involutive automorphism, transposition, geometric charge operator

23.1 Prologue on new concepts

In 1920 Hermann Weyl postulated that the nature of space *admitted every possible metric connection*. Space is dependent on its material content, but in itself it is free and capable of any virtual changes. He called this the *first postulate of freedom* (Weyl 1993, p. 139). We need his postulate if we want to investigate the relation between matter and spacetime that is still not entirely clear. Roy Chisholm who organized the first world conference on Clifford algebras in 1985 demanded that *"there is no distinction between spacetime and the internal interaction space"* (1993, p. 371). This is a great challenge, and if we go that far with our claim we have to show in which way the inner symmetry and dynamics of strong interaction are connected with Lorentz transformations in subnuclear dimensions. Then it is not a trivial question whether and how the stability of baryons is connected with the stability of spacetime. We shall exploit Weyl's first postulate in order to admit new degrees of freedom of motion. The latter involve some special nongrade-preserving rotations, reversions and transpositions that comprise a *reorientation symmetry* (1996), an involutive operation of Clifford algebra which relates to the concept of *polydimensional isotropy*.

It is currently understood that strongly interacting fields are constituted in reflections and transpositions which preserve orientation up to designation of base units, a process rarely observed in high energy physics. There are indeed a few indications and cross references to those facts that can be quoted.[1] Then, William

---

[1] In 1995 Nottale using some results by Feynman derived a spatial fractal dimension of 2 for quantum trajectories. *"The discovery that the typical quantum mechanical paths are continuous but non-*

Pezzaglia Jr. (2000, p. 106) proposed a principle of polydimensional isotropy by saying that *"there is no absolute or preferred direction in the universe to which one can assign the geometry of a vector. Just as which direction you choose to call the z-axis is arbitrary, it is also arbitrary just which geometric element you call the basis vector in z-direction"*. This is in accord with the concept of *local automorphism invariance* of physical laws of motion (Crawford 1994).

A third and the most significant hint originates in the assumption that fermion states formally can be identified with primitive idempotents in Clifford algebra $Cl_{3,1}$ and in the complexified time-space algebra $\mathbb{C} \otimes Cl_{1,3}$ as both algebras contain isomorphic idempotent lattices. Thereby we are following an old idea of Greider and Weiderman(1988), Chisholm and Farewell (1992) and Schmeikal (1996) to construct a tetrahedral symmetry of the four primitives of $Cl_{3,1}$ in order to represent strong interaction and color rotations of quantumchromodynamics (qcd). However, we assume that the tetrahedral rotation stems from an octahedral reorientation symmetry of the idempotents. Thus the multiplets of $SU(3)$ should turn out as real geometric phenomena in flat spacetime. We know that the charge and baryon numbers are conserved in strong interactions. Since in our approach these numbers are now conceived as geometric quantum numbers of algebraic eigenstates, they shall be invariant not only under the usual $SU(3)$ transformations, but also under reorientation of spacetime algebra. We no longer hold the view that at subnuclear level algebraic, topological and metric properties of spacetime are pregiven and are defined from the outset. Instead we may rather speak of local quantum mechanical states of spacetime which are purified by strong interactions.[2]

Just as there are charge- and isospin chains, we may conceive state chains of subnuclear spacetime symmetries such as 4-momenta isospin chains and (hyper)charge chains which turn out as properties of inhomogeneous Lorentz transformations. These somewhat surprising assumptions are, in a way, our start capital. The price we have to pay, at present, is to dispense with the current action principle and Yang–Mills equations (as applied e.g., in Ramond 1990) and to confess at first that it is not yet entirely clear what in this new context has to be regarded as a differential form (Vargas and Torr 2000).

23.2   Algebraic form of physical laws

For reasons of tradition and habit we are modeling motion of electrons and leptons by the Dirac equation. This brings in the Dirac algebra, a geometric algebra generated by matrices over the complex field $\mathbb{C}$. During the last two decades it

---

differentiable and may be characterized by a fractal dimension 2 may be attributed to Feynman (1965, Schweber 1986). Though Feynman evidently did not use the word 'fractal' that was coined in 1975 by Mandelbrot, his description of quantum mechanical paths fully corresponds to this concept".

[2]The concept of a *"dynamical purification of states"* can be comprehended by reading Narnhofer and Thirring 1996.

has been made clear that the algebra over the reals suffices to formulate the Dirac theory and to highlight the geometric meaning of the equations of motion Thus, in the real algebra, the spin S of a fermion can be represented by two γ-matrices in the form of a bivector $-\frac{1}{2}\hbar\gamma_{12}=\frac{1}{2}\hbar\gamma_2\gamma_1$ (Hestenes 1990). Surprisingly, this fact has been rediscovered by physicists several times. The first book I read where the angular 4-momentum was defined by the above bivector was by Harry Lipkin (1965, Ch. 5). Lipkin even identified the γ-matrices as linear combinations of baryon creation- and annihilation operators, for example,

$$\gamma_1 = a_p^+ + a_p, \quad \text{and} \quad \gamma_2 = \mathrm{i}(a_p^+ - a_p), \qquad (2.1)$$

and where p denotes the proton. Angular 4-momenta were $L_{ij} = \frac{1}{2}\gamma_i\gamma_j$ and they obeyed the commutation relations

$$[L_{ij}, L_{jk}] = L_{ik}, \quad i \neq k \qquad (2.2)$$

(see Lipkin 1965, section 5.2: *"Identification of the Lie algebra"*. Today we are sure that even in the Weinberg–Salam model of electroweak interaction, complexification of $C\ell_{1,3}$ was not necessary (Hestenes 1982). It is near at hand, therefore, to pose the question if there is a geometric real version of qcd equations of motion. As soon as we describe subnuclear fermions, the gauge symmetry of the $SU(3)$ comes in. Thus it does not seem so easy at first to show that the imaginary unit is superfluous. The group $SU(3)$ does not seem to have much to do with the spacetime algebra $C\ell_{3,1}$ which is defined over the real field and in addition its representation space is $\mathrm{Mat}(4, \mathbb{R})$, the Majorana algebra. Also the $SU(3)$ is not legitimated by the timespace algebra $C\ell_{1,3}$. But as an algebra it is a subalgebra of the complexified algebras $\mathbb{C} \otimes C\ell_{3,1}$ as well as $\mathbb{C} \otimes C\ell_{1,3}$ in the guise of $SU(4)$ matrices with one vanishing row and column. However, as soon as we become aware of the fact that the emergence of the particle multiplets does not at all rely on complex numbers, but rather on the commutation relations of the Lie algebra we begin to realize that qcd most naturally can be represented in the real algebra $C\ell_{3,1}$. The special unitary group $SU(3)$ is not needed at all. We are thus forced to answer what is a spinor space in the new approach. Formerly we used complex column spinors, now we are free to use real row spinors. We are still conceiving of spinor spaces as minimal ideals. Some may criticize that their basis is heterodimensional, but it was heterodimensional in the Dirac theory too. This can immediately be verified by studying a standard basis (*"erzeugende Einheiten"*) of the standard complex linear spinor space $S = \mathbb{C} \otimes C\ell_{1,3}f$ where f turns out heterodimensional, that is,

$$f = \frac{1}{4}(1 + \gamma_0 + \mathrm{i}\gamma_{12} + \mathrm{i}\gamma_{012}) \qquad (2.3)$$

(see Lounesto 2001, p. 139). Each of the four complex components of one column spinor involves one primitive idempotent of such a form. Last but not least and despite all the mathematics of both $\mathbb{C}h_\chi$ and $\mathrm{Mat}(4, \mathbb{R})$ is easy to handle.

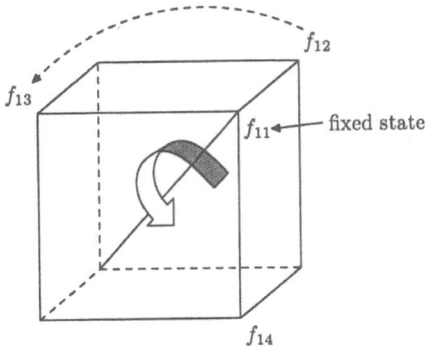

FIGURE 23.1. Recoding by color rotation in $Cl_{3,1}$.

23.3   Recode by motion

The most basic action of intelligence is identification of something. Consider identification of base elements and primitive idempotents in Clifford algebras, erzeugende Einheiten as Hermann Weyl (1931) used to call them. Take a cube with unit vectors or idempotents as corners. A rotation of the cube by an angle of $\frac{2}{3}\pi$ can transpose e_1 onto e_2 and e_2 onto e_3, but leaves the origin unaltered. This can also be achieved by a simple renaming of unit vectors. That is, a geometric operation, being an element of $SO(3)$ at the same time is representing a linguistic operation of recoding of base elements. The whole procedure can be regarded as an involution of the Clifford algebra of $\mathbb{R}^3$ which leaves the orientation of $\mathbb{R}^3$ invariant up to identity of base elements. For idempotents the situation is analogous, but indeed the element of motion is not in the $SO(3)$.

23.4   Transposition in Grassmann monomials

We first define transpositions τ_{ik} abstractly as an exchange of base elements e_i, e_k in Grassmann monomials and more generally in any multivector u of a Clifford algebra $Cl_{p,q}$ with a standard basis $e_1, e_2, \ldots, e_n$, where $n = p + q$, satisfying $e_i^2 = -e_k^2 = 1$, with $1 \leq i \leq p < k \leq n$. It may be that the e_i does occur in the element u in which case it is mapped onto the element e_k. But it may as well be that the e_i is not a factor in any monomial of u. In that case, there is no e_k to be transposed onto e_i. It is therefore that we speak of *transposition* instead of exchange. Depending on the design of the Grassmann monomials constituting the multivector, we can differ between six possible outcomes. Being aware of those, abstracting from indices i, k, we are then able to link the transposition to the other involutions of the Clifford algebra. We use the standard notation^ for the main involution fixing the even subalgebra $Cl_{p,q}^+$. There are three involutions, the grade involution $^\wedge$, the reversion $^\sim$, and the Clifford conjugation $^-$ defined by $u^- = u^{\sim\wedge}$. We add to these the elementary transposition τ_{ik} of any unit vector $\mathbf{e}_i \in Cl_{p,q}$ onto $\mathbf{e}_k$. Any element in $Cl_{p,q}$ is a linear combination of monomials having form

$$X = \mathbf{e}_1^{\eta_1} \mathbf{e}_2^{\eta_2} \ldots \mathbf{e}_n^{\eta_n} = a\mathbf{e}_i^{\eta_i} b\mathbf{e}_k^{\eta_k} c \quad \text{with} \quad \eta_i, \eta_k \in \{0,1\} . \tag{4.1}$$

Transposing $\mathbf{e}_i$ with $\mathbf{e}_k$ brings forth six possible results. We can integrate all of them into

Definition 1.

$$X = a\mathbf{e}_i^{\eta_i} b\mathbf{e}_k^{\eta_k} c \quad \mapsto \quad \tau_{ik}(X) = a\mathbf{e}_k^{\eta_i} b\mathbf{e}_i^{\eta_k} c$$

Depending on the design of Grassmann monomials X we observe the following constitutive properties of transposition τ of base elements $\mathbf{e}_i, \mathbf{e}_k$ in a Grassmann monomial X:

$$\tau_{ik}X = \begin{cases} -X, & \text{for } \eta_i\eta_k = 1; \\ +X, & \text{for } (1-\eta_i)(1-\eta_k) = 1; \\ +ab\mathbf{e}_k c, & \text{for } \eta_i(1-\eta_k) = 1 \text{ and } b \in Cl_{p,q}^+; \\ -ab\mathbf{e}_k c, & \text{for } \eta_i(1-\eta_k) = 1 \text{ and } b \in Cl_{p,q}^-; \\ +a\mathbf{e}_i bc, & \text{for } (1-\eta_i)\eta_k = 1 \text{ and } b \in Cl_{p,q}^+; \\ -a\mathbf{e}_i bc, & \text{for } (1-\eta_i)\eta_k = 1 \text{ and } b \in Cl_{p,q}^-. \end{cases}$$

For base vectors $\mathbf{e}_i, \mathbf{e}_k$ with the positive signature transposition τ_{ik} is a Weyl reflection generated by

$$u_{ik} \equiv \frac{1}{\sqrt{2}}(\mathbf{e}_i - \mathbf{e}_k) \qquad \tau_{ik}(X) = u_{ik} X^\wedge u_{ik}^{-1}. \tag{4.2}$$

In the subspace $Cl_{3,0} \in Cl_{3,1}$ those are representants of the 6 roots from the root lattice A_2 corresponding with the color Lee algebra $su(3, \mathbb{C})$. Those are

$$\{\frac{1}{\sqrt{2}}(\pm\mathbf{e}_1 \mp \mathbf{e}_2), \frac{1}{\sqrt{2}}(\pm\mathbf{e}_2 \mp \mathbf{e}_3), \frac{1}{\sqrt{2}}(\pm\mathbf{e}_1 \mp \mathbf{e}_3)\} .$$

Note, however, that in the Clifford algebra $Cl_{3,1}$ there exists a total of 9 base vectors with signature $+1$, namely the set $\{\mathbf{e}_1, \mathbf{e}_2, \mathbf{e}_3, \mathbf{e}_{14}, \mathbf{e}_{24}, \mathbf{e}_{34}, \mathbf{e}_{124}, \mathbf{e}_{134}, \mathbf{e}_{234}\}$. For base units involving $\mathbf{e}_4$ one cannot construct a 2-component Weyl reflection. One must proceed in a different way. Abstracting from the indices, we use denotation u^τ for the indefinite involutive automorphism $^\tau$ which adds to the equations for involutive operations in $Cl_{p,q}$

	Involutive	Automorphism	Antiautomorphism
Grade involution	$(u^\wedge)^\wedge = u$	$(uv)^\wedge = u^\wedge v^\wedge$	
Reversion	$(u^\sim)^\sim = u$		$(uv)^\sim = v^\sim u^\sim$
Conjugation	$(u^-)^- = u$		$(uv)^- = v^- u^-$
Transposition	$(u^\tau)^\tau = u$	$(uv)^\tau = u^\tau v^\tau$	

Commutation relations: $(u^\wedge)^\tau = (u^\tau)^\wedge$, $(u^\sim)^\tau = (u^\tau)^\sim$.

In accord with equations (4.2) it is legitimate to speak of a transposition $C\ell_{p,q}^\tau$ of the algebra as a Weyl reflection of the grade involuted algebra.

23.5   Transpositions in the basis of Clifford algebra

A generalization to the transposition of base elements in $\mathbb{R}^{p,q}$ are transpositions in the whole basis of the Clifford algebra $C\ell_{p,q}$. We are now especially interested in two things (1) in the representation of transposition operators and (2) in the transposition of primitive idempotents. Generally all primitive idempotents f to be considered in the algebra can be used to form terms $1 - 2f$. Those are transpositions. As can easily be seen transpositions of primitives also involve transpositions in the basis of the algebra. We shall now restrict the rigor to $C\ell_{3,1}$ because this is the most important and entirely real building block for qcd. This is not an artificial restriction, and there is no problem occurring from the difference between spacetime- and timespace algebra.

23.5.1   Transposition in $C\ell_{3,1}$

We define idempotents in a way most of us are perhaps not familiar with. Namely, we introduce six isomorphic color spinor spaces $\mathbb{C}h_\chi$ containing six families of primitives $f_{\chi k}$ with $\chi = 1, \ldots 6$, by minimal basis as follows (Schmeikal 2001).

Definition 2. *Color spaces* $\mathbb{C}h_\chi$ *are defined as:*

$$\mathbb{C}h_1 = \{e_1, e_{24}\}, \quad \mathbb{C}h_2 = \{e_1, e_{34}\}, \quad \mathbb{C}h_3 = \{e_2, e_{34}\},$$
$$\mathbb{C}h_4 = \{e_2, e_{14}\}, \quad \mathbb{C}h_5 = \{e_3, e_{14}\}, \quad \mathbb{C}h_6 = \{e_3, e_{24}\}. \quad (5.1)$$

Each color spinor space is generated by one base unit $e_j, j = 1, 2, 3$ of positive signature and bivector $e_{k4}, k = 1, 2, 3$ and $k \neq j$. Therefore each such space has basis $\{1, e_j, e_{k4}, e_{jk4}\}$. It contains scalar, vector, bivector and trivector parts having signature $\{ +, +, +, + \}$. So we observe $e_j^2 = e_{k4}^2 = e_{jk4}^2 = +1$. The base elements e_j, e_{k4}, e_{jk4} constitute commuting conjugate triples. For arbitrary multivectors ϕ, ψ in any color space $\mathbb{C}h_\chi$ the commutator $[\phi, \psi]$ vanishes. Thus we investigate a very special type of linear space with a commutative geometry embedded into the noncommutative geometry of the Clifford algebra of the Minkowski spacetime. The enormous gain brought on by such a procedure can only be discovered by going deeper into the matter step by step. Let us write down in explicit form the primitives associated with color space $\mathbb{C}h_1$:

$$f_{11} = \tfrac{1}{4}(1 + e_1)(1 + e_{24}), \qquad f_{12} = \tfrac{1}{4}(1 + e_1)(1 - e_{24}),$$
$$f_{13} = \tfrac{1}{4}(1 - e_1)(1 + e_{24}), \qquad f_{14} = \tfrac{1}{4}(1 - e_1)(1 - e_{24}). \quad (5.2)$$

Those mutually annihilating idempotents generate a 16 element lattice of idempotents (Lounesto 2001, p. 227). There is a total of six such lattices which correspond with the six color spaces and are obtained by permuting indices in accord with their definitions. Habitually, we have considered only one such set of primitives, usually $F_1 = \{f_{1k}\}$, because this suffices (a) to decompose the algebra into a sum of minimal left ideals and (b) to define the spinor space in the traditional way. However, it is not enough here, as this would seriously reduce the whole mathematics. We would not be able to construct transposition operators as we are going to do now.

Lemma 1 (Transpositions). *Consider the primitive idempotents $f_{\chi k}$. Then each inhomogeneous linear form $s_{\chi k} = 1 - 2f_{\chi k}$ is a transposition operator. There are 24 such forms.*

This transposes both the 24 primitive idempotents as well as the base units of color spaces $\mathbb{C}h_\chi$ and $C\ell_{3,1}$ in general. We have

$$s_{\chi k}^2 = (1 - 2f_{\chi k})^2 = 1 - 4f_{\chi k} + 4f_{\chi k} = 1 \qquad (5.3)$$

because $f_{\chi k}$ is an idempotent. Thus, the $s_{\chi k}$ are representants of Coxeter reflections[3] with period 2. This is in accordance with their property of being transpositions.

Definition 3 (Transposition of Idempotents). *The action of the transposition operators is somewhat analogous to the action of rotation operators, that is, a primitive $f_{\chi k}$ is transposed by a transposition onto another primitive by forming the term*

$$s_{\beta k}^{-1} f_{\alpha j} s_{\beta k} = s_{\beta k} f_{\alpha j} s_{\beta k} \qquad \text{(notice that } s_{\beta k} \text{ is its own inverse).} \qquad (5.4)$$

A brief calculation shows that for any fixed index χ (fixed color space $\mathbb{C}h_\chi$) the associated primitives $f_{\chi k}$ are invariant under the action of transpositions $s_{\chi k}$. It is therefore that one set of primitives is not sufficient to construct reorientation symmetries. We need all 6 to bring on *qcd*.

Theorem 1 (Reorientation Symmetry of $C\ell_{3,1}$). *The 24 basic transpositions $s_{\chi k}$ of $C\ell_{3,1}$ under Clifford multiplication generate a finite non-abelian group $\Omega(C\ell_{3,1})$ of degree $1152 = 3.2^4.4!$ which implies that the hyperoctahedral subgroup H_4 to dimension 4 belonging to the idempotent hypercubes is contained in $\Omega(C\ell_{3,1})$ with index 3.*

For a treatment of hypercube: see Lounesto 2001, p. 82. The validity of the above theorem has been verified with CLIFFORD, a Maple V package for Clifford algebra computations (see Abłamowicz 1998).

---

[3]For *reflection groups in real Clifford algebras*, see: Morris and Makhool 1992

23.6 Types of action of transpositions in $C\ell_{3,1}$

(i) For a definite family χ, transpositions $s_{\chi k}$ leave F_χ unaltered *pointwise*.

(ii) Given color χ, the sets F_μ with $\mu = \chi+1$, $\chi+3$, $\chi+5 \bmod 6$, are preserved by $s_{\chi k}$, while each is *altered pointwise*. There occurs *intra*-transposition of idempotents within families μ.

(iii) There is a pair of lattices $F_{\chi+2 \bmod 6}$, $F_{\chi+4 \bmod 6}$ between which idempotents are exchanged. In that case there is *inter*-transposition of idempotents. This amounts to an exchange of colors.

These statements will turn out to be of considerable importance for a unified geometric model of quantumchromodynamics.

23.6.1 *Invariance of laws of motion under reorientation*

We showed in 2000 that the set of logic binary connectives is invariant under the whole reorientation group $\Omega(C\ell_{3,1})$ of the algebra. We shall now ask whether invariance with respect to recoding as reorientation plays any role in physics.

Hypothesis 1 (Recoding Invariance). *There are laws of physics which are invariant under recoding, that is, they are unaltered by the action of reorientation symmetries.*

By recoding invariance we may even legitimate:

Hypothesis 2 (Constitution of Equations of Motion). *This means there are laws of motion which derive their dynamic structure from reorientation symmetries. QCD is such a bundle of laws of motion which stems from reorientation.*

23.7 Structure of color spinor spaces $\mathbb{C}h_\chi \subset C\ell_{3,1}$

Consider isospin $\Lambda_3 = \frac{1}{2}(e_{24} - e_{124})$ and the idempotent $f = \frac{1}{2}(1 - e_1)$ both in $\mathbb{C}h_1$. Let $n \in \mathbb{N}$ be a natural number. Then we verify the identities

$$\Lambda_3^{2n} = f \quad \text{and} \quad \Lambda_3^{2n-1} = \Lambda_3. \qquad (7.1)$$

That is, Λ_3 represents a *swap* (or *swop*). The color space $\mathbb{C}h_1$ can be decomposed into two ideals

$$\mathbb{C}h_1 = \mathbb{C}h_1 f \oplus \mathbb{C}h_1 \widehat{f} = \mathcal{G}_1 \oplus \widehat{\mathcal{G}}_1, \qquad (7.2)$$

where $\mathcal{G}_1 = \{f, \Lambda_3\}$ and $\widehat{\mathcal{G}}_1 = \{\widehat{f}, \widehat{\Lambda}_3\}$.

The equations (7.1) imply that spaces $\mathcal{G}_1$ and $\widehat{\mathcal{G}}_1$ are isomorphic with $C\ell_{1,0} = \{1, e_1\} \simeq {}^2\mathbb{R} = \mathbb{R} \oplus \mathbb{R}$ – the double ring of real numbers. Clearly, the color spaces are all isomorphic. Therefore, because of equations (7.2), we end up with a very simple decomposition:

$$\mathbb{C}h_\chi \simeq \mathbb{R} \oplus \mathbb{R} \oplus \mathbb{R} \oplus \mathbb{R} = {}^4\mathbb{R} \quad \text{(the fourfold real ring).} \quad (7.3)$$

In ${}^4\mathbb{R}$ the basis of $\mathbb{C}h_\chi$ is represented as follows:

$$1 = (1,1,1,1), \qquad \mathbf{e}_1 = (1,1,-1,-1),$$
$$\mathbf{e}_{24} = (1,-1,1,-1), \qquad \mathbf{e}_{124} = (1,-1,-1,1). \quad (7.4)$$

By a componentwise multiplication these base elements generate *Klein 4-group*. Therefore we say: the *Dirac group* (Shaw 1995) of color spinor spaces is the *Klein's Vierergruppe*. In the representation (7.4) the idempotents are:

$$f = (0,0,1,1), \qquad \Lambda_3 = (0,0,1,-1) \quad \text{– for } \mathcal{G}_1,$$
$$f_{11} = (1,0,0,0), \qquad f_{12} = (0,1,0,0),$$
$$f_{13} = (0,0,1,0), \qquad f_{14} = (0,0,0,1) \quad \text{– primitives in } C\ell_{3,1}. \quad (7.5)$$

This is a spinorial decomposition into what we formerly called fundamental SU(3)-states where the components f_{12}, f_{13}, f_{14} represent *quarks*. Instead of a column of complex numbers we have now a row of real numbers.

23.8   Quantumchromodynamics

The reorientation group contains six octahedral subgroups $O_{(x)}$ [4], each having a minimal set of two generating transpositions as usual and isomorphic with the symmetric group S_4 having 24 elements. (Schmeikal 1996, p. 90, Schmeikal 2001). Eight of those are *trigonal tetrahedral* and 6 are *tetragonal tetrahedral*. The 8 trigonal operators have the special property that each of them carries out the same flavor rotation in the color space to which it is associated. Consider $\mathbb{C}h_1$, we define such a trigonal operator T by the product of two base reflections, e.g., s_{64}, s_{23}. Multivectors $_{(x)}T_\nu$ have a rather complex form. They are linear combinations of all 16 base elements of the algebra with coefficients equal to $\frac{1}{4}$:

$$_{(1)}T_1 = s_{64}s_{23} = \tfrac{1}{2}(1 + \mathbf{e}_3 + \mathbf{e}_{24} + \mathbf{e}_{234})\tfrac{1}{2}(1 + \mathbf{e}_1 - \mathbf{e}_{34} + \mathbf{e}_{134}) \quad (8.1)$$

An important observation which led us to the concept of color spinor space was the heterodimensional triality of the idempotents $f_{\chi 2}, f_{\chi 3}, f_{\chi 4}$ while fixating a lepton state $f_{\chi 1}$. By the aid of tetrahedral trigonal operators T such triality is mapped onto the base elements $\mathbf{e}_j, \mathbf{e}_{k4}, \mathbf{e}_{jk4} \in C\ell_{3,1}$ of the algebra.[5] While the idempotents representing fermion states are transposed in a three-cycle, that is, flavor-rotated, the base elements of the color spaces are rotated in parallel.

---

[4] An elementary introduction to spacegroups can be found in: Belger and Ehrenberg, 1981, 56.

[5] There are indeed less special and more general ways to represent triality in quantum systems (Fauser 2000).

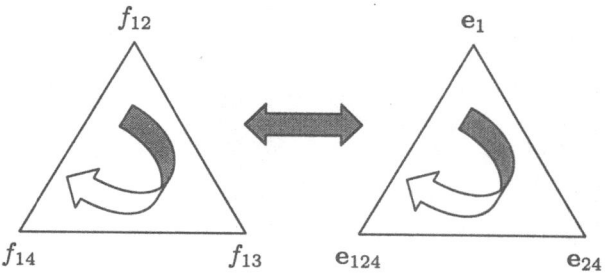

FIGURE 23.2. Quantumchromodynamic transpositions in $C\ell_{3,1}$. Color rotation implies heterodimensional motion.

23.9   The $SU(3)$ symmetry

Classically a trigonal (color- or flavor) rotation is an element of the $SU(3)$. If we complexify the basis and consider $\mathbb{C} \otimes C\ell_{3,1}$ the operator $_{(1)}T_1 = s_{64}s_{23}$ is indeed an element of $SU(3)$ disguised as a $SU(4)$-matrix with only one non-vanishing element, a real unit, in the first row and column. In my earlier works, I investigated the complexified algebras $\mathbb{C} \otimes C\ell_{3,1}$ and $\mathbb{C} \otimes C\ell_{1,3}$. Only lately it turned out that this was not necessary. Yet it was, and still is, helpful to develop the line of argumentation. Namely, there arises the following question almost all by itself:

Question. *If the six fundamental trigonal rotations can be derived from the transpositions of the reorientation symmetry, should it not be natural to derive the generators of the $SU(3)$ from those symmetries too?*

This question was first answered by me in a publication in 1996. In the nowaday context we should put it that way: Take any color spinor space $\mathbb{C}h_1$, then the eight generators of that $su(3)$ which contains trigonal rotations acting on fermions $f_{\chi k}$ are linear combinations with coefficients $\pm\frac{1}{2}$ of only two symmetry elements from some octahedral subgroup of the reorientation group, provided we complexify $C\ell_{3,1}$ (Schmeikal 1996, p. 94) [6]. Each generator is heterodimensional. It contains two base elements of $C\ell_{3,1}$ with different grade. Three of them are to be multiplied by the imaginary unit. Is that unavoidable?

---

[6]Irreducible real representation of O_h in $\mathrm{Mat}(3,\mathbb{R})$ can be found in Petraschen and Trifonow 1969

23.10 Decomplexifying $SU(3) \subset \mathbb{C} \otimes C\ell_{3,1}$ and finding its normal real form in $C\ell_{3,1}$

We know from the old theory that it is sufficient to have eight real operators $t_+, t_-, t_3, u_+, u_-, v_+, v_-, y$ provided they obey the necessary commutation relations. An algebraic representation of the $SU(3)$, it was said, is a correspondence which preserves the below table of commutation relations (Cahn 1984): If the Λ_i

	t_+	t_-	t_3	u_+	u_-	v_+	v_-	y
t_+	0	$2t_3$	$-t_+$	v_+	0	0	$-u_-$	0
t_-	$-2t_3$	0	t_-	0	$-v_-$	u_+	0	0
t_3	t_+	$-t_-$	0	$-u_+/2$	$u_-/2$	$v_+/2$	$-v_-/2$	0
u_+	$-v_+$	0	$u_+/2$	0	$\frac{3}{2}y - t_3$	0	t_-	$-u_+$
u_-	0	v_-	$-u_-/2$	$-\frac{3}{2}y + t_3$	0	$-t_+$	0	u_-
v_+	0	$-u_+$	$-v_+/2$	0	t_+	0	$\frac{3}{2}y + t_3$	$-v_+$
v_-	u_-	0	$v_-/2$	$-t_-$	0	$-\frac{3}{2}y - t_3$	0	v_-
y	0	0	0	u_+	$-u_-$	v_+	$-v_-$	0

TABLE 1. Commutation relations

are defined as follows, the shift operators obey the correct $SU(3)$-commutation relations. Consider at first these elements:

- Real isospin:

$$t_+ = t_1 - t_2, \quad t_- = t_1 + t_2, \quad t_3$$
$$t_1 = \tfrac{1}{2}\Lambda_1, \qquad t_2 = \tfrac{1}{2}\Lambda_2, \qquad t_3 = \tfrac{1}{2}\Lambda_3, \tag{10.1}$$

where $\Lambda_1 = \tfrac{1}{2}(e_{34} - e_{134})$, $\Lambda_2 = \tfrac{1}{2}(-e_{23} + e_{123})$, $\Lambda_3 = \tfrac{1}{2}(e_{24} - e_{124})$.

With equations (10.1) and a little rigor it can be verified that t_+, t_-, t_3 fulfill the isospin Lie algebra $su(2)$ and are therefore generators of the group $SU(2)$. But $\Lambda_1, \Lambda_2, \Lambda_3$ are not. Although $\{\Lambda_1, \Lambda_2, \Lambda_3\}$ are linear combinations of $\{t_\pm, t_3\}$ they obey commutation relations which differ significantly from those of $su(2)$ and $so(3)$: $[\Lambda_1, \Lambda_2] = 2\Lambda_3$, $[\Lambda_2 \Lambda_3] = 2\Lambda_1$, $[\Lambda_1, \Lambda_3] = 2\Lambda_2$. Notice the positive sign of $[\Lambda_1, \Lambda_3]$, which would be negative for $so(3)$. This means that the algebra spans $\mathbf{Spin}_{2,1}^+ \simeq \mathbf{Spin}_{1,2}^+ \simeq Sl(2,\mathbb{R})$. The reason is that shift operators and Λ's live in different spaces; The $t_\pm = \tfrac{1}{4}(\pm e_{23} + e_{34} \mp e_{123} - e_{134})$ involve negative signature and live in an 8-dimensional Clifford algebra. They superimpose such that their linear combinations $\Lambda_1 = t_+ + t_- = \tfrac{1}{2}(e_{34} - e_{134})$ and $\Lambda_2 = t_- - t_+ =$

spans $\mathbf{Spin}_{2,1}^{+} \simeq \mathbf{Spin}_{1,2}^{+} \simeq Sl(2,\mathbb{R})$. The reason is that shift operators and Λ's live in different spaces; The $t_{\pm} = \frac{1}{4}(\pm e_{23} + e_{34} \mp e_{123} - e_{134})$ involve negative signature and live in an 8-dimensional Clifford algebra. They superimpose such that their linear combinations $\Lambda_1 = t_+ + t_- = \frac{1}{2}(e_{34} - e_{134})$ and $\Lambda_2 = t_- - t_+ = \frac{1}{2}(-e_{23} + e_{123})$ decompose that space into two different 4-dimensional subspaces one of which is a color space and isomorphic to $^4\mathbb{R}$, while the other is spanned by $\{1, e_1, e_{23}, e_{123}\}$ and isomorphic to $^2\mathbb{C}$. As a result, in Clifford algebras linear combinations of generators of some group $\mathfrak{G}$ may cover different real or complex forms of an algebra which generates different forms of the group $\mathfrak{G}$. In our case the new form of the isospin subgroup SU(2) is $Sl(2,\mathbb{R})$. Going further now, the other operators have to be:

- Real V-spin:

$$v_+ = v_1 + v_2, \qquad v_- = v_1 - v_2,$$
$$v_1 = \tfrac{1}{2}\Lambda_4, \qquad v_2 = \tfrac{1}{2}\Lambda_5, \qquad (10.2)$$

  where $\Lambda_4 = \frac{1}{2}(e_2 + e_{14})$, $\Lambda_5 = -\frac{1}{2}(e_4 + e_{12})$.

- Real U-spin:

$$u_+ = u_1 + u_2, \qquad u_- = u_1 - u_2,$$
$$u_1 = \tfrac{1}{2}\Lambda_6, \qquad u_2 = \tfrac{1}{2}\Lambda_7, \qquad (10.3)$$

  where $\Lambda_6 = \frac{1}{2}(e_3 + e_{234})$, $\Lambda_7 = -\frac{1}{2}(e_{13} + e_{1234})$.

- Hypercharge:

$$y = \tfrac{1}{2}\Lambda_8 = \tfrac{1}{6}(-2e_1 + e_{24} + e_{124}). \qquad (10.4)$$

If the basis generating the $C\ell_{3,1}$ is chosen to be

$$e_1 = \begin{pmatrix} \sigma_3 & 0 \\ 0 & -\sigma_3 \end{pmatrix}, \qquad e_2 = \begin{pmatrix} -\sigma_{123} & 0 \\ 0 & -\sigma_{123} \end{pmatrix},$$

$$e_3 = \begin{pmatrix} -\sigma_2 & 0 \\ 0 & \sigma_2 \end{pmatrix}, \qquad e_4 = \begin{pmatrix} 0 & \sigma_{123} \\ \sigma_{123} & 0 \end{pmatrix}, \qquad (10.5)$$

the multivectors $\Lambda_1, i\Lambda_2, \Lambda_3, i\Lambda_4, \Lambda_5, i\Lambda_6, \Lambda_7, \Lambda_8$ are essentially the Gell-Mann matrices which constitute the generators in the fundamental representation of the Lee group by hermitian traceless matrices. There are of course many more representations of the Clifford algebra which are leading to no less fundamental representations of the group. But what we are interested in here are the real forms of the algebra su$(3,\mathbb{C})$ which decompose into the same direct sum of root spaces from which the standard classification of particles is done. We apply what we may call a generalization of the *"Weyl unitary trick"* by canceling the imaginary unit factor at the generators $i\Lambda_2, i\Lambda_4, i\Lambda_6$. As all the Lambdas can be conceived as elements of the Majorana algebra Mat$(4,\mathbb{R})$ our trick must result in the *normal real form* of the complex Lie algebra which allows for the typical Cartan decomposition. Clearly, the Λ_i do not span the algebra su$(3,\mathbb{C})$, but one of its noncompact real forms, namely the algebra $sl(3,\mathbb{R})$. Let us see why that must be so.

23.11   Decompositions and real forms of a complex algebra

This section should help us to understand the approach a little better. Consider a Lee algebra g as a linear vector space over a field F, with an antisymmetric product defined by the Lie bracket. The Lie algebra is called real, complex or quaternion in accord with the identity of the number field F. Taking linear combinations of the elements of a real Lie algebra with complex coefficients, a *complexification* of the real algebra is obtained. A real algebra h is a real form of the complex algebra c if c is the complexification of h. By the exponential mapping the algebra g generates a Lie group. Suppose now that g is a compact simple Lie algebra and τ is an involutive automorphism of g such as transposition or conjugation. τ induces a decomposition of g into eigensubspaces $g = k \oplus p$ which correspond with the eigenvalues ± 1 of τ. That is, $\tau(X) = +X$ for $X \in k$ and $\tau(X) = -X$ for $X \in p$. Now, k is a subalgebra, but p is not. If we multiply the elements of p by i - a procedure which we call "*the Hermann Weyl unitary trick*" - we construct a new noncompact algebra $g* = k \oplus ip$. This is a so called *Cartan decomposition*, and k is a *maximal compact subalgebra* of $g*$. Suppose $g = \mathrm{SU}(n, \mathbb{C})$ is the group of unitary complex matrices with determinant $+1$. The algebra of this group consists of complex, traceless, antihermitian matrices. Elements are obtained from the generators X_α by the exponential map $g = e^{c_\alpha X_\alpha}$ with real c_α. Therefore any matrix X in the algebra of $\mathrm{su}(n, \mathbb{C})$ can be written as $X = A + iB$ with A real, skew-symmetric and B real, symmetric, both traceless. Therefore the algebra $g = \mathrm{su}(n, \mathbb{C}) = k \oplus p$ where k is the compact connected subalgebra $\mathrm{so}(n, \mathbb{R})$ of real, skew-symmetric, traceless matrices. p is the subspace of imaginary iB with real, symmetric and traceless B. By the Weyl unitary trick we obtain from g the non-compact algebra $g* = k \oplus ip$ where ip is the subspace of real, symmetric matrices B with zero trace. The new algebra $g* = sl(n, \mathbb{R})$ is the set of traceless $n \times n$ real matrices which generates the linear group of matrices with unit determinant. Referring to the case of $\mathrm{su}(3, \mathbb{C})$ we realize by putting $X_\alpha = \frac{i}{2}\lambda_\alpha$ with Gell-Mann matrices $\lambda_\alpha, \alpha = 1, ..., 8$ that the set X_α under the involution of complex conjugation decomposes into two sets. It splits the compact algebra into $g = \mathrm{su}(3, \mathbb{C}) = k \oplus p$, as p consists of imaginary matrices only. We obtain 3 real matrices $k = \{X_2, X_5, X_7\}$ and $p = \{X_1, X_3, X_4, X_6, X_8\}$. Now k spans the real subalgebra $\mathrm{so}(3, \mathbb{R})$. The Cartan subalgebra $ih_0 = \{X_3, X_8\}$ is hidden in the subspace p.

Any (semi)simple complex algebra g decomposes into a direct sum of root spaces (Sattinger, Weaver 1996) $g = h_0 \oplus \sum_\alpha g_\alpha$ where h_0 is the Cartan subalgebra, in our case $\{t_3, y\}$, and $g_{\pm\alpha} = \{s_{\pm\alpha}\}$ are the sets of shift operators, here $\{t_\pm, u_\pm, v_\pm\}$. There are several real forms

$$\sum_i c_i h_i + \sum_\alpha c_\alpha s_\alpha \qquad c_i, c_\alpha \in \mathbb{C} \qquad (11.1)$$

of the complex Lie algebra which use to be classified according to all the involutive automorphisms τ of g satisfying $\tau^2 = 1$. One of these real forms consists of the subspace where all coefficients are real. This is the *normal real form*. Therefore, consider the above decomposition of $su(3, \mathbb{C})$. We apply the Weyl-trick to get the normal real form $su(3, \mathbb{C})$ which is $sl(3, \mathbb{R})$. It turns out that the k and ip are indeed the subspaces

$$k = \{\frac{(s_\alpha - s_{-\alpha})}{\sqrt{2}}\}, \quad ip = \{h_i, \frac{(s_\alpha + s_{-\alpha})}{\sqrt{2}}\} \tag{11.2}$$

The complex algebra $sl(3, \mathbb{C})$ is characterized by the root lattice A_2 which also belongs to its real forms $su(3, \mathbb{C}), sl(3, \mathbb{R}), su(2, 1; \mathbb{C}), su(1, 2; \mathbb{C})$. There remains the question if the above Cartan decomposition into three plus five generators does only hold true in the special representation (10.5) of the Clifford algebra $Cl_{3,1}$ or if it applies to the multivectors Λ_k as well. The answer is, that the Cartan decomposition into two sets one of which, namely $k = \{\Lambda_2, \Lambda_5, \Lambda_7\}$, spans the compact connected real subalgebra $so(3, \mathbb{R})$, applies independently of the representation of the Clifford algebra. This can be seen from the commutation relations just as well as from an investigation of the exponential map of multivectors which will be carried out in a later section. We conclude that the algebra generated by the Λ_k is a *Clifford algebra normal real form* $sl(3, \mathbb{R})$.

23.12   Original multiplet of Gell-Mann and Zweig

The isospin

$$t_3 = \tfrac{1}{2}\Lambda_3 = \tfrac{1}{4}(e_{24} - e_{124})$$

and the hypercharge

$$y = \tfrac{1}{2}\Lambda_8 = \tfrac{1}{6}(-2e_1 + e_{24} + e_{124})$$

in strong interaction constitute the charge operator as defined by the Gell-Mann–Nishijima relation $q = \frac{y}{2} + t_3$. We thus obtain an algebraic charge operator

$$q = \tfrac{1}{6}(-e_1 + 2e_{24} - e_{124}) = f_{13} + \tfrac{1}{3}(f_{11} - 1) \tag{12.1}$$

which possesses algebraic eigenstates satisfying the eigenvalue equation

$$q\psi = Q\psi \tag{12.2}$$

Using equation (12.1) we can easily calculate the charge Q of fermion states being eigenstates of q :

$$
\begin{aligned}
&qf_{11} = 0, && \text{charge} \quad Q = 0 \quad && \text{lepton}, \\
&qf_{12} = -\tfrac{1}{3}f_{12}, && \text{charge} \quad Q = -\tfrac{1}{3} \quad && \langle s \rangle, \\
&qf_{13} = +\tfrac{2}{3}f_{13}, && \text{charge} \quad Q = +\tfrac{2}{3} \quad && \langle u \rangle, \\
&qf_{14} = -\tfrac{1}{3}f_{14}, && \text{charge} \quad Q = -\tfrac{1}{3} \quad && \langle d \rangle \quad \text{quark.}
\end{aligned}
\tag{12.3}
$$

Demanding strangeness to be

$$s = -f_{12} \qquad (12.4)$$

we obtain $S = -1$ for the isospin singlet f_{12} having hypercharge $Y = -\frac{2}{3}$ but $S = 0$ for the isospin doublet f_{13}, f_{14} as well as for the lepton f_{11} which is consistent. The baryon number operator thus becomes

$$b = y - s = \tfrac{1}{12}(3 - \mathbf{e}_1 - \mathbf{e}_{24} - \mathbf{e}_{124}) \qquad (12.5)$$

On this basis we can calculate the five quantum numbers of the quark triplet as eigenvalues of algebraic operators t_3, y, q, s and b.

	f_{11} $\lvert \nu \rangle$	f_{12} $\lvert s \rangle$	f_{13} $\lvert u \rangle$	f_{14} $\lvert d \rangle$	
t_3	0	0	$+\frac{1}{2}$	$-\frac{1}{2}$	isospin
y	0	$-\frac{2}{3}$	$+\frac{1}{3}$	$+\frac{1}{3}$	hypercharge
q	0	$-\frac{1}{3}$	$+\frac{2}{3}$	$-\frac{1}{3}$	charge
s	0	-1	0	0	strangeness
b	0	$+\frac{1}{3}$	$+\frac{1}{3}$	$+\frac{1}{3}$	baryon number

TABLE 2. Quantum numbers of the fundamental quark triplet

This is indeed the original pattern of the quark triplet as was proposed independently in 1963 by Gell-Mann and Zweig (Fritzsch, Gell-Mann and Leutwyler 1973, Glashow, Iliopoulos and Maiani 1970, Zweig 1964). Involving the other five color spinor spaces other flavors and colors come in.

23.12.1   Invariance of fundamental triplet under reorientation

Claim 1 (Recoding Invariance). *There are laws of physics which are invariant under recoding, that is, they are unaltered by the action of reorientation symmetries. It is now but a matter of rigor to show that the above structure with the fundamental dynamics of qcd is invariant under the reorientation symmetry $\Omega(\mathcal{Cl}_{3,1})$. This means there are laws of motion which derive their dynamic structure from reorientation symmetries. QCD is such a bundle of laws of motion which* *stems from reorientation. Therefore we shall actualize: It is important to consider invariance of physical laws with respect to mere recoding of base elements and erzeugende Einheiten.*

23.13   What is Λ?

Apart from the hypercharge, there is a total of seven real generators Λ_i acting on sixteen possible base elements of the Clifford algebra $C\ell_{3,1}$ in terms of transformations having form

$$e^{-\phi\Lambda_i} e_{j_1 j_2 \ldots j_k} e^{\phi\Lambda_i} \quad \text{with } k \leq 4 \tag{13.1}$$

and thus bringing forth a total of 112 outcomes. Those can be divided into two equivalence classes as follows:

(A1) Generators $\Lambda_1, \Lambda_3, \Lambda_4, \Lambda_6$ involve hyperbolic transformations as relativistic effects on base elements of any grade. They form subspace ip.

(B1) Generators $\Lambda_2, \Lambda_5, \Lambda_7$ involve harmonic transformations as non-relativistic rotations of base elements of any grade: They form the compact real subalgebra $\boldsymbol{k} = \mathrm{so}(3, \mathbb{R})$.

(A2) A flavor rotation T_1 transforms the operators $\Lambda_1, \Lambda_3, \Lambda_4, \Lambda_6$ into an equivalent set which contains $\Lambda_1, \Lambda_4, \Lambda_6$, but projects isospin Λ_3 onto a swap of equivalent form $\Lambda_3^t \in \mathbb{C}h_1$ which is the isospin $\Lambda_3^t = \frac{1}{2}(-e_1 + e_{124})$.

(B2) A flavor rotation $T_1 \in \mathbb{C}h_1$ being a cyclic transposition of generators $\Lambda_2, \Lambda_5, \Lambda_7$ in the compact subspace $\boldsymbol{k}$, at the same time causes color rotations between spaces $\mathbb{C}h_2$ and $\mathbb{C}h_6$. $\frac{1}{2}(e_{34} - e_{134})$ turns into $\frac{1}{2}(-e_3 - e_{234})$. Thereby flavor f_{24} is mapped onto flavor f_{63}.

The essence of all this is (1) that there are two types of generators $\{A, B\}$ with distinguished geometric interpretation and (2) a color rotation never bridges relativistic and non-relativistic modes of dynamics. However, both modalities once observed are heterodimensional.

(C1) Multivectors $\Lambda_1, \Lambda_3, \Lambda_4, \Lambda_6$ are swaps. Squared they give idempotents. Each of them taken alone generates an algebra isomorphic with $C\ell_{1,0}$. From this fact there originates the Lorentz form of transformations. Further they obey equations

$$e^{\phi\Lambda_i} = 1 + \Lambda_i \sinh\phi + \Lambda_i^2(\cosh\phi - 1), \quad i = 1, 3, 4, 6. \tag{13.2}$$

(C2) $\Lambda_2, \Lambda_5, \Lambda_7$ have some more simple geometric property. They are generators of $\mathrm{so}(3, \mathbb{R})$ and can be interpreted as differences between period 4 rotations and their inverses. Though the flavor $\mathrm{SU}(3)$ can be distinguished from the color $\mathrm{SU}(3)$ a flavor rotation in some color space always causes a color rotation in two other color spaces.

23.14   The Λ_i as Lorentz transformations

It was shown that elements $\Lambda_1, \Lambda_2, \Lambda_3$ generate the $Sl(2, \mathbb{R})$. They carry out Lorentz transformations while preserving the base unit e_1 :

- Classical proper orthochronous Lorentz transformations:

$$\Lambda_1 : \begin{cases} e_3' = e_3 \cosh\phi + e_4 \sinh\phi \\ e_4' = e_3 \sinh\phi + e_4 \cosh\phi \end{cases} \qquad \Lambda_2 : \begin{cases} e_2' = e_2 \cos\phi - e_3 \sin\phi \\ e_3' = e_2 \sin\phi + e_3 \cos\phi \end{cases}$$

$$\Lambda_3 : \begin{cases} e_2' = e_2 \cosh\phi + e_4 \sinh\phi \\ e_4' = e_2 \sinh\phi + e_4 \cosh\phi \end{cases} \tag{14.1}$$

Λ_1 acts on a total of 16 base elements of the Clifford algebra only six of which result in a classic form. Thus, Λ_1 preserves all states in $\mathbb{C}h_2$ and classically Lorentz transforms the plane according to (14.1). However, 10 transformations out of 16 are indeed heterodimensional, e.g.:

- Lorentz transforms of bivectors:

$$\Lambda_1 : \begin{cases} e_{13}' = e_{13} \cosh\phi + e_{14} \sinh\phi \\ e_{14}' = e_{13} \sinh\phi + e_{14} \cosh\phi \end{cases} \tag{14.2}$$

Or consider Λ_6 which carries out a heterodimensional transformation of time while preserving the color space $\mathbb{C}h_6$ pointwise

- Lorentz extension of time:

$$\Lambda_6 : \begin{cases} e_4' = e_4 \cosh\phi - e_{34} \sinh\phi \\ e_{34}' = -e_4 \sinh\phi + e_{34} \cosh\phi \end{cases} \tag{14.3}$$

Reader may be interested in the connection with special representations of the Lorentz group. Well, take for example the following matrix representation of the spacetime algebra similar to the one sometimes proposed by Pertti Lounesto (Lounesto 2001, p. 212)

$$e_1 = \begin{pmatrix} \sigma_3 & 0 \\ 0 & -\sigma_3 \end{pmatrix}, \qquad e_2 = \begin{pmatrix} -\sigma_1 & 0 \\ 0 & -\sigma_1 \end{pmatrix},$$

$$e_3 = \begin{pmatrix} 0 & \sigma_3 \\ \sigma_3 & 0 \end{pmatrix}, \qquad e_4 = \begin{pmatrix} i\sigma_2 & 0 \\ 0 & i\sigma_2 \end{pmatrix} \tag{14.4}$$

Then Λ_4 is the following real matrix

$$\Lambda_4 = \tfrac{1}{2}(e_2 + e_{14}) = \begin{pmatrix} 0 & 0 & 0 & 0 \\ 0 & 0 & 0 & 0 \\ 0 & 0 & 0 & 1 \\ 0 & 0 & 1 & 0 \end{pmatrix}.$$

Consider further a classical Lorentz boost in direction 3 which is thus restricted to a 2-dimensional space $\mathbb{R}^{1,1}$, precisely we have a double restricted, proper, orthochronous Lorentz transformation in one time- and one space-dimension:

$$L(\phi) = \begin{pmatrix} 0 & 0 & 0 & 0 \\ 0 & 0 & 0 & 0 \\ 0 & 0 & \cosh\phi & \sinh\phi \\ 0 & 0 & \sinh\phi & \cosh\phi \end{pmatrix}.$$

For small angles we have $L = 1 + \epsilon\Lambda_4$.

An infinitesimal boost in direction of $\mathbf{e}_3$ in our system is given by

$$\mathcal{L} = 1 - \tfrac{1}{2}\epsilon(\mathbf{e}_2 + \mathbf{e}_{14}) \tag{14.5}$$

finite transformations for arbitrary angles ϕ are calculated by the exponential function

$$L(\phi) = e^{\phi\Lambda_4} = \sinh\phi\Lambda_4 + \cosh\phi\Lambda_4 \tag{14.6}$$

The matrix representations Λ_4 and $L(\phi)$ are well known. They can be found already in Neumark's classical work (1963) in chapter III on irreducible linear representations of the Lorentz group (p. 89f.) where they are denoted as b_3 and $b_3(t)$.

Going into details, what is Λ_4 doing? It preserves direction and measure of $\mathbf{e}_2$, $\mathbf{e}_{14}$ and $\mathbf{e}_{124}$, the base elements of $\mathbb{C}h_4$ and it further preserves states f_{11} and f_{14}. It does not preserve $\mathbf{e}_1$, but transforms it along a relativistic heterodimensional path:

- Heterodimensional Lorentz transformation:

$$\mathbf{e}_1' = \mathbf{e}_1 \cosh^2\phi + \mathbf{e}_4 \sinh\phi\cosh\phi + \mathbf{e}_{12}\sinh\phi\cosh\phi - \mathbf{e}_{24}\sinh^2\phi,$$

$$\mathbf{e}_3' = \mathbf{e}_3\cosh\phi - \mathbf{e}_{23}\sinh\phi,$$

$$\mathbf{e}_4' = \mathbf{e}_1\sinh\phi\cosh\phi + \mathbf{e}_4\cosh^2\phi + \mathbf{e}_{12}\sinh^2\phi - \mathbf{e}_{24}\sinh\phi\cosh\phi,$$

$$\mathbf{e}_{12}' = \mathbf{e}_1\sinh\phi\cosh\phi + \mathbf{e}_4\sinh^2\phi + \mathbf{e}_{12}\cosh^2\phi - \mathbf{e}_{24}\sinh\phi\cosh\phi,$$

$$\mathbf{e}_{23}' = -\mathbf{e}_3\sinh\phi + \mathbf{e}_{23}\cosh\phi,$$

$$\mathbf{e}_{24}' = -\mathbf{e}_1\sinh^2\phi - \mathbf{e}_4\sinh\phi\cosh\phi - \mathbf{e}_{12}\sinh\phi\cosh\phi + \mathbf{e}_{24}\cosh^2\phi.$$

Notice, the sum $\mathbf{e}_1' + \mathbf{e}_{24}' = \mathbf{e}_1 + \mathbf{e}_{24}$ is a preserved quantity since $\cosh^2\phi - \sinh^2\phi = 1$. Taking a look at equations (10.2), it is clear that $\tfrac{1}{4}(\mathbf{e}_1 + \mathbf{e}_{24})$ equals the V-spin belonging to some other family χ. It is the V-spin in space 4. Let me stop here because the story becomes more and more complicated and gradually turns into a nonlinear surprise. It is clear that I have avoided any false move. Otherwise you would not have believed me. For charges, in this and other such models, say in $Cl_{6,1}$, do not add. The operators are restricted to fermions. I do not know if that shall turn out to be a hindrance or not. There is one thing that can already be seen very clearly, namely that flavor and color are interconnected and determine some of the qcd.

23.15 Differential operator in color spaces $\mathbb{C}h_\chi$

The decomposition

$$\mathbb{C}h_\chi = \mathbb{C}h_\chi f_{\chi 1} \oplus \mathbb{C}h_\chi f_{\chi 2} \oplus \mathbb{C}h_\chi f_{\chi 3} \oplus \mathbb{C}h_\chi f_{\chi 4} \qquad (15.1)$$

surprisingly takes the simple form

$$\mathbb{C}h_\chi = \xi_1 f_{\chi 1} \oplus \xi_2 f_{\chi 2} \oplus \xi_3 f_{\chi 3} \oplus \xi_4 f_{\chi 4}, \quad \text{with} \quad \xi_k \in \mathbb{R} \qquad (15.2)$$

As the primitive idempotents in the idempotent spaces $\mathbb{C}h_\chi$ are heterodimensional, the Dirac operator D has to be heterodimensional too. Still it has a very simple form

$$D = f_{\chi 1}\frac{\partial}{\partial \xi_1} + f_{\chi 2}\frac{\partial}{\partial \xi_2} + f_{\chi 3}\frac{\partial}{\partial \xi_3} + f_{\chi 4}\frac{\partial}{\partial \xi_4}. \qquad (15.3)$$

In a way, this differential operator can be regarded as a square root of some generalized heterodimensional operator which resembles the Laplacian. However, due to the basis in terms of *erzeugende Einheiten* as Weyl denoted them, its components turn out as decoupled. Since the idempotents are mutually annihilating, the square of form (15.3) gives the homogeneous term

$$D^2 = f_{\chi 1}\frac{\partial^2}{\partial \xi_1} + f_{\chi 2}\frac{\partial^2}{\partial \xi_2} + f_{\chi 3}\frac{\partial^2}{\partial \xi_3} + f_{\chi 4}\frac{\partial^2}{\partial \xi_4}. \qquad (15.4)$$

Notice, we have $\xi_k \in \mathbb{R}$ and $f_{\chi k} \in {}^4\mathbb{R}$ all quantities in real fields. Equation (15.4) gives a most beautiful entry to Clifford-valued differential forms as have been discussed by Vargas and Torr (2001).

REFERENCES

[1] Abłamowicz, R., CLIFFORD - A Maple V Package for Clifford Algebra Computations, http://math.tntech.edu/rafal/cliff5/index.html, released November 18, 1998.

[2] Abłamowicz, R., Lounesto, P., Parra, J.M., Eds.; *Clifford Algebras with Numeric and Symbolic Computations*, Birkhäuser, Boston, 1996.

[3] Abłamowicz, R., Fauser, B., Eds.; *Clifford Algebras and their Application in Mathematical Physics, Vol. 1: Algebra and Physics*, Birkhäuser, Boston, 2000.

[4] Belger, M., Ehrenberg, L.; *Theorie und Anwendung der Symmetriegruppen*, Frankfurt am Main, 1981.

[5] Bergdolt, G.; *Orthonormal basis sets in Clifford algebras*, in "Clifford Algebras with Numeric and Symbolic Computations". Eds. R. Abłamowicz, P. Lounesto, J.M. Parra, Birkhäuser, Boston, 1996, pp. 269–284.

[6] Cahn, R.N.; *Semi-Simple Lie Algebras and Their Representations*, Berkeley, 1984.

[7] Chisholm, J.S.R., Farwell, R.S.; Tetrahedral structure of idempotents of the Clifford algebra $C\ell_{3,1}$, in *Clifford Algebras and their Applications in Mathematical Physics*, A. Micali, et al., eds., Kluwer, Dordrecht, 1992, pp. 27–32.

[8] Chisholm, J.S.R., Farwell, R.S.; Spin gauge theories: Principles and predictions, in: *Clifford Algebras and their Application in Mathematical Physics*, F. Brackx, et al., eds., Dortrecht, 1993, pp. 367–374.

[9] Clifford, W.K.; *The Common Sense of the Exact Sciences*, Ed. K. Pearson, Dover, New York, 1955.

[10] Crawford, J.P.; Local automorphism invariance: Gauge boson mass without a Higgs particle, *J. Math. Phys.* **35**, 1994, 2701–2718.

[11] Fauser, B.; Vertex functions and generalized normal ordering by triple systems in nonlinear spinor field models, *Advances in Applied Clifford Algebras* **10**, 2000, 173–192.

[12] Fritzsch, H., Gell-Mann, M., Leutwyler, H.; Advantages of the color octet gluon picture, *Phys. Lett.* **47B**, 1973, 365–368.

[13] Glashow, S.L., Iliopoulos, J., Maiani, L.; Weak interactions with lepton-hadron symmetry, *Phys. Rev.* **D2**, 1970, 1285–1292.

[14] Greider, K., Weiderman, T.; *Generalised Clifford Algebras as Special Cases of Standard Clifford Algebras*, l'UCD Preprint 16, 1988. (cited after Chisholm 1992), 32.

[15] Hawking, S.W.; *A Brief History of Time*, Bantam, Toronto, 1988.

[16] Hestenes, D.; Space-time structure of weak and electromagnetic interactions, *Found. Physics* **12**, 1982, 153–168.

[17] Hestenes, D.; The Zitterbewegung interpretation of quantum mechanics, *Found. Physics* **10**, 1990, 1213–1232.

[18] Lipkin, H.J.; *Anwendung von Lieschen Gruppen in der Physik*, Mannheim, 1967.

[19] Lounesto, P.; *Clifford Algebras and Spinors*, Cambridge, 2001.

[20] Morris, A.O., Makhool, M.K.; Real projective representations of real Clifford algebras and reflection groups, in *Clifford Algebras and their Application in Mathematical Physics*, A. Micali, et al., eds., Dordrecht, 1992, pp. 27–32.

[21] Narnhofer, H., Thirring, W.; *Why Schrödinger's Cat is Most likely to be Either Alive or Dead*, Preprint ESI 343, Vienna, (1996).

[22] Neumark, M.A.; *Lineare Darstellungen der Lorentzgruppe*, Berlin, 1963.

[23] Petraschen, M.I., Trifonow, E.D.; *Anwendung der Gruppentheorie in der Quantenmechanik*, Leipzig, 1969.

[24] Pezzaglia Jr., W.M.; Dimensionally democratic calculus and principles of polydimensional physics, in *Clifford Algebras and their Application in Mathematical Physics*, R. Abłamowicz and B. Fauser, eds., Birkhäuser, Boston, 2000, pp. 101–124.

[25] Ramond, P.; *Field Theory: a Modern Primer*, Redwood City, 1990.

[26] Sattinger, O.L., Weaver, O.L.; *Lie Groups and Algebras with Applications to Physics, Geometry and Mechanics*, New York, 1986.

[27] Schmeikal, B.; The generative process of space-time and strong interaction - quantum numbers of orientation, in *Clifford Algebras with Numeric and Symbolic Computations*, R. Abłamowicz, P. Lounesto, and J.M. Parra, eds., Birkhäuser, Boston, 1996, pp. 83–100.

[28] Schmeikal, B.; Clifford algebra of quantum logic, in *Clifford Algebras and their Application in Mathematical Physics*, R. Abłamowicz and B. Fauser, eds., Birkhäuser, Boston, 2000, pp. 219–241.

[29] Schmeikal, B.; *On Real Color Spinor Space and* $SU(3)$-*Multiplets of Heterodimensional Lorentz Transformations*, Lecture given at the Institute for Theoretical Physics, Faculty of Physics, Konstanz, Germany, 2000.

[30] Schmeikal, B.; Minimal spin gauge theory, *Advances in Applied Clifford Algebras* **11**, 2001, 63–60.

[31] Schweber, S.S.; Feynman and the visualization of space-time processes, *Rev. Mod. Phys.* **58**, 1986, 449–508.

[32] Shaw, R.; Finite geometry, Dirac groups and the table of real Clifford algebras, in *Clifford Algebras and Spinor Structures*, R. Abłamowicz and P. Lounesto, eds., Kluwer, Dordrecht, 1995, pp. 59–99.

[33] Vargas, J.G., Torr, D.G.; Clifford-valued clifforms: A geometric language for Dirac equations, in *Clifford Algebras and their Application in Mathematical Physics*, R. Abłamowicz and B. Fauser, eds., Birkhäuser, Boston, 2000, pp. 135–154.

[34] Weyl, H.; *Raum-Zeit-Materie*, Heidelberg, 4th edition 1920, reprint: 1993.

[35] Weyl, H.; *Gruppentheorie und Quantenmechanik*, Leipzig, 1931, reprint: Stuttgart 1967.

[36] Zweig,G.; *An SU3 model for strong interaction symmetry and its breaking, 1*, CERN-Reports 8182/TH401, 1964.

[37] Zweig, G.; *An SU3 model for strong interaction symmetry and its breaking, 2*, CERN-Reports 8419/TH412, 1964.

Bernd Schmeikal
Biofield Laboratory
Raiffeisenweg 18, A-4203 Altenberg, Austria
E-mail: bernd.schmeikal@utanet.at

Received: February 4, 2003; Revised: September 11, 2003.

PART IV. PHYSICS

24

The Quantum/Classical Interface: Insights from Clifford's (Geometric) Algebra

William E. Baylis

ABSTRACT Classical relativistic physics in Clifford algebra has a spinorial formulation that is closely related to standard quantum formalism. The algebraic use of spinors and projectors, together with the bilinear relations of spinors to observed currents, gives quantum-mechanical form to many classical results, and the clear geometric content of the algebra makes it an illuminating probe of the quantum/classical interface. This paper extends past efforts to close the conceptual gap between quantum and classical phenomena while highlighting their essential differences. The paravector representation of spacetime in $C\ell_3$ is used in particular to provide insight into spin-$\frac{1}{2}$ systems and their measurement.

Keywords: quantum/classical interface, Lorentz transformations, spin, gauge transformations, eigenspinors, paravectors

24.1 Introduction

Clifford's geometric algebra gives a spinorial approach to classical relativistic mechanics. In the algebra, for example, Lorentz rotations (physical boosts and spatial rotations) are *spin* transformations [1]. Spacetime vectors such as the momentum p are rotated by

$$p \to LpL^\dagger, \tag{1.1}$$

where the *rotor* (transformation element) $L = \exp \frac{1}{2}\mathbf{W}$ is generated by the unit bivector for the plane of rotation. For example, for a spatial rotation in the $\mathbf{e}_2\mathbf{e}_1$ plane in the sense of $\mathbf{e}_1 \to \mathbf{e}_2$ by the angle ϕ,

$$L = \exp\left(\tfrac{1}{2}\mathbf{e}_2\mathbf{e}_1\phi\right), \quad L^\dagger = \exp\left(\tfrac{1}{2}\mathbf{e}_1\mathbf{e}_2\phi\right) = L^{-1}.$$

Acting under the axiomatic rule $\mathbf{vv} = \mathbf{v} \cdot \mathbf{v}$ of the algebra, the real-bilinear form (1.1) of the spin transformation rotates any vector $\mathbf{v} = v^1\mathbf{e}_1 + v^2\mathbf{e}_2 + v^3\mathbf{e}_3$ to

$$\mathbf{v} \to L\mathbf{v}L^\dagger = \left(v^1\cos\phi - v^2\sin\phi\right)\mathbf{e}_1 + \left(v^2\cos\phi + v^1\sin\phi\right)\mathbf{e}_2 + v^3\mathbf{e}_3 \,.$$

AMS Subject Classification: 15A66, 81P15, 81R25, 83A05.

Spin transformations are related to bilinear expressions ("matrix elements") of quantum theory, where the average value of a property for a system in a state ψ, for example, is given by the expectation value of the corresponding observable:

$$\langle p \rangle = \langle \psi \,|p|\, \psi \rangle \,.$$

This relation suggests a closer connection between classical and quantum theories than usually recognized. It is the purpose of this paper to explore the quantum/classical (Q/C) interface, *not* by the more common approaches of looking at limits of high quantum numbers or taking $\hbar \to 0$, but by exploiting the spinorial approach in Clifford's geometric algebra of classical physics.

First we need background concepts and notation. We present a minimal introduction here. More details can be found elsewhere [1–5].

24.2   Paravectors in $C\ell_n$

Consider the geometric algebra $C\ell_n$ of an n-dimensional Euclidean vector space, and denote by $\langle C\ell_n \rangle_k$ the space of k-vectors (multivectors of grade k) in the algebra. The linear space of the full algebra is the 2^n-dimensional space given by the direct sum $\oplus_{k=0}^{n} \langle C\ell_n \rangle_k$ of scalars $\langle C\ell_n \rangle_0$, vectors $\langle C\ell_n \rangle_1$, bivectors $\langle C\ell_n \rangle_2$, and linear subspaces of higher grades. Linear subspaces can of course be formed from direct sums of any of the graded subspaces. Of particular interest is the space of *paravectors* $p = p^0 + \mathbf{p} = \langle p \rangle_0 + \langle p \rangle_1 \in C\ell_n$ and their exterior products:

$$\text{paravector space} = \langle C\ell_n \rangle_{0,1} \,, \ (n+1)\text{-dim}$$
$$\text{biparavector space} = \langle C\ell_n \rangle_{1,2} \,, \ \binom{n+1}{2}\text{-dim}$$
$$\text{triparavector space} = \langle C\ell_n \rangle_{2,3} \,, \ \binom{n+1}{3}\text{-dim} \,.$$

In general, the linear space of grade-0 paravectors is $\langle C\ell_n \rangle_0$, the same as the linear space of grade-0 vectors; the space of the $(n+1)$-grade paravector volume element is the same as that of the volume element of the vector space, namely $\langle C\ell_n \rangle_n$; and the space of k-grade multiparavectors is the sum of $(k-1)$-grade and k-grade vector spaces: $\langle C\ell_n \rangle_{k-1,k} \equiv \langle C\ell_n \rangle_{k-1} \oplus \langle C\ell_n \rangle_k \,, 0 < k < n+1$.

We need two conjugations (anti-automorphisms) and their combination, which act on the paravector p and, more generally, on any product pq of elements according to,

Clifford (bar) conjugation $\bar{p} = p^0 - \mathbf{p}, \qquad \overline{pq} = \bar{q}\bar{p}$

reversal (dagger) conjugation $p^\dagger = p, \qquad (pq)^\dagger = q^\dagger p^\dagger$

grade automorphism (bar-dagger) $\bar{p}^\dagger = \bar{p}, \qquad (\overline{pq})^\dagger = \overline{(pq)^\dagger} = \bar{p}^\dagger \bar{q}^\dagger .$

Since the orthonormal basis vectors of a Euclidean space can all be represented by Hermitian matrices, reversal is the same as Hermitian conjugation. Recognizing this possibility and anticipating its relevance to the Q/C interface, we adopt the dagger notation for reversal.

24.2.1   Paravector basis and metric

If $\{e_1, e_2, \ldots, e_n\}$ is an orthonormal basis of the original Euclidean space, then

$$\langle e_j e_k \rangle_0 = \tfrac{1}{2} \left(e_j e_k + e_k e_j \right) = \langle e_k e_j \rangle_0 = \delta_{jk} , \tag{2.1}$$

and the basis of paravector space can be written $\{e_0, e_1, e_2, \ldots, e_n\}$, where we identify $e_0 \equiv 1$ for convenience in expanding paravectors $p = p^\mu e_\mu$, $\mu = 0, 1, \ldots, n$. The metric of paravector space is determined by the quadratic form: a scalar-valued product of a paravector p with itself or a conjugate. Although p^2 is unsatisfactory because it generally has vector parts, $p\bar{p} = \langle p\bar{p} \rangle_0 = \bar{p}p$ is a scalar. We adopt it as the quadratic form:

$$Q(p) = p\bar{p}.$$

By "polarization" $p \to p + q$ we find the corresponding inner product

$$\langle p, q \rangle = \langle p\bar{q} \rangle_0 = \tfrac{1}{2} \left(p\bar{q} + q\bar{p} \right) = p^\mu q^\nu \langle e_\mu \bar{e}_\nu \rangle_0 \equiv p^\mu q^\nu \eta_{\mu\nu} ,$$

where the natural metric of the paravector space

$$(\eta_{\mu\nu}) = \mathrm{diag}\, (1, -1, -1, \ldots, -1)$$

has the form of a Minkowski (pseudo-Euclidean) metric.

24.2.2   Spacetime as paravector space in $\mathcal{C}\ell_3$

The metric of paravector space makes paravectors in $\mathcal{C}\ell_3$ ideal for modeling vectors in spacetime. We refer to $\mathcal{C}\ell_3$ as the algebra of physical space (APS), emphasizing its roles as the algebra of both multivectors and multiparavectors in physical space. (Extensions to curved spacetimes and higher dimensions are possible, but we restrict ourselves here to flat spacetimes with $n = 3$ and use units with $c = 1$.) The proper velocity of a frame with coordinate velocity $\mathbf{v} = d\mathbf{x}/dt$ is the unit paravector

$$u = \frac{dx}{d\tau} = \gamma \left(1 + \mathbf{v} \right) = u^\mu e_\mu ,$$

where for timelike displacements $dx = dt + d\mathbf{x}$, the proper time τ is the Lorentz scalar related by

$$d\tau^2 = dx d\bar{x} = \eta_{\mu\nu} dx^\mu dx^\nu,$$

and $\gamma = dt/d\tau \geq 1$ is given by $u\bar{u} = 1 = \gamma^2 \left(1 - \mathbf{v}^2\right)$. Other paravectors model the spacetime vectors

$$p = mu = E + \mathbf{p} : \text{paramomentum}$$
$$j = j^\mu \mathbf{e}_\mu = \rho + \mathbf{j} : \text{current density}$$
$$A = A^\mu \mathbf{e}_\mu = \phi + \mathbf{A} : \text{paravector potential}$$
$$\partial = \partial_\mu \mathbf{e}_\nu \eta^{\mu\nu} = \partial_t - \nabla : \text{gradient operator}.$$

Biparavectors represent oriented planes in spacetime. For example the electromagnetic field is

$$\mathbf{F} = \langle \partial \bar{A} \rangle_{1,2} = \tfrac{1}{2} F^{\mu\nu} \langle \mathbf{e}_\mu \bar{\mathbf{e}}_\nu \rangle_{1,2} = \mathbf{E} + i\mathbf{B}$$

with basis biparavectors $\langle \mathbf{e}_\mu \bar{\mathbf{e}}_\nu \rangle_{1,2}$, which generate Lorentz rotations. In APS, $i = \mathbf{e}_1\mathbf{e}_2\mathbf{e}_3 = \mathbf{e}_0\bar{\mathbf{e}}_1\mathbf{e}_2\bar{\mathbf{e}}_3$ is the volume element of both the vector and paravector spaces, and for any element x, $^*x = -ix$ is its (Clifford–Hodge) *dual*. Thus the magnetic field $\mathbf{B}$ is the vector dual to $i\mathbf{B}$, the spatial-plane part of $\mathbf{F}$.

In APS we move seamlessly between covariant quantities (multiparavectors) and their multivector components in the physical space (spatial hypersurface) of the observer.

24.2.3   Spin transformations

Lorentz rotors L are unimodular ($L\bar{L} = 1$) and have the form

$$L = \pm \exp\left(\tfrac{1}{2}\mathbf{W}\right) \in SL(2,\mathbb{C}), \quad \mathbf{W} = \tfrac{1}{2}W^{\mu\nu} \langle \mathbf{e}_\mu \bar{\mathbf{e}}_\nu \rangle_{1,2}.$$

If $\mathbf{W}$ is a vector (a timelike biparavector), then $L = B = B^\dagger$ is a boost, whereas if $\mathbf{W}$ is a bivector (a spacelike biparavector), then $L = R = \bar{R}^\dagger$ is a spatial rotation. Every rotor L can be uniquely factored into the product of a boost and a rotation $L = BR$. The spin transformations of paravectors and their products take the canonical forms

$$p \to LpL^\dagger, \quad \text{odd multiparavector grade}$$
$$\mathbf{F} \to L\mathbf{F}\bar{L}, \quad \text{even multiparavector grade}.$$

24.2.4   Eigenspinors

A Lorentz rotor of particular interest is the *eigenspinor* Λ, which relates the system reference frame to the observer. It transforms as $\Lambda \to L\Lambda$ and is generally a reducible element of the spinor carrier space of Lorentz rotations[1] $L \in \$pin_+(1,3) \simeq SL(2,\mathbb{C})$. This transformation is distinct from that for paravectors and their products and characterizes Λ as a *spinor*. The *eigen* part of its name refers to its association with the system, which in the simplest case is a single particle. Any property of the system in the reference frame is easily transformed by Λ to the lab (the observer's frame). For example, the proper velocity of a massive particle is taken to be $u = e_0 = 1$ in the reference frame; that is, the reference frame at proper time τ is an inertial frame that moves with the particle at that instant. In the lab, the proper velocity is

$$u = \Lambda e_0 \Lambda^\dagger = \Lambda \Lambda^\dagger, \qquad (2.2)$$

which is also the timelike basis paravector of a frame moving with proper velocity u. The complete moving frame is the paravector basis $\left\{ u_\mu = \Lambda e_\mu \Lambda^\dagger \right\}$ with $u \equiv u_0$.[1]

24.2.5 Time evolution

The eigenspinor Λ of an accelerating or rotating system changes in time, with $\Lambda(\tau)$ giving the spacetime rotation at proper time τ. Eigenspinors at different times are related by

$$\Lambda(\tau_2) = L(\tau_2, \tau_1) \Lambda(\tau_1),$$

where by the group property of rotations, the time-evolution operator

$$L(\tau_2, \tau_1) = \Lambda(\tau_2) \bar\Lambda(\tau_1)$$

is another Lorentz rotation. It is found by solving an equation of motion

$$\dot\Lambda = \tfrac{1}{2} \Omega \Lambda = \tfrac{1}{2} \Lambda \Omega_{\text{ref}}, \qquad (2.3)$$

with the *spacetime rotation rate* $\Omega = \dot\Lambda\bar\Lambda - \Lambda\dot{\bar\Lambda} = \Lambda\Omega_{\text{ref}}\bar\Lambda$, where Ω_{ref} is the value of the biparavector Ω in the reference frame. This relation allows us to compute time-rates of change of any property known in the reference frame. For example, the acceleration in the lab is

$$\dot u = \langle \Omega u \rangle_\Re = \left\langle \Lambda \Omega_{\text{ref}} \Lambda^\dagger \right\rangle_\Re = \Lambda \langle \Omega_{\text{ref}} \rangle_\Re \Lambda^\dagger. \qquad (2.4)$$

Here, $\langle x \rangle_\Re = \tfrac{1}{2}\left(x + x^\dagger\right)$ indicates the *real* (grades 0 + 1) part of x. Equation (2.4) is just the covariant Lorentz-force equation if we identify $\Omega = eF/m$. This identification provides a convenient operational definition of the electromagnetic field F.

---

[1]An important distinction between the paravector formalism in APS and the vector formalism of spacetime algebra [5] (STA) is that the basis vectors here give the object frame *relative to the observer* rather than as an independent frame. The Lorentz rotation is the same whether we rotate the system forward (an active transformation) or the observer backward (a passive transformation) or some combination. The space/time split of a property is its expansion into vector grades in the observer's basis $\{e_\mu\}$:

$$p = \langle p \rangle_0 + \langle p \rangle_1 = p^0 + \mathbf{p}$$
$$\mathbf{F} = \langle \mathbf{F} \rangle_1 + \langle \mathbf{F} \rangle_2 = \mathbf{E} + i\mathbf{B}.$$

To find the split for another observer, one expands p in her paravector basis $\{u_\mu\}$ and $\mathbf{F}$ in her biparavector basis $\left\{ \langle u_\mu \bar u_\nu \rangle_{1,2} \right\}$, where $u = u_0$ is her proper velocity.

The equation of motion (2.3) for eigenspinors, especially when coupled to algebraic projectors, often offers a more powerful approach to finding classical trajectories of charges in electromagnetic fields than a direct frontal attack on the Lorentz-force equation, as demonstrated in new analytic solutions for the autoresonant laser accelerator [6]. The eigenspinor gives not only the velocity of the particle, but also its orientation.

24.3   Gauge transformations

As seen in Eq. (2.4), the acceleration (and by integration, the proper velocity and the world line) of a massive particle depends only on the *real (spatial vector) part* of Ω_{ref}. It is independent of "gauge transformations" of the imaginary part:

$$\Omega_{\mathrm{ref}} \rightarrow \Omega_{\mathrm{ref}} - i\omega .$$

Furthermore by Eq. (2.2), the proper velocity u is independent of a rotation $\Lambda \rightarrow \Lambda R_0$ of the reference frame. The reference frame is a *rest frame* for any rotation R_0.

A particular rest frame of interest is the one related to the lab by a pure boost. In other words, $R_0 = \bar{R}$ so that $\Lambda = BR \rightarrow B$. The addition of a reference-frame rotation

$$\Lambda \rightarrow \Lambda \exp_T \left[-\frac{i}{2} \int^\tau \omega \, d\tau \right] \tag{3.1}$$

does not alter the acceleration or velocity of the particle. The notation $\exp_T$ refers to a time-ordered exponential, which is important when ω changes orientation.

Equation (3.1) is a *local gauge transformation*, and we can play the gauge game with our classical eigenspinor. The equation of motion (2.3) is invariant under the gauge transformation if the derivative $d/d\tau$ of Λ is replaced by the *gauge-covariant* derivative

$$\frac{d\Lambda}{d\tau} \rightarrow \frac{d\Lambda}{d\tau} + i\Lambda \mathbf{G} ,$$

and the gauge transformation (3.1) is accompanied by a transformation of the gauge potential $\mathbf{G}$:

$$\mathbf{G} \rightarrow \mathbf{G} + \tfrac{1}{2}\omega .$$

We have the gauge freedom to choose the orientation of the reference frame. In particular, we can choose $\mathbf{e}_3$ to lie along ω in the reference frame: $\omega = \omega \mathbf{e}_3$. In this way, the free (constant momentum) solution $\Lambda(\tau) = \Lambda(0)$ is replaced by

$$\Lambda(\tau) = \Lambda(0) \, e^{-i\theta(\tau)\mathbf{e}_3/2}, \tag{3.2}$$

where the rotation in the reference frame is by the angle

$$\theta(\tau) = \int_0^\tau d\tau' \omega(\tau') . \tag{3.3}$$

Now ω is a scalar that may contain a Lorentz-invariant constant ω_0 as well as an interaction with the 0th component A_{ref}^0 of a paravector field A as viewed in the reference frame:

$$\omega = \omega_0 + 2e\hbar^{-1} A_{\text{ref}}^0 = \omega_0 + 2e\hbar^{-1} \langle A\bar{u} \rangle_0 . \tag{3.4}$$

Here, e is the coupling constant, the factor $\hbar^{-1}$ has been added so that eA_{ref}^0 has units of energy, and we noted that the reference-frame value A_{ref}^0 can be expressed as the Lorentz-invariant scalar product $\langle A\bar{u} \rangle_0$. The potential A is related to the electromagnetic paravector potential below.

24.3.1   Classical de Broglie waves

It is common to model elementary particles classically by point charges. An obvious advantage of point charges is that they are simple and structureless. A disadvantage is that their electromagnetic energy is infinite, requires mass renormalization, and leads to preacceleration.

We proceed by assuming no *a priori* distribution. Instead, a current density $j(x)$ is related by an eigenspinor field $\Lambda(x)$ to the reference-frame density ρ_{ref} :

$$j(x) = \Lambda(x) \rho_{\text{ref}}(x) \Lambda^\dagger(x) . \tag{3.5}$$

This form allows the velocity and orientation to be different at different spacetime positions x. The current density (3.5) can be written in terms of the density-normalized eigenspinor Ψ as

$$j = \Psi\Psi^\dagger, \quad \Psi = \rho_{\text{ref}}^{1/2} \Lambda .$$

Consider first the simplest case: a free particle ($A = 0$) of *fixed momentum* $p = mu$. For such a particle, we can define a proper time τ for the system, and we look for an eigenspinor Ψ_p that depends only on τ. There is a constraint on the reference-frame charge distribution ρ_{ref} in this case since the conservation of charge implies the continuity equation

$$0 = \langle \bar{\partial} j \rangle_0 = \langle (\bar{\partial}\rho_{\text{ref}}) u \rangle_0 = \dot{\rho}_{\text{ref}} ,$$

where we used (3.5) and the fact that $\rho_{\text{ref}}(x)$ is a scalar. The dot indicates the proper-time derivative $d/d\tau = \langle u\bar{\partial} \rangle_0$.

The spacetime dependence of $\Lambda(x)$ and $\Psi(x)$ for in this case is given by (3.2) and (3.3) with $\omega = \omega_0$ and the relation

$$\tau = \langle (x - x_0) \bar{u} \rangle_0 .$$

It is precisely that of a *de Broglie plane wave*

$$\Psi(x) = \Psi_0(x_0) \exp\left(-i\frac{\omega_0 \mathbf{e}_3}{2m} \langle \bar{p}(x - x_0) \rangle_0 \right) \tag{3.6}$$

as long as we identify $\omega_0 = 2m/\hbar$, but here the complex phase angle in the de Broglie wave takes on physical significance as a real rotation angle in the rest frame in the (arbitrary) reference plane $e_2 e_1 = -i e_3$. The wave is seen to satisfy a Klein–Gordon equation

$$\partial \bar{\partial} \Psi = -\left(\frac{\omega_0}{2m}\right)^2 p \bar{p} \Psi = -\left(\frac{\omega_0}{2}\right)^2 \Psi \,,$$

which also suggests the association $\hbar \omega_0 = 2m$. As above, we can take $\omega_0 = e_3 \omega_0$ since R can rotate any reference direction to the direction s of the spin in the lab: $R e_3 = s R$.

24.3.2 Electromagnetic gauge potential

In the presence of a nonvanishing gauge potential A, the rotation angle θ of the eigenspinor $\Psi(\tau) = \Psi(0) e^{-i\theta e_3/2}$ is given by Eqs. (3.3) and (3.4) to be

$$\tfrac{1}{2}\theta = \int_0^\tau d\tau' \left(\tfrac{1}{2}\hbar\omega_0 + e \langle A\bar{u}\rangle_0\right)/\hbar = \left\langle \int_{\bar{x}_0}^{\bar{x}} d\bar{x} \left(\tfrac{1}{2}\hbar\omega_0 u + eA\right)/\hbar \right\rangle_0 \,,$$

where we used $\bar{u} = d\bar{x}/d\tau$ and $u\bar{u} = 1$. If we superimpose solutions along different trajectories, there will be substantial cancellation from the different angles of rotation, and only at extreme values of θ can constructive interference create the caustics that correspond to classical trajectories. However, $\tfrac{1}{2}\theta$ is just the classical action and its extremization gives the classical equations of motion for a charge in the electromagnetic gauge potential A, where $\tfrac{1}{2}\hbar\omega_0 = m$ acts as the particle *mass*. With this identification, the de Broglie wave in the presence of A becomes

$$\Psi(x) = \Psi_0(x_0) \exp\left(-i e_3 \left\langle \int_{\bar{x}_0}^{\bar{x}} d\bar{x}\,(p + eA)/\hbar \right\rangle_0\right). \tag{3.7}$$

24.3.3 Spin and g-factor

As introduced above, a reference frame $\{e_\mu\}$ is transformed to the *moving frame* $\{u_\mu = \Lambda e_\mu \Lambda^\dagger\}$ by the Lorentz transformation Λ, and the moving time axis $u_0 = u = \Lambda \Lambda^\dagger$ is the proper velocity of that frame with respect to the observer. It is natural to ask whether any of the transformed spatial vectors, say $u_3 = \Lambda e_3 \Lambda^\dagger$, carries any physical meaning. By taking e_3 to be the rotation axis in the reference frame, the corresponding unit spacelike paravector u_3 in the lab is associated with the *spin*. We refer to $u_3 = w$ as the *spin paravector* even though the physical spin requires an extra scalar factor of $\hbar/2$ to ensure the correct magnitude and units.

Since e_0 and e_3 are orthogonal, so are u_0 and u_3 and hence u and w :

$$\langle u_3 \bar{u}_0 \rangle_0 = \langle w\bar{u}\rangle_0 = \langle e_3 \bar{e}_0 \rangle_0 = 0,$$

and $u_3 \bar{u}_0 = w\bar{u}$ is a biparavector (a spacetime plane). Its dual is another biparavector

$$\mathbf{S} = \Lambda e_1 \bar{e}_2 \bar{\Lambda} = -i\Lambda e_3 \bar{\Lambda} = -iw\bar{u} \,.$$

The dual of the spin paravector $w = iSu = w^{\dagger}$ is seen to be the *triparavector* $\mathbf{S}u$. This identifies w as the *Pauli–Lubański spin* paravector.

The g-factor determines the precession rate of the particle in a magnetic field in a *rest* (commoving inertial) frame. The reference frame, related to the lab by the eigenspinor Λ, is one rest frame. As with any Lorentz rotor, Λ can always be factored into the product of a real boost $B = B^{\dagger}$ and a unitary rotation $R = \bar{R}^{\dagger}$ namely $\Lambda = BR$. The boost part of the eigenspinor is

$$B = u^{1/2} = \frac{1 + u}{\sqrt{2\langle 1 + u\rangle_S}} = \frac{m + p}{\sqrt{2m\,(m + E)}}.$$

It is convenient to use the *rest* frame related to the lab by a pure boost. It is related to the reference frame by a spatial rotation. The unit spin paravector in this rest frame is the vector

$$\mathbf{s} = R\mathbf{e}_3 R^{\dagger}, \tag{3.8}$$

and it is related to the spin paravector w and the biparavector $\mathbf{S}$ in the lab by $w = B\mathbf{s}B = iSu$. The equation of motion for w is

$$\dot{w} = \frac{d}{d\tau}\left(\Lambda\mathbf{e}_3\Lambda^{\dagger}\right) = \langle\Omega w\rangle_{\Re} = B\langle\Omega_{\text{rest}}\mathbf{s}\rangle_{\Re}B = \langle\Omega_B w\rangle_{\Re} + B\dot{\mathbf{s}}B$$

where $\Omega_{\text{rest}} = \bar{B}\Omega B$ and $\Omega_B \equiv 2\dot{B}\bar{B}$. The precession of $\mathbf{s}$ as viewed in the rest frame is

$$\dot{\mathbf{s}} = \langle\Omega_R\mathbf{s}\rangle_{\Re} = \boldsymbol{\omega}_R \times \mathbf{s} \tag{3.9}$$

with $\Omega_R \equiv 2\dot{R}\bar{R} = -\Omega_R^{\dagger} = -i\boldsymbol{\omega}_R$. If the spacetime rotation rate Ω_{rest} has only a bivector part associated with the spin and an external magnetic field $\mathbf{B}_{\text{rest}}$ times the g-factor,

$$\Omega_{\text{rest}} = -i\boldsymbol{\omega}_{\text{rest}} = -i\left(\omega_0\mathbf{s} - \frac{eg}{2m}\mathbf{B}_{\text{rest}}\right),$$

then the eigenspinor evolution gives

$$\dot{\Lambda} = \tfrac{1}{2}\Omega\Lambda = \tfrac{1}{2}B\Omega_{\text{rest}}R = -\tfrac{1}{2}iB\boldsymbol{\omega}_{\text{rest}}R = \dot{B}R + B\dot{R}.$$

The proper acceleration in this case vanishes:

$$\dot{u} = \dot{\gamma} + \dot{\mathbf{u}} = 2\left\langle\dot{\Lambda}\Lambda^{\dagger}\right\rangle_{\Re} = \langle -iB\boldsymbol{\omega}_{\text{rest}}B\rangle_{\Re} = 0$$

and it follows that $\dot{B} = 0$ and $\boldsymbol{\omega}_R = \boldsymbol{\omega}_{\text{rest}}$. In this case, the precession rate of $\mathbf{s}$ is simply $-\tfrac{1}{2}eg\mathbf{B}_{\text{rest}}/m$. More generally, the presence of a rest-frame electric field adds an acceleration $e\mathbf{E}_{\text{rest}}/m$ to Ω_{rest}:

$$\Omega_{\text{rest}} = \frac{e}{m}\left(\mathbf{E}_{\text{rest}} + i\frac{g}{2}\mathbf{B}_{\text{rest}}\right) - i\omega_0\mathbf{s}.$$

Since

$$\mathbf{E}_{\text{rest}} = \tfrac{1}{2}\left(\mathbf{F}_{\text{rest}} + \mathbf{F}_{\text{rest}}^{\dagger}\right), \quad i\mathbf{B}_{\text{rest}} = \tfrac{1}{2}\left(\mathbf{F}_{\text{rest}} - \mathbf{F}_{\text{rest}}^{\dagger}\right)$$

and $\mathbf{F}_{\text{rest}} = \bar{B}FB$, transformation to the lab gives the rotation rate

$$\Omega = B\Omega_{\text{rest}}\bar{B} = \frac{e}{2m}\left[\mathbf{F} + u\mathbf{F}^{\dagger}\bar{u} + \frac{g}{2}\left(\mathbf{F} - u\mathbf{F}^{\dagger}\bar{u}\right)\right] + \Omega_0$$

$$= \frac{e}{4m}\left[(2+g)\mathbf{F} + (2-g)u\mathbf{F}^{\dagger}\bar{u}\right] + \Omega_0 , \quad (3.10)$$

where $\Omega_0 = -i\omega_0 B s\bar{B} = \omega_0 S$. The equation of motion for w is independent of Ω_0, giving

$$\dot{w} = \langle\Omega w\rangle_{\Re} = \frac{e}{4m}\left[(2+g)\langle\mathbf{F}w\rangle_{\Re} + (2-g)\langle u\mathbf{F}^{\dagger}\bar{u}w\rangle_{\Re}\right] ,$$

which is exactly the algebraic form of the BMT equation [7]. We can also write

$$\Omega_{\text{rest}} = \bar{B}\Omega B = 2\bar{B}\Lambda\bar{\Lambda}B = 2\bar{B}\left(\dot{B}R + B\dot{R}\right)\bar{R}$$

$$= 2\left(\bar{B}\dot{B} + \dot{R}\bar{R}\right) = \Omega_B^{\dagger} + \Omega_R ,$$

the bivector (imaginary) part of which gives the precession rate corrected for Thomas precession:

$$\omega_R = \omega_{\text{rest}} + \omega_{\text{Th}} ,$$

where $\omega_{\text{Th}} = i\langle\Omega_B\rangle_{\Im} = 2i\langle\dot{B}\bar{B}\rangle_{\Im} = -i\langle\Omega_B^{\dagger}\rangle_{\Im}$ is the proper Thomas precession frequency.

If we demand that the eigenspinor evolution (2.3) is linear in Λ, then Ω should not depend on the proper velocity u. Alternatively, we can demand that the particle described is *elementary* in the sense that its motion is described by a single eigenspinor (in contrast to compound particles that require more than a single eigenspinor for the description of their motion), even though it is not necessarily a point particle [8]. In either case, from (3.10), we must have $g = 2$ and the spacetime rotation rate equal to $e\mathbf{F}/m + \omega_0 S$.

24.3.4   Magnetic moment

In a constant rest-frame magnetic field $\mathbf{B}_{\text{rest}}$, with $g = 2$, the classical equation of motion for the eigenspinor $\Lambda = BR$ is

$$\dot{\Lambda} = \tfrac{1}{2}B\Omega_{\text{rest}}R, \quad \Omega_{\text{rest}} = i\left(\frac{e}{m}\mathbf{B}_{\text{rest}} - \omega_0 s\right) ,$$

and it has the solution $\dot{B} = 0$ and

$$\Lambda(\tau) = B\exp\left\{\tfrac{1}{2}i\left(\frac{e}{m}\mathbf{B}_{\text{rest}} - \omega_0 s\right)\tau\right\} R(0) .$$

This describes a rotation rate in the rest frame arising both from the external magnetic field $\mathbf{B}_{\text{rest}}$ and from the intrinsic spin. If we assume a magnetic field

$B_{\text{rest}} \ll m\omega_0/e$ ($\simeq 4.414 \times 10^9$ tesla when $\omega_0 = 2m/\hbar$), the rotation associated with the intrinsic spin dominates:

$$\Lambda(\tau) = B \exp\left\{-\tfrac{1}{2}i\omega_0\mathbf{s}\left(1 - \frac{e}{m\omega_0}\mathbf{sB}_{\text{rest}}\right)\tau\right\} R(0)$$

$$\simeq B \exp\left\{-\tfrac{1}{2}i\omega_0\mathbf{s}'\left(1 - \frac{e}{m\omega_0}\mathbf{s}\cdot\mathbf{B}_{\text{rest}}\right)\tau\right\} R(0)$$

with

$$\mathbf{s}' = R_1\mathbf{s}R_1^\dagger \simeq \mathbf{s}\left(1 - \frac{e}{m\omega_0}\langle\mathbf{sB}_{\text{rest}}\rangle_V\right)$$

$$R_1 = \exp\left(\frac{e}{2m\omega_0}\langle\mathbf{sB}_{\text{rest}}\rangle_V\right).$$

The shift in proper rotation frequency of $\Lambda(\tau)$, together with the association $\hbar\omega_0 = 2m$ of that frequency with the mass of the charge, implies a potential energy that would change the mass. For a charge at rest in a magnetic field $\mathbf{B}_{\text{rest}}$, this potential energy can be expressed

$$-\frac{e\hbar}{2m}\mathbf{s}\cdot\mathbf{B}_{\text{rest}} = -\boldsymbol{\mu}\cdot B_{\text{rest}}$$

with a magnetic dipole moment

$$\boldsymbol{\mu} = \frac{e\hbar}{2m}\mathbf{s} = g\mu_0\frac{\mathbf{s}}{2},$$

where $\hbar\mathbf{s}/2$ is the spin vector, $\mu_0 = e\hbar/(2m)$ is the Bohr magneton, and $g = 2$ is the g factor. This result is consistent with the precession (3.9) of $\mathbf{s}$ about the magnetic field as found above from the equation of motion for Λ.

24.4   Dirac equation

A simple classical equation of motion[8, 9] follows from the Lorentz transformation $p = \Lambda m\Lambda^\dagger$ and the unimodularity of Λ:

$$p\bar{\Lambda}^\dagger = m\Lambda.$$

If as above we put $\Psi = \rho_{\text{ref}}^{1/2}\Lambda$, then Ψ obeys the same equation:

$$p\bar{\Psi}^\dagger = m\Psi.$$

The equation is invariant under gauge rotations $\Lambda \to \Lambda R$ and real linear combinations of solutions are also solutions.

If Ψ is a real linear superposition of classical de Broglie waves (3.7), the momentum p can be replaced by a differential operator:

$$p\bar{\Psi}^\dagger = i\hbar\partial\bar{\Psi}^\dagger\mathbf{e}_3 - eA\bar{\Psi}^\dagger = m\Psi. \tag{4.1}$$

This is a covariant algebraic form of the Dirac equation in APS. To make it easier to solve, we split it into two complementary minimal left ideals with projectors $\mathsf{P}_{\pm 3} \equiv \frac{1}{2}(1 \pm \mathbf{e}_3)$:

$$p\bar{\Psi}^\dagger\mathsf{P}_{+3} = (i\hbar\partial - eA)\,\bar{\Psi}^\dagger\mathsf{P}_{+3} = m\Psi\mathsf{P}_{+3}\,,$$

$$p\bar{\Psi}^\dagger\mathsf{P}_{-3} = (-i\hbar\partial - eA)\,\bar{\Psi}^\dagger\mathsf{P}_{-3} = m\Psi\mathsf{P}_{-3}\,,$$

where we noted the "pacwoman" property $\mathbf{e}_3\mathsf{P}_{\pm 3} = \pm\mathsf{P}_{\pm 3}$. We can bar-dagger conjugate the second equation to project it into the same minimal left ideal of the algebra as the first:

$$\bar{p}\Psi\mathsf{P}_{+3} = (i\hbar\bar{\partial} - e\bar{A})\,\Psi\mathsf{P}_{+3} = m\bar{\Psi}^\dagger\mathsf{P}_{+3}\,.$$

Then, defining

$$\psi^{(W)} = \frac{1}{\sqrt{2}}\begin{pmatrix} \Psi\mathsf{P}_{+3} \\ \bar{\Psi}^\dagger\mathsf{P}_{+3} \end{pmatrix},$$

we obtain the pair of projected equations in the matrix form

$$\begin{pmatrix} 0 & p \\ \bar{p} & 0 \end{pmatrix}\psi^{(W)} = \begin{pmatrix} 0 & i\hbar\partial - eA \\ i\hbar\bar{\partial} - e\bar{A} & 0 \end{pmatrix}\psi^{(W)} = m\psi^{(W)}. \tag{4.2}$$

If the standard matrix representation $\mathbf{e}_\mu \to \underline{\sigma}_\mu$ is used for APS in terms of the Pauli spin matrices $\underline{\sigma}_\mu$, Eq. (4.2) is the traditional form[10] of the Dirac equation $p^\mu\gamma_\mu\psi^{(W)} = m\psi^{(W)}$ with gamma matrices in the Weyl (or spinor) representation. The projection of the algebraic Ψ by $\mathsf{P}_{\pm 3}$ is seen to be equivalent to multiplication of $\psi^{(W)}$ by the traditional chirality projectors $\frac{1}{2}(1 \pm \gamma_5)$ with $\gamma_5 = -i\gamma_0\gamma_1\gamma_2\gamma_3$:

$$\tfrac{1}{2}(1 \pm \gamma_5)\,\psi^{(W)} \Leftrightarrow \Psi\mathsf{P}_{\pm 3}\,.$$

The Dirac bispinor in the Dirac–Pauli (or standard) representation is related by

$$\psi^{(DP)} = \frac{1}{\sqrt{2}}\begin{pmatrix} 1 & 1 \\ 1 & -1 \end{pmatrix}\psi^{(W)} = \begin{pmatrix} \langle\Psi\rangle_+\mathsf{P}_{+3} \\ \langle\Psi\rangle_-\mathsf{P}_{+3} \end{pmatrix},$$

where $\langle\Psi\rangle_\pm = \frac{1}{2}(\Psi \pm \bar{\Psi}^\dagger)$ are the even and odd parts of Ψ and correspond to the *large* and *small* components at low velocities:

$$\langle\Psi\rangle_+ = \rho_{\text{ref}}^{1/2}\langle B\rangle_+ R = \rho_{\text{ref}}^{1/2}\sqrt{\frac{m+E}{2m}}R \simeq \rho_{\text{ref}}^{1/2}R$$

$$\langle\Psi\rangle_- = \rho_{\text{ref}}^{1/2}\langle B\rangle_- R = \frac{\mathbf{p}}{m+E}\langle\Psi\rangle_+\,.$$

The last expression for $\langle\Psi\rangle_+$ on the RHS is the low-velocity approximation. In the rest frame, the small component disappears and the eigenfunction is even. We say the particle has even *intrinsic parity*. These classical expressions correspond closely to their quantum counterparts.

24.4.1 Spin distributions

To study spin distributions, the low-velocity limit

$$\Psi \simeq \langle \Psi \rangle_+ \simeq \rho_{\text{ref}}^{1/2} R \tag{4.3}$$

is sufficient. In terms of Euler angles ϕ, θ, χ, about space-fixed axes, the rotor R can be expressed by

$$R = \exp\left(-\tfrac{1}{2}ie_3\phi\right)\exp\left(-\tfrac{1}{2}ie_2\theta\right)\exp\left(-\tfrac{1}{2}ie_3\chi\right). \tag{4.4}$$

The classical spin direction at position $\mathbf{r}$ in a static system is $\mathbf{s} = Re_3R^\dagger$ [see (3.8) above], where R may be a function of $\mathbf{r}$. A distribution of such spin directions is thus $\rho_{\text{ref}}Re_3R^\dagger$, where the positive scalar $\rho_{\text{ref}} = \rho_{\text{ref}}(\mathbf{r})$ is the density of spins. As seen below, simple measurements of the spin direction give only one component at a time. The distribution of the component of the spin in the direction of an arbitrary unit vector $\mathbf{m}$ is

$$\rho_{\text{ref}}\,\mathbf{s}\cdot\mathbf{m} = \left\langle \rho_{\text{ref}}Re_3R^\dagger\mathbf{m}\right\rangle_0.$$

In terms of the projector $\mathsf{P}_{+3} = (\mathsf{P}_{+3})^2$, since $e_3 = \mathsf{P}_{+3} - \mathsf{P}_{-3}$ and for any elements p, q, $\langle pq \rangle_0 = \langle qp \rangle_0 = \langle \overline{pq} \rangle_0$, the distribution is

$$2\left\langle \rho_{\text{ref}}R\mathsf{P}_{+3}R^\dagger\mathbf{m}\right\rangle_0 = 2\left\langle \mathsf{P}_{+3}R^\dagger\rho_{\text{ref}}^{1/2}\mathbf{m}\rho_{\text{ref}}^{1/2}R\mathsf{P}_{+3}\right\rangle_0 = \text{tr}\left\{\psi^\dagger\mathbf{m}\psi\right\}, \tag{4.5}$$

where by ψ we mean the standard matrix representation of the ideal spinor

$$\psi \equiv \rho_{\text{ref}}^{1/2}R\mathsf{P}_{+3} \equiv e^{-i\chi/2}\rho_{\text{ref}}^{1/2}\begin{pmatrix} e^{-i\phi/2}\cos\tfrac{1}{2}\theta & 0 \\ e^{i\phi/2}\sin\tfrac{1}{2}\theta & 0 \end{pmatrix}.$$

If the column of zeros is dropped, ψ is a two-component spinor, as familiar from the usual nonrelativistic Pauli theory. Such spinors carry an irreducible representation of the rotation group $SU(2)$.[2] The term $\psi^\dagger\mathbf{m}\psi$ is then a scalar and tr can be omitted from (4.5). Although we derived the spin distribution as a classical expression, it has *precisely the quantum form* if we recognize that the matrix representation of the unit vector $\mathbf{m}$, namely $\mathbf{m} = m^j e_j \rightarrow m^1\sigma_x + m^2\sigma_y + m^3\sigma_z$, is traditionally (but misleadingly, since it represents a vector, not a scalar) written $\mathbf{m}\cdot\boldsymbol{\sigma}$. From the definition of ρ_{ref} the spinor ψ satisfies the usual normalization condition,

$$2\int d^3x\,\langle\psi^\dagger\psi\rangle_0 \equiv \langle\psi|\psi\rangle = 1$$

---

[2]More generally, once projected onto a minimal left ideal, any spinor $\Psi = \rho_{\text{ref}}^{1/2}BR$ is equivalent to a dilated spatial rotation, thereby reducing this representation of the noncompact Lorentz group to a scaling factor times the compact group $SU(2)$.

and the average component of the spin in the direction m is

$$2 \int d^3x \, \langle \psi^\dagger \mathbf{m}\psi \rangle_0 \equiv \langle \psi \, |\mathbf{m}| \, \psi \rangle .$$

From expression (4.5) one sees that the real paravector

$$\mathsf{P_s} = R\mathsf{P}_{+3}R^\dagger = \tfrac{1}{2}(1 + \mathbf{s})$$

embodies information about the classical spin state at a given point in space or in a homogeneous ensemble. It is equivalent to the quantum spin density operator ϱ for the pure state of spin s, and it is also a projector that acts as a state filter. The component of the spin in the m direction is $2 \langle \mathsf{P_s \, m} \rangle_0$, whose matrix representation is identical to the usual quantum expression, traditionally written $\tfrac{1}{2}\mathrm{tr}\,\{\varrho\,\mathbf{m}\cdot\boldsymbol{\sigma}\}$. Note that the part of a rotation R around the spin axis becomes a phase factor of $\psi = \varrho_{\mathrm{ref}}^{1/2} R\mathsf{P}_{+3}$, since the pacwoman property gives $e^{-i\chi e_3/2}\mathsf{P}_{+3} = e^{-i\chi/2}\mathsf{P}_{+3}$. Thus, when phase factors of ψ are ignored, information about the axial rotation of R is lost. Good quantum calculations keep track of *relative* phase, and this appears to be the only aspect of the phase (rotation about e_3) that can be determined experimentally.

One way of seeing whether the system is in a given state of spin n is to apply the state filter to the spin density operator ϱ and see what remains:

$$\mathsf{P}_n\varrho\mathsf{P}_n = \left(\mathsf{P}_n\varrho + \bar\varrho\bar{\mathsf{P}}_n\right)\mathsf{P}_n = 2\,\langle \mathsf{P}_n\varrho\rangle_0\,\mathsf{P}_n .$$

The scalar coefficient $2\,\langle \mathsf{P}_n\varrho\rangle_0 = \langle(1+\mathbf{n})\,\varrho\rangle_0$ is the probability of finding the system described by ϱ in the state $\mathbf{n}$. For a system in the pure state $\varrho = \mathsf{P_s} = \tfrac{1}{2}(1+\mathbf{s})$, the probability is

$$2\,\langle \mathsf{P}_n\mathsf{P_s}\rangle_0 = \tfrac{1}{2}\,\langle(1+\mathbf{n})(1+\mathbf{s})\rangle_0 = \tfrac{1}{2}(1+\mathbf{n}\cdot\mathbf{s}). \tag{4.6}$$

This is unity if the system is definitely in the state $\mathbf{n}$, whereas it vanishes if the system is in a state *orthogonal* to $\mathbf{n}$. Thus, $\mathbf{s} = \mathbf{n}$ is required for the states to be the same and $\mathbf{s} = -\mathbf{n}$ for the states to be orthogonal. Note that the same mathematics can describe light polarization [2].

24.4.2   Spin $\tfrac{1}{2}$ and state expansions

The value of $\tfrac{1}{2}$ for the spin of elementary spinors considered here arises in several ways. It is the group-theoretical label for the irreducible spinor representation of the rotation group $SU(2)$ carried by ideal spinors. It is also required by the fact that any rotation can be expressed as a linear superposition of two independent orthogonal rotations defined for any direction in space. The Euler-angle form (4.4) of any rotor R can be rewritten

$$R = \exp\left(-\tfrac{1}{2}i\mathbf{n}\theta\right)\exp\left[-\tfrac{1}{2}ie_3\left(\phi+\chi\right)\right]$$
$$= \left(\cos\tfrac{1}{2}\theta - i\mathbf{n}\sin\tfrac{1}{2}\theta\right)\exp\left[-\tfrac{1}{2}ie_3\left(\phi+\chi\right)\right], \tag{4.7}$$

where $\mathbf{n} = \exp\left(-\frac{1}{2}i\mathbf{e}_3\phi\right)\mathbf{e}_2\exp\left(i\frac{1}{2}\mathbf{e}_3\phi\right)$ is a unit vector in the $\mathbf{e}_1\mathbf{e}_2$ plane. Therefore, any rotor R is a real linear combination $\cos\frac{1}{2}\theta R_\uparrow + \sin\frac{1}{2}\theta R_\downarrow$ of two rotors

$$R_\uparrow = \exp\left[-i\frac{1}{2}\mathbf{e}_3\left(\phi+\chi\right)\right], \quad \text{and} \quad R_\downarrow = -inR_\uparrow = \exp\left(-\frac{1}{2}in\pi\right)R_\uparrow$$

that are mutually orthogonal: $\left\langle R_\uparrow R_\downarrow^\dagger \right\rangle_0 = \langle -in \rangle_0 = 0$.

By projecting the rotors with P_{+3} we obtain the equivalent relation of ideal spinors:

$$R\mathrm{P}_{+3} = \cos\tfrac{1}{2}\theta R_\uparrow \mathrm{P}_{+3} + \sin\tfrac{1}{2}\theta R_\downarrow \mathrm{P}_{+3},$$

$$\psi \equiv \rho_{\text{ref}}^{1/2}R\mathrm{P}_{+3} = \cos\tfrac{1}{2}\theta\psi_\uparrow + \sin\tfrac{1}{2}\theta\psi_\downarrow,$$

$$\psi_\uparrow = \rho_{\text{ref}}^{1/2}e^{-i(\phi+\chi)/2}\mathrm{P}_{+3}, \ \psi_\downarrow = -in\psi_\uparrow.$$

Traditional orthonormality conditions hold:

$$2\left\langle \psi_\uparrow \psi_\downarrow^\dagger \right\rangle_0 = 2\left\langle \psi_\uparrow \psi_\uparrow^\dagger in \right\rangle_0 = 2\rho_{\text{ref}}\left\langle \mathrm{P}_{+3}in \right\rangle_0 = 0$$

$$2\left\langle \psi_\downarrow \psi_\downarrow^\dagger \right\rangle_0 = 2\left\langle \psi_\uparrow \psi_\uparrow^\dagger \right\rangle_0 = 2\rho_{\text{ref}}\left\langle \mathrm{P}_{+3} \right\rangle_0 = \rho_{\text{ref}}.$$

It follows that the amplitudes are

$$\langle\psi_\uparrow|\psi\rangle = 2\int d^3x \left\langle \psi\psi_\uparrow^\dagger \right\rangle_0 = \cos\tfrac{1}{2}\theta, \ \langle\psi_\downarrow|\psi\rangle = 2\int d^3x \left\langle \psi\psi_\downarrow^\dagger \right\rangle_0 = \sin\tfrac{1}{2}\theta$$

giving probabilities as found above in Eq. (4.6).

$$|\langle\psi_\uparrow|\psi\rangle|^2 = \cos^2\tfrac{1}{2}\theta = \tfrac{1}{2}\left(1 + \mathbf{s}\cdot\mathbf{e}_3\right), \ |\langle\psi_\downarrow|\psi\rangle|^2 = \sin^2\tfrac{1}{2}\theta = \tfrac{1}{2}\left(1 - \mathbf{s}\cdot\mathbf{e}_3\right).$$

By tacking on an additional fixed rotation, the treatment can be extended to measurements along an arbitrary axis.

24.4.3   Stern–Gerlach experiment

The basic measurement of spin is that of the Stern–Gerlach experiment [12], in which a beam of ground-state silver atoms is split by a magnetic-field gradient into distinct beams of opposite spin polarization. It is a building block of real and thought experiments in quantum measurement. [11]

Consider a nonrelativistic beam of ground-state atoms that travels with velocity $\mathbf{v} = v_x\mathbf{e}_1$ through a static magnetic field $\mathbf{B}$ that vanishes everywhere except in the vicinity of the Stern–Gerlach magnet, whose net effect on the beam is a vertical force proportional to the z component of the spin s_z :

$$\mu_z\frac{\partial B_z}{\partial z}\mathbf{e}_3 = -2\mu_0 s_z\frac{\partial B_z}{\partial z}\mathbf{e}_3.$$

Now, for the rotors $R_\uparrow$ and $R_\downarrow$, the spin directions are

$$R_\uparrow e_3 R_\uparrow^\dagger = e_3, \ R_\downarrow e_3 R_\downarrow^\dagger = -e_3 \ .$$

These can be re-expressed in a form

$$e_3 R_\uparrow = R_\uparrow e_3, \ e_3 R_\downarrow = -R_\downarrow e_3$$

that becomes an eigenfunction equation when projected onto the P_{+3} ideal:

$$e_3 \psi_\uparrow = \psi_\uparrow, \ e_3 \psi_\downarrow = -\psi_\downarrow \ .$$

Evidently $s_z = \hbar e_3/2$ is the spin operator for ψ. Thus, the full rotor R is split into $R = \cos\frac{1}{2}\theta R_\uparrow + \sin\frac{1}{2}\theta R_\downarrow$, the two parts of which experience opposite forces from the Stern–Gerlach magnet. This splits the incident beam into two isolated branches, analogous to the way a birefringent crystal splits a beam of light into two branches of orthogonal polarization. The fraction of the initial beam in the upper branch is $\cos^2\frac{1}{2}\theta = \frac{1}{2}(1+\cos\theta)$ whereas in the lower branch the fraction is $\sin^2\frac{1}{2}\theta = \frac{1}{2}(1-\cos\theta)$, just as found from the square amplitudes above. The two-valued property of the measurement is a direct result of the decomposition of the rotation operator R into rotations about any two opposite directions, rotations that correspond to "spin-up" and "spin-down" components.

24.5   Conclusions

We have derived many quantum results for a classical system. The quantum/classical interface has become almost transparent in the spinor approach of Clifford's geometric algebra. Quantum effects demand *amplitudes*, and these are provided classically by rotors of the geometric algebra. In particular, the *eigenspinor* is an amplitude closely associated with quantum spinor wave functions. Linear equations for the classical eigenspinor suggest superposition and hence the possibility of quantumlike interference. Projectors onto minimal left ideals of the algebra are needed to reduce the spinor representation of Lorentz rotations in the algebra and to simplify the equation of motion, and by using them, many of the classical results take quantum form. The association helps clarify various quantum phenomena, and extensions to multiparticle systems promise to demystify such quantum phenomena as entanglement. The Q/C relation appears to be much more than superficial. The basic spinor representation gives spin $\frac{1}{2}$, and the linearity of the equation of motion gives $g = 2$.

Of course this is not to say that quantum and classical physics are the same. Indeed, the classical spinor approach of geometric algebra, while removing many superficial differences, highlights an important quantum property that is not (at least not yet) explained classically, namely the basic quantization of particles. Its resolution at the Q/C interface, if there is a resolution, may require a second-quantized extension of the classical eigenspinor equations.

Acknowledgement

Support from the Natural Sciences and Engineering Research Council of Canada is gratefully acknowledged.

REFERENCES

[1] P. Lounesto, *Clifford Algebras and Spinors*, second edition, Cambridge University Press, Cambridge (UK), 2001.

[2] W.E. Baylis, Applications of Clifford Algebras in Physics, Lecture 4 in *Lectures on Clifford Geometric Algebras,* ed. by R. Abłamowicz and G. Sobczyk, Birkhäuser, Boston, 2003.

[3] W.E. Baylis, *Electrodynamics: A Modern Geometric Approach*, Birkhäuser, Boston, 1999.

[4] W.E. Baylis, editor, *Clifford (Geometric) Algebra with Applications to Physics, Mathematics, and Engineering*, Birkhäuser, Boston, 1996.

[5] D. Hestenes, *Spacetime Algebra*, Gordon and Breach, New York, 1966.

[6] W.E. Baylis and Y. Yao, Relativistic Dynamics of Charges in Electromagnetic Fields: An Eigenspinor Approach, *Phys. Rev. A* **60** (1999), 785–795.

[7] V. Bargmann, L. Michel, and V. L. Telegdi, *Phys. Rev. Lett.* **2** (1959), 435.

[8] W.E. Baylis, Classical eigenspinors and the Dirac equation, *Phys. Rev. A* **45** (1992), 4293–4302.

[9] W.E. Baylis, Eigenspinors and electron spin, *Adv. Appl. Clifford Algebras* **7(S)** (1997), 197–213.

[10] V.B. Berestetskii, E.M. Lifshitz, and L.P. Pitaevskii, *Quantum Electrodynamics* (Volume 4 of *Course of Theoretical Physics*), 2nd edn. (transl. from Russian by J. B. Sykes and J. S. Bell), Pergamon Press, Oxford, 1982.

[11] D.Z. Albert, Bohm's alternative to quantum mechanics, *Sci. Am.* **270** (1994, no. 5), 58–67.

[12] L.E. Ballentine, *Quantum Mechanics: a modern development*, World Scientific, Singapore, 1998.

William E. Baylis
Department of Physics, University of Windsor
Windsor, ON, Canada N9B 3P4
E-mail: baylis@uwindsor.ca

Submitted: September 30, 2002; Revised: April 20, 2003.

25

Standard Quantum Spheres

Francesco Bonechi, Nicola Ciccoli, and Marco Tarlini

ABSTRACT We review the deformation of Podleś 2-sphere and of the 4-sphere introduced in [2], which display a quite similar behavior. Special attention is devoted to the development of the local picture of charts.

Keywords: Poisson Lie group, symplectic leaves, cyclic cohomology

25.1   Introduction

In the last two years several examples of deformed spheres have been produced [2, 7]. A particular attention has been devoted to investigate their properties according to the paradigm of noncommutative geometry *à la* Connes [6], mainly their cyclic homology and cohomology. From this point of view we can distinguish two opposite behaviors: while the Hochschild and cyclic homology of the sphere in [7] is isomorphic to the classical one, in the case of [2] all nontrivial content is in zero dimension. This fact can be paraphrased by saying that the Connes–Landi sphere is almost classical, and indeed it fulfills all the axioms of noncommutative geometry of [6], while the quantum space of [2] is very degenerate.

These two examples well represent a general feature of two distinct kind of deformations: the standard deformations associated to the well known Drinfeld–Jimbo quantum groups and the twisted ones which quantize the semiclassical structure defined by a generic element of the (wedge product of) Cartan subalgebra (see [1, 13]).

The prototype of the standard case is the standard Podleś two sphere [11], whose noncommutative geometry was studied in [10]. The 4-sphere of [2] represents a natural generalization. In fact it defines the same C^*-algebra of the 2-sphere and the same cyclic cohomology. This is indeed an extreme case of quantum degeneracy: according to the paradigm of noncommutative geometry, these two quantum spaces, whose classical limit has different dimension, are topologically equivalent.

In this report we review the main properties of the Podleś 2-sphere and of the 4-sphere of [2]. We develop also the local structure, *i.e.*, the quantum charts obtained

AMS Subject Classification: 14D21, 19D55, 20G42, 58B32, 58B34, 81R50.

by "removing" the poles. These charts can be used to define local trivialization for the monopole and instanton vector bundles. In fact a quantum vector bundle is described by deforming the space of sections regarded as a finitely generated projective module; this is a global picture and there is not a clear notion of what are transition functions and the (quantum?) group acting on the fiber. In [9] there is a definition of locally trivial line bundles of an algebra obtained by deformation quantization. The analysis of the local charts can be considered as a first attempt in order to understand if the instanton bundle on the quantum 4-sphere can be studied in this framework.

25.2 Podleś Poisson and quantum 2-sphere

Podleś spheres were introduced in [11] as polynomial algebras together with the universal C^*-algebra associated to them. In [10] their Hochschild and cyclic homology and cohomology were computed. In this section we will review their definition and properties. Let $A(S^2)$ be the algebra of polynomial functions over the two sphere generated by $\alpha \in \mathbb{C}$ and $\tau \in \mathbb{R}$ and the relation $|\alpha|^2 = (\tau + d)(1 - \tau)$, for $d \geq 0$. We define a Poisson structure on S^2 by giving the Poisson bracket

$$\{\alpha, \tau\} = -2\alpha\tau, \qquad \{\alpha, \alpha^*\} = 4\tau^2 + 2(d - 1)\tau.$$

All the points with $\tau = 0$ are degenerate (i.e., 0-dimensional symplectic leaves). If $d > 0$ the set of degenerate points is a circle; the upper and lower hemispheres obtained by removing the circle are symplectic leaves of dimension 2. If $d = 0$ the circle of degenerate points shrinks to the north pole P_N and there is only one two–dimensional symplectic leaf $S^2 \backslash P_N$. In the following we will denote with P_S the south pole, corresponding to $\tau = 1$ in the case with $d = 0$.

Let us describe the local structure in the case $d = 0$. Let $\mathbb{R}_I^2 = S^2 \backslash P_N$ be the first chart and let $z = \alpha/\tau$. The Poisson structure reads

$$\{z, z^*\} = 2(1 + |z|^2).\tag{2.1}$$

On the second chart $\mathbb{R}_{II}^2 = S^2 \backslash P_S$ we define $w = \alpha^*/(1 - \tau)$ and obtain

$$\{w, w^*\} = 2|w|^2(1 + |w|^2).\tag{2.2}$$

Notice that in the intersection $\mathbb{R}_I^2 \cap \mathbb{R}_{II}^2$ we have $w = 1/z$.

The quantization of these Poisson structures is the algebra $A(S_{q,d}^2)$ defined by the following relations ($q \in \mathbb{R}$) :

$$\alpha\tau = q^2\tau\alpha, \quad \alpha^*\alpha = (\tau + d)(1 - \tau), \quad \alpha\alpha^* = (1 - q^2\tau)(d + q^2\tau).$$

This algebra can be obtained as a quantum homogeneous space of the standard $SU_q(2)$ (i.e., the quantum Hopf fibration, [3]). One should expect that unitary irreducible representations correspond to symplectic leaves of the semiclassical structure (for this general correspondence see for example [4]). Let us analyze first the case $d > 0$. There is a family of one dimensional representations corresponding to the zero–dimensional leaves: for $|\mu| = 1$ let $\rho_\mu(\alpha) = -\sqrt{d}\mu$, $\rho_\mu(\tau) = 0$. To the two–dimensional leaves there correspond the representations $\rho_\pm : A(S^2_{d,q}) \rightarrow B(\ell^2(\mathbb{N}))$,

$$\rho_\pm(\tau)|n\rangle = q^{2n}|n\rangle, \qquad \rho_\pm(\alpha)|n\rangle = \omega^\pm_{n-1}|n - 1\rangle, \qquad (2.3)$$

where

$$\omega^-_n = d\sqrt{(1 - q^{2(n+1)})(1 + dq^{2(n+1)})}, \quad \omega^+_n = \sqrt{(1 - q^{2(n+1)})(d + q^{2(n+1)})}.$$

Let us concentrate on the case with $d = 0$, the standard Podleś 2-sphere S^2_q. Irreducible representations consist then of one character ϵ and one infinite–dimensional representation $\sigma = \rho_+$. The universal C^*-algebra generated by $A(S^2_q)$ is $\tilde{K}$, the minimal unitization of compact operators.

We can define a Fredholm module (see [6] for the definition) by using the representation σ. It is given by $(\ell^2(\mathbb{N}) \oplus \ell^2(\mathbb{N}), F)$, where

$$F = \begin{pmatrix} 0 & 1 \\ 1 & 0 \end{pmatrix},$$

and $A(S^2_q)$ acts as

$$\begin{pmatrix} \sigma(a) & 0 \\ 0 & \epsilon(a) \end{pmatrix}.$$

The character of this Fredholm module (i.e., a 0-cyclic cocycle) is

$$\text{tr}_\sigma : A(S^2_q) \rightarrow \mathbb{C},$$

defined by $\text{tr}_\sigma(a) = \text{tr}(\sigma(a) - \epsilon(a))$.

The quantum vector bundle of the q-monopole is defined by giving the following idempotent $e \in M_2(A(S^2_q))$ (see [3, 8]):

$$e = \begin{pmatrix} \tau & \alpha^* \\ \alpha & 1 - q^2\tau \end{pmatrix}$$

whose charge is $\text{tr}_\sigma(\text{Tr}(e)) = \text{tr}(1 + (1 - q^2)\tau) = 1$.

Let us describe the local structure of S^2_q. Let $\mathbb{R}^2_{Iq}$ be the quantum space obtained by removing the north pole from S^2_q; i.e., let us define $A(\mathbb{R}^2_{Iq})$ as the $*$-subalgebra of $A(S^2_q)[\tau^{-1}]$ generated by $z = \tau^{-1}\alpha$. We then have the relation

$$z^*z = q^2zz^* - q^4(1 - q^2). \qquad (2.4)$$

Being $\sigma(\tau)$ invertible, the representation σ of $A(S^2_q)$ gives a representation of $A(S^2_q)[\tau^{-1}]$ and then of $A(\mathbb{R}^2_{Iq})$, which reads

$$\sigma(z)|n\rangle = q^{-(n-1)}(1 - q^{2n})^{1/2}|n - 1\rangle. \qquad (2.5)$$

The localization with respect to the south pole is more problematic (see the Erratum of [3]). Let us define $A(\mathbb{R}^2_{IIq})$ as the $*$-subalgebra of

$$A(S^2_q)[\{(1 - q^{2n}\tau)^{-1}\}_{n \in \mathbb{Z}}]$$

generated by $w = \alpha^*(1 - q^2\tau)^{-1}$. We finally get

$$w^*w - q^2 ww^* = -(1 - q^2)ww^*w^*w, \qquad (2.6)$$

which formally quantizes the Poisson structure in (2.2); this is the algebra considered in [5]. Only the trivial representation ϵ of $A(S^2_q)$ gives a representation of the localization. The following argument indicates that the trivial representation is the unique unitary representation of $\mathbb{R}^2_{IIq}$. Let $N = w^*w$ and $M = ww^*$; we have that $[M, N] = 0$. If $N|\nu\rangle = \nu|\nu\rangle$, then by using (2.6) we find that $M|\nu\rangle = \mu(\nu)|\nu\rangle$, with

$$\mu(\nu) = \frac{\nu}{(q^2 - (1 - q^2)\nu)}.$$

By repeated use of (2.6) it is easy to show that $Nw^{*k}|\nu\rangle = \nu_k w^{*k}|\nu\rangle$ with

$$\nu_k = \frac{\nu}{q^{2k} - \nu(1 - q^{2k})}.$$

But for each $\nu > 0$ there exists $k(\nu)$ such that $\nu_k < 0$ for each $k > k(\nu)$, and this contradicts the hypothesis of unitarity.

25.3   Poisson and quantum 4-sphere

The quantum 4-sphere we are going to describe was introduced in [2]. Let $A(S^4)$ be the algebra of polynomial functions over S^4, generated by $a_1, a_2 \in \mathbb{C}$ and $\tau \in \mathbb{R}$ with the relation $|a_1|^2 + |a_2|^2 = \tau(1 - \tau)$. We define the following Poisson structure on S^4 :

$$\{a_1, a_2\} = a_1 a_2, \quad \{a_i, \tau\} = -2a_i\tau, \quad \{a_1, a_2^*\} = -3a_1 a_2^*,$$
$$\{a_1, a_1^*\} = 2\tau^2 - 2|a_1|^2, \quad \{a_2, a_2^*\} = 2\tau^2 + 2|a_1|^2 - 2|a_2|^2. \qquad (3.1)$$

Symplectic foliation is the same foliation of the Podleś 2-sphere with $d = 0$. In fact there is a zero–dimensional leaf (the north pole P_N, $\tau = 0$) and a symplectic plane $\mathbb{R}^4_I = S^4 \backslash P_N$. In order to describe the symplectic $\mathbb{R}^4_I$ let us define $z_i = a_i/\tau$; after a simple computation we get

$$\{z_1, z_2\} = z_1 z_2, \quad \{z_1, z_2^*\} = z_1 z_2^*, \quad \{z_1, z_1^*\} = 2(1 + |z_1|^2),$$
$$\{z_2, z_2^*\} = 2(1 + |z_1|^2 + |z_2|^2). \qquad (3.2)$$

The Poisson structure on the second chart $\mathbb{R}^4_{II} = S^4 \backslash P_S$ is described by introducing $w_i = a_i^*/(1 - \tau)$. The Poisson brackets between these local coordinates read

$$\{w_1, w_2\} = -w_1 w_2, \quad \{w_1, w_2^*\} = w_1 w_2^*(3 + 4|w_1|^2 + 4|w_2|^2),$$

$$\{w_1, w_1^*\} = 2|w_1|^2(|w_1|^2 + 1) - 2|w_2|^4,$$

$$\{w_2, w_2^*\} = 2|w_2|^2(1 + |w_2|^2) - 2|w_1|^2(|w_1|^2 + 1). \tag{3.3}$$

The algebra $A(\Sigma_q^4)$ quantizes the algebra of polynomial functions on the sphere:

$$a_1 \tau = q^2 \tau a_1, \quad a_2 \tau = q^2 \tau a_2, \quad a_1 a_2 = q^{-1} a_2 a_1, \quad a_1 a_2^* = q^3 a_2^* a_1,$$

$$a_1 a_1^* = q^2 a_1^* a_1 + q^2(1 - q^2)\tau^2, \quad a_2 a_2^* = q^4 a_2^* a_2 + q^2(1 - q^2)\tau,$$

$$a_1^* a_1 + a_2^* a_2 = \tau(1 - \tau). \tag{3.4}$$

Accordingly with the symplectic foliation of the semiclassical limit there are only two irreducible representations: the character $\epsilon : A(\Sigma_q^4) \to \mathbb{C}$ defined by $\epsilon(a_i) = \epsilon(\tau) = 0$, and the infinite–dimensional representation $\sigma : A(\Sigma_q^4) \to B(\ell^2(\mathbb{N})^{\otimes 2})$,

$$\sigma(a_1)|n_1, n_2\rangle = q^{2n_2 + n_1}(1 - q^{2n_1})^{1/2}|n_1 - 1, n_2\rangle,$$

$$\sigma(\tau)|n_1, n_2\rangle = q^{2(n_1 + n_2)}|n_1, n_2\rangle,$$

$$\sigma(a_2)|n_1, n_2\rangle = q^{n_1 + n_2}(1 - q^{2n_2})^{1/2}|n_1, n_2 - 1\rangle. \tag{3.5}$$

Analogously to the 2-sphere, the universal C^*-algebra is the minimal unitization of compacts $\tilde{K}$, giving rise to an unexpected isomorphism between these two quantum topological spaces. We can define a Fredholm module (see [6]) by using the representation σ. It is given by $(\ell^2(\mathbb{N})^{\otimes 2} \oplus \ell^2(\mathbb{N})^{\otimes 2}, F)$, where

$$F = \begin{pmatrix} 0 & 1 \\ 1 & 0 \end{pmatrix},$$

and $A(\Sigma_q^4)$ acts as

$$\begin{pmatrix} \sigma(a) & 0 \\ 0 & \epsilon(a) \end{pmatrix}.$$

The character of this Fredholm module (*i.e.*, a 0-cyclic cocycle) is

$$\mathrm{tr}_\sigma : A(\Sigma_q^4) \to \mathbb{C},$$

defined by $\mathrm{tr}_\sigma(a) = \mathrm{tr}(\sigma(a) - \epsilon(a))$.

The quantum vector bundle of the instanton is defined by giving the following idempotent:

$$G = \begin{pmatrix} q^2 \tau & 0 & a_1 & q a_2^* \\ 0 & q^2 \tau & a_2 & -q^2 a_1^* \\ a_1^* & a_2^* & 1 - \tau & 0 \\ q a_2 & -q^2 a_1 & 0 & 1 - q^4 \tau \end{pmatrix},$$

and its charge is computed as $\text{tr}_\sigma(\text{Tr}\, G) = -1$.

Let us describe the local structure of this quantum sphere. The first chart is $\mathbb{R}^4_{Iq}$; the polynomial functions $A(\mathbb{R}^4_{Iq})$ are defined as the $*$-subalgebra of $A(\Sigma^4_q)[\tau^{-1}]$ generated by $z_i = \tau^{-1}a_i$. After some easy computations we get

$$z_1 z_2 = q^{-1} z_2 z_1, \quad z_1 z_2^* = q^{-1} z_2^* z_1,$$

$$z_1 z_1^* = q^{-2} z_1^* z_1 + q^2(1 - q^2),$$

$$z_2 z_2^* = q^{-2} z_2^* z_2 + q^2(q^{-2} - 1)z_1 z_1^* + q^2(1 - q^2). \tag{3.6}$$

The representation σ of $A(\Sigma^4_q)$ restricts to the following representation of $A(\mathbb{R}^4_{Iq})$,

$$\sigma(z_1)|n_1, n_2\rangle = q^{2-n_1}(1 - q^{2n_1})^{1/2}|n_1 - 1, n_2\rangle,$$

$$\sigma(z_2)|n_1, n_2\rangle = q^{2-n_1-n_2}(1 - q^{2n_2})^{1/2}|n_1, n_2 - 1\rangle. \tag{3.7}$$

The second chart is more difficult to obtain. Analogously to the 2-sphere case, in order to satisfy the Ore condition (see [12]), the localization with respect to the south pole P_S must be defined by inverting the set $\{1 - q^{2n}\tau\}_{n\in\mathbb{Z}}$. We will not explicitly describe the quantization of the Poisson structure on $\mathbb{R}^4_{II}$ given in (3.3): in fact it is purely formal and presents the same problems of the quantization of $\mathbb{R}^2_{II}$ outlined in the previous section.

REFERENCES

[1] P. Aschieri, F. Bonechi, On the noncommutative geometry of twisted spheres. *Lett. Math. Phys.* **59**, (2002), 133–156.

[2] F. Bonechi, N. Ciccoli, M. Tarlini, Noncommutative instantons on the 4-sphere from quantum groups. *Commun. Math. Phys.* **226**, (2002), 419–432.

[3] T. Brzeziński, S. Majid, Quantum group gauge theory on quantum spaces. *Commun. Math. Phys.* **157**, (1993), 591–638. Erratum **167** p. 235 (1995).

[4] V. Chari, A. Pressley, *A Guide To Quantum Groups*, Cambridge University Press, Cambridge, 1995.

[5] C.-S. Chu, P.-M. Ho, B. Zumino, The quantum 2-sphere as a complex quantum manifold. *Z. Phys. C Part. Fields* **70**, 2, (1996), 339–344.

[6] A. Connes, *Noncommutative Geometry*, Academic Press, San Diego, 1994.

[7] A. Connes, G. Landi, Noncommutative manifolds, the instanton algebra and isospectral deformations. *Commun. Math. Phys.* **221**, (2001), 141–159.

[8] P. Hajac, S. Majid, Projective module description of the q-monopole. *Commun. Math. Phys.* **206**, (1999), 247–264.

[9] B. Jurco, P. Schupp, J. Wess, Noncommutative line bundle and Morita equivalence. hep-th/0106110.

[10] T. Masuda, Y. Nakagami, J. Watanabe, Noncommutative differential geometry on the quantum two sphere of Podleś. I: An algebraic viewpoint. *K–theory* **5**, (1991), 151–175.

[11] P. Podleś, Quantum spheres. *Lett. Math. Phys.* **14**, (1987), 193–202.

[12] B. Stenstrom, *Rings of Quotients*, Springer, Berlin, 1975.

[13] J. Varilly, Quantum symmetry groups of noncommutative spheres. *Commun. Math. Phys.* **221**, (2001), 511–523.

Francesco Bonechi
INFN, Sezione di Firenze
Via Sansone 1, Sesto Fiorentino (Firenze), Italy, 50019
E-mail: bonechi@fi.infn.it

Nicola Ciccoli
Dipartimento di Matematica
Università di Perugia, Italy
ciccoli@dipmat.unipg.it

Marco Tarlini
INFN, Sezione di Firenze
Via Sansone 1, Sesto Fiorentino (Firenze), Italy, 50019
E-mail: tarlini@fi.infn.it

Submitted: June 1, 2002.

26

Clifford Algebras, Pure Spinors and the Physics of Fermions

Paolo Budinich

ABSTRACT The equations defining pure spinors are interpreted as equations of motion formulated on the lightcone of momentum space $P = \mathbb{R}^{1,9}$. Most of the equations for fermion multiplets, usually adopted by particle physics, are then naturally obtained and their properties, such as internal symmetries, charges, and families, appear to be due to the correlation of the associated Clifford algebras with the three complex division algebras: complex numbers, quaternions and octonions. Pure spinors could be relevant not only because the underlying momentum space is compact, but also because they may throw some light on several problematic aspects of particle physics.

Keywords: Clifford algebras, bilinear forms, pure spinors

26.1 Introduction

We will adopt the spinor geometry as formulated by its discoverer É. Cartan [1], who specially stressed the mathematical elegance of the spinors he named "simple" (renamed "pure" by C. Chevalley [2]). It has been shown [3, 4] how the role of division algebras, correlated with Clifford algebras, may be set in evidence by dealing algebraically [2] with spinors and to be at the origin of the $SU(3) \otimes SU(2)_L \otimes U(1)$ groups of the standard model. We will show how Cartan's equation defining simple or pure spinors may be identified as the spinor-field equation in ten-dimensional space-time $M = \mathbb{R}^{1,9}$, after restriction to $M = \mathbb{R}^{1,3}$. In this way, we naturally obtain most of the equations adopted traditionally *ad hoc* that describe the behavior of fermion multiplets, presenting internal symmetry, including that of the standard model. The fact that they may be obtained from the traditional ten-dimensional approach after restriction to $M = \mathbb{R}^{1,3}$, could mean that the extra dimensions are redundant, at least for some of the internal symmetry groups (like isospin $SU(2)$ and flavour $SU(3)$).

Properties of Clifford algebras and associated simple or pure spinors may explain some problematic features of elementary particles, such as the origin of strong and electroweak charges and perhaps also of their relative values, as well

AMS Subject Classification: 15A66, 17B37, 20C30, 81R25.

as a correlation of the three families of leptons and baryons with the three imaginary units of quaternions [5].

26.2   Clifford algebras, spinors and Cartan's equations

Here we will summarize some elements of spinor geometry necessary for the following. For more on the subject see [1, 2, 6].

26.2.1   Hints on spinors

Given $V = \mathbb{C}^{2n}$ and the corresponding Clifford algebra $\mathbb{C}\ell(2n)$, generated by γ_a; a spinor ψ is a 2^n-dimensional vector of the endomorphism space S of $\mathbb{C}\ell(2n)$: $\mathbb{C}\ell(2n) = \text{End}\, S$ and $\psi \in S$.

The Cartan's equation defining ψ is:

$$z_a \gamma^a \psi = 0\,, \quad a = 1, 2, \ldots, 2n \tag{2.1}$$

where $z \in V$. Given $\psi \neq 0$ (implying $z_a z^a = 0$), defines the d-dimensional totally null, projective plane $T_d(\psi)$. For $d = n$ (maximal), ψ was named simple (by Cartan [1]) or pure (by Chevalley [2]), and, according to Cartan, it is isomorphic (up to a sign) to $T_n(\psi)$.

Given $\mathbb{C}\ell(2n) = \text{End}\, S$ and $\psi, \phi \in S$, we have [7]:

$$\phi \otimes B\psi = \sum_{j=0}^{n} F_j \tag{2.2}$$

where B is the main automorphism of $\mathbb{C}\ell(2n)$ [6] and:

$$F_j = {}_{[}\gamma_{a_1} \gamma_{a_2} \cdots \gamma_{a_j]} T^{a_1 a_2 \ldots a_j}\,, \tag{2.3}$$

in which the γ matrices are antisymmetrized and the antisymmetric tensor T is given by:

$$T_{a_1 a_2 \ldots a_j} = \frac{1}{2^n} \langle B\psi, {}_{[}\gamma_{a_1} \gamma_{a_2} \cdots \gamma_{a_j]} \phi \rangle\,. \tag{2.4}$$

Proposition 1. *Take $\phi = \psi$ in (2.2); then ϕ is simple or pure iff*

$$F_0 = F_1 = \cdots = F_{n-1} = 0\,, \quad F_n \neq 0 \tag{2.5}$$

while equation (2.2) becomes:

$$\phi \otimes B\phi = F_n \tag{2.6}$$

representing the maximal totally null plane $T_n(\phi)$ to which ϕ is isomorphic (up to a sign) [7].

For $\mathbb{C}\ell(2n)$, with generators $\gamma_1, \gamma_2, \ldots, \gamma_{2n}$ and with the volume element (normalized to 1) $\gamma_{2n+1} := \gamma_1\gamma_2\cdots\gamma_{2n}$, the spinors: $\psi_\pm = \frac{1}{2}(1\pm\gamma_{2n+1})\psi$ are named Weyl spinors, and they are 2^{n-1}-dimensional and associated to the even subalgebra $\mathbb{C}\ell_0(2n)$ of the algebra $\mathbb{C}\ell(2n)$. Weyl spinors may be simple or pure, and they are all so for $n \leq 3$. For $n > 3$ the constraint equations given by equations (2.5) are $1, 10, 66, 364$, for $n = 4, 5, 6, 7$, respectively.

26.2.2   Isomorphisms of Clifford algebras and of spinors

There are the isomorphisms of Clifford algebras:

$$\mathbb{C}\ell(2n) \simeq \mathbb{C}\ell_0(2n+1) \tag{2.7}$$

both central simple. The spinors associated with $\mathbb{C}\ell(2n)$ and $\mathbb{C}\ell_0(2n+1)$ are named Dirac and Pauli spinors and will be indicated with ψ_D and ψ_P respectively. There are also the isomorphisms:

$$\mathbb{C}\ell(2n+1) \simeq \mathbb{C}\ell_0(2n+2) \tag{2.8}$$

both nonsimple. The Weyl spinors associated with $\mathbb{C}\ell_0(2n+2)$ will be indicated with ψ_W.

From the above we get [8, 9] that a 2^n component Dirac or Pauli spinor is isomorphic to a doublet of 2^{n-1} component Dirac, Pauli or Weyl spinors. This isomorphism may be explicitly represented. In fact, let γ_a be the generators of $\mathbb{C}\ell(2n)$ and Γ_A those of $\mathbb{C}\ell(2n+2)$. Then we have for $\Gamma_a^{(j)} = \sigma_j \otimes \gamma_a$:

$$\Psi^{(j)} = \begin{pmatrix} \psi_1^{(j)} \\ \psi_2^{(j)} \end{pmatrix}, \quad a = 1, 2, \ldots, 2n, \quad j = 0, 1, 2, 3, \tag{2.9}$$

where σ_j are Pauli matrices for $j = 1, 2, 3$, while $\sigma_0 := 1_2$, and $\Psi^{(0)}$ is a doublet of Dirac spinors while $\Psi^{(j)}$ is a doublet of Weyl ($j = 1, 2$) or Pauli ($j = 3$) spinors. There are the unitary transformations U_j :

$$U_j = 1 \otimes L + \sigma_j \otimes R = U_j^{-1}, \quad j = 0, 1, 2, 3. \tag{2.10}$$

where $L := \frac{1}{2}(1 + \gamma_{2n+1})$, $R := \frac{1}{2}(1 - \gamma_{2n+1})$. We have [8]:

$$U_j\Gamma_A^{(0)}U_j^{-1} = \Gamma_A^{(j)}, \quad U_j\Psi^{(0)} = \Psi^{(j)} \tag{2.11}$$

for $A = 1, 2, \ldots, 2n+2$, $j = 0, 1, 2, 3$. We may then affirm:

Proposition 2. *Dirac, Pauli, Weyl spinor doublets are isomorphic.*

26.2.3   Cartan's equations

These may be obtained if we multiply equation (2.2) from the left by γ_a and from the right by $\gamma_a\phi$ and set it to zero after summing over a :

$$\gamma_a\phi \otimes B\psi \, \gamma^a\phi = z_a\gamma^a\phi = 0 \tag{2.12}$$

where the vector components z_a are bilinear in the spinors ψ and ϕ :

$$z_a = \frac{1}{2^n} \langle B\psi, \gamma_a \phi \rangle. \tag{2.13}$$

We have [7]:

Proposition 3. *For arbitrary ψ in $z_a = \langle B\psi, \gamma_a \phi \rangle$, $z_a z^a = 0$ if and only if ϕ is simple or pure.*

Following Cartan [1] we know that for $\mathbb{C}\ell(2n)$ only Weyl spinors, or semi-spinors as Cartan called them, may be simple. These are spinors with only 2^{n-1} non-null components of the even subalgebra $\mathbb{C}\ell_0(2n)$. Therefore the equations for 2^n-component simple or pure spinors ϕ may be obtained from $\mathbb{C}\ell(2n+2)$ only.

The above, including Proposition 3, may be easily restricted to the Clifford algebras $\mathbb{C}\ell(1, 2n-1)$ of Lorentzian signature of our interest, by substituting in (2.13) z_a with

$$p_a = \frac{1}{2^n} \psi^\dagger \gamma_0 \gamma_a \phi \tag{2.14}$$

where $\psi^\dagger$ means ψ hermitian conjugate and γ_0 is the timelike generator, and which are real.

26.3   The equations of motion for fermion multiplets, the $U(1)$ of the strong charge

We will now apply the above to the case of the ten-dimensional space-time $M = \mathbb{R}^{1,9}$. Let us then consider $\mathbb{C}\ell(1,9) = \text{End } S$ with generators $\mathcal{G}_\alpha$ whose Dirac spinors $\Phi \in S$ have 32 components. As seen above, we may obtain the Cartan equations for the corresponding simple or pure spinors starting from $\mathbb{C}\ell(1,11) = \text{End } S$ with generators Γ_β, volume element Γ_{13} and 64-components Dirac spinor Ξ :

$$P_\beta \Gamma^\beta (1 \pm \Gamma_{13}) \Xi = 0, \quad \beta = 1, 2, \dots, 12, \tag{3.1}$$

where

$$\Xi = \begin{pmatrix} \Phi_+ \\ \Phi_- \end{pmatrix} \tag{3.2}$$

with $\Phi_\pm$ Weyl spinor. From (3.1) we obtain, for $(1 + \Gamma_{13})$:

$$(P_\alpha \mathcal{G}^\alpha + i P_{11} \mathcal{G}_{11} + P_{12}) \Phi = 0, \quad \alpha = 1, 2, \dots, 10. \tag{3.3}$$

Because of the isomorphisms discussed in section 26.2.2, we can consider the 32-component spinor Φ as a Dirac spinor of $\mathbb{C}\ell(1,9)$. In fact, we have only to take for the generators $\mathcal{G}_\alpha$ and $\mathcal{G}_{11}$ the representation

$$\mathcal{G}_A = 1_2 \otimes G_A, \quad \mathcal{G}_{9,10,11} = i\sigma_{1,2,3} \otimes G_9, \quad A = 1, 2, \dots, 8, \tag{3.4}$$

where G_A and G_9 are the generators and volume element of $\mathbb{C}\ell(1,7)$ and

$$\Phi = \begin{pmatrix} \Theta_1 \\ \Theta_2 \end{pmatrix} \tag{3.5}$$

where Θ_1 and Θ_2 are 16-component Dirac spinors of $\mathbb{C}\ell(1,7)$ and equation (3.3) becomes [9], after defining:

$$(P_9 + iP_{10}) = \rho e^{i\frac{\varphi}{2}}$$

where $\rho = \sqrt{P_9^2 + P_{10}^2}$:

$$(P_A G^A + iP_{11}G_9 + P_{12})e^{i\frac{\varphi}{2}}\Theta_1 + i\rho G_9\Theta_2 = 0,$$
$$(P_A G^A - iP_{11}G_9 + P_{12})\Theta_2 + i\rho G_9 e^{i\frac{\varphi}{2}}\Theta_1 = 0 \tag{3.3'}$$

which manifests the $U(1)$ invariance of the Θ_1 spinor with respect to a phase factor $e^{i\frac{\varphi}{2}}$ corresponding to a rotation of an angle φ in the circle

$$P_9^2 + P_{10}^2 = \rho^2 \tag{3.6}$$

generated in Φ spinor space by $J_{9,10} = \frac{1}{2}[G_9, G_{10}]$. We will interpret it as the origin of the baryon number or strong charge for the fermions represented by Θ_1 of which those of Θ_2 should then be free. Then we conjecture that $\Theta_1 := \Theta_B$ represents baryons and $\Theta_2 := \Theta_L$ leptons. In order to have non-null solutions Φ for equation (3.3) the vector P has to be null:

$$P_\beta P^\beta = 0, \quad \beta = 1, 2, \ldots, 12, \tag{3.7}$$

and if P_β are bilinear in Φ and if these are simple or pure (66-constraint equations [8]) this is identically guaranteed by Proposition 3.

26.4   The baryon quadruplet Θ_B in $M = \mathbb{R}^{1,9}$

Let us now derive from equation (3.3) the equation of motion for Θ_B. Recalling that Φ is a $\mathbb{C}\ell(1,9)$ Dirac spinor we may, exploiting the known isomorphism, adopt the projector $\frac{1}{2}(1 + G_{11})$, and taking into account that:

$$P_{11} = \Phi^\dagger G_0 G_{11}(1 + G_{11})\Phi \equiv 0 \equiv \Phi^\dagger G_0(1 + G_{11})\Phi = P_{12} \tag{4.1}$$

we easily obtain the Cartan equation:

$$P_\alpha G^\alpha (1 + G_{11})\Phi = (P_A G^A + iP_9 G^9 + P_{10})\Theta_B = 0, \quad A = 1, 2, \ldots, 8, \tag{4.2}$$

where Θ_B may be conceived as a $\mathbb{C}\ell(1,7)$ Dirac spinor, apt to represent a quadruplet of $\mathbb{C}\ell(1,3)$-Dirac spinors.

Observe that the 32-component spinor Φ obeying equation (3.3) in $P = \mathbb{R}^{1,11}$ has been reduced to the 16-component spinor Θ_B obeying equations (4.2) in $P = \mathbb{R}^{1,9}$; it might be conceived as a form of "dimensional reduction" which consists in reducing to one half the dimension of spinor space while decoupling two terms from the equations of motion.

26.4.1 Restriction from $M = \mathbb{R}^{1,9}$ to $M = \mathbb{R}^{1,3}$

Let us now consider $M = \mathbb{R}^{1,9}$ and indicate with x_μ the space-time coordinates and with X_j, $(j = 5, 6, \ldots, 10)$, the other six. Then equation (4.2) may be considered as the Fourier transform of the spinor field equation in M:

$$\left(i \frac{\partial}{\partial x_\mu} G^\mu + i \sum_{j=5}^{10} \frac{\partial}{\partial X_j} G^j \right) \Theta_B(x, X) = 0 . \tag{4.3}$$

Let us now operate the restriction of this equation to ordinary space-time $M = \mathbb{R}^{3,1}$ by setting $X_j = 0$ and defining

$$i \frac{\partial}{\partial X_j} \Theta_B(x, X) \Big|_{X_j=0} := P_j(x) \Theta_B(x), \tag{4.4}$$

we obtain:

$$\left[i \frac{\partial}{\partial x_\mu} G^\mu + \sum_{j=5}^{8} P_j(x) G_j + i P_9(x) G_9 + P_{10}(x) \right] \Theta_B(x) = 0 \tag{4.5}$$

which identifies with equation (4.2) if P_μ are considered as generators of Poincaré translations: $P_\mu \to i \frac{\partial}{\partial x_\mu}$ and then the Θ_B spinor and P_j, for $j > 5$, have to be considered x-dependent. We may then conclude that the Cartan equations for Θ_B, may be identified with those defined in the ten-dimensional space-time $M = \mathbb{R}^{1,9}$, after restriction to $M = \mathbb{R}^{1,3}$.

To admit non-null solutions for Θ_B, the vector P has to be null:

$$P_\alpha P^\alpha = 0, \qquad \alpha = 1, 2, \ldots, 10 , \tag{4.6}$$

which is identically satisfied if P_α are bilinear in spinors as in (2.14), and Θ_B is simple. Equation (4.6) defines the invariant mass $\mathcal{M}$:

$$P_\mu P^\mu = P_5^2 + P_6^2 + P_7^2 + P_8^2 + P_9^2 + P_{10}^2 = \mathcal{M}^2 \tag{4.7}$$

defining S^5 presenting an $SU(6)$ group of symmetry orthogonal to the Poincaré group. Therefore the maximal internal symmetry for the quadruplet of fermions represented by Θ_B will be $SU(4)$, covering of $SU(6)$. Since, as seen, dimensional reduction decouples the equations of motion, the invariant mass $\mathcal{M}$ will be larger for larger fermion multiplets.

26.4.2 Flavour $SU(3)$

We know that [9]: $\mathrm{Spin}(1,9) \cong SL(2,o)$ where o stands for octonions, and the isomorphism is valid only if restricted to the Lie algebra. The automorphism group G_2 of octonions has a subgroup $SU(3)$ if one of its seven imaginary units is fixed. It may be shown [8] that the generators $\mathcal{G}_7, \mathcal{G}_8, \mathcal{G}_9$ of $\mathbb{C}\ell(1,9)$ may be expressed in terms of the imaginary units e_1, e_2, e_3, of octonions while $i\mathcal{G}_{11}$ may be identified with e_7. Then the projectors $\frac{1}{2}(1 \pm \mathcal{G}_{11})$, select a particular direction (e_7) in G_2 such that the terms:

$$U_\pm = \tfrac{1}{2}(1 \pm \mathcal{G}_{11}), \quad V_\pm^{(n)} = \tfrac{1}{2}\mathcal{G}_{6+n}(1 \pm \mathcal{G}_{11}), \quad n = 1, 2, 3, \qquad (4.8)$$

represent the so-called complex octonions. They define an $SU(3)$ invariant algebra for which $V_+^{(n)}$ and $V_-^{(n)}$ transform as the (3) and $(\bar{3})$ representations of $SU(3)$ respectively, while U_+ and U_- transform as singlets [10]. Therefore, we write equation (4.2) in the form:

$$P_\alpha \mathcal{G}^\alpha (1 + \mathcal{G}_{11})\Phi = \left(P_a \mathcal{G}^a + \sum_{n=1}^{3} P_{6+n} V_+^{(n)} + P_{10} \right) U_+ \Phi = 0 \qquad (4.9)$$

with

$$P_{6+n} = \Phi^\dagger \mathcal{G}_0 \mathcal{G}_{6+n}(1 - \mathcal{G}_{11})\Phi = \Phi^\dagger G_0 V_-^{(n)}\Phi. \qquad (4.10)$$

The term summed over n is $SU(3)$ covariant and, in fact, $SU(3)$-flavour covariant since $SU(3)$ contains $SU(2)$-isospin as a subgroup, as we will show later. Observe that the $\mathcal{G}_{6+n}$ from which $SU(3)$ derives are reflection operators in the spinor space.

Now eq. (4.10) may not have a direct physical interpretation, however by acting with $V_+^{(n)}$ on the vacuum of a Fock representation of spinor space [7] one could represent, as minimal left ideals, the three spinors representing quarks. Then, the known 3×3 representation of the pseudo-octonions algebra for baryons conceived as 3-quark states (the eight-fold way) may be obtained.

26.4.3   Color $SU(3)$

Complex octonions may also be defined with the first four generators $\mathcal{G}_\mu$ in equation (4.9). They are [8]:

$$V_{\mu\pm}^{(0)} = -\tfrac{1}{2}i\mathcal{G}_\mu^{(0)}(1 \pm \mathcal{G}_{11}), \quad V_{\mu\pm}^{(n)} = -\tfrac{1}{2}i\mathcal{G}_\mu^{(n)}(1 \pm \mathcal{G}_{11}), \quad n = 1, 2, 3, \quad (4.11)$$

where $\mathcal{G}_\mu^{(0)} = e_0 \otimes \Gamma_\mu$, with $e_0 = 1$ and $\mathcal{G}_\mu^{(n)} = e_n \otimes \Gamma_\mu$ where the indices n refer to the isomorphism discussed in section 26.2.2. As before they transform as singlets and triplets for $SU(3)$. We have seen that Θ_B presents a $U(1)$ symmetry in the form of a local phase factor which will impose a covariant derivative, and equation (4.9) takes the form [8]:

$$P_\alpha \mathcal{G}^\alpha (1 + \mathcal{G}_{11})\Phi = \left[i\left(\frac{\partial}{\partial x_\mu} - g A_{(n)}^\mu \right) V_\mu^{(n)} + P_5 \mathcal{G}_5 + P_6 \mathcal{G}_6 \right.$$

$$\left. + \sum_{n=1}^{3} P_{6+n} V_+^{(n)} + P_{10} \right] U_+ \Phi = 0, \quad (4.12)$$

assuming

$$A_\mu^{(n)} = \Phi^\dagger \mathcal{G}_0 \mathcal{G}_\mu^{(n)}(1 - \mathcal{G}_{11})\Phi = 2i\Phi^\dagger \mathcal{G}_0 V_{\mu-}^{(n)}\Phi \qquad (4.13)$$

and then, in the covariant derivative, an $SU(3)$ covariant term appears interpretable as $SU(3)$-color. It could allow one to correlate the three colors with the imaginary units of quaternions (Pauli matrices). Acting on the vacuum of spinor space with $V_\mu^{(n)}$ [7], one could generate colored quarks and the baryons as three colored quark states. This would then mean to deal algebraically with spinors [2] as elements of Clifford algebras. In fact $SU(3)$-color and the other components of the standard group have also been found algebraically [3, 4].

26.4.4 The nucleon doublet, isospin $SU(2)$ and the $U(1)$ of the electric charge

Let us now perform a dimensional reduction with the projector $\frac{1}{2}(1 + G_9)$ which will reduce the spinor Θ_B to an 8-component N :

$$\tfrac{1}{2}(1 + G_9)\Theta_B = N = \begin{pmatrix} \psi_1 \\ \psi_2 \end{pmatrix} \qquad (4.14)$$

and correspondingly, as easily verified:

$$P_9 = \Theta_B^\dagger G_0 G_9(1 + G_9)\Theta_B \equiv 0 \equiv \Theta_B^\dagger G_0(1 + G_9)\Theta_B = P_{10} \qquad (4.15)$$

and the equation of motion for N, will be:

$$\tfrac{1}{2}(1 + G_9)\Theta_B = \left(i\frac{\partial}{\partial x_\mu}\Gamma^\mu + P_5\Gamma_5 + P_6\Gamma_6 + P_7\Gamma_7 + P_8 \right) N = 0. \qquad (4.16)$$

We will now suppose that ψ_1 and ψ_2 in (4.14) represent a $\mathbb{C}\ell(1,3)$ Dirac spinor, by which we have to impose:

$$\Gamma_\mu^{(0)} = 1 \otimes \gamma_\mu \,, \quad \Gamma_{5,6,7}^{(0)} = -i\sigma_{1,2,3} \otimes \gamma_5.$$

Then (4.16) becomes:

$$\left(i\frac{\partial}{\partial x_\mu}\gamma^\mu - i\boldsymbol{\pi}\cdot\boldsymbol{\sigma} \otimes \gamma_5 + P_8 \right) \begin{pmatrix} p \\ n \end{pmatrix} = 0 \qquad (4.17)$$

where

$$\boldsymbol{\pi} = N^\dagger \Gamma_0 \boldsymbol{\sigma} \otimes \gamma_5 N \qquad (4.18)$$

and where we have set $\psi_1 := p$ and $\psi_2 := n$ representing the proton and neutron respectively, since (4.17) well represents the $SU(2)$-isospin covariant nucleon-pion equation of motion. Observe that the pseudoscalar nature of the pion isotriplet $\pi(x)$ derives from having imposed that p and n are $\mathbb{C}\ell(1,3)$-Dirac spinors, because of which Γ_5, Γ_6 and Γ_7 have to contain γ_5, in order to anticommute with Γ_μ.

Isospin $SU(2)$ is generated by $\Gamma_5, \Gamma_6, \Gamma_7$ which are reflection operators in the spinor space N. It is not then the covering of $SO(3)$ in the space spanned by X_5, X_6, X_7 and it could not be otherwise, since equation (4.17) is obtained from equation (4.3) after setting to zero these coordinates. It derives instead from quaternions. In fact $\mathbb{C}\ell(1,7) = \mathbb{H}(8)$ and equation (4.17) is manifestly quaternionic. Then one may affirm that, according to this derivation, quaternions are at the origin of the isospin symmetry of nuclear forces.

We now write explicitly (4.17) for p and n. Defining

$$p_5 \pm i p_6 = \rho^{i \frac{\varepsilon}{2}},$$

we obtain [9] an equation presenting a $U(1)$ phase invariance represented by $e^{i\frac{\varepsilon}{2}}$ and generated by $J_{5,6} = \frac{1}{2}[\Gamma_5, \Gamma_6]$ of the proton p. Since $\varepsilon(x)$ is local it imposes a covariant derivative, and then an electric charge e, for the proton: in fact the resulting equation of motion for the doublet N is:

$$\left\{ \left[i\frac{\partial}{\partial x_\mu} - \frac{e}{2}(1 - \Gamma_5\Gamma_6)A_\mu \right] \gamma^\mu - i\boldsymbol{\pi} \cdot \boldsymbol{\sigma} \otimes \gamma_5 + P_8 \right\} \begin{pmatrix} p \\ n \end{pmatrix} = 0, \qquad (4.19)$$

well representing the nuclear and electromagnetic properties of the nucleon.

26.5   The lepton quadruplet $\Theta_{\mathcal{L}}$, the $U(1)$ of the electroweak charge

The equation for $\Theta_{\mathcal{L}}$ will be:

$$p_\alpha \mathcal{G}^\alpha (1 - \mathcal{G}_{11})\Phi = 0, \quad \alpha = 1, 2, \ldots, 10, \qquad (5.1)$$

from which we will derive the equation in $M = \mathbb{R}^{1,3}$:

$$\left[i\frac{\partial}{\partial x_\mu}G^\mu + \sum_{j=5}^{9} p_j(x)G_j - p_{10}(x) \right] \Theta_{\mathcal{L}} = 0. \qquad (5.2)$$

26.5.1   The equations of motion for $\Theta_{\mathcal{L}}$ and the $U(1)$ for the electroweak charge

Let us now suppose $\Theta_{\mathcal{L}}$ to have the form:

$$\Theta_{\mathcal{L}} = \begin{pmatrix} L_1 \\ L_2 \end{pmatrix} = \begin{pmatrix} \ell_{11} \\ \ell_{12} \\ \ell_{21} \\ \ell_{22} \end{pmatrix} \qquad (5.3)$$

where L_1, L_2 are $\mathbb{C}\ell(1,7)$-Dirac spinors, with the same procedure as in section 26.3 and in section 26.4.4. It is easy to show that the fermion doublet L_1 presents a $U(1)$ phase invariance generated in spinor space by $J_{7,8} = \frac{1}{2}[G_7, G_8]$. We will suppose it at the origin of the electroweak charge for the lepton doublet L_1 from which instead L_2 should be free.

26.5.2   The $SU(2)_L$ phase invariance and the electroweak model

Let us suppose that the four leptons in $\Theta_{\mathcal{L}}$ of equation (5.3) may represent either Dirac or Pauli spinors in space-time $M = \mathbb{R}^{1,3}$. After defining

$$p_5\sigma_1 + p_6\sigma_2 + p_7\sigma_3 := \tfrac{1}{2}\boldsymbol{\omega} \cdot \boldsymbol{\sigma}$$

we obtain from eq. (5.2) [9]:

$$p_\mu\sigma_3 \otimes \gamma^\mu L_{1R} + (p_9 - p_{10})L_{1L} + \left(p_8 + \tfrac{1}{2}i\boldsymbol{\omega} \cdot \boldsymbol{\sigma}\right)L_{2L} = 0,$$
$$p_\mu\sigma_3 \otimes \gamma^\mu L_{2R} - (p_9 + p_{10})L_{2L} - \left(p_8 - \tfrac{1}{2}i\boldsymbol{\omega} \cdot \boldsymbol{\sigma}\right)L_{1L} = 0. \qquad (5.4)$$

Let us now define:

$$\left(p_8 - \tfrac{1}{2}i\boldsymbol{\omega} \cdot \boldsymbol{\sigma}\right) = \rho e^{-\frac{1}{2}i\boldsymbol{\omega}\cdot\boldsymbol{\sigma}} = \rho e^{\frac{q}{2}} \qquad (5.5)$$

where $q = -i\boldsymbol{\sigma} \cdot \boldsymbol{\omega}$ is an imaginary quaternion and

$$\rho = \sqrt{p_5^2 + p_6^2 + p_7^2 + p_8^2}.$$

Equation (5.4) may be easily brought to a form manifesting a quaternionic $SU(2)_L$ phase invariance for L_{1L}. Observe that this $SU(2)$ originates, like the one of isospin in equation (4.17), from $\Gamma_5, \Gamma_6, \Gamma_7$ which are reflection operators building up the Lie algebra $su(2)$. Here, however, the algebra has been exponentiated in equation (5.5) and therefore, this time, $SU(2)$ is the covering of $SO(3)$; the symmetry group of the sphere S_2 of the quaternion q :

$$-q^2 = p_5^2 + p_6^2 + p_7^2. \qquad (5.6)$$

Since the quaternion q is x dependent it will induce a covariant derivative with a non-Abelian gauge interaction: $D_\mu = \partial_\mu - i\boldsymbol{\sigma} \cdot \mathbf{W}_\mu$ for L_{1L} only and then the Dirac equation for $L_1 = \binom{e}{\nu_L}$, where e represents the electron and ν_L the left-handed neutrino is easily obtained [8]:

$$\left(i\frac{\partial}{\partial x_\mu}\gamma^\mu + m\right)\binom{e}{\nu_L} + \boldsymbol{\sigma} \cdot \mathbf{W}_\mu\gamma^\mu\binom{e_L}{\nu_L} + B_\mu\gamma^\mu e_R = 0 \qquad (5.7)$$

where $\mathbf{W}_\mu$ is an isotriplet vector field and B_μ an isosinglet. Equation (5.7) is notoriously the geometrical starting point of the electroweak model.

26.6   Dimensional reduction and families

In section 26.4.4 we adopted the projector $\frac{1}{2}(1+G_9)$ to reduce the 16-component spinor Θ_B to the 8-component one N. Now the first six generators G_a of $\mathbb{C}\ell(1,7)$ are traditionally set in the form:

$$G_a^{(j)} = \sigma_j \otimes \Gamma_a, \quad j = 0,1,2,3, \quad a = 1,2,\dots,6, \tag{6.1}$$

where Γ_a are the generators of $\mathbb{C}\ell(1,5)$ and $\sigma_0 = 1$. The above projector is obtained for $j = 1,2$; therefore we will now indicate $\frac{1}{2}(1+G_9)\Theta_B = \Psi^{(1,2)}$, and we have seen in equation (4.15) that $P_9 \equiv 0 \equiv P_{10}$. Instead for $j = 0$ and $j = 3$ we obtain the projectors

$$\tfrac{1}{2}(1 + iG_7G_8)\Theta_B = \Psi^{(0)} \quad \text{and} \quad \tfrac{1}{2}(1 + iG_8G_9)\Theta_B = \Psi^{(3)} \tag{6.2}$$

respectively and correspondingly: $P_7 \equiv 0 \equiv P_8$ and $P_8 \equiv 0 \equiv P_9$ respectively and the corresponding so reduced 8-dimensional space, for Θ_B simple or pure, is null defining then the invariant masses.

$$P_\mu P^\mu = P_5^2 + P_6^2 + P_7^2 + P_8^2 = M_{(1,2)}^2,$$
$$P_\mu P^\mu = P_5^2 + P_6^2 + P_7^2 + P_{10}^2 = M_{(3)}^2,$$
$$P_\mu P^\mu = P_5^2 + P_6^2 + P_9^2 + P_{10}^2 = M_{(0)}^2. \tag{6.3}$$

Because of the isomorphisms discussed in section 26.2.2 the three spinors and $\Psi^{(0)}$, $\Psi^{(3)}$ and $\Psi^{(1,2)}$ may represent doublets of fermions which, in the limit of absence of external, strongly interacting fields (the pion in equation (4.17)), will obey the Dirac equations. Their masses might be different, and since, in general, the characteristic energy of the phenomena increases with the dimension of the P-space, we could expect $M_0 > M_3 > M_{(1,2)}$.

Repeating the same with the lepton quadruplet $\Theta_{\mathcal{L}}$ we obtain three lepton doublets $L_1^{(1,2)}, L_1^{(3)}, L_1^{(0)}$ representing charged-neutral fermions. Now the strongly interacting fields, the pions, should be absent; if these are represented by p_5, p_6, p_7 (equation (4.17)), the corresponding invariant mass equations become:

$$m_{(1,2)}^2 = p_8^2, \quad m_{(3)}^2 = p_{10}^2, \quad m_{(0)}^2 = p_9^2 + p_{10}^2$$

and then, we could conjecture that $L_1^{(1,2)}, L_1^{(3)}, L_1^{(0)}$ represent the electron-neutrino, muon-neutrino, and tau-neutrino doublets respectively.

There should then be three families of leptons and the corresponding ones of baryons, originating from quaternions. The possible quaternionic origin of three lepton families was also obtained by T. Dray and C. Manogue [5] from octonion division algebra correlated with $\mathbb{C}\ell(1,9)$.

26.7   The chargeless leptons

If we act on $\Theta_{\mathcal{L}}$ in equation (5.3) obeying equation (5.2) with $\frac{1}{2}(1 - G_9)$ or with the other two possible projectors discussed in section 26.6, we will obtain three families of leptons $L_2^{(1,2)}, L_2^{(3)}, L_2^{(0)}$ which however, as seen above, should be free from both strong and electroweak charges. Therefore, we will need a further dimensional reduction through $(1 \pm \Gamma_7)$ which gives [9]:

$$(p_\mu \gamma^\mu + p_5 \gamma_5)\psi = 0, \tag{7.1}$$

for ψ real, which is a candidate for the Majorana equation. A further dimensional reduction through $(1 \pm \gamma_5)$ reduces (7.1) to

$$p_\mu \gamma^\mu (1 \pm \gamma_5)\psi = 0 \tag{7.2}$$

since $p_5 \equiv 0 \equiv p_6$ that is, the Weyl equation for massless neutrinos.

Then the final reduction brings us to Majorana and Weyl spinors. They then naturally appear as the most elementary form of fermions. Should this scheme be realized in nature, they should have been created at the Big Bang, as stable fermions with only gravitational interaction; through which they could have also been created and accumulated during the life of the universe in high energy events connected with supernovae and black holes. They could then be a possible natural source of black matter.

Observe that in this picture it is not Dirac's equation that is the basic equation for fermions, but rather those of Majorana and Weyl. One could then try to start from these elementary spinors the construction of both the fermion miltiplets and the momentum space bilinear in them. It is remarkable that in so doing [8] also the Lorentzian signature of the Minkowski spacetime results unambiguously defined, once more, from the quaternion division algebra, of which then that signature could appear as the image in nature.

26.8   The boson field equations, composite fields

In our approach the boson fields appear at first as external fields as was the case for the electromagnetic potential A_μ and of the pion field π in equation (4.19). However, since the boson fields here are bilinear in spinor fields also their equations of motion should be obtainable from the spinor field equation. In fact it was shown [8] that from Weyl equation (7.2) it is possible to obtain for the electromagnetic tensor, (so named already by Cartan [1]):

$$F_\pm^{\mu\nu} = \tilde{\psi}[\gamma^\mu, \gamma^\nu](1 \pm \gamma_5)\psi \tag{8.1}$$

the equations

$$p_\mu F_+^{\mu\nu} = 0, \quad \varepsilon_{\mu\nu\rho\tau} p^\nu F_-^{\rho\tau} = 0, \tag{8.2}$$

that is Maxwell's equations, and also the non-homogeneous ones [11].

Observe that in equation (8.1) the electromagnetic field $F_\pm^{\mu\nu}$ is bilinear in the Weyl spinor fields, which does not imply that, in the quantized theory, the photon should be conceived as a bound state of neutrinos. In fact "the neutrino theory of light" is only viable within strict limitations [12]. In a similar way it is possible to obtain the equations of motion for the pion field $\pi(x)$, not necessarily as a bound state of nucleon-antinucleon [13]. It is reasonable to expect the possibility to extend this to the other boson fields.[1]

26.9   The role of simplicity

Simplicity, through Proposition 3, mainly imposes the nullness of the momentum-spaces bilinearly constructed with at least one simple spinor. This in turn implies that: to the embedding of 4-dimensional spinor spaces in 8-, 16-, and 32-dimensional ones, the parallel embedding of the corresponding 4-dimensional vector momentum spaces in 6-, 8- and 10-dimensional ones corresponds, which, being null define compact manifolds (spheres) imbedded in each other, where infinity is absent.

Because of this one might expect that simplicity, while not determinant for some features of phenomenology concerning internal symmetry—where one could then expect to deal also with non-pure or generalized pure spinors [15]—instead determines the underlying projective geometry rigorously and may then have deep meanings relevant for some open problems of physics. We will briefly mention some of them.

26.9.1   The masses

If momentum spaces are null or optical, they define invariant masses like equation (4.7), where values increase with the dimension of both the fermion multiplets and of momentum spaces. Before discussing their role in physics we need to recall that the terms appearing in these equations are representing external boson-field interactions with multiplicative coupling constants. Therefore this problem is correlated with the one of determining the values of the charges, after which it might present aspects of relevance for physics.

26.9.2   The charges

The values of the charges are traditionally taken from the experimental data and as such inserted by hand in the equations of motion. In the present approach instead, in which the equations of motion are derived from spinor geometry in momentum

---

[1]A remarkable feature of the present approach based on spinor geometry is that it produces both the present conception of hadrons bi- or multi-linear in quarks (see equation (4.13)) and the old one where the strong nuclear forces were mainly mediated by the pion, bilinear in the nucleon as in equation (4.17) with equation (4.18). Now both conceptions have their own range of validity such that if the present conception of nuclear physics as a collection of phenomenological models [14] should be extended to the whole domain of particle physics, spinor geometry, from which the various models appear to be derivable, could furnish the theoretical background for the phenomenological models.

space, which in turn appears to consist of spheres imbedded in each other, one could expect that everything should be unambiguously defined including charges. In fact the $U(1)$ of the charges are generated by $J_{5,6}$; $J_{7,8}$ and $J_{9,10}$ in spinor space. A suggestion that this might be true and realizable comes from the old Fock equation [16] for the hydrogen atom, written precisely in the sphere S_3 of compactified momentum space, which may be set in the following adimensional form:

$$\psi(\mathbf{u}) = \frac{1}{V(S_3)} \frac{e^2}{\hbar c} \frac{mc}{p_0} \int_{S_3} \frac{\psi(\mathbf{u}')}{(\mathbf{u} - \mathbf{u}')^2} \, d^3\mathbf{u}' \qquad (9.1)$$

where $\mathbf{u}$ is a unit vector of S_3, p_0 a unit of momentum, m the mass of the electron, and where also the electric charge e appears inserted in the adimensional fine structure constant $\alpha^{-1} = \frac{\hbar c}{e^2} = 137,036,\ldots$. Equation (9.1) manifests the $SO(4)$ covariance induced in the two-body system by the Coulomb potential and H-atom eigenfunctions and energy levels are obtained by Fock through harmonic analysis from B_3 to S_3.

One may then conjecture that also the other factor in front of equation (9.1); that is α, (generated by $J_{5,6}$) could be obtained through harmonic analysis from the higher-dimensional momentum space spheres, found above. A preliminary computation (with P. Nurowski) indicates that the strong charge g (generated by $J_{9,10}$) turns out to be larger than the electric charge e [9] simply because the volumes of the unit spheres, which appear in the denominator, decrease after a certain dimension, since

$$V(B_{2k+2})/V(B_{2k}) = \pi/(k+1)$$

which is < 1 for $k > 2$. Then both masses and charges should increase with dimensions, and charges could manifest momentum dependence in dynamical processes.

26.9.3   The constraint equations

While determining underlying geometry they might also have some direct physical meanings. One is triality, arising from the one-constraint equation for 8-component spinors, which could have a role in supersymmetry. Another is the role of the ten-constraint equations for 16-component spinors manifest in the covariant quantization of super strings, even if at the ghost stage [17]. But also the 66-constraint equations for 32-component spinors Φ correlating the baryon quadruplet Θ_B with the lepton one Θ_L could perhaps solve the problem of the often predicted and never found lifetime of the proton, imposing its stability [8].

Conclusions

The most remarkable feature that seems to emerge from this approach is that the vectors of momentum space (bilinear in spinors), where the spinor field equations are formulated, have to be null (because of Proposition 3), and that then the corresponding momentum spaces, consisting of spheres imbedded in each other, are compact. In these momentum spaces, the quantum field theory that results is *a priori* free from ultraviolet divergences.

Our approach is consistent with Cartan's conjecture [1] that simple spinors can be conceived as the elementary constituents of Euclidean geometry whose ordinary vectors are sums (or integrals) of the null-vectors emerging, bilinearly, from simple spinor geometry. In fact this approach may represent a particular realization of that conjecture.

In this framework a striking parallelism appears to emerge between geometry and physics: while the Euclidean geometry of space-time appears to be appropriate for describing the classical mechanics of macroscopic bodies (celestial mechanics), the (quantum) mechanics of the elementary constituents of matter, namely fermions, naturally appears to have to be described with the elementary constituents of this geometry, that is with simple spinors, and in momentum space. It is then reasonable to expect that quantum mechanics may not be properly dealt with, nor properly understood, on the basis of space-time Euclidean geometry, but that in dealing with the elementary constituents of matter, the purely Euclidean concept of point events should be replaced by that of string: a continuous sum, or integral, of null vectors [8].

REFERENCES

[1] É. Cartan, *Leçons sur la theorie des spineurs*, Hermann, Paris, 1937.

[2] C. Chevalley, *The Algebraic Theory of Spinors*, Columbia U.P., New York, 1954.

[3] G. Trayling and W.E. Baylis, *J. Phys. A: Math. Gen.* **34**, 2001, 3309–3323.

[4] G.M. Dixon, *Division Algebras: Octonions, Quaternions and Complex Numbers and the Algebraic Design of Physics*, Kluwer, 1994.

[5] T. Dray and C.A. Manogue, *Mod. Phys. Lett. A* **14**, 1999, 93–95.

[6] P. Budinich and A. Trautman, *The Spinorial Chessboard*, Springer, New York, 1989.

[7] P. Budinich and A. Trautman, *J. Math. Phys.* **30**, 1989, 2125–2131.

[8] P. Budinich, From the geometry of pure spinors with their division algebra to fermion's physics, *Found. Phys.* **32**, 2002, 1347–1398.

[9] P. Budinich, The possible role of pure spinors in some sectors of particle physics, hep-th/0207216, **31**, July 2002.

[10] F. Gürsey and C.H. Tze, *On the Role of Division, Jordan, and Related Algebras in Particle Physics*, World Scientific, Singapore, 1996.

[11] D. Hestenes, *Space-Time Algebra*, Gordon Breach, New York, 1987.

[12] W.A. Perkins, *Lorentz, CPT and Neutrinos*, World Scientific, Singapore, 2000; K. Just, K. Kwong and Z. Oziewicz, hep-th/005263, 27 May (2000).

[13] E. Fermi and C.N. Yang, *Phys. Rev.* **76**, 1949, 1739–1743.

[14] G. Boniolo, C. Petrovich and G. Pisent, *Notes on the Philosophical Status of Nuclear Physics*, Foundations of Science, 2002, 1–28.

[15] A. Trautman and K. Trautman, *J. Geom. Phys.* **15**, 1994, 122–134.

[16] V. Fock, *Zeitsch. f. Phys.* **98**, 1935, 145–154.

[17] M. Matone, L. Mazzucato, I. Oda, D. Sorokin and M. Tonin, The superembedding origin of the Berkovit's pure spinor covariant quantization of superstrings, hep-th/0206104, 12 June 2002.

Paolo Budinich
The Abdus Salam International Centre for Theoretical Physics
Trieste, Italy
E-mail: budinich@ictp.trieste.it, fit@ictp.trieste.it

Submitted: August 2, 2002; Revised: July 29, 2003.

27

Spinor Formulations for Gravitational Energy-Momentum

Chiang-Mei Chen, James M. Nester, and Roh-Suan Tung

ABSTRACT We first describe a class of spinor-curvature identities (SCI) which have gravitational applications. Then we sketch the topic of gravitational energy-momentum, its connection with Hamiltonian boundary terms and the issues of positivity and (quasi)localization. Using certain SCIs, several spinor expressions for the Hamiltonian have been constructed. One SCI leads to the celebrated Witten positive energy proof and the Dougan–Mason quasilocalization. We found two other SCIs which give alternate positive energy proofs and quasilocalizations. In each case the spinor field has a different role. These neat expressions for gravitational energy-momentum have much appeal. However it seems that such spinor formulations just have no room for angular momentum, which leads us to doubt that spinor formulations can really correctly capture the elusive gravitational energy-momentum.

Keywords: gravitation, energy-momentum, positive energy, quasi-local quantity, Hamiltonian

27.1 Introduction

One of the most outstanding results in classical gravitation theory (more specifically we mean in GR: general relativity, Einstein's gravity theory), obtained via Clifford algebra and spinor methods, was Witten's positive energy proof [33]. This seminal work (which was inspired by an analogous result in quantum supergravity [10, 14]) led to many new ideas regarding gravitational energy and its localization. To appreciate this work and its importance we need to recall some facts about one of nature's most elusive quantities: gravitational energy.

A suitable expression which could provide a physically reasonable description of the energy-momentum density for gravitating systems has long been sought. All candidates had several shortcomings. In particular they violated a fundamental theoretical requirement—that gravitational energy should be positive—as well as requirements concerning localization and reference frame independence. Since Witten's positive energy proof (unlike the earlier indirect proof of Schoen and

AMS Subject Classification: 83C05, 83C40, 83C60.

Yau [29]) can be understood in terms of the Hamiltonian, the Hamiltonian density associated with this proof provides a *locally positive localization*—and thus has real promise as a *truly physical* energy-momentum density for gravitational fields. To fulfill this promise certain features need further consideration; an outstanding one concerns the role of—and even the need for—the spinor field, which seemed rather mysterious.

Here we give a survey (further details can be found in the references) of three (classical, commuting, non-supersymmetric) spinor/Clifford algebra formulations of the GR Hamiltonian known to us. We examine the underlying mathematics (spinor-curvature identities), outline their associated positive energy proofs and energy-momentum quasilocalizations, and note the various distinct roles of the spinor field. Although they give good expressions for energy-momentum, these spinor expressions do not seem to have the proper qualities for giving a good description of relativistic angular momentum. Hence they apparently do not succeed in giving a full physical description for the energy-momentum density of asymptotically flat gravitating systems.

The plan of this work is as follows. In section 2 we present our notation and conventions for geometric objects, forms, the Dirac algebra, spinors, and "Clifforms". In section 3 a succinct presentation of a class of *spinor curvature identities* is given; three special cases which have gravitational applications are noted. Then in section 4 the topic of gravitational energy-momentum is discussed; we note the fundamental theoretical requirement of positivity (finally proved over 20 years ago) and the still outstanding issue of the localization (or quasilocalization) of gravitational energy-momentum. In section 5 we explain the Hamiltonian approach to energy-momentum, noting the important roles played by the Hamiltonian boundary term; the standard "ADM" Hamiltonian (albeit in non-standard variables) for GR along with a good choice of boundary term (which necessarily requires reference values) are presented. In section 6 we consider three alternate GR Hamiltonians obtained via certain spinor curvature identities along with suitable associated spinor field reparameterizations. The first leads to the celebrated Witten positive energy proof; it also gives the famous Dougan–Mason energy-momentum quasilocalization. The second alternate Hamiltonian uses SU(2) spinors (i.e., the spinors of 3-dimensional space) in a way similar to, but distinct from, the first case; it yields another proof and quasilocalization. The third alternative is fundamentally quite different; unlike the former cases (where the spinor field was introduced as an alternate reparameterization of the Hamiltonian) now the spinor field enters into the Lagrangian as a dynamic physical field; this case yields yet another positivity proof and quasilocalization. The following section 7 notes that (i) the spinor formulations have the extremely nice property of not requiring any explicit reference values, (ii) in all cases the *Hamiltonian boundary variation principle* reveals the associated boundary conditions, and (iii) the various distinct roles played by the spinor fields in these three formulations. In section 8 consideration is given to the other conserved quantities of an asymptotically flat space: angular momentum and the center-of-mass moment. The key role played by a certain type of term in the conventional Hamiltonian and the absence of this type of term in all of the aforementioned spinor formulations is noted. The concluding section summarizes the virtues of the spinor Hamiltonian expressions; we note, however, that the limitation in connection with angular momentum and especially the center-of-mass moment raises grave doubts as to whether these spinor formulations are really properly representing the physics of gravitating systems.

27.2   Conventions

Geometry (especially with a metric compatible connection, as is assumed here) can be conveniently described using differential forms. The basis one-forms $\vartheta^\mu := e^\mu{}_i dx^i$ (dual to the basis vectors e_ν) are chosen to be orthonormal (and are generally non-holonomic). The metric is then given by $g = g_{\mu\nu}\vartheta^\mu \otimes \vartheta^\nu$ with $g_{\mu\nu}$ constant. (We will not need the coordinate frame metric components $g_{ij} = g_{\mu\nu}e^\mu{}_i e^\nu{}_j$.) The connection is described by the one-form $\Gamma^{\mu\nu} = \Gamma^{\mu\nu}{}_i dx^i$. Because of the metric compatibility condition (and the frame type choice) the connection one-form is antisymmetric: $\Gamma^{\mu\nu} \equiv \Gamma^{[\mu\nu]}$. In general the connection may have torsion, which is neatly described by the 2-form field

$$T^\mu := D\vartheta^\mu := d\vartheta^\mu + \Gamma^\mu{}_\nu \wedge \vartheta^\nu = \tfrac{1}{2}T^\mu{}_{\alpha\beta}\vartheta^\alpha \wedge \vartheta^\beta. \qquad (2.1)$$

The curvature is also described by a 2-form field:

$$R^\alpha{}_\beta := d\Gamma^\alpha{}_\beta + \Gamma^\alpha{}_\gamma \wedge \Gamma^\gamma{}_\beta = \tfrac{1}{2}R^\alpha{}_{\beta\mu\nu}\vartheta^\mu \wedge \vartheta^\nu. \qquad (2.2)$$

The unit volume element (in 4-dimensional spacetime) is given by $\eta := \vartheta^0 \wedge \vartheta^1 \wedge \vartheta^2 \wedge \vartheta^3$. We often use the dual Grassmann basis, which can be constructed from η by contraction (aka the interior product): $\eta_\mu := i_{e_\mu}\eta$, $\eta_{\mu\nu} := i_{e_\mu}\eta_\nu$, $\eta_{\mu\nu\alpha} := i_{e_\mu}\eta_{\nu\alpha}$, and $\eta_{\mu\nu\alpha\beta} := i_{e_\mu}\eta_{\nu\alpha\beta}$, the latter is the totally antisymmetric Levi-Civita tensor. Alternately these objects can be obtained via the Hodge dual: $\eta^{\mu\nu\cdots} := *(\vartheta^\mu \wedge \vartheta^\nu \wedge \cdots)$. A succinct notation, neatly suited to our material, is geometric (Clifford) algebra valued forms, sometimes referred to as *Clifforms* [11, 13, 21]. With the Dirac conventions

$$\gamma_\alpha\gamma_\beta + \gamma_\beta\gamma_\alpha = 2g_{\alpha\beta} = \mathrm{diag}(+1,-1,-1,-1), \qquad (2.3)$$

$$\gamma_{\alpha\beta\dots} := \gamma_{[\alpha}\gamma_\beta \cdots], \quad \gamma := \gamma^0\gamma^1\gamma^2\gamma^3, \qquad (2.4)$$

$$\bar\psi := \psi^\dagger\beta, \qquad \beta \equiv \beta^\dagger, \qquad (\beta\gamma^\mu)^\dagger \equiv \beta\gamma^\mu, \qquad (2.5)$$

we define the *frame* (a vector-valued one-form), the *torsion* (a vector-valued 2-form), the *connection* (a bivector-valued one-form), and the *curvature* (a bivector-valued 2-form), respectively, by

$$\vartheta := \vartheta^\mu\gamma_\mu, \quad T := D\vartheta = d\vartheta + \Gamma \wedge \vartheta + \vartheta \wedge \Gamma = T^\mu\gamma_\mu, \qquad (2.6)$$

$$\Gamma := \tfrac{1}{4}\Gamma^{\mu\nu}\gamma_{\mu\nu}, \quad R := d\Gamma + \Gamma \wedge \Gamma = \tfrac{1}{4}R^{\mu\nu}\gamma_{\mu\nu}. \qquad (2.7)$$

The differentials of Dirac spinors are

$$D\psi := d\psi + \Gamma\psi, \qquad D\bar{\psi} := d\bar{\psi} - \bar{\psi}\Gamma, \qquad (2.8)$$

$$D^2\psi \equiv R \wedge \psi, \qquad D^2\bar{\psi} \equiv -\bar{\psi} \wedge R. \qquad (2.9)$$

27.3   Some spinor curvature identities

Key to our work are certain *spinor-curvature identities* (SCIs) [26]. They readily follow from

$$d[\bar{\psi}A \wedge D(B\psi) - (-1)^a D(\bar{\psi}A) \wedge B\psi] \equiv$$
$$2D(\bar{\psi}A) \wedge D(B\psi) + (-1)^a \bar{\psi}A \wedge D^2(B\psi) - (-1)^a D^2(\bar{\psi}A) \wedge B\psi. \quad (3.1)$$

(Here A and B are Clifford algebra-valued forms of rank a and b.) Using

$$D^2(B\psi) = R \wedge B\psi, \qquad D^2(\bar{\psi}A) = -\bar{\psi}A \wedge R$$

we find the SCIs

$$2D(\bar{\psi}A) \wedge D(B\psi) \equiv 2(-1)^a \bar{\psi}A \wedge R \wedge (B\psi)$$
$$+ d[\bar{\psi}A \wedge D(B\psi) - (-1)^a D(\bar{\psi}A) \wedge B\psi]. \quad (3.2)$$

Qualitatively $(D\psi)^2 \equiv \psi^2 R$ plus a total differential. One can get various linear combinations of the curvature depending on the choice of A, B. We have found three special cases with gravitational applications; they contain, respectively, (i) the Einstein 3-form in four dimensions, (ii) the scalar curvature in three dimensions, and (iii) the scalar curvature in four dimensions.

We know that one can also have identities of the general form (3.2) with the spinor field ψ replaced by a vector or tensor. However we have not found any such *tensor-curvature identities* with the property that the term linear in the curvature reduces to the Einstein or scalar curvature—which is what we need for our gravitational applications. This technical point is apparently the reason why we need to use spin $\frac{1}{2}$ to help clarify certain things about gravity, a field which is fundamentally spin 2.

27.4   The energy-momentum of gravitating systems

Isolated gravitating systems have gravitational fields which are asymptotically flat (very far away the field is essentially Newtonian). For such spaces the total energy-momentum (EM) is well defined [22]. An essential fundamental theoretical requirement (from thermodynamics and stability: otherwise systems could emit an unlimited amount of energy while decaying deeper into ever more negative energy states) is that the energy of gravitating systems should be *positive*. Essentially this means that gravity acts like a purely attractive force. (The total energy is just $E = Mc^2$, with M being the apparent asymptotic Newtonian mass; thus positive energy means $M > 0$, hence an attractive force.) This was finally rigorously proved for GR by Schoen and Yau [29] via an indirect argument. Soon thereafter Witten [33] gave his celebrated direct spinorial positive energy proof (see also [23]).

Although positivity has been settled, the *location* of the energy of gravitating systems has remained an outstanding issue since Einstein's day. Sources (which have a well-defined local EM density) exchange EM with the gravitational field —*locally*— hence it was natural to expect a *local* gravitational EM density. But no suitable expression has been found. Standard techniques (e.g., translation symmetry and Noether's theorem) give only non-covariant (coordinate dependent) *pseudotensor* expressions (for recent discussions see [4, 5]). It was eventually realized that this localization problem is entirely consistent with the *equivalence principle*: 'gravity is not observable at a point' [22]. Nowadays the more popular idea is *quasilocal* EM (i.e., associated with a closed 2 surface) [3].

27.5   The Hamiltonian approach

A good definition of energy is: the value of the Hamiltonian. For both a finite or infinite region the Hamiltonian includes a 3-volume term and a bounding 2-surface integral term:

$$H(N) = \int_V N^\mu \mathcal{H}_\mu + \oint_{S=\partial V} \mathcal{B}(N), \qquad (5.1)$$

here N is the spacetime vector field describing the evolution (timelike displacement) of the spatial volume V.

For gravitating systems it follows from Noether's theorem and local translation (diffeomorphism) symmetry that $\mathcal{H}_\mu \propto$ field eqns (the initial value constraints). Consequently the volume term (although it serves to generate the Hamiltonian equations) has vanishing numerical value. The boundary term plays a doubly important role: it gives the value of the quasilocal quantities, and it also gives the boundary conditions. In addition to its dependence on the dynamic fields and the displacement vector field, the boundary term generally also depends on a choice of reference fields (which determines the "zero" for the quasilocal values). There is yet considerable freedom. Indeed, at least formally, one could say that there are an infinite number of possible choices for the boundary term $\mathcal{B}$; each corresponds to a distinct selection among the infinite number of conceivable choices for the boundary conditions. Thus additional criteria are very much needed. We proposed "covariant-symplectic" boundary conditions; it turns out that there are only two choices that satisfy this property (essentially they correspond to Dirichlet and Neumann boundary conditions) [4, 6–8].

For GR in terms of differential forms the standard "ADM" Hamiltonian is given by the spatial integral of the 3-form

$$\mathcal{H} = N^{\mu} R^{\alpha\beta} \wedge \eta_{\alpha\beta\mu} + i_N \Gamma^{\alpha\beta} D\eta_{\alpha\beta} + d\mathcal{B}(N). \tag{5.2}$$

This is easily verified if one just notes that $R^{\alpha\beta} \wedge \eta_{\alpha\beta\mu} \equiv -2G^{\nu}{}_{\mu}\eta_{\nu}$ is a 3-form version of the Einstein tensor. When (5.2) is integrated over space only the coefficient of η_0 contributes; that coefficient is $2G^0{}_{\mu}$, the well-known covariant components that make up the ADM Hamiltonian, see, e.g., [1, 18, 22].

The total differential, when integrated over a spatial region, yields an integral over the boundary of the region. The weak field limit (which applies asymptotically) fixes the form of $\mathcal{B}$, but only to linear order. Our best choice for the boundary term in general is

$$\mathcal{B}(N) = -\Delta\Gamma^{\alpha\beta} \wedge i_N\eta_{\alpha\beta} - \overset{\circ}{D}{}^{[\beta} \overset{\circ}{N}{}^{\alpha]} \Delta\eta_{\alpha\beta}, \tag{5.3}$$

where $\Delta\Gamma := \Gamma - \overset{\circ}{\Gamma}$, $\Delta\eta := \eta - \overset{\circ}{\eta}$. Here $\overset{\circ}{\Gamma}$, $\overset{\circ}{\eta}$ indicate reference values—usually taken to be the flat space field values. (Note: all of the quasilocal quantities vanish when the dynamic fields take on the reference values on the boundary.) The boundary term (5.3) yields quasilocal values with good limits asymptotically [15, 16] and has good correspondence with other well-established expressions [6].

The significance of the choice of boundary term (5.3) is revealed by the variation of the Hamiltonian:

$$\delta\mathcal{H}(N) = \delta\vartheta^{\alpha} \wedge \frac{\delta\mathcal{H}(N)}{\delta\vartheta^{\alpha}} + \delta\Gamma^{\alpha\beta} \wedge \frac{\delta\mathcal{H}(N)}{\delta\Gamma^{\alpha\beta}} + di_N(\Delta\Gamma^{\alpha\beta} \wedge \delta\vartheta^{\mu} \wedge \eta_{\alpha\beta\mu}). \tag{5.4}$$

In addition to the field equation terms (we do not need their explicit functional form here) we obtain a Hamiltonian boundary variation term. It has a *symplectic structure* [19] which, according to the *boundary variation principle*, reveals which variables are to be held fixed. In this case we should fix (certain projected components of) the orthonormal frame ϑ^{μ}, (geometrically that is equivalent to holding the metric fixed) [6–8].

27.6   Spinor expressions

Using certain special cases of the general spinor curvature identity (3.2), we obtain three alternate spinor formulations for the GR Hamiltonian and its boundary term. From each we get both a positive energy proof and a quasilocal EM expression. We discuss the first case in some detail and then briefly survey the novel features in the other two cases.

27.6.1   The Witten spinor approach

In the above ADM Hamiltonian 3-form (5.2), use a spinor parameterization for the Hamiltonian displacement

$$N^\mu = \overline{\psi}\gamma^\mu\psi. \tag{6.1}$$

With an appropriately adjusted boundary term, a suitable *spinor-curvature identity* then gives the Hamiltonian 3-form associated with the famous Witten positive energy proof:

$$\mathcal{H}_w(\psi) := -4D\overline{\psi} \wedge \gamma\vartheta \wedge D\psi + i_N\Gamma^{\alpha\gamma}\, D\eta_{\alpha\gamma}$$
$$\equiv +N^\mu R^{\alpha\gamma} \wedge \eta_{\alpha\gamma\mu} + i_N\Gamma^{\alpha\gamma}\, D\eta_{\alpha\gamma} + d\mathcal{B}_w\,, \tag{6.2}$$

where

$$\mathcal{B}_w := -2(\overline{\psi}\gamma\vartheta \wedge D\psi + D\overline{\psi} \wedge \gamma\vartheta\psi)\,. \tag{6.3}$$

(Here we are correcting some old sign convention errors.) We stress that this is an acceptable alternate form for (5.2), the GR Hamiltonian [24]. Note that the spinor field can take on almost any value, as long as it is asymptotically constant. For such spinor fields the boundary term (6.3), notwithstanding appearances, actually gives the same asymptotic values as those given by (5.3).

For positive energy, first note that, because of vanishing torsion, one of the Hamiltonian terms vanishes: $D\eta_{\mu\nu} = T^\lambda \wedge \eta_{\mu\nu\lambda} = 0$. The Hamiltonian density is thus

$$\mathcal{H}_w(\psi) = -4\left(D_\alpha\overline{\psi}\gamma\gamma_\lambda D_\beta\psi\right)\vartheta^\alpha \wedge \vartheta^\lambda \wedge \vartheta^\beta$$
$$\equiv -4\left(D_\alpha\overline{\psi}\gamma\gamma_\lambda D_\beta\psi\right)\eta^{\alpha\lambda\beta\mu}\eta_\mu\,. \tag{6.4}$$

Now, using the grade 3 identity

$$\gamma_\lambda\gamma_{\mu\nu} + \gamma_{\mu\nu}\gamma_\lambda \equiv -2\gamma_{\lambda\mu\nu} \equiv 2\eta_{\lambda\mu\nu\kappa}\gamma\gamma^\kappa, \tag{6.5}$$

the Hamiltonian density becomes

$$\mathcal{H}_w(\psi) = 2D_\alpha\overline{\psi}\left(\gamma^{\alpha\beta}\gamma^\mu + \gamma^\mu\gamma^{\alpha\beta}\right)D_\beta\psi\eta_\mu. \tag{6.6}$$

When integrated over space only the $\mu = 0$ term survives; this means that the other two indices must be spatial. Finally, using $\gamma^{ab} \equiv \gamma^a\gamma^b - g^{ab}$, $g^{ab} = -\delta^{ab}$ and the usual type of representation wherein $\beta = \gamma^0$, the Hamiltonian density has the 3+1 (space+time) decomposition

$$\mathcal{H}_w(\psi) \simeq 2D_a\overline{\psi}\left(\gamma^{ab}\gamma^0 + \gamma^0\gamma^{ab}\right)D_b\psi\eta_0 \propto |D_k\psi|^2 - |\gamma^k D_k\psi|^2. \tag{6.7}$$

Now, exploiting the freedom in the choice of ψ, note that the Hamiltonian density (and consequently the energy) is manifestly non-negative, for any ψ solving the (3-dimensional, elliptic) *Witten equation*: $\gamma^k D_k\psi = 0$. (This proof, due to Witten, was later derived directly from the classical, *anti-commuting* spinor, supergravity result [9, 17]; there is an interesting argument that GR has positive energy *because* it admits a supersymmetric extension.)

Moreover we get a 'locally positive localization' for gravitational energy (albeit the *localization* actually depends on the solution of an elliptic equation and hence really depends on the fields globally). Examined in more detail, the argument also shows, (i) that the 4-energy-momentum P^μ is future time-like, and (ii) P^μ vanishes only for Minkowski space.

Altogether these are very beautiful arguments for some nice important results. Nevertheless, more than 20 years later, it is still not clear as to how much the Witten argument really captures the correct physics.

Consider the (static, spherically symmetric) Schwarzschild solution in an isotropic Cartesian frame

$$\vartheta^0 = N dt, \quad \vartheta^i = \varphi^2 dx^i, \quad \text{with} \quad \varphi = 1 + m/2r, \quad N\varphi = 1 - 2m/r.$$

The Witten equation is easily solved: $\psi = \varphi^{-2}\psi_{const}$. One can then substitute this solution into the Hamiltonian and the boundary term, and thereby conclude that, for the Schwarzschild solution, $\frac{1}{8}$ of the energy has been "localized" within the black hole horizon and $\frac{7}{8}$ is outside. We have no physical understanding of this curious distribution.

Note also that for closed spaces, such as an S^3 type cosmology, the spatial hypersurface has no boundary, so it should have vanishing total energy. Hence the Witten positive energy proof (or indeed any other positivity proof) should not go through. We are not yet sure which step in the Witten argument breaks down for such closed spaces.

The Witten spinor Hamiltonian is also important for the quasilocal values it can yield. When integrated over a finite spatial region, the Hamiltonian boundary term $\mathcal{B}_w$ (6.3) defines a quasilocal energy-momentum for any choice of the spinor field on the boundary. This is a popular approach to quasilocal energy. Several similar quasilocal boundary expressions of this type have been investigated. In this case one wants to determine the spinor field quasilocally (i.e., it should depend on the fields only on the boundary 2-surface S). In particular Dougan and Mason [12] take ψ to be "holomorphic" on the boundary. As an alternative one could satisfy the 2-dimensional elliptic equation $\gamma^A \nabla_A \psi = 0$. A nice investigation of the various options for equations to select the value of the spinor field for such quasilocal expressions has been carried out by Szabados [30].

27.6.2   An SU(2) formulation

Here we briefly describe a 3-dimensional alternative, similar to but distinct from the Witten formulation, which uses the SU(2) spinors of the 3-dimensional spatial hypersurface [25]. We begin from the well-known ADM Hamiltonian [18, 22]

$$H(N) = \int d^3x \Big\{ N \big[g^{-\frac{1}{2}} (\pi^{mn} \pi_{mn} - \tfrac{1}{2}\pi^2) - g^{\frac{1}{2}} R \big] + 2\pi^m{}_k \nabla_m N^k \Big\}$$
$$+ \oint dS_k N \delta^{kc}_{am} g^{mb} \Gamma^a{}_{bc}, \quad (6.8)$$

which we have supplemented by boundary term expressions that are valid in asymptotic Cartesian frames. In this case we concern ourselves only with energy (not momentum), so we take the shift N^k to vanish. Within the ADM Hamiltonian density the scalar curvature term and the boundary term do not have a definite sign; the idea is to replace them with alternatives that are more definite. The 3-scalar curvature can be replaced using $N = \varphi^\dagger \varphi$ and the 3-dimensional SCI

$$2\Big[\nabla(\varphi^\dagger i\sigma) \wedge \nabla\varphi - \nabla\varphi^\dagger \wedge \nabla(i\sigma\varphi) \Big] \equiv dB_{su2} - \varphi^\dagger \varphi R^{ab} \wedge \epsilon_{abc}\vartheta^c, \quad (6.9)$$

where $R^{ab} \wedge \epsilon_{abc}\vartheta^c = Rg^{1/2}d^3x$, $\sigma := \sigma_a \vartheta^a$ is a Pauli matrix-valued one-form, and

$$B_{su2} := 2[\nabla\varphi^\dagger \wedge i\sigma\varphi + \varphi^\dagger i\sigma \wedge \nabla\varphi], \quad (6.10)$$

which is a legitimate (since they agree asymptotically) alternative to the boundary term in (6.8).

The Hamiltonian density now takes the form

$$\mathcal{H}(\varphi) = (\varphi^\dagger \varphi) g^{-\frac{1}{2}} \Big[\pi^{mn} \pi_{mn} - \tfrac{1}{2}\pi^2 \Big] d^3x$$
$$+ \Big[\nabla(\varphi^\dagger i\sigma) \wedge \nabla\varphi - \nabla\varphi^\dagger \wedge \nabla(i\sigma\varphi) \Big]. \quad (6.11)$$

By arguments similar to those used in connection with the Witten Hamiltonian, the quadratic $\nabla\varphi$ terms can be diagonalized to the form $|\nabla_k\varphi|^2 - |\sigma^k\nabla_k\varphi|^2$. Consequently the Hamiltonian density is non-negative—and hence the total energy is positive— on *maximal* spacelike hypersurfaces (in this standard time gauge condition the trace of the ADM canonical momentum π vanishes) if the 3-dimensional spinor is chosen to satisfy the 3-dimensional elliptic equation $\sigma^k\nabla_k\varphi = 0$.

Again, the boundary term B_{su2} gives a quasilocal energy for any choice of φ; in particular, as in the previous case, we can use holomorphic spinors or $\sigma^A\nabla_A\varphi = 0$.

27.6.3   The QSL approach

Our last alternative has some similar features to the above two approaches but differs from them in a very fundamental way. In the above the spinor field was introduced into the Hamiltonian as a technical device to aid in obtaining a locally non-negative Hamiltonian density. Instead one can introduce a spinor field into the Lagrangian, then it becomes a basic dynamical gravitational field. The key is another *spinor-curvature* identity, this time involving the 4-dimensional scalar curvature, which led us to a *Quadratic Spinor Lagrangian* (QSL) for GR [27]. (For the relation to teleparallel GR, aka TEGR, $GR_{||}$, see [31]) The Einstein–Hilbert scalar curvature Lagrangian equals (up to an exact differential) the QSL

$$\mathcal{L}_{qs} := 2D\overline{\Psi}\gamma \wedge D\Psi \equiv -R * 1 + d(D\overline{\Psi} \wedge \gamma\Psi + \overline{\Psi}\gamma \wedge D\Psi), \qquad (6.12)$$

where $\Psi = \vartheta\psi$ is a spinor one-form field. The spinor field is to be varied subject to the normalization constraints $\overline{\psi}\psi = 1$, $\overline{\psi}\gamma\psi = 0$ (which can be enforced via Lagrange multipliers).

In order to construct the Hamiltonian it is convenient to work from the corresponding first-order Lagrangian:

$$\mathcal{L}_\Psi := D\overline{\Psi} \wedge P + \overline{P} \wedge D\Psi + \tfrac{1}{2}\overline{P} \wedge \gamma P \qquad (6.13)$$

(which yields the pair of first-order equations: $2D\Psi = -\gamma P$ and its conjugate). From this we find the covariant Hamiltonian 3-form

$$\begin{aligned}
\mathcal{H}_\Psi &:= \overline{P} \wedge \pounds_N \Psi + \pounds_N \overline{\Psi} \wedge P - i_N \mathcal{L}_\Psi \\
&\equiv - i_N(\tfrac{1}{2}\overline{P} \wedge \gamma P) \\
&\quad - \left[i_N \overline{\Psi} DP + D\overline{\Psi} \wedge i_N P + \overline{\Psi} \wedge i_N \omega P - d(i_N \overline{\Psi} P) + \text{c.c.} \right], \quad (6.14)
\end{aligned}$$

(the time derivative here is given by the Lie derivative). The Hamiltonian boundary term

$$\mathcal{B}(N) = i_N \overline{\Psi} P + \overline{P} i_N \Psi, \qquad (6.15)$$

again yields quasilocal values for any choice of spinor field on the boundary.

One distinguishing feature of this approach, in which the spinor field is introduced into the Lagrangian, is that the displacement vector field N remains entirely independent of the spinor field. Another is that the connection does not appear as a primary dynamic variable. Nevertheless it is again possible to arrange that the Hamiltonian is locally non-negative. By arguments similar to those used before, with the lapse-shift choice $N^\mu = (f^2, 0, 0, 0)$, and the spinor field satisfying the (3-dimensional, elliptic) conformal Witten equation: $\gamma^k D_k(f\psi) = 0$, the QSL Hamiltonian density is locally non-negative on maximal slices. This yields another locally positive 'localization' along with a positive energy proof.

For further discussion of additional details and features of these cases see the already cited references as well as the overview in [5].

27.7   Properties of the spinor expressions

The spinor quasilocal expressions have certain common properties. A noteworthy one is that the expressions are essentially algebraic in N^μ; unlike the standard GR expression (5.3) they have no DN terms. (The importance of this feature will be discussed soon.) Another is perhaps the real *beauty* of these spinor formulations: they do not need an *explicit* reference configuration—whereas in the standard GR expression (5.3) the reference configuration is essential. When examined in detail it can be seen that the spinor field *implicitly* determines the reference values [5, 8]. That is one of its jobs.

There are other distinct roles for the spinor fields. The *Hamiltonian variation boundary principle* tells us what must be held fixed on the boundary. This clarifies the role of the fields. Essentially for the spinor formulations we find that (in addition to fixing the usual frame components) the spinor field ψ should be held fixed on the boundary. In each of the cases this has a distinct significance. For the Witten Hamiltonian, fixing the spinor field fixes the displacement N^μ, for the SU(2) spinor alternative it means holding the *lapse* fixed, whereas for the QSL fixed ψ amounts to fixing the observer's orthonormal frame on the boundary. In each case the spinor field has a different role. Altogether we have good examples of what spinors can mean geometrically and physically.

In each case, for displacements corresponding to 4-dimensional spacetime translations, the quasilocal spinor expressions (6.3, 6.10, 6.15) asymptotically agree with a standard one (5.3) and hence each expression gives similar reasonable values for the total energy-momentum.

27.8   Angular momentum and center-of-mass moment

However energy-momentum is not the whole story. For an asymptotically flat space each asymptotic symmetry has its associated conserved quantities. Energy-momentum is associated with spacetime translations. Such spaces also have asymptotic rotation and Lorentz boost symmetry. A proper physical Hamiltonian formalism allows for displacements which have the asymptotic Poincaré form

$$N^\mu = N^\mu_\infty + \lambda^\mu{}_\nu x^\nu. \tag{8.1}$$

Here N^μ_∞ is a constant spacetime translation and the constants $\lambda^{\mu\nu} = \lambda^{[\mu\nu]}$ describe an infinitesimal Lorentz transformation (including rotations and boosts). The value of the Hamiltonian generating the Lorentz displacements then gives six additional conserved quantities [2, 28]. The physical significance of these additional quantities is most easily recognized by recalling that, for a relativistic particle,

$$L^{\mu\nu} := x^\mu p^\nu - x^\nu p^\mu \tag{8.2}$$

includes both *angular momentum*,

$$L^{ij} := x^i p^j - x^j p^i, \quad \text{and} \quad L^{0k} := x^0 p^k - x^k p^0, \tag{8.3}$$

the *center-of-mass moment*.

The conventional variable GR Hamiltonian description of section 5 does yield good values for these additional quantities for gravitating systems. An important contribution here is the $i_N\Gamma \sim DN$ connection-Møller–Komar type term in

(5.3) (such terms have often been overlooked in quasilocal investigations [3]). We found that this term plays a key role in: (i) black hole thermodynamics [6], (ii) certain angular momentum calculations [15, 16, 32] (iii) *all* center-of-mass moment calculations [20]. (Note: from (8.1) it follows that asymptotically the DN term has a contribution proportional to λ.)

Now such terms are completely absent in all three of our spinor formulations. Without them we (i) do not know how to obtain the first law of black hole thermodynamics, (ii) have difficulties in obtaining the angular momentum (within the Witten Hamiltonian (6.2), modifying the displacement to $\bar{\psi}\gamma\gamma^\mu\psi$ will work—but only *if* we take the strange asymptotics $\psi^\dagger\psi \sim r$), (iii) simply *cannot see how* to get the correct center-of-mass moment. Now we have not entirely given up trying, but so far we have not managed to include the effects of such $D^{[\mu}N^{\nu]}$ type terms in the spinor formulations. Presently we are preparing detailed discussions of these crucial issues. Here we can summarize our tentative conclusions but regretfully cannot include any more of the technical supporting details.

27.9   Conclusions

We considered the spinor formulations for the Hamiltonian and the associated positive energy proofs and quasilocal expressions. At first the role of the spinor field seemed mysterious. However the boundary variation principle clarifies the role of the spinor field (and indeed all other variables). Our spinor Hamiltonian expressions illustrate the variety of roles that spinor fields can play. Spinors give beautiful positive energy proofs (especially Witten's) and very neat formulas for quasilocal energy-momentum. It is especially noteworthy that there is no need for extra reference fields on the boundary.

Yet it seems that such spinor formulations have a serious limitation: the present expressions cannot give angular momentum and the center-of-mass moment. Moreover, we do not see how any natural adjustment of these spinor formulations for quasilocal Hamiltonian boundary terms can successfully give expressions for these quantities. Apparently the spinor Hamiltonians cannot give all of the physically conserved quantities of asymptotically flat spacetimes. Hence, although they are deservedly popular and quite good for many purposes, we now believe that such spinor formulations cannot really capture *in a physically correct way* the still elusive gravitational energy-momentum.

Acknowledgments

Our thanks to S. Deser and S. N. Liang for suggestions and corrections. This work was supported by the National Science Council of the Republic of China under the grants NSC90-2112-M-008-041, NSC90-2112-M-002-055.

REFERENCES

[1] R. Arnowitt, S. Deser and C. W. Misner, The dynamics of general relativity in *Gravitation: An Introduction to Current Research*, Ed. L. Witten, Wiley, New York, 1962, pp. 227–265.

[2] R. Beig and N. Ó Murchadha, The Poincaré group is the symmetry group of canonical general relativity, *Ann. Phys.* **174** (1987), 463–498.

[3] J. D. Brown and J. W. York, Jr., Quasilocal energy and conserved charges derived from the gravitational action, *Phys. Rev. D* **47** (1993), 1407–1409.

[4] C.-C. Chang, J. M. Nester and C.-M. Chen, Pseudotensors and quasilocal energy-momentum, *Phys. Rev. Lett.* **83** (1999), 1897–1901; arXiv: gr-qc/9809040.

[5] C.-C. Chang, J. M. Nester and C.-M. Chen, Energy-momentum (quasi-) localization for gravitating systems, in *Gravitation and Astrophysics*, Eds. Liao Liu, Jun Luo, X-Z. Li, J-P. Hsu, World Scientific, Singapore, 2000, pp. 163–173; arXiv: gr-qc/9912058.

[6] C.-M. Chen and J. M. Nester, Quasilocal quantities for general relativity and other gravity theories, *Class. Quantum Grav.* **16** (1999), 1279–1304; arXiv: gr-qc/9809020.

[7] C.-M. Chen and J. M. Nester, A symplectic Hamiltonian derivation of quasilocal energy-momentum for GR, *Gravitation & Cosmology* **6** (2000), 257–270; arXiv: gr-qc/0001088.

[8] C. M. Chen, J. M. Nester and R. S. Tung, Quasilocal energy-momentum for geometric gravity theories, *Phys. Lett.* **A 203** (1995), 5–11.

[9] S. Deser, Positive classical gravitational energy from classical supergravity, *Phys. Rev. D* **27** (1983), 2805–2808.

[10] S. Deser and C. Teitelboim, Supergravity has positive energy, *Phys. Rev. Lett.* **39** (1977), 249–252.

[11] A. Dimakis and F. Müller-Hoissen, Clifform calculus with applications to classical field theories, *Class. Quantum Grav.* **8** (1991), 2093–2132.

[12] A. Dougan and L. Mason, Quasilocal mass constructions with positive energy, *Phys. Rev. Lett.* **67** (1991), 2119–2122.

[13] F. Estabrook, Lagrangians for Ricci flat geometries, *Class. Quantum Grav.* **8** (1991), L151–154.

[14] M. T. Grisaru, Positivity of the energy in Einstein theory, *Phys. Lett.* **73B** (1978), 207–208.

[15] R. D. Hecht and J. M. Nester, A new evaluation of PGT mass and spin, *Phys. Lett. A* **180** (1993), 324–331.

[16] R. D. Hecht and J. M. Nester, An evaluation of mass and spin at null infinity for the PGT and GR gravity theories, *Phys. Lett. A* **217** (1996), 81–89.

[17] G. T. Horowitz and A. Strominger, On Witten's expression for gravitational energy, *Phys. Rev. D* **27** (1983), 2793–2804.

[18] J. Isenberg and J. M. Nester, Canonical gravity, in: *General Relativity and Gravitation: One Hundred Years After the Birth of Albert Einstein*, Vol I., Ed. A. Held, Plenum, New York, 1980, pp. 23–97.

[19] J. Kijowski and W. M. Tulczyjew, *A Symplectic Framework for Field Theories*, Lecture Notes in Physics, Vol. **107**, Springer, Berlin, 1979.

[20] F. F. Meng, Quasilocal Center-of-Mass Moment in GR, MSc. Thesis, National Central University, 2002 (unpublished).

[21] E. W. Mielke, *Geometrodynamics of Gauge Fields—On the Geometry of Yang-Mills and Gravitational Gauge Theories*, Akademie, Berlin, 1987.

[22] C. W. Misner, K. Thorne and J. A. Wheeler, *Gravitation*, Freeman, San Fransisco, 1973.

[23] J. M. Nester, A new gravitational energy expression with a simple positivity proof, *Phys. Lett. A* **83** (1981), 241–242.

[24] J. M. Nester, The gravitational Hamiltonian, in *Asymptotic Behavior of Mass and spacetime Geometry*, Lecture Notes in Physics, Vol. **202**, Ed. F. Flaherty, Springer, Berlin, 1984, pp. 155–163.

[25] J. M. Nester and R. S. Tung, Another positivity proof and gravitational energy localizations, *Phys. Rev. D* **49** (1994), 3958–3962.

[26] J. M. Nester, R. S. Tung and V. Zhytnikov, Some spinor curvature identities, *Class. Quant. Grav.* **11** (1994), 983–987.

[27] J. M. Nester and R. S. Tung, A quadratic spinor Lagrangian for general relativity, *Gen. Rel. Grav.* **27** (1995), 115–119.

[28] T. Regge and C. Teitelboim, Role of surface integrals in the Hamiltonian formulation of general relativity, *Ann. Phys.* **88** (1974), 286–319.

[29] R. Schoen and S. T. Yau, On the proof of the positive mass conjecture in general relativity, *Comm. Math Phys.* **65** (1979), 45–76.

[30] L. Szabados, Two-dimensional Sen connections in general relativity, *Class. Quantum Grav.* **11** (1994), 1833–1847.

[31] R. S. Tung and J. M. Nester, The quadratic spinor Lagrangian is equivalent to the teleparallel theory, *Phys. Rev. D* **60** (1999), 021501.

[32] K. H. Vu, Quasilocal Energy-Momentum and Angular Momentum for Teleparallel Gravity, MSc. Thesis, National Central University, 2000 (unpublished).

[33] E. Witten, A simple proof of the positive energy theorem, *Comm. Math. Phys.* **80** (1981), 381–402.

Chiang-Mei Chen
Department of Physics
National Taiwan University
Taipei, 106, Taiwan, R.O.C.
E-mail: cmchen@phys.ntu.edu.tw

Roh-Suan Tung
Department of Physics
National Central University
Chung-Li, 320, Taiwan, R.O.C.
E-mail: rohtung@phy.ncu.edu.tw

James M. Nester
Department of Physics
National Central University
Chung-Li, 320, Taiwan, R.O.C.
E-mail: nester@phy.ncu.edu.tw

Submitted: September 6, 2002; Revised: November 28, 2002.

28

Chiral Dirac Equations

Claude Daviau

ABSTRACT A chiral relativistic wave equation is proposed for neutrinos. This wave equation allows a mass term. We study this equation in the Clifford algebra of space, and in the frame of the Clifford spacetime algebra, second-order equation, plane waves, Lagrangian formalism, conservative current. Next we extend the wave equation to the complete spacetime algebra and we obtain a chiral wave equation with a mass term and a charge term. We study the relativistic invariance, the Lagrangian formalism, the second-order equation, we solve the equation in the case of the hydrogen atom. We obtain the right number of energy levels and the right energy levels. We study the enlargement of the gauge invariance, with a real matrix formalism. The electric gauge group may be extended to $SO(8)$, a subgroup of $SO(16)$.

Keywords: Clifford algebra, Pauli algebra, Dirac theory, electrons, neutrinos, hydrogen atom, gauge invariance, orthogonal groups

28.1 Different frames to see different things

The Dirac equation [1] comes from the restricted relativity. Dirac wanted a first-order equation to obtain a conservative probability density, and the second-order Klein–Gordon equation. Today, the Dirac equation is considered as the true wave equation for each particle with spin $\frac{1}{2}$: electron, muon, neutrino, quarks, ... First we write this equation with the usual formalism

$$[\gamma^\mu(\partial_\mu + iqA_\mu) + im]\psi = 0, \quad q = \frac{e}{\hbar c}, \quad m = \frac{m_0 c}{\hbar}, \quad (1.1)$$

where the negative e is the electron's charge and A_μ are the covariant components of the exterior electromagnetic potential. We use a $+---$ signature for the spacetime metric, so $A^0 = A_0$ and $A^j = -A_j$, $j = 1, 2, 3$. We use the common Dirac matrices

$$\psi = \begin{pmatrix} \psi_1 \\ \psi_2 \\ \psi_3 \\ \psi_4 \end{pmatrix}, \quad \gamma_0 = \gamma^0 = \begin{pmatrix} I_2 & 0 \\ 0 & -I_2 \end{pmatrix}, \quad \gamma_j = -\gamma^j = \begin{pmatrix} 0 & -\sigma_j \\ \sigma_j & 0 \end{pmatrix}, \quad (1.2)$$

AMS Subject Classification: 15A66, 35Q40, 17B45.

where

$$I_2 = \sigma_0 = \sigma^0 = \begin{pmatrix} 1 & 0 \\ 0 & 1 \end{pmatrix}, \qquad \sigma_1 = -\sigma^1 = \begin{pmatrix} 0 & 1 \\ 1 & 0 \end{pmatrix},$$

$$\sigma_2 = -\sigma^2 = \begin{pmatrix} 0 & -i \\ i & 0 \end{pmatrix}, \qquad \sigma_3 = -\sigma^3 = \begin{pmatrix} 1 & 0 \\ 0 & -1 \end{pmatrix}.$$

We also use $\gamma_{ij} = \gamma_i \gamma_j$, $\gamma_5 = -i\gamma_{0123}$. After the discovery of the maximal parity violation by weak interactions, quantum theory uses the Weyl spinors ξ and η linked to ψ by

$$U\psi = \begin{pmatrix} \xi \\ \eta \end{pmatrix}, \qquad U = U^{-1} = \frac{1}{\sqrt{2}}(\gamma_0 + \gamma_5),$$

$$\xi = \frac{1}{\sqrt{2}} \begin{pmatrix} \psi_1 + \psi_3 \\ \psi_2 + \psi_4 \end{pmatrix}, \qquad \eta = \frac{1}{\sqrt{2}} \begin{pmatrix} \psi_1 - \psi_3 \\ \psi_2 - \psi_4 \end{pmatrix}. \tag{1.3}$$

All experimental results of particle physics are interpreted with the previous frame. For instance neutrinos' waves are purely chiral, with only a ξ or a η wave, and the charge conjugation exchanges the ξ wave with an η wave, and vice-versa. Next, since the mass term of (1.1) connects ξ to η, it is necessary to suppose that the neutrino's mass term is exactly null. But physicists have good reasons to suppose that mass terms are small but not equal to zero.

We will see here that it is possible to obtain chirality with a mass term. But to find this new result, we must first change the usual frame. Three other frames are available.

28.2   The Clifford algebra of space

To obtain space algebra formalism [2–7] we associate to each ψ the $\phi = f(\psi)$ defined in $C\ell_{3,0}$ by

$$
\begin{aligned}
f(\psi) &= \phi \\
&= a_1 + a_2\sigma_{32} + a_3\sigma_{31} + a_4\sigma_{12} + a_5\sigma_{123} + a_6\sigma_1 + a_7\sigma_2 + a_8\sigma_3 \quad (2.1)
\end{aligned}
$$

where the a_j are

$$\psi_1 = a_1 + ia_4, \quad \psi_2 = -a_3 - ia_2, \quad \psi_3 = a_8 + ia_5, \quad \psi_4 = a_6 + ia_7. \tag{2.2}$$

The space algebra is isomorphic to the Pauli algebra generated by the σ_j matrices and their products, $M_2(\mathbb{C})$. But the f isomorphism is only an isomorphism of real vector space, not of complex vector space, because we have

$$f(i\psi) = \phi\sigma_{12}. \tag{2.3}$$

We do not have $f(i\psi) = if(\psi)$: as f is only $\mathbb{R}$-linear, and not $\mathbb{C}$-linear. Consequently we shall use here only the real vector space or real Clifford algebra structures. We write ψ^* the complex conjugate of ψ, and we use reversion:

$$\tilde{\phi} = \phi^\dagger$$
$$= a_1 - a_2\sigma_{32} - a_3\sigma_{31} - a_4\sigma_{12} - a_5\sigma_{123} + a_6\sigma_1 + a_7\sigma_2 + a_8\sigma_3 , \quad (2.4)$$

grade involution:

$$\hat{\phi} = a_1 + a_2\sigma_{32} + a_3\sigma_{31} + a_4\sigma_{12} - a_5\sigma_{123} - a_6\sigma_1 - a_7\sigma_2 - a_8\sigma_3 , \quad (2.5)$$

and conjugation:

$$\bar{\phi} = a_1 - a_2\sigma_{32} - a_3\sigma_{31} - a_4\sigma_{12} + a_5\sigma_{123} - a_6\sigma_1 - a_7\sigma_2 - a_8\sigma_3 . \quad (2.6)$$

For each A and B in space algebra

$$(AB)^\dagger = B^\dagger A^\dagger, \quad \overline{AB} = \bar{B}\,\bar{A}, \quad \hat{\bar{A}} = \bar{\hat{A}}, \quad \widehat{AB} = \hat{A}\hat{B}. \quad (2.7)$$

For each $\phi = f(\psi)$,

$$f(\psi^*) = \sigma_2\hat{\phi}\sigma_2 , \qquad f(\gamma^\mu\psi) = \sigma^\mu\hat{\phi} . \quad (2.8)$$

Using f we obtain the wave equation equivalent to (1.1)

$$f\left([\gamma^\mu(\partial_\mu + iqA_\mu) + im]\psi\right) = 0$$

which gives

$$\sigma^\mu\partial_\mu\hat{\phi} + q\sigma^\mu A_\mu\hat{\phi}\sigma_{12} + m\phi\sigma_{12} = 0 . \quad (2.9)$$

In the space algebra $\nabla = \sigma^\mu\partial_\mu$, $A = \sigma^\mu A_\mu$, and so we have

$$\hat{\nabla} = \partial_0 + \vec{\partial}, \qquad \vec{\partial} = \sigma_1\partial_1 + \sigma_2\partial_2 + \sigma_3\partial_3 ,$$
$$\hat{A} = A^0 - \vec{A}, \qquad \vec{A} = A^1\sigma_1 + A^2\sigma_2 + A^3\sigma_3 . \quad (2.10)$$

The Dirac equation becomes

$$\nabla\hat{\phi} + qA\hat{\phi}\sigma_{12} + m\phi\sigma_{12} = 0 \quad (2.11)$$

or, equivalently,

$$\nabla\hat{\phi}\sigma_{12} = m\phi + qA\hat{\phi} . \quad (2.12)$$

The Dirac theory is not well described by the classical complex formalism. This formalism says that we can construct tensorial densities, and how many. The densities without derivative are

$$\Omega_1 = \bar{\psi}\psi , \qquad\qquad \bar{\psi} = \psi^\dagger\gamma_0 , \quad (2.13)$$
$$J^\mu = \bar{\psi}\gamma^\mu\psi , \quad (2.14)$$
$$S^{\mu\nu} = i\bar{\psi}\gamma^\mu\gamma^\nu\psi , \quad (2.15)$$
$$K^\mu = -\bar{\psi}\gamma_5\gamma^\mu\psi , \quad (2.16)$$
$$\Omega_2 = -i\bar{\psi}\gamma_5\psi , \qquad i\gamma_5 = \gamma_0\gamma_1\gamma_2\gamma_3 , \quad (2.17)$$

where Ω_1 is a scalar, J is a spacetime vector, S is a spacetime bivector, K is a spacetime pseudo-vector, Ω_2 is a spacetime pseudo-scalar. The J^0 component is the density probability

$$J^0 = \psi_1\psi_1{}^* + \psi_2\psi_2{}^* + \psi_3\psi_3{}^* + \psi_4\psi_4{}^* . \tag{2.18}$$

With the classical Dirac formalism, we have 16 real densities. The dimension of the matrix algebra over $\mathbb{C}$ is 16. But if we translate the Dirac equation into space algebra, or spacetime algebra, new tensorial densities appear. With space algebra we obtain

$$J = \phi\phi^\dagger = \sigma^\mu J_\mu , \tag{2.19}$$

$$K = \phi\sigma_3\phi^\dagger = \sigma^\mu K_\mu , \tag{2.20}$$

$$R = \phi\overline{\phi} = \Omega_1 + i\Omega_2 , \tag{2.21}$$

$$S = \phi\sigma_3\overline{\phi} = S^{23}\sigma_1 + S^{31}\sigma_2 + S^{12}\sigma_3 + S^{10}\sigma_{23} + S^{20}\sigma_{31} + S^{30}\sigma_{12} . \tag{2.22}$$

The σ_3 direction seems privileged, and we can see that with the wave equation (2.12) where we have only σ_{12}. The fact that the third direction of the frame seems privileged in the Dirac theory was pointed by de Broglie as early as in 1934 [8]. But we can obtain more densities if we also use σ_1 and σ_2. We obtain two vectors and two bivectors

$$K_{(1)} = \phi\sigma_1\phi^\dagger , \qquad K_{(2)} = \phi\sigma_2\phi^\dagger , \tag{2.23}$$

$$S_{(1)} = \phi\sigma_1\overline{\phi} , \qquad S_{(2)} = \phi\sigma_2\overline{\phi} . \tag{2.24}$$

These tensors are not electric gauge invariant, but their transformation, under a Lorentz rotation, is identical to the classical transformation of tensors: If M is an element of $SL(2,\mathbb{C})$, and V is a spacetime vector $V = \sigma^\mu V_\mu$. The transformation $r : V \mapsto V' = MVM^\dagger$ is a Lorentz rotation, element of the $\mathcal{L}^\uparrow{}_+$ restricted Lorentz group. With

$$\phi' = M\phi , \quad \nabla' = M\nabla M^\dagger , \quad A' = MAM^\dagger , \tag{2.25}$$

the Dirac equation (2.12) gives the same equation

$$\nabla'\widehat{\phi}'\sigma_{12} = m\phi' + qA'\widehat{\phi}' \tag{2.26}$$

because we have $M^\dagger = \widehat{M}^{-1}$ and $\overline{M} = M^{-1}$.

We have also [5]

$$\phi = \sqrt{2}(\xi \quad \sigma_{13}\eta^*) , \qquad \widehat{\phi} = \sqrt{2}(\eta \quad \sigma_{13}\xi^*) . \tag{2.27}$$

The Weyl spinors ξ, η transform as

$$\xi' = M\xi , \qquad \eta' = (M^\dagger)^{-1}\eta = \widehat{M}\eta ,$$

$$\sigma_{13}\eta'^* = M\sigma_{13}\eta^* , \quad \eta'^* = \widehat{M}^*\eta^* = \sigma_2 M\sigma_2\eta^* , \tag{2.28}$$

and so the link between ϕ, $\widehat{\phi}$, ξ and η is Lorentz invariant. The spacetime vectors J, $K_{(3)} = K$, $K_{(1)}$, $K_{(2)}$ become

$$J' = \phi'\phi^{\dagger'} = MJM^{\dagger}, \qquad K'_{(j)} = \phi'\sigma_j\phi^{\dagger'} = MK_{(j)}M^{\dagger}. \qquad (2.29)$$

The scalar Ω_1, the pseudo-scalar Ω_2 and the bivectors $S_{(1)}$, $S_{(2)}$, $S_{(3)} = S$ become

$$\Omega'_1 + i\Omega'_2 = \phi'\overline{\phi}' = M\phi\overline{\phi}M = MM^{-1}\phi\overline{\phi} = \Omega_1 + i\Omega_2, \qquad (2.30)$$

$$S'_{(j)} = \phi'\sigma_j\overline{\phi}' = MS_{(j)}M^{-1}. \qquad (2.31)$$

With the $K_{(j)}$ and $S_{(j)}$ tensors, we have $36 = \binom{8+1}{2}$ densities, and not only 16.

28.3   Chiral wave equation with a mass term

To obtain the maximal parity violation by weak interactions, quantum field theory allocates to neutrinos only chiral waves, neutrinos have only left waves ψ_L and antineutrinos only right waves ψ_R,

$$\psi_L = \tfrac{1}{2}(I_4 - \gamma_5)\psi, \qquad \psi_R = \tfrac{1}{2}(I_4 + \gamma_5)\psi. \qquad (3.1)$$

With the f transformation, we obtain

$$\widehat{\phi}_L = \widehat{\phi}\tfrac{1}{2}(1 + \sigma_3) = \sqrt{2}(\eta \quad 0), \qquad (3.2)$$

$$\phi_R = \phi\tfrac{1}{2}(1 + \sigma_3) = \sqrt{2}(\xi \quad 0). \qquad (3.3)$$

The chiral wave equation is [9]

$$\nabla\widehat{\phi}_L = m\phi_L\sigma_2 \qquad (3.4)$$

or, equivalently,

$$\widehat{\nabla}\phi_L = -m\widehat{\phi}_L\sigma_2 \qquad (3.5)$$

and we obtain to the second order

$$\Box\phi_L = \nabla\widehat{\nabla}\phi_L = -m\nabla\widehat{\phi}_L\sigma_2$$
$$= -m(m\phi_L\sigma_2)\sigma_2 = -m^2\phi_L, \qquad (3.6)$$

$$(\Box + m^2)\phi_L = 0. \qquad (3.7)$$

So ϕ_L is a solution of the Klein–Gordon equation. With the Dirac formalism, it seemed impossible to reconcile chirality with a mass term. So neutrinos and antineutrinos were considered without mass term. Today it is necessary to explain why only a small number of the solar neutrinos are received, and neutrinos must have a small mass term. It may be interesting to have a chiral equation with a mass term.

28.4   The chiral equation with spacetime algebra

To go from the space algebra to the spacetime algebra, we can use the matrix representation (1.2). A spacetime vector $\mathbf{A}$ reads

$$\mathbf{A} = \begin{pmatrix} 0 & \widehat{A} \\ A & 0 \end{pmatrix} = \gamma^\mu A_\mu , \quad \gamma^j = \gamma^j , \quad j = 1, 2, 3, \quad \gamma^0 = \gamma_5 . \tag{4.1}$$

The gradient $\partial = \gamma^\mu \partial_\mu$ is

$$\partial = \begin{pmatrix} 0 & \widehat{\nabla} \\ \nabla & 0 \end{pmatrix} . \tag{4.2}$$

To each $\phi = f(\psi)$ we associate $\Psi = g(\phi)$ by

$$\Psi = \begin{pmatrix} \widehat{\phi} & 0 \\ 0 & \phi \end{pmatrix} =$$

$$a_1 + a_2 \gamma_{23} + a_3 \gamma_{13} + a_4 \gamma_{21} + a_5 \gamma_{0123} + a_6 \gamma_{10} + a_7 \gamma_{20} + a_8 \gamma_{30} , \tag{4.3}$$

and the Dirac equation takes the Hestenes' form [10–19]

$$\partial \Psi \gamma_{21} = m \Psi \gamma_0 + q \mathbf{A} \Psi . \tag{4.4}$$

The invariance under Lorentz rotations becomes

$$\partial \mapsto \partial' = R \partial \tilde{R} , \quad \mathbf{A} \mapsto \mathbf{A}' = R \mathbf{A} \tilde{R} , \quad \Psi \mapsto \Psi' = R \Psi , \tag{4.5}$$

$$R = \begin{pmatrix} \widehat{M} & 0 \\ 0 & M \end{pmatrix} , \quad \tilde{R} = \begin{pmatrix} M^\dagger & 0 \\ 0 & \overline{M} \end{pmatrix} = R^{-1} , \tag{4.6}$$

where $\tilde{R}$ is the reverse, defined by $\tilde{\gamma}_\mu = \gamma_\mu , \widetilde{AB} = \tilde{B}\tilde{A}$. Left and right parts of the waves read now

$$\Psi_L = \Psi \tfrac{1}{2}(1 - \gamma_{30}) , \qquad \Psi_R = \Psi \tfrac{1}{2}(1 + \gamma_{30}) \tag{4.7}$$

and the chiral wave equation (3.4) reads now

$$\partial \Psi_L = m \Psi_L \gamma_2 . \tag{4.8}$$

To obtain plane waves, we set

$$\Psi_L = \cos(p_\mu x^\mu) \Psi_1 + \sin(p_\mu x^\mu) \Psi_2 , \quad \mathbf{p} = \gamma^\mu p_\mu \tag{4.9}$$

where Ψ_1 and Ψ_2 are fixed left terms, $\mathbf{p}$ is the spacetime vector impulse-energy. We obtain

$$-\mathbf{p}\Psi_1 = m \Psi_2 \gamma_2 , \qquad \mathbf{p}\Psi_2 = m \Psi_1 \gamma_2 , \tag{4.10}$$

and so we obtain

$$\Psi_2 = \frac{\mathbf{p}}{m}\Psi_1\boldsymbol{\gamma}_2, \qquad \mathbf{p}^2\Psi_1\boldsymbol{\gamma}_2 = m^2\Psi_1\boldsymbol{\gamma}_2. \qquad (4.11)$$

So non-null plane waves exist only if the relativistic condition $\mathbf{p}^2 = m^2$ is true. In this case Ψ_1 is any fixed left term, and Ψ_2 is given by (4.11).

It is possible to construct a Lagrangian formalism to deduce the chiral wave (4.8). To obtain the Lagrangian density we must consider at the same time a left wave Ψ_L and a right wave Ψ_R. We use here Lasenby's article [20]. $< A >$ is the scalar part of A. The Lagrangian density reads

$$\mathcal{L} = < \boldsymbol{\partial}\Psi_L\boldsymbol{\gamma}_1\tilde{\Psi}_R + m\Psi_L\boldsymbol{\gamma}_{12}\tilde{\Psi}_R > . \qquad (4.12)$$

The Lagrangian equations are

$$\partial_{\tilde{\Psi}_L}\mathcal{L} = \boldsymbol{\partial}(\partial_{\widetilde{\boldsymbol{\partial}\Psi}_L}\mathcal{L}), \qquad (4.13)$$

$$\partial_{\tilde{\Psi}_R}\mathcal{L} = \boldsymbol{\partial}(\partial_{\widetilde{\boldsymbol{\partial}\Psi}_R}\mathcal{L}). \qquad (4.14)$$

We have here

$$\partial_{\tilde{\Psi}_R}\mathcal{L} = \boldsymbol{\partial}\Psi_L\boldsymbol{\gamma}_1 + m\Psi_L\boldsymbol{\gamma}_{12}, \qquad \partial_{\widetilde{\boldsymbol{\partial}\Psi}_R}\mathcal{L} = 0, \qquad (4.15)$$

$$\partial_{\tilde{\Psi}_L}\mathcal{L} = m\Psi_R\boldsymbol{\gamma}_{21}, \qquad \partial_{\widetilde{\boldsymbol{\partial}\Psi}_L}\mathcal{L} = \Psi_R\boldsymbol{\gamma}_1, \qquad (4.16)$$

and the Lagrangian equations give

$$\boldsymbol{\partial}\Psi_L = m\Psi_L\boldsymbol{\gamma}_2, \qquad \boldsymbol{\partial}\Psi_R = m\Psi_R\boldsymbol{\gamma}_2. \qquad (4.17)$$

Parity violation induces us to see Ψ_R as the anti-wave of Ψ_L. The Lagrangian formalism involves equality for the mass of the antineutrino to the mass of the neutrino.

With a left wave Ψ_L, several tensors vanish. For instance the tensors without derivative give

$$\tilde{\Psi}_L = \tfrac{1}{2}(1 + \boldsymbol{\gamma}_{30})\tilde{\Psi}, \qquad (4.18)$$

$$0 = \Psi_L\tilde{\Psi}_L = \Omega_1 + \Omega_2\boldsymbol{\gamma}_{0123}, \qquad (4.19)$$

$$\mathbf{J} = \Psi_L\boldsymbol{\gamma}_0\tilde{\Psi}_L = -\Psi_L\boldsymbol{\gamma}_3\tilde{\Psi}_L = -\mathbf{K}, \qquad (4.20)$$

$$\mathbf{S}_{(3)} = \Psi_L\boldsymbol{\gamma}_{21}\tilde{\Psi}_L = 0, \qquad (4.21)$$

$$\mathbf{S}_{(1)} = \Psi_L\boldsymbol{\gamma}_{32}\tilde{\Psi}_L, \qquad (4.22)$$

$$\mathbf{S}_{(2)} = \Psi_L\boldsymbol{\gamma}_{13}\tilde{\Psi}_L = \boldsymbol{\gamma}_{0123}\mathbf{S}_{(1)}, \qquad (4.23)$$

and so we have only $10 = \binom{4+1}{2}$ independent densities without derivative. The chiral wave Ψ_L is associated to a conservative current, so we have

$$\partial_\mu J^\mu = \boldsymbol{\partial} \cdot \mathbf{J} = 0. \qquad (4.24)$$

To obtain this relation, we may write the components of the current **J**. Using (4.3) and (4.7) we obtain

$$\Psi_L =$$
$$a_1 + a_2\gamma_{23} + a_3\gamma_{13} + a_4\gamma_{21} - a_4\gamma_{0123} + a_3\gamma_{10} + a_2\gamma_{20} - a_1\gamma_{30} . \quad (4.25)$$

Next we obtain

$$J^0 = 2(a_1{}^2 + a_2{}^2 + a_3{}^2 + a_4{}^2) , \quad (4.26)$$
$$J^1 = 4(a_1a_3 + a_2a_4) , \quad (4.27)$$
$$J^2 = 4(a_1a_2 - a_3a_4) , \quad (4.28)$$
$$J^3 = 2(-a_1{}^2 + a_2{}^2 + a_3{}^2 - a_4{}^2) , \quad (4.29)$$

and (4.8) is equivalent to the system

$$\begin{aligned}
\partial_0 a_1 + \partial_1 a_3 + \partial_2 a_2 - \partial_3 a_1 &= ma_2 , \\
\partial_0 a_2 + \partial_1 a_4 + \partial_2 a_1 + \partial_3 a_2 &= -ma_1 , \\
\partial_0 a_3 + \partial_1 a_1 - \partial_2 a_4 + \partial_3 a_3 &= -ma_4 , \\
\partial_0 a_4 + \partial_1 a_2 - \partial_2 a_3 - \partial_3 a_4 &= ma_3 .
\end{aligned} \quad (4.30)$$

The calculation, with the preceding equations, of $\partial_0 J^0$ gives (4.24). **J** is not only conservative, but also isotropic:

$$\mathbf{J} \cdot \mathbf{J} = 0 . \quad (4.31)$$

This result is well known with the Weyl spinors. Chirality, conservative and isotropic current for the probability are in complete agreement with a mass term.

28.5   Chiral equation with charge term

When you come from the Dirac frame to the spacetime algebra, a question inevitably arises: Why does Ψ have only an even value? A preceding study [21] induces the possibility of having a wave equation for a general Ψ term, not only even. For Ψ_L, the restriction to the even subalgebra makes impossible the construction of an equation with charge term, because no term with square -1 commutes with γ_{03} and γ_2. But if we do not restrict Ψ_L to have only even values, we can construct an electric gauge invariance and a charge term. The wave reads

$$\Psi_L = \Psi \tfrac{1}{2}(1 - \gamma_{30}) , \qquad \Psi_L = \Psi_1 + \Psi_2\gamma_2 , \quad (5.1)$$
$$\begin{aligned}
\Psi_1 = a_1 &+ a_2\gamma_{23} + a_3\gamma_{13} + a_4\gamma_{21} \\
&- a_4\gamma_{0123} + a_3\gamma_{10} + a_2\gamma_{20} - a_1\gamma_{30} ,
\end{aligned} \quad (5.2)$$
$$\begin{aligned}
\Psi_2 = a_5 &+ a_6\gamma_{23} + a_7\gamma_{13} + a_8\gamma_{21} \\
&- a_8\gamma_{0123} + a_7\gamma_{10} + a_6\gamma_{20} - a_5\gamma_{30} .
\end{aligned} \quad (5.3)$$

The chiral wave with mass and charge term for an electron is

$$\partial\Psi_L + q\mathbf{A}\Psi_L\gamma_2 = m\Psi_L\gamma_2. \tag{5.4}$$

Relativistic invariance is always (4.5–4.6). This equation is gauge invariant under

$$\Psi_L \mapsto \Psi'_L = \Psi_L e^{a\gamma_2}, \qquad \mathbf{A} \mapsto \mathbf{A}' = \mathbf{A} - \tfrac{1}{q}\partial a \tag{5.5}$$

since we obtain

$$\partial\Psi'_L + q\mathbf{A}'\Psi'_L\gamma_2 = m\Psi'_L\gamma_2. \tag{5.6}$$

Equation (5.4) comes from the Lagrangian density

$$\mathcal{L} = <\partial\Psi_L\gamma_1\tilde{\Psi}_R + m\Psi_L\gamma_{12}\tilde{\Psi}_R + q\mathbf{A}\Psi_L\gamma_{12}\tilde{\Psi}_R > \tag{5.7}$$

which gives with (4.13–4.14) the equation (5.4) and

$$\partial\Psi_R + q\mathbf{A}\Psi_R\gamma_2 = m\Psi_R\gamma_2. \tag{5.8}$$

So if charge conjugation is, in accordance with experimental results, the $L \rightleftarrows R$ exchange, we must see this equation as

$$\partial\Psi_R + (-q)(-\mathbf{A})\Psi_R\gamma_2 = m\Psi_R\gamma_2 \tag{5.9}$$

considering with Ziino [22, 23] that charge conjugation changes the sign of each charge, and must change $\mathbf{A}$ to $-\mathbf{A}$. The antiparticle's mass must be exactly the particle's mass.

At the second order, we use

$$\partial\partial = \Box, \qquad \partial\mathbf{A} = \partial_\mu A^\mu + \mathbf{F},$$
$$\partial(\mathbf{A}\Psi) = (\partial\mathbf{A})\Psi + 2A^\mu\partial_\mu\Psi - \mathbf{A}\partial\Psi \tag{5.10}$$

where $\mathbf{F}$ is the electromagnetic bivector field, and we obtain

$$(\Box + m^2 - q^2\mathbf{A}^2)\Psi_L = q(\mathbf{F} + \partial_\mu A^\mu + 2A^\mu\partial_\mu)\Psi_L\gamma_2. \tag{5.11}$$

This equation seems very close to the second-order equation coming from the Dirac equation (4.4). For the tensors, we also note a high similarity: For the tensors without derivative, we obtain exactly the same number of scalars, vectors, bivectors, giving 16 densities invariant under the electric gauge, and 20 densities varying with the electric gauge. The gauge invariant tensors are

$$\Psi_L\gamma_0\tilde{\Psi}_L = \Omega_1 + \mathbf{J} + \Omega_2\gamma_{0123}, \tag{5.12}$$

$$\Psi_L\gamma_{02}\tilde{\Psi}_L = \mathbf{B} + \mathbf{K}\gamma_{0123}. \tag{5.13}$$

$\mathbf{J}$ is the conservative current density of probability. Ω_1 is scalar, and Ω_2 a pseudo-scalar. $\mathbf{K}$ is a spacetime vector, $\mathbf{B}$ a spacetime bivector. To build the 20 densities rotating under an electric gauge transformation we use grade involution $\widehat{\Psi}_L = \Psi_1 - \Psi_2 \gamma_2$ and conjugation $\overline{\Psi}_L = \widehat{\widetilde{\Psi}}_L$ and we obtain the tensors

$$\Psi_L \gamma_0 \overline{\Psi}_L = \mathbf{v}_1 + \mathbf{b}_1, \qquad \Psi_L \gamma_{02} \overline{\Psi}_L = \mathbf{v}_2 + \mathbf{b}_2, \qquad (5.14)$$

$\mathbf{v}_1$ and $\mathbf{v}_2$ are spacetime vectors, $\mathbf{b}_1$ and $\mathbf{b}_2$ are spacetime bivectors and so we obtain 20 tensorial densities. We can remark that from a Ψ with 2^n real components we obtain $2^{n-1}(2^n + 1) = \binom{2^n+1}{2}$ tensorial densities and 2^{2n-2} are gauge invariant under the electric gauge transformation.

28.6   Solution of the chiral wave for the H atom

Now we solve (5.4) in the case of the hydrogen atom, and prove that the chiral left wave Ψ_L gives the same result as the usual Dirac equation. We use the spherical coordinates

$$x^1 = r \sin\theta \cos\varphi, \quad x^2 = r \sin\theta \sin\varphi, \quad x^3 = r \cos\theta, \qquad (6.1)$$
$$\vec{\sigma}_k = \gamma_{k0}, \quad i_k = \gamma_{0123}\vec{\sigma}_k, \qquad (6.2)$$
$$\vec{r} = x^1 \vec{\sigma}_1 + x^2 \vec{\sigma}_2 + x^3 \vec{\sigma}_3. \qquad (6.3)$$

Then we set, following H. Krüger [24]

$$S = e^{-\frac{\varphi}{2} i_3} e^{-\frac{\theta}{2} i_2}, \qquad (6.4)$$
$$\vec{\partial} = \vec{\sigma}_1 \partial_1 + \vec{\sigma}_2 \partial_2 + \vec{\sigma}_3 \partial_3, \qquad (6.5)$$
$$\Omega = r^{-1}(\sin\theta)^{-\frac{1}{2}} S, \qquad (6.6)$$
$$\vec{\partial}' = \vec{\sigma}_3 \partial_r + \frac{1}{r}\vec{\sigma}_1 \partial_\theta + \frac{1}{r\sin\theta}\vec{\sigma}_2 \partial_\varphi. \qquad (6.7)$$

Then we obtain

$$\vec{\partial} = \Omega \vec{\partial}' \Omega^{-1}. \qquad (6.8)$$

With the hydrogen atom we have

$$q\mathbf{A} = qA^0\gamma_0 = -\frac{\alpha}{r}\gamma_0 \qquad (6.9)$$

where α is the fine structure constant. Multiplying (5.4) by γ_0 from the left, we obtain

$$(\partial_0 + \vec{\partial})\Psi_L - \frac{\alpha}{r}\Psi_L = m\gamma_0\Psi_L\gamma_2. \qquad (6.10)$$

Multiplying by Ω^{-1} from the left, we obtain

$$(\partial_0 + \vec{\partial}\,')(\Omega^{-1}\Psi_L) - \frac{\alpha}{r}\Omega^{-1}\Psi_L = m\gamma_0\Omega^{-1}\Psi_L\gamma_2 \,. \qquad (6.11)$$

Separation of the x^0 and φ variables from the r and θ variables introduces three real constants E, δ and λ, and a ϕ function of r and θ

$$\Omega^{-1}\Psi_L = \phi e^{(\lambda\varphi+\delta-Ex^0)\gamma_2} \,. \qquad (6.12)$$

This way we obtain

$$(E + \frac{\alpha}{r})\phi + \vec{\sigma}_3\partial_r\phi\gamma_2 + \frac{1}{r}\vec{\sigma}_1\partial_\theta\phi\gamma_2 - \frac{\lambda}{r\sin\theta}\vec{\sigma}_2\phi + m\gamma_0\phi = 0 \,. \qquad (6.13)$$

Now we separate the even and odd parts of ϕ :

$$\phi = \phi_1 + \phi_2\gamma_2 \,, \qquad (6.14)$$

$$\phi_1 = b_1 + b_2\gamma_{23} + b_3\gamma_{13} + b_4\gamma_{21} - b_4\gamma_{0123}$$
$$+ b_3\gamma_{10} + b_2\gamma_{20} - b_1\gamma_{30} \,, \qquad (6.15)$$

$$\phi_2 = b_5 + b_6\gamma_{23} + b_7\gamma_{13} + b_8\gamma_{21} - b_8\gamma_{0123}$$
$$+ b_7\gamma_{10} + b_6\gamma_{20} - b_5\gamma_{30} \,, \qquad (6.16)$$

and obtain the system

$$(E + \frac{\alpha}{r})\phi_1 - (\vec{\sigma}_3\partial_r + \frac{1}{r}\vec{\sigma}_1\partial_\theta)\phi_2 - \frac{\lambda}{r\sin\theta}\vec{\sigma}_2\phi_1 + m\gamma_0\phi_2\gamma_2 = 0 \,, \qquad (6.17)$$

$$(E + \frac{\alpha}{r})\phi_2 + (\vec{\sigma}_3\partial_r + \frac{1}{r}\vec{\sigma}_1\partial_\theta)\phi_1 - \frac{\lambda}{r\sin\theta}\vec{\sigma}_2\phi_2 - m\gamma_0\phi_1\gamma_2 = 0 \,. \qquad (6.18)$$

This system is equivalent to

$$(E + \frac{\alpha}{r})b_1 + \partial_r b_5 - \frac{1}{r}\partial_\theta b_7 - \frac{\lambda}{r\sin\theta}b_2 + mb_6 = 0 \,, \qquad (6.19)$$

$$(E + \frac{\alpha}{r})b_2 - \partial_r b_6 - \frac{1}{r}\partial_\theta b_8 - \frac{\lambda}{r\sin\theta}b_1 - mb_5 = 0 \,, \qquad (6.20)$$

$$(E + \frac{\alpha}{r})b_3 - \partial_r b_7 - \frac{1}{r}\partial_\theta b_5 + \frac{\lambda}{r\sin\theta}b_4 - mb_8 = 0 \,, \qquad (6.21)$$

$$(E + \frac{\alpha}{r})b_4 + \partial_r b_8 - \frac{1}{r}\partial_\theta b_6 + \frac{\lambda}{r\sin\theta}b_3 + mb_7 = 0 \,, \qquad (6.22)$$

$$(E + \frac{\alpha}{r})b_5 - \partial_r b_1 + \frac{1}{r}\partial_\theta b_3 - \frac{\lambda}{r\sin\theta}b_6 - mb_2 = 0 \,, \qquad (6.23)$$

$$(E + \frac{\alpha}{r})b_6 + \partial_r b_2 + \frac{1}{r}\partial_\theta b_4 - \frac{\lambda}{r\sin\theta}b_5 + mb_1 = 0 \,, \qquad (6.24)$$

$$(E + \frac{\alpha}{r})b_7 - \partial_r b_3 + \frac{1}{r}\partial_\theta b_1 + \frac{\lambda}{r\sin\theta}b_8 + mb_4 = 0 \,, \qquad (6.25)$$

$$(E + \frac{\alpha}{r})b_8 - \partial_r b_4 + \frac{1}{r}\partial_\theta b_2 + \frac{\lambda}{r\sin\theta}b_7 - mb_3 = 0 \,. \qquad (6.26)$$

To separate the variables r and θ we let

$$b_1 = f_1 A, \quad b_3 = -f_3 B, \quad b_5 = f_4 A, \quad b_7 = f_2 B,$$
$$b_2 = f_2 A, \quad b_4 = -f_4 B, \quad b_6 = f_3 A, \quad b_8 = f_1 B, \tag{6.27}$$

where f_1, f_2, f_3, f_4 are real functions of r and A, B are real functions of θ, and we introduce a constant κ such that

$$A' + \frac{\lambda}{\sin \theta} B = \kappa B, \qquad B' + \frac{\lambda}{\sin \theta} A = -\kappa A. \tag{6.28}$$

System (6.19–6.26) is then equivalent to the radial system

$$(E + \frac{\alpha}{r})f_1 + f_4' + \frac{\kappa}{r}f_2 + mf_3 = 0, \tag{6.29}$$

$$(E + \frac{\alpha}{r})f_2 - f_3' + \frac{\kappa}{r}f_1 - mf_4 = 0, \tag{6.30}$$

$$(E + \frac{\alpha}{r})f_3 + f_2' + \frac{\kappa}{r}f_4 + mf_1 = 0, \tag{6.31}$$

$$(E + \frac{\alpha}{r})f_4 - f_1' + \frac{\kappa}{r}f_3 - mf_2 = 0. \tag{6.32}$$

After adding and subtracting we obtain:

$$0 = (E + \frac{\alpha}{r} - m)(f_2 + f_4) - (f_1 + f_3)' + \frac{\kappa}{r}(f_1 + f_3), \tag{6.33}$$

$$0 = (E + \frac{\alpha}{r} + m)(f_1 + f_3) + (f_2 + f_4)' + \frac{\kappa}{r}(f_2 + f_4), \tag{6.34}$$

$$0 = (E + \frac{\alpha}{r} - m)(f_1 - f_3) - (f_2 - f_4)' + \frac{\kappa}{r}(f_2 - f_4), \tag{6.35}$$

$$0 = (E + \frac{\alpha}{r} + m)(f_2 - f_4) + (f_1 - f_3)' + \frac{\kappa}{r}(f_1 - f_3). \tag{6.36}$$

So the radial system is made of two separate systems, similar to the radial system obtained by Darwin for the solutions of the Dirac equation:

$$(E + \frac{\alpha}{r} - m)f - g' + \frac{\kappa}{r}g = 0, \tag{6.37}$$

$$(E + \frac{\alpha}{r} + m)g + f' + \frac{\kappa}{r}f = 0. \tag{6.38}$$

Then we obtain, with two real constants a and b,

$$f_2 + f_4 = 2af, \quad f_1 + f_3 = 2ag, \quad f_2 - f_4 = 2bg, \quad f_1 - f_3 = 2bf, \tag{6.39}$$

which gives

$$f_1 = ag + bf, \quad f_2 = af + bg, \quad f_3 = ag - bf, \quad f_4 = af - bg. \tag{6.40}$$

To solve the angular system we let

$$A = V - U, \qquad B = U + V \tag{6.41}$$

and obtain

$$U' - \frac{\lambda}{\sin\theta}U = -\kappa V, \qquad V' + \frac{\lambda}{\sin\theta}V = \kappa U. \qquad (6.42)$$

If $\lambda > 0$ we let

$$U = \sin^\lambda\theta[\sin(\tfrac{1}{2}\theta)C' - (\kappa + \tfrac{1}{2} - \lambda)\cos(\tfrac{1}{2}\theta)C], \qquad (6.43)$$

$$V = \sin^\lambda\theta[\cos(\tfrac{1}{2}\theta)C' + (\kappa + \tfrac{1}{2} - \lambda)\sin(\tfrac{1}{2}\theta)C]. \qquad (6.44)$$

If $\lambda < 0$ we let

$$U = \sin^{-\lambda}\theta[\cos(\tfrac{1}{2}\theta)C' + (\kappa + \tfrac{1}{2} + \lambda)\sin(\tfrac{1}{2}\theta)C], \qquad (6.45)$$

$$V = \sin^{-\lambda}\theta[-\sin(\tfrac{1}{2}\theta)C' + (\kappa + \tfrac{1}{2} + \lambda)\cos(\tfrac{1}{2}\theta)C]. \qquad (6.46)$$

U and V are solutions of the angular system (6.42) if and only if C is a solution of

$$C'' + 2|\lambda|\cot(\theta)C' + [(\kappa + \tfrac{1}{2})^2 - \lambda^2]C = 0. \qquad (6.47)$$

It is the differential equation of the Gegenbauer polynomials. With

$$(a)_0 = 1, \qquad (a)_n = a(a+1)\cdots(a+n-1) \qquad (6.48)$$

we obtain:

$$\frac{C(\theta)}{C(0)} = \sum_{n=0}^{\infty} \frac{(|\lambda| - \kappa - \tfrac{1}{2})_n(|\lambda| + \kappa + \tfrac{1}{2})_n}{(\tfrac{1}{2} + |\lambda|)_n n!}\sin^{2n}\tfrac{1}{2}\theta. \qquad (6.49)$$

This sum is finite if κ is an integer and λ is a half-integer. The differential equation (6.47) is a second-order equation, so the general solution has two arbitrary constants. But there is only one polynomial solution, and the other independent solution gives nonintegrable functions, which is physically excluded. Then the polynomial solution has only one arbitrary constant, $C(0)$.

The radial system (6.37–6.38) is classical. The resolution introduces Laguerre's polynomials. The degree n' of these polynomials gives the last quantum number. To obtain the right number of different states, it is necessary to see that there is no integrable solution with $n' = 0$ and $\kappa > 0$. The condition of integrability of the solutions induces the Sommerfeld formula for the energy levels

$$E = \frac{m}{\sqrt{1 + \dfrac{\alpha^2}{(n' + \sqrt{\kappa^2 - \alpha^2})^2}}}. \qquad (6.50)$$

The principal quantum number of the Schrödinger theory is $n = n' + |\kappa|$.

The radial system (6.37–6.38) was derived from the Dirac equation using the kinetic momentum operators, and diagonalizing not only the J^2 and J_3 operators, but also a K ad-hoc operator giving the κ quantum number. We have seen that it is possible to use Krüger's method to separate the variables of the Dirac equation. Then the separation gives, not only the classical solutions, but other solutions which are linear combinations of the classical solutions with opposite values of κ. We must remark that with the chiral equation (5.4), the separation of variables induces directly (6.37–6.38), with only one value for κ. So the classical system (6.37–6.38) seems more convenient for (5.4) than for the Dirac equation itself!

With the chiral equation, we obtain exactly the same number of states, the same quantum numbers n', κ, λ, the same energy levels as with the classical frame. It is not necessary to make a big theory with Hermitian operators, unitary matrix representations, and spin operators to obtain good results for the hydrogen atom. All that is needed is to separate the variables correctly. Integrability of the solution gives for the angular functions κ and λ, for the radial functions the integer n'.

28.7   The real matrix formalism

Quantum theory uses a Hermitian scalar product, which is, in the case of the Dirac equation,

$$< \psi | \psi' > = \iiint \left(\sum_{i=1}^{4} \psi_i^* \psi_i' \right) dv . \tag{7.1}$$

To this Hermitian scalar product is associated the norm

$$\| \psi \|^2 = < \psi | \psi > = \iiint \left(\sum_{i=1}^{4} \psi_i^* \psi_i \right) dv = \iiint J^0 \, dv . \tag{7.2}$$

In the space algebra, or the spacetime algebra, this norm reads

$$\| \Psi \|^2 = \iiint J^0 \, dv = \iiint \left(\sum_{i=1}^{8} a_i^2 \right) dv . \tag{7.3}$$

To this norm we have no reason, within the frame of a real algebra, to associate a Hermitian scalar product. But it is natural to associate a Euclidean scalar product

$$\Psi \cdot \Psi' = \iiint \left(\sum_{i=1}^{8} a_i a_i' \right) dv . \tag{7.4}$$

To use this scalar product and the orthogonal transformation replacing the unitary transformation of the complex theory, we associate to each ψ of the classical formalism, to each Ψ of the even subalgebra of spacetime algebra, the real unicolumn matrix

$$\chi = \begin{pmatrix} a_1 \\ a_2 \\ \vdots \\ a_8 \end{pmatrix} \tag{7.5}$$

and the Euclidean scalar product reads

$$\chi \cdot \chi' = \Psi \cdot \Psi' = \iiint \chi^t \chi' \, dv . \tag{7.6}$$

The Dirac equation (1.1) or (2.11) or (4.4) is equivalent to the matrix equation

$$[\Gamma^\mu(\partial_\mu + qA_\mu P_3) + mP_3]\chi = 0 \tag{7.7}$$

where the Γ_μ and P_j matrices are

$$\Gamma^0 = \Gamma_0 = \begin{pmatrix} I_4 & 0 \\ 0 & -I_4 \end{pmatrix}, \qquad \Gamma^1 = -\Gamma_1 = \begin{pmatrix} 0 & \gamma_{013} \\ \gamma_{013} & 0 \end{pmatrix}, \tag{7.8}$$

$$\Gamma^2 = -\Gamma_2 = \begin{pmatrix} 0 & -\gamma_{03} \\ \gamma_{03} & 0 \end{pmatrix}, \qquad \Gamma^3 = -\Gamma_3 = \begin{pmatrix} 0 & -\gamma_{01} \\ \gamma_{01} & 0 \end{pmatrix}, \tag{7.9}$$

$$P_1 = \begin{pmatrix} \gamma_{013} & 0 \\ 0 & \gamma_{013} \end{pmatrix}, \quad P_2 = \begin{pmatrix} -\gamma_{05} & 0 \\ 0 & \gamma_{05} \end{pmatrix}, \quad P_3 = \begin{pmatrix} -\gamma_{135} & 0 \\ 0 & \gamma_{135} \end{pmatrix}. \tag{7.10}$$

We obtain

$$\Gamma^\mu\Gamma^\nu + \Gamma^\nu\Gamma^\mu = 2g^{\mu\nu}I_8 , \qquad \Gamma^\mu P_j = P_j\Gamma^\mu . \tag{7.11}$$

As with the classical formalism, it is possible to associate to (7.7) an adjoint equation by

$$\overline{\chi} = \chi^t\Gamma_0 , \tag{7.12}$$

$$\overline{\chi}[(\partial_\mu - qA_\mu P_3)\Gamma^\mu - mP_3] = 0 . \tag{7.13}$$

Multiplying (7.7) by $\overline{\chi}$ from the left, (7.13) by χ from the right and adding we obtain the conservative law of the J current

$$\partial_\mu J^\mu = 0 , \qquad J^\mu = \overline{\chi}\Gamma^\mu\chi . \tag{7.14}$$

The left and right parts of the wave are

$$\chi_L = \tfrac{1}{2}(I_8 - \Gamma^{0123}P_3)\chi , \qquad \chi_R = \tfrac{1}{2}(I_8 + \Gamma^{0123}P_3)\chi . \tag{7.15}$$

With the frame of complex matrices, Dirac matrices γ^μ may be replaced by any γ'^μ such that

$$\gamma'^\mu = S\gamma^\mu S^{-1} , \quad S^{-1} = S^\dagger , \quad \psi' = S\psi . \tag{7.16}$$

They are defined only up a unitary matrix. If $\partial_\mu S = 0$, then from (1.1) we obtain

$$[\gamma'^\mu(\partial_\mu + iqA_\mu) + im]\,\psi' = 0 \tag{7.17}$$

and the tensorial densities have the same form as (2.13–2.17). It is possible [25] to gauge (7.16). Gauge groups are essential for modern physics: electro-weak interactions are described with a $U(1) \times SU(2)$ local gauge group [26] and strong interactions are described with $SU(3)$ local gauge group. The gauge group coming from (7.16) is $U(2)$ and is not able to include the $U(1) \times SU(2) \times SU(3)$ of the standard model.

If we use the real formalism, the Dirac equation (7.7) is invariant under

$$\chi \mapsto \chi' = A\chi \quad \text{where} \quad A^{-1} = A^t, \tag{7.18}$$

$$\Gamma^\mu \mapsto \Gamma'^\mu = A\Gamma^\mu A^{-1}, \quad P_3 \mapsto P_3' = AP_3A^{-1}. \tag{7.19}$$

Orthogonal matrices A generate $SO(8)$ orthogonal group, much larger than the $U(2)$ group of the complex frame. Complex formalism and real formalism are not equivalent. For instance the vector space of all complex 4×4 matrices is 16-dimensional above $\mathbb{C}$, and 32-dimensional above $\mathbb{R}$, but the vector space $M_8(\mathbb{R})$ of all real 8×8 real matrices is 64-dimensional above $\mathbb{R}$. A basis of $M_8(\mathbb{R})$ is made of the 16 matrices I_8, Γ_μ, $\Gamma_{\mu\nu}$, $\Gamma_{\mu\nu\rho}$, Γ_{0123}, and the 48 matrices $\Gamma P = P\Gamma$, where Γ is one of the 16 preceding matrices and P is P_1, P_2, P_3. These 64 matrices may be separated into two parts, 36 verify $A^2 = I_8$, $A^t = A$ and give the 36 tensorial densities $\chi^t A\chi$, 28 verify $A^2 = -I_8$, $A^t = -A$ and generate the Lie algebra of $SO(8)$.

Similarly, to any Ψ element of the full spacetime algebra, which reads

$$\Psi = a_1 + a_2\gamma_{23} + a_3\gamma_{13} + a_4\gamma_{21} + a_5\gamma_{0123} + a_6\gamma_{10}$$
$$+ a_7\gamma_{20} + a_8\gamma_{30} + a_9\gamma_0 + a_{10}\gamma_{023} + a_{11}\gamma_{013}$$
$$+ a_{12}\gamma_{021} + a_{13}\gamma_{132} + a_{14}\gamma_1 + a_{15}\gamma_2 + a_{16}\gamma_3 \tag{7.20}$$

we associate the unicolumn matrix

$$X = \begin{pmatrix} a_1 \\ a_2 \\ \vdots \\ a_{16} \end{pmatrix}. \tag{7.21}$$

Scalar product and norm read now

$$X \cdot X' = \iiint X^t X' \, dv, \qquad \|X\|^2 = \iiint X^t X \, dv. \tag{7.22}$$

This scalar product is invariant under the orthogonal transformations

$$X \mapsto X' = MX, \qquad M^{-1} = M^t, \tag{7.23}$$

where M is a 16×16 matrix. The vector space $M_{16}(\mathbb{R})$ includes 16 L matrices which are the matrices of the left multiplication $\Psi \mapsto \gamma\Psi$, and 16 R matrices which are the matrices of the right multiplication $\Psi \mapsto \Psi\gamma$. We note respectively $I_8, L_\mu, L_{\mu\nu}, L_{\mu\nu\rho}, L_{0123}$ the matrices of the left multiplication by $1, \gamma_\mu, \gamma_{\mu\nu}, \gamma_{\mu\nu\rho}, \gamma_{0123}$, and $I_8, R_\mu, R_{\mu\nu}, R_{\mu\nu\rho}, R_{0123}$ the matrices of the right multiplication by $1, \gamma_\mu, \gamma_{\mu\nu}, \gamma_{\mu\nu\rho}, \gamma_{0123}$. We must notice that, changing a right multiplication to a left operation, we obtain $R_{\mu\nu} = R_\nu R_\mu$. The 256 matrices $M = LR = RL$ form a basis of $M_{16}(\mathbb{R})$. As with $M_8(\mathbb{R})$, these 256 matrices split into two parts. 136 matrices verify $M^2 = I_{16}, M^t = M$ and give the 136 tensorial densities $X^t M X$. 120 matrices verify $M^2 = -I_{16}, M^t = -M$ and form a basis of the Lie algebra of $SO(16)$. Each matrix may be computed from

$$L_\mu = \begin{pmatrix} 0 & \Gamma_\mu \\ \Gamma_\mu & 0 \end{pmatrix}, \qquad \mu = 0,1,2,3, \tag{7.24}$$

$$R_0 = \begin{pmatrix} 0 & I_8 \\ I_8 & 0 \end{pmatrix}, \qquad R_j = \begin{pmatrix} 0 & \Gamma_{0123}P_j \\ -\Gamma_{0123}P_j & 0 \end{pmatrix}, \qquad j = 1,2,3. \tag{7.25}$$

The left and right parts of the wave read now

$$X_L = \tfrac{1}{2}(I_{16} - R_{03})X, \qquad X_R = \tfrac{1}{2}(I_{16} + R_{03})X. \tag{7.26}$$

The chiral wave equation (5.4) for the electron reads

$$L^\mu(\partial_\mu + qA_\mu R_2)X_L = mR_2 X_L. \tag{7.27}$$

We set $\overline{X}_L = X^t{}_L L_0$ and obtain the adjoint equation

$$\overline{X}_L(\partial_\mu - qA_\mu R_2)L^\mu = -m\overline{X}_L R_2. \tag{7.28}$$

Multiplying (7.27) by $\overline{X}_L$ from the right and (7.28) by X_L from the right and adding, we obtain the conservative current

$$\partial_\mu J^\mu = 0, \qquad J^\mu = \overline{X}_L L^\mu X_L. \tag{7.29}$$

Equation (7.27) is invariant under the orthogonal transformation

$$X_L \mapsto X'_L = MX_L, \qquad \text{where} \quad M^{-1} = M^t, \tag{7.30}$$

$$X_R \mapsto X_R = MX_R, \qquad L^\mu \mapsto L^{\mu'} = ML^\mu M^{-1}. \tag{7.31}$$

The G group generated by (7.30–7.31) is a subgroup of $SO(16)$ isomorphic to $SO(8)$. The Lie algebra of G is generated by the 28 matrices

$$R_1p_3, L_0R_1p_3, L_{01}R_1p_3, L_{02}R_1p_3, L_{03}R_1p_3, L_{123}R_1p_3, R_2p_3, L_0R_2p_3,$$
$$L_{01}R_2p_3, L_{02}R_2p_3, L_{03}R_2p_3, L_{123}R_2p_3, R_3p_3, L_0R_3p_3, L_{01}R_3p_3, L_{02}R_3p_3,$$
$$L_{03}R_3p_3, L_{123}R_3p_3, L_1p_3, L_2p_3, L_3p_3, L_{12}p_3, L_{23}p_3, L_{31}p_3, L_{012}p_3,$$
$$L_{023}p_3, L_{031}p_3, L_{0123}p_3,$$

where

$$p_3 = \tfrac{1}{2}(I_{16} - R_{03}), \qquad q_3 = \tfrac{1}{2}(I_{16} + R_{03}). \qquad (7.32)$$

Let $N = Mp_3$ be one of the 28 generators. We obtain

$$e^{aN} = I_{16} + \sum_{i=1}^{\infty} \frac{(aMp_3)^n}{n!}$$

$$= p_3 + q_3 + \left(\sum_{i=1}^{\infty} \frac{a^n M^n}{n!}\right)p_3 = q_3 + e^{aM}p_3, \quad (7.33)$$

$$p_3 X_R = 0, \quad q_3 X_R = X_R, \quad q_3 X_L = 0, \quad p_3 X_L = X_L, \qquad (7.34)$$

$$e^{aN} X_R = X_R, \qquad e^{aN} X_L = e^{aM} X_L. \qquad (7.35)$$

With G, the right part of the wave, X_R, is invariant, and G acts only on the left part of the wave, which is transformed into another left term. That is similar to the Weinberg–Salam model for the electro-weak interactions. And since the electric gauge invariance is (5.5), the generator of the electric gauge is $R_2 p_3$, an element of G. So it must be possible to extend the electric gauge invariance so as to integrate the $U(1) \times SU(2)$ model in the frame of (7.30–7.31).

28.8   Concluding remarks

Gauge groups coming naturally from the real formalisms are $SO(8)$ and $SO(16)$. The Lie algebra of $SO(8)$ accommodates the Lie algebra of $U(1) \times SU(2)$ and may be able to describe electro-weak interactions acting only on left waves. The Lie algebra of $SO(16)$ is great enough to accommodate $U(1) \times SU(2) \times SU(3)$.

To each wave equation (5.4) we can associate two different equations, permuting circularly the index $1, 2, 3$. We can consider

$$\Psi_L' = \Psi \tfrac{1}{2}(1 - \gamma_{10}), \qquad (8.1)$$

$$\partial \Psi_L' + q\mathbf{A}\Psi_L'\gamma_3 = m\Psi_L'\gamma_3, \qquad (8.2)$$

and

$$\Psi_L'' = \Psi \tfrac{1}{2}(1 - \gamma_{20}), \qquad (8.3)$$

$$\partial \Psi_L'' + q\mathbf{A}\Psi_L''\gamma_1 = m\Psi_L''\gamma_1. \qquad (8.4)$$

So we have naturally three different kinds of chiral waves, and three different equations, and this fact may be the reason for the existence of three generations of fundamental fermions. These three different generations must be separately treated in the electro-weak interactions, because the gauge groups issued from the three different equations are isomorphic, but not identical.

REFERENCES

[1] P.A.M. Dirac, *Proc. Roy. Soc.* (London) **117**, 610-624, 1928.

[2] W.E. Baylis, Eigenspinors and electron spin, in *The Theory of the Electron*, Advances in Applied Clifford Algebras **7** (S), 1997.

[3] C. Daviau, Sur l'équation de Dirac dans l'algèbre de Pauli, *Ann. Fond. Louis de Broglie* **22**, $n°$ 1, 87–103, 1997.

[4] C. Daviau, Dirac equation in the space Clifford algebra, in *Clifford Algebras and their Application in Mathematical Physics*, Volker Dietrich, Klaus Habetha and Gerhard Jank, eds., Aachen 1996, Kluwer, Dordrecht, 1998, pp. 67–87.

[5] C. Daviau, Sur les tenseurs de la théorie de Dirac en algèbre d'espace, *Ann. Fond. Louis de Broglie* **23**, $n°$ 1, 27–37, 1998.

[6] C. Daviau, Application à la théorie de la lumière de Louis de Broglie d'une réécriture de l'équation de Dirac, *Ann. Fond. Louis de Broglie* **23**, $n°$ 3–4, 121–127, 1998.

[7] C. Daviau, Equations de Dirac et fermions fondamentaux, *Ann. Fond. Louis de Broglie* **24**, $n°$ 1–4, 175–194, 1999, and **25**, $n°$ 1, 93–106, 2000.

[8] L. de Broglie, *L'électron magnétique*, Hermann, Paris, 1934, 138.

[9] C. Daviau, Vers une mécanique quantique sans nombre complexe, *Ann. Fond. Louis de Broglie* **26**, $n°$ 1–3, 149–171, 2001.

[10] D. Hestenes, *Spacetime Algebra*, Gordon & Breach, New York 1966, 1987, 1992.

[11] D. Hestenes, Real spinor fields, *J. Math. Phys.* **8**, $n°$ 4, 798–808, 1967.

[12] D. Hestenes, Local observables in the Dirac theory, *J. Math. Phys.* **14**, $n°$ 7, 893–905, 1973.

[13] D. Hestenes, Proper particle mechanics, *J. Math. Phys.* **15**, $n°$ 10, 1768–1777, 1974.

[14] D. Hestenes, Proper dynamics of a rigid point particle, *J. Math. Phys.* **15**, $n°$ 10, 1778–1786, 1974.

[15] D. Hestenes, Observables, operators, and complex numbers in the Dirac theory, *J. Math. Phys.* **16**, $n°$ 3, 556–572, 1975.

[16] D. Hestenes, A unified language for Mathematics and Physics & Clifford algebra and the interpretation of quantum mechanics in *Clifford Algebras and Their Applications in Mathematics and Physics*, J.S.R. Chisholm & A.K. Common, eds., Reidel, Dordrecht, 1986, pp. 1–23 and pp. 321–346.

[17] R. Boudet, La géométrie des particules du groupe SU(2) et l'algèbre réelle d'espace-temps, *Ann. Fond. Louis de Broglie* **13**, $n°$ 1, 105–137, 1988.

[18] R. Boudet, *Le corpuscule de Louis de Broglie et la géométrie de l'espace-temps*, Courants, Amers, Ecueils en microphysique, Fond. Louis de Broglie, 1993, pp. 77–87.

[19] R. Boudet, The Takabayasi moving frame, from a potential to the Z boson, in *The Present Status of the Quantum Theory of the Light*, S. Jeffers and J.P. Vigier, eds., Kluwer, Dordrecht, 1995, pp. 1–11.

[20] A. Lasenby, C. Doran, S. Gull, A Multivector Derivative Approach to Lagrangian Field Theory, *Found. of Phys.* **23**, $n°$ 10, 1295–1327, 1993.

[21] C. Daviau, Sur une équation d'onde relativiste et ses solutions à symétrie interne, *Ann. Fond. Louis de Broglie* **26**, $n°$ 4, 699–724, 2001.

[22] G. Ziino, Massive chiral fermions: a natural account of chiral phenomenology in the framework of Dirac's fermion theory, *Ann. Fond. L. de Broglie* **14**, $n°$ 4, 427–438, 1989.

[23] G. Ziino, On the true meaning of "maximal parity violation": ordinary mirror symmetry regained from "CP symmetry", *Ann. Fond. L. de Broglie* **16**, $n°$ 3, 343–353, 1991.

[24] H. Krüger, New solutions of the Dirac equation for central fields in *The Electron*, D. Hestenes and A. Weingartshofer, eds., Kluwer, Dordrecht, 1991, pp. 49–81.

[25] R.S. Farwell and J.S.R. Chisholm, Unified spin gauge theory models, in *Clifford Algebras and Their Applications in Mathematics and Physics*, J.S.R. Chisholm & AK Common, eds., Reidel, Dordrecht, 1986, pp. 363–370.

[26] S. Weinberg, A model of leptons, *Phys. Rev. Lett.* **19**, 1264–1290, 1967.

Claude Daviau
La Lande, 44522 Pouillé-les-coteaux, France.
E-mail: daviau.claude@wanadoo.fr

Fondation Louis de Broglie
23 rue Marsoulan, 75012 Paris, France

Submitted January 2002; Revised July 4, 2003.

29

Using Octonions to Describe Fundamental Particles

Tevian Dray and Corinne A. Manogue

ABSTRACT In previous work, the standard 4-dimensional Dirac equation was rewritten in terms of quaternionic 2-component spinors, leading to a formalism which unifies the treatment of massive and massless particles, and which describes the correct particle spectrum to be a generation of leptons, with the correct number of spin/helicity states. Furthermore, precisely three such generations naturally combine into an octonionic description of the 10-dimensional massless Dirac equation. We extend this formalism to 3-component octonionic "spinors", which may lead to a description of fundamental particles in terms of the exceptional Jordan algebra, consisting of 3×3 octonionic Hermitian matrices.

Keywords: Dirac equation, dimensional reduction, quaternions, octonions

29.1 Introduction

We summarize here our previous work, especially [1, 2], which uses the octonions to describe fundamental particles. That work introduced a 2-component octonionic spinor formalism to describe leptons, which we generalize here to a 3-component formalism, along the lines of [3, 4]. We conclude by speculating briefly on whether this formalism might lead to an octonionic description of quarks.

Others have attempted similar descriptions of fundamental particles, using Clifford algebras and related techniques, including the octonions. (See for instance [5–7] and the references cited therein.) We wish to call the reader's attention, in particular, to the work of Trayling and Baylis [8, 9], a summary of which can be found elsewhere in this volume.

29.2 Octonionic formalism

29.2.1 The octonions

The *octonions* $\mathbb{O}$ are the nonassociative, noncommutative, normed division algebra over the reals. The octonions are spanned by the identity element 1 and seven imaginary units, which we label as $\{i, j, k, k\ell, j\ell, i\ell, \ell\}$. Each imaginary

---

AMS Subject Classification: 81R99, 17A35, 17C90, 15A33, 15A18.

unit squares to -1,

$$i^2 = j^2 = k^2 = \cdots = \ell^2 = -1 \qquad (2.1)$$

and the full multiplication table can be conveniently encoded in the 7-point projective plane, as shown in Figure 29.1; each line is to be thought of as a circle.

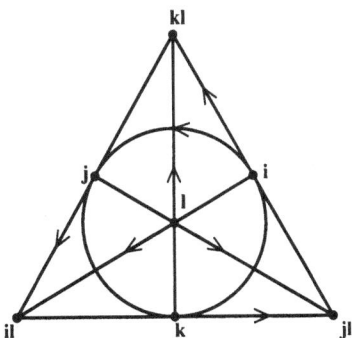

FIGURE 29.1. The representation of the octonionic multiplication table using the 7-point projective plane. Each of the seven oriented lines gives a quaternionic triple.

The octonionic units can be grouped into (the imaginary parts of) quaternionic subalgebras in seven different ways, corresponding to the seven lines in the figure; these will be referred to as quaternionic triples. Within each triple, the arrows give the orientation, so that, e.g., $ij = k = -ji$. Any three imaginary basis units which do not lie in a such a triple anti-associate. Note that any two octonions automatically lie in (at least one) quaternionic triple, so that expressions containing only two independent imaginary octonionic directions do associate. *Octonionic conjugation* is given by reversing the sign of the imaginary basis units, and the norm is just

$$|p| = \sqrt{p\bar{p}} \qquad (2.2)$$

which satisfies the defining property of a normed division algebra, namely

$$|pq| = |p||q| \, . \qquad (2.3)$$

29.2.2   Vectors and spinors

We follow [10, 11] in representing real $(9 + 1)$-dimensional Minkowski space in terms of 2×2 Hermitian octonionic matrices. In analogy with the complex case, a vector field q^μ with $\mu = 0, \ldots, 9$ can be thought of under this representation as a matrix

$$Q = \begin{pmatrix} q^+ & \bar{q} \\ q & q^- \end{pmatrix} \qquad (2.4)$$

where $q^\pm = q^0 \pm q^9 \in \mathbb{R}$ are the components of q^μ in two null directions and $q = q^1 + q^2 i + \cdots + q^8 \ell \in \mathbb{O}$ is an octonion representing the transverse spatial coordinates. Following [12], we define

$$\widetilde{Q} = Q - \mathrm{tr}(Q)I \, . \qquad (2.5)$$

Furthermore, since Q satisfies its characteristic polynomial, we have

$$-Q\widetilde{Q} = -\widetilde{Q}Q = -Q^2 + \mathrm{tr}(Q)Q = \det(Q)I = g_{\mu\nu}q^\mu q^\nu I \qquad (2.6)$$

where $g_{\mu\nu}$ is the Minkowski metric (with signature $(+, -, \ldots, -)$). We can therefore identify the tilde operation with the metric dual, so that $-\widetilde{Q}$ represents the covariant vector field q_μ.

This can be thought of (up to associativity issues) as a Weyl representation of the underlying Clifford algebra $C\ell(9, 1)$ in terms of 4×4 gamma matrices of the form

$$q^\mu \gamma_\mu = \begin{pmatrix} 0 & Q \\ -\widetilde{Q} & 0 \end{pmatrix} \qquad (2.7)$$

which are now octonionic. It is readily checked that

$$\gamma_\mu \gamma_\nu + \gamma_\nu \gamma_\mu = 2 g_{\mu\nu} \qquad (2.8)$$

as desired.

In this language, a Majorana spinor $\Psi = \begin{pmatrix} \psi \\ \chi \end{pmatrix}$ is a 4-component octonionic column, whose chiral projections are the Majorana–Weyl spinors $\begin{pmatrix} \psi \\ 0 \end{pmatrix}$ and $\begin{pmatrix} 0 \\ \chi \end{pmatrix}$, which can be identified with the 2-component octonionic columns ψ and χ, which in turn can be thought of as generalized Penrose spinors. Writing

$$\gamma_\mu = \begin{pmatrix} 0 & \sigma_\mu \\ -\widetilde{\sigma}_\mu & 0 \end{pmatrix} \qquad (2.9)$$

or equivalently

$$Q = q^\mu \sigma_\mu = q_\mu \sigma^\mu \qquad (2.10)$$

defines the *octonionic Pauli matrices* σ_μ. The matrices σ_a, with $a = 1, \ldots, 9$, are the natural generalization of the ordinary Pauli matrices to the octonions, and we define $\sigma_0 = I$. These matrices satisfy

$$\gamma_0 \gamma_\mu = \begin{pmatrix} -\widetilde{\sigma}_\mu & 0 \\ 0 & \sigma_\mu \end{pmatrix}. \qquad (2.11)$$

29.2.3   Octonionic Dirac equation

The momentum-space massless Dirac equation (Weyl equation) in 10 dimensions can be written in the form

$$\gamma_0 \gamma_\mu p^\mu \Psi = 0 \qquad (2.12)$$

Choosing $\Psi = \begin{pmatrix} \psi \\ 0 \end{pmatrix}$ to be a Majorana–Weyl spinor, and using (2.11) and (2.10), (2.12) takes the form

$$\widetilde{P}\psi = 0 \qquad (2.13)$$

which is the octonionic Weyl equation. In matrix notation, it is straightforward to show that the momentum p^μ of a solution of the Weyl equation must be null: (2.13) says that the 2×2 Hermitian matrix P has 0 as one of its eigenvalues, [1] which forces $\det(P) = 0$, which is

$$\widetilde{P}P = 0 \tag{2.14}$$

which in turn is precisely the condition that p^μ be null.

Equations (2.14) and (2.13) are algebraically the same as the octonionic versions of two of the superstring equations of motion, as discussed in [12, 15, 16], and are also the octonionic superparticle equations [17]. As implied by those references, (2.14) implies the existence of a 2-component spinor θ such that

$$P = \pm\theta\theta^\dagger \tag{2.15}$$

where the sign corresponds to the time orientation of P, and the general solution of (2.13) is

$$\psi = \theta\xi \tag{2.16}$$

where $\xi \in \mathbb{O}$ is arbitrary. The components of θ lie in the complex subalgebra of $\mathbb{O}$ determined by P, so that (the components of) θ and ξ (and hence also P) belong to a quaternionic subalgebra of $\mathbb{O}$. Thus, for solutions (2.16), the Weyl equation (2.12) itself becomes quaternionic. Furthermore, it follows immediately from (2.16) that

$$\psi\psi^\dagger = \pm|\xi|^2 P. \tag{2.17}$$

29.3   Leptons

29.3.1   Dimensional reduction

The description in the preceding section of 10-dimensional Minkowski space in terms of Hermitian octonionic matrices is a direct generalization of the usual description of ordinary (4-dimensional) Minkowski space in terms of complex Hermitian matrices. As discussed in [1], if we fix a complex subalgebra $\mathbb{C} \subset \mathbb{O}$, then we single out a 4-dimensional Minkowski subspace of 10-dimensional Minkowski space. The projection of a 10-dimensional null vector onto this subspace is a causal 4-dimensional vector, which is null if and only if the original vector was already contained in the subspace, and timelike otherwise. The time orientation of the projected vector is the same as that of the original, and the induced mass is given by the norm of the remaining six components. Furthermore, the ordinary Lorentz group $SO(3,1)$ clearly sits inside the Lorentz group

---

[1]It is *not* true in general [13, 14] that the determinant of an $n \times n$ Hermitian octonionic matrix is the product of its (real) eigenvalues, unless $n = 2$; however, see also [4].

$SO(9, 1)$ via the identification of their double-covers, the spin groups $\mathrm{Spin}(d, 1)$, namely[2]

$$\mathrm{Spin}(3, 1) = SL(2, \mathbb{C}) \subset SL(2, \mathbb{O}) = \mathrm{Spin}(9, 1) \,. \tag{3.1}$$

Therefore, all it takes to break 10 spacetime dimensions to four is to choose a preferred octonionic unit to play the role of the complex unit. We choose ℓ rather than i to fill this role, preferring to save i, j, k for a (distinguished) quaternionic triple. The projection π from $\mathbb{O}$ to $\mathbb{C}$ is given by

$$\pi(q) = \tfrac{1}{2}(q + \ell q \bar{\ell}) \tag{3.2}$$

and we thus obtain a preferred $SL(2, \mathbb{C})$ subgroup of $SL(2, \mathbb{O})$, corresponding to the "physical" Lorentz group.

29.3.2 Spin

Since we now have a preferred 4-dimensional Lorentz group, we can use its rotation subgroup $SU(2) \subset SL(2, \mathbb{C})$ to define spin. However, care must be taken when constructing the Lie algebra $su(2)$, due to the lack of commutativity.

The action of $M \in SU(2)$ on a Hermitian matrix Q (thought of as a spacetime vector via (2.10)) is given by

$$Q \mapsto MQM^\dagger \,. \tag{3.3}$$

Under this action, we can identify the basis rotations as usual as

$$R_z = \begin{pmatrix} e^{\ell \frac{\phi}{2}} & 0 \\ 0 & e^{-\ell \frac{\phi}{2}} \end{pmatrix}, \qquad R_y = \begin{pmatrix} \cos \frac{\phi}{2} & \sin \frac{\phi}{2} \\ -\sin \frac{\phi}{2} & \cos \frac{\phi}{2} \end{pmatrix},$$

$$R_x = \begin{pmatrix} \cos \frac{\phi}{2} & \ell \sin \frac{\phi}{2} \\ \ell \sin \frac{\phi}{2} & \cos \frac{\phi}{2} \end{pmatrix} \tag{3.4}$$

corresponding to rotations by the angle ϕ about the z, y, and x axes, respectively.

The infinitesimal generators of the Lie algebra $su(2)$ are obtained by differentiating these group elements, via

$$L_a = \left. \frac{dR_a}{d\phi} \right|_{\phi=0} \tag{3.5}$$

where as before $a = x, y, z$. For reasons which will become apparent, we have *not* multiplied these generators by $-\ell$ to obtain Hermitian matrices. We have instead

$$2L_z = \begin{pmatrix} \ell & 0 \\ 0 & -\ell \end{pmatrix}, \quad 2L_y = \begin{pmatrix} 0 & 1 \\ -1 & 0 \end{pmatrix}, \quad 2L_x = \begin{pmatrix} 0 & \ell \\ \ell & 0 \end{pmatrix}, \tag{3.6}$$

---

[2]The last equality is more usually discussed at the Lie algebra level. Manogue and Schray [18] gave an explicit representation using this language of the *finite* Lorentz transformations in 10 spacetime dimensions. For further discussion of the notation $SL(2, \mathbb{O})$, see also [19].

which satisfy the commutation relations

$$[L_a, L_b] = \epsilon_{abc} L_c \tag{3.7}$$

where ϵ is completely antisymmetric and $\epsilon_{xyz} = 1$.

Spin eigenstates are usually obtained as eigenvectors of the Hermitian matrix $-\ell L_z$, with real eigenvalues. Here we must be careful to multiply by ℓ in the correct place. We define

$$\hat{L}_z \psi := -L_z \psi \ell \tag{3.8}$$

which is well defined by alternativity, so that

$$\hat{L}_z = -\ell_R \circ L_z \tag{3.9}$$

where the operator ℓ_R denotes right multiplication by ℓ and where $\circ$ denotes composition. The operators $\hat{L}_a$ are self-adjoint with respect to the inner product

$$\langle \psi, \chi \rangle = \pi(\psi^\dagger \chi) . \tag{3.10}$$

We therefore consider the eigenvalue problem

$$\hat{L}_z \psi = \psi \lambda \tag{3.11}$$

with $\lambda \in \mathbb{R}$. It is straightforward to show that the real eigenvalues are[3] $\lambda_\pm = \pm\frac{1}{2}$ as expected. However, the form of the eigenvectors is a bit more surprising:

$$\psi_+ = \begin{pmatrix} A \\ kD \end{pmatrix} \quad \text{and} \quad \psi_- = \begin{pmatrix} kB \\ C \end{pmatrix} \tag{3.12}$$

where $A, B, C, D \in \mathbb{C}$ are any elements of the preferred complex subalgebra, and k is any imaginary octonionic unit orthogonal to ℓ, so that k and ℓ anticommute. Thus, the components of spin eigenstates are contained in the quaternionic subalgebra $\mathbb{H} \subset \mathbb{O}$ which is generated by ℓ and k.

Therefore, if we wish to consider spin eigenstates, ℓ must be in the quaternionic subalgebra $\mathbb{H}$ defined by the solution. We can further assume without loss of generality that $\mathbb{H}$ takes the Cayley–Dickson form [22, 23]

$$\mathbb{H} = \mathbb{C} + \mathbb{C}k = (\mathbb{R} + \mathbb{R}\ell) + (\mathbb{R} + \mathbb{R}\ell)k . \tag{3.13}$$

Thus, the only possible nonzero components of p_μ are

$$p_t = p_0, \quad p_x = p_1, \quad p_k = p_4, \quad p_{k\ell} = p_5, \quad p_y = p_8, \quad \text{and} \quad p_z = p_9,$$

corresponding to the gamma matrices with components in $\mathbb{H}$. We can further assume (via a rotation in the $(k, k\ell)$-plane if necessary) that $p_5 = 0$, so that

$$P = \pi(P) + m\,\sigma^k \tag{3.14}$$

---

[3] Remarkably, there are also eigenvalues which are not real [20, 21].

where $\pi(P) = p_\alpha \sigma^\alpha \equiv \mathbf{p}$ with $\alpha = 0, 1, 8, 9$ (or equivalently $\alpha = t, x, y, z$) is complex, and corresponds to the 4-dimensional momentum of the particle, with squared mass

$$m^2 = p_\alpha p^\alpha = -\det(\pi(P)) . \tag{3.15}$$

Inserting (3.14) into (2.13), yields

$$(\widetilde{\mathbf{p}} + m\sigma_k)\psi = 0 \tag{3.16}$$

and we see that we have come full circle: Solutions of the *octonionic* Weyl equation (2.12) are described precisely by the *quaternionic* formalism of [2, 3], and the dimensional reduction scheme determines the mass term.

29.3.3 Particles

For each solution ψ of (2.16), the momentum is proportional to $\psi\psi^\dagger$ by (2.17). Up to an overall factor, we can therefore read off the components of the 4-dimensional momentum p_α directly from $\pi(\psi\psi^\dagger)$. We can use a Lorentz transformation to bring a massive particle to rest, or to orient the momentum of a massless particle to be in the z-direction.

If $m \neq 0$, we can distinguish particles from antiparticles by the sign of the term involving m, which is the coefficient of σ_k in P. Equivalently, we have the particle/antiparticle projections (at rest)

$$\Pi_\pm = \tfrac{1}{2}(\sigma_t \pm \sigma_k) . \tag{3.17}$$

If $m = 0$, however, we can only distinguish particles from antiparticles in momentum space by the sign of p^0, as usual; this is the same as the sign in (2.17). Similarly, in this language, the chiral projection operator is constructed from

$$\Upsilon^5 = \sigma^t \sigma^x \sigma^y \sigma^z = -\begin{pmatrix} \ell & 0 \\ 0 & \ell \end{pmatrix} . \tag{3.18}$$

However, as with spin, we must multiply by ℓ in the correct place, obtaining

$$\hat{\Upsilon}^5 = \ell_R \circ \Upsilon^5 . \tag{3.19}$$

As a result, even though Υ^5 is a multiple of the identity, $\hat{\Upsilon}^5$ is not, and the operators $\tfrac{1}{2}(\sigma_t \pm \hat{\Upsilon}^5)$ project $\mathbb{H}^2$ into the Weyl subspaces $\mathbb{C}^2 \oplus \mathbb{C}^2 k$ as desired.

Combining the spin and particle information, over the quaternionic subalgebra $\mathbb{H} \subset \mathbb{O}$ determined by k and ℓ, we thus find one massive spin-$\tfrac{1}{2}$ particle at rest, with two spin states, namely

$$e_\uparrow = \begin{pmatrix} 1 \\ k \end{pmatrix} \quad \text{and} \quad e_\downarrow = \begin{pmatrix} -k \\ 1 \end{pmatrix} \tag{3.20}$$

whose antiparticle is obtained by replacing k by $-k$ (and changing the sign in (2.17)). We also find one massless spin-$\frac{1}{2}$ particle involving k moving in the z-direction, with a single helicity state,

$$\nu_z = \begin{pmatrix} 0 \\ k \end{pmatrix} \tag{3.21}$$

which corresponds, as usual, to both a particle and its antiparticle. It is important to note that $\nu_{-z} = \begin{pmatrix} k \\ 0 \end{pmatrix}$ corresponds to a massless particle with the same helicity moving in the opposite direction, not to a different particle with the opposite helicity. Each of the above states may be multiplied (on the *right*) by an arbitrary complex number.

There is also a single *complex* massless spin-$\frac{1}{2}$ particle, with the opposite helicity, which is given in momentum space by

$$\emptyset_z = \begin{pmatrix} 0 \\ 1 \end{pmatrix} . \tag{3.22}$$

As with the other massless momentum space states, this describes both a particle and an antiparticle. Alone among the particles, this one does not contain k, and hence does not depend on the choice of identification of a particular quaternionic subalgebra $\mathbb{H}$ satisfying $\mathbb{C} \subset \mathbb{H} \subset \mathbb{O}$.

29.3.4 Generations

We have shown how the massless Dirac equation in 10 dimensions reduces to the (massive and massless) Dirac equation in four dimensions when a preferred octonionic unit is chosen. As previously pointed out in in [2, 3], the resulting quaternionic Dirac equation describes one massive particle with two spin states, one massless particle with only one helicity, and their antiparticles. We identify this set of particles with a generation of leptons.

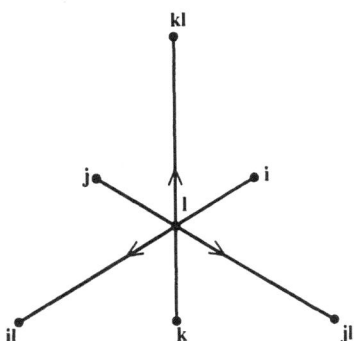

FIGURE 29.2. Each oriented line represents a quaternionic subalgebra of the octonions. The three subalgebras shown each contain the preferred complex subalgebra containing ℓ, but do not otherwise intersect.

Furthermore, as can be seen from Figure 29.2, there is room in the octonions for exactly three such quaternionic descriptions which have only their complex part in common, corresponding to replacing k in turn by i and j. We identify these three quaternionic spaces as describing three generations of leptons.

There is, however, one further massless particle/antiparticle pair, given by (3.22). Being purely complex, it does not belong to any generation, and it has the opposite helicity from the other massless particles. We do not currently have a physical interpretation for this additional particle; if this theory is to correspond to nature, then this additional particle must for some reason not interact much with anything else.

29.4   Quarks?

29.4.1   3-Spinors

We have argued in Section 29.2 that the ordinary momentum-space (massless and massive) Dirac equation in $3 + 1$ dimensions can be obtained via dimensional reduction from the Weyl (massless Dirac) equation in $9 + 1$ dimensions. This latter equation can be written as the eigenvalue equation (2.13) where P is a 2×2 octonionic Hermitian matrix corresponding to the 10-dimensional momentum and tilde again denotes trace reversal. The general solution of this equation is given by (2.15) and (2.16) where θ is a 2-component octonionic vector whose components lie in the same complex subalgebra of $\mathbb{O}$ as do those of P, and where $\xi \in \mathbb{O}$ is arbitrary. (Such a θ must exist since $\det(P) = 0$.)

It is then natural to introduce a 3-component formalism; this approach was used by Schray [17, 31] for the superparticle. Defining $\Psi = \left(\begin{smallmatrix} \theta \\ \xi \end{smallmatrix} \right)$ we have first of all that

$$\mathcal{P} := \Psi\Psi^\dagger = \begin{pmatrix} P & \psi \\ \psi^\dagger & |\xi|^2 \end{pmatrix} \tag{4.1}$$

so that Ψ combines the bosonic and fermionic degrees of freedom. It is tempting to refer to Ψ as a *3-spinor*. Lorentz transformations can be constructed by iterating ("nesting") transformations of the form [18]

$$P \mapsto MPM^\dagger, \tag{4.2}$$
$$\psi \mapsto M\psi \tag{4.3}$$

which can be elegantly combined into the transformation

$$\mathcal{P} \mapsto \mathcal{M}\mathcal{P}\mathcal{M}^\dagger \tag{4.4}$$

with

$$\mathcal{M} = \begin{pmatrix} M & 0 \\ 0 & 1 \end{pmatrix}. \tag{4.5}$$

29.4.2   The Albert algebra

We therefore turn to a 3-component formalism, and consider the *Albert algebra*, which is the exceptional Jordan algebra consisting of 3×3 octonionic Hermitian matrices, which we call *Jordan matrices*. The *Jordan product* [24, 25]

$$A \circ B = \tfrac{1}{2} \left(AB + BA \right) \tag{4.6}$$

of two such matrices is commutative but not associative. We have in particular that

$$A^2 \equiv A \circ A \tag{4.7}$$

and we *define*

$$A^3 := A^2 \circ A \equiv A \circ A^2 \tag{4.8}$$

which differs from the cube of A using ordinary matrix multiplication. Other operations on Jordan matrices are the *trace*, and the *Freudenthal product* [26]

$$A * B = A \circ B - \tfrac{1}{2} \left(A \operatorname{tr}(B) + B \operatorname{tr}(A) \right)$$
$$+ \tfrac{1}{2} \left(\operatorname{tr}(A) \operatorname{tr}(B) - \operatorname{tr}(A \circ B) \right) I \tag{4.9}$$

where I denotes the identity matrix and with the important special case

$$A * A = A^2 - (\operatorname{tr} A) A + \sigma(A) I \tag{4.10}$$

where

$$\sigma(A) = \tfrac{1}{2} \left((\operatorname{tr} A)^2 - \operatorname{tr}(A^2) \right) \equiv \operatorname{tr}(A * A). \tag{4.11}$$

There is also *trace reversal*

$$\widetilde{A} = A - \operatorname{tr}(A) I \equiv -2 I * A \tag{4.12}$$

and, finally, the *determinant*

$$\det(A) = \frac{1}{3} \operatorname{tr} \left((A * A) \circ A \right) \tag{4.13}$$

which can equivalently be defined by

$$\left((A * A) \circ A \right) = (\det A) I. \tag{4.14}$$

Expanding (4.14) using (4.10), we obtain the remarkable result that Jordan matrices satisfy the usual characteristic equation [26]

$$A^3 - (\operatorname{tr} A) A^2 + \sigma(A) A - (\det A) I = 0. \tag{4.15}$$

Explicitly, a Jordan matrix can be written as

$$A = \begin{pmatrix} p & a & \bar{b} \\ \bar{a} & m & c \\ b & \bar{c} & n \end{pmatrix} \tag{4.16}$$

with $p, m, n \in \mathbb{R}$ and $a, b, c \in \mathbb{O}$, where the bar denotes octonionic conjugation. The definitions above then take the concrete form

$$\operatorname{tr} \mathcal{A} = p + m + n,$$
$$\sigma(\mathcal{A}) = pm + mn + pn - |a|^2 - |b|^2 - |c|^2,$$
$$\det \mathcal{A} = pmn + b(ac) + \bar{b}(\bar{a}\bar{c}) - n|a|^2 - m|b|^2 - p|c|^2. \tag{4.17}$$

The *Cayley plane*, also called the *Moufang plane*, consists of those Jordan matrices $\mathcal{V}$ which satisfy the restriction [27, 28]

$$\mathcal{V} \circ \mathcal{V} = \mathcal{V}; \qquad \operatorname{tr} \mathcal{V} = 1. \tag{4.18}$$

As shown in [28], the conditions (4.18) force the components of $\mathcal{V}$ to lie in a *quaternionic* subalgebra of $\mathbb{O}$ (which depends on $\mathcal{V}$). Basic (associative) linear algebra then shows that each element of the Cayley plane is a primitive idempotent (an idempotent which is *not* the sum of other idempotents), and can be written as

$$\mathcal{V} = vv^\dagger \tag{4.19}$$

where v is a 3-component octonionic column vector, whose components lie in the quaternionic subalgebra determined by $\mathcal{V}$, and which is normalized by

$$v^\dagger v = \operatorname{tr} \mathcal{V} = 1. \tag{4.20}$$

Note that v is unique up to a *quaternionic* phase. Furthermore, using (4.10) and its trace (4.11), it is straightforward to show that, for any Jordan matrix $\mathcal{B}$,

$$\mathcal{B} * \mathcal{B} = 0 \iff \mathcal{B} \circ \mathcal{B} = (\operatorname{tr} \mathcal{B}) \mathcal{B} \tag{4.21}$$

which agrees with (4.18) up to normalization, and which is therefore the condition that $\pm\mathcal{B}$ can be written in the form (4.19) (without the restriction (4.20)). Note further that for any Jordan matrix satisfying (4.21), the normalization $\operatorname{tr} \mathcal{B}$ can only be zero if v, and hence $\mathcal{B}$ itself, is zero, so that

$$\mathcal{B} * \mathcal{B} = 0 = \operatorname{tr} \mathcal{B} \iff \mathcal{B} = 0 \tag{4.22}$$

since the converse is obvious.

29.4.3   The Jordan eigenvalue problem

Consider finally the Jordan eigenvalue problem, which is really the *eigenmatrix* problem

$$\mathcal{A} \circ \mathcal{V} = \lambda \mathcal{V} \tag{4.23}$$

for Jordan matrices $\mathcal{A}$ and $\mathcal{V}$, and where λ must be real (since both sides are Hermitian). We summarize here the relevant results from [4].

We restrict $\mathcal{V}$ in (4.23) to the Cayley plane (4.18), which ensures that the eigenmatrices $\mathcal{V}$ are primitive idempotents; they really do correspond to "eigenvectors" v. Recall that this forces the components of $\mathcal{V}$ to lie in a quaternionic subalgebra of $\mathbb{O}$ (which depends on $\mathcal{V}$) even though the components of $\mathcal{A}$ may not.

Restricting the eigenvalues in this way corresponds to the traditional eigenvalue problem in the following sense. If $\mathcal{A}$, $v \neq 0$ lie in a quaternionic subalgebra of the octonions, then the Jordan eigenvalue problem (4.23) together with the restriction (4.18) becomes

$$\mathcal{A}\,vv^\dagger + vv^\dagger\mathcal{A} = 2\lambda\,vv^\dagger. \tag{4.24}$$

Multiplying (4.24) on the right by v and simplifying the result using the trace of (4.24) leads immediately to $Av = \lambda v$ (with $\lambda \in \mathbb{R}$), that is, the Jordan eigenvalue equation implies the ordinary eigenvalue equation in this context. Since the converse is immediate, the Jordan eigenvalue problem (4.23) (with $\mathcal{V}$ restricted to the Cayley plane but $\mathcal{A}$ octonionic) is seen to be a reasonable generalization of the ordinary eigenvalue problem.

We now show how to construct eigenmatrices $\mathcal{V}$ of (4.23), restricted to lie in the Cayley plane. From the definition of the determinant, we have

$$0 = \det(\mathcal{A} - \lambda I) = (\mathcal{A} - \lambda I) \circ \big((\mathcal{A} - \lambda I) * (\mathcal{A} - \lambda I)\big). \tag{4.25}$$

Thus, setting

$$\mathcal{Q}_\lambda = (\mathcal{A} - \lambda I) * (\mathcal{A} - \lambda I) \tag{4.26}$$

we have

$$(\mathcal{A} - \lambda I) \circ \mathcal{Q}_\lambda = 0 \tag{4.27}$$

so that $\mathcal{Q}_\lambda$ is a solution of (4.23).

As shown in [4], we have

$$\mathcal{Q}_\lambda * \mathcal{Q}_\lambda = 0. \tag{4.28}$$

If $\mathcal{Q}_\lambda \neq 0$, we can renormalize $\mathcal{Q}_\lambda$ by defining

$$\mathcal{P}_\lambda = \frac{\mathcal{Q}_\lambda}{\mathrm{tr}(\mathcal{Q}_\lambda)}. \tag{4.29}$$

Each resulting $\mathcal{P}_\lambda$ is in the Cayley plane, and is hence a primitive idempotent. Due to (4.28), we can write $\mathcal{P}_\lambda = v_\lambda v_\lambda^\dagger$ and we call v_λ the (generalized) eigenvector of $\mathcal{A}$ with eigenvalue λ. We have

$$\mathcal{A} \circ v_\lambda v_\lambda^\dagger = \lambda v_\lambda v_\lambda^\dagger \quad \text{as well as} \quad v_\lambda^\dagger v_\lambda = 1.$$

Putting this all together, if there are no repeated eigenvalues, then the eigenmatrix problem leads to the decomposition

$$\mathcal{A} = \sum_{i=1}^{3} \lambda_i \mathcal{P}_{\lambda_i} \tag{4.30}$$

in terms of orthogonal primitive idempotents, which expresses each Jordan matrix $\mathcal{A}$ as a sum of squares of *quaternionic* columns. We emphasize that the components of the eigenmatrices $\mathcal{P}_{\lambda_i}$ need not lie in the same quaternionic subalgebra, and that $\mathcal{A}$ is octonionic. Nonetheless, it is remarkable that $\mathcal{A}$ admits a decomposition in terms of matrices which are, individually, quaternionic.

As shown in [4], decompositions analogous to (4.30) can also be found when there is a repeated eigenvalue, but the terms corresponding to the repeated eigenvalue can not be written in terms of the projections $\mathcal{P}_\lambda$, and of course the decomposition of the corresponding eigenspace is not unique.

29.4.4 Exceptional groups

The octonions are closely related to all five exceptional groups. G_2 is the automorphism group of the octonions, F_4 is the automorphism group of both the Jordan and Freudenthal products, and E_6 is the invariance group of the determinant (4.13), while E_7 and E_8 are related to certain projective spaces over the octonions [29]. We focus here on F_4 and E_6.

Manogue and Schray [18] showed how to view $SL(2, \mathbb{O})$ as (the double cover of) $SO(9, 1)$. The "3-spinor" formalism of Section 29.4.1 then leads to the representation (4.5) of $SO(9, 1)$. Since the transformations of the form (4.4) which preserve the determinant are precisely E_6, this shows how to view $SO(9, 1)$ as a subgroup of E_6. In fact, E_6 is just the union of the three obvious $SO(9, 1)$ subgroups obtained from this one by permuting rows and columns. This shows that E_6 can be thought of as (the double cover of) "$SL(3, \mathbb{O})$". Furthermore, under the above identification, the rotation subgroup $SO(9)$ lies in F_4; F_4 turns out to be the group which preserves the trace of a Jordan matrix. F_4 is thus the rotation subgroup of E_6, which can be thought of as (the double cover of) "$SU(3, \mathbb{O})$".

29.4.5 The Dirac equation revisited

It turns out that the Dirac equation (2.13) is equivalent to

$$\mathcal{P} * \mathcal{P} = 0 \qquad (4.31)$$

which shows both that solutions of the Dirac equation correspond to the Cayley plane and that the Dirac equation admits E_6 as a symmetry group. Using the particle interpretation described above [1, 2] then leads to the interpretation of (part of) the Cayley plane as representing three generations of leptons.

How are we to interpret Jordan matrices which are *not* in the Cayley plane? This is where the decomposition (4.30) comes in. If p is the number of nonzero eigenvalues in the decomposition of a Jordan matrix $\mathcal{A}$, we will refer to $\mathcal{A}$ as a *p-square*. If $\det(\mathcal{A}) \neq 0$, then $\mathcal{A}$ is a 3-square. If $\det(\mathcal{A}) = 0 \neq \sigma(\mathcal{A})$, then $\mathcal{A}$ is a 2-square. Finally, if $\det(\mathcal{A}) = 0 = \sigma(\mathcal{A})$, then $\mathcal{A}$ is a 1-square (unless also $\text{tr}(\mathcal{A}) = 0$, in which case $\mathcal{A} \equiv 0$). It is intriguing that, since E_6 preserves both the determinant and the condition $\sigma(\mathcal{A}) = 0$, E_6 therefore preserves the class of

p-squares for each p. If, as argued above, 1-squares correspond to leptons, is it possible that 2-squares are mesons and 3-squares are baryons?

29.5   Discussion

We have shown how to embed our 2-component octonionic description of leptons into a 3-component formalism, resulting in an equivalent description in which leptons are elements of the Cayley plane. As pointed out in the last section, since any Jordan matrix admits a unique decomposition into at most three elements of the Cayley plane, it is tempting to identify composite matrices as mesons and baryons. The remarkable property which makes this possible is that individual elements of the Cayley plane have components lying in a quaternionic subalgebra of the octonions, which we interpret as a generation structure.

Note finally that there do not appear to be any free quarks in this 3×3 matrix language, as we have used the Cayley plane itself to describe leptons. Whether this reflects the absence of free quarks in nature, or simply the preliminary nature of our model, remains to be determined. In this regard, it would be helpful to have a description of charge in this language, which is our next goal.

REFERENCES

[1] Corinne A. Manogue and Tevian Dray, Dimensional reduction, *Mod. Phys. Lett.* **A14** (1999), 93–97.

[2] Tevian Dray and Corinne A. Manogue, Quaternionic spin, in *Clifford Algebras and their Applications in Mathematical Physics*, Eds. Rafał Abłamowicz and Bertfried Fauser, Birkhäuser, Boston, 2000, pp. 29–46.

[3] Tevian Dray and Corinne A. Manogue, Octonions and fundamental particles, in *A Festschrift in Honor of Paulo Budinich*, Eds. Franco Bradamante and Giuseppe Furlan, Consorzio per l'Incremento degli Studi e delle Ricerche dei Dipartimenti di Fisica dell' Università di Trieste, 2002, pp. 66–87.

[4] Tevian Dray and Corinne A. Manogue, The exceptional Jordan eigenvalue problem, *Internat. J. Theoret. Phys.* **38** (1999), 2901–2916.

[5] Geoffrey M. Dixon, *Division Algebras: Octonions, Quaternions, Complex Numbers and the Algebraic Design of Physics*, Kluwer Academic Publishers, Boston, 1994.

[6] Feza Gürsey and Chia-Hsiung Tze, *On the Role of Division, Jordan, and Related Algebras in Particle Physics*, World Scientific, Singapore, 1996.

[7] S. Okubo, *Introduction to Octonion and Other Non-Associative Algebras in Physics*, Cambridge University Press, Cambridge, 1995.

[8] Gregory J. Trayling, *An Algebraic Basis for the Standard-Model Gauge Group*, Ph.D. dissertation, Department of Physics, University of Windsor, 2000.

[9] Greg Trayling and W.E. Baylis, A geometric basis for the standard-model gauge group, *J. Phys.* **A34** (2001), 3309–3324.

[10] A. Sudbery, Division Algebras, (Pseudo) Orthogonal groups and spinors, *J. Phys.* **A17** (1984), 939–955.

[11] K. W. Chung and A. Sudbery, Octonions and the Lorentz and conformal groups of ten-dimensional space-time, *Phys. Lett.* **B198** (1987), 161–164.

[12] Corinne A. Manogue and Anthony Sudbery, General solutions of covariant superstring equations of motion, *Phys. Rev.* **D40** (1989), 4073–4077.

[13] Tevian Dray and Corinne A. Manogue, The octonionic eigenvalue problem, *Adv. Appl. Clifford Algebras* **8** (1998), 341–364.

[14] Tevian Dray and Corinne A. Manogue, Finding octonionic eigenvectors using *Mathematica*, *Comput. Phys. Comm.* **115** (1998), 536–547.

[15] David B. Fairlie and Corinne A. Manogue, Lorentz invariance and the composite string, *Phys. Rev.* **D34** (1986), 1832–1834.

[16] David B. Fairlie and Corinne A. Manogue, A parameterization of the covariant superstring, *Phys. Rev.* **D36** (1987), 475–479.

[17] Jörg Schray, The general classical solution of the superparticle, *Class. Quant. Grav.* **13** (1996), 27–38.

[18] Corinne A. Manogue and Jörg Schray, Finite Lorentz transformations, automorphisms, and division algebras, *J. Math. Phys.* **34** (1993), 3746–3767.

[19] Corinne A. Manogue and Tevian Dray, Octonionic Möbius transformations, *Mod. Phys. Lett.* **A14** (1999), 1243–1255.

[20] Tevian Dray, Jason Janesky, and Corinne A. Manogue, Octonionic Hermitian matrices with non-real eigenvalues, *Adv. Appl. Clifford Algebras* **10** (2000), 193–216.

[21] Tevian Dray, Jason Janesky, and Corinne A. Manogue, Some properties of 3×3 octonionic Hermitian matrices with non-real eigenvalues, Oregon State University, 2000, 12 pages. (http://xxx.lanl.gov/abs/math/0010255)

[22] L.E. Dickson, On quaternions and their generalization and the history of the eight square theorem, *Ann. Math.* **20** (1919), 155–171.

[23] Richard D. Schafer, *An Introduction to Nonassociative Algebras*, Academic Press, New York, 1966 and Dover, Mineola NY, 1995.

[24] P. Jordan, Über die Multiplikation quantenmechanischer Größen, *Z. Phys.* **80** (1933), 285–291.

[25] P. Jordan, J. von Neumann, and E. Wigner, On an algebraic generalization of the quantum mechanical formalism, *Ann. Math.* **35** (1934), 29–64.

[26] Hans Freudenthal, Zur Ebenen Oktavengeometrie, *Proc. Kon. Ned. Akad. Wet.* **A56** (1953), 195–200.

[27] Hans Freudenthal, Lie groups in the foundations of geometry, *Adv. Math.* **1** (1964), 145–190.

[28] F. Reese Harvey, *Spinors and Calibrations*, Academic Press, Boston, 1990.

[29] John C. Baez, The octonions, *Bull. Amer. Math. Soc.* **39** (2002), 145–205.

[30] Tevian Dray, Corinne A. Manogue, and Susumu Okubo, Orthonormal eigenbases over the octonions, *Algebras Groups Geom.* **19** (2002), 163–180.

[31] Jörg Schray, *Octonions and Supersymmetry*, Ph.D. dissertation, Department of Physics, Oregon State University, 1994.

Tevian Dray[4]
Department of Mathematics
Oregon State University
Corvallis, OR 97331
E-mail: tevian@math.orst.edu

Corinne A. Manogue[4]
Department of Physics
Oregon State University
Corvallis, OR 97331
E-mail: corinne@physics.orst.edu

Submitted: September 3, 2002.

---

[4]Some of this work was done while the authors were the Hutchcroft Visiting Professors of Mathematics at Mount Holyoke College, South Hadley, MA 01075.

30

Applications of Geometric Algebra in Electromagnetism, Quantum Theory and Gravity

Anthony Lasenby, Chris Doran, and Elsa Arcaute

ABSTRACT We review the applications of geometric algebra in electromagnetism, gravitation and multiparticle quantum systems. We discuss a gauge theory formulation of gravity and its implementation in geometric algebra, and apply this to the fermion bound state problem in a black hole background. We show that a discrete energy spectrum arises in an analogous way to the hydrogen atom. A geometric algebra approach to multiparticle quantum systems is given in terms of the multiparticle spacetime algebra. This is applied to quantum information processing, multiparticle wave equations and to conformal geometry. The application to conformal geometry highlight some surprising links between relativistic quantum theory, twistor theory and de Sitter spaces.

Keywords: geometric algebra, quantum theory, multiparticle quantum theory, conformal geometry, wave equations, Dirac equation, scattering, gauge theory, gravitation, de Sitter space, black holes, bound states

30.1 Introduction

The applications of geometric algebra we discuss in this paper are largely to problems in relativistic physics. We start with a brief introduction to the geometric algebra of spacetime; the spacetime algebra or STA. The first application of this is to electromagnetic scattering problems. We describe a general method for solving the full Maxwell equations in the presence of an arbitrarily-shaped conductor. The second application is to the study of the Dirac equation in a black hole background. We show the existence of a spectrum of bound states created by the black hole. Each of these states has an imaginary contribution to its energy, which can be understood in terms of a decay process occurring at the singularity.

The next topic we discuss is the application of geometric algebra to conformal models of space and spacetime. The STA, for example, can also be viewed as

AMS Subject Classification: 83C57, 83C60, 81R25, 81Q05, 78A45.

the conformal algebra of the Euclidean plane. This provides a means of visualising Lorentz transformations which was first explored by Penrose. The conformal model of spacetime is constructed in a 6-dimensional geometric algebra. Rotors in this space encode the full spacetime conformal group and the conformal model provides a unified framework for Lorentzian, de Sitter and anti-de Sitter spaces.

The final topic we discuss is the extension of the STA approach to multiparticle quantum systems. The framework we use for this is the multiparticle spacetime algebra or MSTA. This is the geometric algebra of relativistic configuration space. It provides an ideal setting for studying problems in quantum information theory and gives a new means of encoding quantum entanglement. The MSTA is also the appropriate arena for the study of wave equations for particles with general spin. Lagrangians for multiparticle wave equations are constructed and applied to the case of a spin-0 particle, with surprising consequences for the stress-energy tensor and the coupling of the field to gravity.

On the surface, many of these topics seem quite unrelated. We hope to convince the reader that there are strong links between them, and that geometric algebra is the appropriate tool for understanding and exploiting these relationships. For example, the spacetime conformal model has a natural construction in terms of multiparticle states in the MSTA. This exposes the links between conformal geometry, the MSTA and the twistor programme. Conformal geometry and the MSTA also turn out to provide a framework for constructing supersymmetric models in geometric algebra.

The STA (spacetime algebra) is the geometric algebra of spacetime [4, 5, 13]. It is generated by four vectors $\{\gamma_\mu\}$ which satisfy

$$\gamma_\mu \cdot \gamma_\nu = \tfrac{1}{2}(\gamma_\mu\gamma_\nu + \gamma_\nu\gamma_\mu) = \eta_{\mu\nu} = \text{diag}(+ - - -). \tag{1.1}$$

Throughout, Greek indices run from 0 to 3 and Latin indices run from 1 to 3. We use a signature in which $\gamma_0^2 = -\gamma_i^2 = 1$, and natural units $c = \hbar = G$ are assumed throughout. The reverse operation is denoted by a tilde, as in $\tilde{R}$. The full STA is spanned by

$$
\begin{array}{ccccc}
1 & \{\gamma_\mu\} & \{\gamma_\mu \wedge \gamma_\nu\} & \{I\gamma_\mu\} & I = \gamma_0\gamma_1\gamma_2\gamma_3 \\
1 \text{ scalar} & 4 \text{ vectors} & 6 \text{ bivectors} & 4 \text{ trivectors} & 1 \text{ pseudoscalar}
\end{array} \tag{1.2}
$$

The algebraic properties of the STA are those of the Dirac matrices, but there is never any need to introduce an explicit matrix representation in calculations. As well as quantum theory, the STA has been applied to relativistic mechanics [9, 14, 16], scattering [5], tunnelling [5, 11], and gravitation [20]. Many of these applications are summarised in [8].

Suppose now that we wish to study physics in the rest frame defined by the γ_0 vector. We define

$$\sigma_k = \gamma_k\gamma_0, \tag{1.3}$$

so that

$$\sigma_i\sigma_j + \sigma_j\sigma_i = 2\delta_{ij}. \tag{1.4}$$

The set $\{\sigma_i\}$ therefore generate the geometric algebra of the three-dimensional space defined by the γ_0 frame. We also see that

$$\sigma_1\sigma_2\sigma_3 = \gamma_1\gamma_0\gamma_2\gamma_0\gamma_3\gamma_0 = \gamma_0\gamma_1\gamma_2\gamma_3 = I, \tag{1.5}$$

so relative space and spacetime share the same pseudoscalar. The algebra of space is therefore the *even subalgebra* of the STA. This subalgebra contains the scalars and pseudoscalars, and six (spacetime) bivectors. These bivectors are split into timelike and spacelike bivectors by the chosen velocity vector (γ_0 in this case). This split is conveniently illustrated by the electromagnetic bivector F. We have

$$F = E + IB \tag{1.6}$$

where $E = E^k\sigma_k$ is the electric field and $B = B^k\sigma_k$ the magnetic field. These are recovered from F by forming

$$E = \tfrac{1}{2}(F - \gamma_0 F\gamma_0), \qquad IB = \tfrac{1}{2}(F + \gamma_0 F\gamma_0). \tag{1.7}$$

These expressions clearly show how the split of the (invariant) bivector F into electric and magnetic parts depends on the velocity of the observer. If γ_0 is replaced by a different velocity, new fields are obtained.

30.2   Electromagnetism

The Maxwell equations can be written

$$\nabla\cdot E = \rho, \qquad\qquad \nabla\cdot B = 0,$$
$$\nabla\wedge E = -\partial_t(IB), \qquad \nabla\wedge B = I(J + \partial_t E), \tag{2.1}$$

where the $\wedge$ product takes on its three-dimensional definition. If we now write $\nabla = \gamma^\mu\partial_\mu$, $J = (\rho + J)\gamma_0$ and $F = E + IB$, the Maxwell equations combine into the single, relativistically covariant equation [13]

$$\nabla F = J. \tag{2.2}$$

Here we are interested in monochromatic scattering, which can be treated in a unified manner by introducing a free-space multivector Green's function. The essential geometry of the problem is illustrated in Figure 30.1. The incident field F_i sets up oscillating currents in the object, which generate an outgoing radiation field F_s. The total field is given by

$$F = F_i + F_s. \tag{2.3}$$

For monochromatic waves the time dependence is conveniently expressed as

$$F(x) = F(r)e^{-i\omega t}, \tag{2.4}$$

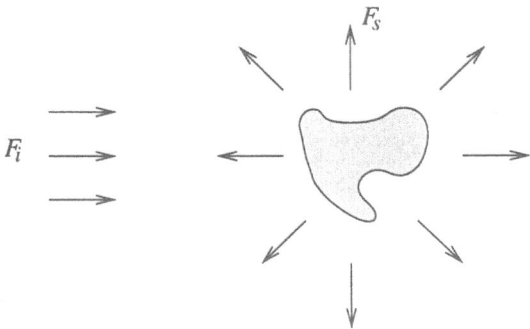

FIGURE 30.1. *Scattering by a localised object.* The incident field F_i sets up oscillating currents in the object, which generate an outgoing radiation field F_s.

so that the Maxwell equations reduce to

$$\nabla F - i\omega F = 0. \tag{2.5}$$

We seek a Green's function satisfying

$$\dot{G}\dot{\nabla} + i\omega G = \delta(\boldsymbol{r}), \tag{2.6}$$

where the overdot denotes the scope of the derivative operator. The solution to this problem is

$$G(\boldsymbol{r}) = \frac{e^{i\omega r}}{4\pi}\left(\frac{i\omega}{r}(1 - \sigma_r) + \frac{\boldsymbol{r}}{r^3}\right), \tag{2.7}$$

where $\sigma_r = \boldsymbol{r}/r$ is the unit vector in the direction of $\boldsymbol{r}$. If S_1 denotes the surface just outside the scatterers (so that no surface currents are present over S_1), then the scattered field can be shown to equal

$$F_s(\boldsymbol{r}) =$$
$$\frac{1}{4\pi}\oint_{S_1} e^{i\omega d}\left(\frac{i\omega}{d} + \frac{i\omega(\boldsymbol{r} - \boldsymbol{r}')}{d^2} - \frac{\boldsymbol{r} - \boldsymbol{r}'}{d^3}\right) n' F_s(\boldsymbol{r}') \, |dS(\boldsymbol{r}')|, \tag{2.8}$$

where

$$d = |\boldsymbol{r} - \boldsymbol{r}'|, \tag{2.9}$$

and n points into the surface. The key remaining problem is to find the fields over the surface of the conductor. This is solved by treating the surface as a series of simplices and explicitly solving for the surface currents over each simplex. A subtlety here is that each part of the conductor creates a field which is seen by every other part of the conductor. The problem of finding a self-consistent solution to these equations can be converted to a matrix inversion problem, which is sufficient to provide the fields over the conductor. Equation (2.8) is then an exact, analytic expression for the fields at any point in space (given the discretised model of the conductor) and the integrals can be computed numerically to any desired accuracy.

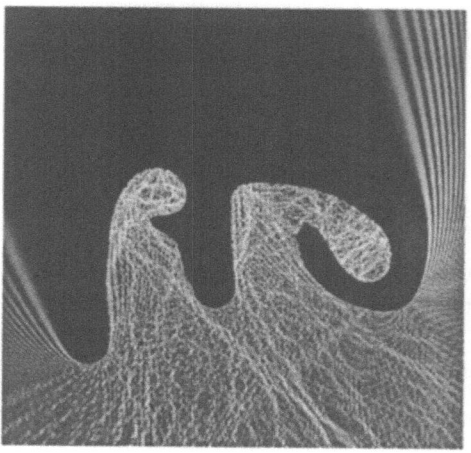

FIGURE 30.2. *Scattering in two dimensions.* The plots show the intensity of the electric field, with higher intensity coloured lighter.

An example of this method is shown in Figure 30.2. The calculations fully incorporate all diffraction effects and polarisations, as well as correctly accounting for obliquity factors. The plots show the intensity of the electric field, with higher intensity coloured lighter. The incident radiation enters from the bottom right of the diagram and scatters off a conductor with complicated surface features. The conductor is closed in the shadow region. Various diffraction effects are clearly visible, as is a complicated pattern of hot and cold regions.

30.3   Single-particle quantum theory and gravity

The non-relativistic wavefunction for a spin-$\frac{1}{2}$ particle is a Pauli spinor. It represents the complex superposition of two possible spin states,

$$|\psi\rangle = \alpha_1|\uparrow\rangle + \alpha_2|\downarrow\rangle, \tag{3.1}$$

where α_1 and α_2 are complex numbers. There are various means of representing $|\psi\rangle$ in the Pauli algebra. One route is via the introduction of idempotents of the form $\frac{1}{2}(1+\sigma_3)$. A more direct route, which has many advantages in computations, is to map the degrees of freedom in $|\psi\rangle$ onto an element of the even subalgebra of the Pauli algebra [4, 5, 19]. The appropriate map is

$$|\psi\rangle = \begin{pmatrix} a^0 + ia^3 \\ -a^2 + ia^1 \end{pmatrix} \leftrightarrow \psi = a^0 + a^k I\sigma_k. \tag{3.2}$$

This map is clearly one-to-one. The spin-up and spin-down basis states are represented by

$$| \uparrow \rangle \leftrightarrow 1, \qquad | \downarrow \rangle \leftrightarrow -I\sigma_2, \tag{3.3}$$

and the action of the Pauli operators on a state is represented by

$$\hat{\sigma}_k |\psi\rangle \leftrightarrow \sigma_k \psi \sigma_3 \quad (k = 1, 2, 3). \tag{3.4}$$

The factor of σ_3 on the right-hand side of ψ ensures that $\sigma_k \psi \sigma_3$ remains in the even subalgebra. Its presence does not break the rotational invariance because rotations are encoded in *rotors* which multiply ψ from the left.

The product of the three Pauli matrices has the same effect as the unit imaginary since

$$\hat{\sigma}_1 \hat{\sigma}_2 \hat{\sigma}_3 = \begin{pmatrix} i & 0 \\ 0 & i \end{pmatrix}. \tag{3.5}$$

It follows that

$$i|\psi\rangle \leftrightarrow \sigma_1 \sigma_2 \sigma_3 \psi (\sigma_3)^3 = \psi I \sigma_3, \tag{3.6}$$

so the action of the unit imaginary on a column spinor has been replaced by the action of a bivector.

Relativistic spin-$\frac{1}{2}$ states are described by Dirac spinors. These have eight real degrees of freedom and can be represented by the even subalgebra of the full STA. The corresponding action of the Dirac matrices and the unit imaginary is then

$$\hat{\gamma}_\mu |\psi\rangle \quad \leftrightarrow \quad \gamma_\mu \psi \gamma_0 \quad (\mu = 0, \dots, 3), \tag{3.7}$$

$$i|\psi\rangle \quad \leftrightarrow \quad \psi I \sigma_3. \tag{3.8}$$

In this representation, the Dirac equation takes the form [4, 5, 13]

$$\nabla \psi I \sigma_3 - eA\psi = m\psi\gamma_0 \tag{3.9}$$

where $A = \gamma^\mu A_\mu$.

We will shortly examine this equation in a gravitational context.

30.3.1   Gravity as a gauge theory

Gravity was first formulated as a gauge theory in the 1960s by Kibble [17]. It turns out that this approach to gravity allows us to develop a theory which fully exploits the advantages of the STA [20]. The theory is constructed in terms of gauge fields in a flat spacetime background. Features of the background space are not gauge invariant, so are not physically measurable. This is the manner in which the gauge theory approach frees us from any notion of an absolute space or time. The theory requires two gauge fields; one for local translations and one for Lorentz transformations. The first of these can be understood by introducing a set of coordinates x^μ, with associated coordinate frame e_μ. In terms of these we have

$$\nabla = e^\mu \partial_\mu, \tag{3.10}$$

and the Lorentzian metric is

$$ds^2 = e_\mu \cdot e_\nu \, dx^\mu \, dx^\nu. \tag{3.11}$$

The gauging step is now simply to replace the e_μ frame with a set of gauge fields g_μ which are no longer tied to a flat-space coordinate frame. That is, we take

$$e_\mu(x) \mapsto g_\mu(x). \tag{3.12}$$

This requires the introduction of 16 gauge degrees of freedom for the four vectors g_μ. The gauge-invariant line element is now

$$ds^2 = g_\mu \cdot g_\nu \, dx^\mu \, dx^\nu, \tag{3.13}$$

from which we can read off that the effective metric is given by

$$g_{\mu\nu} = g_\mu \cdot g_\nu. \tag{3.14}$$

It is possible to develop general relativity in terms of the g_μ alone, since these vectors are sufficient to recover a metric. But quantum theory requires a second gauge field associated with Lorentz transformations (a spin connection), and the entire theoretical framework is considerably simpler if this is included from the start. The connection consists of a set of four bivector fields Ω_μ, which contain 24 degrees of freedom. The full gauge theory is then a first-order theory with 40 degrees of freedom. This is usually easier to work with than the second-order metric theory (with 10 degrees of freedom) because the first-order equations afford better control over the non-linearities in the theory. The final theory is locally equivalent to the Einstein–Cartan–Kibble–Sciama (ECKS) extension of general relativity. But the fact that the gauge theory is constructed on a topologically trivial flat spacetime can have physical consequences, even if these are currently mainly restricted to theoretical discussions.

The minimally-coupled Dirac equation is defined by

$$g^\mu D_\mu \psi I \sigma_3 = m\psi\gamma_0, \tag{3.15}$$

where

$$D_\mu \psi = \left(\partial_\mu + \tfrac{1}{2}\Omega_\mu \right) \psi. \tag{3.16}$$

A consequence of minimal coupling is that, if we form the classical point-particle limit of the Dirac theory, we find that the mass term drops out of the effective equation of motion. So classical particles follow trajectories (geodesics) that do not depend on their mass. This form of the equivalence principle follows directly from minimal coupling in the gauge-theory context.

30.3.2   Schwarzschild black holes

The Schwarzschild solution, written in terms of the time measured by observers in radial free-fall from rest at infinity, is given by [7]

$$ds^2 = dt^2 - \left(dr + \left(\frac{2GM}{r} \right)^{\frac{1}{2}} dt \right)^2 - r^2 d\Omega^2. \tag{3.17}$$

If we let $\{e_r, e_\theta, e_\phi\}$ denote a standard spatial polar-coordinate frame, we can set

$$g_0 = \gamma_0 + \left(\frac{2GM}{r}\right)^{\frac{1}{2}} e_r, \qquad g_i = \gamma_i, \quad (i = r, \theta, \phi). \qquad (3.18)$$

The g_μ vectors are easily shown to reproduce the metric of equation (3.17). The Dirac equation in a black-hole background, with the present gauge choices, takes the simple form [7]

$$\nabla \psi I \sigma_3 - \left(\frac{2GM}{r}\right)^{\frac{1}{2}} \gamma_0 \left(\partial_r \psi + \frac{3}{4r} \psi\right) I \sigma_3 = m \psi \gamma_0. \qquad (3.19)$$

The full, relativistic wave equation for a fermion in a spherically-symmetric black hole background reduces to the free-field equation with a single interaction term H_I, where

$$H_I \psi = \left(\frac{2GM}{r}\right)^{\frac{1}{2}} i\hbar \left(\partial_r \psi + \frac{3}{4r} \psi\right). \qquad (3.20)$$

(Here we have inserted dimensional constants for clarity.) Our choice of gauge has converted the problem to a Hamiltonian form, though with a non-Hermitian interaction Hamiltonian, which satisfies

$$H_I - H_I^\dagger = -i\hbar(2GMr^3)^{1/2}\delta(x). \qquad (3.21)$$

This gauge provides a number of insights for quantum theory in black hole backgrounds. It has recently been used to compute the fermion scattering cross section, providing the gravitational analogue of the Mott formula [7], and also applied to the bound state problem [24, 25].

A gravitational bound state is a state with separable time dependence and which is spatially normalizable. These states form a discrete spectrum in an analogous manner to the Hydrogen atom. But the lack of Hermiticity of H_I implies that the energies also contain an imaginary component, which ensures that the states *decay*. A discrete spectrum is a consequence of the fact that boundary conditions must be simultaneously applied at the horizon and at infinity. For a given complex energy we can simultaneously integrate in from infinity and out from the horizon. These integrated functions will in general not match and so fail to generate a solution. Matching only occurs at specific discrete energy eigenvalues, distributed over the complex plane.

A sample energy spectrum is shown in Figure 30.3, which plots the real part of the energy, in units of mc^2, for the $S_{1/2}$, $P_{1/2}$ and $P_{3/2}$ states. (We follow the standard spectroscopic naming conventions.) The plots are shown as a function of the dimensionless coupling constant α,

$$\alpha = \frac{mM}{m_p^2}, \qquad (3.22)$$

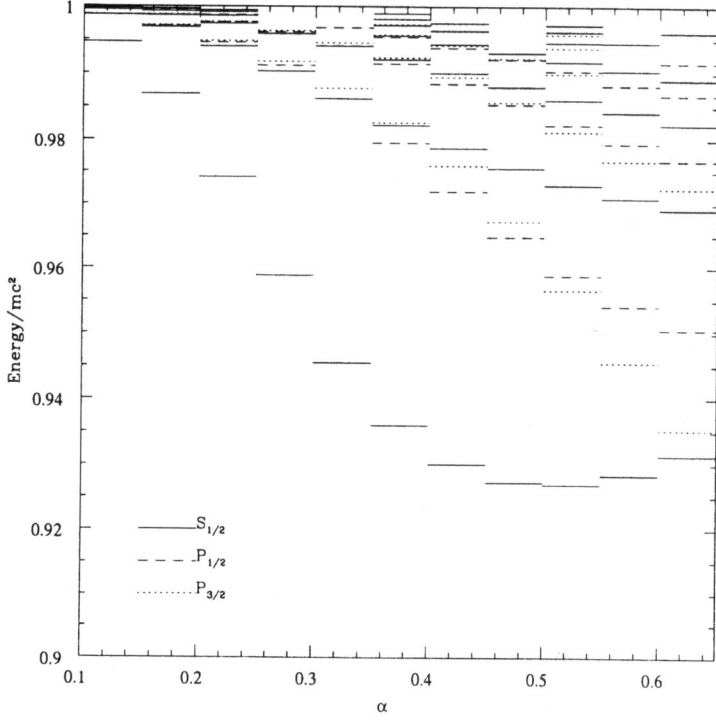

FIGURE 30.3. The real part of the bound state energy, in units of mc^2. The lines represent the value of the energy for the coupling at the left of the line, with α ranging from 0.1 to 0.6 in steps of 0.05.

where m_p is the Planck mass. For m the mass of an electron and $\alpha \approx 1$, then M is in the range appropriate for primordial black holes. The plots need to be extended to much larger values of α to describe solar mass black holes. One feature of the real part of the energy is seen more clearly in Figure 30.4. Beyond a coupling of around $\alpha = 0.65$ the $1S_{1/2}$ is no longer the system ground state, which passes to the $2P_{3/2}$ state. This corresponds to the classical observation that the lowest energy stable circular orbits around a black hole have increasing angular momentum for increasing black hole mass.

A substantial amount of work remains in this area. The spectrum needs to be computed for much larger values of α, and the work also needs to be extended to the Kerr and Reissner–Nordstrom cases. For each eigenstate the real energy is accompanied by a decay rate, which shows that the wavefunction evolution is not unitary. This is essentially because the black hole acts as an open system. One would therefore like to understand the decay rates in terms of a more complete, quantum description of a singularity. So far such a description has proved elusive. Furthermore, a fundamental question to address is whether the energy differences

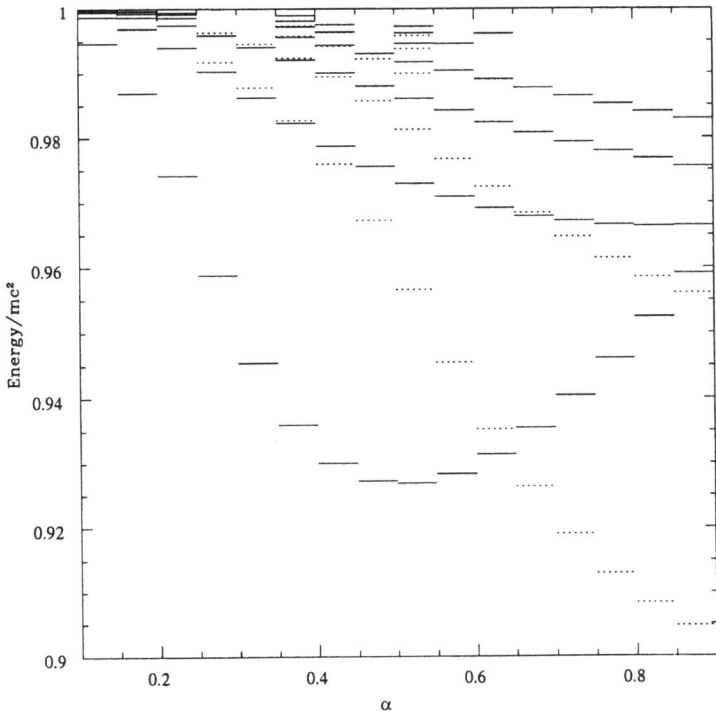

FIGURE 30.4. The energy spectra of the $S_{1/2}$ (solid) and $2P_{3/2}$ (dashed) states. Beyond a coupling of around $\alpha = 0.65$ the $2P_{3/2}$ state takes over as the system's ground state.

between shells corresponds to anything observable. Can quantum states spontaneously jump to lower orbits, with associated radiation? This behaviour is to be expected from the analogy with the Hydrogen atom, but is rather different from the classical picture of stable orbits around a black hole. Finally, we would clearly like to incorporate multiparticle effects in the description of the quantum physics around a black hole. Some ideas in this direction are discussed in section 30.5.

30.4   Conformal geometry

Conformal models of Euclidean geometry are currently the source of much interest in the computer graphics community [6, 22]. Here we aim to illustrate some features of spacetime conformal geometry. A useful starting point for constructing conformal models is the stereographic projection (Figure 30.5). We start with a Euclidean base space $\mathbb{R}^n$ of dimension n, and label points in this space with vectors x from some chosen origin. The stereographic projection maps the entire

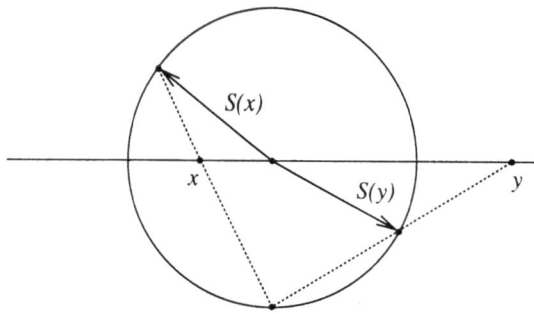

FIGURE 30.5. *The stereographic projection.* Given a point x we form the line through this point and the south pole of the sphere. The point where this line intersects the sphere defines the image of the projection.

space $\mathbb{R}^n$ onto the unit sphere in $\mathbb{R}^{n+1}$. If we let e denote the (new) vector perpendicular to the Euclidean base space, pointing to the south pole of the sphere, then the image on the sphere of the point x is

$$S(x) = \cos\theta\,\hat{x} - \sin\theta\,e, \tag{4.1}$$

where θ is obtained from the distance $r = |x|$ by

$$\cos\theta = \frac{2r}{1 + r^2}, \quad \sin\theta = \frac{1 - r^2}{1 + r^2}. \tag{4.2}$$

The stereographic projection has two important features. The first is that the origin of $\mathbb{R}^n$ is mapped to $-e$ and so is no longer a special point. That is, it is no longer represented by the zero vector. The second is that the point at infinity is mapped to e, so is also no longer algebraically special. Both of these features are valuable in graphics applications. Most computer graphics routines, including OpenGL, employ a projective representation of $\mathbb{R}^n$, as opposed to a stereographic one. The projective representation has the further advantage that it is *homogeneous*. In a homogeneous representation both X and λX represent the *same* point in $\mathbb{R}^n$, even if λ is *negative*. Such a representation is crucial if one wishes to let blades represent geometric objects. If B denotes a blade, then the geometric object associated with B is the solution space of the equation

$$B \wedge X = 0. \tag{4.3}$$

Clearly, if X solves this equation, then so does λX.

To convert the result of the stereographic projection to a homogeneous representation we introduce a further vector $\bar{e}$ with negative norm, $\bar{e}^2 = -1$. We now add this to the result of the stereographic projection to form the vector

$$X = S(x) + \bar{e}. \tag{4.4}$$

This vector is now *null*, $X^2 = 0$. This condition is homogeneous, so we can let X and λX represent the same Euclidean point x. Two important null vectors are provided by

$$n = e + \bar{e}, \qquad \bar{n} = e - \bar{e}, \tag{4.5}$$

which represent the point at infinity (n) and the origin $(\bar{n})$. With a rescaling, we can now write the map from the Euclidean point x to a null vector X as

$$X = F(x) = -(x - e)n(x - e) = (x^2 n + 2x - \bar{n}). \tag{4.6}$$

This clearly results in a null vector, $X^2 = 0$, as it is formed by a reflection of n. Of course, X can be arbitrarily rescaled and still represent the same point, but the representation in terms of $F(x)$ is often convenient in calculations.

30.4.1   Euclidean transformations

If we form the inner product of two conformal vectors, we find that

$$F(x) \cdot F(y) = -2|x - y|^2. \tag{4.7}$$

So conformal geometry encodes distances in a natural way [10, 15]. This is its great advantage over projective geometry as a framework for Euclidean space. Allowing for arbitrary scaling we have

$$|x - y|^2 = -2\frac{X \cdot Y}{X \cdot n\, Y \cdot n}, \tag{4.8}$$

which is manifestly homogeneous in X and Y. This formula returns the dimensionless distance. To introduce dimensions we require a fundamental length scale λ, so that we can then write

$$F(x) = \frac{1}{\lambda^2}(x^2 n + 2\lambda x - \lambda^2 \bar{n}). \tag{4.9}$$

This is simply the conformal representation of x/λ.

The group of Euclidean transformations consists of all transformations of Euclidean space which leave distances invariant. In conformal space these must leave the inner product invariant, so are given by reflections and rotations. Since the South pole represents the point at infinity, we expect that any Euclidean transformation of the base space will leave the n unchanged. This requirement is also clear from equation (4.8). For example, rotations about the origin in $\mathbb{R}^n$ are given by

$$x \mapsto Rx\tilde{R}, \tag{4.10}$$

where R is a rotor. The image of the transformed point, setting $\lambda = 1$, is

$$F(Rx\tilde{R}) = x^2 n + 2Rx\tilde{R} - \bar{n} = R(x^2 n + 2x - \bar{n})\tilde{R} = RX\tilde{R}, \tag{4.11}$$

which follows since $Rn\tilde{R} = n$ and $Rn\tilde{R} = \bar{n}$. Rotations about the origin are therefore given by the same formula in both spaces. Of greater interest are *translations*. We define the rotor T_a by

$$T_a = e^{na/2} = 1 + \tfrac{1}{2}na. \tag{4.12}$$

This satisfies

$$\begin{aligned}
T_a X \tilde{T}_a &= \tfrac{1}{2}(1 + \tfrac{1}{2}na)(x^2 n + 2x - \bar{n})(1 - \tfrac{1}{2}na) \\
&= \tfrac{1}{2}((x+a)^2 n + 2(x+a) - \bar{n}) \\
&= F(x+a).
\end{aligned} \tag{4.13}$$

So the conformal rotor T_a performs a translation in Euclidean space. (This is the origin of the biquaternion approach in three dimensions.) Furthermore, since translations and rotations about the origin are handled by rotors, it is possible to construct simple rotors encoding rotations about arbitrary points. Such operations are essential in graphics routines.

30.4.2   Minkowski and de Sitter spaces

The conformal framework applies straightforwardly to spacetime. One immediate observation is that spacetime can be viewed as a conformal model for the plane. This provides a method of visualising the Lorentz group in terms of conformal (angle-preserving) transformations of the plane, or the 2-sphere. This model has been developed extensively by Penrose & Rindler [29]. Of greater interest here is the conformal model of full spacetime. In this model points in spacetime are described by null vectors in a space of signature $(2, 4)$. The conformal representation is again constructed by equation (4.6), with $x = x^\mu \gamma_\mu$ a point in spacetime.

In conformal geometry, lines and circles are treated in a unified manner as trivector blades. That is, the unique line through A, B and C is encoded by the trivector $L = A \wedge B \wedge C$. If L includes the point at infinity, the line is straight. A circle is the set of all points in a plane a fixed distance from some chosen point (if the line is straight the 'centre' is at infinity). Circles in Lorentzian spaces therefore include hyperbolas with lightlike asymptotes. So trivector blades in the geometric algebra $\mathcal{G}(2, 4)$ can define straight lines, circles and (both branches of a) hyperbola in a single unified manner. A straightforward example of a hyperbola is the trivector $\gamma_0 \gamma_1 \bar{e}$. This encodes the set of all points in the $\gamma_0 \gamma_1$ plane satisfying $x^2 = 1$.

A significant feature of the conformal picture is that it easily generalises to include hyperbolic and spherical geometry [8, 23, 26]. To convert from a conformal model of Euclidean space to spherical space we simply replace the distance formula of equation (4.8) by

$$d(x,y) = \lambda \sin^{-1}\left(-\frac{X \cdot Y}{2X \cdot \bar{e}\, Y \cdot \bar{e}}\right)^{\frac{1}{2}}, \tag{4.14}$$

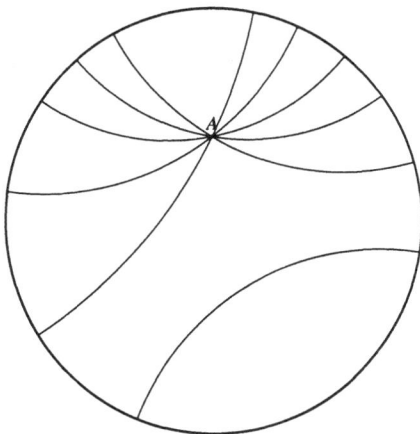

FIGURE 30.6. *The Poincaré disc*. Points inside the disc represent points in a hyperbolic space. A set of *d*-lines are also shown.

so $\bar{e}$ replaces n to convert to spherical geometry. All conformal representations of objects (points, lines etc.) are unchanged, though their geometric interpretation is altered. Similarly, hyperbolic geometry is recovered by imposing the distance measure

$$d(x,y) = \lambda \sinh^{-1}\left(-\frac{X \cdot Y}{2X \cdot e\, Y \cdot e}\right)^{\frac{1}{2}}. \qquad (4.15)$$

The straight line, or geodesic, through two points X and Y in a hyperbolic geometry is defined by the trivector $X \wedge Y \wedge e$. These are called *d*-lines in two dimensions. If we plot hyperbolic points on the Euclidean plane we arrive at the Poincaré disk model of two-dimensional hyperbolic geometry (Figure 30.6). In this picture *d*-lines are (Euclidean) circles which intersect the unit circle at right angles. Figure 30.6 illustrates the hyperbolic version of the parallel postulate. Given a line, and a point A not on the line, one can construct infinitely many lines through A which do not intersect the original line. (For Euclidean geometry only one line exists, and there are none for spherical geometry.)

Similar considerations apply to the accompanying symmetry groups. The Euclidean group is the set of conformal reflections and rotations which leave n invariant. Similarly, spherical transformations keep $\bar{e}$ invariant and hyperbolic transformations keep e invariant. A natural question is how does this generalise to spacetime. The answer is quite straightforward. The geometry obtained by keeping e invariant is that of de Sitter space and for $\bar{e}$ we recover anti-de Sitter space (AdS). The second of these is currently a subject of much interest due to the AdS/CFT correspondence and its relation to the holographic principle [27]. It is therefore gratifying to see that anti-de Sitter space has a very natural encoding in geometric algebra. For example, the geodesic through the points with conformal representation X and Y is encoded in the trivector $X \wedge Y \wedge \bar{e}$, see [18]. These lines can be intersected, reflected, etc. in an extremely straightforward manner. The same operations are considerably harder to formulate if one adopts the tools of differential geometry.

30.5   Multiparticle quantum theory

The STA representation of single-particle quantum states has existed since the 1960s. But it was only in the 1990s that this approach was successfully generalised to multiparticle systems [5]. The essential construction is the geometric algebra of the $4n$-dimensional relativistic configuration space. This is called the multiparticle spacetime algebra or MSTA. With a separate copy of spacetime for each particle present, the MSTA is generated by the $4n$ vectors $\{\gamma_\mu^a\}$, $\mu = 0, \ldots, 3$, $a = 1, \ldots, n$. These satisfy

$$\gamma_\mu^a \cdot \gamma_\nu^b = \tfrac{1}{2}(\gamma_\mu^a \gamma_\nu^b + \gamma_\nu^b \gamma_\mu^a) = \eta_{\mu\nu}\delta^{ab}, \tag{5.1}$$

so generators from distinct spaces all anticommute.

Relative bivectors from separate spaces are constructed in the same way as the single-particle case. Again taking γ_0^a to represent the frame velocity vector, we write

$$\sigma_k^a = \gamma_k^a \gamma_0^a. \tag{5.2}$$

But now we see that for $a \neq b$ the bivectors from different spaces *commute*

$$\sigma_i^a \sigma_j^b = \gamma_i^a \gamma_0^a \gamma_j^b \gamma_0^b = \gamma_i^a \gamma_j^b \gamma_0^b \gamma_0^a = \gamma_j^b \gamma_0^b \gamma_i^a \gamma_0^a = \sigma_j^b \sigma_i^a. \tag{5.3}$$

This provides a natural construction of the tensor product in terms of the geometric product. As such, the MSTA is clearly a natural arena to study multiparticle Hilbert space, which itself is constructed from a tensor product of the individual Hilbert spaces.

The simplest construction of a 2-particle state in the MSTA is as the product of two single particle spinors, $\phi^1 \psi^2$. But this generates a space of 64 real dimensions, whereas we only expect 32 (for the 16-dimensional complex space). We have failed to account for the fact that quantum theory requires a single, global complex structure. We must therefore ensure that post-multiplying the states by either $(I\sigma_3)^1$ or $(I\sigma_3)^2$ results in the same state. That is

$$\psi(I\sigma_3)^1 = \psi(I\sigma_3)^2. \tag{5.4}$$

A simple rearrangement now yields

$$\psi = -\psi(I\sigma_3)^1(I\sigma_3)^2 = \psi\tfrac{1}{2}\big(1 - (I\sigma_3)^1(I\sigma_3)^2\big). \tag{5.5}$$

If we now define

$$E = \tfrac{1}{2}\big(1 - (I\sigma_3)^1(I\sigma_3)^2\big), \tag{5.6}$$

we see that all states must satisfy

$$\psi = \psi E. \tag{5.7}$$

The 2-particle *correlator* E is an idempotent, $E^2 = E$, and so it removes precisely half the degrees of freedom from a general product state, as required. The complex structure is now provided by right multiplication by the multivector J, where

$$J = E(I\sigma_3)^1 = E(I\sigma_3)^2 = \tfrac{1}{2}((I\sigma_3)^1 + (I\sigma_3)^2). \tag{5.8}$$

It follows that $J^2 = -E$.

30.5.1 Entanglement

Entanglement is perhaps the key feature of the quantum world without a classical counterpart. Entangled (pure) states are those which cannot be factored into a direct product of single-particle states. A simple example is provided by the non-relativistic singlet state

$$|\chi\rangle = \frac{1}{\sqrt{2}} \left\{ \begin{pmatrix} 1 \\ 0 \end{pmatrix} \otimes \begin{pmatrix} 0 \\ 1 \end{pmatrix} - \begin{pmatrix} 0 \\ 1 \end{pmatrix} \otimes \begin{pmatrix} 1 \\ 0 \end{pmatrix} \right\}. \tag{5.9}$$

The MSTA version of this is

$$\chi = \frac{1}{\sqrt{2}} \left((I\sigma_2)^1 - (I\sigma_2)^2 \right) E. \tag{5.10}$$

This satisfies the following property:

$$(I\sigma_k)^1 \chi = -(I\sigma_k)^2 \chi, \qquad k = 1\ldots 3. \tag{5.11}$$

So if M is a multivector of the form $M = M^0 + M^k I\sigma_k$, we have

$$M^1 \chi = \tilde{M}^2 \chi. \tag{5.12}$$

This provides a quick proof that the singlet state is rotationally invariant, since under a joint rotation

$$\chi \mapsto R^1 R^2 \chi = R^1 \tilde{R}^1 \chi = \chi, \tag{5.13}$$

where R is a spatial rotor.

A useful parameterisation of a general 2-particle state is provided by the Schmidt decomposition, which enables us to write a non-relativistic state as

$$|\psi\rangle = \cos(\tfrac{1}{2}\eta)e^{i\tau/2} \begin{pmatrix} \cos(\tfrac{1}{2}\phi^1)e^{\frac{1}{2}i\theta^1} \\ \sin(\tfrac{1}{2}\phi^1)e^{-\frac{1}{2}i\theta^1} \end{pmatrix} \otimes \begin{pmatrix} \cos(\tfrac{1}{2}\phi^2)e^{\frac{1}{2}i\theta^2} \\ \sin(\tfrac{1}{2}\phi^2)e^{-\frac{1}{2}i\theta^2} \end{pmatrix}$$
$$+ \sin(\tfrac{1}{2}\eta)e^{-i\tau/2} \begin{pmatrix} \sin(\tfrac{1}{2}\phi^1)e^{-\frac{1}{2}i\theta^1} \\ -\cos(\tfrac{1}{2}\phi^1)e^{\frac{1}{2}i\theta^1} \end{pmatrix} \otimes \begin{pmatrix} \sin(\tfrac{1}{2}\phi^2)e^{-\frac{1}{2}i\theta^2} \\ -\cos(\tfrac{1}{2}\phi^2)e^{\frac{1}{2}i\theta^2} \end{pmatrix}. \tag{5.14}$$

The MSTA version of the Schmidt decomposition is very revealing. We let

$$R = \exp(\tfrac{1}{2}\theta^1 I\sigma_3)\exp(-\tfrac{1}{2}\phi^1 I\sigma_2)\exp(\tfrac{1}{2}\tau I\sigma_3),$$
$$S = \exp(\tfrac{1}{2}\theta^2 I\sigma_3)\exp(-\tfrac{1}{2}\phi^2 I\sigma_2)\exp(\tfrac{1}{2}\tau I\sigma_3) \tag{5.15}$$

so that we can write

$$\psi = R^1 S^2 \big(\cos(\tfrac{1}{2}\eta) + \sin(\tfrac{1}{2}\eta)(I\sigma_2)^1 (I\sigma_2)^2\big) E. \tag{5.16}$$

This neat expression shows how all information regarding the entanglement is contained in a single term on the right of the product state. This gives a form of decomposition which generalises to arbitrary numbers of particles, though the calculations are far from straightforward and much work remains [28]. Geometric algebra brings geometric insight into the nature of the multiparticle Hilbert space and can be used to quantify the nature of entanglement for pure and mixed states. This approach has helped develop new models of decoherence, some of which are currently being tested in nuclear magnetic resonance experiments [12].

30.5.2   Relativistic states and wave equations

We now turn to relativistic multiparticle states. There are many theoretical problems encountered in constructing a wavefunction treatment of interacting particles and we will not discuss these issues here. (See Dolby & Gull for a recent [3] account of the case of non-interacting particles in a classical background electromagnetic field.) Here we wish to demonstrate how the MSTA is used for constructing wave equations for particles with spin other than $\tfrac{1}{2}$. This adapts the work of Bargmann & Wigner [2].

We start with a 2-particle relativistic wavefunction, which is a function of the four spacetime coordinates x^μ, so

$$\psi = \psi(x^\mu)E. \tag{5.17}$$

This satisfies the 2-particle Dirac equation of the form

$$\nabla^1 \psi^1 J\gamma_0^1 = m\psi. \tag{5.18}$$

Any dependence on the particle space label is removed by assigning definite symmetry or antisymmetry to ψ under particle interchange. A spin-1 equation is constructed from a symmetric ψ, and a spin-0 equation from an antisymmetric ψ. It is the latter that interests us here. To simplify this problem we first introduce the relativistic singlet states

$$\epsilon = \big((I\sigma_2)^1 - (I\sigma_2)^2\big)\tfrac{1}{2}(1 + \sigma_3^1)\tfrac{1}{2}(1 + \sigma_3^2)E,$$
$$\bar{\epsilon} = \big((I\sigma_2)^1 - (I\sigma_2)^2\big)\tfrac{1}{2}(1 - \sigma_3^1)\tfrac{1}{2}(1 - \sigma_3^2)E. \tag{5.19}$$

These are both relativistic states because they contain factors of the idempotent $\tfrac{1}{2}(1 - I^1 I^2)$. This is sufficient to ensure that, for a general even multivector M, we have

$$M^1\epsilon = \tilde{M}^2\epsilon, \qquad M^1\bar{\epsilon} = \tilde{M}^2\bar{\epsilon}. \tag{5.20}$$

In particular, if R is a relativistic rotor and we perform a joint Lorentz transformation in both spaces, we see that

$$\epsilon \mapsto R^1 R^2 \epsilon = R^1 \tilde{R}^1 \epsilon = \epsilon, \qquad (5.21)$$

with the same property holding for $\bar{\epsilon}$. The 32 real degrees of freedom in an arbitrary relativistic 2-particle state can be mapped into particle-1 space by writing

$$\psi = M^1 \epsilon + N^1 \bar{\epsilon} + s^1 \epsilon \gamma_0^1 + t^1 \bar{\epsilon} \gamma_0^1, \qquad (5.22)$$

where M and N are general even-grade multivectors, and s and t are odd-grade.

As ϵ and $\bar{\epsilon}$ are antisymmetric under particle interchange, an antisymmetric state must have $M = \tilde{M}$, $N = \tilde{N}$ and $\tilde{t} = s$. The twelve scalar degrees of freedom in a general antisymmetric wavefunction can therefore be written as

$$\psi = (\alpha + I^1 \beta)\epsilon + (\theta + I^1 \eta)\bar{\epsilon} + (u + Iv)^1 \bar{\epsilon} J \gamma_0^1 + (u - Iv)^1 \epsilon J \gamma_0^1, \qquad (5.23)$$

where u and v are vectors, and α, β, θ, η are scalars. A joint rotation of ψ shows that u and v do transform as vectors, since

$$R^1 R^2 (u + Iv)^1 \epsilon = \left(R(u + Iv)\tilde{R}\right)^1 \epsilon. \qquad (5.24)$$

So, while an antisymmetric ψ provides a useful way of generating a spin-0 wave equation, the wavefunction clearly contains some spin-1 terms.

The Dirac equation (5.18) applied to the antisymmetric state ψ leads to a first-order version of the Klein–Gordon equation, in a form similar to that first obtained by Kemmer,

$$\nabla(\alpha + I\beta) = m(u + Iv), \quad \nabla(u + Iv) = -m(\alpha + I\beta). \qquad (5.25)$$

Two degrees of freedom are eliminated by the Dirac equation, which sets $\theta = \alpha$ and $\eta = -\beta$. The equations describe a complex scalar field, with an associated first-order potential of the form of a vector + trivector. The complex structure now arises naturally on the pseudoscalar. The field equations are obtained simply from the standard Dirac Lagrangian, which takes the MSTA form

$$L = \langle \nabla^1 \psi (I\gamma_3)^1 \tilde{\psi} - m\psi\tilde{\psi} \rangle. \qquad (5.26)$$

The canonical stress-energy tensor defined by this Lagrangian is

$$T(a) = \langle a \cdot \nabla^1 \psi (I\gamma_3)^1 \tilde{\psi} \rangle_1^1 \qquad (5.27)$$

where the right-hand side denotes the projection onto the vectors in particle space 1. The surprising feature of this stress-energy tensor is that it is *not* equal to that usually assigned to the Klein–Gordon field. In terms of ψ the conventional stress-energy tensor is

$$T(a) = m\langle a^2 \psi \gamma_0^1 \gamma_0^2 \tilde{\psi} \rangle_1^1. \qquad (5.28)$$

This stress-energy tensor is canonical to a somewhat strange symmetry of the Lagrangian and is not obviously related to translations.

In terms of a complex scalar field ϕ the conventional stress-energy tensor is

$$T(a)_{\mathrm{conv}} = \tfrac{1}{4}\left(\nabla\phi a\nabla\phi^* + \nabla\phi^* a\nabla\phi + 2m^2\phi\phi^*\right), \qquad (5.29)$$

whereas our stress-energy tensor is

$$T(a) = \tfrac{1}{4}\left(\nabla\phi^*\, a\cdot\nabla\phi + \nabla\phi\, a\cdot\nabla\phi^* - \phi^* a\cdot\nabla(\nabla\phi) - \phi a\cdot\nabla(\nabla\phi^*)\right). \qquad (5.30)$$

Both tensors are symmetric and have the same form for plane-wave states. They differ by a total divergence and lead to the same total energy when integrated over a hypersurface. The results of quantum field theory are therefore largely unaffected by the choice of tensor. But gravity is affected by the *local* form of the matter stress-energy tensor, which raises the question; which is the correct stress-energy tensor to use as a source of gravitation? This is a particularly important question to resolve because the stress-energy tensor for the scalar field is a cornerstone of many areas of modern cosmology, including inflation, quintessence, and cosmic strings. The two stress-energy tensors clearly produce different dynamics in a cosmological setting because the conventional tensor includes pressure terms which are absent from our new tensor. Furthermore, the coupling to gravity is different for the two fields, as the antisymmetrised MSTA state couples into the torsion sector of the theory.

30.5.3   Twistors and conformal geometry

In section 30.4.2 we saw that the orthogonal group $O(2,4)$ provides a double-cover representation of the spacetime conformal group. The representation is a double cover because the null vectors X and $-X$ generate the same point in spacetime. For physical applications we are typically interested in the restricted conformal group. This consists of transformations that preserve orientation and time sense, and contains translations, proper orthochronous rotations, dilations and special conformal transformations. The restricted orthogonal group, $SO^+(2,4)$, is a double-cover representation of the restricted conformal group. But the group $SO^+(2,4)$ itself has a double-cover representation provided by the rotor group $\mathrm{spin}^+(2,4)$. So the group of all rotors in conformal space is a four-fold covering of the restricted conformal group. The rotor group $\mathrm{spin}^+(2,4)$ is isomorphic to the Lie group $SU(2,2)$, so the action of the restricted conformal group can be represented in terms of complex linear transformations of four-dimensional vectors in a complex space of signature $(2,2)$. This is the basis of the *twistor* programme, initiated by Roger Penrose [30].

We can establish a simple realisation of a single-particle twistor in the STA. We define the spinor inner product by

$$\langle\tilde{\psi}\phi\rangle_q = \langle\tilde{\psi}\phi\rangle - \langle\tilde{\psi}\phi I\sigma_3\rangle I\sigma_3. \qquad (5.31)$$

This defines a complex space with precisely the required $(2, 2)$ metric. We continue to refer to ψ and ϕ as *spinors*, as they are acted on by a spin representation of the restricted conformal group. To establish a representation of the conformal group we simply need a representation of the bivector algebra of $\mathcal{G}(2, 4)$ in terms of an action on spinors in the STA. A suitable representation is provided by [8]

$$e\gamma_\mu \leftrightarrow \gamma_\mu \psi \gamma_0 I \sigma_3 = \gamma_\mu \psi I \gamma_3,$$
$$\bar{e}\gamma_\mu \leftrightarrow I \gamma_\mu \psi \gamma_0. \tag{5.32}$$

One can now track through to establish representations for the action of members of the conformal group. A spacetime translation by the vector a has the spin representation

$$T_a(\psi) = \psi + a\psi I \gamma_3 \tfrac{1}{2}(1 + \sigma_3). \tag{5.33}$$

It is straightforward to establish that the spinor inner product $\langle \tilde{\psi}\phi \rangle_q$ is invariant under this product.

If we now pick a spinor ϕ to represent the origin in spacetime, we can write

$$\phi = \zeta \tfrac{1}{2}(1 + \sigma_3) - \pi I \sigma_2 \tfrac{1}{2}(1 - \sigma_3), \tag{5.34}$$

where ζ and π are Pauli spinors. The conventional STA representation of the (valence-1) *twistor* $Z^\alpha = (\omega^A, \pi_{A'})$ is obtained by applying a translation of $-r$ [19],

$$Z = \phi - r\phi I \gamma_3 \tfrac{1}{2}(1 + \sigma_3). \tag{5.35}$$

In twistor theory, twistors are viewed as more primitive objects than points, with spacetime points generated by antisymmetrised pairs of twistors. The condition that two twistors generate a real point can be satisfied by setting

$$X = \omega \tfrac{1}{2}(1 - \sigma_3) + r\omega I \gamma_3 \tfrac{1}{2}(1 + \sigma_3),$$
$$Z = \kappa \tfrac{1}{2}(1 - \sigma_3) + r\kappa I \gamma_3 \tfrac{1}{2}(1 + \sigma_3), \tag{5.36}$$

where ω and κ are Pauli spinors. The antisymmetrised product of X and Z can be constructed straightforwardly in the MSTA as

$$\psi_r = (X^1 Z^2 - Z^1 X^2)E. \tag{5.37}$$

If we now apply the decomposition of equation (5.23) we find that

$$\psi_r = (r\cdot r\,\epsilon - (r^1 \epsilon \gamma_0^1 + r^2 \epsilon \gamma_0^2)J - \bar{\epsilon})\langle I\sigma_2 \tilde{\omega}\kappa \rangle_q. \tag{5.38}$$

The singlet state ϵ therefore represents the point at infinity, with $\bar{\epsilon}$, representing the origin ($r = 0$). Furthermore, if ψ_r and ϕ_s are the valence-2 twistors representing the points r and s, we find that

$$-\frac{\langle \tilde{\psi}_r \phi_s \rangle_q}{2\langle \tilde{\psi}_r \epsilon \rangle_q \langle \tilde{\phi}_s \epsilon \rangle_q^*} = (r - s)\cdot(r - s). \tag{5.39}$$

The multiparticle inner product therefore recovers the square of the spacetime distance between points. This establishes the final link between the MSTA, twistor geometry and the conformal representation of distance geometry of equation (4.8). Conformal transformations are applied as joint transformations in both particle spaces. So, for example, a translation is generated by

$$\psi_r \mapsto \psi'_r = T_{a^1} T_{a^2} \psi_r. \tag{5.40}$$

After a little work we establish that

$$\psi'_r = (r + a) \cdot (r + a)\, \epsilon - (r + a)^1 \eta \gamma_0^1 J - \bar{\epsilon}, \tag{5.41}$$

as required (see also [21]).

We have now established a multiparticle quantum representation of conformal space, so can apply the insights of section 30.4.2 to extend the representation to de Sitter and anti-de Sitter spaces. We have seen that the *infinity twistor* of Lorentzian spacetime is represented by the singlet state ϵ. It follows that the infinity twistors for de Sitter and anti-de Sitter spaces are $(\epsilon + \bar{\epsilon})$ and $(\epsilon - \bar{\epsilon})$ respectively [1]. Much work remains in fully elucidating aspects of the twistor programme in the MSTA setup. The links described in this paper should convince the reader that this will prove to be a fruitful line of research. Furthermore, the representation of conformal transformations developed here is naturally related to supersymmetry. This suggests that conformal spacetime algebra may well prove to be the natural setting for supersymmetric field theory.

Acknowledgements

The work reviewed in this paper has been carried out in collaboration with a number of researchers. In particular, we thank Jonathan Pritchard, Alejandro Cáceres, Rachel Parker, Suguru Furuta and Stephen Gull of the University of Cambridge. As ever, we have benefited enormously from discussions with David Hestenes.

REFERENCES

[1] E. Arcaute, A. Lasenby and C. Doran, Twistors and geometric algebra. To be submitted to *J. Math. Phys.*

[2] V. Bargmann and E. Wigner, Group theoretical discussion of relativistic wave equations, *Proc. Nat. Sci.* (USA) **34**, (1948), 211.

[3] C. Dolby and S. Gull, New approach to quantum field theory for arbitrary observers in electromagnetic backgrounds, *Annals. Phys.* **293**, (2001), 189–214.

[4] C.J.L. Doran, A. N. Lasenby and S. F. Gull, States and operators in the spacetime algebra, *Found. Phys.* **23** (9), (1993), 1239–1264.

[5] C. Doran, A. Lasenby, S. Gull, S. Somaroo and A. Challinor, Spacetime algebra and electron physics. In P. W. Hawkes, editor, *Advances in Imaging and Electron Physics* Vol. **95**, Academic Press, 1996, pp. 271–386.

[6] C. Doran, A. Lasenby and J. Lasenby, Conformal geometry, Euclidean space and geometric algebra. In J. Winkler & M. Niranjan editors, *Uncertainty in Geometric Computations*, Kluwer, 2002, pp. 41–58.

[7] C. Doran and A. Lasenby, Perturbation theory calculation of the black hole elastic scattering cross section, *Phys.Rev.* D **66** (2002) 024006.

[8] C. Doran and A. Lasenby, *Geometric Algebra for Physicists*, Cambridge University Press, 2003.

[9] C. Doran and A. Lasenby, Physical Applications of Geometric Algebra. Lecture course presented at the University of Cambridge. Course notes available at http://www.mrao.cam.ac.uk/~clifford/ptIIIcourse/.

[10] A.W.M. Dress and T.F. Havel, Distance geometry and geometric algebra, *Found. Phys.* **23**, (1993), 1357.

[11] S.F. Gull, A.N. Lasenby and C.J.L. Doran, Electron Paths, Tunnelling and diffraction in the spacetime algebra, *Found. Phys.* **23** (10), (1993), 1329–1356.

[12] T. Havel and C. Doran, Geometric algebra in quantum information processing. In S. Lomonaco, ed. *Quantum Computation and Quantum Information Science*, pp. 81–100. AMS Contemporary Math series (2000). quant-ph/0004031

[13] D. Hestenes, *Space-time Algebra*, Gordon and Breach, 1966.

[14] D. Hestenes, *New Foundations for Classical Mechanics (Second Edition)*, Kluwer Academic Publishers, Dordrecht, 1999.

[15] D. Hestenes, Old wine in new bottles: A new algebraic framework for computational geometry. In E. Bayro-Corrochano and G. Sobczyk, eds. *Geometric Algebra with Applications in Science and Engineering*, Birkhäuser, Boston, 2000, pp. 3–17.

[16] D. Hestenes and G. Sobczyk, *Clifford Algebra to Geometric Calculus: A Unified Language for Mathematics and Physics*, D. Reidel, Dordrecht, 1984.

[17] T.W.B. Kibble, Lorentz invariance and the gravitational field, *J. Math. Phys.* **2** (3), (1961), 212.

[18] A. Lasenby, Conformal geometry and the universe. To appear in *Phil. Trans. R. Soc. Lond.* A.

[19] A. Lasenby, C. Doran and S. Gull. 1993. 2-spinors, twistors and supersymmetry in the space-time algebra. In Z. Oziewicz, B. Jancewicz and A. Borowiec, eds., *Spinors, Twistors, Clifford Algebras and Quantum Deformations*, Kluwer Academic, Dordrecht, 233.

[20] A. Lasenby, C. Doran and S. Gull, Gravity, gauge theories and geometric algebra, *Phil. Trans. Roy. Soc. Lond.* A**356**, (1998), 487–582.

[21] A. Lasenby and J. Lasenby, Applications of geometric algebra in physics and links with engineering. In E. Bayro and G. Sobczyk, eds., *Geometric algebra: a geometric approach to computer vision, neural and quantum computing, robotics and engineering*, Birkhäuser, Boston, 2000, pp. 430–457.

[22] A. Lasenby and J. Lasenby, Surface evolution and representation using geometric algebra, In Roberto Cipolla and Ralph Martin, eds., *The Mathematics of Surfaces IX: Proceedings of the ninth IMA conference on the mathematics of surfaces*, Springer, London, 2001, pp. 144–168.

[23] A.N. Lasenby, J. Lasenby and R. Wareham, A covariant approach to geometry and its applications in computer graphics, 2003. Submitted to *ACM Transactions on Graphics*.

[24] A. Lasenby and C. Doran, Geometric algebra, Dirac wavefunctions and black holes. In *Advances in the Interplay Between Quantum and Gravity Physics*, P.G. Bergmann and V. de Sabbata, eds., Kluwer 2002, pp. 251–283.

[25] A. Lasenby, C. Doran, J. Pritchard and A. Cáceres, Bound states and decay times of fermions in a Schwarzschild black hole background, gr-qc/0209090.

[26] H. Li, Hyperbolic geometry with Clifford algebra, *Acta Appl. Math.* **48**, (1997), 317–358.

[27] J. Maldacena, The large N limit of superconformal field theories and supergravity, *Adv. Theor. Math. Phys.* **2**, (1998), 231–252.

[28] R. Parker and C. Doran, Analysis of 1 and 2 particle quantum systems using geometric algebra. In *Applied Geometrical Algebras in Computer Science and Engineering*, C. Doran, L. Dorst and J. Lasenby, eds., AGACSE 2001, Birkhäuser, 2001, 215. quant-ph/0106055

[29] R. Penrose and W. Rindler, *Spinors and Space-time, Volume 1, Two-Spinor Calculus and Relativistic Fields*, Cambridge University Press 1984.

[30] R. Penrose and W. Rindler, *Spinors and Space-time, Volume 2, Spinor and Twistor Methods in Space-time Geometry*, Cambridge University Press 1986.

Anthony Lasenby
Astrophysics Group, Cavendish Laboratory
Madingley Road, Cambridge, CB3 0HE, UK
E-mail: a.n.lasenby@mrao.cam.ac.uk

Chris Doran
Astrophysics Group, Cavendish Laboratory
Madingley Road, Cambridge, CB3 0HE, UK
E-mail: c.doran@mrao.cam.ac.uk

Elsa Arcaute
Astrophysics Group, Cavendish Laboratory
Madingley Road, Cambridge, CB3 0HE, UK
E-mail: ea235@mrao.cam.ac.uk

Submitted: January 16, 2003.

31

Noncommutative Physics on Lie Algebras, $(\mathbb{Z}_2)^n$ Lattices and Clifford Algebras

Shahn Majid

ABSTRACT We survey noncommutative spacetimes with coordinates being enveloping algebras of Lie algebras. We also explain how to do differential geometry on noncommutative spaces that are obtained from commutative ones via a Moyal-product type cocycle twist, such as the noncommutative torus, θ-spaces and Clifford algebras. The latter are noncommutative deformations of the finite lattice $(\mathbb{Z}_2)^n$ and we compute their noncommutative de Rham cohomology and moduli of solutions of Maxwell's equations. We exactly quantize noncommutative $U(1)$-Yang–Mills theory on $\mathbb{Z}_2 \times \mathbb{Z}_2$ in a path integral approach.

Keywords: quantum groups, noncommutative geometry, quantum field theory, quantum gravity, finite lattice, twisting

31.1   Introduction

Noncommutative geometry has made rapid progress in the past two decades and is arguably now mature enough to be fully computable and relevant to real physical experiments. The benefit of noncommutative geometry is that whereas the usual coordinates on a space are commutative and most 'nice' commutative algebras could be viewed that way, noncommutative geometry, by relaxing the commutativity assumption, allows practically any algebra to be viewed geometrically. This opens up a brave new world of possibilities for model-building in which our favorite noncommutative algebras, such as matrices, angular momentum operators, Clifford algebras, can be viewed as noncommutative coordinates with all of the geometry that that entails. And why should we need such a possibility? First of all, we already encounter such noncommutative geometries when we quantise a system. Recall that classical mechanics is described by classical symplectic geometry, Hamilton–Jacobi equations of motion, etc. What happens to all this geometry when we quantise, is it all just thrown away after we have got the right commu-

AMS Subject Classification: 58B32, 58B34, 20C05, 81T75. This article is in final form and no version of it will be submitted for publication elsewhere.

tators? The correspondence principle in quantum mechanics says that it should not be thrown away, that classical variables such as angular momentum should have their analogues in the quantum system with analogous properties. But this is a rather vague statement. Analogues of what? It seems likely that the quantum system should be as rich if not richer in structure than the classical one and should have among other things analogues of all of the geometry of phase space, Hamiltonian flows etc., if only we had the 'eyes' to see it. Noncommutative geometry can in principle provide those eyes.

In this article we will not apply noncommutative geometry to quantum phase space exactly, though that is an interesting direction that deserves more study. Instead we will look at the even more radical proposal that spacetime itself may have noncommutative coordinates, an as yet undiscovered but potentially new physical phenomenon if it were ever verified. It could be called 'cogravity' for reasons explained in Section 31.2. The idea is not new [1], but we shall focus on three simple examples based on Lie algebras viewed as noncommutative spacetimes, with some new observations for the θ-spacetime case.

After the preliminary Section 31.2, we shall move to new results. The first, in Section 31.3 is an application of noncommutative geometry to quantize $U(1)$-Yang–Mills theory on the finite set $\mathbb{Z}_2 \times \mathbb{Z}_2$ all the way up to Wilson loop vacuum expectation values. Section 31.4 covers the electromagnetic theory on Clifford algebras $\mathrm{Cliff}(n)$ viewed as noncommutative coordinates. This is not only for fun; Clifford algebras can be viewed as 'Moyal-product' cocycle type quantizations of $\mathbb{Z}_2^n$ in the same spirit as recently popular proposals for θ-spacetime in string theory, but now in a discrete version. Section 31.5 is a more mathematical section for experts and is the general theory behind such cocycle twist quantisations. This theory depends on Drinfeld's quantum groups work but its full impact remains not so well known in the operator theory approach to noncommutative geometry of Connes [2] and others. We point out that cocycle twisting means by the Majid–Oeckl twisting theorem [3] that the differential geometry also twists (or gets 'quantised') by the same cocycle and that this covers the algebraic (but not functional analytic) aspects of many famous examples including the celebrated noncommutative torus A_θ.

Next, a few words about the relation between the operator theory and quantum groups approaches to noncommutative geometry. From a mathematical point of view the former began in the 1940s with the theorem of Gelfand and Naimark characterising functions on locally compact spaces as commutative C^* algebras and hence proposing any noncommutative C^* algebra as a 'noncommutative space'. Vector bundles were characterised in the 1960s by the Serre–Swann theorem as finitely generated projective modules, a notion again working in the noncommutative case. Operator K-theory led in the 1980s to cyclic cohomology and ultimately to Connes' formalisation of the Dirac operator on a spin manifold in operator algebra terms as a 'spectral triple' [2]. At the same time in the 1980s there emerged from the theory of integrable systems a large class of examples of noncommutative algebras with manifest 'geometrical' content, namely quantum groups $\mathbb{C}_q[G]$ deforming the usual coordinate ring of a simple Lie group G. Also at the same time in the 1980s there emerged a different 'bicrossproduct' class of quantum groups [4] again built on Lie groups but this time from a self-duality approach to quantum gravity and Planck-scale physics. Moreover, just as Lie groups and their homogeneous spaces provide key examples of classical differential geometry, so quantum groups *should* provide key examples of noncommutative geometry. This 'quantum groups approach' to noncommutative geometry starts with differential structures on quantum groups [5], gauge theory with quantum group fiber [6] and eventually a notion of a 'quantum manifold' as an algebra equipped with a quantum group frame bundle [7]. It is important for us here, however, that whereas the q-deformation literature and examples like $\mathbb{C}_q[SU_2]$ played an important role in developing the right axioms, the final theory, like the operator algebras approach, applies in principle to any algebra including ones that have nothing whatever to do with q-deformations. It should be said also that the operator theory approach tends to be mathematically more sophisticated while the quantum groups one tends to be more computational and examples-led. One of the aims of Section 31.5 is to promote convergence between the two approaches.

We will need some precise notation. Let M be a unital algebra over $\mathbb{C}$ (say). We regard M as 'coordinate algebra' on a space, even though it need not be commutative. By 'differential calculus' we mean (Ω^1, d) where Ω^1 is an $M - M$ bimodule, $\mathrm{d} : M \to \Omega^1$ obeys the Leibniz rule

$$\mathrm{d}(fg) = (\mathrm{d}f)g + f\,\mathrm{d}g, \quad \forall f, g \in M \tag{1.1}$$

and elements of the form $f\,\mathrm{d}g$ span Ω^1. We also want in practice that the kernel of d is spanned by 1, the constants. Note that we only need Ω^1 to be a bimodule— indeed, assuming $f\,\mathrm{d}g = (\mathrm{d}g)f$ for all f, g would imply $\mathrm{d}(fg - gf) = 0$ which would make d trivial on a very noncommutative algebra. So we do not assume this, but only the weaker bimodule condition $f((\mathrm{d}g)h) = (f\,\mathrm{d}g)h$ for all f, g, h. This is the starting point of all approaches to noncommutative geometry. After this, one can extend to an entire exterior algebra of differential forms Ω, d with d a super-derivation of degree 1 with $\mathrm{d}^2 = 0$. How this is done is different in the two approaches; in the quantum groups one we use considerations of a background symmetry to narrow down the possible extensions with the result that in most cases there is a reasonably natural choice. In the Connes approach it is defined by choosing an operator to be the 'Dirac' operator. Another ingredient one needs is a Hodge $\star$ operator $\Omega^m \to \Omega^{n-m}$ where n is the top degree (which we assume exists, and which is called the volume-dimension of the noncommutative space).

With these few ingredients one can already do a lot of physics, in particular most of electromagnetism. The simplest version of this which we call 'Maxwell theory' is to think of a gauge field $A \in \Omega^1$ modulo exact forms, with curvature $F = \mathrm{d}A$. Next there is a nonlinear version which we call $U(1)$-Yang–Mills theory where we consider $A \in \Omega^1$ modulo the transformation

$$A \mapsto uAu^{-1} + u\,\mathrm{d}u^{-1} \tag{1.2}$$

with u an invertible element of M. Here the curvature is $F = \mathrm{d}A + A \wedge A$ and transforms by conjugation. Such a theory has all the flavour of a non-Abelian gauge theory but not because of a non-Abelian gauge group, rather because differentials and functions do not commute. One can go on and do non-Abelian gauge theory, Riemannian geometry and construct a Dirac operator $\displaystyle{\not}D$. This would be beyond our present scope but we refer to concrete models such as in [7–9]. It should be mentioned that the results do not usually fit precisely the axioms of a spectral triple, but perhaps something like them.

31.2   Noncommutative spacetime and cogravity

This section is by way of 'warm up' for the reader to get comfortable with the methods in a more familiar setting before applying it to other algebras. We look particularly at models of noncommutative spacetime where the coordinate algebra is the enveloping algebra $U(\mathfrak{g})$ and $\mathfrak{g}$ is a Lie algebra, with structure constants $c_{ij}{}^{k}$ say. Putting in a parameter λ, it means commutation relations

$$[x_i, x_j] = \lambda\, c_{ij}{}^{k} x_k . \tag{2.1}$$

This is a well-studied system in mathematical physics, namely $U(\mathfrak{g})$ is the standard quantisation of the Kirillov–Kostant Poisson bracket on $\mathfrak{g}^{*}$ defined canonically by the Lie algebra structure. The symplectic leaves for this are the coadjoint orbits and their quantisation is given by setting the Casimir of $U(\mathfrak{g})$ to a fixed number.

   What if actual space or spacetime coordinates x_i, t might be elements of such a noncommutative algebra instead of numbers? As in quantum mechanics it means of course that they cannot be simultaneously diagonalised so that there will be some uncertainty or order-of-measurement dependence in the precise location of an event. But there would be many more effects as well, depending on the noncommutative algebra used. We nevertheless propose:

Claim 1 *(A) Noncommutative x_i, t would be a new (as yet undiscovered) physical effect 'cogravity' (B) Even if originating in quantum gravity corrections of Planck scale order, the effect could in principle be tested by experiments today.*

   We will look at the second claim in Section 31.2.1 by way of our first example, the Majid–Ruegg κ-spacetime. Here we look briefly at Claim (A). The first thing to note is that such spacetimes $U(\mathfrak{g})$ are noncommutative analogues of *flat* space. This is because they are Hopf algebras with a trivial additive coproduct and counit

$$\Delta 1 = 1 \otimes 1, \quad \epsilon 1 = 1, \quad \Delta \xi = \xi \otimes 1 + 1 \otimes \xi, \quad \epsilon \xi = 0, \quad \forall\, \xi \in \mathfrak{g} \tag{2.2}$$

extended multiplicatively, which is to say an additive Abelian 'group' structure on the noncommutative space. On the other hand, if these were coordinates of spacetime, then the dual Hopf algebra should be the momentum coordinate algebra. Here it is the usual commutative coordinate ring $\mathbb{C}[G]$ of the underlying Lie group G, which has a non-Abelian multiplicative coproduct for group multiplication. So momentum space is a non-Abelian group manifold with curvature

	Position	Momentum
Gravity	Curved $\sum_i x_i^2 = 1$	Noncommutative $[p_i, p_j] = \imath\epsilon_{ijk}p_k$
Cogravity	Noncommutative $[x_i, x_j] = \imath\epsilon_{ijk}x_k$	Curved $\sum_i p_i^2 = 1$
Qua.Mech.	$[x_i, p_j] = \imath\hbar\delta_{ij}$	

FIGURE 31.1. Gravity and cogravity related by non-Abelian Fourier transform

(at least in compact cases like $SU_2 = S^3$), but an ordinary (commutative) space. Moreover, there is a non-Abelian Fourier transform

$$\mathcal{F} : U(\mathfrak{g}) \to \mathbb{C}[G] \tag{2.3}$$

or more precisely on a completion of these algebras to allow functions with good decay properties. We see that the physical meaning of noncommutative space co-ordinates is equivalent under Fourier transform to curvature or gravity in momentum space. The reader is probably more familiar with the flipped situation in which position space is classical and curved, such as a non-Abelian group $SU(2)$, and its Fourier dual is a noncommutative space with noncommuting momenta or covariant derivatives, such as $[p_i, p_j] = \imath\epsilon_{ijk}p_k$. The idea of noncommutative space or spacetime is mathematically just the same but with the roles of position and momentum swapped. We summarise a typical situation in Table 1 to give some idea of the physical meaning of the mathematics that we propose here. It makes it clear that this reversed possibility of noncommutative position space and curved momentum space, is potentially new physical effect dual to gravity. Associated to it is a new dimensionful parameter, say λ controlling the noncommutativity of spacetime or curvature of momentum space. It is independent of usual curvature in position space, which is associated to the Newton constant parameter, and to Planck's constant which controls noncommutativity between the position and momentum sectors. This point of view was explained in detail in [1].

Turning to technical details, we use the following general construction which comes out of the analysis of translation-invariant differential structures on quantum groups. If $M = U(\mathfrak{g})$, a differential calculus is specified by a right ideal $\mathcal{I} \subset U(\mathfrak{g})^+$ where $U(\mathfrak{g})^+$ denotes expressions in the generators with no constant term. Let $\Lambda^1 = U(\mathfrak{g})^+/\mathcal{I}$ be the quotient and $\pi : U(\mathfrak{g})^+ \to \Lambda^1$ be the projection map. Let $\tilde{\pi}$ be the projection from $U(\mathfrak{g})$ where we first apply $\mathrm{id} - 1\epsilon$ which projects to $U(\mathfrak{g})^+$, and then π. We have, $\forall f \in U(\mathfrak{g})$, $\omega \in \Lambda^1$,

$$\Omega^1 = U(\mathfrak{g})\Lambda^1, \qquad \mathrm{d}f = (\mathrm{id} \otimes \tilde{\pi})\Delta f, \qquad \omega f = f_{(1)}\pi(\tilde{\omega}.f_{(2)}), \tag{2.4}$$

where $\tilde{\omega}$ is a representative projecting onto ω and $\Delta f = f_{(1)} \otimes f_{(2)}$ is a notation. The elements of Λ^1 are the 'basic 1-forms' and others are given by these with 'functional' coefficients from $U(\mathfrak{g})$ on the left. Note that 1-forms and such 'functions' do not commute. The wedge product for the entire exterior algebra is given in the present class of examples simply by elements of Λ^1 anticommuting among themselves. The Hodge $\star$ operator is similarly the usual one among such basic one-forms. Finally, given a basis $\{e_\mu\}$ of Λ^1, we define partial derivatives ∂^μ by (summation understood):

$$\mathrm{d}f = (\partial^\mu f)e_\mu, \quad \forall f \in U(\mathfrak{g}). \tag{2.5}$$

A natural way to specify the ideal $\mathcal{I}$ is as the kernel of a representation $\rho : \mathfrak{g} \to \mathrm{End}(V)$ extended as a representation of $U(\mathfrak{g})^+$. Then we can identify Λ^1 as the image of ρ and have the formulae

$$\mathrm{d}\xi = \rho(\xi), \qquad [\omega, \xi] = \omega.\rho(\xi), \quad \forall \xi \in \mathfrak{g}, \ \omega \in \mathrm{im}\,(\rho), \tag{2.6}$$

where we take the matrix product on the right.

31.2.1 λ-spacetime and gamma-ray bursts

We start with one of the most accessible noncommutative spacetime in this family [11]

$$[t, x_i] = \imath\lambda x_i, \qquad [x_i, x_j] = 0 \tag{2.7}$$

where λ has time dimension. If the effect is generated by quantum gravity corrections, we might expect $\lambda \sim 10^{-44}s$, the Planck time. One can also work with $\kappa = \lambda^{-1}$. Such a spacetime in two dimensions (i.e., the enveloping algebra of $U(b_+)$ where $b_+ \subset su_2$) was first proposed in [10] with an additional q parameter which one may set to 1. The first thing the reader will be concerned about is that this proposal manifestly breaks Lorentz invariance, so cannot be correct. What was shown in [11] was that there *is* not only a Lorentz but a full Poincaré invariance, but under a quantum group

$$U(so_{1,3}) \blacktriangleright\!\!\triangleleft \mathbb{C}[\mathbb{R}^{1,3}]. \tag{2.8}$$

This was shown also to be isomorphic to a 'κ-Poincaré' quantum group that had been proposed from another point of view (that of contraction from $U_q(so_{2,3})$) by J. Lukierski et al. in [12] but without a noncommutative spacetime on which to act.

What is important is that all of the geometrical consequences are not ad hoc but naturally follow within our approach to noncommutative geometry from the choice (2.7) of algebra. The first step is to compute the natural differential structure. There is not much choice and we take the 4-dimensional representation

$$\rho(x_\mu) = \imath\lambda \begin{pmatrix} 0 \\ e_\mu \end{pmatrix} \tag{2.9}$$

where the $e_\mu = (0 \cdots 1 \cdots 0)$ has 1 in the $\mu + 1$-th position and $\mu = 0, 1, 2, 3$. We write $x_0 \equiv t$. The basic 1-forms are provided by the $\mathrm{d}x_\mu$ as certain matrices, and these span the image of ρ, since the exponentiation of the Lie algebra has a similar form in this representation. The result from the general theory above is then

$$(\mathrm{d}x_j)x_\mu = x_\mu \, \mathrm{d}x_j, \qquad (\mathrm{d}t)x_\mu - x_\mu \, \mathrm{d}t = \imath\lambda \, \mathrm{d}x_\mu. \tag{2.10}$$

The form for d then implies

$$\partial^i : f(x,t) :=: \frac{\partial}{\partial x_i} f(x,t) :, \quad \partial^0 : f(x,t) :=: \frac{f(x,t+\imath\lambda) - f(x,t)}{\imath\lambda} : \tag{2.11}$$

for normal ordered polynomial functions. We use such normal ordered functions, with t to the right, to describe a general function in the spacetime. Under this identification we can extend all formulae to formal power series. Note that we see the effect of the noncommutative spacetime as forcing a lattice-like finite difference for the time derivative, and that this is actually by an imaginary time displacement. This is similar to the $+\imath\epsilon$ prescription in quantum field theory where operations more naturally take place in Euclidean space and must be Wick rotated back to the Minkowski picture. Note also that the noncommutativity of functions and the time direction generates the exterior derivative in the sense

$$[\mathrm{d}t, f] = \imath\lambda \, \mathrm{d}f, \quad \forall f \tag{2.12}$$

which is a typical feature of many noncommutative geometries but has no classical analogue.

Next, at least formally, we have eigenfunctions of the ∂^μ given by

$$\psi_{k,\omega} = e^{\imath k x}e^{\imath\omega t}, \quad \psi_{k,\omega}\psi_{k',\omega'} = \psi_{k+e^{-\lambda\omega}k',\omega+\omega'},$$

$$(\psi_{k,\omega})^{-1} = \psi_{-ke^{\lambda\omega},-\omega} \tag{2.13}$$

where we also show the product and inversion of such functions. We see from the latter that the Fourier dual or momentum space is the non Abelian Lie group $\mathbb{R}{\bowtie}_\lambda\mathbb{R}^3$. The invariant integration is the usual one on normal ordered functions and hence, allowing for the required ordering, the Fourier transform is given by

$$\mathcal{F}(: f(x,t) :)(k,\omega) = \int \psi_{k,\omega} : f(x,t) := \int \mathrm{d}x\,\mathrm{d}t\, e^{\imath k \cdot x}e^{\imath\omega t}f(e^{-\lambda\omega}x,t)$$

$$= e^{\lambda\omega}\mathcal{F}_{\text{usual}}(f)(e^{\lambda\omega}k,\omega), \tag{2.14}$$

i.e., reduces to a usual Fourier transform. We also compute the scalar wave operator from $\star\,\mathrm{d}\star\,\mathrm{d}$ and obtain the usual form $(\partial^0)^2 - \sum_i(\partial^i)^2$, which now has massless modes given by plane waves with

$$\frac{2}{\lambda^2}(\cosh(\lambda\omega) - 1) - k \cdot ke^{\lambda\omega} = 0. \tag{2.15}$$

This is a straightforward application of the Fourier theory on non-Abelian enveloping algebras introduced in [1].

More details and in particular a physical analysis, appeared in [13]. Critically, one has to make a postulate for how the mathematics shall be related to experimental numbers. Here, given the solvable Lie algebra structure, we proposed that expressions shall be identified only when normal ordered. In effect, one measures t first in any experiment. Under such an assumption one can analyse the wave-velocity of the above plane waves and argue that the dispersion relation has the classical form. Both of these steps are needed for any meaning to predictions from the theory. We can then find for the massless wave speed:

$$\left|\frac{\mathrm{d}\omega}{\mathrm{d}k}\right| = e^{-\lambda\omega} \tag{2.16}$$

in units where 1 is the usual speed of light. We assumed that light propagation has the same features as our analysis for massless fields, in which case the physical prediction is that the speed of light depends on energy.

One may, for example, plug in numbers from gamma-ray burst data as follows. These gamma-ray bursts have been shown in some cases to travel cosmological distances before arriving on Earth, and have a spread of frequencies from 0.1–100 MeV in energy terms. According to the above, the relative time delay Δ_t on travelling distance L for frequencies ω, $\omega + \Delta_\omega$ is

$$\Delta_t \sim \lambda\Delta_\omega \frac{L}{c} \sim 10^{-44}\mathrm{s} \times 100\mathrm{MeV} \times 10^{10}\mathrm{y} \sim 1\ \mathrm{ms} \tag{2.17}$$

where we put in the worst case for λ, namely the Planck time. We see that arrival times would be spread by the order of milliseconds, which is in principle observable! To observe it would need a statistical analysis of many gamma-ray burst events, to look for an effect that was proportional to distance travelled (since little is known about the initial creation profile of any one burst). This in turn would require accumulation of distance-data for each event by astronomers, such as has been achieved in some cases by coordination between the (now lost) BEPPO-SAX satellite to detect the gamma-ray burst and the Hubble telescope to lock in on the host galaxy during the afterglow period. With the design and implementation of such experiments and statistical analysis, we see that one might in principle observe the effect even if it originates in quantum gravity.

Let us mention finally that there are many other effects of noncommutative spacetime, some of which might be measured in earthbound experiments. For example, the LIGO/VIRGO gravitational wave interferometer project, although intended to detect gravitational waves, could also detect the above variable speed of light effect; a detailed theoretical model has yet to be built, but some initial speculations are in [14]. Similarly, reversal of momentum in our theory is done by group inversion, which means $(k, \omega) \to (-ke^{\lambda\omega}, -\omega)$, a modification perhaps detectable as CPT violation in neutral Kaon resonances [13]. The problem of interpretation in scattering theory is still open, however: what is the meaning of non-Abelian momentum and how might one detect it? Let us not forget also that the usual Lorentz and Poincaré group covariance is modified to a certain quantum group. It contains the usual $U(so_{1,3})$ as a sub-Hopf algebra but acting in a modified non-linear way, which means that special relativity effects are slightly modified. This is another source of potential observability. In short, we have indicated the reasons for Claim (B) above, but much needs to be done by way of physical interpretation and experimental design.

Finally, we comment on the rest of the geometry. The exterior algebra and coho-mology holds no surprises (the latter is trivial). Indeed, the space is geometrically as trivial as $\mathbb{R}^{1,3}$. This is consistent with our view that the above predictions have nothing to do with gravity, it is an independent effect. Thus, the curvature in the Maxwell theory of a gauge field $A = A^\mu \, \mathrm{d}x_\mu$ is

$$F = \mathrm{d}A = \partial^\mu A^\nu \, \mathrm{d}x_\mu \wedge \mathrm{d}x_\nu \qquad (2.18)$$

and its components have the usual antisymmetric form because the basic forms anticommute as usual. Because the Hodge $\star$ operations on them are also as usual when we keep all differentials to the right, and because the partial derivatives commute, the Maxwell operator $\star \, \mathrm{d} \, \star \, \mathrm{d}$ on 1-forms has the same form as the usual one, namely the scalar wave operator as above if we take A in Lorentz gauge $\partial^\mu A_\mu = 0$. This is why Maxwell light propagation is as in the scalar field case as assumed above. If we take a static electric source $J = \rho(x) \, \mathrm{d}t$, then the scalar potential and electric flux are as usual, since the spatial derivatives are as usual. Magnetostatic solutions likewise have the same form. Mixed equations with time dependence have the usual form but with ∂^0 for the time derivative. The $U(1)$-Yang–Mills theory appears more complicated but has similar features to the Maxwell one. Now the curvature is

$$F = (1 + \imath \lambda A^0) \, \mathrm{d}A + [A^0, A^i] \, \mathrm{d}t \wedge \mathrm{d}x_i + \tfrac{1}{2}[A^i, A^j] \, \mathrm{d}x_i \wedge \mathrm{d}x_j \qquad (2.19)$$

where the extra terms are from $A \wedge A$ using the relations (2.12) between functions and 1-forms. A gauge transformation is

$$\begin{aligned}
A^i &\mapsto uA^iu^{-1} + u(1 + \imath \lambda A^0)\partial^i u^{-1}, \\
A^0 &\mapsto uA^0u^{-1} + u(1 + \imath \lambda A^0)\partial^0 u^{-1}.
\end{aligned} \qquad (2.20)$$

The Dirac operator and spinor theory requires more machinery and has not yet been worked out in any meaningful (not ad-hoc) manner. Likewise, quantum field theory on the noncommutative Minkowski space is possible, starting with the Fourier transform above, but has not been fully worked out.

31.2.2   Angular momentum space and fuzzy spheres

Next we look at angular momentum operators but now regarded in a reversed role as noncommutative position space, which means the Lie algebra su_2 with relations

$$[x_i, x_j] = \imath \lambda \epsilon_{ijk} x_k. \qquad (2.21)$$

One may add a commutative time coordinate if desired, but the first remarkable discovery is that this is not required: when one makes the analysis of differential calculi there is only one natural choice and it is already four, not three dimensional! The algebra itself needs no introduction, but some aspects have been studied under the heading 'fuzzy spheres'. More precisely, these are finite-dimensional matrix algebras viewed as the image of (2.21) in a fixed spin representation, in which case one is seeing effectively the quotient where the Casimir $x \cdot x$ is equal to a constant. We are not taking this point of view here but working directly with the infinite-dimensional coordinate algebra (2.21) itself. This model of noncommutative geometry appeared recently in [15] and we give only a brief synopsis of a few aspects. First of all, the reader may ask about the Euclidean group invariance. This is preserved, but again as a quantum group

$$U(su_2) \bowtie \mathbb{C}[SU_2] \tag{2.22}$$

where SU_2 is a curved momentum space (as promised above). This is an example of a Drinfeld quantum double as well as a partially trivial bicrossproduct. As $\lambda \to \infty$ it becomes an S^3 of infinite radius, i.e., flat $\mathbb{R}^3$ acting by usual translations as it should. The $U(su_2)$ acts by the adjoint action which becomes usual rotations in the limit. We see that nontrivial quantum groups arise in very basic physics wherever noncommutative operators obeying the angular momentum relations are present.

For the differential geometry, we take $\rho(x_i) = \frac{1}{2}\lambda\sigma_i$ the usual Pauli-matrix representation and the basic 1-forms $\Lambda^1 = M_2$, the space of 2×2 matrices since the image of ρ in this case is everything. Then

$$\mathrm{d}x_i = \tfrac{1}{2}\lambda\sigma_i, \quad \text{and}$$

$$(\,\mathrm{d}x_i)x_j - x_j\,\mathrm{d}x_i = \tfrac{1}{2}\imath\lambda\epsilon_{ijk}\,\mathrm{d}x_k + \tfrac{1}{4}\lambda\delta_{ij}e_0, \quad e_0 x_i - x_i e_0 = \lambda\,\mathrm{d}x_i, \tag{2.23}$$

where e_0 is the 2×2 identity matrix which, together with the Pauli matrices σ_i completes the basis of basic 1-forms. It provides a natural time direction, even though there is no time coordinate. Indeed, the first cohomology is nontrivial and spanned by e_0, i.e., it is a closed 1-form which is not d of anything (it is denoted by θ in [15]). Nevertheless, like $\mathrm{d}t$ in the previous section, we see from (2.23) that e_0 generates the exterior derivative by commutator,

$$[e_0, f] = \lambda\,\mathrm{d}f, \quad \forall f \in U(su_2). \tag{2.24}$$

The partial derivatives defined by

$$\mathrm{d}f = (\partial^i f)\,\mathrm{d}x_i + (\partial^0 f)e_0$$

for all f are hard to write down explicitly; they are given in [15]. Nevertheless, the formal group elements

$$\psi_k = e^{\imath k.x} \tag{2.25}$$

are the plane waves and eigenfunctions for the partial derivatives. For the same reasons as in our previous model in Section 31.2.1, the scaler Laplacian comes out as $(\partial^0)^2 - \sum_i (\partial^i)^2$ when we take a local Minkowski metric. Its value on plane waves is

$$\frac{1}{\lambda^2}\left((\cos(\frac{\lambda|k|}{2}) - 1)^2 + 4\sin^2(\frac{\lambda|k|}{2}) \right). \tag{2.26}$$

The Maxwell theory may likewise be worked out and has similarities with the usual one due to the fourth dimension provided by e_0, except that we will only have static solutions since we have no time variable. This time the coordinates are fully 'tangled up' by the relations (2.21) and solutions are rather hard to write down explicitly. One solution, for a uniform electric charge density and spherical boundary conditions at infinity (i.e., constructed as a series of concentric shells) has timelike source $J = e_0$, scalar potential $x \cdot x$ and electric field proportional to x, i.e., radial, see [15]. There is similarly a magnetic solution for a uniform current density. Non-uniform solutions have yet to be worked out due only to their algebraic complexity. The Dirac operator is known and given in [15] also, as are coherent states in which the noncommutative coordinates behave as close as possible to classical. The explicit form of ∂^0 is also interesting and takes the form

$$\partial^0 f = \frac{\lambda}{8} \sum_i (\partial^i)^2 f + O(\lambda^2) \qquad (2.27)$$

which is the free particle Hamiltonian to lowest order. This is a general feature of many noncommutative algebras, that there is an extra cotangent direction e_0 induced by the noncommutative geometry as generating d by commutator, and one could even say that this is the 'origin of time evolution' if one defines the corresponding energy as ∂^0 and asks that it be $\frac{\partial}{\partial t}$ in the algebra with a variable t adjoined. Details of such a philosophy will appear elsewhere.

Finally, we note that there are a couple of physical models in which this kind of noncommutative space could appear naturally. One is $2 + 1$ quantum gravity in a Euclidean version based on an $iso(3)$-Chern–Simons theory. There one finds [16] that the quantum states have a quantum group symmetry, namely the double (2.22), suggesting that our above model should provide a description of the relevant effective geometry. The other is with a certain form of ansatz for matrix models in string theory, under which the theory reduces to one on a fuzzy sphere. On the other hand, the problem of formulating the physical consequences of the above noncommutative geometry is independent of the underlying theory of which it may be an effective model.

31.2.3 θ-space and the Heisenberg algebra

The first proposal for spacetime was probably made by Snyder [17] in the 1940s even before the modern machinery of noncommutative geometry, and took the form

$$[x_\mu, x_\nu] = \imath \theta_{\mu\nu}, \qquad \theta_{\mu\nu} = -\theta_{\nu\mu} \in \mathbb{C} \qquad (2.28)$$

where $\theta_{\mu\nu}$ were operators with further properties arranged in such a way as to preserve Lorentz covariance. More recently such algebras have been revived by string theorists with θ now a number and called noncommutative 'θ-space'. This is no longer any kind of noncommutative Euclidean or Minkowski space since it does not appear to have any (pseudo)orthogonal group or quantum group appropriate to that. Rather, it is just the usual Heisenberg algebra of quantum mechanics under another context and has a symplectic character. It can also be viewed as a noncommutative torus in an unexponentiated form. For our present treatment we assume that the space is $2n$-dimensional and $\theta_{\mu\nu}$ nondegenerate. We take the latter in normal form and replaced by a single central variable, say $t = x_0$, i.e., we take

$$[x_i, x_j] = 0, \qquad [x_i, x_{-j}] = \imath\lambda\delta_{ij}t, \qquad [x_{-i}, x_{-j}] = 0 \qquad (2.29)$$

in terms of new variables grouped as positive and negative index. This is now of our enveloping algebra form generated by a $2n + 1$-dimensional Heisenberg Lie algebra. We will also give a different treatment of (2.28) by twisting theory in Section 31.5.

For the calculus we take the standard representation

$$\rho(x_i) = \imath\lambda \begin{pmatrix} 0 & e_i & 0 \\ 0 & 0 & 0 \\ 0 & 0 & 0 \end{pmatrix}, \qquad \rho(x_{-i}) = \imath\lambda \begin{pmatrix} 0 & 0 & 0 \\ 0 & 0 & e_i^t \\ 0 & 0 & 0 \end{pmatrix},$$

$$\rho(t) = \imath\lambda \begin{pmatrix} 0 & 0 & 1 \\ 0 & 0 & 0 \\ 0 & 0 & 0 \end{pmatrix} \qquad (2.30)$$

where e_i is a row vector with 1 in the i-th position and e_i^t is its transpose. The general construction then gives the basic 1-forms $\mathrm{d}x_0$, $\mathrm{d}x_i$, $\mathrm{d}x_{-i}$ as certain matrices. They span the image of ρ since the Heisenberg Lie algebra exponentiates to a similar form in this representation. We obtain

$$(\mathrm{d}x_{-i})x_{\pm j} = x_{\pm j}\,\mathrm{d}x_{-i}, \qquad (\mathrm{d}x_i)x_j = x_j\,\mathrm{d}x_i,$$
$$(\mathrm{d}x_i)x_{-j} - x_{-j}\,\mathrm{d}x_i = \imath\lambda\delta_{ij}\,\mathrm{d}t \qquad (2.31)$$

and $\mathrm{d}t$ central. The partial derivatives defined by

$$\mathrm{d}f = (\partial^i f)\,\mathrm{d}x_i + (\partial^{-i}f)\,\mathrm{d}x_{-i} + (\partial^0 f)\,\mathrm{d}t$$

then turn out to be just the usual derivatives on functions provided these are normal ordered with all x_{-i} to the left of all x_j. This is even simpler than our first example above.

The plane wave eigenfunctions of the partial derivatives are the group elements

$$\psi_{k_-,k_+,\omega} = e^{\imath k_- \cdot x_-} e^{\imath k_+ \cdot x_+} e^{\imath\omega t},$$
$$\psi_{k_-,k_+,\omega}\psi_{k'_-,k'_+,\omega'} = \psi_{k_-+k'_-,k_++k'_+,\omega+\omega'+\lambda k_+\cdot k'_-} \qquad (2.32)$$

for the Heisenberg group. Integration is the usual one on normal ordered functions and hence, allowing for this, the Fourier transform is given by

$$\mathcal{F}(: f :)(k_-, k_+, \omega)$$

$$= \int \psi_{k_-, k_+, \omega} : f :$$

$$= \int \mathrm{d}x_- \, \mathrm{d}x_+ \, \mathrm{d}t \, e^{\imath k_- \cdot x_-} e^{\imath k_+ \cdot x_+} e^{\imath \omega t} f(x_- - \lambda k_+ t, x_+, t)$$

$$= \mathcal{F}_{\text{usual}}(f)(k_-, k_+, \omega + \lambda k_- \cdot k_+), \tag{2.33}$$

i.e., reduces to a usual Fourier transform.

For the Laplacian, because the algebra does not have a Euclidean covariance it is not very natural to take the usual metric and Hodge $\star$ operator on forms, but if one does this one would have the usual $(\partial^0)^2 - (\partial^-)^2 - (\partial^+)^2$ etc. Since the derivatives are the usual ones on normal ordered expressions, in momentum space it becomes $\omega^2 - k_-^2 - k_+^2$, etc. without any modifications. The same applies in the Maxwell theory; the only subtlety is to keep all expressions normal ordered. On the other hand it is not clear how logical such a metric is since there is no relevant background symmetry of orthogonal type. Probably more natural is a Laplacian like $(\partial^0)^2 - \partial^- \cdot \partial^+$. On the other hand, the $U(1)$-Yang–Mills theory, as with the Chern–Simons theory if one wants it, involves commutation relations between functions and forms and begins to show a difference.

Finally, if we want the original (2.28) we should quotient by $t = 1$. Then the calculus is also quotiented by $\mathrm{d}t = 0$ and we see that the calculus becomes totally commutative in the sense $[\mathrm{d}x_\mu, x_\nu] = 0$ as per the classical case. This is the case relevant to the string theory literature where one finds that field theory etc. at an algebraic level has just the same form as classically. This is also the conclusion in another approach based on symmetric categories [18]. Although a bit trivial noncommutative-geometrically, the model is still physically interesting and we refer to the physics literature [19]. As well as D-branes, there are applications to the physics of motion in background fields and the quantum Hall effect. Moreover, the situation changes when one considers non-Abelian gauge theory and/or questions of analysis, where there appear instantons. From the quantum groups point of view one should consider frame bundles and spin connections with a frame group or quantum group of symplectic type using the formalism of [7]. One can also use the automorphisms provided by the quantum double of the Heisenberg algebra. These are some topics for further development of this model.

31.3 Quantum $U(1)$-Yang–Mills theory on the $\mathbb{Z}_2 \times \mathbb{Z}_2$ lattice

In this section we move to a different class of examples of noncommutative geometry, where the algebra is that of functions on a group lattice. It is important that the machinery is identical to the one above, just applied to a different algebra and a corresponding analysis of its differential calculus. In this sense it is part of one noncommutative 'universe' in a functorial sense and not an ad-hoc construction specific to lattices. In this case we gain meaningful answers even for a finite lattice. In usual lattice theory this would make no sense because constructions are justified only in the limit of zero lattice spacing, other aspects are errors; in noncommutative geometry the finite lattice or finite group is an exact geometry in its own right.

Specifically, we look at functions $\mathbb{C}[G]$ on a finite group G. In this case the invariant calculi are described by ad-stable subsets $\mathcal{C} \subset G$ not containing the identity. The elements of the subset are the allowed 'directions' by which we may move by right translation from one point to another on the group. Hence the elements of $\mathcal{C}$ label the basis of invariant 1-forms $\{e_a\}$. The differentials are

$$\Omega^1 = \mathbb{C}[G].\mathcal{C}\mathcal{C}, \quad f = \sum_{a \in \mathcal{C}} (\partial^a f) e_a, \quad \partial^a = R_a - \mathrm{id}, \quad e_a f = R_a(f) e_a, \quad (3.1)$$

where $R_a(f) = f((\ )a)$ is right translation. There is a standard construction for the wedge product of basic forms as well. This setup is an immediate corollary of the analysis of [5] but has been emphasised by many authors, such as [20] or more recently [21, 22]. Moreover, one can take a metric $\delta_{a,b^{-1}}$ in the $\{e_a\}$ basis, which leads to a canonical Hodge $\star$ operation if the exterior algebra is finite dimensional. Then one may proceed to Maxwell and Yang–Mills theory as well as gravity.

We now demonstrate some of these ideas in the simplest case $G = \mathbb{Z}_2 \times \mathbb{Z}_2$. It is a baby version of the treatment already given for S_3 in [21] but we take it further to the complete quantum theory in a path integral approach. We take $\mathcal{C} = \{x = (1,0), y = (0,1)\}$ as the two allowed directions from each point, i.e., our spacetime consists of a square with the allowed directions being edges. The corresponding basic 1-forms are e_x, e_y and we have

$$\mathrm{d}f = (\partial^x f) e_x + (\partial^y f) e_y, \quad \mathrm{d}e_x = \mathrm{d}e_y = 0, \quad e_x^2 = e_y^2 = \{e_x, e_y\} = 0.$$

The top form is $e_x \wedge e_y$ and the Hodge $\star$ is

$$\star 1 = e_x \wedge e_y, \quad \star e_x = e_y, \quad \star e_y = -e_x, \quad \star(e_x \wedge e_y) = 1.$$

We write a $U(1)$ gauge field as a 1-form $A = A^x e_x + A^y e_y$. Its curvature

$$F = \mathrm{d}A + A \wedge A = F^{xy} e_x \wedge e_y, \quad F^{xy} = \partial^x A^y - \partial^y A^x + A^x R_x A^y$$

is covariant as $F \mapsto uFu^{-1}$ under

$$A \mapsto uAu^{-1} + u\,\mathrm{d}u^{-1}, \quad A^a \mapsto \frac{u}{R_a(u)} A^a + u\partial^a u^{-1}$$

for any unitary u (any function of modulus 1). We specify also the reality condition $A^* = A$ where the basic forms are self-adjoint in the sense $e_i^* = e_i$. This translates in terms of components as

$$\bar{A}^a = R_a A^a, \quad \bar{F}^{xy} = -R_{xy}(F^{xy}) \quad (3.2)$$

under complex conjugation and implies that $F^* = F$. Such reality should also be imposed in the examples in the previous section if one wants to discuss Lagrangians. Finally, we change variables by

$$A_x + 1 = \lambda_x e^{\imath \theta_x}, \quad A_y + 1 = \lambda_y e^{\imath \theta_y}, \tag{3.3}$$

where the $\lambda_x, \lambda_y \geq 0$ are real and θ_x, θ_y are angles. The reality condition means in our case that a connection is determined by real numbers

$$\lambda_1 = \lambda_x(x), \quad \lambda_2 = \lambda_y(x), \quad \lambda_3 = \lambda_x(y), \quad \lambda_4 = \lambda_y(y) \tag{3.4}$$

and similarly $\theta_1, \ldots, \theta_4$.

With these ingredients the Yang–Mills Lagrangian $\mathcal{L} e_x \wedge e_y = -\frac{1}{2} F^* \wedge \star F$, along the same lines as for any finite group, takes the form (discarding total derivatives),

$$\mathcal{L} = \tfrac{1}{2}|F^{xy}|^2 = \tfrac{1}{2}(\lambda_x^2 \partial^x \lambda_y^2 + \lambda_y^2 \partial^y \lambda_x^2) + \lambda_x^2 \lambda_y^2 - \lambda_x \lambda_y R_y(\lambda_x) R_x(\lambda_y) w_1 \tag{3.5}$$

where the Wilson loop

$$w_1 = \Re\, e^{\imath \theta_x} e^{\imath R_x \theta_y} e^{-\imath R_y \theta_x} e^{-\imath \theta_y} \tag{3.6}$$

is the real part of the holonomy around the square where we displace by x, then by y, then back by x and then back by y, as explained in [21]. When we sum over all points on the lattice, we have the action

$$S = \sum \mathcal{L} = (\lambda_1^2 + \lambda_3^2)(\lambda_2^2 + \lambda_4^2) - 4\lambda_1 \lambda_2 \lambda_3 \lambda_4 \cos(\theta_1 - \theta_2 + \theta_3 - \theta_4). \tag{3.7}$$

We now quantise this theory in a path integral approach. We 'factorise' the partition function into one where we first hold the λ_i variables fixed and do the θ_i integrals, and then the λ_i integrals. The first step is therefore a lattice $U(1)$-type theory with fixed λ_i. The latter are somewhat like a background choice of 'lengths' associated to our allowed edges. We refer to [21] for a discussion of the interpretation. We assume the usual measure on the gauge fields before making the polar transformation, then

$$\mathcal{Z} = \int d^4 \lambda \, d^4 \theta \, \lambda_1 \lambda_2 \lambda_3 \lambda_4 \, e^{-\alpha S}$$

$$= \int_0^\infty d^4 \lambda \, \lambda_1 \lambda_2 \lambda_3 \lambda_4 \, e^{-\alpha(\lambda_1^2 + \lambda_3^2)(\lambda_2^2 + \lambda_4^2)} \mathcal{Z}_\lambda \tag{3.8}$$

where $\alpha > 0$ is a coupling constant and

$$\mathcal{Z}_\lambda = \int_0^{2\pi} d^4 \theta \, e^{4\alpha \lambda_1 \lambda_2 \lambda_3 \lambda_4 \cos(\theta_1 - \theta_2 + \theta_3 - \theta_4)}$$

$$= \int_0^{2\pi} d\theta_2 \, d\theta_3 \, d\theta_4 \int_{-\theta_2 + \theta_3 - \theta_4}^{2\pi - \theta_2 + \theta_3 - \theta_4} d\theta \, e^{\beta \cos(\theta)}$$

$$= (2\pi)^3 \int_0^{2\pi} d\theta \, e^{\beta \cos(\theta)} = (2\pi)^4 I_0(\beta) \tag{3.9}$$

where $\beta = 4\alpha\lambda_1\lambda_2\lambda_3\lambda_4$ and I_0 is a Bessel function, and we changed variables to $\theta = \theta_1 - \theta_2 + \theta_3 - \theta_4$. For the expectation values of Wilson loops in this $U(1)$ part of the theory we take the real part of the holonomy $w_n = \cos(n\theta)$ along a loop that winds around our square n times. Then similarly to the above, we have

$$\langle w_n \rangle_\lambda = \langle \cos(n\theta) \rangle_\lambda = \frac{I_n(\beta)}{I_0(\beta)}. \tag{3.10}$$

In other words the usual Fourier transform on a circle becomes under quantization a Bessel transform,

$$\langle \text{Fourier Transform} \rangle_\lambda \cdot \mathcal{Z}_\lambda = \text{Bessel transform}$$

$$\left\langle \sum_n a_n \cos(n\theta) \right\rangle_\lambda \cdot \mathcal{Z}_\lambda = (2\pi)^4 \sum_n a_n I_n(\beta)$$

as the effect of taking the vacuum expectation value. The left-hand side here is the unnormalised expectation value. The same results apply in a 'Minkowski' theory where α is replaced by $\imath\alpha$, with Bessel J functions instead. Of course, the appearance of Bessel functions is endemic to lattice theory; here it is not an approximation error but a clean feature of the $\mathbb{Z}_2 \times \mathbb{Z}_2$ theory.

At the other extreme, we consider the reverse order in which the θ_i integrals are deferred. We change variables by

$$\lambda_1 = \tfrac{1}{2}\sqrt{x}(a + \sqrt{2 - a^2}), \qquad \lambda_2 = \tfrac{1}{2}\sqrt{y}(b + \sqrt{2 - b^2}),$$

$$\lambda_3 = \tfrac{1}{2}\sqrt{x}(-a + \sqrt{2 - a^2}), \qquad \lambda_4 = \tfrac{1}{2}\sqrt{y}(-b + \sqrt{2 - b^2}),$$

where $a, b \in [-1, 1]$ and $x, y > 0$. Then

$$\lambda_1^2 + \lambda_3^2 = x, \quad \lambda_2^2 + \lambda_4^2 = y, \quad \lambda_1\lambda_2\lambda_3\lambda_4 = \tfrac{1}{4}xy(1 - a^2)(1 - b^2)$$

$$\mathcal{Z} = (2\pi)^3 \int_0^{2\pi} d\theta\, \mathcal{Z}_\theta,$$

$$\mathcal{Z}_\theta = \frac{1}{16} \int_0^\infty dx\, dy \int_0^1 da\, db\, e^{-\alpha xy(1 - \cos(\theta))(1 - a^2)(1 - b^2))} \frac{xy(1 - a^2)(1 - b^2)}{\sqrt{(2 - a^2)(2 - b^2)}}.$$

Only the product $z = xy$ is relevant here and moving to this and the ratio x/y as variables, the latter gives a logarithmically divergent constant factor which we discard, leaving the z-integral, which we do. We let

$$A(a, b) = (1 - a^2)(1 - b^2).$$

Then up to an overall factor

$$\mathcal{Z}_\theta = \int_0^\infty dz \int_0^1 da\, db\, \frac{Aze^{-\alpha z(1 - \cos(\theta)A)}}{\sqrt{(2 - a^2)(2 - b^2)}}$$

$$= \frac{1}{\alpha^2} \int_0^1 da\, db \frac{A}{(1 - \cos(\theta)A)^2 \sqrt{(2 - a^2)(2 - b^2)}}. \tag{3.11}$$

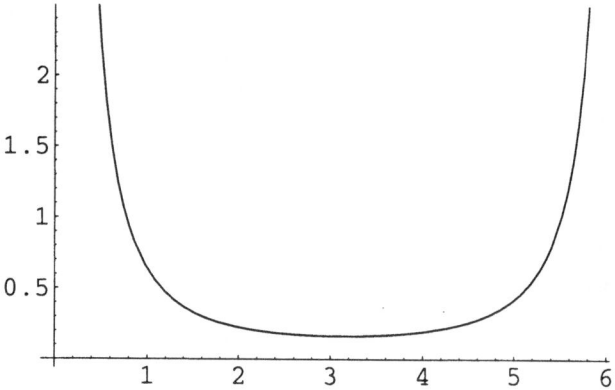

FIGURE 31.2. Vacuum expectation values $\langle \lambda_1 \lambda_2 \lambda_3 \lambda_4 \rangle_\theta$ plotted for $\theta \in [0, 2\pi]$ and $\alpha = 1$

This integral is convergent for all $\theta \neq 0$. The unnormalised expectation of the 'scale Wilson loop' is similarly

$$\langle \lambda_1 \lambda_2 \lambda_3 \lambda_4 \rangle_\theta \cdot \mathcal{Z}_\theta = \frac{1}{2\alpha^3} \int_0^1 da \, db \frac{A^2}{(1 - \cos(\theta)A)^3 \sqrt{(2 - a^2)(2 - b^2)}}$$

which is again convergent. If $\theta = 0$ we have divergent integrals but can regularise the theory by, for instance, doing the a-integrals from $\epsilon > 0$, giving $1/\epsilon^2$ and $1/\epsilon^4$ divergences respectively. We plot $\langle \lambda_1 \lambda_2 \lambda_3 \lambda_4 \rangle_\theta$ in Figure 2 with $\alpha = 1$. There is a similar but sharper appearance to $\mathcal{Z}_\theta$ itself. The point $\theta = \frac{\pi}{2}$ is special and has values

$$\mathcal{Z}_{\frac{\pi}{2}} = \frac{1}{4}, \quad \langle \lambda_1 \lambda_2 \lambda_3 \lambda_4 \rangle_{\frac{\pi}{2}} = \frac{\pi^2}{32},$$

while the minima occur at $\theta = \pi$ with small positive value.

Finally, we look at the full theory. From our first point of view, for the expectation of the $U(1)$-Wilson loops in the full theory, we need integrals

$$\langle (\lambda_1 \lambda_2 \lambda_3 \lambda_4)^m w_n \rangle \cdot \mathcal{Z}$$
$$= (2\pi)^4 \int_0^\infty d^4\lambda \, (\lambda_1 \lambda_2 \lambda_3 \lambda_4)^{m+1} e^{-\alpha(\lambda_1^2 + \lambda_3^2)(\lambda_2^2 + \lambda_4^2)} I_n(4\alpha \lambda_1 \lambda_2 \lambda_3 \lambda_4).$$

We change the λ_i variables to z, a, b and discard a log divergent factor as before. Then up to an overall constant,

$$\langle (\lambda_1 \lambda_2 \lambda_3 \lambda_4)^m w_n \rangle \cdot \mathcal{Z} =$$
$$\int_0^\infty dz \int_0^1 da \, db \frac{z^{m+1} A^{m+1} e^{-\alpha z} I_n(\alpha z A)}{4^{m+1} \sqrt{(2 - a^2)(2 - b^2)}}. \quad (3.12)$$

For example, we have

$$\mathcal{Z} = \frac{1}{4\alpha^2} \int_0^1 da\,db\, \frac{A}{(1-A^2)^{\frac{3}{2}}\sqrt{(2-a^2)(2-b^2)}},$$

$$\langle \lambda_1\lambda_2\lambda_3\lambda_4 \rangle \cdot \mathcal{Z} = \frac{1}{16\alpha^3} \int_0^1 da\,db\, \frac{A(2+A^2)}{(1-A^2)^{\frac{5}{2}}\sqrt{(2-a^2)(2-b^2)}},$$

$$\langle w_1 \rangle \cdot \mathcal{Z} = \frac{1}{4\alpha^2} \int_0^1 da\,db\, \frac{A^2}{(1-A^2)^{\frac{3}{2}}\sqrt{(2-a^2)(2-b^2)}}.$$

These are all divergent, with $Z \sim 1/\alpha^2\epsilon$ and so on. The physical reason is the singular contribution from configurations in the functional integral where $\theta = 0$ as seen above. Moreover, if we regularise by doing the a-integral (say) from $\epsilon > 0$, then

$$\langle (\lambda_1\lambda_2\lambda_2\lambda_4)^m \rangle \sim \frac{1}{\alpha^m\epsilon^{2m}} \sim \langle (\lambda_1\lambda_2\lambda_2\lambda_4)^m w_1 \rangle. \tag{3.13}$$

In particular, we have a finite answer $\langle w_1 \rangle = 1$.

In summary, for the full theory we see that some of the physical vacuum expectation values connected with the phase part of the noncommutative gauge theory, such as the Wilson loop $\langle w_1 \rangle$, are finite but not necessarily interesting. However, if one makes α itself a divergent function of the regulator (such as $\alpha = 1/\epsilon^2$), one can render all the $\langle (\lambda_1\lambda_2\lambda_3\lambda_4)^m \rangle$ and $\langle (\lambda_1\lambda_2\lambda_3\lambda_4)^m w_1 \rangle$ finite as well and indeed one obtains nontrivial answers. A systematic treatment will not be attempted here (one should choose the regulator more physically, among other things) but it does appear that the theory is largely renormalisable.

31.4   Clifford algebras as noncommutative spaces

An idea for 'quantisation' in physics is the Moyal product where we work with the same vector space of functions on $\mathbb{R}^n$ but modify the product to a noncommutative one. Exactly the same multiplication-alteration idea can be used to construct Clifford algebras on the vector space $\mathbb{C}[(\mathbb{Z}_2)^n]$ but modifying its product. The formulae are similar but instead of real-valued vectors $x \in \mathbb{R}^n$ we work with $\mathbb{Z}_2$-valued vectors, so that Clifford algebras are in a precise sense discrete 'quantisations' of the $(\mathbb{Z}_2)^n$ lattice. We will explain this a bit more from the Drinfeld-twist point of view in [23, 24], but for the present purposes the general idea of building Clifford algebras by multiplication-alteration factors is well known. It is mentioned in [25] for example, as well as in the mathematics literature where Clifford algebras were constructed as twisted $(\mathbb{Z}_2)^n$ group rings in [26]. On the other hand, our point of view tells us immediately how to do the differential calculus etc. on such spaces and that one has the same cohomology, eigenvalues of the Laplace operator etc. as the 'classical' untwisted case. Before doing this, Section 31.4.1 finishes the differential geometry for the 'classical' case $(\mathbb{Z}_2)^n$ as in Section 31.3, but now for general n.

31.4.1   Cohomology and Maxwell theory on $(\mathbb{Z}_2)^n$

We use the setting for finite groups as in Section 31.3, but now functions are on $(\mathbb{Z}_2)^n$. For $\mathcal{C}$ we take the n directions where we step $+1$ in each of the $\mathbb{Z}_2$ directions, leaving the others unchanged. The differentials are

$$(\partial^a f)(x) = f(x + (0, \ldots, 1, \ldots, 0)) - f(x), \qquad (4.1)$$

where 1 is in the a-th position. We denote by e_a the basic 1-forms, so that

$$\mathrm{d}f = \sum_{a=1}^{n} (\partial^a f) e_a, \quad e_a f = R_a(f) e_a, \quad e_a^2 = 0, \quad \{e_a, e_b\} = 0, \quad \mathrm{d}e_a = 0 \qquad (4.2)$$

where $R_a(f) = f(x + (0, \ldots, 1, \ldots, 0))$ in the a-th place. Because of the cyclic nature of the group $\mathbb{Z}_2$ it is easy to see that no function can obey $\partial^1 f = 1$ (or similarly in any other direction). From this kind of argument one can find that the noncommutative de Rham cohomology (the closed forms modulo the exact ones) is represented by the Grassmann algebra Λ of basic forms generated by the $\{e_a\}$ with the anticommutativity relations in (4.2),

$$H^{\cdot}(\mathbb{C}[(\mathbb{Z}_2)^n]) = \Lambda. \qquad (4.3)$$

Here $H^1 = \mathbb{C}e_1 \oplus \cdots \oplus \mathbb{C}e_n$ has the same form as for a classical torus $S^1 \times \cdots \times S^1$ and for the same reason (and the same holds for $(\mathbb{Z}_m)^n$ for all m). In particular, the top form is $e_1 \cdots e_n$ and is not exact. Moreover, because the exterior algebra among the basic forms is just the usual one, the Hodge $\star$ is the usual one given by the totally antisymmetric epsilon tensor. We take the Euclidean metric δ_{ab} on the basic forms. Then the spin zero wave operator is

$$\square = \frac{1}{4} \star \mathrm{d} \star \mathrm{d} = \frac{1}{4} \sum_a \partial^a \partial^a = -\frac{1}{2} \sum_a \partial^a. \qquad (4.4)$$

The plane wave eigenfunctions of the partial derivatives ∂^a are

$$\psi_k(x) = (-1)^{k \cdot x}, \qquad \partial^a \psi_k = -2 k_a \psi_k \qquad (4.5)$$

labelled by momenta $k \in (\mathbb{Z}_2)^n$. These diagonalise the wave operator, which has eigenvalue

$$\square \psi_k = |k| \psi_k, \qquad |k| = \sum_a k_a = k \cdot k \qquad (4.6)$$

since each $k_a \in \{0, 1\}$. The eigenvalues range from $0, \ldots, n$ and there are $\binom{n}{m}$ eigenfunctions for eigenvalue m. In particular, there are no massless modes other than the constant function 1.

For the spin 1 equation, since the formulae among basic 1-forms are as for $\mathbb{R}^n$, the Maxwell operator $\star \mathrm{d} \star \mathrm{d}$ on 1-forms $A = \sum_a A^a e_a$ is given by the above $\square$ on each component function if A is in Lorentz gauge $\partial \cdot A = 0$. Its eigenvalues therefore also range from $0, \ldots, n$ according to the degree of all components. The Maxwell operator on the $(n-1)2^n + 1$ dimensional space of gauge fields A such that $\partial \cdot A = 0$ can be fully diagonalised by the following modes. There are (i) $n2^{n-1}$ modes of the form

$$A = \psi_k e_a \tag{4.7}$$

of degree $|k| = 0, \ldots, n-1$, where $a = 1, \ldots, n$ and k is such that its a-component k_a vanishes. In addition there are (ii) $\binom{n}{m}(m-1)$ modes of degree $m = 2, \ldots, n$ of the form

$$A = \psi_k(\mu_1 e_{a_1} + \cdots + \mu_m e_{a_m}), \quad \mu_1 + \cdots \mu_m = 0, \tag{4.8}$$

where the momentum vector k has components $k_{a_1} = \cdots = k_{a_m} = 1$ and the rest zero. Adding up the modes of type (ii) gives $n2^{n-1} + 1 - 2^n$, which with the modes of type (i) gives a total of $(n-1)2^n + 1$, which is the required number since the space of A is $n2^n$-dimensional while the image of d, i.e., the number of exact forms, is $2^n - 1$-dimensional. Note that the gauge fixing by $\partial \cdot A$ still leaves possible $A \mapsto A + \mathrm{d}f$ where $\Box f = 0$, but then $f = 1$ by the above, and $\mathrm{d}f = 0$, i.e., the gauge-fixing is fully effective. These eigenfunctions A with degree > 0 are also the allowed sources, being a basis of solutions of $\partial \cdot J = 0$ in the image of the Maxwell wave operator. Hence we can solve fully the Maxwell equations. One could also rework these results with a Minkowski metric in the e_a basis.

31.4.2 Noncommutative geometry on Clifford algebras

In this section we look at the standard Clifford algebra $\mathrm{Cliff}(n)$ with generators γ_a, $a = 1, \ldots, n$ and relations

$$\{\gamma_a, \gamma_b\} = \delta_{a,b}. \tag{4.9}$$

In [24, 26] this is presented as a twist of the group algebra of $(\mathbb{Z}_2)^n$. This is the momentum group for our classical space $\mathbb{C}[(\mathbb{Z}_2)^n]$ above, with elements the plane waves ψ_k. The 'quantisation' procedure consists of changing the product between such basis vectors to a new one

$$\psi_k \bullet \psi_m = F(k,m)\psi_{k+m}, \quad F(k,m) = (-1)^{\sum_{j<i} k_i m_j} \tag{4.10}$$

where F is a cocycle on the group $(\mathbb{Z}_2)^n$ just because it is bilinear. These relations are such that if we write

$$\psi_a = \psi_{(0,\ldots,1,\ldots,0)}, \quad \psi_a(x) = \begin{cases} -1 & \text{if } x_a = 1 \\ 0 & \text{else} \end{cases}, \tag{4.11}$$

where 1 in the momentum vector is in the a-th position, then

$$\psi_a \bullet \psi_b + \psi_b \bullet \psi_a = \delta_{a,b}$$

so that we can identify $\gamma_a = \psi_a$. We similarly define

$$\gamma_k = \gamma_1^{k_1} \cdots \gamma_n^{k_n} \tag{4.12}$$

for any $k \in (\mathbb{Z}_2)^n$ and identify $\gamma_k = \psi_k$. In this way the whole basis of $\mathbb{C}[(\mathbb{Z}_2)^n]$ is identified with the vector space of the Clifford algebra. This approach makes $\mathrm{Cliff}(n)$ a braided-commutative algebra in a symmetric monoidal category, see [24].

We can similarly identify all normal ordered expressions where the γ_a occur in increasing order from left to right with the identical 'classical' expression in terms of the commutative ψ_a. Moreover, because the Clifford algebra is obtained by a cocycle twist, it follows from the general theory in the next section that we may likewise identify the differential calculi. Indeed, the entire exterior algebra $\Omega(\mathrm{Cliff}(n))$ is a twist by the same cocycle F but now of the exterior algebra of differential forms on $\mathbb{C}[(Z_2)^n]$ from Section 31.4.1, with the same cohomology and gauge theory. Explicitly,

$$\Omega(\mathrm{Cliff}(n)) = \mathrm{Cliff}(n).\Lambda, \tag{4.13}$$

where Λ is our previous Grassmann algebra (4.2) generated by the anticommuting e_a. We have

$$\mathrm{d}f = \sum_a (\partial^a f) e_a, \quad \partial^a : f :=: \partial^a f :, \tag{4.14}$$

where $: f :$ is the normal ordered element defined by writing f in terms of the ψ_a and on the right we use the partial derivatives from the previous section. Similarly for the noncommutation relations between functions and 1-forms. Explicitly,

$$\partial^a \gamma_k = -2k_a \gamma_k, \qquad e_a \gamma_b = \begin{cases} -\gamma_b e_a & \text{if } a = b \\ \gamma_b e_a & \text{else} \end{cases}. \tag{4.15}$$

Taking the δ_{ab} metric on the e_a basis as before, and the corresponding Hodge $\star$ etc., we have the Laplace operator $\frac{1}{4} \star \mathrm{d} \star \mathrm{d}$ on plane waves

$$\Box \gamma_k = |k| \gamma_k \tag{4.16}$$

so that the total degree function on the Clifford algebra has a nice geometrical interpretation as the eigenvalue of the Laplacian. Similarly the diagonalisation of the wave operator in spin 1 has the same form as in Section 31.4.1. For example, in $\mathrm{Cliff}(2)$ there are five eigenfunctions with $\partial \cdot A = 0$, namely

$$e_1, \ e_2, \ \gamma_2 e_1, \ \gamma_1 e_2, \ \gamma_1 \gamma_2 (e_1 - e_2).$$

The allowed sources in the Maxwell theory are spanned by the latter three. The corresponding solutions for A and their curvatures are respectively

$$A = \gamma_2 e_1, \ \gamma_1 e_2, \ \tfrac{1}{2}\gamma_1\gamma_2(e_1 - e_2),$$
$$F = \mathrm{d}A = 2\gamma_2 e_1 \wedge e_2, \ -2\gamma_1 e_1 \wedge e_2, \ 2\gamma_1\gamma_2 e_1 \wedge e_2.$$

One also has a $U(1)$-Yang–Mills theory on the Clifford algebra, where the curvature is $dA + A \wedge A$ and gauge transform is (1.2) under an invertible element of the Clifford algebra. The subgroup of such gauge transformations restricted to preserve the generating space $\{\gamma_a\}$ under the Clifford adjoint operation is the Clifford group, giving a geometrical interpretation of that.

31.5   Twisting of differentials and the noncommutative torus A_θ.

In this more technical section we explain the theorem for the systematic twisting of differentials on quantum groups and a small corollary of it which covers our above treatment of differential calculus on Clifford algebras as well as on the noncommutative torus and other spaces. The theory works at the level of a Hopf algebra H, but for the purposes of the present article this could just be the coordinate algebra $\mathbb{C}[G]$ of an algebraic group or the group algebra of a discrete group; Hopf algebras are a nice way to treat the continuum and finite theories on an identical footing, as well as allowing one to generalise to quantum groups if one wants q-deformations etc.

A 2-cocycle on the group in our terms is an element $F : H \otimes H \to \mathbb{C}$ obeying the condition

$$F(b_{(1)} \otimes c_{(1)})F(a \otimes b_{(2)}c_{(2)}) = F(a_{(1)} \otimes b_{(1)})F(a_{(2)}b_{(2)} \otimes c),$$
$$F(1 \otimes a) = \epsilon(a) \tag{5.1}$$

for all $a, b, c \in H$, where $\Delta a = a_{(1)} \otimes a_{(2)} \in H \otimes H$ is our notation for the coproduct. We will similarly denote

$$(\mathrm{id} \otimes \Delta)\Delta a = (\Delta \otimes \mathrm{id})\Delta a = a_{(1)} \otimes a_{(2)} \otimes a_{(3)}$$

for the iterated coproduct. On a group manifold, the counit ϵ simply evaluates at the group identity. It is a construction going back to Drinfeld that in this case there is a new Hopf algebra H_F with product

$$a \bullet b = F(a_{(1)} \otimes b_{(1)})a_{(2)}b_{(2)}F^{-1}(a_{(3)} \otimes b_{(3)}), \quad \forall a, b \in H \tag{5.2}$$

where F^{-1} is the assumed convolution inverse. See [27] for details and proofs. Another fact is that for a bicovariant calculus on a Hopf algebra H, the exterior algebra $\Omega(H)$ is a super-Hopf algebra [29]. Then:

Theorem 2 (*S.M. & R. Oeckl* [3]) *Let F be a 2-cocycle on a Hopf algebra H with bicovariant calculus $\Omega(H)$. Extend F to a super-cocycle on $\Omega(H)$ by zero on degree > 0. Then $\Omega(H_F) = \Omega(H)_F$ is a bicovariant calculus on H_F.*

This took care of how calculi on Hopf algebras themselves behave under twisting. We need a slight but immediate extension of this. Let A be an algebra on which H coacts by an algebra homomorphism $\Delta_L : A \to H \otimes A$. We will use the notation $\Delta_L a = a^{(\bar{1})} \otimes a^{(\bar{\infty})}$. In the case when H is functions on a group, given any group element we evaluate the H part of the output of Δ_L on the element and have a map $A \to A$, i.e., a coaction is just the same thing as a group action but expressed in terms of the coordinates on the group. A fundamental theorem [28] is that when we twist H to H_F we must also twist A to a new algebra A_F for it to be H_F-covariant; its product is

$$a \bullet b = F(a^{(\bar{1})} \otimes b^{(\bar{1})}) a^{(\bar{\infty})} b^{(\bar{\infty})}, \quad \forall a, b \in A. \tag{5.3}$$

There is obviously the same theorem in the category of super algebras coacted upon by super-Hopf algebras, which is trivial to spell out (one just puts the signed super-transposition in place of the usual transposition in all constructions). Now suppose that A is covariant under H and is equipped with a covariant differential calculus such that the coaction extends to all of the exterior super-algebra $\Omega(A)$ of (possibly noncommutative) differential forms as a comodule algebra under H. So we can apply (5.3) to $\Omega(A)$ in place of A. Clearly,

Corollary 3 *If $\Omega(A)$ is H-covariant on an H-covariant algebra A and if F is a cocycle on H, then $\Omega(A_F) = \Omega(A)_F$ is an H_F-covariant calculus on A_F.*

One could also view $\Omega^1(A)$ as an $\Omega^1(H)$-super comodule algebra and then using Theorem 2 and the super-version of (5.3) applied to $\Omega(A)$ we see in fact that $\Omega(A_F)$ is an $\Omega(H_F)$-super comodule algebra.

Since any Hopf algebra coacts covariantly on itself via $\Delta_L = \Delta$, a canonical example is to take $A = H$ with this coaction. Then for any cocycle F we have a new algebra A_F via (5.3) covariant under H_F obtained via (5.2). This was the theory behind [23, 24]. Now suppose that $\Omega(H)$ is a bicovariant calculus on H, then it is in particular left covariant. So $\Omega(A) = \Omega(H)$ is left H-covariant as induced by Δ_L on functions. We can therefore apply Corollary 3 to obtain $\Omega(A_F)$ also. In fact [5] $\Omega(A) \cong A \otimes \Lambda$ where Λ is the algebra of left-invariant forms generated by $\{e_a\}$ say. The coaction on these forms is trivial and hence when we apply (5.3) to $\Omega(A)$ we find simply

$$e_a \wedge_\bullet e_b = F(e_a{}^{(\bar{1})} \otimes e_b{}^{(\bar{1})}) e_a{}^{(\bar{\infty})} \wedge e_b{}^{(\bar{\infty})} = e_a \wedge e_b$$
$$a \bullet e_a = F(a^{(\bar{1})} \otimes e_a{}^{(\bar{1})}) a^{(\bar{\infty})} e_a{}^{(\bar{\infty})} = a e_a$$
$$e_a \bullet a = F(e_a{}^{(\bar{1})} \otimes a^{(\bar{1})}) e_a{}^{(\bar{\infty})} a^{(\bar{\infty})} = e_a a, \quad \forall a \in A. \tag{5.4}$$

So the form relations are unchanged in this basis $\{e_a\}$. We get the same result in the super-twisting point of view: the supercoproduct in $\Omega(H)$ looks like $\underline{\Delta} e_a = \underline{\Delta}_R e_a + 1 \otimes e_a$ where $\underline{\Delta}_R e_a \subset \Lambda^1 \otimes H$ is a certain right coaction. Here Λ^1 is spanned by the $\{e_a\}$. Since the cocycle is trivial on 1 in the sense in (5.1) and similarly for $F(a \otimes 1)$, and since extended by zero on the e_a, we see that the super version of (5.3) with $\Omega^1(A) = \Omega^1(H)$ and supercoaction $\underline{\Delta}_L = \underline{\Delta}$ gives the same result that all of these products involving 1-forms are unchanged.

Let us see how all of this works for a discrete group G. Here $H = \mathbb{C}G$ is the group algebra spanned by group elements, with coproduct $\Delta g = g \otimes g$ for all $g \in G$. The equation (5.1) then reduces to a group cocycle. Yet, because H is cocommutative, the twisting (5.2) has no effect and $H_F = H$ is unchanged for all F. On the other hand, we can take $A = \mathbb{C}G$ as a left-covariant algebra under $\Delta_L = \Delta$. This time (5.3) means a new algebra A_F with product

$$g \bullet h = F(g, h)gh, \quad \forall g, h \in G, \tag{5.5}$$

extended linearly. This is the special case used in Section 31.4.

We now look at the initial differential structure. To keep the picture simple we assume G is Abelian (this is not necessary). Then under Fourier transform $\mathbb{C}G \cong \mathbb{C}[\hat{G}]$, where $\hat{G}$ is the group of characters (if G is infinite then this will be compact and $\mathbb{C}[\hat{G}]$ is the algebraic coordinate ring). In the compact case we use indeed the classical calculus on this 'position space' $\hat{G}$ while in the finite case we use the setup (3.1) for a chosen conjugacy class. That is the geometrical picture, but we do not have to actually do the Fourier transform to the coordinate ring picture, we rather work directly with the 'momentum group' G dual to the position space. Then

$$\Omega(\mathbb{C}G) = (\mathbb{C}G)\Lambda \tag{5.6}$$

where Λ here is the usual Grassmann algebra of basic forms $\{e_a\}$. The differential and relations have the form

$$dg = \sum_a (\chi_a(g) - 1)ge_a, \quad e_a g = \chi_a(g)ge_a, \quad \forall g \in G \tag{5.7}$$

and the super-coproduct is

$$\underline{\Delta} g = g \otimes g, \quad \underline{\Delta} e_a = e_a \otimes 1 + 1 \otimes e_a. \tag{5.8}$$

Here $\{\chi_a\}$ are some subset of characters, the allowed directions in $\hat{G}$ that define the calculus. This is equivalent to our treatment in Section 31.4.1 for $G = (\mathbb{Z}_2)^n$, with g in the role of plane waves ψ_k.

Next we take $A = \mathbb{C}G = H$ another copy of the same algebra, F a group cocycle and $\Delta_L = \Delta$ to get our twisted version $A_F = (\mathbb{C}G)_F$ as in (5.5). And we take $\Omega(A) = (\mathbb{C}G)\Lambda = \Omega(H)$ as above. Then from (5.4) we know that $\Omega((\mathbb{C}G)_F)$ has the identical form to the untwisted case. The only part that changes is the algebra $(\mathbb{C}G)_F$ itself which becomes twisted and typically (depending on F) noncommutative. This is the reason for exactly the same form of calculus on $\mathrm{Cliff}(n)$ in Section 31.4.2 as for $(\mathbb{Z}_2)^n$.

Now let us cover the celebrated noncommutative torus A_θ in the same way. This has the relations $vu = e^{\imath\theta}uv$ where θ is an angle. For our initial situation we take $G = \mathbb{Z} \times \mathbb{Z}$ in (5.7), with free commuting generators u, v. Here the character group is $S^1 \times S^1$ and the calculus is defined by two characters

$$\chi_{1,\phi}(u^m v^n) = e^{\imath\phi m}, \quad \chi_{2,\phi}(u^m v^n) = e^{\imath\phi n}$$

with ϕ as a parameter. The only difference is that we rescale d and then take the limit $\phi \to 0$ of this family of calculi, giving

$$d(u^m v^n) = \lim_{\phi \to 0} \frac{1}{\imath\phi} \left((\chi_{1,\phi}(u^m v^n) - 1) u^m v^n e_1 + (\chi_{2,\phi}(u^m v^n) - 1) u^m v^n e_2 \right)$$

$$= u^m v^n (m e_1 + n e_2),$$

$$e_a u^m v^n = u^m v^n e_a.$$

This is nothing but the usual classical differential calculus on $\mathbb{C}[S^1 \times S^1]$, but written algebraically and in momentum space. Next, on $\mathbb{Z} \times \mathbb{Z}$ we take cocycle

$$F(u^m v^n, u^s v^t) = e^{\imath \theta n s}, \tag{5.9}$$

where θ denotes a fixed parameter. Then the formula (5.5) gives

$$v \bullet u = e^{\imath\theta} uv = e^{\imath\theta} u \bullet v \tag{5.10}$$

which is a noncommutative torus from our algebraic point of view. That its product is a twisting is known to experts from another point of view [31] and also for more general θ-spaces [32] from the twisting point of view in recent work [33, 34]. That one automatically gets the differential geometry as a twist seems to be less well known. Thus, for the differential structure we obtain

$$du = u e_1, \quad dv = v e_2, \quad e_a u = u e_a, \quad e_a v = v e_a \tag{5.11}$$

since this is the same as the classical form. Likewise the wedge products are as per the classical form in which the e_a anticommute. Note then that

$$(dv)u = v e_1 u = v \bullet u e_1 = e^{\imath\theta} u \bullet v e_1 = e^{\imath\theta} u\, dv,$$

$$dv \wedge du = v e_2 \wedge u e_1 = e_2 \wedge v \bullet u e_1 = e^{\imath\theta} e_2 \wedge u \bullet v e_1 = -e^{\imath\theta}\, du \wedge dv$$

which is how the calculus is usually presented, as noncommutation between du, v etc., but we see that it expresses nothing more than the noncommutativity of A_θ itself. This goes some way towards explaining why one can develop Yang–Mills and other geometry on a noncommutative torus [30] so much like on a usual torus, with two commuting derivations as vector fields, etc. For a general noncommutative algebra one does not expect many derivations.

We can just as easily apply this theory to $H = A = \mathbb{C}[\mathbb{R}^n]$. Here H has the linear coproduct $\Delta x_\mu = x_\mu \otimes 1 + 1 \otimes x_\mu$ and on it we take the cocycle

$$F(f \otimes g) = e^{\imath \sum \theta_{\mu\nu} \partial^\mu \otimes \partial^\nu} (f \otimes g)(0) \tag{5.12}$$

where we apply differential operators on functions f, g and evaluate at zero. Then the algebra $\mathbb{C}[\mathbb{R}^n]_F$ has product

$$f \bullet g = \left(\sum_{m=0}^{\infty} \frac{(\sum \theta_{\mu\nu} \partial^\mu \otimes \partial^\nu)^m}{m!} \Big|_0 (f_{(1)} \otimes g_{(1)}) \right) f_{(2)} g_{(2)}$$

$$= \cdot e^{\imath \sum_{\mu,\nu} \theta_{\mu\nu} \partial^\mu \otimes \partial^\nu} (f \otimes g) \tag{5.13}$$

where $(\partial^m|_0 f_{(1)}) f_{(2)} = \partial^m f$. This is the Moyal product and this point of view was explored in [3] in the context of generalising the cocycle to a non-Abelian group rather than $\mathbb{R}^n$. It was picked up and discussed explicitly in [18] as well as by other authors. In particular, the coordinate functions x_μ with the $\bullet$ product obey the algebra (2.28) studied in Section 31.2.3. From this point of view the bilinear form that defines the Clifford algebra twisting (4.10) is like a finite difference matrix. Meanwhile the cochain that similarly twists $(\mathbb{Z}_2)^n$ to the octonions [23] has an additional cubic term that is responsible for their nonassociativity, i.e., more like a discrete 'Chern–Simons' theory as remarked there. Note that when F is only a cochain our Corollary 3 induces not a usual associative exterior algebra on the octonions but a superquasialgebra in the same sense as the octonions are a quasialgebra (i.e., associative when viewed in a monoidal category).

Further afield, other θ-manifolds have been of interest of late, see [35] and references therein. It is clear that these too are twistings as noncommutative algebras and therefore that one can recover the algebraic side of their noncommutative geometry by twisting the classical geometry as above. It is assumed that M is a classical manifold admitting a group action by a compact Lie group $K \supseteq S^1 \times S^1$. We assume an algebraic description exists (otherwise one needs to do some analysis), so that we have classical coordinate Hopf algebras $\mathbb{C}[K] \to \mathbb{C}[S^1 \times S^1]$ and a coaction $\mathbb{C}[M] \to \mathbb{C}[K] \otimes \mathbb{C}[M]$. We suppose that this extends to the classical exterior algebra of M also. The action of the subgroup here corresponds to a coaction $\mathbb{C}[M] \to \mathbb{C}[S^1 \times S^1] \otimes \mathbb{C}[M]$. Then with the same cocycle (5.9), we obtain a new algebra $\mathbb{C}[M]_F$ by the comodule algebra twisting theorem (5.3), and we obtain a differential calculus $\Omega(\mathbb{C}[M]_F)$ on it by Corollary 3, and ultimately an entire twisted noncommutative geometry with a parameter θ. Equivalently, we can obviously pull back F as a cocycle $F : \mathbb{C}[K] \otimes \mathbb{C}[K] \to \mathbb{C}$ and do all of the above with the $\mathbb{C}[K]$-coaction directly (the result is the same). On the other hand, whereas $\mathbb{C}[S^1 \times S^1]$ does not itself twist as a Hopf algebra, now (5.2) gives a new Hopf algebra $\mathbb{C}[K]_F$ and this coacts on $\Omega(\mathbb{C}[M]_F)$ by Corollary 3. By Theorem 2 we also have $\Omega(\mathbb{C}[K]_F)$ and this supercoacts. It seems that these elementary deductions fit with observations from a different point of view (not via the twisting theory) in [35]. We see only an 'easy' algebraic part of that theory but it does seem to indicate some useful convergence between that operator theory approach and the quantum groups one.

REFERENCES

[1] S. Majid. Duality principle and braided geometry. Springer Lec. Notes in Phys. **447**, 125–144, 1995.

[2] A. Connes. *Noncommutative Geometry*. Academic Press, 1994.

[3] S. Majid and R. Oeckl. Twisting of quantum differentials and the Planck scale Hopf algebra. *Commun. Math. Phys.* **205**, 617–655, 1999.

[4] S. Majid. Hopf algebras for physics at the Planck scale. *J. Classical and Quantum Gravity* **5**, 1587–1606, 1988.

[5] S.L. Woronowicz. Differential calculus on compact matrix pseudogroups (quantum groups). *Commun. Math. Phys.* **122**, 125–170, 1989.

[6] T. Brzeziński and S. Majid. Quantum group gauge theory on quantum spaces. *Commun. Math. Phys.* **157**, 591–638, 1993. Erratum 167:235, 1995.

[7] S. Majid. Riemannian geometry of quantum groups and finite groups with nonuniversal differentials. *Commun. Math. Phys.* **225**, 131–170, 2002.

[8] F. Ngakeu, S. Majid and D. Lambert. Noncommutative Riemannian geometry of the alternating group A_4. *J. Geom. Phys.* **42**, 259–282, 2002.

[9] S. Majid. Ricci tensor and Dirac operator on $\mathbb{C}_q[SL_2]$ at roots of unity. *Lett. Math. Phys.* **63**, 39–54, 2003.

[10] S. Majid. On q-Regularization. *Int. J. Mod. Phys. A* **5**, 4689–4696, 1990

[11] S. Majid and H. Ruegg. Bicrossproduct structure of the κ-Poincaré group and noncommutative geometry. *Phys. Lett. B* **334**, 348–354, 1994.

[12] J. Lukierski, A. Nowicki, H. Ruegg and V.N. Tolstoy. q-Deformation of Poincaré algebra. *Phys. Lett. B* **271**, p. 321, 1991.

[13] G. Amelino-Camelia and S. Majid. Waves on noncommutative spacetime and gamma-ray bursts. *Int. J. Mod. Phys. A* **15**, 4301–4323, 2000.

[14] G. Amelino-Camelia. Gravity-wave interferometers as quantum-gravity detectors *Nature* **398**, p. 216, 1999.

[15] E. Batista and S. Majid. Noncommutative geometry of angular momentum space $U(su_2)$. *J. Math. Phys.* **44**, 107–137, 2003.

[16] B.J. Schroer. Combinatorial quantization of Euclidean gravity in three dimensions, *preprint* math.QA/0006228.

[17] S. Snyder. Quantized space-time *Phys. Rev.* **71**, 38–41, 1947.

[18] R. Oeckl. Untwisting noncommutative $\mathbb{R}^d$ and the equivalence of quantum field theories. *Nucl.Phys. B* **581**, 559–574, 2000.

[19] N. Seiberg and E. Witten. String theory and noncommutative geometry. *J. High En. Phys.*, 9909:032, 1999.

[20] K. Bresser, F. Mueller-Hoissen, A. Dimakis, A. Sitarz. Noncommutative geometry of finite groups *J. Phys. A* **29**, 2705–2736, 1996.

[21] S. Majid and E. Raineri. Electromagnetism and gauge theory on the permutation group S_3. *J. Geom. Phys.* **44**, 129–155, 2002.

[22] S. Majid and T. Schucker. $\mathbb{Z}_2 \times \mathbb{Z}_2$ lattice as a Connes–Lott–quantum group model. *J. Geom. Phys.* **43**, 1–26, 2002.

[23] H. Albuquerque and S. Majid. Quasialgebra structure of the Octonions. *J. Algebra* **220**, 188–224, 1999.

[24] H. Albuquerque and S. Majid. Clifford algebras obtained by twisting of group algebras. *J. Pure Applied Algebra* **171**, 133–148, 2002.

[25] G. Dixon. *Division Algebras: Octonions, Quaternions, Complex Numbers and the Algebraic Design of Physics*. Kluwer, 1994.

[26] S. Caenepeel, F. Van Oystaeyen. A note on generalized Clifford algebras and representations. *Comm. in Algebra* **17**, 93–102, 1989.

[27] S. Majid. *Foundations of Quantum Group Theory*. Cambridge University Press, 1995.

[28] S. Majid. q-Euclidean space and quantum Wick rotation by twisting. *J. Math. Phys.* **35**, 5025–5034, 1994.

[29] T. Brzeziński. Remarks on bicovariant differential calculi and exterior Hopf algebras. *Lett. Math. Phys.* **27**, p. 287, 1993.

[30] A. Connes and M.A. Rieffel. Yang–Mills for noncommutative two-tori. *AMS Contemp. Math. Series* **62**, 237–266, 1987.

[31] M. Rieffel. Deformation quantization for actions of $\mathbb{R}^d$. *Memoirs AMS* **106**, 1993.

[32] A. Connes and G. Landi. Noncommutative manifolds, the instanton algebra and isospectral deformations. *Commun. Math. Phys.* **221**, 141–159, 2001.

[33] A. Sitarz. Twist and spectral triples for isospectral deformations. *Lett. Math. Phys.* **58**, 69–79, 2001.

[34] J. Varilly. Quantum symmetry groups of noncommutative spheres. *Commun. Math. Phys.* **221**, 511–523, 2001.

[35] A. Connes and M. Dubois-Violette. Noncommutative finite-dimensional manifolds, I: Spherical manifolds and related examples. *Commun. Math. Phys.* **230**, 539–579, 2002.

Shahn Majid
School of Mathematical Sciences
Queen Mary, University of London
327 Mile End Rd, London E1 4NS, UK.
E-mail: s.majid@qmw.ac.uk

Submitted: May 20, 2002.

32

Dirac Operator on Quantum Homogeneous Spaces and Noncommutative Geometry

Robert M. Owczarek

ABSTRACT Spectral triples constitute basic structures of Connes' noncommutative geometry. In order to unify noncommutative geometry with quantum group theory, it is necessary to provide a proper description of spectral triples on quantum groups and, in a wider context, on their homogeneous spaces. Thus, a way to define Dirac-like operators on such quantum spaces has to be found. This paper is a brief summary on some problems we are facing in searching for such a way. We suggest that some modifications of either noncommutative geometry or quantum group theory, or both, are inevitable.

Differential calculi and their relationship to the Dirac operators are the key concepts to understand for defining the spectral triples in question. Since there is no canonical way to construct such calculi on the quantum groups, and since there are two basic kinds, left-(right-) covariant and bicovariant, there are a number of issues to be resolved. We discuss these difficulties as well as difficulties implicit in the definition of the Dirac operator, which are present already at the classical manifolds level.

Definition of quantum homogeneous spaces constitutes another problem. This is also briefly reviewed in the paper. Then a basic example associated with the quantum 2-sphere of Podleś, as a homogeneous space of the quantum $SU(2)$ group of Woronowicz, is discussed. A naturally defined Dirac operator on such a 2-sphere does not satisfy one of Connes' axioms.

Keywords: Dirac operator, noncommutative geometry, homogeneous spaces

32.1   Introduction

In this section the motivation behind the study of quantum homogeneous spaces and Dirac operators on them is presented. A gap exists between axioms of noncommutative geometry [3] and examples of quantum spaces supplied by the theory of quantum groups and quantum homogeneous spaces (it was mentioned, for example, in the discussion following the talk by Prof. Varilly at the confer-

AMS Subject Classification: 46L87, 58B34, 58B32.

ence). Noncommutative geometry theorists are inclined to ignore these spaces. They argue, for example, that some topological invariants are trivial for them. As they claim, nonvanishing of the invariants is crucial for a noncommutative (as well as commutative) space to be genuine. Therefore, vanishing of the invariants is the rationale to ignore such a space. On the other hand, examples of quantum spaces in noncommutative geometry are basically concentrated only around noncommutative tori. Even examples of noncommutative spherical spaces built within noncommutative geometry are based on this apparently fundamental example (see, e.g., [5]). This approach was recently summarized in a paper by Connes and Dubois–Violette [4]. The reason for such limited choice is that the spectral axioms for the Dirac operator are the easiest satisfied in the case when the "quantization" preserves spectrum. This is quite a strong condition. It is arguable if it should be imposed. At the current stage of research it is too early for proposing an alternative.

The spectrum preservation requirement excludes the important examples of quantum spaces provided by the theory of quantum groups from studies in noncommutative geometry. Such exclusion looks quite artificial from the point of view of the theory. One would expect that all spaces described within the C^*-algebraic framework should be included. Quantum groups and homogeneous quantum spaces began their lives in C^*-algebraic framework with the work of Woronowicz [18, 19] and Podleś [11].

Actually, we expect that reconciliation of the quantum group inspired spaces with noncommutative geometry is inevitable. Almost for sure it will lead to some reformulation of the Connes axioms. In the meantime the studies towards this reconciliation should lead to exciting and fruitful research. Of course, there are a number of problems to be solved this way. Some of them are discussed below. In this paper some already existing results are discussed from the new perspective of quantum spaces. The aim of such discussion is to present the current stage of the subject, point out the open questions, and pave the way for further study in this direction.

There are numerous problems to be solved for a proper formulation of the Dirac operators on spaces. Let us mention here only some of them. First, we discuss the problem that while some authors claim they construct the Dirac operators, they build a kind of a Dirac–Kähler operator instead. The operators are different, even at the level of classical differential geometry. Next, the very definition of quantum homogeneous spaces is discussed. Finally, the differential calculi on quantum homogeneous spaces are briefly discussed.

Let us mention that the definition of Clifford algebras in the quantum case is not discussed here despite its importance in the general theory of Dirac operators on noncommutative spaces. The reason is that our basic example could be constructed using a classical Clifford algebra. However, proper reformulation of the notion of a Clifford algebra is indispensable for a complete theory of the Dirac operator in noncommutative geometry.

The example of the Dirac operator on a quantum sphere of Podleś [11] will be also briefly discussed. The spectrum of this operator has asymptotic behavior inconsistent with Connes' axioms.

32.2   Standard formulation of the Dirac operator and its reformulation for noncommutative geometry

Before discussing construction of Dirac operators in the noncommutative geometry context, let us recall the standard approach to defining the spin structures and Dirac operators on classical Riemannian manifolds. This material can be found in many places, e.g., in [2]. The first ingredient is the double cover of the special orthogonal group, which is the structure group for the principal bundle of oriented orthonormal frames, by the spin group,

$$1 \to \mathbb{Z}_2 \to \mathbf{Spin}(n) \overset{\rho_n}{\to} SO(n) \to 1 \,. \tag{2.1}$$

An orientable, smooth, Riemannian manifold (M, g) is equipped with the $SO(n)$ principal frame bundle, where $n = \dim M$. It actually makes sense, and is quite useful in the case of homogeneous Riemannian spaces, to consider orthonormal frame bundles with structure groups, which are proper subgroups of the $SO(n)$ group. Then, among Riemannian manifolds we distinguish the spin manifolds. These are the Riemannian manifolds, with a spin structure defined on them; that is a principal frame bundle structure with $\mathbf{Spin}(n)$ or appropriate subgroup of $\mathbf{Spin}(n)$ as the structure group, "double covering" the orthonormal bundle of M. In other words, the following diagram is commutative:

$$
\begin{array}{ccc}
\tilde{F} \times \mathbf{Spin}(n) & \overset{\eta \times \rho_n}{\longrightarrow} & F \times SO(n) \\
\downarrow & & \downarrow \\
\tilde{F} & \overset{\eta}{\longrightarrow} & F \\
\searrow_{\tilde{\pi}} & & \swarrow_{\pi} \\
 & M &
\end{array}
\tag{2.2}
$$

where F, $\tilde{F}$ are total spaces of the orthonormal frame bundle and the spin structure principal bundle, respectively, η is the covering bundle morphism, $\tilde{\pi}$, π are respective bundle projections. A connected, orientable Riemannian manifold (M, g) admits a spin structure iff $w_2(M) = 0$. Since the $\mathbf{Spin}(n)$ group is contained in the corresponding Clifford algebra Cl_n, the standard representation $\gamma_n : Cl_n \to \operatorname{End} S_n$ induces the representation $\gamma_n : \mathbf{Spin}(n) \to \operatorname{End} S_n$, called the spinor representation. The spinor bundle S is then defined as a vector bundle associated with the principal bundle of the spin structure, $\tilde{F}$.

For a Riemannian, oriented manifold (M, g) there exists a unique connection 1-form ω on the orthonormal frame bundle F, with values in the Lie algebra $so(n)$. One can then define a connection 1-form $\tilde{\omega}$ on $\tilde{F}$, taking values in the Lie algebra $spin(n)$ by the formula

$$\tilde{\omega} := \rho_n^{-1}{}_* \eta^* \omega, \tag{2.3}$$

where $\rho_n^{-1}{}_*$ is the push-forward operation associated to ρ_n^{-1}, and η^* is the pull-back operation associated to η.

This 1-form is called the spin connection. This connection form determines a covariant derivative ∇ acting on sections of an arbitrary bundle associated to the principal bundle, including the spinor bundle $\mathcal{S}$.

If $\{e_a(x) : a \leq n, x \in U \text{ (open in M)}\}$ is a local section of F, a local Dirac operator is defined as

$$D : \Gamma_U^\infty(\mathcal{S}) \to \Gamma_U^\infty(\mathcal{S}) : \Psi \mapsto \sum_a \gamma^a \nabla_a \Psi . \tag{2.4}$$

Here, $\Gamma_U^\infty(\mathcal{S})$ is the space of smooth sections of $\mathcal{S}$ over

$$U, \gamma^a := \gamma(\alpha(e_*^a)),$$

$\alpha : \mathbb{R}^{n*} \to \mathbb{R}^n$ is the isomorphism induced by the scalar product g_0, $e_a : a \leq n$ is an orthonormal basis of $\mathbb{R}^n$, $e_a^* : a \leq n$ is the dual basis of $\mathbb{R}^{n*}$, and $\nabla_a = \nabla_{e_a}$ are covariant derivatives w.r.t. $e_a(x)$.

One can check that D does not depend on the section, so D is globally defined. It is the Dirac operator.

The same construction can be pursued in another, equivalent, way, which is better suited for the purpose of developing noncommutative geometry (see, e.g., [17]). Of course the starting point is the same. One deals with an n-dimensional orientable Riemannian manifold (M, g). Then one builds a Clifford algebra bundle $C\ell(M)$ over M. The fibers of the bundle are Clifford algebras. More exactly, for $n = 2m$,

$$C\ell_x(M) = C\ell(T_x M, g_x)^{\mathbb{C}} \simeq M_{2^m}(\mathbb{C}).$$

For $n = 2m + 1$,

$$C\ell(T_x M, g_x)^{\mathbb{C}} \simeq M_{2^m}(\mathbb{C}) \oplus M_{2^m}(\mathbb{C}).$$

Hence $C\ell_x(M)$ is taken to be the even part of the Clifford algebra,

$$C\ell_x(M) := C\ell^{even}(T_x M, g_x)^{\mathbb{C}} \simeq M_{2^m}(\mathbb{C}).$$

As a result, the Clifford bundles are graded in the even-dimensional case, but not in the odd-dimensional case.

The fibers of this bundle are to be considered as algebras of endomorphisms of a vector space. These fibers give rise to a vector bundle. As a result of this construction one considers instead of spin structures, $spin^C$ structures. The $spin^C$ structures are defined using the algebras

$$A := C_0(M), \quad B := C_0(M, C\ell(M)).$$

The spinor bundle is defined as a $B - A$-equivalence module $\mathcal{S}$. The $spin^C$ structure itself is defined as a pair $(\epsilon, \mathcal{S})$, where ϵ is an orientation on TM.

Let us mention that existence of $spin^C$ structures is guaranteed if the condition $w_3(TM) = 0$ is satisfied. The $\mathcal{S}$ denotes the bundle of smooth sections of the vector bundle of spinors. Therefore, $\mathcal{S}$ is $\Gamma^\infty(S)$, which defines S. We can introduce the Hilbert space of square-integrable spinors $\mathcal{H} = L^2(M, S)$. In case M is even-dimensional, there is a $\mathbb{Z}_2$ grading in the space $\mathcal{S}$, coming from the $\mathbb{Z}_2$ grading of the Clifford algebra bundle. This grading implies in its turn $\mathbb{Z}_2$ grading of $\mathcal{H}$, $\mathcal{H} = \mathcal{H}^+ \oplus \mathcal{H}^-$, where $\mathcal{H}^+$, $\mathcal{H}^-$ are (± 1)-eigenspaces of the grading operator called χ; $\chi^2 = 1$. The χ's correspond to γ^5 matrix known in physics.

As the last step, the Dirac operator is defined in a very similar way as in the standard case, by means of spin connection, covariant derivative, and the representation γ. This formulation, essentially equivalent to the standard one, allowed Connes to state his set of axioms for spectral triples. It leads, when applied incorrectly, to one of the problems addressed below.

32.3   Potential difficulty with the definition of Dirac operator using Clifford algebras

Following the standard way of defining the Dirac operator on noncommutative spaces may lead, when done carelessly, to some problems. These were very clearly explained already in a preprint by Trautman [16], but since the same problematic constructions keep appearing in some papers, let us recall the content of this not widely known preprint here.

Let (V, g) be a finite-dimensional real or complex vector space, with scalar product g. Let $\bigwedge V, C\ell(V, g)$ denote the Grassmann and the Clifford algebra over V, respectively. Then, one defines two maps $d, \delta : V \to \mathrm{End}(\bigwedge V)$, by

$$d(u)(w) := u \wedge w \quad \text{and} \quad \delta(u)(w) := \langle g(u)|w \rangle, \quad u \in V, w \in \bigwedge V,$$

where $g(u) \in V^*$ acts on $w \in \mathrm{End}(\bigwedge V)$ as a graded derivation of degree -1. It extends the standard definition of $g(u)$, which is

$$\langle g(u)|v \rangle = g(u, v), v \in V.$$

These definitions imply $d(u)^2 = 0$, $\delta(u)^2 = 0$, where $u \in V$ is arbitrary, and

$$(d(u) + \delta(v))^2 = g(u, v) \, \mathrm{id}_{\bigwedge V} .$$

Therefore, $d + \delta : V \to \mathrm{End}(\bigwedge V)$ enjoys the Clifford property. Thus, by universality property of the Clifford algebras, $d + \delta$ extends to a homomorphism of algebras with unity

$$d + \delta : C\ell(V, g) \to \mathrm{End}(\bigwedge V).$$

This representation is decomposable. The decomposition is particularly easy to describe in case (V, g) is neutral (in particular even dimensional, but could be either real or complex). Namely, there is a decomposition $V = N \oplus P$ into totally null subspaces of the same dimension, equal to half the dimension of V. Then one can define $S := \bigwedge N \subset \bigwedge V$, and

$$\gamma : V \to \text{End } S, \quad \gamma(u) := \sqrt{2}(d(n) + \delta(p))|_S,$$

where $u = n + p$. As a result of this definition, $\gamma(u)^2 = g(u, u)\, \text{id}_S$. Again, by the universality property of Clifford algebras, γ can be extended to an algebra homomorphism $\gamma : C\ell(V, g) \to \text{End } S$, which is an irreducible and faithful representation of $C\ell(V, g)$. Spinors defined this way are called Chevalley spinors or algebraic spinors.

At first it looks like the algebraic spinors are equivalent to the standard ones, also when considered on Riemannian manifolds. However, they are not the same. Namely, manifolds admitting algebraic spinor structures must be almost complex. There are many examples of Riemannian manifolds admitting spin structures (spin manifolds), which are not almost complex. The Dirac operator, which is a curved space version of the $d + \delta$ operator is also not the same as the Dirac operator, and is usually called the Dirac–Kähler operator.

This problem should be taken into account also when constructing a Dirac-like operator on a noncommutative space.

32.4   Problem of definition of quantum homogeneous space

There is a plethora of definitions of quantum homogeneous spaces. Many definitions of the same notion means lack of a proper one. A bit more detailed discussion of this problem is given below.

First, let us recall a few definitions of the quantum homogeneous spaces. They differ only by minor editing from their originals. The first definition reads:

Definition 1 (Oeckl [10]). *Let $\pi : P \to H$ be a surjection of Hopf algebras. Then the left P-comodule algebra*

$$B := P^H = \{\, p \in P : p_{(1)} \otimes \pi(p_{(2)}) = p \otimes 1 \,\}$$

is called a quantum homogeneous space

Let us recall another definition, which is also quite algebraic in flavor. In fact there are a number of very similar definitions to this one.

Definition 2 (Hermisson [7]). *Let χ be a subalgebra of $\mathcal{H}$ ($\mathcal{H}$ is a Hopf algebra over $\mathbb{C}$) and a left coideal, i.e., $\delta(\chi) \subset \mathcal{H} \otimes \chi$, which is what we call a function algebra of a quantum homogeneous space.*

In our opinion, a definition of homogeneous quantum spaces should involve, besides purely algebraic, also some functional analytic conditions. However, such definitions as a rule become quite complicated, as shown by two different definitions of quantum homogeneous spaces given by Podleś. The reasons for these complications lie, besides other issues, in the splitting of classically equivalent notions related to homogeneous spaces . We cite here some results by Podleś concerning this issue.

Let us begin with the first definition by Podleś.

Definition 3 (Podleś [11]). *Let X be a quantum space and $G = (A, u)$ be a compact matrix pseudogroup ($A = C(G)$ in our notation). We say that a C^*-homomorphism*

$$\Gamma : C(X) \to C(X) \otimes C(G)$$

is an action of G on X if

$$(\mathrm{id} \otimes \Phi_G)\Gamma = (\Gamma \otimes \mathrm{id})\Gamma,$$

where Φ_G is [the coproduct of $C(G)$]. If the finite-dimensional vector subspace $W \subset C(X)$ is Γ-invariant (i.e., $\Gamma W \subset W \otimes C(G)$), then Γ defines a representation d of G acting on W by the formula $\hat{d} = \Gamma |_W$ (...). If d is irreducible, then we say that $\mathbf{d}$ (the class of representations equivalent to d) enters the spectrum Γ. Moreover, if for a fixed $\mathbf{d}$, the subspace W is only one, then we say that $\mathbf{d}$ has multiplicity 1. We say that (X, Γ) is homogeneous if the trivial representation enters the spectrum Γ with multiplicity 1.

Now, let us recall another definition of the quantum homogeneous spaces. It consists actually of a chain of definitions and theorems.

Definition 4 (Definition 1.4 in Podleś [14]). *Let X be a quantum space and G a quantum group. We say that a C^*-homomorphism $\Gamma : C(X) \to C(X) \otimes C(G)$ is an action of G on X if*

(a) $(\Gamma \otimes \mathrm{id})\Gamma = (\mathrm{id} \otimes \Phi_G)\Gamma$,

(b) $\langle (I \otimes y)\Gamma x : x \in C(X), y \in C(G) \rangle = C(X) \otimes C(G)$.

Definition 5 (Definition 1.3 in Podleś [14]). *We say that a quantum group $H = (B, v)$ is a (compact) subgroup of a quantum group $G = (A, u)$ if $\dim v = \dim u$ and there exists a C^*-homomorphism $\theta_{HG} : A \to B$ s.t. $\theta_{HG}(u_{ij}) = v_{ij}, i, j = 1, 2, \ldots, \dim u$. The quotient space $H \backslash G$ is defined by*

$$C(H \backslash G) = \{ x \in C(G) : (\theta_{HG} \otimes \mathrm{id})\Phi_G x = I \otimes x \}.$$

Theorem 1 (Theorem 1.7 in Podleś [14]). *Let H be a subgroup of quantum group G. Then*

$$\Gamma_{H \backslash G} = \Phi_G |_{C(H \backslash G)} : C(H \backslash G) \to C(H \backslash G) \otimes C(G)$$

is an action of G on $H \backslash G$.

Definition 6 (Definition 1.8 in Podleś [14]). *All pairs* $(H\backslash G, \Gamma_{H\backslash G})$ *obtained in the above way (and the pairs isomorphic to them) are called quotients. We say that a pair* (X, Γ) *is embeddable if* $C(X) \neq \{0\}$ *and there exists a faithful* C^*-*homomorphism* $\psi : C(X) \to C(G)$ *such that* $\Phi_G \psi = (\psi \otimes \text{id}) \Gamma$ (...). *We say that* (X, Γ) *is homogeneous if* $c_0 = 1$. (*c_0 is defined as the multiplicity of the trivial representation in the decomposition of* $\theta_{HG}(u_0)$, *where* u_0 *is also trivial, into irreducible components*).

Proposition 1 (Proposition 1.9 in Podleś [14]). *Let* Γ *be an action of a quantum group* G *on a quantum space* X. *Then*

(a) (X, Γ) *is a quotient* $\Rightarrow$ (X, Γ) *is embeddable* $\Rightarrow$ (X, Γ) *is homogeneous.*

(b) *In the classical case* (X, Γ) *is a quotient* $\iff$ (X, Γ) *is embeddable* $\iff$ (X, Γ) *is homogeneous.*

There is a family of quantum spaces, which for different values of parameters defining particular members of the family fits one of the kinds of quantum homogeneous spaces. It is also of our special interest in this paper. Namely, it is the family of Podleś spheres, S^2_{qc}, $q \in (-1, 1)\backslash\{0\}$, $c \in [0, \infty]$, with the action Γ of the quantum groups $S_q U(2)$ denoted by σ_{qc}.

The following proposition is stated and proven in [14]. Here,

$$c(i) = \frac{-q^{2i}}{(1 + q^{2i})^2}, \quad i = 1, 2, \ldots.$$

Proposition 2 (Proposition 2.6 in Podleś [14]). *Let* $q \in (-1, 1)\backslash\{0\}$, $c \in \{c(1), c(2), \ldots\} \cup [0, \infty]$. *Then*

(a) (S^2_{qc}, σ_{qc}) *is a quotient* $\iff$ $c = c(1)$ *or* $c = 0$.

(b) (S^2_{qc}, σ_{qc}) *is embeddable* $\iff$ $c = c(1)$ *or* $c \in [0, \infty]$.

(c) (S^2_{qc}, σ_{qc}) *is homogeneous.*

The above review of definitions of quantum homogeneous spaces shows the degree of complication we are facing in solving this problem.

32.5   Problems with differential calculi on quantum homogeneous spaces

Differential calculi on quantum spaces are a necessary ingredient in constructing a Dirac operator on the spaces. In this section we shortly discuss the problems connected with the definition of the differential calculi.

The differential calculi are, as a rule, constructed from the differential calculi on quantum groups, the homogeneous spaces of which we consider. This results in the ambiguity of the definition. Namely, the differential calculi on quantum groups are not uniquely defined. They are basically of two kinds, bicovariant and left- (or right-) covariant. A beautiful theory of such calculi was created by Woronowicz [18, 20]. In particular, the different calculi may have different dimensions. Moreover, the dimension may, in contrast to the classical space case, be different from the dimension of the space (the definition of the latter is also problematic, but we do not touch this subject here).

In particular, Podleś [12, 13] defined both $3D$ and $2D$ calculi on quantum spheres (though, e.g., $2D$ calculi exist only for $c = 0$).

A deeper thought leads to the conclusion that this dimension ambiguity of differential calculi should be resolved simultaneously with solving the ambiguity of definition of "Clifford algebras" and their "spinor representations".

32.6 Some (not completely satisfactory) solutions

In this section we present a solution to some of the above problems we found in the case of the Podleś spheres, with $c = 0$.

Let us briefly recall different approaches to defining quantum spheres. A number of authors proposed such definitions. Within axioms of noncommutative geometry such spaces were defined, for example, by Connes, Landi, Sitarz, Dąbrowski, Bonechi, Aschieri, Dubois–Violette, Varilly. Among the axioms are conditions that impose strong constraints on the asymptotic behavior of the spectrum of the Dirac operators. Therefore, as a rule, the Dirac operators for the quantum spheres defined this way have spectra identical to their classical counterparts. This may arguably be considered a weakness of this approach.

Another set of quantum spheres was defined within the framework of the theory of quantum groups. In this set important examples were defined by Podleś. The example discussed below is built around these very spaces. The interesting result concerning the spaces (see e.g., [9]) is that a natural Dirac operator on them is incompatible with Connes' axioms from noncommutative geometry .

Let us mention that Dirac operators were defined also by other authors, e.g., Pinzul and Stern [15] though no spectrum was given there, Bibikov and Kulish [1] though the Dirac operator and its spectrum seem to have serious problems, Durdevich [6] whose results on the Dirac operator and its spectrum coincide with ours (though his claim that the Dirac operator is structured according to the general theory developed by him is actually not true.)

Let us recall that the spectrum of the Dirac operator constructed in [9] for the family of quantum spheres S^2_{q0} (which in Podleś classification are quotient quantum spaces) has the form

$$\frac{q^{j+\frac{1}{2}} - q^{-(j+\frac{1}{2})}}{q - q^{-1}}$$

where $j = \frac{1}{2}, \frac{3}{2}, \ldots$. One obvious observation about this spectrum is that it is q-dependent. The second, which requires a little work and was presented in [9], is the asymptotic behavior of the Dirac operator with respect to the number of the eigenvalues. It is

$$\lambda_N \sim q^{-\sqrt{\frac{N}{2}}}.$$

This is obviously incompatible with Connes' axioms. This result shows that more work towards understanding noncommutative geometry of quantum spaces is needed.

32.7   Conclusions and outlook

In this paper we showed why a proper definition of Dirac operators on quantum homogeneous spaces is both a valuable task from the point of view of general theory of noncommutative geometry and quantum groups, but also a very complicated one.

Let us make some guesses about the future developments. One interesting development concerns differential calculi on quantum groups. Recently a reformulation of the $3D$ calculus of Woronowicz over the quantum $SU(2)$ group has been given [8]. The notion of twisted graded traces was introduced there. It is an ingredient which substitutes the graded trace of Connes in building a consistent differential geometry theory on quantum groups. As a result even now the axioms of Connes should be modified.

It is not obvious that this change would influence axioms concerning asymptotic behavior of Dirac operators, but it is the author's belief. This modification should result from further studies of such operators on quantum homogeneous spaces. The only result of the author in this area concerns the Podleś sphere with the parameter $c = 0$. This is the case called by Podleś a quotient space. However, in his classification of quantum spheres there are also more irregular cases. It is reasonable to expect that the spectra of Dirac operators in these cases are even more exotic.

The case considered in this paper did not apply the notion of quantum Clifford algebras. Some other cases however would need this notion. Currently the author is considering the case of the quantum $SU(2)$ group of Woronowicz. Already there the "structure group" of the natural "spinor bundle" is not classical and leads to consideration of a quantum Clifford algebra problem. It is another major step to be taken in order to develop the general theory of Dirac operators on quantum spaces.

Acknowledgements

I would like to thank the organizers of the conference, in particular Giovanni Landi for inviting me, Rafał Abłamowicz and Bertfried Fauser for everything they did for the smooth organization of the conference.

REFERENCES

[1] P.N. Bibikov, P.P. Kulish, Dirac operators on quantum $SU(2)$ group and quantum sphere, preprint, arXiv: q-alg/9608012, (1996).

[2] M. Cahen, S. Gutt, Spin structures on compact simply connected Riemannian symmetric spaces, in *Proceedings of the Workshop on Clifford Algebra, Clifford Analysis and their Applications in Mathematical Physics, Ghent, 1988*, Simon Stevin **62** (1988), pp. 209–242.

[3] A. Connes, Gravity coupled with matter and the foundation of non-commutative geometry, *Comm. Math. Phys.* **182** (1996), 155–176.

[4] A. Connes, M. Dubois–Violette, Noncommutative finite-dimensional manifolds. I. Spherical manifolds and related examples [R1], preprint, arXiv: math.QA/0107070, 2001.

[5] A. Connes, G. Landi, Noncommutative manifolds, the instanton algebra and isospectral deformations, *Comm. Math. Phys.* **221** (2001), 141–159.

[6] M. Durdevich, Quantum spinor structures for quantum spaces, *Intern. Journ. Theor. Phys.* **40** (2001), 115–138.

[7] U. Hermisson, Construction of covariant differential calculi on quantum homogeneous spaces, *Lett. Math. Phys.* **46** (1998), 313–322.

[8] J. Kustermans, G.J. Murphy, L. Tuset, Differential calculi over quantum groups and twisted cyclic cocycles, preprint, arXiv: math.QA/0110199, 2001.

[9] R. Owczarek, Dirac operator on the Podleś sphere, *Intern. Journ. Theor. Phys.* **40** (2001), 163–170.

[10] R. Oeckl, Structure theorem for covariant bundles on quantum homogeneous spaces, *Czech. Journ. Phys.* **51** (2001), 1401–1406.

[11] P. Podleś, Quantum spheres, *Lett. Math. Phys.* **14** (1987), 193–202.

[12] P. Podleś, Differential calculus on quantum spheres, *Lett. Math. Phys.* **18** (1989), 107–119.

[13] P. Podleś, The classification of differential structures on quantum 2-spheres, *Comm. Math. Phys.* **150** (1992), 167–179.

[14] P. Podleś, Symmetries of quantum spaces. Subgroups and quotient spaces of quantum SU(2) and SO(3) groups, *Comm. Math. Phys.* **170** (1995), 1–20.

[15] A. Pinzul, A. Stern, Dirac operator on the quantum sphere, *Phys. Lett. B* **512** (2001), 217–224.

[16] A. Trautman, Dirac and Chevalley spinors: A comparison, SISSA preprint, 1987.

[17] J.C. Varilly, An introduction to noncommutative geometry, in *Proceedings of the EMS Summer School on Noncommutative Geometry and Applications*, Monsoraz and Lisboa, September 1997, P. Almeida ed.; arXiv: physics/9709045, 1997.

[18] S.L. Woronowicz, Twisted SU(2) group : An example of a non-commutative differential calculus, *Publ. RIMS*, Kyoto Univ. **23** (1987), 117–181.

[19] S.L. Woronowicz, Compact matrix pseudogroups, *Comm. Math. Phys.* **111** (1987), 613–665.

[20] S.L. Woronowicz, Differential calculus on compact matrix pseudogroups (quantum groups), *Comm. Math. Phys.* **122** (1989), 125–170.

Robert Owczarek
On leave from Los Alamos National Laboratory
RRES-CH, LANL
MS J594, Los Alamos NM 87545, USA
E-mail: rmo@lanl.gov

On leave from
Institute of Fundamental Technological Research
Polish Academy of Sciences
Świętokrzyska 21
00-041 Warsaw, Poland

Submitted: August 20, 2002; Revised: June 28, 2003.

33

r-Fold Multivectors and Superenergy

Jose M. Pozo and Josep M. Parra

ABSTRACT A general structure combining Grassmann, Clifford and tensor products is presented. The r-fold multivectors provide the basis for the natural extension of Grassmann and Clifford algebras when several geometric entities are multilinearly related. Any tensor can be organized and understood as an r-fold multivector when its antisymmetries are taken into account. The r-fold Clifford algebra is contrasted with the multiparticle geometric algebra. The application of r-fold Clifford algebra to the study of superenergy tensors in physics is shown to provide their simplest definition. In addition, it constitutes a most efficient tool for obtaining and proving their essential properties, such as dominant positivity and conditions for their conservation.

Keywords: multivector, Grassmann, Clifford, tensor product, superenergy, dominant positivity property

33.1 Introduction

Sometimes the direct or tensor product, the exterior or Grassmann product, and the geometric or Clifford product are seen as competing structures in mathematical physics. However, this is not our point of view. In this contribution we present a structure that integrates all three products. We start by following Grassmann and Clifford themselves in considering the exterior product as the generator of the most fundamental geometrical entities, the multivectors, which constitute the *Grassmann space* Λ of the vector space E. Far from considering the Grassmann algebra as the Clifford algebra for a totally degenerated metric space E, we understand the Grassmann space Λ as the common linear space for both the Grassmann algebra and the Clifford algebra. We take Λ, instead of the vector space E, as the basic building block for constructing tensors. An element of the tensor product of r Grassmann spaces is an *r-fold multivector*, which retains the geometrical content of each block. We note that any tensor over E can be so *structured* by taking into account its antisymmetries, which in fact reflects its geometrical content. Now any operation on the Grassmann space, such as involutions, duality or the

---

AMS Subject Classification: 15A66, 15A69, 15A75, 83C40.

exterior and the Clifford products, is naturally extended to the r-fold Grassmann space.

We will consider the r-fold extension of the *Clifford algebra* (rfCA). This structure is compared to the *multiparticle geometric algebra* [1]. These are seen as two different algebras on the same linear space, and they are distinguished as the *tensor product* and the *graded tensor product* of Clifford algebras. We pay attention to the orthogonal transformations in both frameworks and consider the different interpretations and applications of each. To finish section 33.2, we briefly discuss different r-fold extensions of the nabla differential operator that depend on the block in which the Clifford-product action of ∇ is performed.

As a successful application of r-fold Clifford algebra, we present in section 33.3 the main results obtained by us [2] in the domain of *superenergy (s-e) tensors* for the gravitational field and other matter fields. Our contribution is based on the general algebraic construction defined by Senovilla [3], which generalizes the classical Bel–Robinson and Bel s-e tensors [4] to s-e tensors $T\{A\}$ generated by any field A in spacetimes of arbitrary dimension. Their relevance in many mathematical and physical contexts is due to the fact that they always satisfy the *dominant positivity property*, which is a generalization of the dominant energy condition usually required for a physically acceptable energy-momentum tensor.

The definition of the s-e of a field A requires structuring it as an r-fold multivector. The use of rfCA simplifies drastically both the expression of the s-e tensors and the proof of the dominant positivity property. It also enables us to formulate a fairly general sufficient condition on A that guarantees local conservation of the s-e tensor $T\{A\}$. The section starts with a short historical motivation of s-e tensors and closes with examples in some relevant physical fields.

33.2 r-fold multivectors

33.2.1   The Grassmann structure of r-fold multivectors

Let us consider a vector space E of dimension d over the field $\mathbb{K}$ of real or complex numbers. Via the exterior product, E generates the *Grassmann* or *exterior space*

$$\Lambda \equiv \bigwedge(E) = \Lambda^0 \oplus \Lambda^1 \oplus \cdots \oplus \Lambda^d,$$

which is a graded space where $\Lambda^0 \equiv \mathbb{K}$ contains the scalars, $\Lambda^1 \equiv E$ the vectors, $\Lambda^2 \equiv E \wedge E$ the bivectors and in general $\Lambda^r \equiv \bigwedge^r E$ contains the r-vectors. A general element $A \in \Lambda$ (not necessarily homogeneous) of the Grassmann space is called a *multivector*.

Understood as a generating product, the Grassmann product defines a structured linear space, over which the Grassmann algebra is not the only natural structure. To clarify this point of view, we can consider the Grassmann construction as a geometrically operational way to get the space of fully antisymmetric tensors over E. When E is endowed with additional structure, other products on

Λ can naturally arise. For instance, providing E with a volume element allows the definition of the shuffle Cayley product and the corresponding Grassmann–Cayley (bi)algebra. If E is endowed with a non-degenerate metric of signature (p, q), we can define the Clifford geometric product and the corresponding Clifford (bi)algebra $C\ell_{p,q}$.

Accordingly, with the graded structure of the Grassmann space, the r-*degree projection* $\langle A \rangle_r$ is defined on any multivector $A \in \Lambda$, as its projection into the subspace Λ^r,

$$\forall A \in \Lambda, \qquad A = \sum_{r=0}^{d} \langle A \rangle_r, \qquad \text{where} \quad \langle A \rangle_r \in \Lambda^r.$$

There is a natural four-element involution group that can be defined solely from the graded structure of Λ. Its elements are the identity, the main involution *, the reversion $^\sim$, and the 'Clifford'[1] conjugation $^-$. For a homogeneous r-vector $A \in \Lambda^r$ they give

$$A^* = (-1)^r A, \qquad \widetilde{A} = (-1)^{\frac{r(r-1)}{2}} A, \qquad \overline{A} = (-1)^{\frac{r(r+1)}{2}} A.$$

These four involutions are the automorphisms or antiautomorphisms of the Grassmann product that extend the identity, $\mathbf{1}$, and the additive inversion, $-\mathbf{1}$, from E to Λ:

$$\widetilde{a} = a = -a^* = -\overline{a}, \qquad \forall a \in \Lambda^1,$$
$$(A \wedge B)^* = A^* \wedge B^*, \qquad \widetilde{A \wedge B} = \widetilde{B} \wedge \widetilde{A}, \qquad \overline{A \wedge B} = \overline{B} \wedge \overline{A}.$$

A complete basis of the linear space Λ is denoted $\{e_I\}$, where Latin capital letters are used for antisymmetric multi-indices. Then any multivector $A \in \Lambda$ can be expressed, using the Einstein summation convention, by $A = A^I e_I$.

By means of the tensor product of r copies of the Grassmann space[2] we get the r-*fold Grassmann space*,

$$^r\Lambda \equiv \bigotimes^r \Lambda = \overbrace{\Lambda \otimes \Lambda \otimes \cdots \otimes \Lambda}^{r} = T_0^r(\Lambda).$$

An element of $^r\Lambda$ is a rank-r tensor on the Grassmann space and will be called an r-*fold multivector*. Each factor Λ will be called a *block*. Its expression in a basis, using multi-indices, is

$$A = A^{I_1 I_2 \cdots I_r} e_{I_1} \otimes e_{I_2} \otimes \cdots \otimes e_{I_r}. \tag{2.1}$$

---

[1]The name 'Clifford' is used here following the common usage, but the transformation is defined independently of any metric structure.

[2]A trivial generalization would be considering a collection of r different Grassmann spaces. But this generalization will not be developed here.

An r-fold multivector A that is homogeneous at each block is said to be *degree-defined*.

It is important to observe that an r-fold multivector can have terms whose content in some blocks is a scalar. Accordingly, terms of the form $e_{I_1} \otimes 1 \otimes e_{I_3}$ are possible and are strictly different from $e_{I_1} \otimes e_{I_3}$. However, we can define a canonical $r \hookrightarrow s$ *left-immersion* and a canonical $r \hookrightarrow s$ *right-immersion*, in order to relate ${}^r\Lambda \hookrightarrow {}^s\Lambda$, $r < s$, by completing respectively the right-hand side or the left-hand side of $A \in {}^r\Lambda$ with $r - s$ blocks of unity:

$$li_{r \hookrightarrow s} : A^{J_1 \cdots J_r} e_{J_1} \otimes \cdots \otimes e_{J_r} \mapsto A^{J_1 \cdots J_r} e_{J_1} \otimes \cdots \otimes e_{J_r} \otimes 1 \otimes \cdots \otimes 1$$

$$ri_{r \hookrightarrow s} : A^{J_1 \cdots J_r} e_{J_1} \otimes \cdots \otimes e_{J_r} \mapsto 1 \otimes \cdots \otimes 1 \otimes A^{J_1 \cdots J_r} e_{J_1} \otimes \cdots \otimes e_{J_r}.$$

A complete basis of the linear space ${}^r\Lambda$ is denoted by $\{e_{\underline{I}}\}$, where underlined capital letters denote r-*fold multi indices*, $\underline{I} \equiv \{I_1, \ldots, I_s\}$. Then an r-fold multivector $A \in {}^r\Lambda$ can be expressed as $A = A^{\underline{I}} e_{\underline{I}}$.

The degree projection and the operations of the involution group, defined on Λ, can be now extended to ${}^r\Lambda$. The r-*fold degree projection*, and the r-*fold reversion, main involution and Clifford conjugation* apply the corresponding operation independently to every block. In this case, when the same transformation is applied to all blocks, the same symbol as with simple multivectors is used:

$$\langle A \rangle_{s_1, \ldots, s_r} = A^{I_1 \cdots I_r} \langle e_{I_1} \rangle_{s_1} \otimes \cdots \otimes \langle e_{I_r} \rangle_{s_r}, \qquad \langle A \rangle_s \equiv \langle A \rangle_{s, \ldots, s},$$

$$\widetilde{A} = A^{I_1 \cdots I_r} \widetilde{e_{I_1}} \otimes \cdots \otimes \widetilde{e_{I_r}}, \qquad \overline{A} = A^{I_1 \cdots I_r} \overline{e_{I_1}} \otimes \cdots \otimes \overline{e_{I_r}},$$

$$A^* = A^{I_1 \cdots I_r} e_{I_1}{}^* \otimes \cdots \otimes e_{I_r}{}^*.$$

Evidently, it is possible to apply a different involution on each block, but the combined involution will not be, in general, automorphic or antiautomorphic for the exterior product. Although this kind of transformation is not expected to be of frequent usage, a suggested notation would be

$$A^{\overset{1,4\ \ 3}{\sim\ *}} = A^{I_1 I_2 I_3 I_4 \cdots I_r} \widetilde{e_{I_1}} \otimes e_{I_2} \otimes e_{I_3}{}^* \otimes \widetilde{e_{I_4}} \otimes \cdots \otimes e_{I_r}.$$

The general applicability of r-fold multivectors is manifest from the observation that any tensor

$$\hat{A} = \hat{A}^{\mu_1 \cdots \mu_s} e_{\mu_1} \otimes \cdots \otimes e_{\mu_s} \in T^s_0(E)$$

corresponds to an r-fold multivector, A, obtained by reordering and grouping antisymmetric indices in separate blocks.

$$A \in \Lambda^{n_1} \otimes \Lambda^{n_2} \otimes \cdots \otimes \Lambda^{n_r} \subset {}^r\Lambda$$

where $n_1 + n_2 + \cdots + n_r = s$. Thus, the *reordered tensor* A will be an r-*fold* $(n_1, n_2, \ldots, n_r)$-*vector*.

33.2.2   *r*-fold Clifford and Clifford–Grassmann algebras

When E is endowed with a metric g of signature (p, q), besides the exterior product, another associative [5] product is defined on the multivector space Λ. This product, named after Clifford [6], who called it *geometric*, will be called the *Clifford geometric product*. Although it is usually denoted by mere juxtaposition, when some ambiguity can arise in the expressions, the use of a specific symbol seems appropriate. We suggest the readily available symbol ©, which contains the initial of Clifford. The Grassmann space Λ endowed with © constitutes the *Clifford geometric algebra* $C\ell_{p,q} \equiv (\Lambda, ©)$. As a simple case, the product of a general multivector $A \in \Lambda$ with a vector $b \in \Lambda^1$ can be expanded into inner and exterior products:

$$Ab \equiv A©b = A \llcorner b + A \wedge b \qquad \text{and} \qquad bA \equiv b©A = b \lrcorner A + b \wedge A.$$

The Grassmann product of two multivectors is always contained in its Clifford product and can be extracted by means of the degree projectors. In spite of this fact, it is natural to consider the Grassmann–Clifford bialgebra $(\Lambda, \wedge, ©)$ if we want to keep track of the fact that the exterior product is metric-independent.

Both Grassmann and Clifford products can be naturally extended to the r-fold Grassmann space $^r\Lambda$. This extension involves an independent Grassmann or Clifford product in each block. Note that a different product can be chosen for each block. We will mainly be concerned with the *r-fold Clifford product*, defined as

$$AB \equiv A©B = \left(A^{I_1 \cdots I_r} e_{I_1} \otimes \cdots \otimes e_{I_r} \right) \left(B^{J_1 \cdots J_r} e_{J_1} \otimes \cdots \otimes e_{J_r} \right)$$
$$\equiv A^{I_1 \cdots I_r} B^{J_1 \cdots J_r} (e_{I_1} e_{J_1}) \otimes \cdots \otimes (e_{I_r} e_{J_r}).$$

The space $^r\Lambda$ endowed with this product constitutes the *r-fold Clifford algebra* (*r*fCA):

$$(^r\Lambda, ©) = {}^rC\ell_{p,q} \equiv \overset{r}{\bigotimes} C\ell_{p,q}.$$

The 2-fold Clifford–Grassmann algebra $(^2\Lambda, ©\wedge)$, involving the geometric product in the first block and the exterior product in the second one, has been previously used in [7–9], and its elements are sometimes called *Clifforms*.

We will also need to consider the product between an r-fold multivector A and an s-fold multivector B with $r \neq s$. If for instance $r < s$, then its definition involves the previous application of the left immersion $li_{r \hookrightarrow s}$ to the shortest factor, A.

$$AB \equiv li_{r \hookrightarrow s}(A)©B$$
$$= A^{I_1 \cdots I_r} B^{J_1 \cdots J_s} (e_{I_1} e_{J_1}) \otimes \cdots \otimes (e_{I_r} e_{J_r}) \otimes e_{J_{r+1}} \otimes \cdots \otimes e_{J_s}.$$

The election of one of the two unitary volumes fixes an orientation in the vector space E. The product with this *unit volume* ϵ implements the *Hodge duality*[3]

$$* : \Lambda \to \Lambda, \qquad A \mapsto *A \equiv \epsilon A$$

which satisfies $*\Lambda^k = \Lambda^{d-k}$. Note that this operator is metric dependent.

The Hodge duality can be extended to r-fold multivectors. As it can be applied independently to each block, a symbol must be used to specify the transformed block:

$$\underset{s}{*}A \equiv A^{I_1 \cdots I_r} e_{I_1} \otimes \cdots \otimes *e_{I_s} \otimes \cdots \otimes e_{I_r} = (1 \otimes \overset{s-1}{\cdots} \otimes 1 \otimes \epsilon) \copyright A \, .$$

Evidently we can apply the Hodge duality to any number of blocks. In this case we can use the notation

$$\underset{s_1, s_2, \ldots, s_m}{*} \equiv \underset{s_1}{*} \underset{s_2}{*} \cdots \underset{s_m}{*} \, .$$

A pair of REDUCE files CGAgen.red and r-foldCGA.red, available at http://ffn.ub.es/~jmparra, is being developed. They currently implement all the structures described above for r-fold multivectors: Grassmann and Clifford products, the involution group, the degree projectors, the tensor product and the left immersion. The program operates on algebraic expressions (components are variables and the bases are explicit but arbitrary). Obviously, the standard Clifford algebra is included as the simplest case. The number of blocks and the complexity of expressions are only subject to the computer limitations. The dimension of the space E is limited to 10.

33.2.3   Tensor product versus graded tensor product of Clifford algebras

An interesting issue to consider is the relationship between the rfCA and a multiparticle geometric algebra (MPGA) introduced by the Geometric Algebra Research Group of Cambridge (UK) [1, 10, 11]. To compare the two structures, it is convenient to express r-fold multivectors in an alternative notation more similar to the one used in MPGA. This change of perspective is also helpful in a more complete understanding of the r-fold multivectors structure.

Let us consider representing the position of each block, counted from the left, by a superscript label. For the simplest case of 2-fold multivectors we have:

$$A \otimes 1 \mapsto \overset{(1)}{A}, \qquad 1 \otimes B \mapsto \overset{(2)}{B}, \qquad A \otimes B \mapsto \overset{(1)}{A}\overset{(2)}{B} = \overset{(1)}{A} \copyright \overset{(2)}{B}.$$

---

[3]This definition differs in some sign conventions from the standard definition of the Hodge duality. However our definition is more natural in the Clifford algebra framework and it encompasses the whole of the Grassmann space under a unique expression, so that the square of the operator is $*^2 = \epsilon^2$, a sign independent of the degree of the argument.

Observe that this label ordering is coherent with the left-immersion. With this notation the r-fold Clifford product implies the commutation of factors with different block-labels:

$$(\overset{(1)(2)}{AB}) \textcircled{\tiny\raisebox{0.2ex}{c}} (\overset{(1)(2)}{CD}) = \overset{(1)(2)(1)(2)}{AB\,CD} = (\overset{(1)}{AC})(\overset{(2)}{BD}). \tag{2.2}$$

The MPGA is the Clifford algebra generated by the direct sum of r copies of the vector space E. Each copy of E is given a label indicating its sequential ordering. Vectors with different labels are orthogonal, and hence they anticommute. It is this property that establishes the essential difference between the MPGA and the rfCA:

$$C\ell(\textstyle\bigoplus_{i=1}^{r} \overset{(i)}{E}) \neq \textstyle\bigotimes_{i=1}^{r} C\ell(\overset{(i)}{E}).$$

The multiparticle geometric product has been implemented in the MAPLE package CLIFFORD by Abłamowicz in a notation similar to the r-fold notation presented here. There, the structure is defined as the *graded* tensor product of Clifford algebras,

$$C\ell(\textstyle\bigoplus_{i=1}^{r} \overset{(i)}{E}) = \textstyle\bigotimes_{\mathrm{grad}}{}_{i=1}^{r} C\ell(\overset{(i)}{E}).$$

To better appreciate the difference between the two products, let us denote the multiparticle geometric product by the symbol $\textcircled{\tiny\raisebox{0.2ex}{c}}_{\mathrm{grad}}$, and consider the expression analogous to (2.2) for homogeneous factors $B \in \Lambda^r$ and $C \in \Lambda^s$:

$$(\overset{(1)(2)}{AB}) \underset{\mathrm{grad}}{\textcircled{\tiny\raisebox{0.2ex}{c}}} (\overset{(1)(2)}{CD}) = (-1)^{rs} \overset{(1)}{A}\overset{(1)}{C}\overset{(2)}{B}\overset{(2)}{D} = (-1)^{rs} (\overset{(1)}{AC})(\overset{(2)}{BD}).$$

Observe that, in fact, we have again two different products defined on the same linear space $\overset{r}{\Lambda}$. Thus, they define the two algebras $(\overset{r}{\Lambda}, \textcircled{\tiny\raisebox{0.2ex}{c}})$ and $(\overset{r}{\Lambda}, \underset{\mathrm{grad}}{\textcircled{\tiny\raisebox{0.2ex}{c}}})$. The adjective 'graded' for tensor product algebras does not define a different linear space but only a different product on it.

While the MPGA is by definition the Clifford algebra $C\ell_{rp,rq}$, the rfCA $^rC\ell_{p,q}$ is not in general a Clifford algebra. It is a Clifford algebra only when dim E is even (or $r = 1$). The proof in the affirmative sense is based on the periodicity theorems for Clifford algebras. The proof in the negative sense is based on the existence of more than two linearly independent elements in the center of $^rC\ell_{p,q}$.

One of the most significant differences between the algebras concerns extensions of the Spin isometry group of the underlying metric space E. If we take seriously the view that the $r \times d$ dimensional space $\bigoplus_{i=1}^{r} \overset{(i)}{E}$ is the generating space of the MPGA, then its Spin group, $\mathrm{Spin}_{rp,rq}$, strictly includes rotations mixing differently labeled vectors. These mixing transformations are naturally outside the scope of the r-fold multivector approach. Indeed its implementation would be rather artificial and cumbersome. The maximal natural extension of $\mathrm{Spin}_{p,q}$ to r-fold multivectors is the group

$$^r\mathrm{Spin}_{p,q} \equiv \{R_1 \otimes \cdots \otimes R_r \mid R_i \in \mathrm{Spin}_{p,q}\},$$

which implements independent orthogonal transformations on each block,

$$A \mapsto RA\widetilde{R}, \quad \text{where} \quad A \in {}^r\Lambda, \quad R \in {}^r\mathrm{Spin}_{p,q}. \tag{2.3}$$

Note that this group has a dimension $r\binom{d}{2}$, and is a small subgroup of $\text{Spin}_{rp,rq}$, which has a dimension $\binom{rd}{2}$. However, the known past uses of the MPGA have been limited to non-mixing transformations, that is, to elements of $^r\text{Spin}_{p,q}$. Moreover, the elements of this subgroup are even in each block, hence they behave in the same way for both products, $\copyright$ and $\underset{\text{grad}}{\copyright}$.

We conclude that both approaches can be used equally well for current common applications of MPGA. Nevertheless, new non-trivial mathematical and physical results may appear from the consideration of the whole group $\text{Spin}_{rp,rq}$ in the MPGA. For instance, in its primary intended application to multiparticle quantum mechanics, this group includes interchanges between particles, possibly interpretable as exchange interactions. However, in the interpretation of r-fold multivectors as fields not related to different particles, our view seems more natural as it does not introduce additional anticommutation rules and unnecessary structures. In fact, for the main application presented here, namely, the superenergy tensors, the non-commutativity of different blocks would cause grave difficulties for expressions involving inhomogeneous r-fold multivectors.

33.2.4   The ∇ differential operator on r-fold multivectors

When we consider the tangent vector space $E = T(\mathcal{M})$ on some metric manifold $\mathcal{M}$ with an affine connection, we can perform derivatives of r-fold multivectors. The basic differential operator $\nabla \equiv e^\mu \nabla_\mu$ is algebraically a vector. Thus, it combines a derivative with a vector structure. While the derivative of an r-fold multivector $A \in {}^r\Lambda$ poses no problem following the standard Leibniz rule, the vector content of ∇ can be combined in different ways with the r-fold multivector structure.

A first possibility is to consider Clifford multiplication applied to a specific block. Following the left-hand multiplication rule of an r-fold multivector with an s-fold multivector, $\nabla A \equiv \nabla \copyright A$ naturally means performing the Clifford product with the first block:

$$\nabla A = \nabla_\mu A^{I_1 \cdots I_r} \left(e_\mu e_{I_1} \right) \otimes e_{I_2} \otimes \cdots \otimes e_{I_r}.$$

However, we may be interested in performing Clifford multiplication of the vector part of ∇ with the sth block:

$$\overset{(s)}{\nabla} A = \nabla_\mu A^{I_1 \cdots I_r} e_{I_1} \otimes \cdots \otimes \left(e_\mu e_{I_s} \right) \otimes \cdots \otimes e_{I_r}.$$

Here, the product involved also involves Clifford multiplication, and we have used the block-label notation to denote the operator

$$\overset{(s)}{\nabla} = 1 \otimes \overset{s-1}{\cdots} \otimes 1 \otimes \nabla.$$

A second possibility is to consider the tensor product instead of the Clifford product:

$$\nabla \otimes A = \nabla_\mu A^{I_1 \cdots I_r} e_\mu \otimes e_{I_1} \otimes e_{I_2} \otimes \cdots \otimes e_{I_r}.$$

Note that this operation increases by one the number of blocks. It is equivalent to $\nabla \copyright (1 \otimes A)$.

33.3   Application: Superenergy

33.3.1   Motivation

The name 'superenergy' (s-e) was first applied to the Bel–Robinson (BR) and Bel tensors defined from the Weyl conformal tensor and the Riemann tensor respectively [4]. The motivation for this name is that although they are 4-index tensors, they share many properties with energy-momentum tensors: positivity, conservation in vacuum, and some symmetry properties. Considering the positivity property as the most fundamental one, Senovilla [3, 12] obtained a general algebraic construction of a positive s-e tensor $T\{A\}$ from an arbitrary *seed* tensor A. The construction unifies in a single procedure the BR tensor, the Bel tensor, and many energy-momentum tensors of different physical fields as well .

The physical meaning of the classical gravitational s-e is still under debate [4, 13–18]. However, s-e tensors (recently renamed causal tensors) have several physical and mathematical applications. The interaction between gravity and other fields has been approached through the interchange of their *superenergy* [3, 15, 19]. A simple geometric criterion for assuring the causal propagation of a physical field has been formulated in terms of the divergence of its s-e tensor [3]. Mathematically, the positivity property of the BR tensor is essential in the proof of the global nonlinear stability of Minkowski spacetime [20]. More recently, the null-cone preserving maps have been completely classified by means of the s-e tensors [21]. Using this classification, a causal hierarchy of Lorentzian manifolds and a revised concept of isocausality, which includes the case of conformally equivalent spacetimes, have been obtained [22].

The construction of the s-e tensor $T\{A\}$ is well defined in Lorentzian manifolds of arbitrary dimension, retaining its positivity properties. This fact makes these tensors interesting objects in almost any higher-dimensional theory such as supergravity, string theory or M-theory [23, 24].

The original definition of $T\{A\}$ was framed in standard tensor formalism while taking the seed tensor A as an r-fold multivector. This definition involves 2^r terms, making its manipulation and the proof of its positivity property rather cumbersome. A reformulation by means of 2-spinors made by Bergqvist [25] leads to a simple and elegant proof of the positivity. However, this spinorial formalism is tied to dimension four. In [26] we introduced an alternative formulation for s-e tensors using r-fold Clifford algebra. This formulation was used in [2, 26] to obtain a proof of the positivity of $T\{A\}$, which is as elegant as the 2-spinorial one, and it is valid for any dimension. Additionally, it may be considered a paradigmatic example of the deep connection between some spinorial uses and geometrically significant applications of real Clifford algebra.

The r-fold Cliffordian formulation of $T\{A\}$ is in fact a generalization, because it can be applied to seed tensors of undefined degree. This generalization is not trivial since new nonvanishing mixed terms appear that do not satisfy the same symmetries [2]. Furthermore, by using this formulation, the authors [2] found a simple sufficient condition on the seed tensor that guarantees the conservation of its s-e $T\{A\}$.

33.3.2   Ingredients

The broad applicability of superenergy tensors $T\{A\}$ is a consequence of the few prerequisites needed for the definition of both the tensor $T\{A\}$ and its positivity property. The first ingredient is a vector space of arbitrary dimension $d = p + 1$, endowed with a metric of hyperbolic signature, $p - 1$. In particular, we can consider the tangent space of a differential manifold with one time-like dimension and p space-like dimensions. The choice of the opposite signature only introduces a few sign changes into some expressions.

The second ingredient is an arbitrary tensor field

$$\hat{A} = \hat{A}^{\mu_1 \cdots \mu_s} \, e_{\mu_1} \otimes \cdots \otimes e_{\mu_s},$$

organized as an r-fold multivector:

$$\otimes^s E \ni \hat{A}^{\mu_1 \cdots \mu_s} \, e_{\mu_1} \otimes \cdots \otimes e_{\mu_s} \mapsto A^{[\mu_{11} \cdots \mu_{1n_1}] \cdots [\mu_{r1} \cdots \mu_{rn_r}]}$$

$$\mapsto A \equiv A^{I_1 \cdots I_r} \, e_{I_1} \otimes \cdots \otimes e_{I_r} \in \Lambda^{n_1} \otimes \cdots \otimes \Lambda^{n_r}$$

where $n_1 + \cdots + n_r = s$.

33.3.3   Standard tensorial definition

The tensor definition of $T\{A\}$ requires us to consider the collection of all possible 2^r combinations of Hodge duals:

$$T\{A\} = \tfrac{1}{2} \sum_{\mathcal{P} \subset \{1, \ldots, r\}} \underset{\mathcal{P}}{*} A \times \underset{\mathcal{P}}{*} A.$$

The special cross product used in the definition above is defined by

$$(A \times A)_{\mu_1 \nu_1 \cdots \mu_r \nu_r} \equiv$$

$$\left(\prod_{i=1}^{r} \frac{1}{(n_i - 1)!} \right) A_{\mu_1 \, \lambda_{12} \cdots \lambda_{1n_1} \cdots \mu_r \, \lambda_{r2} \cdots \lambda_{rn_r}} \, A_{\nu_1}{}^{\lambda_{12} \cdots \lambda_{1n_1}}{}_{\cdots}{}^{\lambda_{r2} \cdots \lambda_{rn_r}}{}_{\nu_r}$$

Observe that $T\{A\}$ then has r pairs of vector indices, one pair proceeding from each block, and that by definition the tensor is symmetric in the two indices of each pair:

$$T\{A\}_{\mu_1 \nu_1 \cdots \mu_s \nu_s \cdots \mu_r \nu_r} = T\{A\}_{\mu_1 \nu_1 \cdots (\mu_s \nu_s) \cdots \mu_r \nu_r}.$$

33.3.4 *r-fold Clifford algebra formulation*

The *r*fCA formulation is inspired by the Clifford algebra expression [27, 28] of the standard electromagnetic energy-momentum tensor:

$$T(u) = -\tfrac{1}{2} F u \overline{F} \in \Lambda^1, \quad \forall u \in \Lambda^1. \tag{3.1}$$

Equivalently, we can obtain its components in a basis:

$$T(v, u) = -\tfrac{1}{2} \langle v F u \overline{F} \rangle_0 \quad \text{and} \quad T_{\mu\nu} = -\tfrac{1}{2} \langle e_\mu F e_\nu \overline{F} \rangle_0.$$

The generalization of expression (3.1) to an *r*-fold multivector A in place of the electromagnetic field, leads us directly to the definition of the s-e tensor as an endomorphism:

$$T\{A\} : \bigotimes\nolimits^r \Lambda^1 \to \bigotimes\nolimits^r \Lambda^1$$

$$T\{A\}(u_1 \otimes u_2 \otimes \cdots \otimes u_r) \equiv (-1)^r \tfrac{1}{2} \langle A\,(u_1 \otimes \cdots \otimes u_r)\,\overline{A}\,\rangle_1. \tag{3.2}$$

Equivalently, applied to *r* pairs of vectors we have

$$T\{A\}(v_1 \otimes \cdots \otimes v_r,\ u_1 \otimes \cdots \otimes u_r)$$
$$= (-1)^r \tfrac{1}{2} \langle (v_1 \otimes \cdots \otimes v_r) A\,(u_1 \otimes \cdots \otimes u_r)\overline{A}\,\rangle_0.$$

Finally, we can obtain its components in a basis $\{e_\mu\}$ by

$$T\{A\}_{\mu_1 \nu_1 \cdots \mu_r \nu_r} = (-1)^r \tfrac{1}{2} \langle (e_{\mu_1} \otimes \cdots \otimes e_{\mu_r}) A\,(e_{\nu_1} \otimes \cdots \otimes e_{\nu_r}) \overline{A}\,\rangle_0.$$

In order to compare the standard-tensorial and *r*fCA definitions of $T\{A\}$ we must first consider the objects A to which each of them can be applied. While the standard-tensorial definition applies to degree-defined *r*-fold multivectors, $A \in \Lambda^{n_1} \otimes \Lambda^{n_2} \otimes \cdots \otimes \Lambda^{n_r}$, the Clifford-algebra definition applies to general *r*-fold multivectors, $A \in \bigotimes^r \Lambda \equiv {}^r\Lambda$. Both definitions are equivalent for degree-defined *r*-fold multivectors. The straightforward extension of the tensorial definition to non-degree-defined *r*-fold multivectors:

$$T\Big\{ \sum_{k_1, \ldots, k_r} \langle A \rangle_{k_1, \ldots, k_r} \Big\} = \sum_{k_1, \ldots, k_r} T\{\langle A \rangle_{k_1, \ldots, k_r}\},$$

does not coincide with the corresponding superenergy in the Clifford algebra formulation. For the case of a single multivector seed $A = \sum_{k=0}^d A_{[k]}$, where $A_{[k]} \in \Lambda^k$, we would have

$$T\{A\}(u, v) = \sum_k T\{A_{[k]}\}(u, v) + \sum_k \langle \langle A_{[k+2]} \widetilde{A_{[k]}} \rangle_2 (v \wedge u) \rangle_0.$$

In general, the difference consists of the appearance of additional *interaction-like* cross terms. In contrast to ones proceeding from a homogeneous seed tensor, these terms are not symmetric in each of the index pairs, but they are antisymmetric in the pair in which two different degrees interact. Hence, they are necessarily excluded from the tensorial definition. Thus, the Clifford formulation is properly a generalization. On the one hand, these antisymmetric terms in the s-e tensor are unwanted in order to obtain conserved quantities [19]. On the other hand, the full expression for non-degree-defined seed tensors allows us to consider any local Lorentz transformation as the s-e tensor of the corresponding rotor field, and this, in turn, will be essential in the general proof of the positivity of $T\{A\}$.

33.3.5   The dominant positivity property

An orthochronous proper Lorentz transformation of a vector $u \in \Lambda^1$ is performed by a *rotor* $R \in \mathrm{Spin}_{p,1}$, by means of the expression $\mathcal{R}(u) = Ru\bar{R}$. Similarly, for an *r*-fold 1-vector $\underline{u} = u_1 \otimes \cdots \otimes u_r \in \bigotimes^r \Lambda^1$, an *r-fold rotor*, $\underline{R} \in {}^r\mathrm{Spin}_{p,1}$, implements an independent Lorentz transformation in each block. The formula (2.3) can now be seen as an instance of the superenergy tensor,

$$\mathcal{R}(\underline{u}) = \underline{R}\,\underline{u}\,\overline{\underline{R}} = (-1)^r\,2T\{\underline{R}\}(\underline{u}).$$

The group ${}^r\mathrm{Spin}_{p,1}$ can be extended to include dilations by means of an additional scalar factor:

$${}^r\mathrm{DSpin}_{p,1}^+ \equiv \{\,\lambda R \mid \lambda \in \mathbf{R}^+,\ R \in {}^r\mathrm{Spin}_{p,1}^+\,\},\quad u \mapsto (\lambda R)u\overline{(\lambda R)} = \lambda^2 \mathcal{R}(u).$$

Definition 33.3.1. *A tensor $M \in \bigotimes^s E$ on a Lorentzian metric space E is said to satisfy the **dominant positivity property** DPP if for all collections $\{u_i\}$ of causal and future-pointing vectors $M(u_1, \ldots, u_s) \geq 0$.*

Theorem 33.3.2. $T\{A\}$ *satisfies the DPP for all $A \in {}^r\Lambda$.*

A detailed proof can be found in [2]. Here we present an outline.

Outline of Proof for Theorem 33.3.2. A future-pointing vector u can always be expressed as the result of applying a local Lorentz transformation and a dilation to a chosen unitary time-like future-pointing vector e_0. Then, for any collection of time-like future-pointing vectors grouped as an *r*-fold vector, $\underline{u} = u_1 \otimes u_2 \otimes \cdots \otimes u_r$, we can write

$$\underline{u} = R_{\underline{u}}\,\underline{e_0}\,\overline{R_{\underline{u}}},\quad \text{for some } R_{\underline{u}} \in {}^r\mathrm{DSpin}_{p,1}^+,\quad \text{where } \underline{e_0} = e_0 \otimes \overset{r}{\cdots} \otimes e_0.$$

The key point is that applying the cyclic property of the scalar part of the geometric product, we can now write the general expression

$$2T\{A\}(\underline{u})(\underline{v}) = (-1)^r \langle A\underline{u}\overline{A}\underline{v}\rangle_0$$

$$= (-1)^r \langle (\overline{R_{\underline{v}}}AR_{\underline{u}})\,\underline{e_0}\,\overline{(\overline{R_{\underline{v}}}AR_{\underline{u}})}\,\underline{e_0}\rangle_0 = 2T\{A'\}(\underline{e_0})(\underline{e_0})$$

as the $\{0,...,0\}$ component of a new s-e tensor of seed $A' \equiv \overline{R_u} A R_u$. It only remains to prove that $T\{A'\}(\underline{e_0})(\underline{e_0}) \geq 0$ for any $A' \in {}^r Cl_{p,1}$. This is easily accomplished by splitting A' into parts orthogonal and parallel to the direction e_0 in each block. The result is a summation of 2^r positively defined squares.   □

33.3.6   Conserved superenergy tensors

The energy-momentum tensor of any isolated physical field (or the total energy-momentum, otherwise) is expected to satisfy the local conservation law $\nabla^\mu T_{\mu\nu} = 0$. Its extension to general s-e tensors is simply $\nabla^{\mu_1} T_{\mu_1\nu_1\cdots\mu_r\nu_r} = 0$. For $T\{A\}$, this condition corresponds to

$$0 = (-1)^r \nabla \cdot T\{A\}(\underline{u}) = \tfrac{1}{2}\langle \dot{\nabla} A \underline{u} \dot{\bar{A}} \rangle_{0,1,...,1}$$
$$= \tfrac{1}{2}(\langle \nabla A\, \underline{u}\, \bar{A} \rangle_{0,1,...,1} - \langle A\, \underline{u}\, \overline{\nabla A} \rangle_{0,1,...,1}) = \langle \nabla A\, \underline{u}\, \bar{A} \rangle_{0,1,...,1},$$

where the overdots indicate the elements that must be differentiated by $\dot{\nabla}$, and the Leibniz rule and the cyclic property have been used.

While $T\{A\}$ satisfies the DPP for all $A \in {}^r \Lambda$, the conservation of the s-e tensor depends on the dynamics of the field A involved. The expression obtained above leads to the following sufficient condition for the conservation of $T\{A\}$.

Theorem 33.3.3. *If $A \in {}^r \Lambda$ satisfies $\nabla A = \lambda A$, with $\lambda \in \mathbb{R}$, then $T\{A\}$ is divergence-free.*

Proof. From the condition $\nabla A = \lambda A$ we obtain

$$\nabla \cdot T\{A\}(\underline{u}) = (-1)^r \lambda \langle A\, \underline{u}\, \bar{A} \rangle_{0,1,...,1}$$

which vanishes due to the properties of the Clifford conjugation:

$$\langle A\, \underline{u}\, \bar{A} \rangle_{0,1,...,1} = (-1)^{r-1}\overline{\langle A\, \underline{u}\, \bar{A} \rangle}_{0,1,...,1} = -\langle A\, \underline{u}\, \bar{A} \rangle_{0,1,...,1}.$$

The relevance of this sufficient condition can be seen in the following examples.

Example 1. *The Maxwell equation for the electromagnetic field is $\nabla F = j$. Hence, in vacuum, the electromagnetic energy-momentum tensor $T\{F\}$ is conserved. In addition, Chevreton's electromagnetic s-e tensor* [15],

$$T\{\nabla \otimes F + \dot{F} \otimes \dot{\nabla}\},$$

is also conserved in vacuum flat spacetime, since

$$\nabla F = 0 \Rightarrow \nabla(\nabla \otimes F) = 1 \otimes \nabla^2 F = 0 \quad and \quad \nabla(\dot{F} \otimes \dot{\nabla}) = (\dot{\nabla}F) \otimes \dot{\nabla} = 0.$$

Example 2. *The Klein–Gordon scalar field is defined by the equation* $\nabla^2\phi = m^2\phi$. *Its energy-momentum tensor is given by*

$$T\{(\nabla + m)\phi\} = \nabla\phi \otimes \nabla\phi - \tfrac{1}{2}((\nabla\phi)^2 + m^2\phi^2)g.$$

Its conservation follows from

$$\nabla^2\phi = m^2\phi \;\Rightarrow\; \nabla(\nabla + m)\phi = m(\nabla + m)\phi.$$

Teyssandier's order-4 SE tensor [19, 29] *is* $T\{(\nabla + m) \otimes (\nabla + m)\phi\}$, *which is also conserved in flat spacetime, since*

$$\nabla(\nabla + m) \otimes (\nabla + m)\phi = m(\nabla + m) \otimes (\nabla + m)\phi.$$

Example 3. *The Riemann and the Weyl tensor satisfy in Einstein spaces (vacuum with cosmological constant) of any dimension the equations:*

$$\nabla R = \nabla C = 0.$$

Thus, the generalization of the Bel–Robinson tensor, $T\{C\}$, *and the Bel tensor,* $T\{R\}$, *are conserved.*

Acknowledgments

The authors gratefully acknowledge the financial support of the 6th ICCAAMP organizers in Cookeville. This work has been partially supported by the contracts BFM2000-0604 from the Spanish Ministry of Education, and 2000SGR/23 from the DGR of the Generalitat de Catalunya.

REFERENCES

[1] C.J.L. Doran, *Geometric Algebra and its Application to Mathematical Physics.* Ph.D. thesis, University of Cambridge, 1994.

[2] J.M. Pozo and J.M. Parra, Positivity and conservation of superenergy tensors. *Class. Quantum Grav.* **19**, 967–983, 2002.

[3] J.M.M. Senovilla, Super-energy tensors. *Class. Quantum Grav.* **17**, 2799–2842, 2000. *Preprint* gr-qc/9906087.

[4] L. Bel, Les états de radiation et le problème de l'énergie en relativité générale. *Cahiers de Physique* **16**, 59–80, 1962. English translation: *Gen. Rel. Grav.* **32**, 2047–78 (2000). And independently, I. Robinson *unpublished* 1959 (see also I. Robinson, On the Bel–Robinson tensor, *Class. Quantum Grav.* **14**, A331–A333, 1997)

[5] M. Riesz, *Clifford Numbers and Spinors.* The Inst. for Fluid Dynamics and Applied Math., Lecture Series **38**. Univ. of Maryland, University Park, 1958. (Reprinted as facsimile (Dordrecht: Kluwer 1993) ed. E.F. Bolinder and P. Lounesto).

[6] W.K. Clifford, Applications of Grassmann's extensive algebra. *Am. J. Math.* **1**, 350–358, 1878.

[7] F.J. Chinea, A Clifford algebra approach to general relativity. *Gen. Rel. Grav.* **21** (1), 21–44, 1989.

[8] A. Dimakis and F. Müller-Hoissen, Clifform calculus with applications to classical field theories. *Class. Quantum Grav.* **8**, 2093–2132, 1991.

[9] J.M. Parra and J.M. Pozo, Einstein Equations with Spacetime Geometric Clifford Algebra, 194–197. In Bona et al. [30], 1998.

[10] A. Lasenby, S. Gull and C. Doran, STA and the interpretation of quantum mechanics, In *Clifford (Geometric) Algebras with Applications to Physics, Mathematics, and Engineering.* W.E. Baylis (Editor), Birkhäuser, Boston, 1996, pp. 147–169. 1996.

[11] S.S. Somaroo, A.N. Lasenby, and C.J.L. Doran, Geometric algebra and the causal approach to multiparticle quantum mechanics. *J. Math. Phys.* **40**, 3327–3340, 1999.

[12] J.M.M. Senovilla, Remarks on superenergy tensors, 175–182. In Martín et al. [31], 1999. (Senovilla J.M.M. 1999 *Preprint* gr-qc/9901019).

[13] G.T. Horowitz and B.G. Schmidt, *Proc. R. Soc.* A **381**, 215, 1982.

[14] M.A.G. Bonilla and J.M.M. Senovilla, Some properties of the Bel and Bel–Robinson tensors. *Gen. Rel. Grav.* **29**, 91–116, 1997.

[15] M. Chevreton, Sur le tenseur de superénergie du champ électromagnétique. *Nuovo Cimento* **34**, 901–913, 1964.

[16] G. Bergqvist and M. Ludvigsen, Quasi-local momentum near a point. *Class. Quantum Grav.* **4**, L29–L32, 1987.

[17] L.B. Szabados, On certain quasi-local spin-angular momentum expressions for small spheres. *Class. Quantum Grav.* **16**, 2889–2904, 1999.

[18] J.D. Brown, S.R. Lau, and J.W. York, Canonical quasilocal energy and small spheres. *Phys. Rev. D* **59**, 064028, 1999.

[19] J.M.M. Senovilla, (super)^n-energy for arbitrary fields and its interchange: conserved quantities. *Mod. Phys. Lett.* **15**, 159–166, 2000. *Preprint* gr-qc/9905057.

[20] D. Christodoulou and S. Klainerman, *The Global Nonlinear Stability of the Minkowski Space.* Princeton Univ. Press, Princeton, 1993.

[21] G. Bergqvist and J.M.M. Senovilla, Null cone preserving maps, causal tensors and algebraic Rainich theory. *Class. Quantum Grav.* **18**, 5299–5325, 2001.

[22] A. Garcia-Parrado and J.M.M. Senovilla, Causal relationship: a new tool for the causal characterization of Lorentzian manifolds. *Class. Quant. Grav.* **20**, 625-664, 2003.

[23] G.W. Gibbons, Quantum gravity/strings/m-theory as we approach the 3rd millennium. In N Dadhich and J Narlikar, editors, *Gravitation and Relativity: At the turn of the Millenium,* (Proc. of the GR-15 Conference), pp. 259–280, Pune, India, 1998. IUCAA. gr-qc/9803065.

[24] S. Deser, The immortal Bel–Robinson tensor, 35–43. In Martín et al. [31], 1999.

[25] G. Bergqvist, Positivity of general superenergy tensors. *Commun. Math. Phys.* **207**, 467–479, 1999.

[26] J.M. Pozo and J.M. Parra, Clifford algebra approach to superenergy tensors, 283–287. In Ibáñez [32], 2000. *Preprint* gr-qc/9911041).

[27] M. Riesz, Sur certain notions fondamentales en théorie quantique relativiste. In *C. R. 10ᵉ Congrès Math. Scandinaves (Copenhagen)*, pp. 123–148, 1946. *Marcel Riesz Collected Papers* ed. L. Gårding and L. Hörmander (Berlin: Springer-Verlag 1988), 545–570.

[28] D. Hestenes, *Space-Time Algebra*. Gordon & Breach, New York, 1966.

[29] P. Teyssandier, Superenergy tensors for a massive scalar field, 319–324. In Ibáñez [32], 2000. (see also Teyssandier P 1999 *Preprint* gr-qc/9905080).

[30] C. Bona, J. Carot, L. Mas, and J. Stela, editors, *Analytical and Numerical Approaches to Relativity: Sources of Gravitational Radiation, Proc. of the Spanish Relativity Meeting, ERE97 (Palma de Mallorca)*. Universitat de les Illes Balears, 1998.

[31] J. Martín, E. Ruiz, F. Atrio, and A. Molina, editors, *Relativity and Gravitation in General, Proc. of the Spanish Relativity Meeting in Honour of the 65ᵗʰ Birthday of L. Bel, ERE98 (Salamanca)*. World Scientific, Singapore, 1999.

[32] J. Ibáñez, editor, *Recent Developments in Gravitation, Proc. of the Spanish Relativity Meeting, ERE99 (Bilbao)*. Universidad del Pais Vasco, 2000.

José María Pozo
Departament de Física Fonamental Universitat de Barcelona
Diagonal 647, E-08028 Barcelona, Spain
E-mail: jpozo@ffn.ub.es

Josep Manel Parra
Departament de Física Fonamental
Universitat de Barcelona
Diagonal 647, E-08028 Barcelona, Spain
E-mail: jmparra@ffn.ub.es

Submitted: September 20, 2002. Edited: August 9, 2003

34

The $C\ell_7$ Approach to the Standard Model

Greg Trayling and William E. Baylis

ABSTRACT A recent geometric approach to the standard model in terms of the Clifford algebra $C\ell_7$ is summarized. The complete gauge group of the standard model is shown to arise naturally and uniquely by considering all rotations in seven-dimensional space that (1) conserve the spacetime components of the particle and antiparticle currents and (2) do not couple the right-chiral neutrino. The spinor mediates a physical coupling of Poincaré and isotopic symmetries within the restrictions of the Coleman–Mandula theorem. The four extra spacelike dimensions in the model form a basis for the Higgs isodoublet field. The charge assignments of both the fundamental fermions and the Higgs boson are produced exactly.

Keywords: standard model, gauge groups, unification

34.1   Introduction

One of the most fundamental problems in particle physics is to explain the seemingly disparate gauge symmetries underlying the electromagnetic, weak and strong interactions. The prevailing wisdom is to embed the generators of these groups into some larger master group, but this leads to numerous theoretical problems. Larger groups contain extra gauge generators that should materialize as extra particle forces, and the present lack of any experimental evidence along these lines demands complicated theoretical remedies to explain their absence. Furthermore, invoking some master group does not address the question of the origin of gauge symmetries; one is still left with a large group inexplicably cast in some abstract space.

This paper demonstrates how the exact gauge symmetries

$$U(1)_Y \otimes SU(2)_L \otimes SU(3)_C$$

of the minimal standard model arise as the rotational symmetries of a reducible representation of the Poincaré group in a linear space with only four extra spacelike dimensions. Rather than embed the gauge groups into some master group,

AMS Subject Classification: 15A66, 81V22, 81T13.

we infix the Dirac algebra into the more mathematically uniform Clifford algebra $C\ell_7$. This is used to construct an algebraic approach that builds on a previous formulation [1] in geometric algebra of the Dirac theory. In our geometric model the gauge symmetries are not imposed but arise naturally from the algebra itself as unique symmetry groups of the current. Our approach of studying rotational symmetries in a higher-dimensional space may be viewed as an extension of the well-known association of spin with spatial rotations and the treatment of charge symmetry as a rotational symmetry in isospin space.

Section 2 summarizes the conventions adopted, as they vary throughout the Clifford algebra community. Section 3 develops the notion of spinors in $C\ell_7$. It shows how spinors representing all the fermions of a single generation can be combined into a single algebraic spinor, how the currents are calculated from such spinors, and how the contributions from individual fermions can be projected out. In section 4, we study the rotational symmetries of these spinors and show that they give exactly the gauge symmetries of the standard model with the correct weak hypercharge assignments. Section 5 shows how the four extra spatial dimensions and their transformation properties are precisely what is needed for the four components of a minimal Higgs field.

34.2   Algebraic foundations

In the real Clifford algebra $C\ell_7$, the unit vectors $e_1, e_2, \ldots, e_7$ are chosen to represent orthogonal spacelike directions in the tangent space of a seven-dimensional manifold. We further choose e_1, e_2, e_3 to represent the three observed directions in physical space, and relegate e_4, e_5, e_6, e_7 to four hidden dimensions that are orthogonal to physical space. We will be primarily concerned with rotations in the tangent space, so the question of whether these extra dimensions are compact or infinite in extent will be tactfully avoided. It will later be shown how the particle current in these directions is equivalent to the Higgs field, so their presence is ultimately manifested as the mass of the particles.

The product of any number of vectors is completely determined by the anticommutator $e_j e_k + e_k e_j = 2\delta_{jk}$, $j, k = 1, \ldots, 7$. All elements of the algebra can be reduced to real linear combinations of $2^7 = 128$ basis forms, each one representing a geometric object. There are in general $\binom{7}{k}$ independent k-vectors in $C\ell_7$. Two basic antiautomorphic involutions are used. The reversion of $K \in C\ell_7$, denoted $K^\dagger$, reverses the order of appearance of all vector elements within K. Clifford conjugation, denoted by $\bar{K}$, both reverses the order and negates all vector elements of K. The algebra $C\ell_7$ is appealing in that the volume element of the algebra commutes with all elements and squares to -1. It can therefore be associated identically with the unit imaginary

$$i \equiv e_1 e_2 e_3 e_4 e_5 e_6 e_7 \tag{2.1}$$

and used to reduce products of real vectors to elements of a complex space with

64 basis forms. This fortuitous circumstance occurs for every $C\ell_{3+4n}$ with non-negative integer n, and $C\ell_7$ is the smallest of the series that contains the Dirac algebra as a subalgebra.

The formalism used here builds on the physical applications of the algebra of physical space, in particular the use of *paravectors* [2, 3] to model spacetime vectors. Paravectors are sums of scalars and vectors such as

$$V = V^0 + V^1\mathbf{e}_1 + V^2\mathbf{e}_2 + V^3\mathbf{e}_3 \equiv V^\mu\mathbf{e}_\mu,$$

where for notational convenience we denote the unit scalar by e_0. The linear space of paravectors in physical space has a Minkowski spacetime metric $\eta_{\mu\nu}$ with signature $(1, 3)$. The metric arises from the square norm of paravectors $V\bar{V} = V^\mu V_\mu$ as $\eta_{\mu\nu} = \langle \mathbf{e}_\mu\bar{\mathbf{e}}_\nu \rangle_S$. Here, $\langle \cdots \rangle_S$ means the scalar part of the enclosed expression, and we adopt the summation convention for repeated indices, with lower-case Greek indices taking integer values $0\ldots 3$. The algebra generated by products of paravectors in physical space is just $C\ell_3$, which admits a covariant formulation of relativity and has also been shown to provide a natural formulation of the single-particle Dirac theory [1].

Proper and orthochronous Lorentz transformations of spacetime vectors are effected by bilinear transformations of the form [4]

$$V \to LVL^\dagger \tag{2.2}$$

where $L \in C\ell_3$ is any unimodular element: $L\bar{L} = 1$, and $C\ell_3$ is understood here to refer specifically to the algebra of physical space, the space spanned by $\{\mathbf{e}_1, \mathbf{e}_2, \mathbf{e}_3\}$. Every such L can be expressed as the product

$$L = \exp\left(w/2\right)\exp\left(\theta/2\right)$$

of a spatial rotation $L_R = \exp\left(\theta/2\right)$ in the plane of the bivector $\theta = \frac{1}{2}\theta^{jk}\mathbf{e}_j\bar{\mathbf{e}}_k$ and a pure boost $L_B = \exp\left(w/2\right)$ in the direction of the rapidity $w = w^j\mathbf{e}_j$. An advantage of the formalism is that the generators of the transformations have direct physical significance. We will consider more general rotations in $C\ell_7$ that have the form of equation (2.2) but are generated by bivectors that are not restricted to the three spatial planes of physical space.

An immediate consequence of this formalism is that vectors in the higher dimensions are Lorentz invariant. If z is any linear combination of $\mathbf{e}_4, \mathbf{e}_5, \mathbf{e}_6, \mathbf{e}_7$, its product with any $K \in C\ell_3$ satisfies

$$zK = \bar{K}^\dagger z. \tag{2.3}$$

It follows that z is invariant under any Lorentz transformation (2.2) with $L \in C\ell_3: z \to LzL^\dagger = L\bar{L}z = z$.

34.3   Algebraic spinors and currents

Algebraic spinors may be defined as entities that transform under the restricted Lorentz group according to the rule

$$\Psi \to L\Psi. \tag{3.1}$$

Spinors are thus elements of the carrier space of a representation (generally a reducible representation) of the Poincaré group. To describe one generation of the standard model, we use the algebraic spinor $\Psi \in C\ell_7$. It contains the spinors for the leptons as well as for three colors of quarks and all their antiparticles. The transformations (3.1) are preserved by multiplication from the right by Lorentz-invariant factors, in particular by hermitian idempotent elements (projectors) that project the spinor Ψ onto left ideals of $C\ell_7$. In particular, there are eight independent primitive idempotents that can each be used to reduce Ψ to a spinor representing a fermion doublet.

In constructing an algebraic expression for the particle current J, we seek a form that is bilinear in the spinors, transforms as a paravector, and satisfies $J^\dagger = J$. The simplest solution to this, and the one that we adopt here, has the same form as found for the Dirac theory in $C\ell_3$:

$$J = \Psi\Psi^\dagger \to L\Psi\Psi^\dagger L^\dagger. \tag{3.2}$$

A specific component of J may be extracted by contracting it with its associated direction through

$$J_\mu = \langle \Psi\Psi^\dagger \bar{e}_\mu \rangle_S = \langle \Psi^\dagger \bar{e}_\mu \Psi \rangle_S. \tag{3.3}$$

It is useful to distinguish transformations acting on the left from others that act on the right. Those on the left include Lorentz transformations and rotations in the space of the extra four dimensions. Since they operate on orthogonal subspaces, rotations in the space spanned by $\{e_4, e_5, e_6, e_7\}$ commute with the Lorentz transformations. They are applied to the spinor after the particles have been given the motion and orientation described by Ψ and will be called "exterior" transformations to represent their position, as in equation (3.2), in transformations of the current J. Transformations applied from the right will similarly be called "interior". They are applied to the particles in their reference frame, before they acquire the motion and orientation implied by the spinor. Note that exterior transformations are not synonymous with *external* transformations, since the extra four dimensions may relate to properties that are commonly considered to be *internal*. Exterior transformations mix the components within a single pair of fermions, whereas interior transformations mix different pairs together.

Primitive idempotents $P(n)$ needed to isolate fermion doublets of Ψ can be constructed from interior products of three pairs of simple projectors $P_\pm = P_\pm^\dagger = P_\pm^2 = \bar{P}_\mp$, where $P_+ + P_- = 1$ and $P_+ P_- = 0$. From among several equivalent choices, we use the three mutually commuting projector pairs

$$P_{\pm 3} \equiv \tfrac{1}{2}(1 \pm e_3), \quad P_{\pm\alpha} \equiv \tfrac{1}{2}(1 \pm ie_4 e_5), \quad P_{\pm\beta} \equiv \tfrac{1}{2}(1 \pm ie_6 e_7). \tag{3.4}$$

The primitive projectors $P(n)$ are given as products of these simple commuting projectors by

$$
\begin{aligned}
&P(1) = P_{+3}P_{+\alpha}P_{-\beta} = \bar{P}(6), \qquad && P(5) = P_{+3}P_{-\alpha}P_{+\beta} = \bar{P}(2), \\
&P(2) = P_{-3}P_{+\alpha}P_{-\beta} = \bar{P}(5), \qquad && P(6) = P_{-3}P_{-\alpha}P_{+\beta} = \bar{P}(1), \\
&P(3) = P_{-3}P_{-\alpha}P_{-\beta} = \bar{P}(8), \qquad && P(7) = P_{-3}P_{+\alpha}P_{+\beta} = \bar{P}(4), \\
&P(4) = P_{+3}P_{-\alpha}P_{-\beta} = \bar{P}(7), \qquad && P(8) = P_{+3}P_{+\alpha}P_{+\beta} = \bar{P}(3),
\end{aligned}
\tag{3.5}
$$

and $\sum_n P(n) = 1$. Each of the eight primitive projectors $P(n)$, applied from the right, projects Ψ onto one of eight minimal left ideals of $C\ell_7$. The nth $\Psi P(n)$ is identified with a distinct pair of fermions and forms current elements in equation (3.2) only with itself.

One pair of simple projectors, applied from the right, can be taken to separate particles from antiparticles. We let this be $P_{\pm 3}$, with P_{+3} designated for particles and P_{-3} for antiparticles. This choice is completely arbitrary and will be generalized below. Each column holds the spinors for a fermion doublet, and the projectors for the two isotopic-spin components are taken to be $P_{\pm \beta}$ applied as an *exterior* operator. The P_{+3} (particle) spinors can be factored explicitly as in Table 1. We have chosen component labels that correspond to the Weyl matrix representation [5], where each four-component spinor in Ψ is further split into two-component spinors of right and left chirality. The P_{-3} spinors have a similar form but have been excluded for brevity.

Lower spinor
$P_{+\beta}\Psi P(1) = \sqrt{8}(\psi_R \mathbf{e}_6 \mathbf{e}_5 + \psi_L \mathbf{e}_1 \mathbf{e}_6)P(1)$
$P_{+\beta}\Psi P(4) = \sqrt{8}(\psi_R \mathbf{e}_6 \mathbf{e}_1 + \psi_L \mathbf{e}_6 \mathbf{e}_5)P(4)$
$P_{+\beta}\Psi P(5) = \sqrt{8}(\psi_R + \psi_L \mathbf{e}_5 \mathbf{e}_1)P(5)$
$P_{+\beta}\Psi P(8) = \sqrt{8}(\psi_R \mathbf{e}_1 \mathbf{e}_5 + \psi_L)P(8)$
Upper spinor
$P_{-\beta}\Psi P(1) = \sqrt{8}(\psi_R + \psi_L \mathbf{e}_1 \mathbf{e}_5)P(1)$
$P_{-\beta}\Psi P(4) = \sqrt{8}(\psi_R \mathbf{e}_5 \mathbf{e}_1 + \psi_L)P(4)$
$P_{-\beta}\Psi P(5) = \sqrt{8}(\psi_R \mathbf{e}_5 \mathbf{e}_6 + \psi_L \mathbf{e}_1 \mathbf{e}_6)P(5)$
$P_{-\beta}\Psi P(8) = \sqrt{8}(\psi_R \mathbf{e}_6 \mathbf{e}_1 + \psi_L \mathbf{e}_5 \mathbf{e}_6)P(8)$

TABLE 1. The algebraic P_{+3} (particle) spinors, where the two-component Weyl spinors are *algebraic* elements defined by $\psi_R = \psi_0 + \psi_1 \mathbf{e}_1$, and $\psi_L = \psi_3 - \psi_2 \mathbf{e}_1$ for each particle with Dirac spinor components $\psi_0, \psi_1, \psi_2, \psi_3$.

The chiral projectors for all fermions in the Weyl representation are the mutually annihilating exterior operators

$$
P_{R/L} = P_{\bar{L}/\bar{R}} = \tfrac{1}{2}(1 \pm \mathbf{e}_4 \mathbf{e}_5 \mathbf{e}_6 \mathbf{e}_7).
\tag{3.6}
$$

Note that the chirality projectors $P_{R/L}$ commute with all elements of the sub-algebra $C\ell_3$ as well as with $P_{\pm 3}, P_{\pm \alpha}, P_{\pm \beta}$ and therefore with all the primitive idempotents $P(n)$. The chirality of Ψ can be flipped by the transformation

$$\Psi \to -e_1 e_2 e_3 e_4 \Psi \tag{3.7}$$

which has the effect of reversing the vector components of the current (3.2) in the span of $\{e_1, e_2, e_3, e_4\}$ while leaving the components in the span of $\{e_0, e_5, e_6, e_7\}$ invariant.

Charge conjugation is realized by the algebraic operation

$$\Psi \to \Psi_C = i e_4 \overline{\Psi}^\dagger. \tag{3.8}$$

The conjugate of $P_{-\beta} \Psi P(1)$, for example, is

$$P_{+\beta} \Psi_C P(6) = i e_4 \sqrt{8} (\bar{\psi}_R^\dagger + \bar{\psi}_L^\dagger e_1 e_5) P(6) \tag{3.9}$$

where $\bar{\psi}_R^\dagger = \psi_0^* - \psi_1^* e_1$ and $\bar{\psi}_L^\dagger = \psi_3^* + \psi_2^* e_1$. The identification (3.8), together with the relation (2.3) and transformation rule (3.1), ensures that spinors and their charge conjugates transform in the same way under the Lorentz group:

$$\Psi_C \to i e_4 \bar{L}^\dagger \bar{\Psi}^\dagger = L i e_4 \bar{\Psi}^\dagger = L \Psi_C. \tag{3.10}$$

In a matrix representation, charge conjugation (3.8) is equivalent to defining the conventional charge conjugates through $\lceil \psi_C = i\gamma^2 \psi^* \rfloor$ and interchanging both upper and lower spinors in a fermion doublet.

Geometrically, charge conjugation transforms the particle current as

$$J \equiv \Psi \Psi^\dagger \to e_4 \bar{\Psi}^\dagger \bar{\Psi} e_4 = e_4 \bar{J} e_4 \tag{3.11}$$

and has the effect of negating the e_4 component while leaving all other directions invariant. This is a discrete symmetry of the higher-dimensional directions that is not accessible by a rotation. The negation of two or four directions can be achieved by rotations, and to negate three directions one simply reverses one direction followed by reversing another two or four. The choice of e_4 and the phase introduced in equation (3.8) are merely convenient choices for the representation used.

The total current obtained by simply adding all the left ideal doublets into a single element Ψ is then

$$J = \Psi \sum_n P(n) \Psi^\dagger = \sum_{a=1}^{16} \lceil J_{(a)}^\mu \rfloor e_\mu + \text{(higher-dim. terms)}. \tag{3.12}$$

The sum here runs over the 16 four-component spinors assigned to the upper and lower halves of the eight minimal left ideals, each of which is ascribed to a distinct fermion. The residual part of the current involves cross-current terms between the upper and lower fermions of the same ideal as well as mass-like terms of the form $\lceil \bar{\psi}\psi \rfloor$, all projected onto the higher dimensions $\mathbf{e}_k$, $k = 4, 5, 6, 7$.

The main idea of this section has simply been that, instead of writing a separate term for each of the particle currents, we can consolidate them into a single expression that accommodates a number of spinorial representations. The advantage of the algebraic formalism becomes evident when we enumerate all the possible rotational symmetries of this current.

34.4 Gauge symmetries

We now show that rotational transformations that leave both the physical spacetime components of the particle and antiparticle currents and the right-chiral neutrino (and left-chiral antineutrino) sterile lead exactly to the standard-model gauge symmetries. This involves generators acting from both the left and right of the algebraic spinor, as these generators usually do not commute with Ψ. By combining the fermion currents into the single form (3.2), we uncover relationships among the fermions that in most other models are simply imposed on abstract spaces.

We begin by considering exterior rotations $\Psi \rightarrow \exp(\theta T)\Psi$ that leave the physical spacetime components of $\Psi\Psi^\dagger$ invariant, where T generates rotations in one or more planes of seven-dimensional space. From the infinitesimal form

$$J \rightarrow (1 + \theta T)\Psi\Psi^\dagger(1 + \theta T^\dagger) \tag{4.1}$$

it is clear from the invariance of J_μ (3.3) for $\mu = 0, 1, 2, 3$, that $\mathbf{e}_\mu T = -T^\dagger \mathbf{e}_\mu = T\mathbf{e}_\mu$. Thus, to leave the spatial components of J invariant, T must commute with $\mathbf{e}_k$, $k = 1, 2, 3$. This reduces the choices for T to linear combinations of the six bivectors $\mathbf{e}_j\mathbf{e}_k$: $(j, k) \in \{4, 5, 6, 7\}$, $j > k$. The projectors $P_{R/L}$ split this $so(4)$ algebra into two independent copies of $su(2)$ with generators of the form $\mathbf{e}_j\mathbf{e}_k P_{L/R}$.

The generators of $SU(2)_L$ may be written in the form

$$T_1 = \tfrac{1}{4}(\mathbf{e}_6\mathbf{e}_4 + \mathbf{e}_5\mathbf{e}_7), \quad T_2 = \tfrac{1}{4}(\mathbf{e}_7\mathbf{e}_4 + \mathbf{e}_6\mathbf{e}_5), \quad T_3 = \tfrac{1}{4}(\mathbf{e}_5\mathbf{e}_4 + \mathbf{e}_7\mathbf{e}_6), \tag{4.2}$$

that implicitly contains the left-chiral projector (3.6) and therefore acts only on left-chiral particles and right-chiral antiparticles. The three generators (4.2) induce simultaneous rotations in a pair of commuting planes and satisfy $[T_a, T_b] = \varepsilon_{abc}T_c$, with the fully antisymmetric structure constants ε_{abc} where $\varepsilon_{123} = 1$. The conventional presence of the unit imaginary in front of T_c has been absorbed into the antihermitian property of the bivectors. The effect of the transformation $\Psi \rightarrow \exp(\theta_a T_a)\Psi$ is identical to that of the prevailing $SU(2)$ prescriptions

$$\begin{pmatrix} \nu_L & u_L & -\bar{e}_R & -\bar{d}_R \\ e_L & d_L & \bar{\nu}_R & \bar{u}_R \end{pmatrix} \rightarrow \exp(-\tfrac{1}{2}i\theta_a\sigma_a)\begin{pmatrix} \nu_L & u_L & -\bar{e}_R & -\bar{d}_R \\ e_L & d_L & \bar{\nu}_R & \bar{u}_R \end{pmatrix}$$

as is readily verified by a suitable matrix representation of the generators [6]. The remaining $SU(2)_R$ generators would couple with ν_R and its conjugate and are therefore omitted.

Now let us look at the possible *interior* rotations $\Psi \rightarrow \Psi \exp(\theta T')$. To emphasize the fact that they act on the *right* side of Ψ, the interior transformations and generators are denoted here with a prime. Any interior unitary transformation leaves $J = \Psi\Psi^\dagger$ invariant, but we want a stronger condition: we demand that the spacetime components of the particle and antiparticle currents be separately invariant. Mathematically, this is equivalent to splitting the current in two using the $\Psi P_{\pm 3}$ spinors

$$J = \tfrac{1}{2}\Psi(1 + e_3)\Psi^\dagger + \tfrac{1}{2}\Psi(1 - e_3)\Psi^\dagger \equiv J_{+3} + J_{-3} \tag{4.3}$$

and requiring each part to be invariant. Generators and other elements placed between Ψ and $\Psi^\dagger$ are Lorentz invariant. Thus, we may involve the elements e_1, e_2, e_3 in the interior symmetries while satisfying the Coleman–Mandula theorem [7], which prohibits any non-trivial combination of the Poincaré and isotopic groups. Under the infinitesimal interior transformation $\Psi \rightarrow \Psi(1+\theta T')$, we have

$$J_{\pm 3} \rightarrow \tfrac{1}{2}\Psi(1 + \theta T')(1 \pm e_3)(1 + \theta T'^\dagger)\Psi^\dagger \tag{4.4}$$

which may be viewed as a transformation of the central $P_{\pm 3}$ projector. We see that the space of available bivector generators that leave e_3 invariant is now spanned by the larger set of 15 bivectors $e_j e_k : (j, k) \in \{1, 2, 4, 5, 6, 7\}$, $j < k$. Insulating the right-chiral neutrino from interior transformations in a similar manner as before now requires that both lepton doublets ($P(1)$ and $P(6)$ in the representation adopted) be avoided. This reduces the number of independent generators to eight, all of which couple quarks of different color charges:

$$T_1' = \tfrac{1}{4}(e_1 e_7 + e_6 e_2), \quad T_2' = \tfrac{1}{4}(e_1 e_6 + e_2 e_7), \quad T_3' = \tfrac{1}{4}(e_1 e_2 + e_7 e_6)$$

$$T_4' = \tfrac{1}{4}(e_6 e_4 + e_5 e_7), \quad T_5' = \tfrac{1}{4}(e_4 e_7 + e_5 e_6), \quad T_6' = \tfrac{1}{4}(e_4 e_1 + e_2 e_5)$$

$$T_7' = \tfrac{1}{4}(e_1 e_5 + e_2 e_4), \quad T_8' = \tfrac{1}{4\sqrt{3}}(e_2 e_1 + 2 e_5 e_4 + e_7 e_6). \tag{4.5}$$

The interior generators have been arranged to give the conventional $SU(3)$ structure constants [5]

$$[T_a', T_b'] = -f_{abc}T_c'. \tag{4.6}$$

The transformation $\Psi \rightarrow \Psi \exp(\theta_a T_a')$ is identical in its effect on the P_{+3} spinor components to

$$(q_{\text{red}}, q_{\text{grn}}, q_{\text{blu}}) \rightarrow (q_{\text{red}}, q_{\text{grn}}, q_{\text{blu}}) \exp(-\tfrac{1}{2} i \theta_a \lambda_a^*) \tag{4.7}$$

where λ_a are the Gell-Mann matrices. This is equivalent to the more familiar

$$\begin{pmatrix} q_{\text{red}} \\ q_{\text{grn}} \\ q_{\text{blu}} \end{pmatrix} \rightarrow \exp(-\tfrac{1}{2} i \theta_a \lambda_a) \begin{pmatrix} q_{\text{red}} \\ q_{\text{grn}} \\ q_{\text{blu}} \end{pmatrix}. \tag{4.8}$$

Under the same algebraic operation, the effect on the conjugate spinors

$$(-\overline{q_{\text{grn}}}, \overline{q_{\text{blu}}}, -\overline{q_{\text{red}}})$$

is equivalent to

$$(\overline{q}_{\text{red}}, \overline{q}_{\text{grn}}, \overline{q}_{\text{blu}}) \rightarrow (\overline{q}_{\text{red}}, \overline{q}_{\text{grn}}, \overline{q}_{\text{blu}}) \exp(\tfrac{1}{2}i\theta_a\lambda_a) \qquad (4.9)$$

which is the correct transformation. The fact that the doublets can be written in the same representation by using either the column (u, d) or the column $(-\bar{d}, \bar{u})$ is a special property of $SU(2)$. Such a construction is not possible for the $SU(3)$ triplet, but the geometric symmetries here provide a separate set of $SU(3)$ sub-matrices, one in terms of $-\lambda_a^*$ and the other in terms of λ_a, operating on the two carrier spaces. It is an advantage of having the conjugate spinors in separate columns of Ψ, that the same algebraic symmetry applies to both particles and antiparticles.

There remains one additional possible symmetry. We need to consider a synchronized double-sided rotation that conspires to cancel out in the case of the right-chiral neutrino. As this rotation is to represent a distinct symmetry, its left- and right-side generators must commute with all $SU(2)$ and $SU(3)$ generators, respectively. The surviving bivector candidates are $(\mathbf{e}_4\mathbf{e}_5 + \mathbf{e}_7\mathbf{e}_6)$ acting from the left, and $(\mathbf{e}_1\mathbf{e}_2 + \mathbf{e}_5\mathbf{e}_4 + \mathbf{e}_6\mathbf{e}_7)$ operating from the right. One may verify with the infinitesimal operator

$$\Psi \rightarrow (1 + \theta_0 T_0)\Psi(1 + \theta_0 T_0') \qquad (4.10)$$

that the solution for which there is no change to the right-chiral neutrino can be normalized to

$$T_0 = \tfrac{1}{2}(\mathbf{e}_4\mathbf{e}_5 + \mathbf{e}_7\mathbf{e}_6), \qquad T_0' = \tfrac{1}{3}(\mathbf{e}_1\mathbf{e}_2 + \mathbf{e}_5\mathbf{e}_4 + \mathbf{e}_6\mathbf{e}_7). \qquad (4.11)$$

Applying this operation to each spinor in turn proves to be identical to the $U(1)_Y$ transformation $\psi_{(j)} \rightarrow \exp(-i\theta_0 Y_{(j)})\psi_{(j)}$ with the weak hypercharge assignments

$$Y(\nu_R, \nu_L, e_R, e_L) = (0, -1, -2, -1) = -Y(\bar{\nu}_L, \bar{\nu}_R, \bar{e}_L, \bar{e}_R), \qquad (4.12)$$

$$Y(u_R, u_L, d_R, d_L) = (\tfrac{4}{3}, \tfrac{1}{3}, -\tfrac{2}{3}, \tfrac{1}{3}) = -Y(\bar{u}_L, \bar{u}_R, \bar{d}_L, \bar{d}_R).$$

It produces the conventional weak hypercharge assignments for both leptons and quarks. This exhausts the rotational gauge symmetries.

The double-sided transformations may be locally gauged by introducing twelve gauge fields $B, W_a, G_a \in C\ell_3$ that transform according to

$$\bar{B} \rightarrow \bar{B} + \frac{2}{g'}\bar{\partial}\theta_0, \quad \bar{W}_a \rightarrow \bar{W}_a + \frac{1}{g}\bar{\partial}\theta_a + \varepsilon_{abc}\theta_b\bar{W}_c, \quad a \in \{1, 2, 3\},$$

$$\bar{G}_a \rightarrow \bar{G}_a + \frac{1}{g_s}\bar{\partial}\theta_a' + f_{abc}\theta_b'\bar{G}_c, \quad a \in \{1, 2, \ldots, 8\}. \qquad (4.13)$$

These enter in the Lagrangian derivative terms

$$\mathcal{L}_\partial = \langle \Psi^\dagger i\partial\Psi \rangle_S - \tfrac{1}{2}g'\langle \Psi^\dagger i\bar{B}(T_0\Psi + \Psi T_0') \rangle_S$$

$$- g\langle \Psi^\dagger i\bar{W}_a T_a \Psi \rangle_S - g_s\langle \Psi^\dagger i\bar{G}_a \Psi T_a' \rangle_S \quad (4.14)$$

where the algebraic derivative operator is defined by [2]

$$\bar{\partial} \equiv \partial_0 + \partial_1 \mathbf{e}_1 + \partial_2 \mathbf{e}_2 + \partial_3 \mathbf{e}_3. \tag{4.15}$$

Although the above terms are similar to the conventional forms, it should be emphasized that all of the currents are simultaneously handled in the same expression using the algebraic spinor Ψ, whose gauge symmetries arise naturally from the geometry of the model.

The $U(1)_Y \otimes SU(2)_L \otimes SU(3)_C$ result here is a general consequence of the algebra for rotations that conserve the particle and antiparticle currents and do not couple ν_R. They are not specific to the ΨP_{+3} spinors. An arbitrary fitting of the doublets into some orthogonal linear combination of the columns is achievable by shuffling the P_{+3} spinors through a transformation $\Psi \rightarrow \Psi S$ where $SS^\dagger = 1$. Furthermore, it can be shown [6] that if we relax the condition that the transformations of Ψ be rotations to see whether generators other than bivectors might play a role, only linear combinations of pairs of commuting bivectors persist. This framework then gives a geometric basis for the gauge group of the standard model, which arises unambiguously through the various rotational symmetries of the algebraic current in seven-dimensional space.

34.5   Higgs field

When looking at the exterior invariances of the current, we previously disregarded the higher-dimensional vector components and allowed them to freely rotate among each other. This Lorentz-invariant vector space is then a carrier space for the set of exterior gauge transformations and affords a natural inclusion of the minimal Higgs field [8]. One can verify that by formulating the complex scalar isodoublet H and conjugate Higgs $H_c = \bar{H}^\dagger$ as

$$H = (-\phi_1 \mathbf{e}_6 + \phi_2 \mathbf{e}_7)P_{-\alpha} + (\phi_3 \mathbf{e}_5 - \phi_4 \mathbf{e}_4)P_{+\beta} \sim \left[\begin{pmatrix} \phi_1 + i\phi_2 \\ \phi_3 + i\phi_4 \end{pmatrix} \right],$$

$$H_c = (\phi_1 \mathbf{e}_6 - \phi_2 \mathbf{e}_7)P_{+\alpha} - (\phi_3 \mathbf{e}_5 - \phi_4 \mathbf{e}_4)P_{-\beta} \sim \left[\begin{pmatrix} \phi_3 - i\phi_4 \\ -\phi_1 + i\phi_2 \end{pmatrix} \right], \tag{5.1}$$

where the ϕ_j are real scalars, the expression

$$\mathcal{L}_M = \frac{1}{\sqrt{2}} \langle \Psi^\dagger G_e H \Psi P_\ell + \Psi^\dagger (G_d H + G_u H_c) \Psi P_q \rangle_S \tag{5.2}$$

is identical to the conventional Higgs-coupling Lagrangian term with coupling strengths $G_{e,d,u}$. The projectors $P_\ell = P(1)$ and $P_q = P(4) + P(5) + P(8)$ are used to separate the lepton and quark currents. The transformation required for gauge invariance,

$$H \rightarrow \exp(\theta_0 T_0 + \theta_a T_a) H \exp(-\theta_0 T_0 - \theta_b T_b) \tag{5.3}$$

is equivalent to the conventional notation

$$\begin{pmatrix} \phi^+ \\ \phi^0 \end{pmatrix} \rightarrow \exp(-iY\theta_0 - \tfrac{1}{2}i\theta_a\sigma_a)\begin{pmatrix} \phi^+ \\ \phi^0 \end{pmatrix} \tag{5.4}$$

where $\phi^+ \equiv \phi_1 + i\phi_2$ and $\phi^0 = \phi_3 + i\phi_4$. The weak hypercharge assignment of $Y = 1$ ($Y = -1$) for the Higgs field (conjugate field) has been recovered naturally from the double-sided algebraic transformation. The Higgs field—one of the least understood aspects of the standard model—thus arises here simply as a coupling to the higher-dimensional vector components of the current.

34.6   Conclusion

We began by formulating a generalized current expression in the Clifford algebra $C\ell_7$ of seven-dimensional space. The addition of four spacelike dimensions to those of physical space is the minimum necessary to incorporate all the fermions of one generation, both particles and antiparticles, into a single spinorial element. By examining all possible rotations of the generalized current that leave the right-chiral neutrino (and left-chiral antineutrino) sterile and conserve the spacetime components of the particle and antiparticle currents, we found that they are precisely those of the gauge group of the standard model. These gauge symmetries are thus seen to be local rotation groups in the tangent space of a manifold with *only four* extra spacelike dimensions. The fact that this is fewer than the minimum of seven extra dimensions required in the Kaluza–Klein type of approach [9] stems both from the availability of double-sided transformations on algebraic spinor elements and from the existence of higher-dimensional multivector subspaces in $C\ell_7$. Finally, the four extra dimensions, together with their exterior transformation properties, are precisely what is needed for the four components of a minimal scalar Higgs field with the correct weak hypercharge assignment.

Many features of the standard model thus flow from the relatively simple geometry of seven-dimensional Euclidean space, but there are many other features that call for explanation, such as the origin of the three generations and the mass spectra. Work is continuing on these and other aspects of the standard model within the framework of our model.

REFERENCES

[1] Baylis, W.E., Classical eigenspinors and the Dirac equation, 1992, *Phys. Rev.* A **45**, 4293–4302.

[2] Baylis, W.E., 1999, *Electrodynamics: A Modern Geometric Approach*, Birkhäuser, Boston.

[3] Baylis, W.E., 2000, Multiparavector subspaces in $C\ell_n$: Theorems and applications in *Clifford Algebras and their Applications in Mathematical Physics*, eds. R. Abłamowicz and B. Fauser, Vol. 1, Birkhäuser, Boston, pp. 3–20.

[4] Baylis, W.E. and Jones, G., The Pauli algebra approach to special relativity, 1989, *J. Phys. A* **22**, 1–15.

[5] Kaku, M., 1993, *Quantum Field Theory*, Oxford University Press, New York.

[6] Trayling, G., and Baylis, W.E., A geometric basis for the standard-model gauge group, 2001, *J. Phys. A: Math.* **34**, 3309–3324.

[7] Coleman, S. and Mandula, J., All possible symmetries of the S matrix, 1967, *Phys. Rev.* **159**, 1251–1256.

[8] Higgs, P.W., Spontaneous symmetry breakdown without massless bosons, 1966, *Phys. Rev.* **145**, 1156–1163.

[9] Witten, E., 1981, Search for a realistic Kaluza–Klein theory, *Nucl. Phys.* B **186**, 412–428.

William E. Baylis
Physics Department, University of Windsor
Windsor, Ontario, Canada N9B 3P4
E-mail: baylis@uwindsor.ca

Greg Trayling
Physics Department, University of Windsor
Windsor, Ontario, Canada N9B 3P4
E-mail: traylin@uwindsor.ca

Submitted: Aug 30, 2002, Revised: November 30, 2002.

PART V. APPLICATIONS IN ENGINEERING

35

Implementation of a Clifford Algebra Co-Processor Design on a Field Programmable Gate Array

Christian Perwass, Christian Gebken, and Gerald Sommer

ABSTRACT We present the design of a Clifford algebra co-processor and its implementation on a Field Programmable Gate Array (FPGA). To the best of our knowledge this is the first such design developed. The design is scalable in both the Clifford algebra dimension and the bit width of the numerical factors. Both aspects are only limited by the hardware resources. Furthermore, the signature of the underlying vector space can be changed without reconfiguring the FPGA. High calculation speeds are achieved through a pipeline architecture.

Keywords: Clifford co-processor, FPGA

35.1 Introduction

Clifford algebra has been applied to many different fields of research, as for example quantum mechanics, theories of gravity, automated geometric reasoning, computer vision and robotics. Clifford algebra is a powerful mathematical tool for symbolic calculations. In order to perform numerical calculations with Clifford algebra, multivectors in $C\ell_n$ have in general to be treated as 2^n dimensional vectors. Therefore, to evaluate the geometric product of two multivectors, 2^{2n} product operations and $2^n(2^n - 1)$ additions have to be performed with the multivector elements in the worst case.

Matrix multiplication has a similar computational complexity. Due to the need for very fast matrix multiplications, as for example in computer graphics, hardware implementations of the matrix product were developed. Since Clifford algebra is increasingly used in applied fields where computational speed is of importance, a hardware implementation of the geometric product and associated operations is of great interest.

AMS Subject Classification: 15-04, 15A66, 65K05.

We present the design of a Clifford algebra co-processor and its implementation on a Field Programmable Gate Array (FPGA). To the best of our knowledge this is the first such design developed. The design is scalable in both the Clifford algebra dimension and the bit width of the numerical factors. Both aspects are only limited by the hardware resources. Furthermore, the signature of the underlying vector space can be changed without reconfiguring the FPGA. High calculation speeds are achieved through a pipeline architecture.

The difference between the symbolic power of Clifford algebra (CA) and its high computational complexity has long been noted by researchers in this field. In order to numerically evaluate, or solve, symbolically powerful CA equations on a computer, one can often translate them into matrix equations which may then be solved or evaluated with standard matrix libraries. This has the advantage that one can use readily available matrix software packages and specialized matrix processing hardware. However, it may not always be obvious how to express a CA equation as a matrix equation. Furthermore, if one always has to translate CA equations into matrix equations, then the symbolic advantage we have gained by using CA is somewhat lost.

Therefore, many software packages have been developed recently, to evaluate and solve CA equations directly. There are packages for the symbolic computer algebra systems Maple [1, 2] and Mathematica [3], a package for the numerical mathematics program MatLab called GABLE [5], the C++ software libraries CLU [14], GluCat [9], the C++ software library generator Gaigen [6], a Java library [4] and stand alone programs CLUCalc, CLUit [14] and CLICAL [11], to name just a few. For researchers who are interested in the geometric aspects of CA, GABLE, CLUCalc and CLUit also visualize the geometric interpretation of multivectors in particular spaces.

When working with matrices one can take advantage of hardware accelerated matrix multiplications. Our goal was to see to what extent a hardware implementation of CA operations is indeed feasible and will speed up the evaluation process.

In this paper we can only give an overview of the design of the co-processor. Only some aspects are discussed in more detail. For a complete, detailed report (in German) see the Diploma thesis of Christian Gebken [15].

We will not give an introduction to Clifford algebra here. Introductory material can be found for example in [7, 10, 12, 16]. However, we give a short introduction to the geometric product, since this is the main Clifford algebra operation to be implemented by the co-processor. Let the basis of a universal Clifford algebra $C\ell_n$ be given by the set $\mathcal{B}_n = \{E_i\}$, which consists of 2^n basis blades. A multivector $A \in C\ell_n$ is then given by $A = \sum_{i=1}^{2^n} \alpha^i E_i$, with $\alpha^i \in \mathbb{R}$, $\forall i \in \{1, \ldots, 2^n\}$. If we agree on a particular basis $\mathcal{B}_n$ of $C\ell_n$, we can therefore represent the multivector A by the vector $(\alpha^1, \alpha^2, \ldots, \alpha^{2^n})$.

The geometric product of two elements of $\mathcal{B}_n$ results again in an element of $\mathcal{B}_n$, up to a sign. The relationship can be expressed with a tensor as follows: $E_i E_j = \sum_{k=1}^{2^n} g_{ij}{}^k E_k$, where the entries of $g_{ij}{}^k$ can be 1, -1 or zero. If $\mathcal{B}_n$ is the basis of a universal Clifford algebra, which we assumed, then the geometric product of basis blades is invertible. Let the three multivectors $A = \sum_{i=1}^{2^n} \alpha^i E_i$, $B = \sum_{i=1}^{2^n} \beta^i E_i$ and $C = \sum_{i=1}^{2^n} \gamma^i E_i$ be related by the geometric product $AB = C$. Then the relationship between their scalar components is given by $\gamma^k = \sum_{i=1}^{2^n} \sum_{j=1}^{2^n} \alpha^i \beta^j g_{ij}{}^k$. This relationship can be used to solve multivector equations [13]. Many software packages use pre-calculated multiplication tables, i.e., the $g_{ij}{}^k$, to implement the geometric, inner and outer product. As will be discussed later, the co-processor represents multivectors as lists of basis blades with associated scalar factors. The geometric product of basis blades will be evaluated explicitly, without a multiplication table, which is more effective in this case.

An explicit evaluation of the geometric product of two blades is done as follows. Given an orthonormal basis $\{e_1, e_2, \ldots, e_n\}$ of a vector space $\mathbb{R}^n$, two blades of the Clifford algebra $C\ell_n$ over $\mathbb{R}^n$ may, for example, be given by $a = \alpha\, e_1 e_2 e_4$ and $b = \beta\, e_1 e_3 e_4$, with $\alpha, \beta \in \mathbb{R}$. Their geometric product may then be evaluated as follows. First of all, the scalar factors of the blades can be multiplied separately, i.e., $ab = (\alpha\beta)\,(e_1 e_2 e_4 e_1 e_3 e_4)$. The resultant blade component can be evaluated by applying the associativity of the geometric product and the rules $e_i e_i = 1$ and $e_i e_j = -e_j e_i$, $i \neq j$. Note that $e_i e_i = -1$ is also possible, depending on the signature of the vector space. The resultant blade component therefore is $-e_2 e_3$, and thus $ab = -(\alpha\beta)\, e_2 e_3$.

35.2   Hardware designs

The goal is to numerically evaluate CA operations like the geometric product with a digital circuit design. Before we discuss the evaluation of a CA product itself, we have to decide on what the Clifford co-processor should be able to do. This, of course, influences the design of the whole chip. Since we want to build a co-processor on an FPGA which is external to the CPU, we do not want to constantly transfer data between the CPU and the FPGA. Therefore, the co-processor should have a list of operations and multivectors which it can evaluate independently from the CPU. The possible operations should allow us to evaluate most CA expressions. Therefore, we want the processor to have the following features. 1. Evaluation of the geometric, inner and outer product, addition and subtraction. 2. Operations between given multivector and previous result. 3. Operations between two previous results. 4. Choice of order of multivectors in operation.

This list of features makes certain demands on the arithmetic logic unit (ALU) of the co-processor. Before we discuss possible ALU designs, we will give a short overview of how an FPGA actually works.

The FPGA we used was a Xilinx XC4085XLA-0.9. This FPGA contains 3136 configurable logic blocks (CLBs) which are connected in a configurable array. Each CLB contains two delay flip-flops (DFF), two look up tables (LUT) and a fast carry logic for arithmetic operations. A DFF can be used as a register to store information and an LUT allows logic functions up to 4 bit to be implemented, like AND, OR, XOR, etc. and combinations of these. Using the CLBs and the configurable connection array, any digital circuit design can be implemented.

Note that this FPGA does not contain any predefined arithmetic units for multiplication or addition. These have to be implemented by a designer. A single 32×32 bit floating point multiplier would already use up most, if not all, available CLBs of this FPGA.

Although the internal clock of the FPGA can run at more than 100 MHz, the maximal achievable frequency for a particular design may be much lower. In our case there was an additional problem due to the FPGA's external RAM on the FPGA board, which only allowed simultaneous read/write access at 20 MHz.

We programmed the FPGA design using the C++ hardware description language (CHDL) [8]. CHDL is a C++ software library which allows the programmer to define a digital circuit using C++ objects and operators. This simplifies the design process compared to the standard hardware description language VHDL.

In general the clock frequencies of FPGAs cannot be as high as those of non-configurable ICs, due to the FPGA's configurable connections between CLBs.

Despite these problems, FPGAs are certainly a good choice to build and test a Clifford co-processor prototype. Furthermore, in certain applications we might want to use different digital circuit designs on the same FPGA consecutively, to solve different aspects of a problem. Here the Clifford co-processor may be a design which could follow some pre-processing operations.

We will now discuss two possible ALU designs.

35.2.1   Direct computation

This design expects as input two complete multivectors. Each operation between basis blades is hardwired, so that all necessary operators exist on the chip simultaneously. This allows for pipelined, parallel processing, but it also needs a large amount of resources.

As an example take the geometric product. For each multiplication between basis blades there exists a separate multiplier which is hardwired to the appropriate multivector elements. Its output is connected to an adder which collects all blade multiplications which result in the same blade. In $C\ell_n$ we would need 2^{2n} multipliers. Since there are 2^n different basis blade combinations whose products result in the same basis blade, each of the 2^n resultant basis blades is the sum of 2^n values. These have to be added in a cascade of $\sum_{i=1}^{n-1} 2^i$ adders. Therefore, we have a total of $2^n \sum_{i=1}^{n-1} 2^i$ adders. For 3d-space this gives 64 multipliers and 48 adders. For 5d-space we already have 1024 multipliers and 960 adders. Even the most advanced FPGAs available today do not have enough capacity to deal with the 5d case. Furthermore, a change of signature or dimension would mean that we have to reconfigure the chip.

Although this would be the simplest and fastest implementation, we did not follow this approach due to its enormous need of resources and its inflexibility.

35.2.2   Basis blade pipeline design

A much more flexible design is that of basis blade pipelines. Here we have a number of pipelines which each deal with an operation between two basis blades. The number of pipelines per operation depends on the amount of resources available. Of course, there has to be additional logic which distributes the different combinations of basis blades between the different pipelines and collects the results appropriately. This does in fact cause some non-trivial problems, which will be discussed later on. Note that at each clock tick (the moment when the clock changes from low to high) we can push a new basis blade combination into every pipeline. This means that if a basis blade pair needs n clock cycles to be processed by a pipeline, we initially have to wait n clock cycles for the first result to appear. However, after this time we obtain a new result after each clock cycle.

In this setup, the geometric product could be evaluated using two different methods: a multiplication table or an explicit calculation. For software packages a multiplication table is the easiest and also a very efficient solution. Here this is not the case, since a multiplication table would have to be stored in memory. On the FPGA we used the memory could only be accessed serially, which would not allow any parallel processing. Furthermore, other parts of the design need to access the memory at the same time, which would have slowed down the processing even further. Therefore, we decided on evaluating the geometric product explicitly. This is discussed in some detail later on.

For example, to evaluate the geometric product each pipeline only needs a single pipelined multiplier and a single adder to add the result to the appropriate result blade. Depending on the amount of resources available we can vary the number of pipelines working in parallel. In fact, we could even have a number of ALUs working in parallel, although then we might run into additional trouble due to interdependencies of the operations.

Due to its flexibility and extensibility we chose this ALU design for the Clifford co-processor.

35.2.3   Co-processor design

Figure 35.1 shows the overall data flow of the co-processor. We assume here that the FPGA which implements the co-processor sits on a PCI card with its own memory and is accessible through the PCI bus. The RAM Access Controller (RAM-AC) has two main objectives. First of all it is used to transfer data from the main board memory to the FPGA memory and vice versa. Secondly, it allows the Central Control Unit (CCU) to access the instructions and previous results. Furthermore, it allows the ALU to write evaluation results to the appropriate addresses of the on-board RAM.

The CCU takes an instruction from the RAM, and feeds the appropriate evaluation pipelines of the ALU with basis blade pairs. The ALU accumulates the evaluation results in a result area in the RAM. More details are given in the next section.

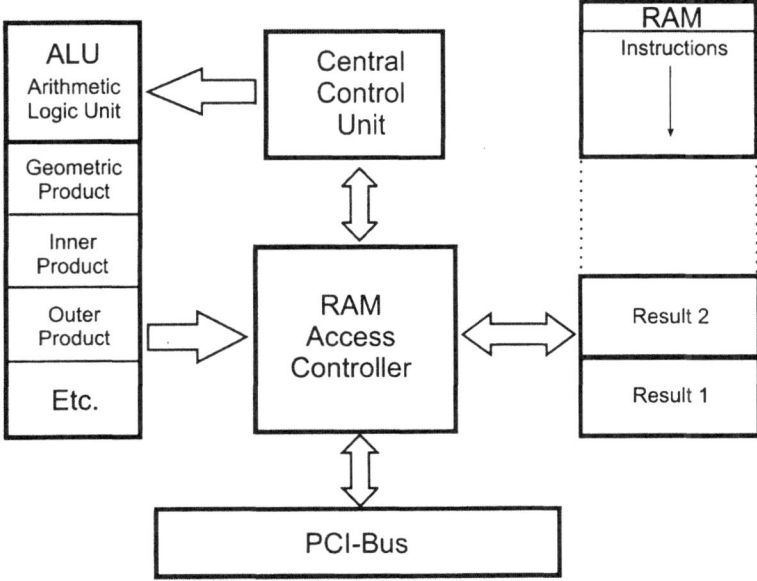

FIGURE 35.1. General data flow in CA co-processor.

35.3   Implementation

The FPGA available for the design implementation was rather small, which meant that we had to make some restrictions.

1. We could only implement a single basis blade pipeline,

2. Scalar factors of basis blades are 24 bit integer numbers,

3. Clifford algebras of up to 8d-vector spaces can be used,

4. Only the geometric product has been implemented.

35.3.1   Basis blades and instructions

Each basis blade consists of a scalar factor and an algebra component. This is shown in Figure 35.2. The scalar component is a 23 bit integer value plus a sign

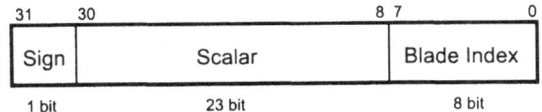

FIGURE 35.2. Structure of a basis blade.

bit. The blade index follows a well-known method used to express basis blades in binary code. Each bit in the blade index stands for a basis vector. If a bit is high, the corresponding basis vector exists in the blade, otherwise it does not. In this way all basis blades of a Clifford algebra can be represented. For example, in $C\ell_8$ the basis blade $e_1e_3e_8$ is represented by the binary code 10000101. Each blade has a unique blade index, so that the latter can be used as a memory offset address. Each operation instruction consists of three instruction words (32 bits each) and the corresponding multivectors, as shown in Figure 35.3. Each

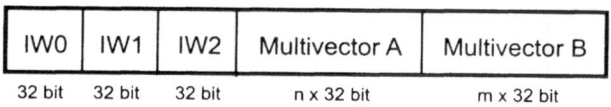

IW0	IW1	IW2	Multivector A	Multivector B
32 bit	32 bit	32 bit	n x 32 bit	m x 32 bit

FIGURE 35.3. Structure of an instruction

multivector consists of a collection of basis blades as shown in Figure 35.2. The number of elements within the multivectors is contained in the instruction words IW1 and IW2, respectively. Furthermore, the operand multivectors do not have to follow the instruction words. The operands can also be result multivectors from previous operations. The structure of the instruction words is as follows.

IW0 Bit	Description
0-9	Offset to the start of the next instruction
11-14	ID of operation type
15-18	Flags
19-31	Reserved

IW1 Bit	Description
0-22	Address of A-operand (first multivector)
13-31	Number of basis blades in A-operand

IW2 Bit	Description
0-22	Address of B-operand (second multivector)
13-31	Number of basis blades in B-operand

The elements of IW0 are fairly self explanatory. The first 10 bits contain the offset to the next instruction. This can be variable, since the number of elements in the multivectors passed with the instructions is variable. There are four bits for the operation type ID, which allows for 16 different operations. Currently only the geometric product has been implemented. The next four bits contain flags which indicate whether this is a proper instruction or the end of the instruction list. Furthermore, there is a flag which tells the CCU whether each element of operand A is multiplied with all elements of operand B or vice versa. That is, the roles of A and B can be switched.

Note that the addresses of the A and B operands are not given as relative address offsets but as absolute values. This has been done to simplify the design somewhat. However, it also means that the multivectors passed with the instructions and the result multivectors have to be at previously known addresses. This can be achieved as follows.

When preparing an instruction list, we know how many result multivectors we will obtain. Since each result multivector can have at most $2^8 = 256$ basis blades, we know how much memory we have to reserve for each ($256 \cdot 4$ bytes $= 1024$ bytes). By placing the nth result multivector memory block at the highest memory address minus $n \cdot 1024$ bytes, we have a simple way of knowing the address of a result multivector when preparing the instruction list. Note that the instruction list itself starts at the lowest memory address, as indicated in Figure 35.1.

35.3.2   The geometric product pipeline

Due to size restrictions of the FPGA we used, only one geometric product pipeline could be implemented. At this point we assume that the CCU has fed the pipeline with an appropriate pair of basis blades and the resultant basis blade will be added to the correct result multivector.

In order to evaluate the geometric product of two basis blades, we have to perform two main operations: the scalar parts of the two basis blades have to be multiplied with a standard multiplier and the blade parts (the blade indices) have to be multiplied with the geometric product operation. If we disregard the sign for a moment, the geometric product of blade indices is simply an XOR operation: if both basis blades contain the same basis vector (e.g., $e_1 e_2$ and $e_1 e_3$ both contain e_1), then this basis vector (e_1 in this case) squares to unity, otherwise the basis vector remains (e_2 and e_3 in this case). This is shown in figure 35.4.

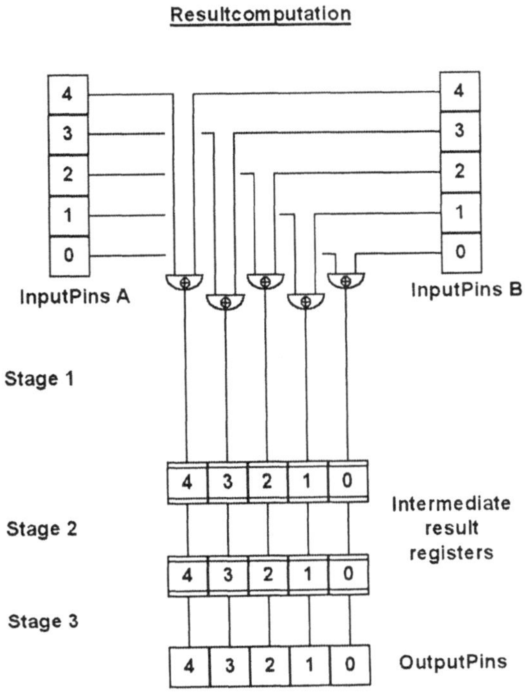

FIGURE 35.4. Geometric product without sign contributions.

For clarity, we have only drawn the digital circuit for the lower five bits of a basis blade index. Empty half circles closed by a straight line symbolize AND gates and if they contain an encircled plus they represent XOR gates. The square boxes represent registers. Registers basically load the value at their input at each clock tick and store it until the next. In this way one can realize a pipeline design.

Due to hardware restrictions, only a limited number of logic gates can be evaluated consecutively within a clock cycle. In order to have a synchronized, pipelined design, registers have to be inserted after a certain number of consecutive logic gates.

In Figure 35.4 we have shown a design with three stages. That is, we need three clock cycles to process the data applied at the input pins. However, at each clock tick a new set of data can be applied to the input pins since the intermediate results of the previous data sets are stored in the registers.

Evaluating the appropriate sign is what makes the geometric product somewhat more complicated. There are three contributions to the sign: the sign of the scalar factors, the sign due to the signature of the basis vectors and the sign due to the swapping of basis vectors. The sign due to the scalar factors is taken care of by the scalar multiplier, which we will not discuss.

Signature

The sign due to the signature of two multivectors is evaluated as shown in Figure 35.5. The co-processor has an 8 bit register which stores the signature of the basis vectors, and which can be set before an instruction set is executed. If a bit in the sign register is high, then the corresponding basis blade squares to minus one. Otherwise it squares to plus one. Hence, the circuit in Figure 35.5. If both basis indices have a common basis vector, i.e., a common high bit, then the squaring of these basis vectors contributes a minus if the corresponding signature bit is high. This is achieved through the AND gates. The XOR gates combine all the separate minus signs. They return low (plus) if there is an even number of minus signs and otherwise high (minus).

Swapping

The digital circuit evaluating the sign due to the swapping of basis vectors is shown in Figure 35.6. For example, if we want to evaluate $(e_1 e_2)(e_1)$ we write $e_1 e_2 e_1 = -e_2 e_1 e_1 = -e_2$, where we have made one swap, exchanging e_1 and e_2. The number of swaps necessary can be evaluated by an XOR cascade. Input pin 0 stands for e_1, input pin 1 for e_2, and so on. If in both operands input pin 4 is high, then both operands have an e_5 component. Before we can square these we potentially have to swap e_5 of operand A with e_1, e_2, e_3 and e_4 of operand B. If an odd number of these are actually present in operand B, then a minus is introduced, otherwise not. This is achieved by the cascade of XORs. The combination of the different swap signs is done in stages 2 and 3.

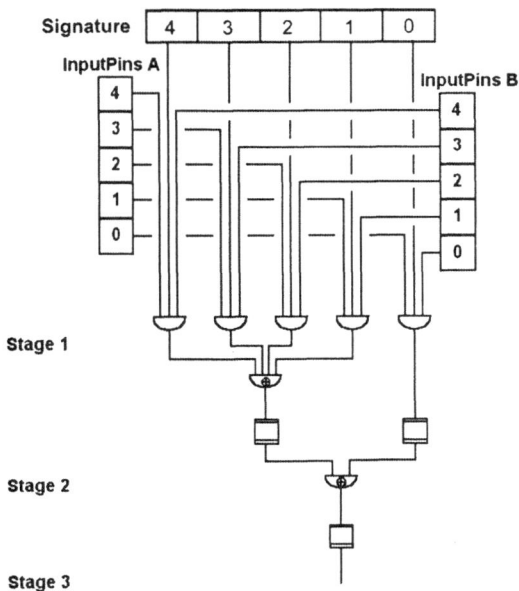

FIGURE 35.5. Evaluation of sign due to signature.

Sign combination

The different sign contributions due to the geometric product are finally combined in stage 3 as shown in Figure 35.7. This circuit shows the final stage of the circuits drawn in Figures 35.5 and 35.6. Of course, this resultant sign has to be XORed again with the resultant sign of the scalar multiplication.

35.3.3   Inner and outer product

The inner and outer product can be evaluated in a very similar manner to the geometric product. In fact, they become the geometric product if certain conditions are satisfied. An implementation of these products would therefore only introduce an additional step, where these conditions are checked and then either zero is returned directly or the geometric product is evaluated.

The inner product of two basis blades is only non-zero if one basis blade is completely contained within the other. We can check this by evaluating operand A AND operand B, which has to be equal either to operand A or to operand B. If this is the case, then the inner product of the two operands is simply the geometric product.

The outer product of two basis blades is only non-zero if they have no basis vector in common. We can check this with a simple AND operation. Again the outer product becomes the geometric product if the two operands have no element in common.

Currently, we have not implemented the inner and outer product.

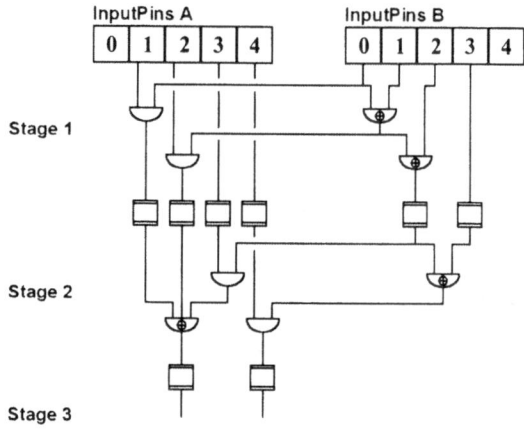

FIGURE 35.6. Sign evaluation due to basis vector swaps.

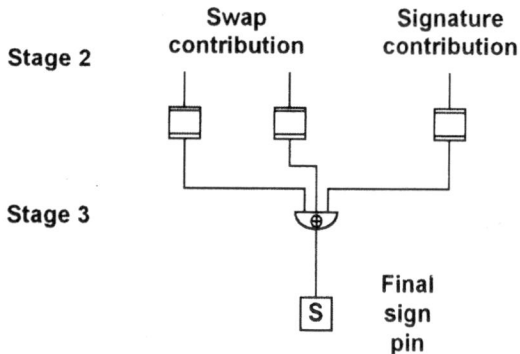

FIGURE 35.7. Combination of signs due to geometric product.

35.3.4   Geometric product evaluation

In the following we discuss how the geometric product of two multivectors is evaluated by the co-processor. The three main steps are:

1. The CCU takes the current set of instruction words and initializes counters with the number of A and B operands (basis blades).

2. For each A operand the CCU loops through all B operands and passes each pair to the ALU. If an operand is zero, the A and B operands are still passed to the ALU but they are marked as invalid, which means that they will not be added to the result multivector.

3. The blade indices of the resultant basis blades that exit the ALU pipeline are used to load the appropriate basis blade from the result multivector memory block. The two basis blades are then added and written back to memory.

Steps 2 and 3 are executed pipelined. That is, at each clock tick a new pair of basis blades is passed to the ALU. If a valid result is available at the ALU output it is added to the result multivector memory block in parallel to the ALU execution.

There are a number of difficulties which mainly have to do with the pipeline synchronization, which we cannot discuss here in detail. However, there is one issue which should be mentioned. For a fixed A operand the geometric product with each B operand will create a different basis blade. Although loading and writing basis blades from and to the result multivector memory block takes a couple of clock cycles, this only introduces a constant delay. That is, we have to store the results of the ALU in a FIFO until the corresponding basis blades from the result multivector memory block arrive. They are then added and written back to memory.

However, if we are at the end of the B operand loop, the A operand changes. This means that, by chance, the new combination of A and B operands might result in the same basis blade as the last combination of operands from the previous B operand loop. Therefore, the CCU would load the same basis blade from the result multivector memory block a second time, before the first basis blade could have been added to it and written back. This is a so called *read after write* conflict: we read obsolete data before it has been updated.

The design we implemented recognizes such conflicts. Any blade pair that causes a conflict is stored in a FIFO and is executed at the end of the B operand loop.

35.4   Evaluation and conclusions

In order to test the evaluation speed of the co-processor and the software packages CLU and Gaigen, we evaluated the geometric product of multivectors that contained all basis blades of the respective Clifford algebra. Of course, this does not test how the skipping of zero basis blades in multivectors is handled by the co-processor and the software packages. However, if we wanted to test this, the question would be what is a realistic distribution of basis blades in multivectors. The benchmark we used is nonetheless one indicator of the performance of the different packages.

Due to hardware restrictions we could only run the FPGA at 20MHz. CLU and Gaigen were tested on 1.5 GHz machines. In real terms, both software packages were much faster than our implementation of the co-processor. However, in order to see whether the hardware implementation does offer an advantage in principle, we have to compare the results at the same frequency. The results are shown in Figure 35.8, where the x-axis gives the dimension of the vector space over which the Clifford algebra is formed and the y-axis gives the number of geometric operations per second (GOPS), i.e., the number of geometric products between multivectors. It can be seen that the optimized code generated by Gaigen is nearly as fast as the co-processor. The co-processor is about 1.5 times faster than Gaigen.

Since, so far, the co-processor does not make any use of parallel pipelines, and modern CPUs also use pipeline processing, this shows that programming a modern CPU can be about as efficient as a pipelined hardware implementation.

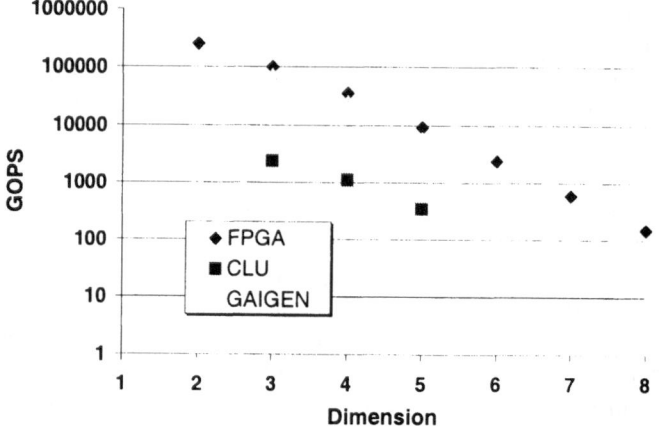

FIGURE 35.8. Evaluation speed benchmarks. GOPS stands for Geometric Operations Per Second.

The main advantages of a hardware implementation of Clifford operations are the following.

- If the co-processor is implemented on an application specific integrated circuit (ASIC) or FPGA, it can run in parallel to the main CPU. That is, the CPU can deal with other things, while the Clifford operations are evaluated.

- If enough resources are available, a number of evaluation pipelines can work in parallel. There could even be a number of parallel instruction pipelines. This could offset the low FPGA clock frequencies.

- A Clifford co-processor could be integrated into the main CPU, just as multimedia operations have been (e.g., MMX).

In conclusion we can say that a Clifford co-processor implementation on even the most modern FPGAs might only be about as fast as an optimized software implementation of Clifford operations. This is mainly due to the comparatively low clock frequencies and a lack of direct support of basic arithmetic operations like floating point multiplication and addition, which are essential for many applications.

Of course, it would be most desirable to have Clifford operations available as part of a standard CPU. However, this will only occur if Clifford operations offer a functionality that is needed by many (profitable) applications, which cannot easily be provided through other mathematical tools, like matrices.

REFERENCES

[1] R. Abłamowicz, Clifford algebra computations with Maple, in *Clifford (Geometric) Algebras*, Banff, Alberta Canada, 1995, Ed. W.E. Baylis, Birkhäuser, Boston, 1996, pp. 463–501.

[2] R. Abłamowicz, B. Fauser, The CLIFFORD Home Page, `math.tntech.edu/rafal/cliff5/index.html`, last visited 15. Sept. 2003.

[3] J. Browne, The GrassmannAlgebra Book Home Page, `www.ses.swin.edu.au/homes/browne/grassmannalgebra/book/`, last visited 15. Sept. 2003.

[4] A. Differ, The Clados Home Page, `sourceforge.net/projects/clados/`, last visited 15. Sept. 2003.

[5] L. Dorst, The GABLE Home Page, `carol.wins.uva.nl/~leo/GABLE/`, last visited 15. Sept. 2003.

[6] D. Fontijne, The Gaigen Home Page, `carol.wins.uva.nl/~fontijne/gaigen/`, last visited 15. Sept. 2003.

[7] D. Hestenes, G. Sobczyk. *Clifford Algebra to Geometric Calculus: A Unified Language for Mathematics and Physics*. Dordrecht, 1984.

[8] K. Kornmesser. The CHDL Home Page, `www-li5.ti.uni-mannheim.de/fpga/?chdl/`, last visited 15. Sept. 2003.

[9] P. Leopardi, The GluCat Home Page, `glucat.sourceforge.net/`, last visited 15. Sept. 2003.

[10] P. Lounesto. *Clifford Algebra and Spinors*. Cambridge University Press, 1997.

[11] P. Lounesto, The CLICAL Home Page, `www.helsinki.fi/~lounesto/CLICAL.htm`, last visited 15. Sept. 2003.

[12] C.B.U. Perwass, *Applications of Geometric Algebra in Computer Vision*. PhD thesis, Cambridge University, 2000.

[13] C.B.U. Perwass, G. Sommer, Numerical evaluation of versors with Clifford algebra, in *Applications of Geometric Algebra in Computer Science and Engineering*, Eds. Leo Dorst, Chris Doran, Joan Lasenby, Birkhäuser, 2002, pp. 341–349.

[14] C.B.U. Perwass, The CLU Home Page, `www.perwass.de/cbup/clu.html`, last visited 15. Sept. 2003.

[15] C. Gebken, *Implementierung eines Koprozessors für geometrische Algebra auf einem FPGA*, Diploma thesis, Christian-Albrechts-University Kiel, 2003.

[16] G. Sommer, editor. *Geometric Computing with Clifford Algebra*. Springer Verlag, 2001.

Christian Perwass
Institut für Informatik
Christian-Albrechts-Unversität Kiel
24105 Kiel, Germany
E-mail: chp@ks.informatik.uni-kiel.de

Christian Gebken
Institut für Informatik
Christian-Albrechts-Unversität Kiel
24105 Kiel, Germany
E-mail: chg@ks.informatik.uni-kiel.de

Gerald Sommer
Institut für Informatik
Christian-Albrechts-Unversität Kiel
24105 Kiel, Germany
E-mail: gs@ks.informatik.uni-kiel.de

Submitted: August 31, 2002.

36

Image Space

Jan J. Koenderink

ABSTRACT "Image processing" is a purely syntactical discipline that transforms images into other images. As such it is distinct from such fields as "image recognition," "feature detection," and so forth, which are semantically oriented. If semantics plays a role, the "meaning" is relative to a user's world model. In the case of purely syntactical operations there is some hope to put the discipline on a foundation of first principles. Image processing in its present state is more of a bag of hat tricks though, and many of its methods are easily shown to be inconsistent. This contribution is an attempt to put image processing on a solid geometrical basis. In order to do so the (currently only vaguely defined though much used) concept of "image space" is formalized. Departing from a few commonly acknowledged properties one arrives at a three-dimensional Cayley–Klein space with a single isotropic dimension that can be understood as a limit of either Euclidean or Minkowskian three-space. In this geometry, planes containing an isotropic line are isomorphic with the dual number plane, whereas any other plane is isomorphic with the conventional complex number plane. The elements of the group of similarities are readily identified with image transformations that "don't change the image," thus an "image" can be defined as an invariant under similarities. The differential invariants are a substitute for the "feature operators" of current image processing. Their formal structure is rather simpler than the corresponding Euclidean differential invariants, but the differential geometry is just as rich as that of conventional Euclidean space.

Keywords: image processing, image space, images, feature detectors, Cayley–Klein spaces, singly isotropic space, dual number plane, Clifford shifts, Clifford planes, Euclidean

36.1 The argument

"Image Space" seems an apt name for an ill-defined entity commonly implied in the (engineering) field of "image processing". Technically an "image" is typically a square array of nonnegative integers, limited by some maximum value (e.g., a 512 x 512 array of integers in the range $[0 \ldots 255]$ ("bytes") is typical). Each entry is known as a "pixel" ("picture element") and the geometry is implied by the rectangular array, which allows one to define neighborhood relations. As technology progresses the pixel density and the range of pixel values increase.

AMS Subject Classification: 51N25, 53B99, 68U10.

Conceptually, pixellation and intensity discretization are irrelevant (mere artifacts of the limited technology). I will ignore them and conceive of "ideal" images as a field of nonnegative real numbers on the Euclidean plane. When people refer to "image space" they implicitly take the intensity domain as a third Euclidean dimension, thus image space would be a three-dimensional Euclidean space. An "image" then is a surface in image space, which may be described by way of the classical, Euclidean differential geometry of surfaces. Indeed, many "feature detectors", and so forth, in common use in image processing, are nothing more than such differential invariants in disguise.

My argument for this contribution is that these conventional methods are quite misplaced and inconsistent with the physical meaning of images. This has potentially important consequences in view of the fact that millions of images per annum are processed (quite likely with inconsistent methodology) in a field like medical imaging alone.

The reason that the conventional practice is inconsistent with basic principles of physics is because the dimension of the picture plane and the intensity domain are incommensurable. Thus one can attach no meaning to any but very special Euclidean movements. For instance, consider a rotation about an axis parallel to the picture plane: This would "mix" in intensity with distance in the image plane, clearly inconceivable. But if the Euclidean group of movements doesn't apply, the corresponding differential invariants can possess no meaning either. And these are the target of operations in image processing. Thus the whole approach is doomed to failure.

In order to put things right (which should be possible in view of the fact that people are perfectly happy with the results of their inconsistent methods), we need to fix the following problems:

- The intensity dimension should be fitted with a physically reasonable metric (for instance, it ought not to matter whether radiance is measured in this or that system of units);

- Image space should be fitted out with its proper group of movements and similarities;

- A geometrical theory of differential invariants should be developed under the action of this group [9]: These will be the *correct* "features" (and so forth) of "image processing done right".

This may look like quite a program. However, as I will show, its implementation is comparatively straightforward. Simple as it may be, it has potentially very important consequences for image processing.

36.2   The "intensity" dimension

To fit the intensity dimension with a suitable structure is a problem of physics. It involves an analysis of the basic operation of intensity measurement. Here I will focus on the case of the radiance of electromagnetic radiation (e.g., photography). Depending upon the type of detector the radiance will be measured as radiant power per area and per solid angle (thermal detectors), or as photon number flux per area and per solid angle. Suppose one uses a CCD array; then it is natural to measure photon count per photoelement per temporal aperture (using either an electronic or mechanical shutter). This photon flux (the result of an elementary measurement) is a nonnegative integer. Knowing the effective area of the photosensitive elements and the exposure time, one may proceed to estimate the irradiance (a nonnegative real number with the dimension number of photons per area and per unit of time). The estimate will vary from trial to trial because of the quantum statistics of photon absorptions; the number detected is known to be Poisson distributed. In order to estimate the radiance from these uncertain data one needs to use Bayesian inference, and in order to use Bayesian inference one needs to assume a prior probability density function for the radiance.

Consider the following experiment: One performs a single measurement and observes n photons (say). From this one predicts the radiance λ (say). Knowing the radiance one may predict the most likely result of another measurement. Physical insight (at least mine) reveals that this prediction should be n, for in the absence of prior information there can be no reason why the prediction should deviate from the observation. But this has important consequences. For instance, if one assumes a uniform prior, it is easy to show that the expected number would be $n + 1$. Thus the assumption allows us to discard the uniform prior offhand. In fact, common statistical wisdom is that one should assume a *hyperbolic* prior (one easily checks that this immediately leads to the correct prediction). We arrive at the hyperbolic prior via a simple physical reasoning [8]: Assume that two different observers use different clocks. Their times are related by a fixed shift (epoch) and a fixed factor (thus both clocks run correctly), say $t_A = \mu t_B + \tau$. The intensities estimated by the observers A and B will be related as $\lambda_A \, dt_A = \lambda_B \, dt_B$. Let the observers assign priors $f_A(\lambda_A) \, d\lambda_A$ and $f_B(\lambda_B) \, d\lambda_B$. I assume that the observers share equal prior knowledge (none); then it must be the case that they assume the same prior, for otherwise the prior would not be objectively determined. Thus mutual consistency implies that $f_A(\lambda_A) \, d\lambda_A = f_B(\lambda_B) \, d\lambda_B$. This leads to the functional equation $f(\lambda) = \mu f(\mu \lambda)$ of which the hyperbolic prior $d\lambda/\lambda$ is the unique solution. Thus the hyperbolic prior is forced on us because of elementary considerations of physics.

Since $d\lambda/\lambda = d \log \lambda$, the hyperbolic prior is uniform on the logarithmic radiance axis. Thus the log radiance axis is the natural structure for the intensity domain. It cannot be assigned a natural origin because a change of physical units would shift the origin. One needs to set

$$z(x, y) = \log \frac{I(x, y)}{I_0}$$

in order to make the relation dimensionally consistent. ($I(x, y)$ denotes the intensity as a function of the Cartesian coordinates of the picture plane.) Thus the intensity dimension is the (full) affine line. In this paper the third coordinate of image space (the first two parameterize the picture plane) will be denoted z. Then $z(x, y)$ is a Monge parameterization of the "intensity surface".

Notice that the Monge parameterization is indeed very natural since only a single intensity is assigned to any point in the picture plane. In particular any generic plane can be parameterized as $z(x, y) = z_0 + (g_x x + g_y y)$, where $\{g_x, g_y\} = \nabla z$. Here a "generic plane" is one that does not contain a pixel (see below). Planes that contain a pixel are special, I will refer to them as "normal planes" for reasons to be explained later.

36.3   The group of proper motions and similarities of image space

Image space is the picture plane (a Euclidean plane) with the intensity dimension (an affine line) attached to each of its points, a trivial fiber bundle. Images are sections of the bundle. We have obtained a simple, consistent structure. In order to arrive at the geometry of image space I will introduce two natural assumptions:

- Image space is a homogeneous space, i.e., it allows of a group of congruences. This assumption is implicit in virtually all existing techniques of image processing (although I have not seen it explicitly mentioned) and hardly needs further justification;

- The intensity domain and the dimensions of the picture plane never "mix" under action of such congruences. This is hardly an assumption since it is enforced by the physics. It has important consequences. For instance, it rules out the Euclidean group as the group of proper movements, thus is inconsistent with current practice.

I will introduce some convenient conventions at this point: A "pixel" will denote the fiber at the "trace" of that pixel (the base point of the fiber). Thus pixels are straight lines parallel to the intensity dimension. Then the second assumption says that the group of movements conserves a bundle of parallel lines [12] (the pixels). Together with the first assumption this fixes the geometry, for the unique three-dimensional homogeneous space that conserves a bundle of parallel lines is one of the twenty seven (three–dimensional) Cayley–Klein geometries [3, 10], the one with a single isotropic direction. Here we are in luck because the geometry and differential geometry of this particular space have been developed to considerable detail by Strubecker and Sachs [13, 14, 16–19]. The important geometry in planes containing a pixel has been developed by these authors too; moreover there exists an English translation of the book by Jaglom [7] on this geometry. Thus I could stop at this point and leave it to the reader to obtain the necessary details from the literature. Instead (since this is a contribution in the section on applied mathematics) I will spell out some of the details as they apply to image processing.

The absolute conic is degenerated to a pair of lines $x \pm iy = 0$ in the plane $z = \infty$, with intersection $\{0, 0, \infty\}$. The lines parallel to the z-axis are conserved by the collinearity

$$x' = e^h(x \cos\varphi - y \sin\varphi) + t_x,$$
$$y' = e^h(x \sin\varphi + y \cos\varphi) + t_y,$$
$$z' = e^\delta z + \alpha_x x + \alpha_y y + \zeta.$$

This is the eight-parameter group G_8 of similarities. Setting $h = \delta = 0$ we obtain the six-parameter subgroup of proper motions G_6. Notice that similarities of image space appear as Euclidean similarities in the image plane if we restrict attention to the traces (projection on the picture plane). The parameter h is the modulus of these similarities "of the first kind". The parameter δ is the modulus of similarities of the second kind. A general similarity is characterized by a *pair* of moduli, quite different from the Euclidean case. The parameter ζ is evidently connected with translations in the intensity domain, whereas $\{\alpha_x, \alpha_y\}$ has to do with rotations in normal planes (see below). The rotations in the picture plane are of the familiar Euclidean type.

In order to obtain some intuitions for the nature of these movements it is convenient to consider the actions of one-parameter subgroups or a representation of their orbits (see Figure 36.1). There exist a number of distinct types, some appear as rotations (Figure 36.2) in traces in the picture plane, others as translations (Figures 36.3), a third type as identities (see Figures 36.4 and 36.5). In Figure 36.6, I show the effect of a similarity of the second kind, that appears as an identity in the picture plane. It corresponds to a contrast change, or "gamma transformation". Geometrically it scales the isotropic angles.

36.4   The structure of image space

This Cayley–Klein geometry is characterized through a single isotropic direction. Let the basis vectors e_x, e_y span the picture plane, e_z the intensity domain. Then we have $e_x e_x = e_y e_y = 1$, $e_z e_z = 0$ (and $e_u e_v = 0$ for $u \neq v$). We form the bivectors $b_x = e_y \wedge e_z$, and so forth, and the trivector $t = e_x \wedge e_y \wedge e_z$. Because of the fact that the geometry has an isotropic vector, the full multiplication table contains a quarter of zeroes. This has important geometrical consequences of which I will review a number of characteristic ones.

First consider the metric. The metric tensor is $g = e_x \otimes e_x + e_y \otimes e_y$, or, equivalently, the classical "line element" is $ds^2 = dx^2 + dy^2$. Thus we see that the metric is "the average" between the metric of Euclidean space and Minkowski space: $ds^2 = dx^2 + dy^2 + \mu dz^2$. For $\mu = +1$ we obtain Euclidean space, for $\mu = -1$ Minkowski space. When we take the limit $\mu \downarrow 0$, the unit sphere (gauge figure) becomes a prolate ellipsoid of revolution, and finally a right circular cylinder. The generators are pixels and the trace the unit circle of the picture plane. Thus image

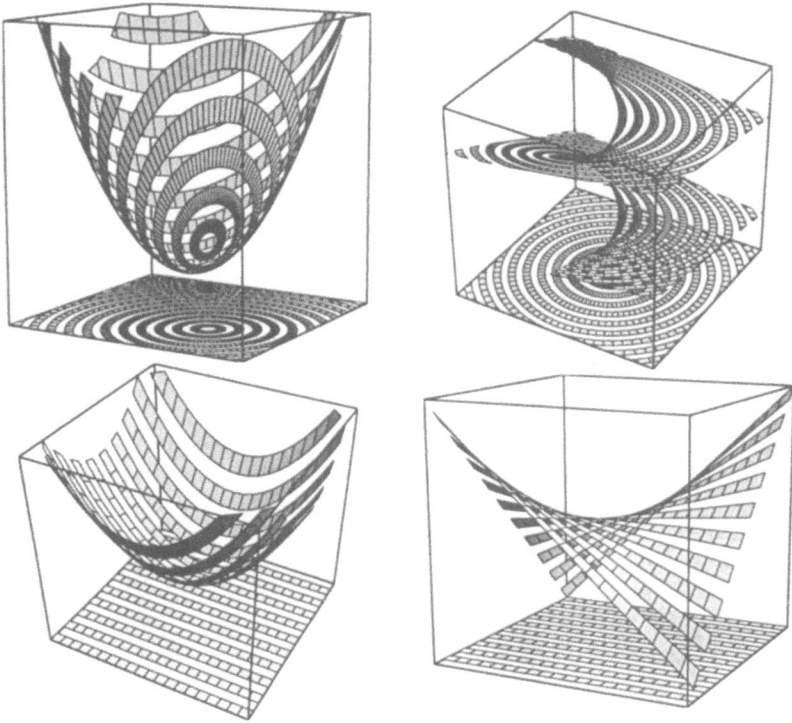

FIGURE 36.1. Some orbits of one-parameter subgroups. On the top row movements that appear as Euclidean rotations in the picture plane, on the bottom row movements that appear as translations. (There also exist nontrivial movements that appear as the identity in the picture plane.) The movement at the bottom right is a Clifford shift (see below). The orbits are Clifford parallels (see below). The "translation" at the bottom left is a parabolic rotation in image space. The orbits drawn are parallel circles on the surface of a sphere. Notice that the "rotation" at top right is not a periodic movement, whereas that on the top left is.

space appears as the infinitesimal neighborhood of the plane $z = 0$. This is indeed a useful interpretation for we see immediately that we cannot rotate pixels into the picture plane, the rotations have the pixels as orbits. When we consider the limit $\mu \uparrow 0$, the gauge figure of Minkowski space, which is a right circular cone with semitopangle $\pi/4$, the "light cone" of relativistic physics, becomes progressively narrower. In the limit the semitopangle becomes zero and the light cone collapses into a pixel. This yields another useful insight: We may consider the pixels as degenerated light cones, thus "pixels have no windows" but are forever isolated from each other. (In physics one would say "elsewhere".) On a single pixel we have a natural order of "lighter" or "darker" (the future and past of relativity) but there is no such order for distinct pixels. In fact (as I will show below) there exist proper movements that may equate the z-components of any two points in distinct

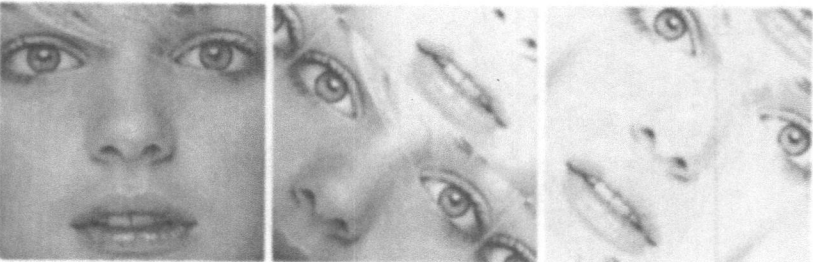

FIGURE 36.2. Two transformations that appear as rotations in the picture plane. At the left the original image. At the center the image has been subjected to a skew, periodic rotation (orbits shown in Figure 36.1 top left), at the right to a screw, non-periodic rotation (orbits shown in Figure 36.1 top right). In order to show a large extent I have tiled the plane with the original image.

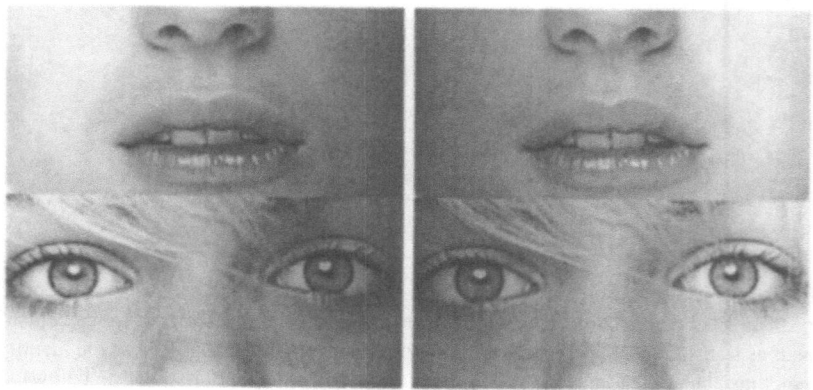

FIGURE 36.3. A Clifford left and a Clifford right shift. These are translations in the picture plane (orbits shown in Figure 36.1 bottom right). In order to show a large extent I have tiled the plane with the original image (Figure 36.2 left).

pixels.

Next consider the projection of a point on a plane. We represent the point by a vector $\mathbf{p}$ (say) and the plane by a bivector $\mathbf{b}$ (say). Let the plane be generic, i.e., not contain a pixel. Then you have

$$\mathbf{p}_\perp = \frac{\mathbf{p} \wedge \mathbf{b}}{\mathbf{b}} (= (\mathbf{p} \wedge \mathbf{b})\mathbf{b}^{-1},$$

where the sequence matters because multiplication is not commutative. Let $\mathbf{p} = x\mathbf{e_x} + y\mathbf{e_y} + z\mathbf{e_z}$ and $\mathbf{b} = u\mathbf{b_x} + v\mathbf{b_y} + w\mathbf{b_z}$, then elementary algebra yields $\mathbf{p}_\perp = -(xu + yv + zw)/w\,\mathbf{e_z}$. Thus the plumb line from a point to a (generic) plane of *any* orientation is parallel to the z-axis! This is highly remarkable because it means that the normals to all planar elements are identical. You cannot use

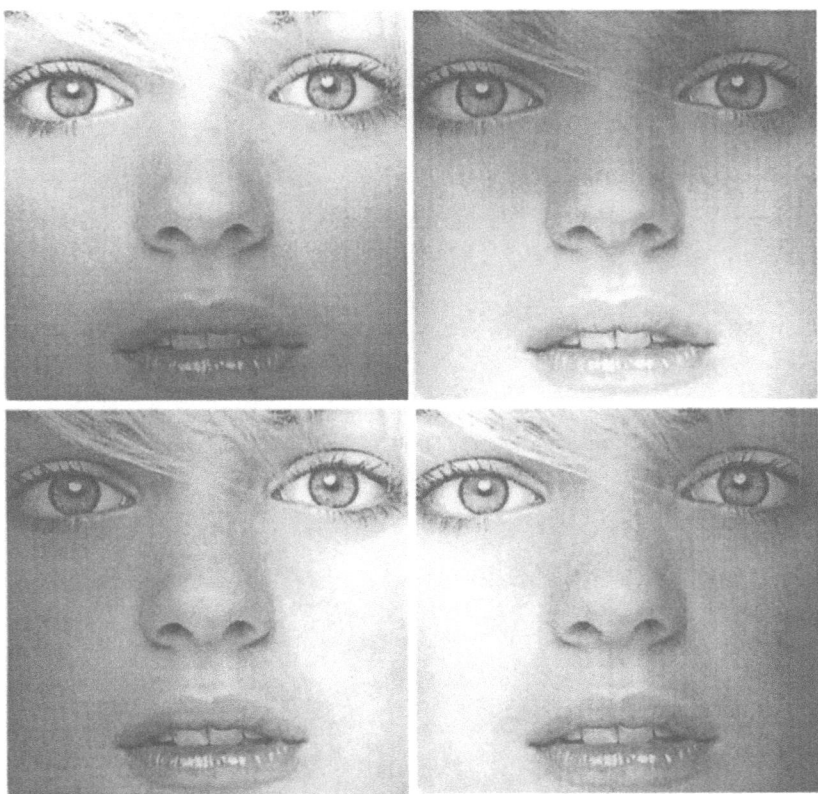

FIGURE 36.4. Rotations that appear as identities in the picture plane, different senses of rotation about horizontal and vertical axes. (Original image Figure 36.2 left.)

the normal direction to indicate the spatial attitude of a plane. For instance, the Gaussian "spherical image" which is central to classical differential geometry is not particularly useful in image space! This strange behavior is the reason for my use of "normal lines" for pixels and "normal planes" for planes containing a pixel.

Consider the rotations generated by bivectors. You have

$$e^{-\frac{\varphi}{2}\mathbf{b_z}}(x\mathbf{e_x} + y\mathbf{e_y} + z\mathbf{e_z})e^{+\frac{\varphi}{2}\mathbf{b_z}}$$
$$= (x\cos\varphi - y\sin\varphi)\mathbf{e_x} + (\mathbf{x}\sin\varphi + \mathbf{y}\cos\varphi)\mathbf{e_y} + \mathbf{z}\mathbf{e_z},$$

as expected, but

$$e^{-\frac{\vartheta}{2}\mathbf{b_x}}(x\mathbf{e_x} + y\mathbf{e_y} + z\mathbf{e_z})e^{+\frac{\vartheta}{2}\mathbf{b_x}} = x\mathbf{e_x} + y\mathbf{e_y} + (z + \vartheta y)\mathbf{e_z},$$

(similar for $\mathbf{b_y}$) which clearly has to be a rotation, though it looks perhaps more like a shear to the Euclidean eye. The Euclidean angle φ is periodic (the angle measure in the picture plane is elliptic, the distance measure parabolic) whereas

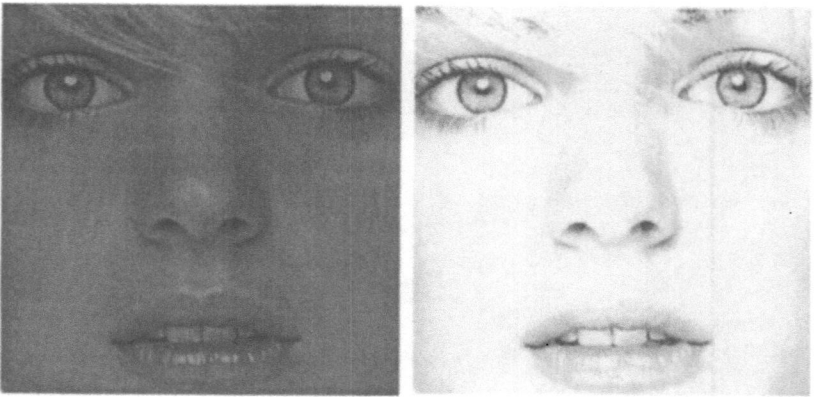

FIGURE 36.5. Full isotropic translations. These are identities in the picture plane. (Original image Figure 36.2 left.)

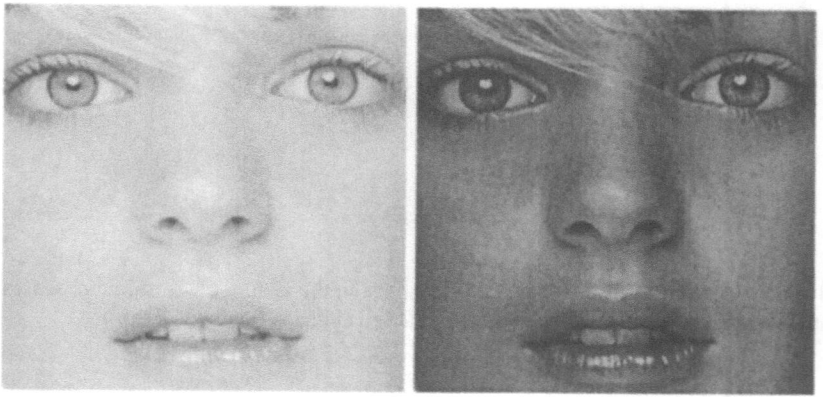

FIGURE 36.6. Similarities of the 2nd kind. These are identities in the picture plane. (Original image Figure 36.2 left.)

the isotropic angle ϑ is not: The angle measure in normal planes is parabolic, just like the distance measure. The isotropic angles take values on the full real axis, the Euclidean angles on the periodically repeated interval $[0, 2\pi)$. Now it becomes clear why the similarities of image space have two moduli: One scales distances, the other angles. This works because both the distance and the angle measures are parabolic. In Euclidean space you can scale only the distances, but you cannot scale the angles because the angle measure is elliptic. That's why the group of similarities of image space is an eight parameter group, instead (as in Euclidean space) a seven parameter group.

36.4.1   The geometry of normal planes

Much of the special nature of singly isotropic geometry is already apparent in the simple case of the geometry of the normal planes. Consider the plane spanned by e_x and e_z. A generic point p can be written $u e_x + v e_z$. Notice that $e_x p = u + \varepsilon v$, where $\varepsilon = e_x \wedge e_z = -b_y$, thus $\varepsilon^2 = 0$. Thus there exists a rather direct relation between the normal plane and the plane of (complex) *dual* numbers with as imaginary unit the nilpotent ε. The dual numbers were studied by Cayley and their use in geometry was initiated by Study. A good reference is Jaglom [6]. The use of dual numbers to represent points of the normal planes is natural because of the structure of the linear transformation $p' = ap + b$. Indeed, splitting real and dual parts in

$$u' + \varepsilon v' = (a_1 + \varepsilon a_2)(u + \varepsilon v) + (b_1 + \varepsilon b_2)$$

yields $u' = a_1 u + b_1$, $v' = a_2 u + a_1 v + b_2$, that is a similarity or movement (in the case $a_1 = 1$) in the normal plane. Thus the algebra of the dual number plane conforms automatically to the defining group of image space.

The dual numbers are in most respects like the familiar imaginary numbers, with a few exceptions. Clearly the dual numbers admit of divisors of zero. When you try to augment the dual number plane with points at infinity you have to distinguish two types of "infinity", namely $\infty = 1/0$ and $\omega = 1/\varepsilon$. Thus you close the dual number plane with a line at infinity to obtain the inversive number plane. Inverse stereographical projection yields the dual number cylinder, the analogue of the familiar complex number sphere.

The nilpotency of the dual unit has many important (and convenient!) consequences. For instance, consider the Taylor expansion

$$f(u + \varepsilon v) = f(u) + \varepsilon f'(u) v$$

in which all orders higher than the first vanish identically. For instance

$$\exp \varepsilon v = 1 + \varepsilon v, \quad \sin \varepsilon v = \varepsilon v, \quad \cos \varepsilon v = 1.$$

Thus any dual number $z = u + \varepsilon v$ can be written in the polar form $\rho \exp \varepsilon \vartheta$, where the modulus $|z| = u$ and the argument $\arg z = v/u$. Notice that the modulus is a *signed* number and that the argument can be an arbitrary real number. Apparently the angle is defined as v/u. This can be understood when you consider the unit circle ("of the first kind") $u^2 = 1$. The points $1 + \varepsilon v_1$ and $1 + \varepsilon v_2$ span an arclength $\varepsilon(v_2 - v_1)$ (which is invariant with respect to rotations), thus the angle can be identified with the arclength along the circumference of the unit circle, just as in the Euclidean case. The unit circle of the first kind $u = \pm 1$ acts as a convenient "protractor" in numerous synthetic constructions. Notice that it has a *line* of centers ($u = 0$).

Consider a curve $u + \varepsilon v(u)$. The first derivative with respect to u is the unit tangent $1 + \varepsilon v'(u)$, thus the curve is parameterized by arclength. The derivative of the tangent with respect to arclength is $\varepsilon v''(u)$, which is $v''(u)$ times the unit normal ε. (As we have seen already the normal is a constant vector.) The number

$v''(u)$ is the *curvature*. Thus a constant curvature curve is of the form $(u - u_0) + \frac{1}{2}\varepsilon\kappa(u-u_0)^2$. These are the "circles of the second kind". They can be transformed in themselves by a movement and indeed behave like Euclidean circles in almost all respects.

When you use them to define angles in the conventional way you obtain the same result as was obtained above. An analytic function in dual numbers can be written [15] as

$$f(u + \varepsilon v) = h(u) + \varepsilon(vh'(u) + g(u)),$$

where $h(u)$ and $g(u)$ are real functions. The analytic functions define *conformal maps* in the sense of the geometry of normal planes. With $h(u) = u$ we obtain the conformal maps $f(u + \varepsilon v) = u + \varepsilon(v + g(u))$ which involve local shifts in the intensity direction (of amount $g(u)$) combined with local rotations (of amount $g'(u)$) and thus are clearly conformal, depending on the arbitrary function $g(u)$.

Consider two points $z_{1,2} = u_{1,2} + \varepsilon v_{1,2}$. A movement $z' = az + b$ with $a_1 = 1$ transforms them into $z'_{1,2} = u_{1,2} + b_1 + \varepsilon(a_2 u_{1,2} + v_{1,2} + b_2)$, thus $(z_2 - z_1)' = (u_2 - u_1) + \varepsilon(a_2(u_2 - u_1) + (v_2 - v_1))$, thus $u'_2 - u'_1 = u_2 - u_1$. Hence $u_2 - u_1$ is an invariant. However, $u_2 - u_1 = 0$ does by no means imply $z_2 = z_1$! Attractive as it might seem to consider $u_2 - u_1$ "the distance" between z_1 and z_2, this will not work. When $u_2 - u_1 = 0$ but $z_1 \neq z_2$, I call the points "parallel". Consider two parallel points $z_{1,2}$. You have $(z_2 - z_1)' = (z_2 - z_1) = \varepsilon(v_2 - v_1)$. Thus $v_2 - v_1$ is an invariant when $u_2 - u_1 = 0$. Then it is reasonable to define the distance between two points $z_{1,2}$ as $u_2 - u_1$ in the generic case, and as $v_2 - v_1$ when the points are parallel.

Although the notion of "parallel points" might sound odd, this is due to the fact that there exists a perfect *metrical* duality between points and lines in the normal planes. Thus the notion of "parallel lines" (which somehow doesn't sound odd) implies the existence of "parallel points". Once you get used to the idea, the duality is a marvelous property that makes the geometry of normal planes in many respects more attractive than that of the Euclidean plane. Thus the distance between two lines is defined as either the (isotropic) angle subtended by them (in the generic case) or the gap (in the isotropic direction) between them if they prove to be parallel. The expression $ux = y - v$ defines lines and points ($\{x, y\}$ point coordinates, $\{u, v\}$ line coordinates) such that all the points $\{x, y\}$ lie on the line $\{u, v\}$ and all lines $\{u, v\}$ pass through the point $\{x, y\}$.

The distance between a point and a line is naturally defined to be zero if the point is on the line. If the point is not on the line you drop a plumb line from the point on the line (in the isotropic direction of course!) and define the separation in the isotropic direction as the distance. It is obviously an invariant under isotropic movements.

Consider the transformation $u' + \varepsilon v' = k_1 u + \varepsilon k_1 k_2 v$. The bilocal vector $z_2 - z_1$ is transformed into $k_1(u_2 - u_1) + \varepsilon k_1 k_2(v_2 - v_1))$, which has modulus $|z'_2 - z'_1| = k_1|z_2 - z_1|$ and argument $arg(z_2 - z_1)' = k_2 arg(z_2 - z_1)$. Thus we obtain two kinds of "similarities". The similarities of the first kind, with modulus k_1 and $k_2 = 1$, scale the (proper) distances, whereas the similarities of the second kind, with modulus k_2 and $k_1 = 1$, scale the angles. The general similarity scales both angles and distances of course.

36.4.2 The geometry of image space proper

Once the geometry of the normal planes is well understood, the geometry of image space should pose no problems. Here we have a perfect metrical duality between lines and planes. The distances between various entities (points, lines, planes) are easily defined and the geometry is not difficult to construct once you have grasped the geometry of the normal planes. Here I will merely point out a few things that are not immediately obvious.

From the perspective of image processing the normal subgroup of "modular limit movements"

$$x' = a + x, \quad y' = b + y, \quad z' = c + c_1 x + c_2 y + c_3 z$$

is important since the Euclidean movements of the image plane are not of immediate interest. Here the coefficient c_3 describes a similarity of the second kind, for $c_3 = 1$ we obtain the important group B_5 of unimodular limit movements. This group has an interesting structure that we can study by considering the isotropic Clifford left and right shifts

$$S_3^{(l)} \begin{cases} x' = \alpha + x, \\ y' = \beta + y, \\ z' = \gamma - \beta x + \alpha y + z, \end{cases} \qquad S_3^{(r)} \begin{cases} x' = A + x, \\ y' = B + y, \\ z' = C + Bx - Ay + z. \end{cases}$$

The group of unimodular limit movements has the Clifford shifts as three-parameter simple transitive subgroups and it can be written as a commutative product of right and left shifts. The Clifford shifts intersect in the full isotropic shift group $x' = x$, $y' = y$, $z' = z + \gamma$. The Clifford shifts commute with the full isotropic shifts. It is easily shown that each unimodular limit movement can be written uniquely as the product of a right and a left Clifford shift up to a full isotropic shift.

The so called "Clifford planes" are hyperbolic paraboloids with full isotropic diameters (e.g., $(x^2 - y^2)/2$). They are covered with two families of orbits of Clifford shifts which are straight lines. The lines from one family are pushed forwards by a shift along the orbits of the other family. The traces of both families are bundles of parallel lines in the picture plane. These so called "Clifford parallels" have many of the properties of families of parallel lines in Euclidean space, and since the surfaces are covered by two such families they are aptly denoted "Clifford *planes*".

The spheres of the second kind of image space (those of the first kind are right circular cylinders with isotropic generators) are the surfaces

$$z(x, y) = z_0 + \tfrac{1}{2}k((x - x_0)^2 + (y - y_0)^2).$$

These play an important role in differential geometry, in particular the unit sphere $z(x, y) = (x^2 + y^2)/2$ will be used to define the isotropic spherical image of surfaces. The spheres can be moved in themselves through isotropic rotations.

The geometry of image space may be expected to be reflected in the conventional image operations used by professionals in various fields (think of photographers [1], TV engineers, computer scientists in the field of image processing or presentation [4]), and so it is. There has arisen a praxis that considers certain transformations of images as of a mere "cosmetic" nature, that is to say, such transformations are supposed to leave the "image" (in some transcendent sense) invariant. Examples abound: If you take a roll of photographic negatives for printing to various laboratories, you will receive literally different but "equivalent" prints in return. People watching the same program on different TV sets see literally different but "equivalent" shows. Computer graphics presented on various CRTs, printouts, beamer images, and so forth are not identical, but merely "equivalent". Many of such transformations have received conventional names and are easily identified as congruences or similarities of image space:

- "gamma transformations" are similarities of the second kind;

- "additive planes" or "gradients" are rotations in normal planes;

- "brightening" and "darkening" are full isotropic shifts.

Many transformations that are considered to change the image without introducing artifacts ("dodging" and "burning", "negative printing") for instance can be identified as certain inversions in planes and spheres. Even several rather drastic changes that *do* change the image can often be so interpreted. For instance, "sandwich printing" corresponds to a conformal transformation involving an arbitrary function (another negative "sandwiched" with the fiducial one). On the other hand, the transformations generally frowned upon (as "not artistic", etc.) such as "flashing" turn out not to be among the similarities, movements, Möbius transformations or conformal mappings. Apparently the present formalism captures the actual praxis quite well.

36.5   Differential geometry of images

36.5.1   The computation of derivative images

It is evident that real images cannot be differentiated. For one thing they are discrete entities. But even in the ideal case there is no guarantee of differentiability. Moreover, real images are the result of observation and are perturbed by various kinds of noise. Thus the notion of differentiation only makes sense in the context of a scale of observation. We use the Schwartz theory of tempered distributions in the form of the "scale space" representation that is conventionally used in image processing. This allows one to compute differential invariants at any given scale coarser than the pixel scale exactly. It is in this way that the differential geometry of surfaces in simply isotropic space can be used as a framework for image processing that is implementable without restriction whatsoever. In a sense it turns image processing from an art into a science. There exists ample literature on the topic of scale space [2].

36.5.2   "Images" as invariants

A "shape" is an invariant under similarities or movements. Thus an "image" in the sense of an independent geometrical structure is an invariant under the similarities and movements of image space. The formal study of such entities is the differential geometry of surfaces in image space.

Here we meet with the initial problem that in image space all normals are parallel, as for a plane in Euclidean space. (Indeed the intrinsic curvature of any surface is zero since I can map any surface *isometrically* on the picture plane!) Thus the classical spherical image (introduced by Gauss [5]) is useless. However, as I will show below, the Hessian of the intensity can naturally be identified as the isotropic Weingarten map. It captures the shape (the changes modulo movements) exactly as in the classical (Euclidean) differential geometry of surfaces. Thus the spectral analysis of the Hessian of the log–intensity yields the isotropic curvature theory of surfaces.

36.5.3   Curvature

The Gaussian and the mean curvatures are the determinant and the trace of the Jacobean of the "surface attitude image", that is the map $\{x, y\} \mapsto \{z_x, z_y\}$ that maps points on the surface $\{x, y, z(x, y)\}$ to the attitude image (often called "gradient space"). The attitude image is the equivalent of the spherical image (or Gauss map) in the Euclidean case. Because all normals are equal in image space, they cannot be used to construct the equivalent of the classical spherical image. We can map points on the image surface to points on the unit sphere $z(x, y) = (x^2 + y^2)/2$ by parallel tangent planes though. When we do a stereographical projection of the unit sphere on its tangent plane at $x = y = 0$ we obtain the attitude image (or gradient space) referred to above. As in Euclidean geometry the stereographical map is conformal, however, in the case of image space it is even an isometry! For you have

$$\{x, y, z(x, y) = \tfrac{1}{2}(x^2 + y^2)\} \mapsto \{\frac{\partial z(x, y)}{\partial x}, \frac{\partial z(x, y)}{\partial y}\} = \{x, y\},$$

thus the trace of the unit sphere (which is the stereographical projection of it from the "north pole" which is the point $\{0, 0, \infty\}$) is indeed equal to gradient space. Hence the stereographical projection is a true image of the unit sphere and more convenient. The attitude image of a surface suffers (Euclidean) movements when the surface is subjected to movements in image space and Euclidean similarities if the surface is subjected to similarities of the second kind. Thus the Jacobean of the attitude map can be identified as the isotropic Weingarten map.

Since the zeroth and first-order terms in a Taylor expansion of the image at any of its points can be removed through the application of a proper movement, the second order is the lowest of interest. It describes the curvature at the image point. We have

$$z(x, y) = \tfrac{1}{2}(z_{xx}x^2 + 2z_{xy}xy + z_{yy}y^2).$$

The invariants $K = z_{xx}z_{yy} - z_{xy}^2$ and $H = (z_{xx} + z_{yy})/2$, the Gaussian and the mean curvature respectively, have a similar interpretation as they have in the classical (Euclidean) theory of surfaces.

Notice that

$$z(\cos \varphi, \sin \varphi) = \tfrac{1}{2}(\tfrac{1}{2}(z_{xx} + z_{yy}) + \tfrac{1}{2}(z_{xx} - z_{yy}) \cos 2\varphi + z_{xy} \sin 2\varphi).$$

We see that $\tfrac{1}{2}(z_{xx} + z_{yy})$ does not depend upon the direction (it is the mean curvature) whereas the terms containing z_{xy} and $\tfrac{1}{2}(z_{xx} - z_{yy})$ act in unison, leading to the invariant $z_{xy}^2 + \tfrac{1}{4}(z_{xx} - z_{yy})^2$ (which can be written as $H^2 - K$), whereas the principal direction is given by $\tfrac{1}{2} \arctan 2z_{xy}/(z_{xx} - z_{yy})$.

After suitable rotation in the image plane, $z(x, y)$ may be written as

$$z(u, v) = \tfrac{1}{2}(z_{uu}u^2 + z_{vv}v^2) = \tfrac{1}{2}(\kappa_1 u^2 + \kappa_2 v^2),$$

with $\kappa_1 \geq \kappa_2$ (in two ways, differing by a rotation over π). The "principal curvatures" $\kappa_{1,2}$ are simply the second-order directional derivatives in the direction of the largest principal curvature (henceforth called "principal direction") and orthogonal to it. A similarity of the second kind merely scales the principal curvatures and may be said to leave the "shape" invariant. We see that the shape is fully characterized by a parameter such as ϑ and the "size" by λ :

$$\tan \vartheta = \frac{\kappa_1 + \kappa_2}{\kappa_1 - \kappa_2}, \qquad \lambda = \sqrt{\kappa_1^2 + \kappa_2^2},$$

where we take λ to be nonnegative. The parameter ϑ takes values in the interval $[-\tfrac{\pi}{2}, \tfrac{\pi}{2}]$; it is undefined for "flat points" ($\kappa_1 = \kappa_2 = 0$) which have no shape. The parameters λ and ϑ may be called "curvedness" and "shape index" (the conventional "shape index" $s = \tfrac{2}{\pi}\vartheta$ takes values on $[-1, +1]$). They are more immediately descriptive of "shape" than the usual differential invariants $K = \kappa_1\kappa_2$ (the "Gaussian curvature") and $2H = \kappa_1 + \kappa_2$ (the "mean curvature"). The curvedness can be interpreted as the isotropic strain energy density (as in the theory of thin plate splines).

The above considerations enable us to obtain a notion regarding the relative abundances of the various shapes. I will give an example in the next subsection as a simple application of the formalism. Although I have heard remarks on (empirical observations of) the relative abundances of shapes, I have never seen a formal treatment attempted.

36.5.4   The relative abundances of shape types and curvedness of an isotropic random Gaussian surface

Consider an isotropic, random Gaussian image $\{x, y, z(x, y)\}$ where Z is normally distributed with some power spectrum $E(k)$. For any isotropic surface the parameters

$$p = \tfrac{1}{2}(z_{xx} + z_{yy}), \quad q = \tfrac{1}{2}(z_{xx} - z_{yy}) \quad \text{and} \quad r = z_{xy}$$

are statistically independent. This is the case because, firstly,

$$\langle pq \rangle = \langle z_{xx}^2 \rangle - \langle z_{yy}^2 \rangle = 0$$

by virtue of isotropy and secondly,

$$\langle qr \rangle = \langle z_{xx} z_{xy} \rangle - \langle z_{yy} z_{xy} \rangle = 0$$

(again, by virtue of isotropy). The same reasoning applies to $\langle pr \rangle$.

For the Gaussian surface we have [11]

$$\langle p^2 \rangle = \tfrac{1}{2} M_4, \quad \langle q^2 \rangle = \langle r^2 \rangle = \tfrac{1}{8} M_4,$$

where the "circular moments" are defined as

$$M_n = 2\pi \int_0^\infty k^n E(k) \, k \, dk.$$

The result will depend somewhat on the shape of the power spectrum $E(\mathbf{k})$ (or—equivalently—the autocorrelation function). Here I take an autocorrelation function of Gaussian shape, say $R(\rho) = \exp(-\rho^2/(2\sigma^2))$. Then we have $M_4 = 16\pi/\sigma^4$. The corresponding normal distribution in parameter space is

$$P(p, q, r) \, dp \, dq \, dr = \frac{\sigma^6}{8\pi^3 \sqrt{2}} e^{-\frac{\sigma^4}{4\pi}(\frac{p^2}{2} + q^2 + r^2)} \, dp \, dq \, dr,$$

thus the isodensity surfaces in $\{p, q, r\}$ space are prolate ellipsoids of revolution with the p-axis as major axis. Notice that

$$p^2 + q^2 + r^2 = \tfrac{1}{2}(z_{xx}^2 + 2z_{xy}^2 + z_{yy}^2) = \tfrac{1}{2}\lambda^2,$$

the "total variation" (an invariant because it equals $2H^2 - K$). Thus the loci of constant curvedness in $\{p, q, r\}$ space are spheres concentric with the origin. The loci of constant shape index are right circular cones about the p-axis. Thus we obtain a system of spherical coordinates of $\{p, q, r\}$ space such that the curvedness is the radius, the shape index the polar distance, and (twice) the orientation of the principal direction the azimuth. (By the way, this is by far the most convenient representation of the second-order structure, it is much less known than it deserves.) This makes it very simple to find the joint histogram of shape index and curvedness values:

$$h(\lambda, \vartheta)\, d\vartheta\, d\lambda = \frac{\lambda^2 \sigma^6 \cos \vartheta}{8\pi^3 \sqrt{2}} e^{-\frac{\sigma^4 \lambda^2}{16\pi}(3+\cos 2\vartheta)}\, d\vartheta\, d\lambda.$$

Integrated over the full curvedness range we have

$$h(\vartheta)\, d\vartheta = \frac{2\sqrt{2} \cos \vartheta}{(3 + \cos 2\vartheta)^{3/2}},$$

and integrated over the full shape index range (erfi(z) denotes the imaginary error function erf(iz))

$$h(\lambda)\, d\lambda = \frac{\lambda \sigma^4}{2\pi} \mathrm{erfi}\left(\frac{\lambda \sigma^2}{2\sqrt{2\pi}}\right) \exp^{-\frac{\lambda^2 \sigma^4}{4\pi}}.$$

The histogram (see Figure 36.7) is zero at the extremes of the shape index scale. The extremes are the umbilics and apparently these are rare. Indeed, since the umbilics are represented by points (the poles), rather than areas, on a sphere of constant curvedness, it is evident that they only occur at isolated points. The histogram is unimodal when $\lambda \sigma^2 < 2\sqrt{\pi}$, then the maximum is arrived at for the isotropic shapes (zero mean curvature). Otherwise the histogram is bimodal, the peaks are located at $\pm \arccos 2\sqrt{\pi}/(\lambda \sigma^2)$. For realistic parameter values the peaks of the histogram lie close to the parabolic shapes. This is indeed what one observes for the overwhelming majority of natural texture in images (see Figures 36.8 and 36.9). For the histogram averaged over the full curvedness range, the extrema are exactly at the parabolic shapes.

36.6   Conclusions

The formalism presented here appears to describe the conventional praxis as exercised by professionals in the image presentation business. It has the virtue of being consistent from the perspective of physics (the image as an array of elementary observations of the radiance). Moreover, it yields a computational framework that allows one to implement differential geometry without any compromise. This indeed raises image processing from the state of a mere art into that of a science. Many of the conventional methods of image processing can easily be interpreted in geometrical terms, and—where necessary—corrected. There are two types of flaws in many conventional methods: One involves the use of ad hoc "hacks", the other the unwarranted use of Euclidean invariants. The latter can be corrected right away by substituting invariants under the correct group. The former can often be interpreted in geometrical terms and given a rational basis, then a principled method can be substituted for the ad hoc one. To do this is quite a program; in this paper I have merely touched upon the basics.

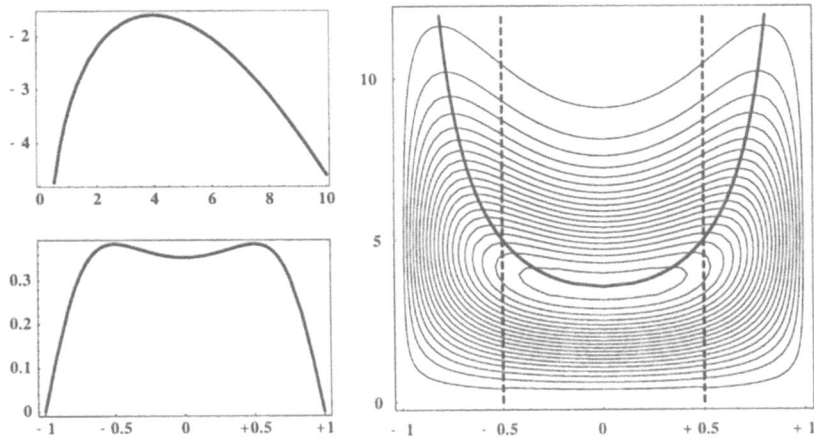

FIGURE 36.7. Prediction of the joint curvature and shape index histogram (right). The parabolic shapes have been indicated with dashed vertical lines. The position of the modes of the distribution is drawn as a thick line, the thin (closed) curves are loci of equal probability density. On the left the predicted histograms for the log-curvedness (top-left) and shape index (bottom-left).

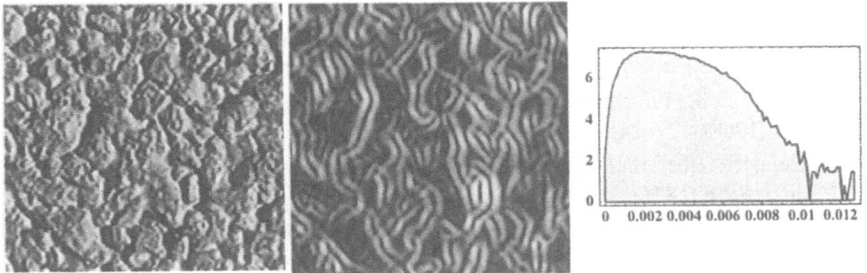

FIGURE 36.8. An image (left) and the distribution of curvedness (center). On the right the histogram of log-curvedness. This is an image of rough plaster, lit from the side. It is by no means a Gaussian random field, yet the prediction applies (at least qualitatively) quite well.

REFERENCES

[1] Adams, A.: *The Print: Contact Printing and Enlarging*. Basic Photo 3. Morgan and Lester, New York, 1950.

[2] Florack, L.: *Image Structure*. Kluwer, Dordrecht, 1997.

[3] Clifford, W. K., Preliminary sketch of biquaternions. *Proc. Lond. Math. Soc.* (1873), 381–395.

[4] Foley, J. D., Dam. A. van, Feiner, S. K. and Hughes, J. F.: *Computer Graphics, Principles and Practice*. Addison-Wesley Publishing Company, Reading, Massachusetts, 1990.

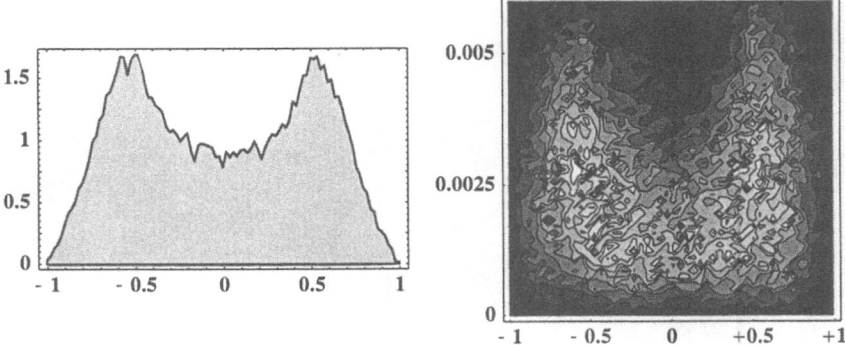

FIGURE 36.9. At the left the histogram of shape indices for the rough plaster image. On the right the joint histogram of curvedness and shape index.

[5] Gauss C. F.: *Algemeine Flächentheorie.* (German translation of the *Disquisitiones generales circa Superficies Curvas*), Hrsg. A. Wangering, Ostwalds Klassiker der exakten Wissenschaften **5**. Engelmann, Leipzig, 1889 (orig. 1827)

[6] Jaglom, I. M.: *Complex Numbers in Geometry.* Transl. E. J. F. Primrose, Academic Paperbacks, New York, 1968.

[7] Jaglom, I. M.: *A Simple Non-Euclidean Geometry and Its Physical Basis: An Elementary Account of Galilean Geometry and the Galilean Principle of Relativity.* Transl. A. Shenitzer, ed. ass. B. Gordon. Springer, New York, 1979.

[8] Jaynes, E. T.: Prior probabilities. *IEEE Trans. on System Science and Cybernetics* **4**(3) (1968), 227–241.

[9] Klein, F.: Über die sogenannte nichteuclidische Geometrie. *Mathematische Annalen Bd.* **6** (1871), 112–145.

[10] Klein, F.: *Vergleichende Betrachtingen über neue geometrische Forchungen* (The "Erlangen Program"). Program zu Eintritt in die philosophische Fakultät und den Senat der Universität zu Erlangen. Deichert, Erlangen, 1872.

[11] Longuet-Higgins, M. S., (1956), *Phil. Trans. R. Soc.* A 249, 321–64.

[12] Pottmann, H., Opitz, K.: Curvature analysis and visualization for functions defined on Euclidean spaces or surfaces. Computer Aided Geometric Design **11** (1994), 655–674.

[13] Sachs, H.: *Ebene Isotrope Geometrie.* Vieweg, Braunscheig/Wiesbaden, 1987.

[14] Sachs, H.: *Isotrope Geometrie des Raumes.* Vieweg, Braunscheig/Wiesbaden, 1990.

[15] Scheffer, G.: Verallgemeinerung der Grundlagen der gewöhnlichen komplexen Funktionen. Sitz. ber. Sächs. Ges. Wiss., *Math.-phys. Klasse*, Bnd. **45** (1893), 828–842.

[16] Strubecker, K.: Differentialgeometrie des isotropen Raumes I. *Sitzungsberichte der Akademie der Wissenschaften Wien* **150** (1941), 1–43.

[17] Strubecker, K.: Differentialgeometrie des isotropen Raumes II. *Math. Z.* **47** (1942), 743–777.

[18] Strubecker, K.: Differentialgeometrie des isotropen Raumes III. *Math. Z.* **48** (1943), 369–427.

[19] Strubecker, K.: Differentialgeometrie des isotropen Raumes IV. *Math. Z.* **50** (1945), 1–92.

Jan J. Koenderink
Helmholtz Instituut
Universiteit Utrecht
Princetonplein 5, 3584CC Utrecht, The Netherlands
E-mail: j.j.koenderink@phys.uu.nl

Submitted: August 9, 2002.

37

Pose Estimation of Cycloidal Curves by using Twist Representations

Bodo Rosenhahn and Gerald Sommer

ABSTRACT This work concerns the 2D-3D pose estimation problem of cycloidal curves. Pose estimation means to estimate the relative position and orientation of a 3D object to a reference camera system. The 3D object features are in this work cycloidal curves, as extensions to classical 3D point or 3D line concepts. This means, we assume knowledge of a 3D cycloidal curve and observe it in an image of a calibrated camera. The aim is to estimate the rotation $\mathbf{R}$ and translation $\mathbf{t}$ to get a best fit of the transformed 3D object model to the observed 2D image data. Furthermore, other concepts such as 3D cycloidal surfaces and the numerical problems of estimating the pose parameters are discussed.

Keywords: 2D-3D pose estimation, cycloidal curves, twists

37.1   Introduction

In this paper we study the 2D-3D pose estimation problem of 3D cycloidal curves. Cycloidal curves are a special case of algebraic curves. In general a cycloidal curve is generated by a circle rolling on a circle or line without slipping [11]. This leads to epitrochoids, hypotrochoids and trochoids as special classes of entities, we will use in the context of 2D-3D pose estimation.

37.1.1   The pose estimation problem

Pose estimation itself is a basic visual task [6] and several approaches for monocular pose estimation exist that relate the position of a 3D object to a reference camera coordinate system [6, 9, 19]. Our preliminary work also concerned the 2D-3D pose estimation problem of rigid objects and kinematic chains [16]. Instead of using invariance as an explicit formulation of geometry, as often has been done in projective geometry, we are using implicit formulations and use constraints to describe the pose estimation problem. The formulas in [15] result in compact constraint equations for pose estimation of rigid objects containing different entities (points, lines, planes). The entities we are now interested in are cycloidal curves.

AMS Subject Classification: 65D17, 68U10.

This means curves we can generate by one, two or more coupled twists. Examples are 3D circles, ellipses, cardioids, cycloids, archimedic spirals, spheres, quadrics, etc. In this paper we will use conformal geometric algebra (CGA) [10] to describe schemes for cycloidal curves and their coupling with the 2D-3D pose estimation problem.

37.1.2   Preliminary work

Our recent work [15, 16] is summarized in Figure 37.1: There we assume as object features 3D points, 3D lines, 3D spheres, 3D circles or kinematic chain segments of a reference model. Further, we assume corresponding extracted features in an image of a calibrated camera. The aim is to find the rotation $\mathbf{R}$ and translation $\mathbf{t}$ of the object, which leads to the best fit of the reference model with the actual extracted entities. To relate 2D image information to 3D entities we reconstruct an extracted image entity to an entity with one dimension higher gained by back-projection in the space. This idea will be used to formulate the problem as a pure kinematic problem.

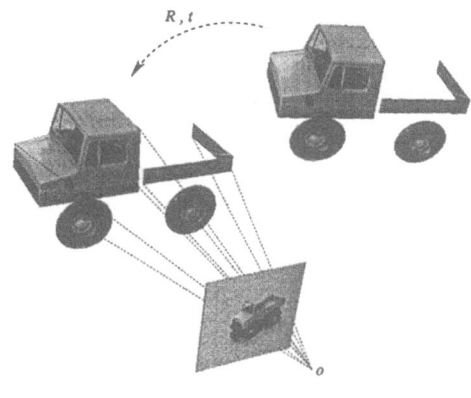

FIGURE 37.1. The scenario. The assumptions are the projective camera model, the model of the object (consisting of points, lines, circles and kinematic chains) and corresponding extracted entities on the image plane. The dashed line describes the pose of the model, which leads to the best fit of the object with the actual extracted entities.

During the last few years we studied several algebras (e.g., the algebras for pure projective [14] or pure kinematic geometry [1]), but the conformal geometric algebra is the most general one that describes the problems involved in pose estimation. The mathematical spaces involved in the pose estimation problem are the projective plane, projective space, kinematic space and Euclidean space. Indeed all geometric algebras for modeling the spaces involved in the pose estimation problem are represented as sub-algebras of the conformal geometric algebra. So we are able to formalize all aspects of the pose estimation problem in one framework.

The motivation for this work is to generalize the already studied entities for pose estimation to more *free-form*-like contours and surfaces. Though points, lines, planes, circles and spheres cover a large range to model objects, we are also interested in, e.g., ellipses to model the shape of eyes etc. This motivates a search for a more general class of entities, which can be used in the context of pose estimation. We now give a very brief summary of algebraic curves and their characterizations collected by Lee in [11]. More detailed information about algebraic curves can also be found in [4]. In our work we will concentrate on a

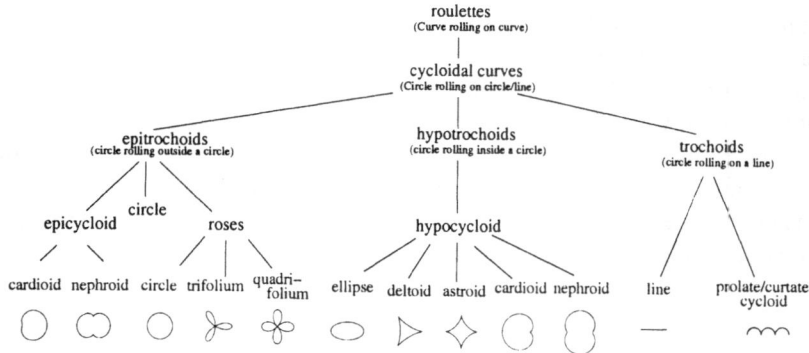

FIGURE 37.2. Tree of algebraic curves.

subclass of *roulettes*, the *cycloidal curves*, which are circles rolling on circles or lines. Figure 37.2 shows a subtree of the family of algebraic curves. Cycloidal curves can be classified as epitrochoids, hypotrochoids and trochoids, which split into further subclasses. Figure 37.2 also shows examples of these curves.

These curves are mostly defined in the 2D plane. For our scenario of pose estimation, we will extend these curves to plane curves in the 3D space.

37.2   Cycloidal curves in conformal geometric algebra

In this section we will introduce the conformal geometric algebra with respect to the basic notation and the description of cycloidal curves in the language of parameterized twist transformations.

37.2.1   Introduction to conformal geometric algebra

In this section we introduce the main properties of the conformal geometric algebra [10]. The aim is to clarify the notation, a more detailed introduction concerning geometric algebras can be found in [18].

A geometric algebra $\mathcal{G}_{p,q,r}$ is built from a vector space $\mathbb{R}^n$, endowed with the signature (p, q, r), $n = p + q + r$, by application of a geometric product. The product defining a geometric algebra is called *geometric product* and is denoted by juxtaposition, e.g., uv for two multivectors u and v. The geometric product consists of an outer ($\wedge$) and an inner ($\cdot$) product, whose roles are to increase or to decrease the order of the algebraic entities, respectively. For later use we introduce the commutator $\times$ and anticommutator $\overline{\times}$ products, respectively for any two multivectors,

$$AB = \tfrac{1}{2}(AB + BA) + \tfrac{1}{2}(AB - BA) =: A\overline{\times}B + A\underline{\times}B.$$

To introduce the conformal geometric algebra (CGA) we follow [10] and start with the *Minkowski plane* $\mathbb{R}^{1,1}$, which has an orthonormal basis $\{e_+, e_-\}$, defined by the properties

$$e_+^2 = 1, \quad e_-^2 = -1, \quad \text{and} \quad e_+ \cdot e_- = 0.$$

A *null basis* can now be introduced by the vectors

$$e_0 := \tfrac{1}{2}(e_- - e_+) \quad \text{and} \quad e := e_- + e_+$$

with $e_0^2 = e^2 = 0$. The vector e_0 can be interpreted as the origin and the vector e as a point at infinity. Furthermore we define $E := e \wedge e_0 = e_+ \wedge e_-$.

In the case of working in an n-dimensional vector space $\mathbb{R}^n$ we couple an additional vector space $\mathbb{R}^{1,1}$, which defines a null space to gain $\mathbb{R}^n \oplus \mathbb{R}^{1,1} = \mathbb{R}^{n+1,1}$. From that vector space we can derive the conformal geometric algebra (CGA) $\mathcal{G}_{n+1,1}$ as a linear space of dimension 2^{n+2}.

The algebras $\mathcal{G}_{3,1}$ and $\mathcal{G}_{3,0}$ are suited to represent the projective and Euclidean space, respectively [8]. Since

$$\mathcal{G}_{4,1} \supseteq \mathcal{G}_{3,1} \supseteq \mathcal{G}_{3,0},$$

both algebras for the projective and Euclidean space constitute subspaces of the linear space of the CGA. It is possible to use operators to relate the different algebras and to guarantee the mapping between the algebraic properties, see [17].

The basis entities of the 3D conformal space are spheres $\underline{s}$, containing the center p and the radius ρ, $\underline{s} = p + \tfrac{1}{2}(p^2 - \rho^2)e + e_0$. Thus, a sphere is a 1-blade. The dual form for a sphere is $\underline{s}^*$. The advantage of the dual form is that $\underline{s}^*$ can be calculated directly from points on the sphere: For four points on the sphere, $\underline{s}^*$ can be written as $\underline{s}^* = \underline{a} \wedge \underline{b} \wedge \underline{c} \wedge \underline{d}$ and is therefore a 4-blade.

A point $\underline{x} = x + \tfrac{1}{2}x^2 e + e_0$ is nothing more than a degenerate sphere with radius $\rho = 0$, which can easily be seen from the representation of a sphere. A point $\underline{x}$ lies on a sphere $\underline{s}$ iff $\underline{x} \wedge \underline{s}^* = 0$, or $\underline{x} \cdot \underline{s} = 0$.

Circles $\underline{z}$ can be described by the intersection of two spheres. The dual representation of circles leads to the outer product of three points on the circle, $\underline{z}^* = \underline{a} \wedge \underline{b} \wedge \underline{c}$.

From the dual form of circles and spheres, affine points $\underline{X}^*$, lines $\underline{L}^*$ and planes $\underline{P}^*$ can easily be represented as degenerate cases, just by involving the point at infinity,

$$\underline{X}^* = e \wedge \underline{x}, \qquad \underline{L}^* = e \wedge \underline{a} \wedge \underline{b}, \qquad \underline{P}^* = e \wedge \underline{a} \wedge \underline{b} \wedge \underline{c}.$$

Note that since we will only work with the entities in their dual representation in the next sections, we neglect the $\star$-sign in the further formulas.

37.2.2   Rigid motions and twists in CGA

As in $\mathcal{G}_{3,0}$, rotations in $\mathcal{G}_{4,1}$ are represented by rotors $\mathbf{R} = \exp\left(-\frac{\theta}{2}l\right)$. The components of the rotor $\mathbf{R}$ are the unit bivector l, which represents the dual of the rotation axis and the angle θ, which represents the amount of the rotation. If we want to translate an entity with respect to a translation vector $\mathbf{t} \in \mathcal{G}_{3,0}$, we can use a so-called *translator*, $\mathbf{T} = (1 + \frac{\mathbf{et}}{2}) = \exp\left(\frac{\mathbf{et}}{2}\right)$, which is a special rotor. Rotations and translations can be estimated by applying rotors and translators as *versor products*, e.g., $\underline{\mathbf{X}}' = \mathbf{R}\underline{\mathbf{X}}\widetilde{\mathbf{R}}$, or $\underline{\mathbf{X}}'' = \mathbf{T}\underline{\mathbf{X}}\widetilde{\mathbf{T}}$. Note, that we use the $\sim$-sign on a multivector to denote its reverse. To express a rigid body motion, we can apply multiplied rotors and translators consecutively. We denote such an operator as a motor M [1]. The rigid body motion of, e.g., a point $\underline{\mathbf{X}}$ can be written as $\underline{\mathbf{X}}' = M\underline{\mathbf{X}}\widetilde{M}$, see also [7].

Following e.g., [12], a rigid body motion of points can be expressed by a rotation around a line in space followed by a translation along this line. This results from the fact, that for every $g \in SE(3)$ there exists a $\xi \in se(3)$ and a $\theta \in \mathbb{R}$ such that $g = \exp(\xi\theta)$. Such transformations are also called *twist* transformations. The Lie algebra element $\xi \in se(3)$ is a twist, and its Lie group element, the exponential $g = \exp(\xi\theta) \in SE(3)$ describes a rigid body motion [5]. A motor describing a twist transformation can be written as

$$M = \exp\left(-\tfrac{1}{2}\theta\left(l + \mathbf{e}m\right)\right) = \exp\left(-\tfrac{1}{2}\theta\Psi\right).$$

A twist can be seen as an infinitesimal version of a screw motion and describes a line in space with an angle θ and a *pitch* h, the ratio of translation to rotation. If the pitch h is zero, the resulting motion is a rotation of an entity (e.g., a point $\underline{\mathbf{X}}$) around a line $\underline{\mathbf{L}}^\star$ in the space. To gain a twist representation, the general idea is to translate both, the entity and the line to the origin, to perform a rotation and to translate back the transformed entity.

The motor M can be interpreted as the exponential of a twist, with the form

$$M = \mathbf{T}\mathbf{R}\widetilde{\mathbf{T}} = \exp\left(-\tfrac{1}{2}\theta\left(l + \mathbf{e}(\mathbf{t} \cdot l)\right)\right).$$

The motion of a point can then be decomposed as

$$\underline{\mathbf{X}}' = M\underline{\mathbf{X}}\widetilde{M} = (\mathbf{T}\mathbf{R}\widetilde{\mathbf{T}})\underline{\mathbf{X}}(\mathbf{T}\widetilde{\mathbf{R}}\widetilde{\mathbf{T}}).$$

We call such a transformation a *general rotation*. Whereas in Euclidean geometry, Lie algebras and Lie groups are only applied to point concepts, motors can also be applied to other entities, like lines, planes, circles, spheres, etc.

37.2.3   Cycloidal curves in conformal geometric algebra

As previously explained, cycloidal curves are circles rolling on circles or lines. In this section we will explain how to generate such curves in conformal geometric algebra. E.g., ellipses are not entities which can be directly described in conformal

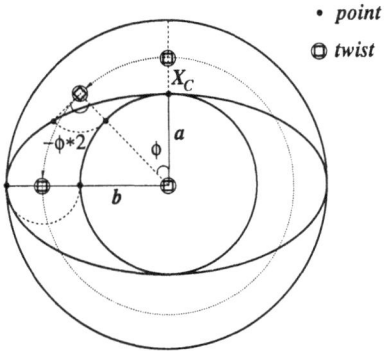

FIGURE 37.3. An ellipse generated by two coupled twists.

geometric algebra. The idea for modeling an ellipse is visualized in Figure 37.3: We assume two parallel twists (modeling general rotations) in the 3D space and a 3D point on the ellipse, and we transform the point around the two twists in a fixed and dependent manner. In this case, we use two coupled parallel (not collinear) twists, rotate the point by -2ϕ around the first twist and by ϕ around the second one. The set of all points for $\phi \in [0, \ldots, 2\pi]$ generates an ellipse as the orbit of the coupled group actions. In general, every cycloidal curve is generated by a

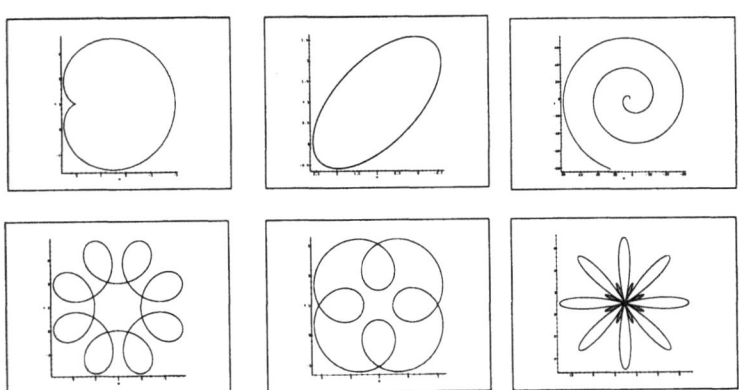

FIGURE 37.4. 3D 2-twist generated curves.

set of twists ξ_i with frequencies λ_i acting on one point $\underline{X}$ on the curve. Since twists can be used to describe general rotations in the 2D plane or 3D space, we call the generated curves nD-$m$twist curves. With nD-$m$twist curves we mean n dimensional curves, generated by m twists with $n, m \in \mathbf{N}$. In the context of the 2D-3D pose estimation problem we use the cycloidal curves as 3D object entities. So we mean 3D-$m$twist curves, if we speak of just $m$twist curves.

We will start with very simple curves, and the simplest one consists of one point (a point on the curve) and one twist. Rotating the point around the twist leads to the parameterized generation of a circle: The corresponding twist transformation can be expressed as a suitable motor M_ϕ and an arbitrary 3D point, $\underline{X}_Z$, on the circle. The 3D orbit of all points on the circle is simply given by

$$\underline{X}_Z^\phi = M_\phi \underline{X}_Z \widetilde{M}_\phi, \qquad \phi \in [0, \ldots, 2\pi].$$

We call a circle also a *1twist* generated curve since it is generated by one twist.

Now we can continue and wrap a second twist around the first one. If we make the amount of rotation of each twist dependent on each other, we gain a 3D curve in general. This curve is firstly dependent on the relative positions and orientation of the twists with respect to each other, the (starting) point on the curve and the ratio of angular frequencies.

The general form of a *2-twist* generated curve is

$$\underline{X}_C^\phi = M_{\lambda_2\phi}^2 M_{\lambda_1\phi}^1 \underline{X}_C \widetilde{M}_{\lambda_1\phi}^1 \widetilde{M}_{\lambda_2\phi}^2, \quad \lambda_1, \lambda_2 \in \mathbb{R}, \quad \phi \in [\alpha_1, \ldots, \alpha_2].$$

The motors M^i are the exponentials of twists, the scalars $\lambda_i \in \mathbb{R}$ determine the ratio of angular frequencies between the twists and $\underline{X}_C$ is a point on the curve. The values α_i define an interval for the boundaries of the curve. For closed curves they are usually $\alpha_1 = 0$ and $\alpha_1 = 2\pi$, but indeed it is also possible to define curve segments. The case $\lambda_1 = \lambda_2 = 1$ leads to cardioids, $\lambda_1 = 2, \lambda_2 = 1$ leads

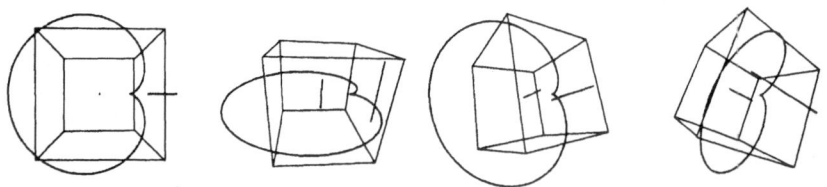

FIGURE 37.5. Perspective views of a 3D 2-twist generated curve. The 2-twist curve and the twists axes are visualized.

to nephroids and $\lambda_1 = 3, \lambda_2 = 1$ leads to deltoids, which can be transformed (by moving the second twist) to a trifolium, etc. Ellipses can easily be described by $\lambda_1 = -2, \lambda_2 = 1$.

Figure 37.4 shows further examples from curves, which can be very easily generated by two coupled twists. Note: Also the archimedic spiral is a 2-twist generated curve. To gain an archimedic spiral, one twist has to be a translator. Since an archimedic spiral is no algebraic curve, the 2-twist generated curves are more general than the cycloidal curves.

All these curves are given in a 3D space. In Figure 37.4 only projections are shown. Figure 37.5 shows a 3D 2-twist generated curve of different projective views.

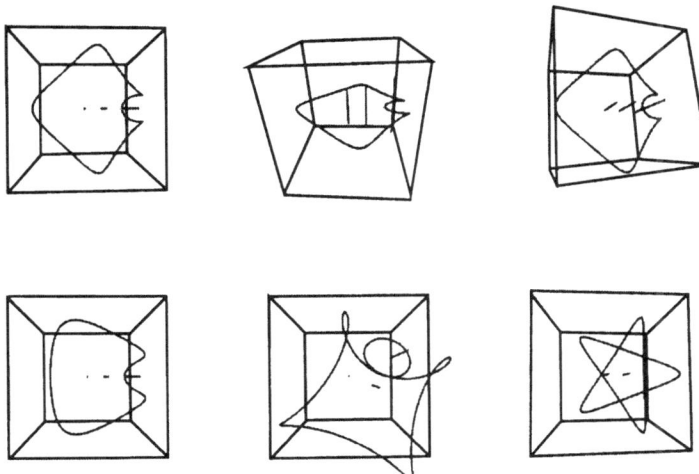

FIGURE 37.6. 3D 3-twist generated curves. The first row shows three different perspective views of a 3-twist generated curve. The other row shows other 3-twist generated curves, generated with different frequencies and twist positions. The three twists are also visualized.

It is also possible to generate 3-twist curves. For this, a point is transformed with respect to three twists. Examples of planar curves are shown in Figure 37.6.

So far, we have only formalized 3D curves. Surfaces can be modeled by rotating, e.g., the 2-twist generated curves around a third twist with a second independent angle ϕ_2, leading to 3-twist generated surfaces. Examples are shown in Figure 37.7. Note: If there is only one variable angle ϕ, the resulting entity is always a 3D curve in the 3D space, whereas the case of two variable angles ϕ_1 and ϕ_2 leads to a 3D surface in the 3D space. The case of three variable angles leads to volumes as highest structures in the 3D space.

The general form of 3-twist generated surfaces is

$$\underline{X}^{\phi_1,\phi_2} = M^3_{\lambda_3\phi_2} M^2_{\lambda_2\phi_1} M^1_{\lambda_1\phi_1} \underline{X} \widetilde{M}^1_{\lambda_1\phi_1} \widetilde{M}^2_{\lambda_2\phi_1} \widetilde{M}^3_{\lambda_3\phi_2},$$

with $\lambda_i \in \mathbb{R}$ and $\phi_1, \phi_2 \in [\alpha_{i_1}, \ldots, \alpha_{i_2}]$. An ellipsoid, for example, is nothing more than a (special) rotated ellipse ($\lambda_3 = \lambda_2 = 1, \lambda_1 = -2$). Its parameterized equation can be written as

$$\underline{X}_Q^{\phi_1,\phi_2} = M^3_{\phi_2} M^2_{\phi_1} M^1_{-2\phi_1} \underline{X}_Q \widetilde{M}^1_{-2\phi_1} \widetilde{M}^2_{\phi_1} \widetilde{M}^3_{\phi_2}, \quad \phi_1, \phi_2 \in [0, \ldots, 2\pi].$$

This is visualized in the first image of Figure 37.7. The second image of Figure 37.7 shows a horizontally rotated cardioid.

The third and fourth images in the second row show rotated hypocycloids. The last row shows rotated spirals, leading to surfaces, comparable to a flower. These surfaces are very easy to generate and can be represented by just a few coupled

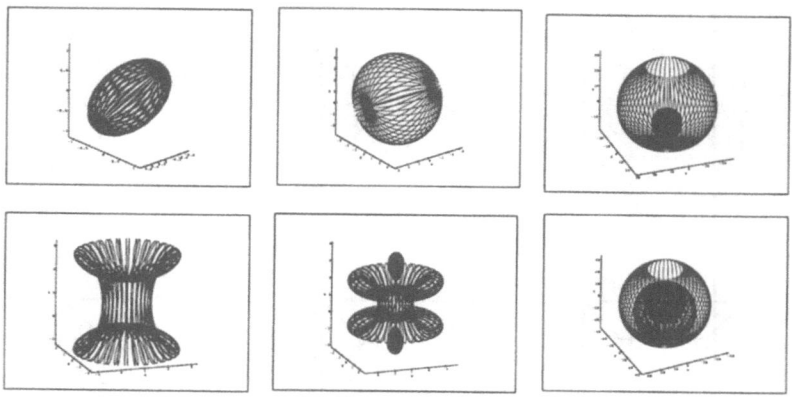

FIGURE 37.7. 3D 3-twist generated surfaces

twists. In Figure 37.7 no grid-plot is shown. Instead the rotated cycloidal curves are shown to visualize the geometric generation of the surface. Lines and planes also fall into the definition of 2-twist and 3-twist generated curves and surfaces. Examples for well-known entities, interpreted as cycloidal curves or surfaces, are points, lines, circles, ellipses, line segments, spirals, spheres, ellipsoids, planes, cylinders, cones, etc.

The rigid body motions of these entities can easily be estimated, just by transforming the generating twists. The transformation of an m twist generated curve can be performed by transforming the m twists (which are just lines in the space), and the point on the curve.

The description of these curves is compact, and rigid transformations can be estimated very quickly. Cycloidal curves and surfaces extend already studied entities to a more general class of entities without losing the advantages of our previous work. Indeed, we build up a hierarchy of entities and further work will concentrate on enlarging these kinds of entities to approximate free-form curves and surfaces. So far the hierarchy of entities consists of the following entities:

points, lines	$\subseteq$	circles	$\subseteq$	cycloidal curves	$\subseteq$	...
planes	$\subseteq$	spheres	$\subseteq$	cycloidal surfaces	$\subseteq$	...

37.3   Pose estimation in conformal geometric algebra

Now we will formalize the 2D-3D pose estimation problem, that is, *a transformed object entity must lie on a projective reconstructed image entity.* Let $\underline{X}$ be an object point and $\underline{L}$ be an object line, given in CGA. The (unknown) transformation of the entities can be written as $M\underline{X}\widetilde{M}$ and $M\underline{L}\widetilde{M}$.

Let x be an image point and l be an image line on a projective plane. The projective reconstruction from an image point in CGA can be written as $\underline{L}_x = \text{e} \wedge O \wedge x$. This leads to a reconstructed projection ray, containing the optical center O of the camera, see Figure 37.1, the image point x and the vector e as the point at infinity. Similarly, $\underline{P}_l = \text{e} \wedge O \wedge l$ leads to a reconstructed projection plane in CGA.

Collinearity and coplanarity can be expressed by the commutator and anticommutator products, respectively. Thus, the constraint equations of pose estimation from image points read

$$
\underbrace{(M \underbrace{X}_{\text{object point}} \widetilde{M})}_{\substack{\text{rigid motion of the object point}}} \underline{\times} \underbrace{\text{e} \wedge (O \wedge x)}_{\substack{\text{projection ray,} \\ \text{reconstructed from the image point}}} = 0,
$$

$$
\underbrace{}_{\substack{\text{collinearity of the transformed object} \\ \text{point with the reconstructed line}}}
$$

Constraint equations to relate 2D image lines to 3D object points, or 2D image lines to 3D object lines, can also be expressed in a similar manner.

The constraint equations implicitly represent a distance measure which has to be zero. Such compact equations subsume the pose estimation problem at hand: find the best motor M which satisfies the constraint.

37.3.1   The pose estimation problem of cycloidal curves

Now we can combine the last two subsections to formalize the pose estimation problem for cycloidal curves: We assume a 3D cycloidal curve, e.g.,

$$
\underline{X}_Z^\phi = M_{\lambda_1\phi}^2 M_{\lambda_2\phi}^1 \underline{X} \widetilde{M}_{\lambda_2\phi}^1 \widetilde{M}_{\lambda_1\phi}^2, \quad \lambda_1, \lambda_2 \in \mathbb{R}, \ \phi \in [0, \ldots, 2\pi].
$$

The rigid motion of this curve incident on a projection ray can be expressed as

$$
\left(M(M_{\lambda_1\phi}^2 M_{\lambda_2\phi}^1 \underline{X} \widetilde{M}_{\lambda_2\phi}^1 \widetilde{M}_{\lambda_1\phi}^2) \widetilde{M} \right) \underline{\times} (\text{e} \wedge (O \wedge x)) = 0.
$$

Since every aspect of the 2D-3D pose estimation problem of cycloidal curves is formalized in CGA, the constraint equation describing the pose problem is compact and easy to interpret: The inner parenthesis contains the parameterized representation of the cycloidal curve with one unknown angle ϕ. The outer parenthesis contains the unknown motor M, describing the rigid body motion of the 3D cycloidal curve. This is the pose we are interested in. The expression is then combined, via the commutator product, with the reconstructed projection ray and has to be zero. This describes the incidence of the transformed curve to a projection ray. The unknowns are the six parameters of the rigid motion M and the angle ϕ for each point correspondence.

37.3.2   The pose estimation problem of cycloidal surfaces

Similar to the previous section, it is possible to formalize constraint equations for incidence of cycloidal surfaces to projection rays,

$$
\left(M(M^3_{\lambda_3\phi_2} M^2_{\lambda_2\phi_1} M^1_{\lambda_1\phi_1} \underline{X} \widetilde{M}^1_{\lambda_1\phi_1} \widetilde{M}^2_{\lambda_2\phi_1} \widetilde{M}^3_{\lambda_3\phi_2}) \widetilde{M} \right)
$$

$$
\underline{\times} \, (\mathrm{e} \wedge (O \wedge x)) = 0.
$$

But this would not cover all geometric aspects of the surface in the pose problem. It is more efficient to build constraints on the surface contour in the image and to model also tangentiality within the constraints. Therefore we use the surface tangential plane $\underline{P_x}$ at each point $\underline{X}$ and claim incidence of the tangential plane $\underline{P_x}$ with each projection ray,

$$
\left(M(M^3_{\lambda_3\phi_2} M^2_{\lambda_2\phi_1} M^1_{\lambda_1\phi_1} \underline{P_x} \widetilde{M}^1_{\lambda_1\phi_1} \widetilde{M}^2_{\lambda_2\phi_1} \widetilde{M}^3_{\lambda_3\phi_2}) \widetilde{M} \right)
$$

$$
\overline{\times} \, (\mathrm{e} \wedge (O \wedge x)) = 0.
$$

The unknowns of these constraint equations are the rigid motion M and the angles ϕ_1 and ϕ_2.

37.4   Experiments

This section shows experimental results.

37.4.1   Estimation of twist parameters

To solve the constraint equations for simple objects, partially represented by cycloidal curves, we linearize the equations with respect to the motors and use the first order Taylor series expression for approximation.

This leads to a mapping of the above mentioned global transformation to a twist representation, which enables incremental changes of pose. This means, we do not search for the parameters of the Lie group $SE(3)$ to describe the rigid body motion [5], but for the parameters which generate their Lie algebra $se(3)$. This idea is taken from [12]. From the resulting linear equations in the unknown 3D rigid body motion we get an error function to be minimized. The main problem of pose estimation of cycloidal curves is that they are in general not convex. This results in the problem of getting trapped in local minima. To avoid global minima, in our first approach we use local search [2] strategies to handle these problems and to find a global minimum.

37.4.2   Pose estimation of convex objects

We will continue with pose estimation results of convex objects. For these examples the gradient method converges directly. The pose, for these types of curves, can be estimated in nearly video real-time (10 frames per second) on a SUN Ultra 10.

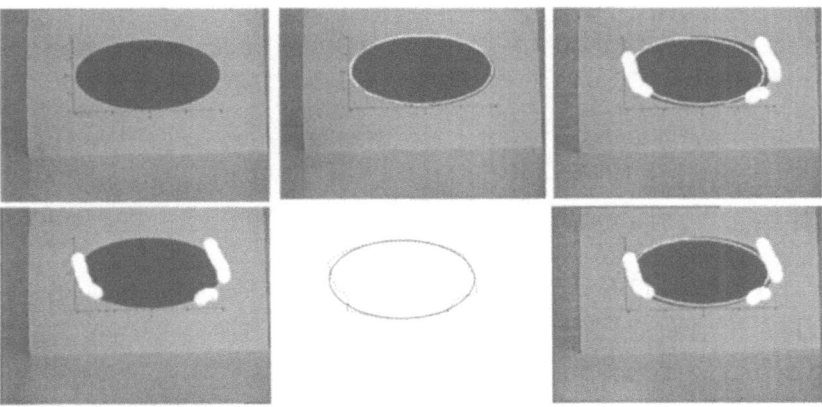

FIGURE 37.8. Pose estimation of a 3D ellipse by using undistorted, distorted and interpolated data.

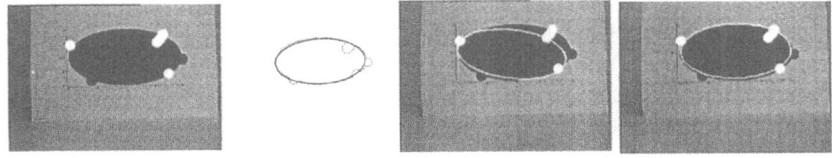

FIGURE 37.9. Pose estimation of a 3D ellipse by using distorted and interpolated data.

Figures 37.8 and 37.9 address the self regularizing properties of image contours: In the first column of Figure 37.8 an undistorted and a distorted image ellipse are shown. The second column shows a pose estimation result found by taking the undistorted image data and the extracted contour of the distorted image. There are two possibilities to deal with the contour points:

On the one hand it is possible to use them directly, on the other hand it is possible to interpolate them to an image ellipse and use the interpolated data. In [20] several approaches for ellipse fitting are presented. To interpolate our contour points, we re-implemented and used the LLS (linear least squares) approach in our scenario. Though the LLS approach is not the best algorithm discussed in [20], it is easy to implement and very fast. The pose results for the raw contour points on the one hand, and the interpolated ellipse points on the other hand, are shown in the third column of Figure 37.8. The upper image shows the pose result achieved by using the pure points, the lower image shows the result achieved by using the interpolated data. Indeed, using the pure points leads to worse results in

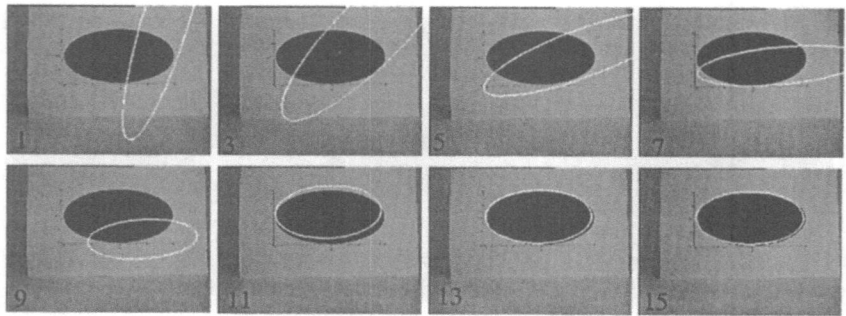

FIGURE 37.10. Convergence behavior of the algorithm during the iteration.

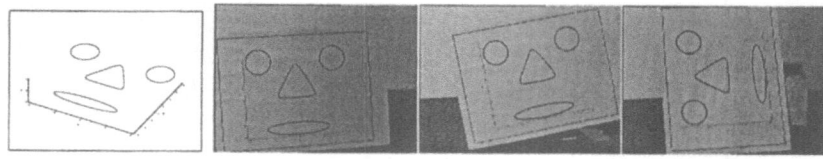

FIGURE 37.11. Pose estimation of an object containing one ellipse, two circles and one deltoid.

comparison to using the interpolated data. To visualize the pose quality the 3D object model is transformed and projected into the image. Figure 37.9 shows a more extreme case of disturbed image data. It can be seen, that the use of interpolated image data leads to more stable results than using the raw contour points.

Figure 37.10 shows the convergence behavior of our algorithm during the iterations. Since the rigid body motion to estimate is very large, the algorithm needs several iterations to converge. If only small movements are observed, the number of iterations (and therefore the computing time) can be reduced significantly.

Figure 37.11 shows pose estimation results of a second 3D object model. This object contains two circles, one ellipse and one deltoid. The left image shows the 3D object model. The other images show the transformed projected object model to visualize the quality of the pose.

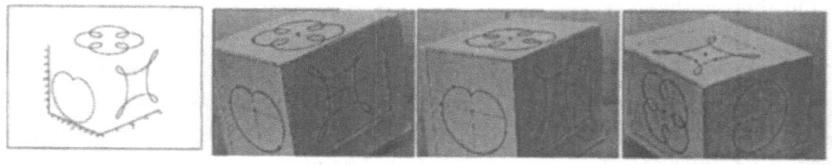

FIGURE 37.12. Pose estimation of an object containing a cardioid and two cycloids.

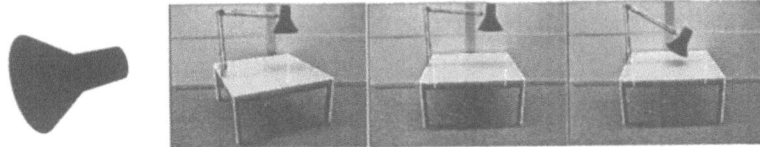

FIGURE 37.13. Pose estimation of 2-twist surface, connected via a 2 d.o.f. kinematic chain to a table.

37.4.3   Pose estimation of non-convex objects

Now we look at pose estimation of non-convex objects. Figure 37.12 shows experimental results for an object, containing three cycloidal non-convex curves. We use all information simultaneously to solve for the pose parameters. Here we have to encapsulate the gradient method for pose estimation within a heuristic, since the entities are not convex any more. In the first image, the object model used is shown. The other three images show pose estimation results for the object. Again we mark the extracted point features in the image and the transformed projected points. Furthermore, the transformed projected cycloidal curves are shown. Since we measured the size of the object model by hand, the pose estimation result is reasonably accurate.

The combination of the gradient method with the heuristic mentioned leads to slow algorithms. While the pose for convex object models can be estimated in 10 frames per second, the estimation of the object presented in Figure 37.12 takes up to 5 minutes to converge to a global minimum. But indeed, it is possible to estimate the global minimum.

Figure 37.13 shows experimental results from a 2-twist surface. The object model is a lamp-shade connected via a 2 d.o.f. kinematic chain to a table. In this experiment, the pose parameters and the angles of the kinematic chain are estimated. The lamp-shade itself is modeled by two conical parts. Since both surface parts are convex, the gradient method converges directly and we do not have to apply a heuristic to estimate the pose.

37.5   Discussion

In this contribution we present a new approach to estimate the pose of 2-twist and 3-twist curves. These cycloidal curves extend the classical pose estimation problems in an interesting and useful way. We believe that for many applications cycloidal curves are better suited to represent natural objects than the often used tessellated surfaces. The representation of these entities is very compact and geometric manipulations, like rigid body motions of the curves, can easily be performed, just by transforming the generating elements of the curve.

The main difficulties that occur in the context of 2D-3D pose estimation of general space curves or surfaces are the local minima in the error functions to be minimized. This occurs naturally in the case of non-convex curves. If the objects are convex (e.g., Figure 37.11) on the other hand, the gradient method for solving the pose parameters converges to the global minimum and the pose can be estimated in (nearly) video real-time (10 fps) on a SUN Ultra 10. But if the curves are non-convex, we have to deal with local minima within our error function. In this case we combine the previously used gradient method with a heuristic and then we are able to estimate the global minimum. But so far, the algorithm is too slow for practical applications. One possibility to deal with this problem is to separate the non-convex curve in a set of curves containing the convex parts. This was done in the experiment of Figure 37.13.

There the pose problem reduces to the simple case of convex object features, but it requires more a priori knowledge about the scenario. Future research will concentrate on faster algorithms to solve such constraint equations for non-convex curves. Furthermore, we will concentrate on free-form contours and surfaces. First results are presented in [17].

37.5.1   Acknowledgments

We would like to thank Christian Perwass and Jon Selig for the fruitful discussions and hints performing this work. Figures 37.5 and 37.6 are generated with the CLU draw programming package [3]. This work has been supported by DFG Graduiertenkolleg No. 357 and by EC Grant IST-2001-3422 (VISATEC).

REFERENCES

[1] Bayro-Corrochano, E., The geometry and algebra of kinematics. In [18], 2001, 457–472.

[2] Beveridge, J.R., Local search algorithms for geometric object recognition: Optimal correspondence and pose. *Ph.D. Thesis, Computer Science Department, University of Massachusetts, Amherst*, 1993. Available as Technical Report CS 93–5.

[3] CLU Library, A C++ Library for Clifford Algebra.   Available at http://www.perwass.de/CLU   *Lehrstuhl für kognitive Systeme, University Kiel*, 2001.

[4] O'Connor, J.J., and Robertson, E.F.,   Famous Curves Index   http://www-history.mcs.st-andrews.ac.uk/history/Curves/Curves.html

[5] Gallier, J., *Geometric Methods and Applications. For Computer Science and Engineering.* Springer Verlag, New York Inc. 2001

[6] Grimson, W. E. L., *Object Recognition by Computer.* The MIT Press, Cambridge, MA, 1990.

[7] Hestenes, D., Li, H., and Rockwood, A., New algebraic tools for classical geometry. In [18], 2001, pp. 3–23.

[8] Hestenes, D., and Ziegler, R., Projective geometry with Clifford algebra. *Acta Applicandae Mathematicae*, Vol. **23**, 1991, 25–63.

[9] Horaud, R., Phong, T.Q., and Tao, P.D., Object pose from 2-d to 3-d point and line correspondences. *International Journal of Computer Vision*, Vol. **15**, 1995, 225–243.

[10] Li, H., Hestenes, D., and Rockwood, A., Generalized homogeneous coordinates for computational geometry. In [18], 2001, pp. 27–52.

[11] Lee, X.,   A   Visual   Dictionary   of   Special   Plane   Curves. http://xahlee.org/SpecialPlaneCurves_dir/specialPlaneCurves.html

[12] Murray, R.M., Li, Z., and Sastry, S.S., *A Mathematical Introduction to Robotic Manipulation*. CRC Press, 1994.

[13] Needham, T., *Visual Complex Analysis*. Oxford University Press, 1997.

[14] Perwass, C., and Lasenby, L., A unified description of multiple view geometry. In [18], 2001, pp. 337–369.

[15] Rosenhahn, B., Zhang, Y., and Sommer, G., Pose estimation in the language of kinematics. In: *Second International Workshop, Algebraic Frames for the Perception-Action Cycle*, AFPAC 2000, Sommer G. and Zeevi Y.Y. (Eds.), LNCS 1888, Springer-Verlag, Heidelberg, 2000, pp. 284–293.

[16] Rosenhahn, B., Granert, O., and Sommer, G., Monocular pose estimation of kinematic chains. In *Applied Geometric Algebras for Computer Science and Engineering*, Birkhäuser Verlag, Dorst L., Doran C. and Lasenby J. (Eds.), pp. 373–383, 2001.

[17] Rosenhahn, B., Perwass, C., and Sommer, G., Pose Estimation of 3D Free-form Contours. Technical Report 0207, University Kiel, 2002.

[18] Sommer, G., editor. *Geometric Computing with Clifford Algebra*. Springer Verlag, 2001.

[19] Walker, M.W., and Shao, L., Estimating 3-d location parameters using dual number quaternions. *CVGIP: Image Understanding*, Vol. **54**, No. 3, 1991, 358–367.

[20] Zhang, Z., Parameter estimation techniques: a tutorial with application to conic fitting. *Image and Vision Computing*, Vol. **15**, 1997, 59–76.

Bodo Rosenhahn
Institut für Informatik und Praktische Mathematik
Christian-Albrechts-Universität zu Kiel
Olshausenstr. 40, 24098 Kiel
Germany
E-mail: bro@ks.informatik.uni-kiel.de

Gerald Sommer
Institut für Informatik und Praktische Mathematik
Christian-Albrechts-Universität zu Kiel
Olshausenstr. 40, 24098 Kiel
Germany
E-mail: gs@ks.informatik.uni-kiel.de

Submitted: August 1, 2002.

Index

Progress in Mathematical Physics

Progress in Mathematical Physics is a book series encompassing all areas of mathematical physics. It is intended for mathematicians, physicists and other scientists, as well as graduate students in the above related areas.

This distinguished collection of books includes authored monographs and textbooks, the latter primarily at the senior undergraduate and graduate levels. Edited collections of articles on important research developments or expositions of particular subject areas may also be included.

This series is reasonably priced and is easily accessible to all channels and individuals through international distribution facilities.

Preparation of manuscripts is preferable in LATEX. The publisher will supply a macro package and examples of implementation for all types of manuscripts.

Proposals should be sent directly to the series editors:

Anne Boutet de Monvel
Mathématiques, case 7012
Université Paris VII Denis Diderot
2, place Jussieu
F-75251 Paris Cedex 05
France

Gerald Kaiser
The Virginia Center for Signals and Waves
1921 Kings Road
Glen Allen, VA 23059
U.S.A.

or to the Publisher:

Birkhäuser Boston
675 Massachusetts Avenue
Cambridge, MA 02139
U.S.A.
Attn: Ann Kostant

Birkhäuser Verlag
40-44 Viadukstrasse
CH-4010 Basel
Switzerland
Attn: Thomas Hempfling

1   COLLET/ECKMANN. Iterated Maps on the Interval as Dynamical Systems
     ISBN 3-7643-3510-6
2   JAFFE/TAUBES. Vortices and Monopoles, Structure of Static Gauge Theories
     ISBN 3-7643-3025-2
3   MANIN. Mathematics and Physics
     ISBN 3-7643-3027-9
4   ATWOOD/BJORKEN/BRODSKY/STROYNOWSKI. Lectures on Lepton Nucleon
     Scattering and Quantum Chromodynamics
     ISBN 3-7643-3079-1
5   DITA/GEORGESCU/PURICE. Gauge Theories: Fundamental Interactions
     and Rigorous Results
     ISBN 3-7643-3095-3